CAMBRIDGE STUDIES IN ADVANCED MATHEMATICS

---

45 Positive Harmonic Functions and Diffusion

Already published

1 W. M. L. Holcombe *Algebraic automata theory*
2 K. Petersen *Ergodic theory*
3 P. T. Johnstone *Stone spaces*
4 W. H. Schikhof *Ultrametric calculus*
5 J.-P. Kahane *Some random series of functions, 2nd edition*
6 H. Cohn *Introduction to the construction of class fields*
7 J. Lambek & P. J. Scott *Introduction to higher-order categorical logic*
8 H. Matsumura *Commutative ring theory*
9 C. B. Thomas *Characteristic classes and the cohomology of finite groups*
10 M. Aschbacher *Finite group theory*
11 J. L. Alperin *Local representation theory*
12 P. Koosis *The logarithmic integral I*
13 A. Pietsch *Eigenvalues and s-numbers*
14 S. J. Patterson *An introduction to the theory of the Riemann zeta-function*
15 H. J. Baues *Algebraic homotopy*
16 V. S. Varadarajan *Introduction to harmonic analysis on semisimple Lie groups*
17 W. Dicks & M. Dunwoody *Groups acting on graphs*
18 L. J. Corwin & F. P. Greenleaf *Representations of nilpotent Lie groups and their applications*
19 R. Fritsch & R. Piccinini *Cellular structures in topology*
20 H. Klingen *Introductory lectures on Siegel modular forms*
22 M. J. Collins *Representations and characters of finite groups*
24 H. Kunita *Stochastic flows and stochastic differential equations*
25 P. Wojtaszczyk *Banach spaces for analysts*
26 J. E. Gilbert & M. A. M. Murray *Clifford algebras and Dirac operators in harmonic analysis*
27 A. Frohlich & M. J. Taylor *Algebraic number theory*
28 K. Goebel & W. A. Kirk *Topics in metric fixed point theory*
29 J. F. Humphreys *Reflection groups and Coxeter groups*
30 D. J. Benson *Representations and cohomology I*
31 D. J. Benson *Representations and cohomology II*
32 C. Allday & V. Puppe *Cohomological methods in transformation groups*
33 C. Soulé *et al. Lectures on Arakelov geometry*
34 A. Ambrosetti & G. Prodi *A primer of nonlinear analysis*
35 J. Palis & F. Takens *Hyperbolicity, stability and chaos at homoclinic bifurcations*
37 Y. Meyer *Wavelets and operators*
38 C. Weibel *An introduction to homological algebra*
39 W. Bruns & J. Herzog *Cohen–Macaulay rings*
45 R. Pinsky *Positive harmonic functions and diffusion*

ROSS G. PINSKY
Department of Mathematics
Technion – Israel Institute of Technology

Positive Harmonic Functions and Diffusion

CAMBRIDGE
UNIVERSITY PRESS

CAMBRIDGE UNIVERSITY PRESS
Cambridge, New York, Melbourne, Madrid, Cape Town, Singapore, São Paulo

Cambridge University Press
The Edinburgh Building, Cambridge CB2 8RU, UK

Published in the United States of America by Cambridge University Press, New York

www.cambridge.org
Information on this title: www.cambridge.org/9780521470148

First published 1995
This digitally printed version 2008

A catalogue record for this publication is available from the British Library

ISBN 978-0-521-47014-8 hardback
ISBN 978-0-521-05983-1 paperback

To my parents

What is the difference between process and event? A process happens regularly, following a relatively permanent pattern; an event is extraordinary, irregular. A process may be continuous, steady, uniform; events happen suddenly, intermittently, occasionally. Processes are typical; events are unique. A process follows a law, events create a precedent.

Being human is not a solid structure or a string of predictable facts, but an incalculable series of moments and facts. As a process, man may be described biologically; as an event he can only be understood creatively, dramatically.

Abraham Joshua Heschel

Contents

Preface

Over the past 50 years, the body of results concerning questions of existence and characterization of positive harmonic functions for second-order elliptic operators has been nourished by two distinct sources – one rich in analysis, the other less well endowed analytically but amply compensated by generous heapings of probability theory (more specifically, martingales and stopping times). For example, the results appearing in the first seven sections of Chapter 4, which have been developed for the most part over the past decade, have been proved with nary a word about probability, while the results appearing in Chapter 6 have traditionally been formulated and proved using a probabilistic approach. On the other hand, the Martin boundary theory of Chapters 7 and 8 have long been studied by distinct probabilistic and analytic methods. My original intention was to write a monograph which would provide an integrated probabilistic and analytic approach to a host of results and ideas related, at least indirectly, to the existence and/or characterization of positive harmonic functions. When the undertaking was still in its inchoate stages, it became apparent that such a monograph, if executed appropriately, might serve as a graduate text for students working in diffusion processes. This direction also seemed appealing. Indeed, too numerous have been the occasions on which I explained a result or a 'meta-result' to a student or colleague and then found myself at a loss when it came to suggesting a reference text. I hope this book might ground some of the folklore. In the end, then, the book has been written with two intentions in mind.

I have endeavoured to keep the book as self-contained as the dictates of good taste permit. In particular, the book ought to be accessible to the graduate student who is familiar with the standard first-year fare in analysis, probability and partial differential equations at an American university. It also ought to be accessible to analysts who come to the subject without a probabilistic pedigree – if they are willing to do a little spade work in the first two chapters; I hope it might tempt them to explore the probabilistic tack a bit. Monroe Donsker, my first teacher of probability, was fond of saying that a whole class of problems may be formulated and proved in two distinct fashions – one that the probabilists will understand and one that the analysts will understand. He would then add that S. R. S. Varadhan, my thesis adviser, was the practitioner par excellence of the dual approach. The outlook of these two has been a beacon for me; I hope the book reflects that source adequately.

A brief word on the contents is in order. Chapters 1 and 2 provide the necessary background on diffusion processes. These processes are constructed in Chapter 1 and their properties are investigated in Chapter 2. The reader who is familiar with diffusion processes will probably find it useful, none the less, to scan Chapter 1, Sections 1.10–1.13, where the martingale problem is introduced and then quickly superseded by what I call the generalized martingale problem, which turns out to be very convenient for the considerations of this work. These pages will also familiarize the reader with notation that is used throughout the book.

The first part of Chapter 3 culls a few basic results from functional analysis and PDEs. The second part treats an elliptic operator $L = \frac{1}{2}\nabla \cdot a\nabla + b \cdot \nabla + V$ on a domain D in the case where D is bounded and possesses a smooth boundary, and where L is uniformly elliptic with coefficients which are smooth up to the boundary. In this setting, a rigorous treatment of the construction of the corresponding semigroup and generator is given; this in turn leads to the theory of the principal eigenvalue.

In Chapter 4, the notion of principal eigenvalue is extended to operators L on arbitrary domains $D \subseteq R^d$ under the assumption that L is uniformly elliptic with smooth coefficients on compact subsets of D. This extension can be characterized in terms of positive harmonic functions and Green's functions, and leads naturally to the classification of operators as subcritical, critical or supercritical. The results given in Sections 4.1–4.7, developed around this theme, are what I call *criticality theory*.

In Sections 4.8 and 4.9, classical results concerning invariant measures and ergodic behavior for diffusion processes are recast in the language of criticality theory. In Section 4.10, the connection between criticality theory and classical spectral theory is made in the case where L is symmetric with respect to some reference measure. Section 4.11 sketches the extension of the results to the case where D is a manifold.

In the one-dimensional case, and in the radially symmetric multi-dimensional case, many calculations may be worked out explicitly; such calculations are carried out in Chapter 5.

Throughout Chapter 6, it is assumed that $V \equiv 0$. In this case, the notions of criticality and subcriticality reduce to those of recurrence and transience, respectively, for diffusion processes. Various techniques are developed to establish criteria for recurrence or transience.

A supercritical operator possesses no positive harmonic functions, while a critical operator possesses a unique positive harmonic function (up to constant multiples). In the subcritical case, the cone of positive harmonic functions may be more interesting. Chapters 7 and 8 treat the structure of the cone $C_L(D)$ of positive harmonic functions for a subcritical operator L on a domain $D \subset R^d$. Chapter 7 begins with a review of the classical analytic theory of the Martin boundary, from which follows the structure of $C_L(D)$, and of the probabilistic theory of h-transforms and conditioned diffusions. The rest of the chapter is devoted to the development of an alternative probabilistic approach to the Martin boundary in terms of what I dub the *exterior harmonic measure boundary*. In Chapter 8, the Martin boundary is calculated explicitly for several classes of operators; in several cases, the calculation is based on the probabilistic approach developed in Chapter 7.

In the case where $V \equiv 0$, an important connection exists between diffusion processes and bounded harmonic functions. This connection is explored in Chapter 9. No such exposition on bounded harmonic functions would be complete without some discussion of Brownian motion on a manifold of negative curvature. However, this topic alone could be the subject of a book. Thus, in the second half of Chapter 9, some familiarity with some rudimentary definitions in differential geometry is assumed. Furthermore, a fundamental comparison result from differential geometry is stated without proof. With the above caveats, complete proofs are presented for the results in this chapter.

Exercises appear at the end of each chapter. In quite a number of them, the reader is asked to supply a proof or a step of a proof that I deemed judicious to omit from the text. Perhaps for that reason, I have been, I confess, too generous with the hints. Historical notes and references also appear at the end of each chapter.

I am indebted to the late Monroe Donsker, and, especially to S. R. S. Varadhan, for their contribution to my mathematics education. I also owe thanks to Rick Durrett, whose stimulating working seminar at UCLA during the 1983–84 academic year first piqued my interest in positive harmonic functions. It is a pleasure to thank Yehuda Pinchover, who has been a patient listener and an unending source of ideas during the writing of this book. Indeed, it was his work on criticality theory, carried out in a purely analytical manner, that crystalized my resolve to write this book. This work was partially supported by the Fund for the Promotion of Research at the Technion. I would like to thank Sylvia Schur and Debbie Miller for their efficient word processing, which spelled a welcome relief for me. Finally, I thank my wife, Jeanette, for her moral support.

Ross G. Pinsky
Haifa, Israel

Notation

We use a hierarchical numbering system for equations and statements. A theorem, corollary, lemma, proposition, or example is a 'statement'. The kth equation in section j of chapter i is labeled $(j.k)$ at the place where it occurs and is cited as $(j.k)$ within chapter i, but as $(i.j.k)$ outside chapter i. The kth statement in section j of chapter i is labeled $j.k$ at the place where it occurs and is cited as STATEMENT $j.k$ within chapter i, but as STATEMENT $i.j.k$ outside chapter i.

1
Existence and uniqueness for diffusion processes

1.1 Stochastic processes and filtrations

Let $(Y, \mathcal{M}, P)$ be a probability space and denote points in Y by q. A Borel measurable function $\eta\colon Y \to R^d$ is called an *R^d-valued random variable* on $(Y, \mathcal{M})$. The *expectation* $E\eta$ of η with respect to the probability measure P is defined by $E\eta = \int_Y \eta \, dP$. The R^d-valued random variable η on $(Y, \mathcal{M}, P)$ induces a probability measure ν on R^d defined by $\nu(B) = P(\eta \in B)$, for $B \in \mathcal{B}_d$, where $\mathcal{B}_d$ is the Borel σ-algebra on R^d; ν is called the *distribution of* η.

A family $\mathcal{M}_t$, $t \geqslant 0$, of σ-algebras satisfying $\mathcal{M}_{t_1} \subseteq \mathcal{M}_{t_2}$ for $t_1 \leqslant t_2$ and $\mathcal{M}_t \subseteq \mathcal{M}$ for all $t \geqslant 0$ is called a *filtration* on $(Y, \mathcal{M})$. If for all $t \geqslant 0$, $\mathcal{M}_t$ includes all sets of P-measure zero, the filtration is called *complete* on $(Y, \mathcal{M}, P)$. If $\mathcal{M}'_t$ is the smallest σ-algebra containing $\mathcal{M}_t$ and all sets of P-measure zero, then $\mathcal{M}'_t$ is called the *completion* of $\mathcal{M}_t$.

A measurable map $\zeta\colon [0, \infty) \times Y \to R^d$ is called an *R^d-valued stochastic process* on $(Y, \mathcal{M}, P)$. We frequently suppress the variable $q \in Y$ and write $\zeta(t)$ for $\zeta(t, q)$. The stochastic process $\zeta(t)$ on $(Y, \mathcal{M}, P)$ induces a probability measure ν on $\mathcal{B}([0, \infty), R^d)$, the space of measurable functions from $[0, \infty)$ to R^d with the sup-norm topology, defined by $\nu(B) = P(\zeta(\cdot) \in B)$, for Borel sets B in $\mathcal{B}([0, \infty), R^d)$. The measure ν is called the *distribution* of $\zeta(\cdot)$. The distribution ν is uniquely determined by the finite-dimensional distributions of $\zeta(t)$, that is, by the distributions of $(\zeta(t_1), \zeta(t_2), \ldots, \zeta(t_n))$, for all $0 \leqslant t_1 < t_2 < \ldots < t_n$ and all $n \geqslant 1$. If $\zeta(t)$ and $\hat{\zeta}(t)$ are two stochastic processes on $(Y, \mathcal{M}, P)$ and $P(\zeta(t) = \hat{\zeta}(t)) = 1$ for each $t \geqslant 0$, then, using a monotone class theorem, it is not difficult to show that $\zeta(\cdot)$ and $\hat{\zeta}(\cdot)$ possess a common distribution on $\mathcal{B}([0, \infty), R^d)$; the stochastic processes $\zeta(t)$ and $\hat{\zeta}(t)$ are called *versions* of one another. The stochastic process $\zeta(t)$ is called *progressively measurable* with respect to $\mathcal{M}_t$ if $\zeta(t)$ is $\mathcal{M}_t$-measurable for each $t \geqslant 0$.

1.2 Conditional expectation, martingales and stopping times

If $\mathcal{G} \subset \mathcal{M}$ is itself a σ-algebra and η is a random variable satisfying $E|\eta| < \infty$, then the conditional expectation $E(\eta|\mathcal{G})$ of η given $\mathcal{G}$ is defined by the following two conditions:

(i) $E(\eta|\mathcal{G})$ is $\mathcal{G}$-measurable;
(ii) $\int_A \eta \, dP = \int_A E(\eta|\mathcal{G}) \, dP$, for all $A \in \mathcal{G}$.

A direct application of the Radon–Nikodym theorem shows that the conditional expectation exists and is unique up to sets of P-measure zero. The basic properties of the conditional expectation are as follows:

1. $E(E(\eta|\mathcal{G})) = E\eta$.
2. $E(\eta_1 + \eta_2|\mathcal{G}) = E(\eta_1|\mathcal{G}) + E(\eta_2|\mathcal{G})$.
3. If $\eta \geq 0$, then $E(\eta|\mathcal{G}) \geq 0$.
4. $|E(\eta|\mathcal{G})| \leq E(|\eta|\,|\mathcal{G})$.
5. $E(\eta|\mathcal{G}) = \eta$ if η is $\mathcal{G}$ measurable.
6. $E(\eta|\mathcal{G}) = E\eta$ if η is independent of $\mathcal{G}$, that is, if $P(\{\eta \in B\} \cap A) = P(\eta \in B)P(A)$ for all $B \in \mathcal{B}_d$ and $A \in \mathcal{G}$.
7. $E(\eta_1\eta_2|\mathcal{G}) = \eta_1 E(\eta_2|\mathcal{G})$ if η_1 is $\mathcal{G}$-measurable.
8. (Bounded convergence theorem for conditional expectation) If $\{\eta_j\}_{j=1}^{\infty}$ is bounded and $\eta = \lim_{j\to\infty} \eta_j$ a.s., then $E(\eta|\mathcal{G}) = \lim_{j\to\infty} E(\eta_j|\mathcal{G})$.
9. (Dominated convergence theorem for conditional expectation) If $\eta = \lim_{j\to\infty} \eta_j$ a.s. and $|\eta_j| \leq Z$ for all $j = 1, 2, \ldots$, where $EZ < \infty$, then $E(\eta|\mathcal{G}) = \lim_{j\to\infty} E(\eta_j|\mathcal{G})$.
10. (Jenson's inequality for conditional expectation) If ϕ is convex and $E|\phi(\eta)| < \infty$, then $\phi(E(\eta|\mathcal{G})) \leq E(\phi(\eta)|\mathcal{G})$.

Of course, all the equalities and inequalities given above are to be taken in the P-almost sure sense.

For $B \in \mathcal{B}_d$, let I_B denote the characteristic function of the set B, that is,

$$I_B(x) = \begin{cases} 1, & \text{if } x \in B \\ 0, & \text{if } x \notin B \end{cases}.$$

We will usually denote $E(I_B(\eta)|\mathcal{G})$ by $P(\eta \in B|\mathcal{G})$. If Z is a random variable, we define the σ-algebra generated by Z as the smallest σ-algebra on Y which contains all sets of the form $\{Z \in B\}$ for $B \in \mathcal{B}_d$. This σ-algebra is denoted by $\sigma(Z)$. We will denote $E(\eta|\sigma(Z))$ and $P(\eta \in B|\sigma(Z))$ by $E(\eta|Z)$ and $P(\eta \in B|Z)$ respectively.

A real-valued progressively measurable stochastic process $\zeta(t)$ is called a *martingale* with respect to $(Y, \mathcal{M}, \mathcal{M}_t, P)$ if $E|\zeta(t)| < \infty$, for all $t \geqslant 0$, and $E(\zeta(t)|\mathcal{M}_s) = \zeta(s)$, for all $0 \leqslant s < t < \infty$. It is called a *submartingale* (*supermartingale*) with respect to $(Y, \mathcal{M}, \mathcal{M}_t, P)$ if $E|\zeta(t)| < \infty$, for all $t \geqslant 0$, and $E(\zeta(t)|\mathcal{M}_s) \geqslant \zeta(s)$ $(E(\zeta(t)|\mathcal{M}_s) \leqslant \zeta(s))$, for all $0 \leqslant s < t < \infty$.

An extended real-valued non-negative random variable $\tau: Y \to [0, \infty]$ is called a *stopping time* with respect to $\mathcal{M}_t$ (or an $\mathcal{M}_t$-stopping time) if $\{\tau \leqslant t\} \in \mathcal{M}_t$, for all $t \geqslant 0$. The following paragraph gives a summary of the basic properties of stopping times.

The *σ-algebra 'up to time τ'* is defined by $\mathcal{M}_\tau = \{A \in \mathcal{M}: A \cap \{\tau \leqslant t\} \in \mathcal{M}_t \text{ for all } t \geqslant 0\}$. Of course, one must verify that $\mathcal{M}_\tau$ as defined is in fact a σ-algebra. If $\tau \equiv t$, then $\mathcal{M}_\tau$ reduces to $\mathcal{M}_t$. Furthermore, if $\sigma \leqslant \tau$ are two stopping times, then $\mathcal{M}_\sigma \subseteq \mathcal{M}_\tau$. Thus, the notation $\mathcal{M}_\tau$ and its designation as the σ-algebra 'up to time τ' are appropriate. If σ and τ are two stopping times, then $\sigma \wedge \tau$ is also a stopping time. Every stopping time τ is $\mathcal{M}_\tau$-measurable. If $\zeta(t)$ is progressively measurable with respect to $\mathcal{M}_t$ and τ is finite-valued, then $\zeta(\tau)$ is $\mathcal{M}_\tau$-measurable.

The principal connection between stopping times and martingales is contained in *Doob's optional sampling theorem*. A version of that theorem which is appropriate for our needs is as follows:

Theorem 2.1 (Doob's optional sampling theorem). *Let $\zeta(t)$ be a martingale (submartingale, supermartingale) and let τ be a stopping time with respect to $(Y, \mathcal{M}, \mathcal{M}_t, P)$. Then $Z(t) \equiv \zeta(t \wedge \tau)$ is also a martingale (submartingale, supermartingale) with respect to $(Y, \mathcal{M}, \mathcal{M}_t, P)$.*

The following two martingale inequalities will be useful in the sequel:

Theorem 2.2. *If $\zeta(t)$ is a right continuous submartingale, then*

$$P(\sup_{0 \leqslant s \leqslant t} \zeta(s) \geqslant \lambda) \leqslant \frac{1}{\lambda} E\zeta^+(t),$$

for all $\lambda > 0$ and $t > 0$, where $\zeta^+(t) = \zeta(t) \vee 0$.

Theorem 2.3. *If $\zeta(t)$ is a right continuous martingale or a non-negative, right continuous submartingale, then for any $\alpha > 1$,*

$$E[\sup_{0 \leqslant s \leqslant t} |\zeta(s)|^\alpha] \leqslant \left(\frac{\alpha}{\alpha - 1}\right)^\alpha E|\zeta(t)|^\alpha.$$

In Chapter 9 we will need the *martingale convergence* theorem.

Theorem 2.4. *Let $\zeta(t)$ be a submartingale with respect to $(Y, \mathcal{M}, \mathcal{M}_t, P)$. Assume that $E|\zeta(t)|$ is bounded in t. Then* $\lim_{t\to\infty} \zeta(t)$ exists a.s.

1.3 Markov processes and semigroups

An R^d-valued stochastic process $\zeta(t)$ on $(Y, \mathcal{M}, \mathcal{M}_t, P)$ which satisfies $P(\zeta(t) \in B|\mathcal{M}_s) = P(\zeta(t) \in B|\zeta(s))$ for $0 \leqslant s < t < \infty$ and $B \in \mathcal{B}_d$ is called an R^d-valued *Markov process* on $(Y, \mathcal{M}, \mathcal{M}_t, P)$. Suppose that for each $0 \leqslant s < t$ and each $x \in R^d$ there exists a probability measure $p(s, x, t, dy)$ on R^d satisfying

$$\left.\begin{aligned} &\text{(i) } p(s, x, t, B) \text{ is measurable as a function of } x \text{ for each } 0 \leqslant s < t \\ &\quad \text{and } B \in \mathcal{B}_d; \\ &\text{(ii) } p(s, x, t, B) = \int_{R^d} p(s, x, u, dy)p(u, y, t, B) \text{ for all } s < u < t \\ &\quad \text{and all } B \in \mathcal{B}_d. \end{aligned}\right\} \tag{3.1}$$

Condition (ii) in (3.1) is known as the *Chapman–Kolmogorov equation*. Suppose, further, that $P(\zeta(t) \in B|\mathcal{M}_s) = P(\zeta(t) \in B|\zeta(s)) = p(s, \zeta(s), t, B)$ for all $0 \leqslant s < t$ and all $B \in \mathcal{B}_d$. Then $p(s, x, t, dy)$ is called the *transition probability function* for the Markov process $\zeta(t)$. A Markov process with a transition probability function $p(s, x, t, dy)$ actually satisfies the following condition:

$$P(\zeta(t + \cdot) \in B|\mathcal{M}_t) = P(\zeta(t + \cdot) \in B|\zeta(t)), \text{ for } t \geqslant 0$$
$$\text{and Borel sets } B \text{ in } \mathcal{B}([0, \infty), R^d). \tag{3.2}$$

The proof is left as exercise (1.2).

In the special case that $p(s, x, t, dy) = p(0, x, t - s, dy)$, for all $0 \leqslant s < t$ and all $x \in R^d$, we will use the notation $p(t, x, dy) = p(s, x, s + t, dy)$. In this case, $P(\zeta(t) \in B|\mathcal{M}_s) = p(t - s, \zeta(s), B)$ depends only on the position of the process $\zeta(s)$ and on the difference $t - s$. Such a Markov process is called *time-homogeneous*. All the Markov processes we consider in this book will be time-homogeneous.

Given a time-homogeneous transition probability function $p(t, x, dy)$, the Kolmogorov construction (see, for example, [Breiman (1968) or (1992)]) guarantees the existence of a measurable space $(Y, \mathcal{M})$ with filtration $\mathcal{M}_t$ and a progressively measurable stochastic process $\zeta(t)$: $[0, \infty) \times Y \to R^d$ such that for each probability measure μ on R^d, there exists a probability measure P_μ on $(Y, \mathcal{M})$ satisfying

(i) $P_\mu(\zeta(0) \in B) = \mu(B)$ for $B \in \mathcal{B}_d$;
(ii) $P_\mu(\zeta(t) \in B|\mathcal{M}_s) = p(t - s, \zeta(s), B)$, for $0 \leqslant s < t$ and $B \in \mathcal{B}_d$.

The process $\zeta(t)$ is then a Markov process on $(Y, \mathcal{M}, \mathcal{M}_t, P_\mu)$. Its distribution is uniquely determined by the transition probability function $p(t, x, \mathrm{d}y)$ and the initial distribution μ. In the special case where $\mu = \delta_x$ is the atom at some $x \in R^d$, we use the notation P_x. Of course, $P_\mu = \int_{R^d} P_x \mu(\mathrm{d}x)$. The expectation corresponding to P_μ is denoted by E_μ.

Let $\mathcal{B}_b(R^d)$ denote the space of real-valued bounded measurable functions on R^d and define the operator $T_t: \mathcal{B}_b(R^d) \to \mathcal{B}_b(R^d)$ by

$$T_t f(x) = E_x f(\zeta(t)) = \int_{R^d} f(y) p(t, x, \mathrm{d}y).$$

Since

$$\begin{aligned} T_{t+s} f(x) = E_x f(\zeta(t+s)) &= E_x(E_x f(\zeta(t+s)) | \mathcal{F}_t) \\ &= E_x E_x(f(\zeta(t+s)) | \zeta(t)) \\ &= E_x E_{\zeta(t)} f(\zeta(s)) \\ &= E_x \int_{R^d} f(y) p(s, \zeta(t), \mathrm{d}y) \\ &= \int_{R^d} \int_{R^d} f(y) p(s, z, \mathrm{d}y) p(t, x, \mathrm{d}z), \end{aligned}$$

it follows from the Chapman–Kolmogorov equations that $T_{t+s} = T_t T_s$ for $0 \leqslant s < t$; therefore T_t defines a *semigroup*. Furthermore, letting $\|\cdot\|$ denote the sup-norm on $\mathcal{B}_b(R^d)$, it is clear that T_t is a *contraction semigroup* on $\mathcal{B}_b(R^d)$, that is, $\|T_t f\| \leqslant \|f\|$ for $t \geqslant 0$. (Additional material on semigroups may be found in Chapter 3, Section 3.4 and in exercise 3.3.)

A Markov process $\zeta(t)$ on $(Y, \mathcal{M}, \mathcal{M}_t, P)$ with transition probability function $p(t, x, \mathrm{d}y)$ is called a *strong Markov process* if $P(\zeta(\tau + t) \in B | \mathcal{F}_\tau) = p(t, \zeta(\tau), B)$ on $\{\tau < \infty\}$ for all $t \geqslant 0$, all $B \in \mathcal{B}_d$ and all stopping times τ. The Markov process is called a *Feller process* if its semigroup T_t leaves invariant $C_b(R^d)$, the space of bounded continuous functions on R^d. In other words, $\zeta(t)$ is a Feller process if $E_x f(\zeta(t))$ is continuous in $x \in R^d$ for each $t \geqslant 0$ and each $f \in C_b(R^d)$. Using the Markov property, it is not difficult to show that the Feller property is equivalent to the apparently stronger requirement that $x \to E_x \Phi(\zeta(\cdot))$ be continuous for all bounded continuous $\Phi: \mathcal{B}([0, \infty), R^d) \to R$. Thus, in terms of weak continuity of measures (see Chapter 7, Section 3), the Feller property may be defined as the weak continuity of P_x as a function of x. In fact, the Feller property can be shown to be equivalent to a yet stronger condition as follows.

Theorem 3.1. *If $\zeta(t)$ is a Feller process, then $x \to E_x\Phi(\zeta(\cdot))$ is continuous for each bounded $\Phi: \mathcal{B}([0, \infty), R^d) \to R$ which is ν – a.s. continuous, where ν is the distribution of $\zeta(t)$ on $\mathcal{B}([0, \infty), R^d)$.*

1.4 Brownian motion

Consider the transition probability function on R^d defined by $p(t, x, dy) = p(t, x, y)\,dy$ with

$$p(t, x, y) = (2\pi t)^{-d/2} \exp\left(\frac{-|x - y|^2}{2t}\right).$$

Then for $t > 0$ and $x \in R^d$, $p(t, x, y)$ is the d-dimensional joint Gaussian density with mean zero and covariance tI. The Markov process with this transition probability function and initial distribution μ is called a d-dimensional *Brownian motion* with initial distribution μ. We will denote this process on $(Y, \mathcal{M}, \mathcal{M}_t, P)$ by $B(t) = B(t, q)$. (We suppress the dependence on μ.) An equivalent definition of Brownian motion with initial distribution μ is this:

$B(t)$ is a Markov process on $(Y, \mathcal{M}, \mathcal{M}_t, P)$ satisfying the following three conditions:

$$\left.\begin{array}{ll} \text{(i)} & B(0) \text{ has distribution } \mu. \\ \text{(ii)} & B(t) - B(s) \text{ and } B(s) \text{ are independent for } 0 \le s < t < \infty. \\ \text{(iii)} & B(t) - B(s) \text{ has the joint Gaussian distribution with mean} \\ & \text{zero and covariance } (t - s)I. \end{array}\right\} \quad (4.1)$$

As previously noted, any R^d-valued stochastic process induces a measure on $\mathcal{B}([0, \infty), R^d)$ which is called the distribution of the process. We want to show that the distribution of Brownian motion is in fact supported on $C([0, \infty), R^d)$, the space of continuous functions from $[0, \infty)$ to R^d. One must be a bit careful because $C([0, \infty), R^d)$ is not a Borel-measurable subset of $\mathcal{B}([0, \infty), R^d)$. There are many ways to proceed, one of which is as follows. Denote elements of $\mathcal{B}([0, \infty), R^d)$ by ω. Let $\{t_k\}_{k=1}^{\infty}$ be a dense, countable set in $[0, \infty)$ and consider the Borel measurable set

$$\mathcal{U} = \{\omega \in \mathcal{B}([0, \infty), R^d) : \omega \text{ is uniformly continuous on } \{t_k\}_{k=1}^{\infty}\}.$$

One can show that the distribution ν of the Brownian motion satisfies $\nu(\mathcal{U}) = 1$. From this it is easy to show that there exists a continuous version of the Brownian motion. See [Breiman (1968) or (1992)] for a particularly simple derivation or, for example, [Karatzas and Shreve (1988)], where the proof of continuity actually reveals that almost every path is Hölder continuous with exponent γ for any $\gamma < \frac{1}{2}$. We shall always assume that our Brownian motions are continuous. The probability measure on $C([0, \infty), R^d)$ corresponding to Brownian motion in the case where the initial distribution μ is the atom at $x = 0$ is called the (d-dimensional) *Wiener measure*.

Brownian motion is a Feller process and a strong Markov process (exercises 1.3 and 1.4).

We note the following simple lemma, which will be required in Section 7.

Lemma 4.1. *Let $B(t)$ be a Brownian motion on $(Y, \mathcal{M}, \mathcal{M}_t, P)$. Then for each $t \geq 0$, the σ-*algebras $\mathcal{M}_t$ and $\sigma(B(s) - B(t); s \geq t)$ *are independent.*

Proof. By appealing to a standard monotone class theorem, it is enough to show that $P(D|\mathcal{M}_t) = P(D)$, for D of the form $D = \{(B(s_j) - B(t)) \in A_j,\ j = 1, 2, \ldots, n)\}$, where $t < s_1 < s_2 < \ldots < s_n$ and $A_j \in \mathcal{B}_d$, $j = 1, 2, \ldots, n$. Using the Markov property and the fact that $B(t)$ is independent of $B(s) - B(t)$, for $s \geq t$, we have $P(D|\mathcal{M}_t) = P(D|B(t)) = P(D)$. □

1.5 Itô processes

Let $B(t)$ be a d-dimensional Brownian motion on $(Y, \mathcal{M}, \mathcal{M}_t, P)$. For $\theta \in R^d$, define

$$\mathrm{e}_\theta(x) = \exp(i\langle\theta, x\rangle),\ x \in R^d.$$

By property (ii) of (4.1), properties (6) and (7) of the conditional expectation, and the Markov property, we have for $0 \leq t_1 < t_2$,

$$\begin{aligned} E(\mathrm{e}_\theta(B(t_2))|\mathcal{M}_{t_1}) &= E(\exp(\mathrm{i}\langle\theta, B(t_1) + (B(t_2) - B(t_1))\rangle)|\mathcal{M}_{t_1}) \\ &= \mathrm{e}_\theta(B(t_1))E(\mathrm{e}_\theta(B(t_2) - B(t_1))|\mathcal{M}_{t_1}) \\ &= \mathrm{e}_\theta(B(t_1))E(\mathrm{e}_\theta(B(t_2) - B(t_1))|B(t_1)) \\ &= \mathrm{e}_\theta(B(t_1))E\mathrm{e}_\theta(B(t_2) - B(t_1)). \end{aligned}$$

By property (iii) of (4.1),

$$E\mathrm{e}_\theta(B(t_2) - B(t_1)) = \exp\left(\frac{-|\theta|^2}{2}(t_2 - t_1)\right).$$

We conclude that

$$E(e_\theta(B(t_2))|\mathcal{M}_{t_1}) = e_\theta(B(t_1))\exp\left(\frac{-|\theta|^2}{2}(t_2 - t_1)\right),$$
$$0 \leqslant t_1 < t_2,\ \theta \in R^d. \quad (5.1)$$

Note that (5.1) uniquely determines the distribution of the process $B(t)$ up to its initial distribution (exercise 1.5). Another way of stating (5.1) is that $X_{i\theta}(t)$ is a martingale relative to $(Y, \mathcal{M}, \mathcal{M}_t, P)$ for all $\theta \in R^d$, where $X_{i\theta}(t) = \exp(i\langle \theta, B(t) - B(0)\rangle + \frac{1}{2}|\theta|^2 t)$.

Now, from (5.1), we have

$$\frac{d}{dt}E(e_\theta(B(t))|\mathcal{M}_{t_1}) = \frac{-|\theta|^2}{2} e_\theta(B(t_1))\exp\left(\frac{-|\theta|^2}{2}(t - t_1)\right),$$
$$\text{for } 0 \leqslant t_1 < t.$$

Thus, for $t_2 > t_1$,

$$\begin{aligned} E(e_\theta(B(t_2)) &- e_\theta(B(t_1))|\mathcal{M}_{t_1}) \\ &= \frac{-|\theta|^2}{2}\int_{t_1}^{t_2} e_\theta(B(t_1))\exp\left(\frac{-|\theta|^2}{2}(t - t_1)\right) dt \\ &= \frac{-|\theta|^2}{2}\int_{t_1}^{t_2} E(e_\theta(B(t))|\mathcal{M}_{t_1})\, dt \\ &= E\left(\int_{t_1}^{t_2}\tfrac{1}{2}\Delta\, e_\theta(B(t))\, dt|\mathcal{M}_{t_1}\right). \end{aligned}$$

That is,

$$e_\theta(B(t)) - \int_0^t \tfrac{1}{2}\Delta\, e_\theta(B(s))\, ds \text{ is a martingale with respect to}$$
$$(Y, \mathcal{M}, \mathcal{M}_t, P), \quad \text{for all } \theta \in R^d. \quad (5.2)$$

Since any $f \in C_0^\infty(R^d)$ can be represented by

$$f(x) = \int_{-\infty}^{\infty} \exp(i\langle \theta, x\rangle)\phi(\theta)\, d\theta,$$

for some rapidly decreasing ϕ, it follows easily fom (5.2) and exercise 1.1, that

$$f(B(t)) - \int_0^t \tfrac{1}{2}\Delta f(B(s))\, ds \text{ is a martingale with respect to}$$
$$(Y, \mathcal{M}, \mathcal{M}_t, P), \quad \text{for all } f \in C_b^2(R^d), \quad (5.3)$$

where $C_b^2(R^d)$ denotes the space of functions on R^d, all of whose partial derivatives up to order 2 are continuous and bounded. Conversely, starting with (5.3), one can obtain (5.1).

Now let

$$\zeta(t) = \sigma B(t) + bt, \tag{5.4}$$

where σ is a constant $d \times d$ matrix and b is a constant d-vector. Letting $a = \sigma\sigma^T$, a calculation similar to the one just carried out reveals that

$$X_{i\theta}(t) \text{ is a martingale with respect to } (Y, \mathcal{M}, \mathcal{M}_t, P),$$
$$\text{where } X_{i\theta}(t) = \exp\left(i\langle \theta, \zeta(t) - \zeta(0) - bt\rangle + \tfrac{1}{2}\langle \theta, a\theta\rangle\right), \tag{5.5}$$

and that

$$f(\zeta(t)) - \int_0^t Lf(\zeta(s))\,\mathrm{d}s \text{ is a martingale with respect to } (Y, \mathcal{M}, \mathcal{M}_t, P),$$
$$\text{for all } f \in C_b^2(R^d), \tag{5.6}$$

where

$$L = \tfrac{1}{2}\sum_{i,j=1}^{d} a_{ij}\frac{\partial^2}{\partial x_i \partial x_j} + \sum_{i=1}^{d} b_i \frac{\partial}{\partial x_i}.$$

Furthermore, (5.5) and (5.6) are equivalent.

We now want to construct a Markov process $\zeta(t)$ satisfying (5.6) in the case where

$$L = \tfrac{1}{2}\sum_{i,j=1}^{d} a_{ij}(x)\frac{\partial^2}{\partial x_i \partial x_j} + \sum_{i=1}^{d} b_i(x) \frac{\partial}{\partial x_i}$$

is a variable coefficient elliptic operator. In the light of (5.4), it is tempting to try to arrive at $\zeta(t)$ by solving the 'differential equation'

$$\mathrm{d}\zeta(t) = \sigma(\zeta(t))\,\mathrm{d}B(t) + b(\zeta(t))\,\mathrm{d}t, \tag{5.7}$$

where $\sigma\sigma^{\mathrm{T}}(x) = a(x)$. The problem with this is that $B(t)$ is not of bounded variation; we must develop a suitable integration theory with respect to $\mathrm{d}B(t)$.

It will actually be expedient to develop an integration theory with respect to processes more general than Brownian motion. To this end, let a: $[0, \infty) \times Y \to S_d$ and b: $[0, \infty) \times Y \to R^d$ be progressively measurable on $(Y, \mathcal{M}, \mathcal{M}_t, P)$, where S_d denotes the space of real symmetric non-negative definite $d \times d$ matrices. Let

$$L_t = \tfrac{1}{2}\sum_{i,j=1}^{\mathrm{d}} a_{ij}(t, q)\frac{\partial^2}{\partial x_i \partial x_j} + \sum_{i=1}^{\mathrm{d}} b_i(t, q) \frac{\partial}{\partial x_i}. \tag{5.8}$$

The following theorem generalizes the equivalence of (5.5) and (5.6).

Theorem 5.1. *Let L_t satisfy (5.8) and assume that a_{ij} and b_i are bounded. Let $\zeta(t) = \zeta(t, q)$ be an almost surely continuous process on $(Y, \mathcal{M}, \mathcal{M}_t, P)$.*

(a) *The following four conditions are equivalent:*

(i) *$X_\theta(t)$ is a martingale with respect to $(Y, \mathcal{M}, \mathcal{M}_t, P)$ for all $\theta \in R^d$, where*

$$X_\theta(t) = \exp\left(\left\langle \theta, \zeta(t) - \zeta(0) - \int_0^t b(s)\,ds \right\rangle - \tfrac{1}{2}\int_0^t \langle \theta, a(s)\theta \rangle\,ds\right);$$

(ii) *$X_{i\theta}(t)$ is a martingale with respect to $(Y, \mathcal{M}, \mathcal{M}_t, P)$ for all $\theta \in R^d$, where*

$$X_{i\theta}(t) = \exp\left(i\left\langle \theta, \zeta(t) - \zeta(0) - \int_0^t b(s)\,ds \right\rangle + \tfrac{1}{2}\int_0^t \langle \theta, a(s)\theta \rangle\,ds\right);$$

(iii) *$f(\zeta(t)) - \int_0^t (L_s f)(\zeta(s))\,ds$ is a martingale with respect to $(Y, \mathcal{M}, \mathcal{M}_t, P)$ for all $f \in C_b^2(R^d)$;*

(iv)
$$f(t, \zeta(t)) - \int_0^t \left(\frac{\partial}{\partial s} + L_s\right) f(s, \zeta(s))\,ds$$
is a martingale with respect to $(Y, \mathcal{M}, \mathcal{M}_t, P)$ for all $f \in C_b^{1,2}([0, \infty) \times R^d)$.

(b) *If $E\exp(\lambda|\zeta(0)|) < \infty$, for all $\lambda \in R$, then $E\exp(\lambda|\zeta(t)|) < \infty$, for all $\lambda \in R$ and all $t \geq 0$. In this case, $f(t, \zeta(t)) - \int_0^t ((\partial/\partial s) + L_s) f(s, \zeta(s))\,ds$ is a martingale for all $f \in C^{1,2}([0, \infty) \times R^d)$ for which there exist constants c_t and k_t for each $t > 0$ such that*

$$\left|\frac{\partial^\alpha f}{\partial x^\alpha}(s, x)\right| \leq c_t \exp(k_t|x|) \text{ and } \left|\frac{\partial f}{\partial s}(s, x)\right| \leq c_t \exp(k_t|x|),$$

for all $x \in R^d$, $0 \leq s \leq t$, $t > 0$ and $|\alpha| \leq 2$.

In particular,

$$\left\langle \theta, \zeta(t) - \int_0^t b(s)\,ds \right\rangle \text{ is a martingale with respect to } (Y, \mathcal{M}, \mathcal{M}_t, P), \tag{5.9}$$

and

$$\left\langle \theta, \zeta(t) - \int_0^t b(s)\,ds \right\rangle^2 - \int_0^t \langle \theta, a(s)\theta \rangle\,ds$$
is a martingale with respect to $(Y, \mathcal{M}, \mathcal{M}_t, P)$. (5.10)

To prove the theorem, we need two results from Chapter 2, so we postpone the proof until Section 5 of Chapter 2.

If a_{ij} and b_i are bounded and $\zeta(t)$ is an almost surely continuous process satisfying the equivalent conditions of Theorem 5.1, we will call $\zeta(t)$ an *Itô process* with *covariance* a and *drift* b. We will write $\zeta(t) \sim I_d(a, b)$.

Note that if $\zeta(t) \sim I_d(I, 0)$, then $\zeta(t)$ is necessarily a d-dimensional Brownian motion with some arbitrary initial distribution.

Return now for a moment to the paragraph following (5.6). There we broached the possibility of constructing a Markov process $\zeta(t)$ which solves (5.6) in the case where L has variable coefficients via a 'differential equation'. Assuming that a and b are bounded, we can now reformulate this in terms of Itô processes. If $\zeta(t)$ solves (5.6), then $\zeta(t)$ is an Itô process with covariance $a(\zeta(t, q))$ and drift $b(\zeta(t, q))$. Thus we are looking for a Markov process $\zeta(t)$ which is an $I_d(a(\zeta(\cdot)), b(\zeta(\cdot)))$-process.

1.6 Stochastic integrals

The integration theory that we present is in respect of Itô processes. Note that from Theorem 5.1(a)–(i), it follows that if $\zeta(t) \sim I_d(a, b)$, then $\tilde{\zeta}(t) \equiv \zeta(t) - \int_0^t b(s)\,ds \sim I_d(a, 0)$. Thus, it suffices to develop an integration theory for $I_d(a, 0)$ processes. Indeed, if $\zeta(t) \sim I_d(a, b)$, then writing $\zeta(t) = \tilde{\zeta}(t) + \int_0^t b(s)\,ds$, where $\tilde{\zeta} \sim I_d(a, 0)$, we can define $d\zeta(t) = d\tilde{\zeta}(t) + b(t)\,dt$, where dt is the Lebesgue measure.

Let $\zeta(t) \sim I_d(a, 0)$ and let $\mathscr{H}_a^d$ denote the space of functions g: $[0, \infty) \times Y \to R^d$ which are progressively measurable with respect to $\mathscr{M}_t$ and which satisfy $E\int_0^t \langle g(s), a(s)g(s)\rangle\,ds < \infty$, for all $t > 0$. Call a function $g \in \mathscr{H}_a^d$ simple if it is bounded and for some positive integer n,

$$g(t) = g\left(\frac{[nt]}{n}\right) \text{ for all } t > 0.$$

For simple functions g, define

$$\int_0^t \langle g(s), d\zeta(s)\rangle = \sum_{k=0}^{\infty} \left\langle g\left(\frac{k}{n}\right), \zeta\left(\frac{k+1}{n} \wedge t\right) - \zeta\left(\frac{k}{n} \wedge t\right)\right\rangle.$$

Obviously, if g_1 and g_2 are simple and λ_1 and λ_2 are constants, then $\lambda_1 g_1 + \lambda_2 g_2$ is also simple and

$$\int_0^t \langle \lambda_1 g_1(s) + \lambda_2 g_2(s), d\zeta(s)\rangle = \lambda_1 \int_0^t \langle g_1(s), d\zeta(s)\rangle + \lambda_2 \int_0^t \langle g_2(s), d\zeta(s)\rangle.$$

Clearly, $\int_0^t \langle g(s), d\zeta(s)\rangle$ is progressively measurable. We will now prove that

(i) $$\exp\left(\theta\int_0^t\langle g(s), \mathrm{d}\zeta(s)\rangle - \frac{\theta^2}{2}\int_0^t\langle g(s), a(s)g(s)\rangle\,\mathrm{d}s\right)$$

is a martingale with respect to $(Y, \mathcal{M}, \mathcal{M}_t, P)$. (6.1)

From the definition of an Itô process, this is equivalent to

$$\int_0^t\langle g(s), \mathrm{d}\zeta(s)\rangle \sim I_1(\tilde{a}, 0), \text{ where } \tilde{a}(t) = \langle g(t), a(t)g(t)\rangle.$$

From Theorem 5.1 it will then follow that

(ii) $\int_0^t\langle g(s), \mathrm{d}\zeta(s)\rangle$ is a martingale with respect to $(Y, \mathcal{M}, \mathcal{M}_t, P)$

(6.1)

and

(iii) $$\left(\int_0^t\langle g(s), \mathrm{d}\zeta(s)\rangle\right)^2 - \int_0^t\langle g(s), a(s)g(s)\rangle\,\mathrm{d}s$$

is a martingale with respect to $(Y, \mathcal{M}, \mathcal{M}_t, P)$. (6.1)

To prove (6.1 (i)), assume that there exists a k_0 such that

$$\frac{k_0}{n} \leqslant t_1 < t_2 \leqslant \frac{k_0+1}{n}.$$

We will show that

$$E\left(\exp\left(\theta\int_0^{t_2}\langle g(s), \mathrm{d}\zeta(s)\rangle - \frac{\theta^2}{2}\int_0^{t_2}\langle g(s), a(s)g(s)\rangle\,\mathrm{d}s\right)\Big|\mathcal{M}_{t_1}\right)$$
$$= \exp\left(\theta\int_0^{t_1}\langle g(s), \mathrm{d}\zeta(s)\rangle - \frac{\theta^2}{2}\int_0^{t_1}\langle g(s), a(s)g(s)\rangle\,\mathrm{d}s\right) \quad \text{for all } \theta \in R. \tag{6.2}$$

This is enough since, by successive conditioning, one then obtains (6.2) for all $0 \leqslant t_1 < t_2 < \infty$. We have

$$E\left(\exp\left(\theta\int_0^{t_2}\langle g(s), \mathrm{d}\zeta(s)\rangle - \frac{\theta^2}{2}\int_0^{t_2}\langle g(s), a(s)g(s)\rangle\,\mathrm{d}s\right)\Big|\mathcal{M}_{t_1}\right)$$
$$= E\left(\exp\left(\theta\int_0^{t_1}\langle g(s), \mathrm{d}\zeta(s)\rangle - \frac{\theta^2}{2}\int_0^{t_1}\langle g(s), a(s)g(s)\rangle\,\mathrm{d}s\right)\right)$$
$$\times E\left(\exp\left(\theta g\left(\frac{k_0}{n}\right)(\zeta(t_2)-\zeta(t_1)) - \frac{\theta^2}{2}\int_{t_1}^{t_2}\left\langle g\left(\frac{k_0}{n}\right),\right.\right.\right.$$
$$\left.\left.\left. a(s)g\left(\frac{k_0}{n}\right)\right\rangle\,\mathrm{d}s\right)\Big|\mathcal{M}_{t_1}\right).$$

To complete the proof, we must show that

$$E\left(\exp\left(\theta g\left(\frac{k_0}{n}\right)(\zeta(t_2)-\zeta(t_1))-\frac{\theta^2}{2}\int_{t_1}^{t_2}\left\langle g\left(\frac{k_0}{n}\right), a(s)g\left(\frac{k_0}{n}\right)\right\rangle \mathrm{d}s\right)\Big|\mathscr{M}_{t_1}\right) = 1. \quad (6.3)$$

Note that if

$$g\left(\frac{k_0}{n}\right)$$

is constant, then (6.3) follows from exercise 1.6. In the general case, although

$$g\left(\frac{k_0}{n}\right)$$

is not constant, it is $\mathscr{M}_{t_1}$-measurable. Thus, (6.3) follows from exercises 1.6 and 1.7.

By a standard mollification procedure, any $g \in \mathscr{H}_a^d$ may be approximated by a sequence of simple functions $\{g_n\}_{n=1}^{\infty} \in \mathscr{H}_a^d$ such that

$$\lim_{n\to\infty} E\int_0^t \langle (g-g_n)(s), a(s)(g-g_n)(s)\rangle \,\mathrm{d}s = 0, \quad \text{for all } t>0. \quad (6.4)$$

Furthermore, if $\sup_{t,q}|g(t,q)| \leqslant c$, then the sequence $\{g_n\}_{n=1}^{\infty}$ may be chosen so that

$$\sup_{t,q}|g_n(t,q)| \leqslant c, \quad \text{for all } n \geqslant 1. \quad (6.5)$$

From (6.1 (iii)), it follows in particular that the martingale

$$\int_0^t \langle g_n(s), \mathrm{d}\zeta(s)\rangle - \int_0^t \langle g_m(s), \mathrm{d}\zeta(s)\rangle$$

satisfies

$$E\left(\int_0^t \langle g_n(s), \mathrm{d}\zeta(s)\rangle - \int_0^t \langle g_m(s), \mathrm{d}\zeta(s)\rangle\right)^2$$
$$= E\int_0^t \langle (g_n-g_m)(s), a(s)(g_n-g_m)(s)\rangle \,\mathrm{d}s.$$

Using this and (6.4) in conjunction with Theorem 2.3, we obtain

$$\lim_{m,n\to\infty} E\sup_{0\leqslant u\leqslant t}\left|\int_0^u \langle g_n(s), \mathrm{d}\zeta(s)\rangle - \int_0^u \langle g_m(s), \mathrm{d}\zeta(s)\rangle\right|^2$$
$$\leqslant \lim_{m,n\to\infty} 4E\int_0^t \langle (g_n-g_m)(s), a(s)(g_n-g_m)(s)\rangle \,\mathrm{d}s = 0.$$

Thus, there exists a subsequence $\int_0^t \langle g_{n'}(s), \mathrm{d}\zeta(s)\rangle$, which converges uniformly on all finite t-intervals for all q outside a set N of P-measure zero. We can define

$$\int_0^t \langle g(s), \mathrm{d}\zeta(s)\rangle = \lim_{n'\to\infty} \int_0^t \langle g_{n'}(s), \mathrm{d}\zeta(s)\rangle$$

on the complement of N and, say, $\int_0^t \langle g(s), \mathrm{d}\zeta(s)\rangle \equiv 0$ on N. Then $\int_0^t \langle g(s), \mathrm{d}\zeta(s)\rangle$ is continuous and, assuming that the filtration $\mathcal{M}_t$ is complete with respect to P, it is also progressively measurable. Finally, it is easy to show that up to sets of measure zero, the definition of $\int_0^t \langle g(s), \mathrm{d}\zeta(s)\rangle$ is independent of the particular approximating sequence $\{g_n\}_{n=1}^{\infty}$. (Given two such sequences, $\{g_n\}_{n=1}^{\infty}$ and $\{h_n\}_{n=1}^{\infty}$, consider the sequence $g_1, h_1, g_2, h_2, \ldots$.) By working a little harder, in fact, it is possible to pick a version of $\int_0^t \langle g(s), \mathrm{d}\zeta(s)\rangle$, which is progressively measurable with respect to $\mathcal{M}_t$, even if $\mathcal{M}_t$ is not complete with respect to P [Stroock and Varadhan (1979), p. 97].

We have shown above that $\int_0^t \langle g_n(s), \mathrm{d}\zeta(s)\rangle$ converges to $\int_0^t \langle g(s), \mathrm{d}\zeta(s)\rangle$ in $L^2(Y, \mathrm{d}P)$; thus, by exercise 1.1, it follows that (6.1(ii)) and (iii) still hold for $\int_0^t \langle g(s), \mathrm{d}\zeta(s)\rangle$. Assuming that

$$\sup_{\substack{q\in Y\\ t>0}} |g(t,q)| \leq c,$$

we now show that (6.1(i)) holds. By (6.5) we may assume that

$$\sup_{\substack{q\in Y\\ t>0}} |g_n(t,q)| \leq c.$$

By Exercise 1.1, it is enough to show that, for each $t > 0$,

$$\exp\left(\theta \int_0^t \langle g_n(s), \mathrm{d}\zeta(s)\rangle - \frac{\theta^2}{2}\int_0^t \langle g_n(s), a(s) g_n(s)\rangle\, \mathrm{d}s\right)$$

converges in $L^1(Y, \mathrm{d}P)$ to

$$\exp\left(\theta \int_0^t \langle g(s), \mathrm{d}\zeta(s)\rangle - \frac{\theta^2}{2}\int_0^t \langle g(s), a(s), a(s) g(s)\rangle\, \mathrm{d}s\right).$$

To do this, we will show that

$$\exp\left(\theta \int_0^t \langle g_n(s), \mathrm{d}\zeta(s)\rangle - \frac{\theta^2}{2}\int_0^t \langle g_n(s), a(s) g_n(s)\rangle\, \mathrm{d}s\right)$$

is uniformly integrable. In particular, we will show that

$$E \exp\left[2\left(\theta \int_0^t \langle g_n(s), \mathrm{d}\zeta(s)\rangle - \frac{\theta^2}{2}\int_0^t \langle g_n(s), a(s) g_n(s)\rangle\, \mathrm{d}s\right)\right] \leq M < \infty,$$

for M independent of n. We have

$$E\exp\left[2\left(\theta\int_0^t\langle g_n(s), d\zeta(s)\rangle - \frac{\theta^2}{2}\int_0^t\langle g_n(s), a(s)g_n(s)\rangle\, ds\right)\right]$$

$$= E\exp\left((2\theta)\int_0^t\langle g_n(s), d\zeta(s)\rangle - \frac{(2\theta)^2}{2}\int_0^t\langle g_n(s), a(s)g_n(s)\rangle\, ds\right.$$

$$\left. + \theta^2\int_0^t\langle g_n(s), a(s)g_n(s)\rangle\, ds\right)$$

$$\leqslant \exp\left(\sup_{q,n}\theta^2\int_0^t\langle g_n(s), a(s)g_n(s)\rangle\, ds\right)$$

$$\times E\exp\left((2\theta)\int_0^t\langle g_n(s), d\zeta(s)\rangle - \frac{(2\theta)^2}{2}\int_0^t\langle g_n(s), a(s)g_n(s)\rangle\, ds\right)$$

$$= \exp\left(\sup_{q,n}\theta^2\int_0^t\langle g_n(s), a(s)g_n(s)\rangle\, ds\right) \leqslant \exp\left(c^2\theta^2\sup_q\int_0^t\|a(s)\|\, ds\right),$$

where the last equality follows from (6.1 (i)) with g replaced by g_n and θ replaced by 2θ.

Since (6.1(i)) holds for bounded g, the following proposition holds from the definition of an Itô process.

Proposition 6.1. *Let $g \in \mathcal{H}_a^d$ be bounded and let $\zeta(t) \sim I_d(a, 0)$. Then $\int_0^t\langle g(s), d\zeta(s)\rangle \sim I_1(\tilde{a}, 0)$, where $\tilde{a}(t) = \langle g(t), a(t)g(t)\rangle$.*

For $g \in \mathcal{H}_a^d$, we call $\int_0^t\langle g(s), d\zeta(s)\rangle$ a *stochastic integral*. The stochastic integral can easily be extended to matrix-valued g. Let g: $[0, \infty) \times Y \to R^d \otimes R^d$ be progressively measurable and $d \times d$ matrix-valued. Assume that each row of g is in $\mathcal{H}_a^d$. This is equivalent to the assertion that $E\int_0^t \mathrm{Tr}\,(g(s)a(s)g^{\mathrm{T}}(s))\, ds < \infty$, for all $t > 0$. We can then define the R^d-valued process $\zeta(t) = \int_0^t g(s)\, d\zeta(s)$ in the obvious way. In the case where g is bounded, Proposition 6.1 generalizes as follows.

Theorem 6.2. *Let g: $[0, \infty) \times Y \to R^d \otimes R^d$ be progressively measurable, bounded and $d \times d$ matrix-valued and let $\zeta(t) \sim I_d(a, 0)$. Then $\int_0^t g(s)\, d\zeta(s) \sim I_d(gag^{\mathrm{T}}, 0)$. In particular, $E\int_0^t g(s)\, d\zeta(s) = 0$ and $E|\int_0^t g(s)\, d\zeta(s)|^2 = E\int_0^t \mathrm{Tr}\,(gag^{\mathrm{T}})(s)\, ds$.*

We end this section with *Itô's formula*. Assume that $\zeta(t) \sim I_d(a, b)$ and that $E\exp(\lambda|\zeta(0)| < \infty$, for all $\lambda > 0$, and let L_t be as in (5.8). Then, by Theorem 5.1,

$$f(t, \zeta(t)) - \int_0^t \left(\frac{\partial}{\partial s} + L_s\right) f(s, \zeta(s))\, \mathrm{d}s$$

is a martingale for all $f \in C^{1,2}([0, \infty) \times R^d)$ which satisfy a certain exponential growth condition. Itô's formula identifies this martingale as $\int_0^t \langle (\nabla f)(s, \zeta(s)), \mathrm{d}\bar{\zeta}(s)\rangle$, where $\bar{\zeta}(t) = \zeta(t) - \int_0^t b(s)\, \mathrm{d}s \sim I_d(a, 0)$.

Theorem 6.3 (Itô's formula). *Let $\zeta(t) \sim I_d(a, b)$ on $(Y, \mathcal{M}, \mathcal{M}_t, P)$, where $a(t, q)$ and $b(t, q)$ are bounded, and assume that $E \exp(\lambda|\zeta(0)|) < \infty$, for all $\lambda > 0$. If $f \in C^{1,2}([0, \infty) \times R^d)$ and for each $t > 0$ there exist constants c_t and k_t such that*

$$\left|\frac{\partial^\alpha f}{\partial x^\alpha}(s, x)\right| \leq c_t \exp(k_t|x|) \text{ and } \left|\frac{\partial f}{\partial s}(s, x)\right| \leq c_t \exp(k_t|x|),$$

$$\text{for all } x \in R^d, 0 \leq s \leq t, t > 0, \text{ and } |\alpha| \leq 2,$$

then

$$f(t, \zeta(t)) = f(0, \zeta(0)) + \int_0^t \langle (\nabla f)(s, \zeta(s)), \mathrm{d}\bar{\zeta}(s)\rangle + \int_0^t \left(\frac{\partial f}{\partial s} + L_s f\right)(s, \zeta(s))\, \mathrm{d}s,$$

where

$$\bar{\zeta}(t) = \zeta(t) - \int_0^t b(s)\, \mathrm{d}s \text{ and } L_t = \tfrac{1}{2} \sum_{i,j=1}^d a_{ij}(t, q) \frac{\partial^2}{\partial x_i \partial x_j} + \sum_{i=1}^d b_i(t, q) \frac{\partial}{\partial x_i}.$$

Remark. Itô's formula actually holds for any $f \in C^{1,2}([0, \infty) \times R^d)$ without any growth restriction. However, without the growth restriction in Theorem 6.3, the expectations of the quantities involved may not exist and, furthermore, we must extend our definition of the stochastic integral to more general integrands. Similarly, the exponential moment condition on $\zeta(0)$ is only needed in order to ensure finite expectations.

For a proof of Itô's formula, see, for example [Gihman and Skorohod (1972)] or [Friedman (1975)]. We will use Itô's formula only very sparingly in this book. For the most part, in our applications, it will be enough to know that $E_x f(\zeta(t \wedge \tau)) = E_x \int_0^{t\wedge\tau} L_s f(\zeta(s))\, \mathrm{d}s$, for all $t > 0$ and all stopping times τ. This equality follows from Doob's optional sampling theorem (Theorem 2.1) and the fact that $f(\zeta(t)) - \int_0^t L_s f(\zeta(s))\, \mathrm{d}s$ is a martingale.

1.7 Stochastic differential equations

We can now define rigorously the stochastic differential equation

$$\text{(i)} \qquad d\zeta(t) = \sigma(\zeta(t))\,dB(t) + b(\zeta(t))\,dt, \qquad (7.1)$$

which was suggested in (5.7). Assume that $B(t)$ is a d-dimensional Brownian motion on $(Y, \mathcal{M}, \mathcal{M}_t, P)$ or, in the language of Itô processes, assume that $B(t) \sim I_d(I, 0)$ on $(Y, \mathcal{M}, \mathcal{M}_t, P)$. Add to (7.1(i)) the condition

$$\text{(ii)} \qquad \zeta(0) = x. \qquad (7.1)$$

By a solution to (7.1(i)) and (ii), we mean a progressively measurable $\zeta_x\colon [0, \infty) \times Y \to R^d$ which is almost surely continuous in t and which satisfies P-almost surely

$$\zeta_x(t) = x + \int_0^t \sigma(\zeta_x(s))\,dB(s) + \int_0^t b(\zeta_x(s))\,ds, \quad t \geq 0. \qquad (7.2)$$

Define the norm $\|\cdot\|$ on $d \times d$ matrices by $\|a\| = (\sum_{i,j=1}^d a_{ij}^2)^{1/2}$. We have the following existence and uniqueness theorem. (Until Section 10, we will focus on the case of bounded coefficients. However, we state and prove the following theorem for unbounded coefficients since this is how it is usually stated and since the proof for unbounded coefficients requires virtually no extra work.)

Theorem 7.1. *Assume that* $\sigma\colon R^d \to R^d \otimes R^d$ *and* $b\colon R^d \to R^d$ *satisfy the Lipschitz condition*

$$\left.\begin{aligned} \|\sigma(x) - \sigma(y)\| &\leq k|x - y|, \quad \text{for all } x, y \in R^d, \\ |b(x) - b(y)| &\leq k|x - y|, \quad \text{for all } x, y \in R^d. \end{aligned}\right\} \qquad (7.3)$$

Then for each $x \in R^d$ *there exists a unique (up to sets of* P-measure zero) *solution* $\zeta_x(t)$ *to* (7.2). *In fact,* $\zeta_x(t)$ *is measurable with respect to* $\sigma(B(s), 0 \leq s \leq t)$, *the* σ*-algebra generated by* $B(s), 0 \leq s \leq t$, *and can be obtained by the method of successive approximations. The distribution of* $\zeta_x(t)$ *on* $C([0, \infty), R^d)$ *is uniquely determined by* $\sigma(x)$ *and* $b(x)$. *That is, if* $\zeta_x(t)$ *satisfies* (7.2) *with respect to a different Brownian motion* $\hat{B}(t)$ *on a probability space* $(\hat{Y}, \hat{\mathcal{M}}, \hat{P})$ *with a filtration* $\hat{M}_t$, *then* $\hat{\zeta}_x(\cdot) \overset{\text{dist}}{=} \zeta_x(\cdot)$. *Finally, for each* $t > 0$, *there exists a constant* C_t *such that*

$$E|\zeta_x(t) - \zeta_{x'}(t)|^2 \leq C_t|x - x'|^2, \text{ for } x, x' \in R^d. \qquad (7.4)$$

Proof. We first prove existence. Note that the Lipschitz assumption on a and b guarantees that these coefficients grow no faster than linearly. That is, there exists a constant $c > 0$ such that $\|\sigma(x)\| \leq c(1 + |x|)$ and $|b(x)| \leq c(1 + |x|)$.

Fix $T > 0$. Define $\zeta_0(t) = x$, $0 \leq t \leq T$, and for $n \geq 1$, define

$$\zeta_n(t) = x + \int_0^t \sigma(\zeta_{n-1}(s))\, dB(s) + \int_0^t b(\zeta_{n-1}(s))\, ds, \quad 0 \leq t \leq T.$$

Set $\Delta_n(t) = E|\zeta_n(t) - \zeta_{n-1}(t)|^2$, for $n \geq 1$. We make the inductive assumption that $E\int_0^T |\zeta_n(t)|^2\, dt < \infty$ and that

$$\Delta_k(t) \leq \frac{L^k t^k}{k!}, \text{ for } 0 \leq t \leq T \text{ and } k = 1, 2, \ldots, n,$$

where L is an appropriate constant. Note that the growth condition on $\sigma(x)$ and $b(x)$ and the condition $E\int_0^T |\zeta_n(t)|^2\, dt < \infty$ guarantee that $E\int_0^T \mathrm{Tr}\,(\sigma(\zeta_n(s))\sigma^{\mathrm{T}}(\zeta_n(s)))\, ds < \infty$ and, thus, that $\int_0^T \sigma(\zeta_n(s))\, dB(s)$, $0 \leq t \leq T$, is well defined.

Clearly, $E\int_0^T |\zeta_0(t)|^2\, dt < \infty$. For $0 \leq t \leq T$, we have

$$\begin{aligned}\Delta_1(t) &\leq 2E\left|\int_0^t \sigma(x)\, dB(s)\right|^2 + 2E\left|\int_0^t b(x)\, ds\right|^2 \\ &\leq 4c^2 t(1 + |x|^2) + 4c^2 t^2(1 + |x|^2) \leq 4c^2(1 + T)(1 + |x|^2)t.\end{aligned}$$

For $n \geq 1$ and $0 \leq t \leq T$, we have

$$\begin{aligned}\Delta_{n+1}(t) &\leq 2E\left|\int_0^t (\sigma(\zeta_n(s)) - \sigma(\zeta_{n-1}(s)))\, dB(s)\right|^2 \\ &\quad + 2E\left|\int_0^t (b(\zeta_n(s)) - b(\zeta_{n-1}(s)))\, ds\right|^2 \\ &= 2\int_0^t E(\sigma(\zeta_n(s)) - \sigma(\zeta_{n-1}(s)))(\sigma(\zeta_n(s)) - \sigma(\zeta_{n-1}(s)))^{\mathrm{T}}\, ds \\ &\quad + 2E\left|\int_0^t (b(\zeta_n(s)) - b(\zeta_{n-1}(s)))\, ds\right|^2 \\ &\leq 2k^2\int_0^t \Delta_n(s)\, ds + 2tk^2\int_0^t \Delta_n(s)\, ds \\ &= 2k^2(1 + t)\int_0^t \Delta_n(s)\, ds \leq 2k^2(1 + T)\int_0^t \Delta_n(s)\, ds.\end{aligned}$$

Letting $L = 2(1 + T)\max(k^2, 2c^2(1 + |x|^2))$, we obtain by induction that for all $n \geq 0$, $E\int_0^T |\zeta_n(t)|^2\, dt < \infty$ and

$$\Delta_{n+1}(t) \leqslant \frac{L^{n+1}T^{n+1}}{(n+1)!}, \text{ for } 0 \leqslant t \leqslant T.$$

It follows then that $\lim_{m\to\infty} \sup_{n\geqslant m} E\int_0^T |\zeta_n(t) - \zeta_m(t)|^2\,dt = 0$. Thus there exists a $\hat{\zeta}_x: [0, \infty) \times Y \to R^d$ such that $\lim_{n\to\infty} E\int_0^T |\hat{\zeta}_x(t) - \zeta_n(t)|^2\,dt = 0$, for all $T > 0$. Since ζ_n converges to $\hat{\zeta}_x$ in $dt \times P$-measure on $[0, T] \times Y$ for each $T > 0$, there exists a subsequence $\{\zeta_{n_k}\}_{k=1}^{\infty}$ such that $\hat{\zeta}_x(t, q) = \lim_{k\to\infty} \zeta_{n_k}(t, q)$ a.s. $dt \times P$ on $[0, \infty) \times Y$. Thus, if we define

$$\tilde{\zeta}_x(t, q) = \begin{cases} \lim\limits_{k\to\infty} \zeta_{n_k}(t, q), \text{ if the limit exists} \\ 0, \text{ otherwise,} \end{cases}$$

then $\tilde{\zeta}_x = \hat{\zeta}_x$ a.s. $dt \times P$ on $[0, \infty) \times Y$ and, for each $t \geqslant 0$, $\tilde{\zeta}_x(t)$ is measurable with respect to $\sigma(B(s), 0 \leqslant s \leqslant t)$ since $\zeta_n(t)$ is clearly measurable with respect to $\sigma(B(s), 0 \leqslant s \leqslant t)$.

Define

$$\zeta_x(t) = x + \int_0^t \sigma(\tilde{\zeta}_x(s))\,dB(s) + \int_0^t b(\tilde{\zeta}_x(s))\,ds, \quad t \geqslant 0. \tag{7.5}$$

We will show that

$$\lim_{n\to\infty} E\int_0^t |\zeta_x(s) - \zeta_n(s)|^2\,ds = 0, \quad \text{for all } t > 0, \tag{7.6}$$

from which it will follow that $\zeta_x = \tilde{\zeta}_x$ a.s. $dt \times dP$ on $[0, \infty) \times Y$. This allows us to replace $\tilde{\zeta}_x$ by ζ_x in the righthand side of (7.5), which proves that ζ_x solves (7.2). To show (7.6), we have for $0 \leqslant s \leqslant t$,

$$\begin{aligned} E|\zeta_x(s) - \zeta_n(s)|^2 &\leqslant 2E\left|\int_0^s (\sigma(\tilde{\zeta}(u)) - \sigma(\zeta_{n-1}(u)))\,dB(u)\right|^2 \\ &\quad + 2E\left|\int_0^s (b(\tilde{\zeta}(u)) - b(\zeta_{n-1}(u)))\,du\right|^2 \\ &\leqslant 2k^2 E\int_0^s |\tilde{\zeta}(u) - \zeta_{n-1}(u)|^2\,du \\ &\quad + 2k^2 sE\int_0^s |\tilde{\zeta}(u) - \zeta_{n-1}(u)|^2\,du \\ &\leqslant 2k^2(1+t)E\int_0^t |\tilde{\zeta}(u) - \zeta_{n-1}(u)|^2\,du. \end{aligned}$$

Since $\lim_{n\to\infty} E\int_0^t |\tilde{\zeta}(u) - \zeta_{n-1}(u)|^2\,du = 0$, (7.6) follows.

Now assume that $\zeta_x(t)$ and $\zeta'_x(t)$ are solutions to (7.2) for x and x' respectively. Then

$$E \sup_{0\le s\le t} |\zeta_x(s) - \zeta_{x'}(s)|^2$$

$$\le 3|x - x'|^2 + 3E \sup_{0\le s\le t}\left|\int_0^s (\sigma(\zeta_x(u)) - \sigma(\zeta_{x'}(u)))\,\mathrm{d}B(u)\right|^2$$

$$+ 3E \sup_{0\le s\le t}\left|\int_0^s (b(\zeta_x(u)) - b(\zeta_{x'}(u)))\,\mathrm{d}u\right|^2$$

$$\le 3|x - x'|^2 + 12E\left|\int_0^t (\sigma(\zeta_x(u)) - \sigma(\zeta_{x'}(u)))\,\mathrm{d}B(u)\right|^2$$

$$+ 3tE\int_0^t |b(\zeta_x(u)) - b(\zeta_{x'}(u))|^2\,\mathrm{d}u$$

$$\le 3|x - x'|^2 + 12k^2 E\int_0^t |\zeta_x(u) - \zeta_{x'}(u)|^2\,\mathrm{d}u$$

$$+ 3tk^2 E\int_0^t |\zeta_x(u) - \zeta_{x'}(u)|^2\,\mathrm{d}u,$$

where we have used Theorem 2.3 in the second inequality. Letting

$$H(t) = E \sup_{0\le s\le t} |\zeta_x(s) - \zeta_{x'}(s)|^2,$$

we have

$$H(s) \le 3|x - x'|^2 + (12k^2 + 3k^2t)\int_0^s H(u)\,\mathrm{d}u, \quad \text{for } 0 \le s \le t.$$

By Gronwall's inequality, we conclude that

$$H(t) \le 3|x - x'|^2 \exp[(12k^2 + 3k^2t)t].$$

This proves the uniqueness of the solution to (7.2) and also proves (7.4). It is clear that the distribution of each ζ_n is independent of the particular Brownian motion $B(t)$ and depends only on σ, b and the distribution of Brownian motion. Thus the same is true of $\zeta(t)$. □

We will now show that $\zeta_x(t)$ is a Markov process. We will denote $P(\zeta_z(t) \in B)|_{z=\zeta_x(s,q)}$, which depends on $q \in Y$, by $P(\zeta_{\zeta_x(s)}(t) \in B)$. Written out explicitly, we have

$$P(\zeta_{\zeta_x(s)}(t) \in B) = P(\zeta_{\zeta_x(s,q)}(t) \in B) = P(q' \in Y\colon \zeta_{\zeta_x(s,q)}(t, q') \in B).$$

Fix $s \ge 0$. To prove the Markov property, we will prove that for $B \in \mathcal{B}_d$, $P(\zeta_x(s + t) \in B|\mathcal{M}_s)$ is measurable with respect to the σ-algebra generated by $\zeta_x(s)$. Let $\hat{B}(t) = B(t + s) - B(s)$, and let $\hat{\zeta}_x(t)$ denote the solution to the stochastic differential equation with the Brownian motion $\hat{B}(t)$ in place of $B(t)$. We can represent $\zeta_x(s + t)$ as

$$\zeta_x(s+t) = \zeta_x(s) + \int_s^{s+t} \sigma(\zeta_x(u))\,dB(u) + \int_s^{s+t} b(\zeta_x(u))\,du$$
$$= \zeta_x(s) + \int_0^t \sigma(\zeta_x(s+u))\,d\hat{B}(u) + \int_0^t b(\zeta_x(s+u))\,du.$$

From this and the pathwise uniqueness of solutions, it follows that $\zeta_x(s+t) = \hat{\zeta}_{\zeta_x(s)}(t)$ a.s. Now $\hat{\zeta}_x(t)$ is measurable with respect to $\sigma(B(t) - B(s),\ t \geq s)$. Thus, by Lemma 4.1, $P(\hat{\zeta}_x(t) \in B | \mathcal{M}_s) = P(\hat{\zeta}_x(t) \in B)$, for any $B \in \mathcal{B}_d$ and $t \geq 0$. Then, by exercise 1.7, $P(\hat{\zeta}_{\zeta_x(s)}(t) \in B | \mathcal{M}_s) = P(\hat{\zeta}_{\zeta_x(s)}(t) \in B)$. This completes the proof since the righthand side above is a function of $\zeta_x(s)$.

From (7.4), it follows that $\zeta_x(t)$ is a Feller process. Then using exercise 1.4, we conclude that $\zeta_x(t)$ is a strong Markov process and that for any finite stopping time τ,

$$P(\zeta_x(\tau+t) \in B | \mathcal{M}_\tau) = P(\zeta_x(\tau+t) \in B | \zeta_x(\tau)) = P(\zeta_{\zeta_x(\tau)}(t) \in B),$$
$$\text{for } B \in \mathcal{B}_d \text{ and } 0 \leq t < \infty. \quad (7.7)$$

We record this as follows.

Theorem 7.2. *The solution $\zeta_x(t)$ of* (7.2) *is a strong Markov process and a Feller process and satisfies* (7.7).

Remark. Note that the proof of the Markov property relied on the pathwise uniqueness of the solution to the stochastic differential equation.

We now want to show that if a and b are bounded, then in fact $\zeta_x(t)$ is an $I_d(a(\zeta_x(\cdot)), b(\zeta_x(\cdot)))$-process. At this stage we need the following lemma.

Lemma 7.3. *Let $a: R^d \to S_d$ satisfy $\langle v, a(x)v\rangle \geq \alpha|v|^2$, for all v, $x \in R^d$, where $\alpha > 0$. If $\|a(x) - a(y)\| \leq C|x-y|$ for all $x, y \in R^d$, then $a^{1/2}$, the unique positive definite square root of a, satisfies*

$$\|a^{1/2}(x) - a^{1/2}(y)\| \leq \frac{D}{2\alpha^{1/2}}|x-y|, \quad \textit{for all } x, y \in R^d.$$

Proof. It is enough to show that if $a(x)$ is a C^1-function from R to S_d such that $\langle v, a(x)v\rangle \geq \alpha|v|^2$, for $x \in R^1$ and $v \in R^d$, then

$$\|(a^{1/2}(x))'\| \leq \frac{1}{2\alpha^{1/2}}\|a'(x)\|.$$

We may assume that $a(\cdot)$ is diagonal at the point x in question. We will prove below that if a is C^1, then so is $a^{1/2}$. Assuming this for the moment, we have

$$a'(x) = a^{1/2}(x)(a^{1/2}(x))' + (a^{1/2}(x))'a^{1/2}(x)$$

and thus, since $a(\cdot)$ is diagonal at x,

$$(a^{1/2}(x))'_{ij} = \frac{(a_{ij}(x))'}{a_{ii}^{1/2}(x) + a_{jj}^{1/2}(x)},$$

which gives the desired inequality.

Let $U \subset\subset R^d$ and let A satisfy $\langle v, a(x)v\rangle \leq A|v|^2$ for $x \in U$ and $v \in R^d$. Then it is easy to check that

$$a^{1/2}(x) = A^{1/2}\sum_{n=0}^{\infty} c_n\left(I - \frac{a(x)}{A}\right)^n, \text{ for } x \in U,$$

where c_n is the nth coefficient in the power series expansion of $(1-z)^{1/2}$ around $z=0$. From this, it follows that if a is C^1, then so is $a^{1/2}$. □

Now let $a\colon R^d \to S_d$ and $b\colon R^d \to R^d$ be bounded and satisfy

$$\left.\begin{array}{l}\text{(i) } \langle v, a(x)v\rangle \geq \alpha|v|^2, \text{ for } x, v \in R^d \text{ and some } \alpha > 0,\\ \text{(ii) } \|a(x) - a(y)\| \leq c|x-y|, \text{ for } x, y \in R^d;\end{array}\right\} \tag{7.8}$$

$$|b(x) - b(y)| \leq c|x-y|, \text{ for } x, y \in R^d. \tag{7.9}$$

Let $\sigma = a^{1/2}$. Then, by Lemma 7.3, σ and b satisfy (7.3). Let $\zeta_x(t)$ denote the solution to (7.2) which is guaranteed by Theorem 7.1. It follows from Theorem 6.2 that $\int_0^t \sigma(\zeta_x(s))\,dB(s) \sim I_d(a(\zeta_x(\cdot)), 0)$ and then, from the definition of an Itô process, it follows that $\zeta_x(t) \sim I_d(a(\zeta_x(\cdot)), b(\zeta_x(\cdot)))$. From Theorem 5.1 we have

$$\left.\begin{array}{l}\text{(i) } \zeta_x(0) = x;\\ \text{(ii) } f(\zeta_x(t)) - \int_0^t Lf(\zeta_x(s))\,ds \text{ is a martingale with respect to } (Y, \mathcal{M}, \mathcal{M}_t, P)\\ \text{for all } f \in C_b^2(R^d), \text{ where } L = \frac{1}{2}\sum_{i,j=1}^d a_{ij}(x)\frac{\partial^2}{\partial x_i \partial x_j} + \sum_{i=1}^d b_i(x)\frac{\partial}{\partial x_i}.\end{array}\right\} \tag{7.10}$$

We conclude that the solution $\zeta_x(t)$ of the stochastic differential equation gives a solution to the problem raised in the final paragraph of Section 5 in the case where a and b are bounded and satisfy (7.8) and (7.9). (In fact, the solution $\zeta_x(t)$ satisfies (7.10) even if a and b are not bounded (see exercise 1.8).) We now consider the question of uniqueness in distribution for Itô processes under the assumption that a and b are bounded and satisfy (7.8) and (7.9). That is, does the fact that $\zeta_x(t)$ solves (7.10) or, equivalently, that $\zeta_x(t)$ solves (7.10(i)) and

$$\zeta_x(t) \sim I_d(a(\zeta_x(\cdot)), b(\zeta_x(\cdot))),$$

uniquely determine its distribution? To continue our study, we will need the following 'chain rule' for stochastic integrals which is easy to prove and is left to the reader.

Lemma 7.4. *Let $\zeta(t) \sim I_d(a, 0)$ and let η, ρ: $[0, \infty) \times Y \to R^d \otimes R^d$ be bounded progressively measurable $d \times d$ matrix-valued functions. Define $\hat{\zeta}(t) = \int_0^t \rho(s)\,\mathrm{d}\zeta(s)$. Then $\int_0^t \eta(s)\,\mathrm{d}\hat{\zeta}(s) = \int_0^t \eta(s)\rho(s)\,\mathrm{d}\zeta(s)$.*

Now assume that $\eta_x(t)$ is an almost surely continuous stochastic process on $(Y, \mathcal{M}, \mathcal{M}_t, P)$ which satisfies (7.10), where the coefficients a and b of L are bounded and satisfy (7.8) and (7.9). We want to show that its distribution is uniquely determined. Letting $\sigma = a^{1/2}$, it follows from Lemma 7.3 that σ and b satisfy (7.3). By definition, $\eta_x(t) \sim I_d(a(\eta_x(\cdot)), b(\eta_x(\cdot)))$ and thus

$$\hat{\eta}_x(t) \equiv \eta_x(t) - \int_0^t b(\eta_x(s))\,\mathrm{d}s \sim I_d(a(\eta_x(\cdot)), 0).$$

Define $\hat{B}(t) = \int_0^t \sigma^{-1}(\eta_x(s))\,\mathrm{d}\hat{\eta}_x(s)$. Then from Theorem 6.2, $\hat{B}(t) \sim I_d(I, 0)$, that is, $\hat{B}(t)$ is a d-dimensional Brownian motion with respect to $(Y, \mathcal{M}, \mathcal{M}_t, P)$. By the chain rule (Lemma 7.4), $\hat{\eta}_x(t) - \hat{\eta}_x(0) = \hat{\eta}_x(t) - x = \int_0^t \sigma(\eta_x(s))\,\mathrm{d}\hat{B}(s)$ and thus $\eta_x(t) = \hat{\eta}_x(t) + \int_0^t b(\eta_x(s))\,\mathrm{d}s = x + \int_0^t \sigma(\eta_x(s))\,\mathrm{d}\hat{B}(s) + \int_0^t b(\eta_x(s))\,\mathrm{d}s$. That is, $\eta_x(t)$ is the unique solution to (7.2) with $B(t)$ replaced by $\hat{B}(t)$. Thus, the conclusions of Theorems 7.1 and 7.2 hold for $\eta_x(t)$. This gives us the following result.

Theorem 7.5. *Assume that a and b are bounded and satisfy (7.8) and (7.9). Then on a suitable space $(Y, \mathcal{M}, P)$ with a filtration $\mathcal{M}_t$, one can define for each $x \in R^d$ an Itô process $\zeta_x(t)$ satisfying $P(\zeta_x(0) = x) = 1$ and $\zeta_x(t) \sim I_d(a(\zeta_x(\cdot)), b(\zeta_x(\cdot)))$. Its distribution is uniquely determined. Furthermore, $\zeta_x(t)$ is a strong Markov process on $(Y, \mathcal{M}, \mathcal{M}_t, P)$ and for any finite stopping time τ, $P(\zeta_x(\tau + t) \in B | \mathcal{M}_\tau) = P(\zeta_{\zeta_x(\tau)}(t) \in B)$ for $B \in \mathcal{B}_d$ and $0 \leq t < \infty$. Finally, for each $t > 0$, there exists a constant $c_t > 0$ such that*

$$E|\zeta_x(t) - \zeta_{x'}(t)|^2 \leq c_t|x - x'|^2, \text{ for } x, x' \in R^d. \qquad (7.11)$$

In particular, then, $\zeta_x(t)$ is a Feller process.

The analysis leading up to Theorem 7.5 also gives a proof of the following theorem.

Theorem 7.6. *Let $\zeta_x(t) \sim I_d(a(\zeta_x(\cdot)), b(\zeta_x(\cdot)))$ on a probability space $(Y, \mathcal{M}, P)$ with filtration $\mathcal{M}_t$ and assume that a and b are bounded and satisfy (7.8) and (7.9). Then there exists a Brownian motion $\hat{B}(t)$ on $(Y, \mathcal{M}, \mathcal{M}_t, P)$ such that $\zeta_x(t)$ is the unique solution to the stochastic differential equation given by (7.2) with the Brownian motion $B(t)$ replaced by $\hat{B}(t)$.*

We conclude this section with an important localization result.

Theorem 7.7. *Let σ_i and b_i, $i = 1, 2$, satisfy (7.3) and assume that $\zeta_x^{(i)}(t)$, $i = 1, 2$, satisfies (7.2) with σ_i and b_i in place of σ and b. Let $U \subset R^d$ be open and assume that $\sigma_1 = \sigma_2$ and $b_1 = b_2$ on U. Define*

$$\tau_i = \inf\{t \geq 0\colon \zeta_x^{(i)}(t) \notin U\},\ i = 1, 2.$$

Then $P(\tau_1 = \tau_2) = 1$ and $P(\sup_{\sigma \leq s \leq \tau_1} |\zeta_1(s) - \zeta_2(s)| = 0) = 1$.

Proof. Let $\phi_i(t) = 1$, if $\zeta_x^{(i)}(s) \in U$ for all $0 \leq s \leq t$, and $\phi_i(t) = 0$ otherwise. Then

$$\begin{aligned}\phi_1(t)(\zeta_x^{(1)}(t) - \zeta_x^{(2)}(t)) = {} & \phi_1(t) \int_0^t (b_1(\zeta_x^{(1)}(s)) - b_2(\zeta_x^{(1)}(s)))\,\mathrm{d}s \\ & + \phi_1(t) \int_0^t (b_2(\zeta_x^{(1)}(s)) - b_2(\zeta_x^{(2)}(s)))\,\mathrm{d}s \\ & + \phi_1(t) \int_0^t (\sigma_1(\zeta_x^{(1)}(s)) - \sigma_2(\zeta_x^{(1)}(s)))\,\mathrm{d}B(s) \\ & + \phi_1(t) \int_0^t (\sigma_2(\zeta_x^{(1)}(s)) - \sigma_2(\zeta_x^{(2)}(s)))\,\mathrm{d}B(s).\end{aligned} \tag{7.12}$$

If $\phi_1(t) = 1$, then $b_1(\zeta_x^{(1)}(s)) = b_2(\zeta_x^{(1)}(s))$, for $0 \leq s \leq t$; thus the first term on the righthand side of (7.12) is identically zero. Also, if $\phi_1(t) = 1$, then $\sigma_1(\zeta_x^{(1)}(s)) = \sigma_2(\zeta_x^{(1)}(s))$, for $0 \leq s \leq t$; thus, by Exercise 1.9, the third term on the righthand side of (7.12) is identically zero. Therefore,

$$\begin{aligned}\phi_1(t)|\zeta_x^{(1)}(t) - \zeta_x^{(2)}(t)|^2 \leq {} & 2\phi_1(t)\left|\int_0^t (b_2(\zeta_x^{(1)}(s)) - b_2(\zeta_x^{(2)}(s)))\,\mathrm{d}s\right|^2 \\ & + 2\phi_1(t)\left|\int_0^t (\sigma_2(\zeta_x^{(1)}(s)) - \sigma_2(\zeta_x^{(2)}(s)))\,\mathrm{d}B(s)\right|^2.\end{aligned} \tag{7.13}$$

Since $\phi_1(t)$ is non-increasing,

$$\phi_1(t)\left|\int_0^t (b_2(\zeta_x^{(1)}(s)) - b_2(\zeta_x^{(2)}(s)))\,ds\right|^2$$
$$\leqslant d\left(\int_0^t \phi_1(s)|b_2(\zeta_x^{(1)}(s)) - b_2(\zeta_x^{(2)}(s))|\,ds\right)^2. \quad (7.14)$$

Next, we show that

$$\theta_1(t)\left|\int_0^t (\sigma_2(\zeta_x^{(1)}(s)) - \sigma_2(\zeta_x^{(2)}(s)))\,dB(s)\right|^2$$
$$\leqslant \left|\int_0^t \phi_1(s)(\sigma_2(\zeta_x^{(1)}(s)) - \sigma_2(\zeta_x^{(2)}(s)))\,dB(s)\right|^2. \quad (7.15)$$

Let $A = \{\phi_1(t) = 1\}$ and $B = \{\phi_1(t) = 0\}$. If $q \in B$, then the lefthand side of (7.15) vanishes; thus (7.15) holds in this case. If $q \in A$, then $\phi_1(s)(\sigma_2(\zeta_x^{(1)}(s)) - \sigma_2(\zeta_x^{(2)}(s))) = (\sigma_2(\zeta_x^{(1)}(s)) - \sigma_2(\zeta_x^{(2)}(s)))$, for $0 \leqslant s \leqslant t$, and, consequently, by exercise 1.9, equality holds in (7.15).

Using (7.14) and (7.15) in (7.13), we obtain

$$\phi_1(t)|\zeta_x^{(1)}(t) - \zeta_x^{(2)}(t)|^2 \leqslant 2d\left(\int_0^t \phi_1(s)|b_2(\zeta_x^{(1)}(s)) - b_2(\zeta_x^{(2)}(s))|\,ds\right)^2$$
$$+ 2\left|\int_0^t \sigma_1(s)(\sigma_2(\zeta_x^{(1)}(s)) - \sigma_2(\zeta_x^{(2)}(s)))\,dB(s)\right|^2. \quad (7.16)$$

Since σ_2 and b_2 satisfy (7.3), it follows from (7.16) that there exists a constant c such that

$$E\phi_1(t)|\zeta_x^{(1)}(t) - \zeta_x^{(2)}(t)|^2 \leqslant c\int_0^t E\phi_1(s)|\zeta_x^{(1)}(s) - \zeta_x^{(2)}(s)|^2\,ds.$$

By Gronwall's inequality, this implies that $E\phi_1(t)|\zeta_x^{(1)}(t) - \zeta_x^{(2)}(t)|^2 = 0$. Thus, by the continuity of $\zeta_x^{(1)}(t)$ and $\zeta_x^{(2)}(t)$, it follows that

$$\sup_{0\leqslant t<\infty} \phi_1(t)|\zeta_x^{(1)}(t) - \zeta_x^{(2)}(t)| = 0 \text{ a.s.}$$

Therefore, $\zeta_x^{(1)}(t,q) = \zeta_x^{(2)}(t,q)$, for all $0 \leqslant t \leqslant \tau_1(q)$, a.s. Consequently, $P(\tau_2 \geqslant \tau_1) = 1$. Reversing the roles of $\zeta_x^{(1)}(t)$ and $\zeta_x^{(2)}(t)$ gives $P(\tau_1 \geqslant \tau_2) = 1$. □

1.8 The formulation of the martingale problem on R^d

In order to extend the existence and uniqueness in distribution result of Theorem 7.5 to coefficients a and b with weaker smoothness requirements, we need to introduce the martingale formulation more explicitly. Let $\Omega = C([0, \infty), R^d)$ denote the space of continuous functions ω from $[0, \infty)$ to R^d with the topology of uniform convergence on bounded intervals. Let $X(t) = X(t, \omega)$ denote the position of ω at time t. It is easy to check that

$$m(\omega, \omega') = \sum_{n=0}^{\infty} \frac{1}{2^n} \frac{\sup_{n \leq t < n+1} |X(t, \omega) - X(t, \omega')|}{1 + \sup_{n \leq t < n+1} |X(t, \omega) - X(t, \omega')|} \tag{8.1}$$

defines a compatable metric on Ω and that (Ω, m) is a complete separable metric space. It is well known that the Borel σ-algebra $\mathcal{F}$ on Ω is given by $\sigma(X(t), 0 \leq t < \infty)$. Define the filtration $\mathcal{F}_t = \sigma(X(s), 0 \leq s \leq t)$. Let

$$L = \tfrac{1}{2} \sum_{i,j=1}^{d} a_{ij}(x) \frac{\partial^2}{\partial x_i \partial x_j} + \sum_{i=1}^{d} b_i(x) \frac{\partial}{\partial x_i}$$

and define the martingale problem for L as follows.

The martingale problem for L on R^d. For each $x \in R^d$, find a probability measure P_x on $(\Omega, \mathcal{F})$ such that

(i) $P_x(X(0) = x) = 1$;
(ii) $f(X(t)) - \int_0^t (Lf)(X(s))\, ds$ is a martingale with respect to $(\Omega, \mathcal{F}, \mathcal{F}_t, P_x)$, for all $f \in C_0^2(R^d)$.

Note that if $\{P_x, x \in R^d\}$ solves the martingale problem for L and the coefficients $a(x)$ and $b(x)$ of L are bounded, then by Theorem 5.1, for each $x \in R^d$, $X(t) = X(t, \omega)$ is an Itô process on $(\Omega, \mathcal{F}, \mathcal{F}_t, P_x)$ with covariance $a(X(\cdot))$ and drift $b(X(\cdot))$. Conversely, assume that $a(x)$ and $b(x)$ are bounded and that for each $x \in R^d$, $\zeta_x(t)\colon [0, \infty) \times Y \to R^d$ is an $I_d(a(\zeta(\cdot)), b(\zeta(\cdot)))$-process on $(Y, \mathcal{M}, \mathcal{M}_t, P)$ satisfying $P(\zeta_x(0) = x) = 1$. Denote the distribution of $\zeta_x(t)$ on Ω by ν_x. Then, again by Theorem 5.1, $\{\nu_x, x \in R^d\}$ solves the martingale problem for

$$L = \tfrac{1}{2} \sum_{i,j=1}^{d} a_{ij}(x) \frac{\partial^2}{\partial x_i \partial x_j} + \sum_{i=1}^{d} b_i(x) \frac{\partial}{\partial x_i}.$$

This gives us the following reformulation of Theorem 7.5 into the language of the martingale problem.

Theorem 8.1. *Assume that $a(x)$ and $b(x)$ are bounded and satisfy* (7.8) *and* (7.9). *Then there exists a unique solution $\{P_x, x \in R^d\}$ to the martingale problem for*

$$L = \tfrac{1}{2}\sum_{i,j=1}^{d} a_{ij}(x)\frac{\partial^2}{\partial x_i \partial x_j} + \sum_{i=1}^{d} b_i(x)\frac{\partial}{\partial x_i}.$$

The family $\{P_x, x \in R^d\}$ *possesses the Feller property and the strong Markov property. For any finite stopping time* τ,

$$P_x(X(\tau + t) \in B | \mathcal{F}_\tau) = P_{X(\tau)}(X(t) \in B), \text{ for } B \in \mathcal{B}_d \text{ and } 0 \leqslant t < \infty.$$

Remark. In the sequel, the expectation with respect to a probability measure P_x will be denoted by E_x.

1.9 The Cameron–Martin–Girsanov transformation and the martingale problem on R^d

To extend existence and uniqueness for the martingale problem to more general drifts b, the following transformation of drift formula is very useful. It is known as the *Cameron–Martin–Girsanov formula*

Theorem 9.1. *Let* $a: R^d \to S_d$ *and* b, $c: R^d \to R^d$ *be bounded and measurable, and let*

$$L = \tfrac{1}{2}\sum_{i,j=1}^{d} a_{ij}(x)\frac{\partial^2}{\partial x_i \, \mathrm{d}x_j} + \sum_{i=1}^{d} b_i(x)\frac{\partial}{\partial x_i}$$

and

$$L_1 = L + \sum_{i=1}^{d} (ac)_i(x)\frac{\partial}{\partial x_i}.$$

Assume that $\{P_x, x \in R^d\}$ *solves the martingale problem for* L. *For* $t \geqslant 0$, *define* $N(t): \Omega \to [0, \infty)$ *by*

$$N(t) = \exp\left(\int_0^t \langle c(X(s)), \mathrm{d}\bar{X}(s)\rangle - \tfrac{1}{2}\int_0^t \langle c(X(s)), ac(X(s))\rangle \, \mathrm{d}s\right),$$

where $\bar{X}(t) = X(t) - \int_0^t b(X(s))\,\mathrm{d}s$. *Then, for each* $x \in R^d$ *there exists a unique probability measure*

$$Q_x \text{ on } (\Omega, \mathcal{F}) \text{ such that, for all } t \geqslant 0, \left.\frac{\mathrm{d}Q_x}{\mathrm{d}P_x}\right|_{\mathcal{F}_t} = N(t).$$

The family $\{Q_x, x \in R^d\}$ *solves the martingale problem for* L_1.

Remark. Note that the stochastic integral appearing in $N(t)$ is well defined. Indeed, the process $X(t)$ is an $I_d(a(X(\cdot)), b(X(\cdot)))$-process with respect to $(\Omega, \mathcal{F}, \mathcal{F}_t, P_x)$ and thus $\bar{X}(t) \sim I_d(a(X(\cdot)), 0)$.

Proof. Define Q_x^t on $\mathcal{F}_t$ by $Q_x^t(A) = E_x(N(t), A)$, $A \in \mathcal{F}_t$. Since $\bar{X}(t) \sim I_d(a(X(\cdot)), 0)$ with respect to $(\Omega, \mathcal{F}, \mathcal{F}_t, P_x)$, it follows from Theorem 6.2 and the definition of an Itô process that $N(t)$ is a martingale with respect to $(\Omega, \mathcal{F}, \mathcal{F}_t, P_x)$. Therefore, $\{Q_x^t, t \geq 0\}$ forms a consistent family of probability measures. (That is, for $t > s$, $Q_x^t = Q_x^s$ on $\mathcal{F}_s$.) It then follows from the Kolmogorov construction that there exists a unique Q_x on $(\Omega, \mathcal{F})$ which satisfies $Q_x = Q_x^t$ on $\mathcal{F}_t$. Now let $t > s$ and let $A \in \mathcal{F}_s$. Denote the expectation with respect to Q_x by E^{Q_x}. Since $\bar{X}(t) \sim I_d(a(X(\cdot)), 0)$ with respect to $(\Omega, \mathcal{F}, \mathcal{F}_t, P_x)$, it follows again from Theorem 6.2 and the definition of an Itô process that if $A \in \mathcal{F}_s$, then

$$
\begin{aligned}
&E^{Q_x}\left(\exp\left(\left\langle \theta, X(t) - X(0) - \int_0^t (b + ac)(X(u))\,\mathrm{d}u \right\rangle \right.\right.\\
&\qquad \left.\left. - \tfrac{1}{2}\int_0^t \langle \theta, a(X(u))\theta\rangle\,\mathrm{d}u\right); A\right)\\
&= E_x\left(\exp\left(\int_0^t \langle \theta + c(X(u)), \mathrm{d}\bar{X}(u)\rangle \right.\right.\\
&\qquad \left.\left. - \tfrac{1}{2}\int_0^t \langle \theta + c(X(u)), a(X(u))(\theta + c(X(u)))\rangle\,\mathrm{d}u\right); A\right)\\
&= E_x\left(\exp\left(\int_0^s \langle \theta + c(X(u)), \mathrm{d}\bar{X}(u)\rangle \right.\right.\\
&\qquad \left.\left. - \tfrac{1}{2}\int_0^s \langle \theta + c(X(u)), a(X(u))(\theta + c(X(u)))\rangle\,\mathrm{d}u\right); A\right)\\
&= E^{Q_x}\left(\exp\left(\left\langle \theta, X(s) - X(0) - \int_0^s (b + ac)(X(u))\,\mathrm{d}u \right\rangle \right.\right.\\
&\qquad \left.\left. - \tfrac{1}{2}\int_0^s \langle \theta, a(X(u))\theta\rangle\,\mathrm{d}u\right); A\right).
\end{aligned}
$$

From this we conclude that $X(\cdot) \sim I_d(a(X(\cdot)), (b + ac)(X(\cdot)))$ with respect to $(\Omega, \mathcal{F}, \mathcal{F}_t, Q_x)$; thus $\{Q_x, x \in R^d\}$ solves the martingale problem for L_1.

The theorem below follows directly from the Cameron–Martin–Girsanov formula.

***Theorem* 9.2.** *Let a, b, c, L, L_1 and $N(t)$ be as in Theorem* 9.1. *Then there exists a one-to-one correspondence between solutions of the martingale problem for L and L_1. The correspondence is given by $P_x \to Q_x$ where $Q_x \ll P_x$ and $\mathrm{d}Q_x/\mathrm{d}P_x|_{\mathcal{F}_t} = N(t)$, $t \geq 0$.*

The follow corollary is an immediate consequence of Theorem 9.2.

***Corollary* 9.3.** *Under the assumptions of Theorem* 9.2, *there exists at least one (exactly one) solution to the martingale problem for L if and only if there exists at least one (exactly one) solution to the martingale problem for L_1.*

We have now arrived at an improved existence and uniqueness theorem.

***Theorem* 9.4.** *Assume that a satisfies* (7.8) *and that $b\colon R^d \to R^d$ is bounded and measurable. Then there exists a unique solution $\{P_x, x \in R^d\}$ to the martingale problem for*

$$L = \tfrac{1}{2}\sum_{i,j=1}^{d} a_{ij}(x)\frac{\partial^2}{\partial x_i \partial x_j} + \sum_{i=1}^{d} b_i(x)\frac{\partial}{\partial x_i}.$$

Proof. There exists a unique solution to the martingale problem for

$$L_0 = \tfrac{1}{2}\sum_{i,j=1}^{d} a_{ij}(x)\frac{\partial^2}{\partial x_i \partial x_j}$$

by Theorem 8.1. Let $c = a^{-1}b$. By (7.8(i)), c is bounded. Now invoke Corollary 9.3. □

1.10 The Stroock–Varadhan solution to the martingale problem on R^d

The martingale problem for an operator L with bounded coefficients is in fact a local problem. This is concretized in the following theorem.

***Theorem* 10.1.** *Let $a\colon R^d \to S_d$ and $b\colon R^d \to R^d$ be bounded and measurable. Suppose that for each $x \in R^d$ there exist an open set $U \ni x$ and bounded measurable coefficients $\tilde{a}$ and $\tilde{b}$ such that*

(i) *$\tilde{a}$ equals a and $\tilde{b}$ equals b on U;*
(ii) *There exists a unique solution to the martingale problem for $\tilde{L}$ with coefficients $\tilde{a}$ and $\tilde{b}$.*

Then there exists a unique solution to the martingale problem for L with coefficients a and b.

For a proof of this result, see [Stroock and Varadhan (1979), Theorem 6.6.1]. Using this result, Theorem 9.4, which required that a satisfy (7.8), may be strengthened by replacing (7.8(i)) by

$$\sum_{i,j=1}^{d} a_{ij}(x) v_i v_j > 0, \text{ for all } x \in R^d \text{ and } v \in R^d - \{0\}. \qquad (10.1)$$

With regard to uniqueness for the martingale problem, in the case of bounded coefficients, this is about as far as we can go by purely probabilistic methods. Relying on certain L^p-estimates for singular integral operators, and on Theorem 10.1, Stroock and Varadhan have proved the following important theorem [Stroock and Varadhan (1979), Chapter 7].

Theorem 10.2. *Let $a: R^d \to S_d$ be bounded and continuous and satisfy* (10.1) *and let $b: R^d \to R^d$ be bounded and measurable. Then there exists a unique solution $\{P_x, x \in R^d\}$ to the martingale problem for*

$$L = \tfrac{1}{2}\sum_{i,j=1}^{d} a_{ij}(x)\frac{\partial^2}{\partial x_i \partial x_j} + \sum_{i=1}^{d} b_i(x)\frac{\partial}{\partial x_i}.$$

The probability measure $P_x(X(t) \in dy)$ possesses a measurable density $p(t, x, y)$. The family $\{P_x, x \in R^d\}$ possesses the Feller property and the strong Markov property. For any finite stopping time τ,

$$P_x(X(\tau + t) \in B|\mathcal{F}_\tau) = P_{X(\tau)}(X(t) \in B) = \int_B p(t, X(\tau), y)\,dy,$$

for $B \in \mathcal{B}_d$ and $t \geq 0$.

Remark. It is important to point out that the Markov property and the strong Markov property are consequences of the uniqueness. To understand this, see [Stroock and Varadhan (1979), Theorem 1.2.10 and Theorem 6.1.3]. Compare this with the remark following Theorem 7.2.

We now discuss the possibility of extending Theorem 10.2 to locally bounded coefficients a and b. For $U \subset R^d$, define the *first exit time* from U by $\tau_U = \inf\{t \geq 0: X(t) \notin U\}$. If U is open, then it is easy to check that τ_U is an $\mathcal{F}_t$-stopping time (exercise 1.10). We begin by noting the following localization theorem [Stroock and Varadhan (1979), Theorem 10.1.1], which extends Theorem 7.7 to the case of more general coefficients.

Theorem 10.3. *Assume that a, $\bar{a}: R^d \to S_d$ and b, $\bar{b}: R^d \to R^d$ are locally bounded and measurable and let*

$$L = \tfrac{1}{2}\sum_{i,j=1}^{d} a_{ij}(x)\frac{\partial^2}{\partial x_i \partial x_j} + \sum_{i=1}^{d} b_i(x)\frac{\partial}{\partial x_i}$$

and

$$\bar{L} = \tfrac{1}{2}\sum_{i,j=1}^{d} \bar{a}_{ij}(x)\frac{\partial^2}{\partial x_i \partial x_j} + \sum_{i=1}^{d} \bar{b}_i(x)\frac{\partial}{\partial x_i}.$$

Assume that the martingale problem for L *has a unique solution* $\{P_x, x \in R^d\}$ *and assume that* $a = \bar{a}$ *and* $b = \bar{b}$ *on some open set* $U \subset R^d$. *Then any solution* $(\bar{P}_x, x \in R^d)$ *to the martingale problem for* $\bar{L}$ *satisfies* $\bar{P}_x|_{\mathcal{F}_{\tau_U}} = P_x|_{\mathcal{F}_{\tau_U}}$, *where* $\tau_U = \inf(t \geq 0\colon X(t) \notin U)$.

In the light of Theorem 10.3, the natural way to approach the martingale problem with locally bounded coefficients is as follows. Assume that $a(x)$ is continuous and satisfies (10.1). Let $\{D_n\}_{n=1}^{\infty}$ be an increasing sequence of domains such that $D_n \subset\subset D_{n+1}$ and $\bigcup_{n=1}^{\infty} D_n = R^d$, and let $\psi_n\colon R^d \to R$ be a C^∞-function satisfying $\psi_n(x) \equiv 1$ on D_n, $\psi_n(x) \equiv 0$ on D_{n+1}^c and $0 \leq \psi_n \leq 1$, for all $x \in R^d$. (The notation $D_n \subset\subset D_{n+1}$ means that D_n is bounded and $\bar{D}_n \subset D_{n+1}$.) Define

$$\left.\begin{aligned} a_n &= \psi_n a + (1 - \psi_n)I, \\ b_n &= \psi_n b. \end{aligned}\right\} \tag{10.2}$$

Let L_n denote the operator with coefficients a_n and b_n and let $\{P_x^{(n)}, x \in R^d\}$ denote the unique solution to the martingale problem for L_n. From Theorem 10.3, it is clear that $P_x^{(m)} = P_x^{(n)}$ on $\mathcal{F}_{\tau_{D_n}}$, for $m > n$, and that if a solution $\{P_x, x \in R^d\}$ to the martingale problem for L exists, then $P_x = P_x^{(n)}$ on $\mathcal{F}_{\tau_{D_n}}$, for all $n = 0, 1, 2, \ldots$, where $\tau_{D_n} = \inf\{t \geq 0\colon X(t) \notin D_n\}$.

Now fix $x \in R^d$ and assume that Q_x is a probability measure on $(\Omega, \mathcal{F})$ which satisfies

$$Q_x = P_x^{(n)} \text{ on } \mathcal{F}_{\tau_{D_n}}, \quad n = 1, 2, \ldots. \tag{10.3}$$

We will show that

(i) $\lim_{n\to\infty} P_x^{(n)}(\tau_{D_n} \leq t) = 0$, for all $t \geq 0$;
(ii) (10.3) uniquely determines Q_x;
(iii) Q_x solves the martingale problem for L on R^d starting from x.

To prove (i), note that since Q_x is a probability measure on $(\Omega, \mathcal{F})$, $\lim_{n\to\infty} Q_x(\tau_{D_n} \leq t) = 0$, for all $t \geq 0$. Now (i) follows, since by (10.3), $Q_x(\tau_{D_n} \leq t) = P_x^{(n)}(\tau_{D_n} \leq t)$. To prove (ii), assume that $B \in \mathcal{F}_t$. Then for any

$$n = 1, 2, \ldots, Q_x(B) = Q_x(B \cap \{\tau_{D_n} \leq t\}) + Q_x(B \cap \{\tau_{D_n} > t\}).$$

Thus

$$
\begin{aligned}
Q_x(B) &= \lim_{n\to\infty} Q_x(B \cap \{\tau_{D_n} \leq t\}) + \lim_{n\to\infty} Q_x(B \cap \{\tau_{D_n} > t\}) \\
&= \lim_{n\to\infty} P_x^{(n)}(B \cap \{\tau_{D_n} > t\}),
\end{aligned}
$$

where the last equality follows from (10.3(i)), and the fact that $B \cap \{\tau_{D_n} \leq t\}$ and $B \cap \{\tau_{D_n} > t\}$ are $\mathcal{F}_t$-measurable. We conclude that Q_x is uniquely determined on $\mathcal{F}_t$ and thus on $\mathcal{F}$. To prove (iii), note that for $f \in C_0^2(R^d)$, $0 \leq s \leq t$, and $A \in \mathcal{F}_s$, we have

$$
\begin{aligned}
&E^{Q_x}\left(f(X(t \wedge \tau_{D_n})) - \int_0^{t\wedge\tau_{D_n}} Lf(X(u))\,du;\ A\right) \\
&\qquad = E^{Q_x}\left(f(X(s \wedge \tau_{D_n})) - \int_0^{s\wedge\tau_{D_n}} Lf(X(u))\,du;\ A\right).
\end{aligned}
$$

This follows from Theorem 2.1, (10.3) and the fact that $L = L_n$ on $\bar{D}_n$. Letting $n \to \infty$ and using (i) and the bounded convergence theorem, we obtain

$$
\begin{aligned}
&E^{Q_x}\left(f(X(t)) - \int_0^{t} Lf(X(u))\,du;\ A\right) \\
&\qquad = E^{Q_x}\left(f(X(s)) - \int_0^{s} Lf(X(u))\,du;\ A\right).
\end{aligned}
$$

This proves (iii).

Our analysis above has shown that if the sequence $P_x^{(n)}$ on $(\Omega, \mathcal{F}_{\tau_{D_n}})$ is extendable to a probability measure on $(\Omega, \mathcal{F})$, then this measure solves the martingale problem for L on R^d starting from x. It has also shown that such a measure, if it exists, is unique, and that the condition $\lim_{n\to\infty} P_x^{(n)}(\tau_{D_n} \leq t) = 0$, for all $t \geq 0$, is a necessary condition for its existence. Our aim now is to prove that this condition is in fact sufficient. Except for the statement concerning the strong Markov property for which the reader is referred to the remark following Theorem 10.2, this will give us a proof of the following theorem.

Theorem 10.4. *Let $a\colon R^d \to S_d$ and $b\colon R^d \to R^d$ be measurable and locally bounded and assume that a is continuous and satisfies* (10.1). *Then there exists at most one solution to the martingale problem for L on R^d with coefficients a and b. Denoting by $\{P_x^{(n)}, x \in R^d\}$ the solution to the martingale problem for L_n with coefficients a_n and b_n as in* (10.2), *then there exists a solution to the martingale problem for a and b if and only if*

$$\lim_{n\to\infty} P_x^{(n)}(\tau_{D_n} \leq t) = 0, \quad \textit{for all } t > 0 \textit{ and all } x \in R^d, \quad (10.4)$$

where $\tau_{D_n} = \inf\{t \geq 0: X(t) \notin D_n\}$. *If* (10.4) *holds, then the unique solution* $\{P_x, x \in R^d\}$ *to the martingale problem for L satisfies*

$$P_x = P_x^{(n)} \textit{ on } \mathcal{F}_{\tau_{D_n}}, \text{ for } n = 1, 2, \ldots, \textit{and } x \in R^d.$$

The family $\{P_x, x \in R^d\}$ *possesses the strong Markov property. For any finite stopping time* τ,

$$P_x(X(\tau + t) \in B|\mathcal{F}_\tau) = P_{X(\tau)}(X(t) \in B).$$

To complete the proof of Theorem 10.4, we must show that (10.4) guarantees the existence of a probability measure P_x on $(\Omega, \mathcal{F})$ which satisfies $P_x = P_x^{(n)}$ on $\mathcal{F}_{\tau_{D_n}}$, for $n = 1, 2, \ldots$, and $x \in R^d$. We will use the following variation of the Tulcea extension theorem.

Theorem 10.5. *Let* $(Y, \mathcal{M})$ *be a measurable space and* $\{\mathcal{M}_n, n \geq 0\}$ *a non-decreasing sequence of* σ*-algebras such that* $\mathcal{M} = \sigma(\bigcup_{n=0}^\infty \mathcal{M}_n)$. *For* $q \in Y$, *define* $A_n(q) = \cap\{B: B \in \mathcal{M}_n \textit{ and } q \in B\}$. *For* $n \geq 1$, *let* $\pi^n(q, \mathrm{d}q')$ *be a transition function from* $(Y, \mathcal{M}_{n-1})$ *to* $(Y, \mathcal{M}_n)$ *satisfying* $\pi^n(q, B) = 0$, *for all* $B \in \mathcal{M}_n$ *such that* $A_{n-1}(q) \cap B = \emptyset$. *Assume also that the following condition holds:*

$$\textit{Every sequence } \{q_n\}_{n=1}^\infty \subset Y \textit{ which satisfies } \bigcap_{n=0}^N A_n(q_n) \neq \emptyset,$$

$$\textit{for every } N, \textit{also satisfies } \bigcap_{n=0}^\infty A_n(q_n) \neq \emptyset. \quad (10.5)$$

Then, given any probability measure P_0 *on* $\mathcal{M}_0$, *there exists a unique probability measure P on* $(Y, \mathcal{M})$ *such that* $P = P_0$ *on* $\mathcal{M}_0$ *and for every n*,

$$P(B) = \int_Y \pi^n(q, B)P(\mathrm{d}q), \text{ for } B \in \mathcal{M}_n.$$

For a proof of Theorem 10.5, see [Stroock and Varadhan (1979), Theorem 1.1.9].

In order to apply Theorem 10.5, we need a construction which will also prove useful in subsequent sections. Let Δ denote the point at infinity in R^d and define $\hat{R}^d = R^d \cup \{\Delta\}$. Endow $\hat{R}^d$ with the one-point compactification topology and let $\hat{\mathcal{B}}_d$ denote the Borel sets on $\hat{R}^d$. Define a metric $\rho(\cdot, \cdot)$ on $\hat{R}^d$ by identifying it with S^d via sterographic projection and using the standard Riemannian metric on

S^d. Let $C([0, \infty), \hat{R}^d)$ denote the space of continuous functions ω from $[0, \infty)$ to $\hat{R}^d$ with the topology of uniform convergence (with respect to the metric on $\hat{R}^d$) on bounded intervals. Let $X(t, \omega)$ denote the position of ω at time t. Replacing $|X(t, \omega) - X(t, \omega')|$ in (8.1) by $\rho(X(t, \omega), X(t, \omega'))$, it follows that the resulting expression defines a compatible metric for $C([0, \infty), \hat{R}^d)$ under which $C([0, \infty), \hat{R}^d)$ becomes a complete separable metric space. Define $\tau_{D_n}(\omega)$ as before and let $\tau_{R^d} \equiv \lim_{n\to\infty} \tau_{D_n}$.

Now define

$$\hat{\Omega}_{R^d} = \{\omega \in C([0, \infty), \hat{R}^d)\text{: either } \tau_{R^d} = \infty \text{ or } \tau_{R^d} < \infty$$
$$\text{and } X(\tau_{R^d} + t) = \Delta, \text{ for all } t \geq 0\}. \quad (10.6)$$

That is, $\hat{\Omega}_{R^d}$ consists of those functions in $C([0, \infty), \hat{R}^d)$ which either never reach Δ or which, upon reaching Δ, remain there forever; $\hat{\Omega}_{R^d}$ is a complete separable metric space since it is a closed subset of $C([0, \infty), \hat{R}^d)$.

Let $\hat{\mathcal{F}}$ denote the Borel σ-algebra on $\hat{\Omega}_{R^d}$ and define the filtration

$$\hat{\mathcal{F}}_t^{R^d} = \sigma(X(s), 0 \leq s \leq t).$$

It is well known that $\hat{\mathcal{F}}^{R^d} = \sigma(\hat{\mathcal{F}}_t^{R^d}, 0 \leq t < \infty)$. Since all functions in $\hat{\Omega}_{R^d}$ which reach Δ remain there forever, we also have $\hat{\mathcal{F}}^{R^d} = \sigma(\hat{\mathcal{F}}_{\tau_{D_n}}^{R^d}, n = 0, 1, 2, \ldots)$, where, by convention, $\tau_{D_0} \equiv 0$.

Now define $\tilde{\Omega}_{R^d}$ to be the space of trajectories from $[0, \infty)$ to $\hat{R}^d$ which either belong to $\hat{\Omega}_{R^d}$ or else satisfy all of the following four conditions: (i) $\tau_{R^d}(\omega) \in (0, \infty)$, (ii) $\omega \in C([0, \tau_{R^d}(\omega)), R^d)$, (iii) $\liminf_{t\to\tau_{R^d}(\omega)} |X(t, \omega)| < \infty$, and (iv) $X(t, \omega) = \Delta$, for $t \geq \tau_{R^d}(\omega)$. Endow $\tilde{\Omega}_{R^d}$ with the topology of uniform convergence (with respect to the metric on $\hat{R}^d$) on bounded intervals. Let $\tilde{\mathcal{F}}^{R^d}$ denote the Borel σ-algebra on $\tilde{\Omega}_{R^d}$ and let $\tilde{\mathcal{F}}_t^{R^d} = \sigma(X(s), 0 \leq s \leq t)$. Analogous to the situation with $\hat{\Omega}_{R^d}$, we have $\tilde{\mathcal{F}}^{R^d} = \sigma(\tilde{\mathcal{F}}_t^{R^d}, 0 \leq t < \infty) = \sigma(\tilde{\mathcal{F}}_{\tau_{D_n}}^{R^d}, n = 0, 1, 2, \ldots)$.

To prove Theorem 10.4, we will apply Theorem 10.5 to the space $(\tilde{\Omega}_{R_d}, \tilde{\mathcal{F}}^{R^d})$ with the non-decreasing sequence of σ-algebras $\{\tilde{\mathcal{F}}_{\tau_{D_n}}^{R^d}\}_{n=0}^{\infty}$. The reason we need the space $(\tilde{\Omega}_{R^d}, \tilde{\mathcal{F}}^{R^d})$ is that (10.5) does not hold for the space $(\hat{\Omega}_{R^d}, \hat{\mathcal{F}}^{R^d})$ with the non-decreasing sequence of σ-algebras $\{\hat{\mathcal{F}}_{\tau_{D_n}}^{R^d}\}$ (exercise 1.11).

Proof of Theorem 10.4. We apply Theorem 10.5 to $(\tilde{\Omega}_{R^d}, \tilde{\mathcal{F}}^{R^d})$ and $\{\tilde{\mathcal{F}}_{\tau_{D_n}}^{R^d}\}_{n=0}^{\infty}$. Clearly,

$$A_n(\omega) = \{\omega' \in \tilde{\Omega}_{R^d}\text{: } \omega'(t), \text{ for } 0 \leq t \leq \tau_{D_n}(\omega)\}. \quad (10.7)$$

It is easy to see that (10.5) holds (exercise 1.11). Fix $x \in R^d$ and define P_0 on $\tilde{\mathcal{F}}_0^{R^d}$ by $P_0(X(0) = x) = 1$. In the sequel, we identify $\mathcal{F}_{\tau_{D_n}}$ with $\tilde{\mathcal{F}}_{\tau_{D_n}}^{R^d}$ in the obvious way. It is easy to check that for each $\omega \in \tilde{\Omega}_{R^d}$ and each $n \geq 1$, there exists a unique probability measure $\delta_\omega \otimes_{\tau_{D_{n-1}}} P^{(n)}$ on τ_{D_n}) which satisfies

$$(\delta_\omega \otimes_{\tau_{D_{n-1}}} P^{(n)})(\omega' \in \tilde{\Omega}^{R^d} : \omega'(s) = \omega(s), 0 \leq s \leq \tau_{D_{n-1}}(\omega)) = 1$$

and

$$(\delta_\omega \otimes_{\tau_{D_{n-1}}} P^{(n)})(B) = P^{(n)}_{X(\tau_{D_{n-1}}(\omega),\omega)}(\omega' : \omega'(\cdot + \tau_{D_{n-1}}(\omega)) \in B),$$

for $B \in \sigma(X((\tau_{D_{n-1}} + t) \wedge \tau_{D_n})), 0 \leq t < \infty$.

For $n \geq 1$, define the transition function from $(\tilde{\Omega}_{R^d}, \tilde{\mathcal{F}}^{R^d}_{\tau_{D_{n-1}}})$ to $(\tilde{\Omega}_{R^d}, \tilde{\mathcal{F}}^{R^d}_{\tau_{D_n}})$ by

$$\pi^n(\omega, B) = (\delta_\omega \otimes_{\tau_{D_{n-1}}} P^{(n)})(B), \text{ for } B \in \tilde{\mathcal{F}}^{R^d}_{\tau_{D_n}}.$$

From (10.7), it is clear that $\pi^n(\omega, B) = 0$, if $A_{n-1}(\omega) \cap B = \emptyset$. We are now in a position to apply Theorem 10.5. There exists a unique measure P_x on $(\tilde{\Omega}_{R^d}, \tilde{\mathcal{F}}^{R^d})$ such that $P_x(X(0) = x) = 1$ and $P_x(B) = \int_{\tilde{\Omega}} \pi^n(\omega, B) P_x(d\omega)$, for $B \in \tilde{\mathcal{F}}^{R^d}_{\tau_{D_n}}$. Note that $\pi^n(\omega, B)$ is $\tilde{\mathcal{F}}^{R^d}_{\tau_{D_{n-1}}}$-measurable. Thus, by induction and the definition of π^n, it follows that $P_x|_{\tilde{\mathcal{F}}^{R^d}_{\tau_{D_n}}} = P^{(n)}|_{\mathcal{F}_{\tau_{D_n}}}$. To complete the proof of Theorem 10.4, we will show that $\operatorname{supp}(P_x) \subseteq \Omega$. For this, it is enough to show that $P_x(\lim_{n\to\infty} \tau_{D_n} < \infty) = 0$ or, equivalently, that $\lim_{n\to\infty} P_x(\tau_{D_n} < t) = 0$, for all $t \geq 0$. But $\lim_{n\to\infty} P_x(\tau_{D_n} \leq t) = \lim_{n\to\infty} P_x^{(n)}(\tau_{D_n} \leq t) = 0$ by (10.4). □

The estimates of Stroock and Varadhan, used to prove Theorem 10.2, show that under the conditions of that theorem, at each $x \in R^d$ the Feller property holds with a certain modulus of continuity which depends only on

$$\inf_{\theta \in R^d - \{0\}} \frac{\langle \theta, a(x)\theta \rangle}{|\theta|^2}, \quad \sup_{\theta \in R^d - \{0\}} \frac{\langle \theta, a(x)\theta \rangle}{|\theta|^2}, \quad \sup_{|x-y| \leq 1} |b(y)|$$

and on the modulus of continuity of a at x. Similarly, certain L^p-estimates for the density $p(t, x, y)$ on compact sets are shown to hold with constants that depend only on quantities such as these with x varying over compact sets [Stroock and Varadhan (1979), Theorem 7.2.4 and Corollary 9.1.10]. From these facts it is easy to deduce that the Feller property holds and that a density exists in the case of locally bounded coefficients. We record this as follows.

Theorem 10.6. *Let $\{P_x, x \in R^d\}$ denote the solution to the martingale problem for L in R^d under the conditions of Theorem 10.4. Then $\{P_x, x \in R^d\}$ possesses the Feller property. Furthermore, the probability measure $P_x(X(t) \in \mathrm{d}y)$ possesses a density $p(t, x, y)$ and for any finite stopping time τ,*

$$P_x(X(\tau + t) \in B|\mathcal{F}_\tau) = P_{X(\tau)}(X(t) \in B) = \int_B p(t, X(\tau), y)\,\mathrm{d}y,$$

for all $t > 0$ and $B \in \mathcal{B}_d$.

1.11 The generalized martingale problem on R^d

If (10.4) does not hold, then the process corresponding to a and b that one attempts to construct via a_n and b_n will *explode*. That is, for some (and in fact, therefore, for all) $x \in R^d$, $P_x(\tau_{R^d} < \infty) > 0$ (see Section 15).

To understand heuristically how this may occur, consider the operator

$$L = \tfrac{1}{2}\frac{\mathrm{d}^2}{\mathrm{d}x^2} + |x|^\delta \frac{\mathrm{d}}{\mathrm{d}x} \text{ on } R, \text{ where } \delta \geq 0.$$

Now, if a solution $\{P_x, x \in R\}$ to the martingale problem were to exist, then recalling the stochastic differential equation of Section 7, one would expect that under P_x, $X(t)$ could be represented as $X(t) = x + B(t) + \int_0^t |X(s)|^\delta \,\mathrm{d}s$, where $B(t)$ is a Brownian motion. Let $y(t)$ be the deterministic solution of the equation obtained by deleting $B(t)$ from the above equation; then $y(t) = x + \int_0^t |y(s)|^\delta \,\mathrm{d}s$ or, equivalently, $y'(t) = |y(t)|^\delta$ and $y(0) = x$. Assume that $x > 0$. Solving the equation, we obtain $y(t) = (x^{1-\delta} + (1-\delta)t)^{(1/1-\delta)}$, if $\delta \neq 1$ and $y(t) = xe^t$, if $\delta = 1$. In particular, if $\delta \leq 1$, then $y(t)$ is defined for all time, while if $\delta > 1$, then $y(t)$ blows up in finite time. It is well known that Brownian motion $B(t)$ satisfies

$$\lim_{t\to 0} \frac{B(t)}{t^{1/2+\varepsilon}} = 0 \text{ a.s., for any } \varepsilon > 0.$$

Thus, if $\delta > 1$, it is clear intuitively that $X(t)$ should also blow up in finite time. For a rigorous proof of this, see Example 5.1.7. We note that in Section 7, the existence of a solution existing for all time to the stochastic differential equation (7.2) was proved in the case of uniformly Lipschitz coefficients; this condition holds for the above operator only if $\delta \leq 1$.

If the coefficients a and b satisfy the conditions of Theorem 10.4, but (10.4) does not hold, then although there is no solution to the martingale problem as we have formulated it in Section 8, there will be a unique solution to the following generalized martingale problem. Recall the definition of $\hat{\Omega}_{R^d}$ given in (10.6).

The generalized martingale problem for L on R^d. Let $\{D_n\}_{n=1}^{\infty}$ be an increasing sequence of domains such that $D_n \subset\subset D_{n+1}$ and $\bigcup_{n=1}^{\infty} D_n = R^d$. For each $x \in \hat{R}^d$, the one-pont compactification of R^d, find a probability measure P_x on $(\hat{\Omega}_{R^d}, \hat{\mathcal{F}}^{R^d})$ such that

(i) $P_x(X(0) = x) = 1$;
(ii) $f(X(t \wedge \tau_{D_n})) - \int_0^{t\wedge\tau_{D_n}} (Lf)(X(s))\,ds$ is a martingale with respect to $(\hat{\Omega}_{R^d}, \hat{\mathcal{F}}^{R^d}, \hat{\mathcal{F}}_t^{R^d}, P_x)$, for all $n = 1, 2, \ldots$, and all $f \in C^2(R^d)$, where $\tau_{D_n} = \inf\{t \geq 0: X(t) \notin D_n\}$.

The proof of Theorem 10.4 revealed that for each $x \in R^d$, there exists a unique probability measure P_x on $(\tilde{\Omega}_{R^d}, \tilde{\mathcal{F}}^{R^d})$ which satisfies

$$P_x|_{\tilde{\mathcal{F}}^{R^d}_{\tau_{D_n}}} = P_x^{(n)}|_{\mathcal{F}_{\tau_{D_n}}}.$$

(Again, we make the natural identification between $\mathcal{F}_{\tau_{D_n}}$ and $\tilde{\mathcal{F}}^{R^d}_{\tau_{D_n}}$.) Clearly, then, P_x satisfies (i) and (ii) above. To complete the proof of existence and uniqueness for the generalized martingale problem on R^d, it remains to show that P_x is in fact supported on $\hat{\Omega}$. That is, we must show that $P_x(\tau_{R^d} < \infty, \liminf_{t\to\tau_{R^d}} |X(t)| < \infty) = 0$. It is enough to show that for any $t_0 > 0$ and any integer $r_0 \geq 1$,

$$P_x(\tau_{R^d} \leq t_0, \liminf_{t\to\tau_{R^d}} |X(t)| < r_0) = 0. \tag{11.1}$$

Define

$$\eta_1 = \inf\{t \geq 0: |X(t)| \leq r_0\},\ \eta_{2n} = \inf\{t \geq \eta_{2n-1}: |X(t)| \geq r_0 + 1\}$$

and

$$\eta_{2n+1} = \inf\{t \geq \eta_{2n}: |X(t)| \leq r_0\}.$$

We have

$$P_x(\tau_{R^d} \leq t_0, \liminf_{t\to\tau_{R^d}} |X(t)| < r_0) \leq P_x\left(\bigcap_{n=1}^{\infty} \{\eta_n \leq t_0\}\right). \tag{11.2}$$

In the calculations below, we will use the notation $E_x^{(n)}$ for the expectation with respect to $P_x^{(n)}$ and E_x for the expectation with respect to P_x. Also, let $\tau_m = \inf\{t \geq 0: |X(t)| \geq m\}$ and let n_0 be such that $\{x: |x| < r_0 + 1\} \subset D_{n_0}$. For any positive integer N, we have

$$
\begin{aligned}
P_x\left(\bigcap_{n=1}^{2N}\{\eta_n \le t_0\}\right) &= E_x \prod_{n=1}^{2N} I_{\{\eta_n \le t_0\}} \\
&= E_x \prod_{n=1}^{2N-1} I_{\{\eta_n \le t_0\}} E_x(I_{\{\eta_{2N}\le t_0\}} | \tilde{\mathcal{F}}^{R^d}_{\eta_{2N-1}}) \\
&\le E_x \prod_{n=1}^{2N-1} I_{\{\eta_n \le t_0\}} E_{X(\eta_{2N-1})} I_{\{\eta_2 \le t_0\}} \\
&= E_x \prod_{n=1}^{2N-1} I_{\{\eta_n \le t_0\}} E_{X(\eta_{2N-1})} I_{\{\tau_{r_0+1} \le t_0\}} \\
&= E_x \prod_{n=1}^{2N-1} I_{\{\eta_n \le t_0\}} E^{(n_0)}_{X(\eta_{2N-1})} I_{\{\tau_{r_0+1} \le t_0\}}, \qquad (11.3)
\end{aligned}
$$

where the penultimate equality follows from the fact that $|X(\eta_{2N-1})| \le r_0$ and that $\eta_2 = \tau_{r_0+1}$ a.s. P_z, for $|z| \le r_0$, and the last equality follows from the fact that $P_x|^{R^d}_{\mathcal{F}_{\tau_{r_0+1}}} = P^{(n_0)}_x|_{\mathcal{F}_{\tau_{r_0+1}}}$. Since $X(t)$ possesses a density under $P^{(n_0)}_x$, we have $P^{(n_0)}_x(\tau_{r_0+1} = t_0) = 0$. From this and Theorem 2.3.3(iv) of Chapter 2, it follows that $I_{\{\tau_{r_0+1} \le t_0\}}$ is a $P^{(n_0)}_x$ – a.s. continuous function on Ω. Thus by Theorem 3.1, $P^{(n_0)}_z(\tau_{r_0+1} \le t_0) = E^{(n_0)}_z I_{\{\tau_{r_0+1} \le t_0\}}$ is a continuous function of z. Let $\gamma = \sup_{|z|\le r_0} P^{(n_0)}_z(\tau_{r_0+1} \le t_0)$. We claim that $\gamma < 1$. To see this, note that if $\gamma = 1$, then by continuity there exists a z_0 with $|z_0| \le r_0$ such that $P^{(n_0)}_{z_0}(\tau_{r_0+1} \le t_0) = 1$. But this is impossible by the Stroock–Varadhan support theorem (Theorem 2.6.1 in Chapter 2). Using this in (11.3), we obtain

$$
\begin{aligned}
P_x\left(\bigcap_{n=1}^{2N}\{\eta_n \le t_0\}\right) &\le \gamma E_x \prod_{n=1}^{2N-1} I_{\{\eta_n \le t_0\}} \\
&= \gamma P_x\left(\bigcap_{n=1}^{2N-1}\{\eta_n \le t_0\}\right) \le \gamma P_x\left(\bigcap_{n=1}^{2N-2}\{\eta_n \le t_0\}\right).
\end{aligned}
$$

By induction, $P_x(\bigcap_{n=1}^{2N}\{\eta_n \le t_0\}) \le \gamma^N$ and thus $P_x(\bigcap_{n=1}^{\infty}\{\eta_n \le t_0\}) = 0$. Now (11.1) follows from (11.2).

We have proved that there exists a unique solution to the generalized martingale problem for L on R^d. The Feller property for $\{P_x, x \in R^d\}$ and the existence of a density again follow from the estimates noted prior to the statement of Theorem 10.6. The density will be a subprobability density on R^d; there may be an atom at $x = \Delta$ which accounts for the rest of the probability. The strong Markov property is again a consequence of uniqueness (see the remark following Theorem 10.2). We record all this as follows.

Theorem 11.1. *Let a and b be as in Theorem* 10.4. *Then there exists a unique solution* $\{P_x, x \in \hat{R}^d\}$ *to the generalized martingale problem for* L *on* R^d *with coefficients a and b. Denoting by* $\{P_x^{(n)}, x \in R^d\}$ *the solution to the martingale problem for* L_n *with coefficients* a_n *and* b_n *as defined in* (10.2), *then* $P_x|_{\hat{\mathcal{F}}_{\tau_{D_n}}^{R^d}} = P_x^{(n)}|_{\mathcal{F}_{\tau_{D_n}}}$, *for* $n = 1, 2, \ldots$, *and* $x \in R^d$ *(where* $\mathcal{F}_{\tau_{D_n}}$ *has been identified with* $\hat{\mathcal{F}}_{\tau_{D_n}}^{R^d}$ *in the obvious way). The subprobability measure* $P_x(X(t) \in dy)$ *on* R^d *possesses a measurable density* $p(t, x, y)$. *The family* $\{P_x, x \in R^d\}$ *possesses the Feller property and the family* $\{P_x, x \in \hat{R}^d\}$ *possesses the strong Markov property. For any finite stopping time* τ,

$$P_x(X(\tau + t) \in B | \hat{\mathcal{F}}_\tau^{R^d}) = P_{X(\tau)}(X(t) \in B) = \int_B p(t, X(\tau), y)\, dy,$$

for all $t \geq 0$ *and* $B \in \mathcal{B}_d$. *(If* (10.4) *holds, then for all* $x \in R^d$, P_x *is in fact supported on* Ω *and* $\{P_x, x \in R^d\}$ *solves the martingale problem for* L *on* R^d.)

Remark. The Feller property does not necessarily hold at $x = \Delta$ (see the remark in Section 4 of Chapter 8, following the definition of *explosion inward from the boundary*).

1.12 The martingale problem on $D \subseteq R^d$

For our applications in Chapter 4, the following set-up will be very important. Let

$$L = \tfrac{1}{2} \sum_{i,j=1}^{d} a_{ij}(x) \frac{\partial^2}{\partial x_i \partial x_j} + \sum_{i=1}^{d} b_i(x) \frac{\partial}{\partial x_i}$$

be defined on an arbitrary domain $D \subseteq R^d$, where $a: D \to S_d$ and $b: D \to R^d$ are locally bounded on D.

The martingale problem for L on D. For each $x \in D$, find a probability measure P_x on $(\Omega, \mathcal{F})$ such that

(i) $P_x(X(0) = x) = 1$;
(ii) $P_x(X(t) \in D \text{ for all } t \geq 0) = 1$;
(iii) $f(X(t)) - \int_0^t (Lf)(X(s))\, ds$ is a martingale with respect to $(\Omega, \mathcal{F}, \mathcal{F}_t, P_x)$, for all $f \in C_0^2(D)$.

Remark. From Theorem 10.3 and from Theorem 2.6.1 of Chapter 2, if $\bar{D} \neq R^d$, then (ii) cannot hold if a is uniformly positive definite and a and b are bounded in D. As we shall see in Chapter 4, the martingale problem in D will have a solution in certain instances if a is bounded on D and the drift $b(x)$ satisfies $\lim_{x\to\partial D} |b(x)| = \infty$.

To treat existence and uniqueness for the solution to the martingale problem on D, choose a sequence $\{D_n\}_{n=1}^{\infty}$ of increasing domains such that $D_n \subset\subset D_{n+1}$ and $D = \bigcup_{n=1}^{\infty} D_n$. For each $n = 1, 2, \ldots$, let ψ_n: $R^d \to R$ be a C^∞-function satisfying $\psi_n(x) = 1$, for $x \in D_n$, $\psi_n = 0$, for $x \notin D_{n+1}$, and $0 \leq \psi_n \leq 1$. Define a_n: $R^d \to S_d$ and b_n: $R^d \to R^d$ by

$$\left.\begin{aligned} a_n &= \psi_n a + (1 - \psi_n)I, \\ b_n &= \psi_n b. \end{aligned}\right\} \tag{12.1}$$

We now make a construction analogous to that used to prove Theorems 10.4 and 11.1. Let $\hat{D} = D \cup \{\Delta\}$ denote the one-point compactification of D. Thus Δ is identified with the boundary ∂D if D is bounded and with ∂D and the point at infinity if D is unbounded. Let $\mathcal{B}(\hat{D})$ denote the Borel sets of $\hat{D}$. Define a metric on $\hat{D}$ as follows. In Section 10, we defined a metric ρ on $\hat{R}^d$ by identifying it with S^d via stereographic projection and using the standard Riemannian metric on S^d. We can then define on $\hat{D}$ the metric

$$\rho_D(x, y) = \begin{cases} \inf_{z\in\partial D\cup\{\Delta\}} \rho(x, z), \text{ if } x \in D \text{ and } y = \Delta \\ \min(\rho(x, y), \rho_D(x, \Delta) + \rho_D(y, \Delta)), \text{ if } x, y \in D. \end{cases}$$

Let $C([0, \infty), \hat{D})$ denote the space of continuous functions ω from $[0, \infty)$ to $\hat{D}$ with the topology of uniform convergence (with respect to the metric on $\hat{D}$) on bounded intervals. Let $X(t, \omega)$ denote the position of ω at time t. Replacing $|X(t, \omega) - X(t,\omega')|$ in (8.1) by $\rho_D(X(t, \omega), X(t, \omega'))$, it follows that the resulting expression defines a compatible metric for $C([0, \infty), \hat{D})$ under which $C([0, \infty), \hat{D})$ becomes a complete separable metric space. Define $\tau_{D_n}(\omega) = \inf\{t \geq 0$: $X(t, \omega) \notin D_n\}$ and let $\tau_D \equiv \lim_{n\to\infty} \tau_{D_n}$. Now define

$$\hat{\Omega}_D = \left\{\begin{aligned} \omega \in C([0, \infty), \hat{D})\text{: either } \tau_D = \infty \text{ or } \tau_D < \infty \text{ and} \\ X(\tau_D + t) = \Delta, \text{ for all } t > 0 \end{aligned}\right\}. \tag{12.2}$$

$\hat{\Omega}_D$ is a complete separable metric space since it is a closed subset of $C([0, \infty), \hat{D})$. Let $\hat{\mathcal{F}}^D$ denote the Borel σ-algebra on $\hat{\Omega}_D$ and define the filtration $\hat{\mathcal{F}}_t^D = \sigma(X(s), 0 \leq s \leq t)$. Then analogous to the case with $\hat{\Omega}_{R^d}$, we have $\hat{\mathcal{F}}^D = \sigma(\hat{\mathcal{F}}_t^D, 0 \leq t < \infty) = \sigma(\hat{\mathcal{F}}_{\tau_{D_n}}^D, n = 1, \ldots)$.

Now define $\tilde{\Omega}_D$ to be the space of trajectories $\omega = X(\cdot, \omega)$ from

$[0, \infty)$ to $\hat{D}$ which either belong to $\hat{\Omega}_D$ or else satisfy all of the following four conditions: (i) $\tau_D(\omega) = (0, \infty)$, (ii) $\omega \in C([0, \tau_D(\omega)), D)$, (iii) $\limsup_{t \to \tau_D(\omega)} \rho_D(X(t, \omega), \Delta) > 0$ and (iv) $X(t, \omega) = \Delta$ for $t \geq \tau_D(\omega)$. Endow $\tilde{\Omega}_D$ with the topology of uniform convergence (with respect to the metric on $\hat{D}$) on bounded intervals. Let $\tilde{\mathcal{F}}^D$ denote the Borel σ-algebra on $\tilde{\Omega}_D$ and let $\tilde{\mathcal{F}}_t^D = \sigma(X(s), 0 \leq s \leq t)$. We have $\tilde{\mathcal{F}}^D = \sigma(\tilde{\mathcal{F}}_t^D, 0 \leq t < \infty) = \sigma(\tilde{\mathcal{F}}_{\tau_{D_n}}^D, n = 1, 2, \ldots)$.

With the above set-up, we may proceed exactly as we did in the proof of Theorem 10.4. This gives us a proof of the following theorem, with the exception of the statement concerning the existence of a density. For the existence of a density, see exercise 1.12.

Theorem 12.1. *Let $a: D \to S_d$ and $b: D \to R^d$ be locally bounded and measurable on D and assume that a is continuous on D and that*

$$\sum_{i,j=1}^{d} a_{ij}(x) v_i v_j > 0, \text{ for } x \in D \text{ and } v \in R^d - \{0\}.$$

Then there exists at most one solution to the martingale problem for L on D with coefficients a and b. Letting $\{P_x^{(n)}, x \in R\}$ denote the unique solution to the martingale problem on R^d for L_n with coefficients a_n and b_n as in (12.1), then there exists a solution $\{P_x, x \in D\}$ to the martingale problem for L on D if and only if

$$\lim_{n \to \infty} P_x^{(n)}(\tau_{D_n} \leq t) = 0, \text{ for all } t > 0 \text{ and } x \in D, \tag{12.3}$$

where $\tau_{D_n} = \inf\{t \geq 0: X(t) \notin D_n\}$. If (12.3) holds, then the solution $\{P_x, x \in D\}$ satisfies

$$P_x = P_x^{(n)} \text{ on } \mathcal{F}_{\tau_{D_n}}, n = 1, 2, \ldots, x \in D.$$

The family $\{P_x, x \in D\}$ possesses the Feller property and the strong Markov property. Furthermore, the probability measure $P_x(X(t) \in dy)$ possesses a density $p(t, x, y)$ and for any finite stopping time τ,

$$P_x(X(\tau + t) \in B | \mathcal{F}_\tau) = P_{X(\tau)}(X(t) \in B) = \int_B p(t, X(\tau), y) \, dy,$$

for all $t \geq 0$ and $B \in \mathcal{B}_d$.

1.13 The generalized martingale problem on $D \subseteq R^d$

Just as we defined the generalized martingale problem on R^d, it will be convenient for our applications in the sequel to define the generalized martingale problem for L on D. Recall the definition of $\hat{\Omega}_D$ given in (12.2).

The generalized martingale problem for L on D. Let $\{D_n\}_{n=1}^{\infty}$ be an increasing sequence of domains such that $D_n \subset\subset D_{n+1}$ and $\bigcup_{n=1}^{\infty} D_n = D$. For each $x \in \hat{D}$, the one-point compactification of D, find a probability measure P_x on $(\hat{\Omega}_D, \hat{\mathcal{F}}^D)$ such that

(i) $P_x(X(0) = x) = 1$;
(ii) $f(X(t \wedge \tau_{D_n})) - \int_0^{t \wedge \tau_{D_n}} (Lf)(X(s))\, ds$ is a martingale with respect to $(\hat{\Omega}_D, \hat{\mathcal{F}}^D, \hat{\mathcal{F}}_t^D, P_x)$, for all $n = 1, 2, \ldots$ and all $f \in C^2(D)$.

The extension from the martingale problem on D to the generalized martingale problem on D is essentially identical to the extension from the martingale problem on R^d to the generalized martingale problem on R^d. We record this extension as follows.

Theorem 13.1. *Let the coefficients a and b of L be as in Theorem* 12.1. *Then there exists a unique solution $\{P_x, x \in \hat{D}\}$ to the generalized martingale problem for L on D. Letting $\{P_x^{(n)}, x \in R^d\}$ denote the unique solution to the martingale problem on R^d for L_n with coefficients a_n and b_n as in* (12.1), *then $P_x|_{\hat{\mathcal{F}}^D_{\tau_{D_n}}} = P_x^{(n)}|_{\mathcal{F}_{\tau_{D_n}}}$, for $n = 1, 2, \ldots$, and $x \in D$ (where $\mathcal{F}_{\tau_{D_n}}$ has been identified with $\hat{\mathcal{F}}^D_{\tau_{D_n}}$ in the obvious way). The family $\{P_x, x \in D\}$ possesses the Feller property and the family $\{P_x, x \in \hat{D}\}$ possesses the strong Markov property. Furthermore, the sub-probability measure $P_x(X(t) \in dy)$ on D possesses a density $p(t, x, y)$ and for any finite stopping time τ,*

$$P_x(X(\tau + t) \in B | \hat{\mathcal{F}}_\tau^D) = P_{X(\tau)}(X(t) \in B), \text{ for all } t \geq 0 \text{ and } B \in \mathcal{B}_d.$$

If (12.3) *holds, then $\{P_x, x \in D\}$ in fact solves the martingale problem for L on D.*

Remark 1. As noted in the remark following Theorem 11.1, the Feller property does not necessarily hold at $x = \Delta$.

Remark 2. Note that the generalized martingale problem for L on D includes the other three martingale problems as particular cases.

Theorem 10.3 extends easily to the context of the generalized martingale problem. We record this as follows.

Theorem 13.2. *Let $\{P_x, x \in \hat{D}_1\}$ solve the generalized martingale problem on $D_1 \subseteq R^d$ for L_1 and let $\{Q_x, x \in \hat{D}_2\}$ solve the generalized* *martingale problem on* $D_2 \subseteq R^d$ *for* L_2. *Let* $U \subset D_1 \cap D_2$ *be a domain on which the coefficients of* L_1 *and* L_2 *coincide. Assume that these coefficients, when restricted to* U, *satisfy the conditions of* Theorem 13.1. *Then for all* $x \in U$, $P_x|_{\mathcal{F}_{\tau_U}} = Q_x|_{\mathcal{F}_{\tau_U}}$. ($\mathcal{F}_{\tau_U}$ has been identified with $\hat{\mathcal{F}}^{D_i}_{\tau_U}$, $i = 1, 2$, in the obvious way.)

1.14 The canonical construction and the definition of a diffusion process

Up until Section 8, we considered processes $\zeta(t)$ on a space $(Y, \mathcal{M}, \mathcal{P})$ and, in particular, in Sections 5–7 we studied Itô processes. Starting in Section 8, we changed tack and began considering measures on the spaces $(\Omega, \mathcal{F})$ and $(\hat{\Omega}_D, \hat{\mathcal{F}}^D)$. At this stage, we want to point out how the set-up in Sections 8–13 can be translated to the framework of Sections 1–7. It is sufficient to consider the generalized martingale problem on $D \subseteq R^d$ since, as noted, this case contains the three other martingale problems.

Assume that the family $\{P_x, x \in \hat{D}\}$ of probability measures on $(\hat{\Omega}_D, \hat{\mathcal{F}}^D)$ solves the generalized martingale problem for L on D. Let $(Y, \mathcal{M}) = (\hat{\Omega}_D, \hat{\mathcal{F}}^D)$ and let $\zeta: [0, \infty) \times Y \to R^d$ be defined by $\zeta(t, \omega) = X(t, \omega)$; that is, ζ is the identity function. Then for each $x \in \hat{D}$, $\zeta = \zeta(t)$ is a stochastic process on $(Y, \mathcal{M}, P_x)$ with distribution P_x on $(\hat{\Omega}_D, \hat{\mathcal{F}}^D)$. This construction is called the *canonical construction* of a stochastic process with distribution P_x on $(\hat{\Omega}_D, \hat{\mathcal{F}}^D)$, and $\zeta(t, \omega)$ is called the *canonical process*. Since $\zeta(t, \omega) = X(t, \omega)$, in the sequel we will just write $X(t, \omega)$.

We now give our definition of a diffusion process. A *diffusion process* is the stochastic process corresponding to a solution to the generalized martingale problem for a second-order elliptic operator L.

1.15 The lifetime of a diffusion process and the definition of explosion

Let $\{P_x, x \in \hat{D}\}$ be the solution to the generalized martingale problem for an operator L on $D \subseteq R^d$. Recall that $\tau_D = \lim_{n\to\infty} \tau_{D_n}$, where $\tau_{D_n} = \inf\{t \geq 0: X(t) \notin D_n\}$ and $\{D_n\}_{n=1}^{\infty}$ is a sequence of domains satisfying $D_n \subset\subset D_{n+1}$ and $\bigcup_{n=1}^{\infty} D_n = D$. Clearly, τ_D is independent of the choice of $\{D_n\}_{n=1}^{\infty}$. We will call τ_D the *lifetime* of the diffusion process $X(t)$. This definition is a bit misleading since $X(\cdot)$ is an element of $(\hat{\Omega}_D, \hat{\mathcal{F}}_D)$ and is thus defined for all $t \in [0, \infty)$. Of course, $X(t) = \Delta$, for all $t \geq \tau_D$, where Δ is the point at infinity in the one-point compactification of D. In order to justify calling τ_D the lifetime of the process, Δ is sometimes called the *cemetery state*.

If $P_x\ (\tau_D < \infty) > 0$ for some (and thus, by the theorem below, for all) $x \in D$, then the diffusion process $X(t)$ is said to *explode on D*.

We have the following theorem.

Theorem 15.1. *Let the coefficients of L be as in Theorem 12.1, and let $\{P_x, x \in \hat{D}\}$ denote the solution to the generalized martingale problem for L on D. If $P_x(\tau_D < \infty) > 0$, for some $x \in D$, then $P_x(\tau_D < \infty) > 0$, for all $x \in D$.*

Proof. The proof is an application of Harnack's inequality (Theorem 4.0.1 in Chapter 4) and is left as exercise 4.15. □

Exercises

1.1 Let $\zeta_n(t)$, $n = 1, 2, \ldots$ and $\zeta(t)$ be stochastic processes on $(Y, \mathcal{M}, \mathcal{M}_t, P)$. Assume that $\zeta_n(t)$ is a martingale with respect to $(Y, \mathcal{M}, \mathcal{M}_t, P)$, for each $n = 1, 2, \ldots,$ and that $\zeta_n(t) \to \zeta(t)$ in $L^1(Y, P)$, for each $t \geq 0$. Then $\zeta(t)$ is also a martingale with respect to $(Y, \mathcal{M}, \mathcal{M}_t, P)$. (Hint: Use the fact that a process $\eta(t)$ is a martingale with respect to $(Y, \mathcal{M}, \mathcal{M}_t, P)$ if and only if $\int_A E(\eta(t)|\mathcal{M}_s)\,dP = \int_A \eta(s)\,dP$, for all $0 \leq s \leq t < \infty$ and $A \in \mathcal{M}_s$.)

1.2 Let $\zeta(t)$ be an R^d-valued Markov process on $(Y, \mathcal{M}, \mathcal{M}_t, P)$ with transition probability function $p(s, x, t, dy)$. Show that (3.2) holds. (Hint: Use induction to show that (3.2) holds for Borel sets B in $\mathcal{B}([0, \infty), R^d)$, which are finite-dimensional rectangles and then appeal to a monotone class theorem.)

1.3 Show that Brownian motion is a Feller process.

1.4 Show that a continuous R^d-valued Feller process $\zeta(t)$ on $(Y, \mathcal{M}, \mathcal{M}_t, P)$ satisfies the strong Markov property and that, in particular, if $p(t, x, y)$ is the transition probability function, then for any finite stopping time τ, $P(\zeta(\tau + t) \in B|\mathcal{M}_\tau) = \int_B p(t, \zeta(\tau), dy)$, for $B \in \mathcal{B}_d$ and $0 \leq t < \infty$. (Hint: First do it for a stopping time with a discrete range. Then approximate the general finite stopping time by a non-increasing sequence $\{\tau_n\}_{n=1}^\infty$ of stopping times with discrete ranges which satisfy $\tau_n \geq \tau$ and $\lim_{n\to\infty} \tau_n = \tau$.)

1.5 Show that (5.1) implies (4.1).

1.6 Let $\zeta(t) \sim I_d(a, b)$ on $(Y, \mathcal{M}, \mathcal{M}_t, P)$. Then

$$E\left(\exp\left(\left\langle \theta, \zeta(t_2) - \zeta(t_1) - \int_{t_1}^{t_2} b(s)\,ds \right\rangle - \tfrac{1}{2}\int_{t_1}^{t_2} \langle \theta, a(s)\theta\rangle\,ds\right) \Big| \mathcal{M}_{t_1}\right) = 1 \text{ a.s.},$$

for all $0 \leq t_1 < t_2 < \infty$ and $\theta \in R^d$.

1.7 Let $(Y, \mathcal{M}, P)$ be a probability space and let $\mathcal{G} \subset \mathcal{M}$ be a sub σ-algebra. Let $(X, \mathcal{B})$ be a measurable space and let $F: Y \times X \to R$ be measurable with respect to $\mathcal{M} \times \mathcal{B}$ and satisfy $\sup_{x \in X} E|F(\cdot, x)| < \infty$. Show that a version $G(q, x)$ of $E(F(\cdot, x)|\mathcal{G})$ can be chosen so that $G(\cdot, \cdot)$ is $\mathcal{M} \times \mathcal{B}$-measurable. Use this to show that if $f: Y \to X$ is $\mathcal{G}$-measurable and $E|F(\cdot, f(\cdot))| < \infty$, then $E(F(\cdot, f(\cdot))|\mathcal{G}) = G(\cdot, f(\cdot))$.

1.8 Use Theorems 7.5 and 7.7 to show that under the conditions of Theorem 7.1, the solution $\zeta_x(t)$ of (7.2) satisfies (7.10) even if a and b are not bounded.

1.9 Let $\zeta(t) \sim I_d(a, b)$ on $(Y, \mathcal{M}, \mathcal{M}_t, P)$ and let $f, g \in \mathcal{H}_a^d$ ($\mathcal{H}_a^d$ is defined in Section 6). If $f(t, q) = g(t, q)$ for all $0 \leq t \leq T$ and all $q \in \mathcal{A} \subset Y$, then $\int_0^t f(s)\, d\zeta(s) = \int_0^t g(s)\, d\zeta(s)$, for all $0 \leq t \leq T$ and $q \in \mathcal{A}$.

1.10 On $(\Omega, \mathcal{F}, \mathcal{F}_t)$ $((\hat{\Omega}_D, \hat{\mathcal{F}}^D, \hat{\mathcal{F}}_t^D))$, define $\tau_U = \inf\{t \geq 0: X(t) \notin U\}$, where $U \subseteq R^d (U \subseteq D)$. If U is open, show that τ_U is an $\mathcal{F}_t$-stopping time ($\hat{\mathcal{F}}_t^D$-stopping time).

1.11 The notation in this exercise is as in Section 10. Consider the spaces $(\tilde{\Omega}_{R^d}, \tilde{\mathcal{F}}^{R^d})$ and $(\hat{\Omega}_{R^d}, \hat{\mathcal{F}}^{R^d})$ with the non-decreasing sequences of σ-algebras $\{\tilde{\mathcal{F}}_{\tau_{D_n}}^{R^d}\}_{n=1}^{\infty}$ and $\{\hat{\mathcal{F}}_{\tau_{D_n}}^{R^d}\}_{n=1}^{\infty}$ respectively. As in Theorem 10.5, define

$$\tilde{A}_n(\omega) = \bigcap\{B: B \in \tilde{\mathcal{F}}_{\tau_{D_n}}^{R^d}$$
$$\text{and } \omega \in B\} \text{ and } \hat{A}_n(\omega) = \bigcap\{B: B \subset \hat{\mathcal{F}}_{\tau_{D_n}}^{R^d} \text{ and } \omega \in B\}.$$

(a) Show that $\tilde{A}_n(\omega) = \{\omega' \in \tilde{\Omega}_{R^d}: X(t, \omega') = X(t, \omega),$ for $0 \leq t \leq \tau_{D_n}(\omega)\}$ and that $\hat{A}_n(\omega) = \{\omega' \in \hat{\Omega}_{R^d}: X(t, \omega') = X(t, \omega),$ for $0 \leq t \leq \tau_{D_n}(\omega)\}$.

(b) Show that condition (10.5) holds for $\tilde{A}_n(\omega)$ but not for $\hat{A}_n(\omega)$.

1.12 Prove the existence of a density in Theorem 12.1 using the result on the existence of a density given in Theorem 10.6. (Hint: For any $B \subset\subset D$, write $P_x(X(t) \in B) = \lim_{n\to\infty} P_x(X(t) \in B, \tau_{D_n} > t)$. Now use the localization result in Theorem 10.3.)

1.13 Let $D = D_1 \times D_2$, where $D_1 \subset R^m$, $0 < m < d$, and $D_2 \subset R^{d-m}$. Denote points in D by $x = (x_1, x_2)$ with $x_i \in D_i$, $i = 1, 2$. Let $L = L_1 + L_2$, where L_1 is an operator on D_1 and where L_2 is an operator on D which satisfies $L_2 f = 0$, for all f depending only on x_1. That is, every term of L_2 involves differentiation in x_2 (and possibly x_1 as well). Assume that there exists a unique solution $\{P_x, x \in \hat{D}\}$ to the generalized martingale problem for L on $D \subseteq R^d$, and denote the canonical process by $X(t) = (X_1(t), X_2(t))$. Assume that there exists a unique solution $\{P_x^{(1)}, x \in \hat{D}_1\}$ to the generalized martingale problem for L_1 on D_1. Let $\{D_{1,n}\}_{n=1}^{\infty}$ be a sequence of domains satisfying $D_{1,n} \subset\subset D_{1,n+1}$ and $\bigcup_{n=1}^{\infty} D_{1,n} = D_1$, let $\tau_{D_{1,n}} = \inf\{t \geq 0: X_1(t) \notin D_{1,n}\}$, and let $\tau_{D_1} = \lim_{n\to\infty} \tau_{D_{1,n}}$. Let τ_D denote the lifetime of $X(t)$ and assume that $\tau_D = \tau_{D_1}$ a.s. P_x, $x \in D$. Then the distribution of $X_1(\cdot)$ under P_{x_1,x_2} is equal to $P_{x_1}^{(1)}$, for all $x_2 \in D_2$. In particular, then, $X_1(t)$ under P_{x_1,x_2} is a Markov process. (Hint: Apply the martingale formulation to functions which depend only on x_1.)

1.14 Let $\{P_x, x \in \hat{D}\}$ solve the generalized martingale problem for L on $\hat{D}$. Consider the domain $D \times D \subset R^d \times R^d$ and denote points $x \in D \times D$ by $x = (x_1, x_2)$, with $x_1, x_2 \in D$. Denote the one-point compactification of $D \times D$ by $(D \times D)\hat{}$. Let L_{x_i}, $i = 1, 2$, denote the operator L in the variable x_i and let $\bar{L} = L_{x_1} + L_{x_2}$ on $D \times D$. Show that $\{P_{x_1} \times P_{x_2}, (x_1, x_2) \in (D \times D)\hat{}\}$ solves the generalized martingale problem for $\bar{L}$ on $D \times D$. Furthermore, if $\{P_x, x \in D\}$ solves the martingale problem for L on D, then $\{P_{x_1} \times P_{x_2}, (x_1, x_2) \in D \times D\}$ solves the martingale problem for $\bar{L}$ on $D \times D$.

1.15 Let L and D be as in Theorem 13.1 and let $\{P_x, x \in \hat{D}\}$ denote the solution to the generalized martingale problem for L on D. Let $\phi: R^d \to (0, \infty)$ be measurable and locally bounded. Define $\sigma: [0, \infty) \times \hat{\Omega}_D \to [0, \infty)$ by $\int_0^{\sigma(t,\omega)} ds/\phi(X(s)) = t$, if $\int_0^{\tau_D} ds/\phi(X(s)) > t$, and let $\sigma(t, \omega) = \tau_D$, if $\int_0^{\tau_D} ds/\phi(X(s)) \leq t$. Define $S: \hat{\Omega}_D \to \hat{\Omega}_D$ by $X(t, S\omega) = X(\sigma(t, \omega), \omega)$. Let $Q_x = P_x S^{-1}$, $x \in \hat{D}$. Then $\{Q_x, x \in \hat{D}\}$ is the solution to the generalized martingale problem for ϕL on D. The diffusion corresponding to Q_x is called a *time change* of the diffusion corresponding to P_x.

Notes

The material in the first two sections is developed in any number of texts; see, for example [Breiman (1968 or 1992)], [Chung (1974)], [Durrett (1991)] or [D. Williams (1991)]. An exposition on Markov processes from the perspective of semigroups can be found in [Ethier and Kurtz (1986)] and in [Lamperti (1977)]. The basic properties of Brownian motion can be found, for example, in [Breiman (1968 or 1992)], [Durrett (1991)] or [Karatzas and Shreve (1988)]. Any number of texts treat stochastic integrals and stochastic differential equations. For a basic treatment, see [Chung and Williams (1990)] (this book only treats stochastic integrals), [Durrett (1984)], [Ethier and Kurtz (1986)], [Friedman (1975)], [Gihman and Skorohod (1972)]; for a more sophisticated treatment, see [Elliot (1982)], [Ikeda and Watanabe (1981)], [Karatzas and Shreve (1988)], [Revuz and Yor (1991)], [Rogers and Williams (1987)]. The martingale formulation in Sections 8–10 is due to Stroock and Varadhan and may be found in [Stroock and Varadhan (1979)]. To the best of my knowledge, the definition and construction of the 'generalized martingale problem' does not appear in the literature.

2
The basic properties of diffusion processes

2.0 Introduction

Throughout this chapter, the following assumption will be in force.

Assumption A.

$$L = \tfrac{1}{2}\sum_{i,j=1}^{d} a_{ij}(x)\frac{\partial^2}{\partial x_i \partial x_j} + \sum_{i=1}^{d} b_i(x)\frac{\partial}{\partial x_i}$$

on a domain $D \subseteq R^d$ satisfies

(i) *$a = \{a_{ij}\}$ and $b = \{b_i\}$ are locally bounded and measurable on D;*
(ii) *a is continuous on D;*
(iii) *$\sum_{i,j=1}^{d} a_{ij}(x)v_i v_j > 0$, for $x \in D$ and $v \in R^d - \{0\}$.*

Note that if Assumption A holds, then by Theorem 1.13.1, there exists a unique solution to the generalized martingale problem for L on D.

The following result will be used frequently in the sequel without reference.

Proposition 0.1. *Let L satisfy Assumption A on a domain $D \subseteq R^d$ and let $\{P_x, x \in \hat{D}\}$ denote the solution to the generalized martingale problem for L on D. Assume that $u \in C^2(U) \cap C(\bar{U})$, for some domain $U \subset\subset D$ and that Lu is bounded on U. Then for $x \in D$,*

$$u(X(t \wedge \tau_U)) - \int_0^{t \wedge \tau_U} Lu(X(s))\,ds$$

is a martingale with respect to $(\hat{\Omega}_D, \hat{\mathcal{F}}^D, \hat{\mathcal{F}}_t^D, P_x)$. If $\{P_x, x \in D\}$ solves the martingale problem for L on D, and if $u \in C^2(D)$ and u and Lu are bounded on D, then $u(X(t)) - \int_0^t Lu(X(s))\,ds$ is a martingale with respect to $(\Omega, \mathcal{F}, \mathcal{F}_t, P_x)$, for $x \in D$.

Proof. We will prove the second statement of the proposition; the first one is proved similarly. Let $\{D_n\}_{n=1}^{\infty}$ be a sequence of domains satisfying $D_n \subset\subset D_{n+1}$ and $\bigcup_{n=1}^{\infty} D_n = D$. Let $v_n \in C_0^2(D)$ satisfy $v_n = u$ on D_n. Since $\{P_x, x \in D\}$ solves the martingale problem for L on D, for each $n = 1, 2, \ldots$, $v_n(X(t)) - \int_0^t Lv_n(X(s))\,ds$ is a martingale with respect to $(\Omega, \mathcal{F}, \mathcal{F}_t, P_x)$. By Doob's optional sampling theorem (Theorem 1.2.1),

$$v_n(X(t \wedge \tau_{D_n})) - \int_0^{t\wedge\tau_{D_n}} Lv_n(X(s))\,ds$$

is also a martingale with respect to $(\Omega, \mathcal{F}, \mathcal{F}_t, P_x)$. Since $v_n = u$ on $\bar{D}_n$, it follows that $u(X(t \wedge \tau_{D_n})) - \int_0^{t\wedge\tau_{D_n}} Lu(X(s))\,ds$ is a martingale with respect to $(\Omega\, \mathcal{F}, \mathcal{F}_t, P_x)$, for all $x \in D_n$. Equivalently, for $x \in D_n$,

$$E_x\left(u(X(t \wedge \tau_{D_n})) - \int_0^{t\wedge\tau_{D_n}} Lu(X(r))\,dr;\, A\right)$$
$$= E_x\left(u(X(s \wedge \tau_{D_n})) - \int_0^{s\wedge\tau_{D_n}} Lu(X(r))\,dr;\, A\right), \quad (0.1)$$

for $0 \leq s < t$ and $A \in \mathcal{F}_s$.

Since $\lim_{n\to\infty} t \wedge \tau_{D_n} = t$ a.s. P_x, for $x \in D$, and since u and Lu are bounded on D, it follows from (0.1) and the bounded convergence theorem that

$$E\left(u(X(t)) - \int_0^t Lu(X(r))\,dr;\, A\right) = E\left(u(X(s)) - \int_0^s Lu(X(r))\,dr;\, A\right),$$

for $x \in D$, $0 \leq s < t$ and $A \in \mathcal{F}_s$. That is, $u(X(t)) - \int_0^t Lu(X(s))\,ds$ is a martingale with respect to $(\Omega, \mathcal{F}, \mathcal{F}_t, P_x)$. □

2.1 Stochastic representation for solutions of elliptic equations and elliptic inequalities

We will establish a stochastic representation for solutions of elliptic equations and a stochastic inequality for functions which satisfy certain elliptic inequalities. We begin with the following important property of diffusions.

Theorem 1.1. *Let L satisfy Assumption A on a domain $D \subseteq R^d$ and let $\{P_x, x \in \hat{D}\}$ be the solution to the generalized martingale problem for L on D. Let $U \subset\subset D$. Then there exists a constant c depending only on*

$$\inf_{x\in U} \inf_{|v|=1} \sum_{i,j=1}^{d} a_{ij}(x) v_i v_j,\ \sup_{x\in U} |b(x)|$$

and the diameter of U such that

$$\sup_{x \in U} E_x \tau_U < c.$$

Proof. Let $u(x) = \exp(\lambda x_1)$, where x_1 denotes the first coordinate of $x \in D$. Then $Lu(x) = (\frac{1}{2}a_{11}(x)\lambda^2 + b_1(x)\lambda)\exp(\lambda x_1)$ and it follows that there exists an $\varepsilon > 0$ and a λ which depend only on

$$\inf_{y \in U} \inf_{|v|=1} \sum_{i,j=1}^{d} a_{ij}(y) v_i v_j \text{ and } \sup_{y \in U} |b(y)|$$

such that $Lu(x) \geqslant \varepsilon \inf_{y \in U} u(y)$, for all $x \in \bar{U}$. Since $u(X(t \wedge \tau_U)) - \int_0^{t \wedge \tau_U} Lu(X(s))\,ds$ is a martingale,

$$\begin{aligned} u(x) &= E_x u(X(t \wedge \tau_U)) - E_x \int_0^{t \wedge \tau_U} Lu(X(s))\,ds \\ &\leqslant \sup_{y \in U} u(y) - \varepsilon(\inf_{y \in U} u(y)) E_x t \wedge \tau_U, \text{ for } x \in U. \end{aligned}$$

Letting $t \to \infty$, we obtain

$$\sup_{x \in U} E_x \tau_U \leqslant \frac{1}{\varepsilon}[(\sup_{x \in U} \exp(\lambda x_1) / \inf_{x \in U} \exp(\lambda x_1)) - 1].$$

□

Theorem 1.2. *Let L satisfy Assumption A on a domain $D \subset\subset R^d$ and let $\{P_x, x \in \hat{D}\}$ denote the solution to the generalized martingale problem for L on D. Let $U \subset\subset D$ be a domain.*

(i) *If $u \in C^2(U) \cap C(\bar{U})$ satisfies*

$$Lu = -f \text{ in } U,$$
$$u = \phi \text{ on } \partial U,$$

where $\phi \in C(\partial U)$ and f is bounded and measurable, then

$$u(x) = E_x \int_0^{\tau_U} f(X(s))\,ds + E_x \phi(X(\tau_U)).$$

(ii) *If $u \in C^2(U) \cap C(U)$ satisfies*

$$Lu \leqslant -f \text{ in } U,$$
$$u \geqslant \phi \text{ on } \partial U,$$

where $\phi \in C(\partial U)$ and f is bounded and measurable, then

$$u(x) \geqslant E_x \int_0^{\tau_U} f(X(s))\,ds + E_x \phi(X(\tau_U)).$$

Proof. We will prove part (i) and leave part (ii) to the reader. Under the assumption of part (i), $u(X(t \wedge \tau_U)) + \int_0^{t\wedge\tau_U} f(X(s))\,ds$ is a martingale. Thus $E_x u(X(t \wedge \tau_U)) + E_x\int_0^{t\wedge\tau_U} f(X(s))\,ds = u(x)$. Since $u \in C(\bar{U})$ and since, by Theorem 1.1, $P_x(\tau_U < \infty) = 1$, it follows that

$$\lim_{t\to\infty} E_x u(X(t \wedge \tau_U)) = E_x u(X(\tau_U)) = E_x \phi(X(\tau_U)).$$

Since f is bounded and since $E_x\tau_U < \infty$ by Theorem 1.1, we obtain from the dominated convergence theorem that

$$\lim_{t\to\infty} E_x \int_0^{t\wedge\tau_U} f(X(s))\,ds = E_x \int_0^{\tau_U} f(X(s))\,ds.$$

The theorem now follows. □

2.2 The behavior of the exit time from a small ball

Let $B_r(x) \subset R^d$ denote the ball of radius r centered at $x \in R^d$. In this section we study the behavior of the exit time from such balls.

Theorem 2.1. *Let L satisfy Assumption A on a domain $D \subseteq R^d$ and let $\{P_x, x \in \hat{D}\}$ denote the solution to the generalized martingale problem for L on D. Let $B_x(r)$ denote the ball of radius r centered at $x \in R^d$.*

(i) *For each $x \in D$ and $\varepsilon > 0$, there exists an $r_0 > 0$ such that*

$$\sup_{y\in B_{r0}(x)} E_y \tau_{B_{r0}(x)} < \varepsilon.$$

(ii) *For each $x \in D$ and $M > 0$, there exists an $r_0 > 0$ such that*

$$\sup_{y\in B_{r0}(x)} E_y \exp(M\tau_{B_{r0}(x)}) < \infty.$$

Proof. Without loss of generality, assume that $x = 0$ and, by localization (Theorem 1.13.2), that $D = R^d$.

(i) Pick $\alpha > 0$ and $R_0 > 0$ such that $\sum_{i,j=1}^d a_{ij}(y)v_i v_j \geqslant \alpha|v|^2$, for all $v \in R^d$ and $|y| \leqslant R_0$, and pick N such that $|b(y)| \leqslant N$, for $|y| \leqslant R_0$. For $r \in (0, R_0)$, define

$$v_r(y) = \frac{r^2 - |y|^2}{d\alpha}, \text{ for } |y| \leqslant r.$$

Then

$$\frac{\alpha}{2}\Delta v_r = -1, \text{ for } |y| < r, \text{ and } v_r(y) = 0, \text{ for } |y| = r.$$

Note that if A and V are non-negative definite $d \times d$ matrices, then $\mathrm{Tr}\, AV \geqslant 0$. Now

$$\tfrac{1}{2}\sum_{i,j=1}^{d} a_{ij}(y)\frac{\partial^2 v_r}{\partial y_i \partial y_j}(y) = \tfrac{1}{2}\mathrm{Tr}\,(aD^2 v_r)(y),$$

where

$$D^2 v_r = \left\{\frac{\partial^2 v_r}{\partial y_i \partial y_j}\right\}$$

is the Hessian of v_r. Since $-D^2 v_r(y)$ is non-negative definite for $|y| \leqslant r$, and since $a(y) \geqslant \alpha I$, for $|y| \leqslant R_0$, it follows that

$$0 \leqslant \tfrac{1}{2}\mathrm{Tr}\,(\alpha I - a(y))D^2 v_r(y) = -1 - \tfrac{1}{2}\sum_{i,j=1}^{d} a_{ij}(y)\frac{\partial^2 v_r}{\partial y_i \partial y_j}(y), \text{ for } |y| \leqslant r. \tag{2.1}$$

Since $|b \cdot \nabla v_r(y)| \leqslant (2N/d\alpha)|y|$, it follows from (2.1) that there exists an $r_0 \leqslant R_0$ such that

$$Lv_r(y) \leqslant -\tfrac{1}{2}, \text{ if } |y| \leqslant r \leqslant r_0.$$

Thus, from Theorem 1.2,

$$v_r(y) \geqslant \tfrac{1}{2}E_y \tau_{B_r(0)}, \text{ if } |y| \leqslant r \leqslant r_0.$$

Part (i) now follows since $\lim_{r\to 0}\sup_{|y|<r} v_r(y) = 0$.

(ii) Let $\varepsilon > 0$ and pick $r_0 = r_0(\varepsilon)$ as in part (i). By Chebyshev's inequality,

$$P_y(\tau_{B_{r_0}(0)} \geqslant 1) \leqslant \varepsilon, \text{ for all } |y| \leqslant r_0. \tag{2.2}$$

By the strong Markov property and (2.2) we have for $|z| \leqslant r_0$ and $n = 2, 3, \ldots,$

$$\begin{aligned}
P_z(\tau_{B_{r_0}(0)} \geqslant n) &= E_z I_{\{\tau_{B_{r_0}(0)} \geqslant n\}} = E_z I_{\{\tau_{B_{r_0}(0)} \geqslant n\}} I_{\{\tau_{B_{r_0}(0)} \geqslant 1\}} \\
&= E_z E_z(I_{\{\tau_{B_{r_0}(0)} \geqslant n\}} I_{\{\tau_{B_{r_0}(0)} \geqslant 1\}} | \mathcal{F}_1) \\
&= E_z I_{\{\tau_{B_{r_0}(0)} \geqslant 1\}} E_z(I_{\{\tau_{B_{r_0}(0)} \geqslant n\}} | \mathcal{F}_1) \\
&= E_z I_{\{\tau_{B_{r_0}(0)} \geqslant 1\}} E_{X(1)} I_{\{\tau_{B_{r_0}(0)} \geqslant n-1\}} \\
&\leqslant \sup_{|y| \leqslant r_0} E_y I_{\{\tau_{B_{r_0}(0)} \geqslant n-1\}} E_z I_{\{\tau_{B_{r_0}(0)} \geqslant 1\}} \\
&= \sup_{|y| \leqslant r_0} P_y(\tau_{B_{r_0}(0)} \geqslant n-1) P_z(\tau_{B_{r_0}(0)} \geqslant 1) \\
&\leqslant \varepsilon \sup_{|y| \leqslant r_0} P_y(\tau_{B_{r_0}(0)} \geqslant n-1).
\end{aligned}$$

Thus,

$$\sup_{|y|\leq r_0} P_y(\tau_{B_{r_0}(0)} \geq n) \leq \varepsilon \sup_{|y|\leq r_0} P_y(\tau_{B_{r_0}(0)} \geq n-1)$$

and, by induction,

$$\sup_{|y|\leq r_0} P_y(\tau_{B_{r_0}(0)} \geq n) \leq \varepsilon^n. \tag{2.3}$$

Thus, if we pick ε such that $\varepsilon \exp(M) < 1$, it follows from (2.3) that

$$\sup_{|y|\leq r} E_y \exp(M\tau_{B_{r_0}(0)}) < \infty. \qquad \square$$

The following theorem concerns the behavior of a diffusion process on a small time interval. We make the convention $|x - \Delta| = \infty$, for $x \in D$, where Δ is the point at infinity in the one-point compactification $\hat{D}$ of D.

Theorem 2.2. *Let L satisfy Assumption A on a domain $D \subseteq R^d$ and let $\{P_x, x \in \hat{D}\}$ denote the solution to the generalized martingale problem for L on D.*

(i) *For all $x \in D$ and $\varepsilon > 0$,*

$$\lim_{t\to 0} P_x(\sup_{0\leq s\leq t} |X(s) - x| \geq \varepsilon) = 0,$$

and the rate of convergence depends only on

$$\sup_{\substack{|y-x|\leq\varepsilon\\ y\in D}} \sup_{|v|=1} \langle v, a(y)v\rangle \text{ and } \sup_{\substack{|y-x|\leq\varepsilon\\ y\in D}} |b(y)|.$$

(ii) *If a and b are bounded and $D = R^d$, then*

$$P_x(\sup_{0\leq s\leq t} |X(s) - x - \int_0^s b(X(u))\,du| \geq \lambda) \leq 2d \exp\left(\frac{-\lambda^2}{2A\,dt}\right),$$

for all $x \in R^d$, $\lambda > 0$, and $t > 0$, where

$$A = \sup_{x\in R^d} \sup_{|v|=1} \langle v, a(x)v\rangle.$$

Proof. (i) Since the event $\{\sup_{0\leq s\leq t} |X(s) - x| \geq \varepsilon\}$ is measurable with respect to $\hat{\mathcal{F}}^D_{\tau_{B_\varepsilon(x)}}$, it follows from Theorem 1.13.2 that we may assume $D = R^d$ and that the coefficients a and b of L are bounded and satisfy

$$\sup_{y\in R^d} \sup_{|v|=1} \langle v, a(y)v\rangle = \sup_{|y-x|<\varepsilon} \sup_{|v|=1} \langle v, a(y)v\rangle \text{ and}$$

$$\sup_{y\in R^d} |b(y)| = \sup_{|y-x|<\varepsilon} |b(y)|.$$

Part (i) now follows from part (ii) and the boundedness of b.

(ii) Let $\theta \in R^d$ satisfy $|\theta| = 1$ and let $\rho > 0$. By Theorem 1.5.1(i),

$$X_{\rho\theta}(t) =$$

$$\exp\left(\left\langle \rho\theta, X(t) - X(0) - \int_0^t b(X(s))\,\mathrm{d}s \right\rangle - \tfrac{1}{2}\int_0^t \langle \rho\theta, a(X(s))\rho\theta\rangle\,\mathrm{d}s\right)$$

is a martingale. Since $E_x X_{\rho\theta}(t) = 1$, for all $t \geq 0$, using Theorem 1.2.2 for the second inequality below, we have

$$P_x\left(\sup_{0\leq s\leq t}\left\langle \theta, X(s) - x - \int_0^s b(X(u))\,\mathrm{d}u \right\rangle \geq \lambda\right)$$

$$\leq P_x\left(\sup_{0\leq s\leq t} X_{\rho\theta}(s) \geq \exp\left(\frac{-At\rho^2}{2} + \rho\lambda\right)\right) \leq \exp\left(-\rho\lambda + \frac{\rho^2 At}{2}\right).$$

Taking

$$\rho = \frac{\lambda}{At},$$

we obtain

$$P_x\left(\sup_{0\leq s\leq t}\left\langle \theta, X(s) - x - \int_0^s b(X(u))\,\mathrm{d}u \right\rangle \geq \lambda\right) \leq \exp\left(\frac{-\lambda^2}{2At}\right).$$

Replacing θ by $-\theta$, we obtain a similar bound for the infimum and combining the two bounds gives

$$P_x\left(\sup_{0\leq s\leq t}\left|X(s) - x - \int_0^s b(X(u))\,\mathrm{d}u\right| \geq \lambda\right) \leq 2d\exp\left(\frac{-\lambda^2}{2A\,\mathrm{d}t}\right),$$

since

$$P_x\left(\sup_{0\leq s\leq t}\left|X(s) - x - \int_0^s b(X(u))\,\mathrm{d}u\right| \geq \lambda\right)$$

$$\leq \sum_{j=1}^d P_x\left(\sup_{0\leq s\leq t}\left|X_j(s) - x_j - \int_0^s b_j(X(u))\,\mathrm{d}u\right| \geq \frac{\lambda}{\sqrt{d}}\right). \quad \square$$

2.3 The Blumenthal 0–1 law and the regularity of boundary points

If L satisfies Assumption A on a domain $D \subseteq R^d$ and if $\{P_x, x \in D\}$ solves the martingale problem for L on D, then $X(t)$ is a Markov process with respect to $(\Omega, \mathcal{F}, \mathcal{F}_t, P_x)$ and $f(X(t)) - \int_0^t Lf(X(s))\,\mathrm{d}s$ is a martingale with respect to $(\Omega, \mathcal{F}, \mathcal{F}_t, P_x)$, for $f \in C_0^2(D)$. Now define $\mathcal{F}_{t^+} = \bigcap_{s>t} \mathcal{F}_s$ and consider the filtration $\{\mathcal{F}_{t^+}\}_{t=0}^{\infty}$ on $(\Omega, \mathcal{F}, P_x)$. For $s > 0$, define the *shift operator* $\theta_s: \Omega \to \Omega$ (or $\theta_s: \hat{\Omega}_D \to \hat{\Omega}_D$) by $X(t, \theta_s \omega) = X(t + s, \omega)$.

Theorem 3.1. *Let L satisfy Assumption A on a domain $D \subseteq R^d$ and assume that $\{P_x, x \in D\}$ solves the martingale problem for L on D. Then*

(i) *$X(t)$ is a Markov process on $(\Omega, \mathcal{F}, \mathcal{F}_{t^+}, P_x)$ and*

$$P_x(X(\cdot, \theta_t \omega) \in B | \mathcal{F}_{t^+}) = P_x(X(\cdot, \theta_t \omega) \in B | \mathcal{F}_t)$$
$$= P_{X(t)}(X(\cdot) \in B), \text{ for } B \subseteq \mathcal{F};$$

(ii) *$f(X(t)) - \int_0^t Lf(X(s))\,ds$ is a martingale with respect to $(\Omega, \mathcal{F}, \mathcal{F}_{t^+}, P_x)$, for all $f \in C_0^2(D)$.*

Proof. We begin with a remark on notation. In expressions such as $P_x(X(t) \in B)$ or $E_x f(X(t))$, $X(t)$ is actually a dummy variable of integration. In the proof below, we will need to consider not only P_x and E_x, but also $P_{X(t)}$ and $E_{X(t)}$, where $X(t)$ is a random variable. Thus, when considering probabilities $P_{X(t)}$ or expectations $E_{X(t)}$, the dummy variable of integration will be denoted by $X'(\cdot)$.

We will show that for each $x \in R^d$, each $s > 0$ and each $B \subseteq \mathcal{B}_d$,

$$P_x(X(t + s) \in B | \mathcal{F}_{t^+}) = P_{X(t)}(X'(s) \in B). \tag{3.1}$$

Then, by exercise 1.2, it will follow that

$$P_x(X(\cdot, \theta_t \omega) \in B | \mathcal{F}_{t^+}) = P_{X(t)}(X'(\cdot) \in B), \text{ for all } B \in \mathcal{F}. \tag{3.2}$$

This gives part (i); part (ii) follows from part (i) and the fact that $f(X(t)) - \int_0^t Lf(X(s))\,ds$ is a martingale with respect to $(\Omega, \mathcal{F}, \mathcal{F}_t, P_x)$. It therefore remains to prove (3.1). Let $f \in C_0(D)$ and let $\varepsilon > 0$ and $s > 0$. Then from Theorem 1.12.1,

$$E_x(f(X(t + s + \varepsilon)) | \mathcal{F}_{t+\varepsilon}) = E_{X(t+\varepsilon)} f(X'(s)).$$

Taking conditional expectations with respect to $\mathcal{F}_{t^+}$ gives

$$E_x(f(X(t + s + \varepsilon)) | \mathcal{F}_{t^+}) = E_x(E_{X(t+\varepsilon)}(f(X'(s))) | \mathcal{F}_{t^+})). \tag{3.3}$$

By the continuity of f, $\lim_{\varepsilon \to 0} f(X(t + s + \varepsilon)) = f(X(t + s))$ and, by the Feller property, $\lim_{\varepsilon \to 0} E_{X(t+\varepsilon)} f(X'(s)) = E_{X(t)} f(X'(s))$. Thus, applying the bounded convergence theorem for conditional expectations to (3.3), we obtain $E_x(f(X(t + s)) | \mathcal{F}_{t^+}) = E_{X(t)} f(X'(s))$. Equation (3.1) now follows since $C_0(D)$ is a separating class for probability measures on D. □

The next result is known as the *Blumenthal 0–1 law*. It will prove to be useful in the sequel for the study of regular boundary points.

Theorem 3.2. *Let L satisfy Assumption A on a domain $D \subseteq R^d$ and assume that $\{P_x, x \in D\}$ solves the martingale problem for L on D. Then, for all $x \in D$, $P_x(A) = 0$ or 1, for any $A \in \mathcal{F}_{0^+}$.*

Proof. From Theorem 3.1, it follows that for $0 < t_1 < t_2 < \ldots < t_m$ and $B = (B_1, B_2, \ldots, B_m)$, where $B_i \in \mathcal{B}_d$,

$$\begin{aligned} P_x((X(t_1), X(t_2), \ldots, X(t_m)) \in B \mid \mathcal{F}_{0^+}) \\ &= P_x((X(t_1), X(t_2), \ldots, X(t_m)) \in B \mid \mathcal{F}_0) \\ &= P_x((X(t_1), X(t_2), \ldots, X(t_m)) \in B). \end{aligned}$$

Thus,

$$P_x(A \mid \mathcal{F}_{0^+}) = P_x(A) \text{ a.s. } P_x, \tag{3.4}$$

for any A of the form $A = \{(X(t_1), \ldots, X(t_m)) \in B\}$. Using the bounded convergence theorem for conditional expectations and the fact that sets of the form A generate $\mathcal{F}$, it follows that (3.4) holds for any $A \in \mathcal{F}$. In particular, if $A \in \mathcal{F}_{0^+}$, then from (3.4), $P_x(A \mid \mathcal{F}_{0^+}) = P_x(A)$ a.s. P_x. However, $P_x(A \mid \mathcal{F}_{0^+}) = 1_A$, the characteristic function of the set A. So $1_A = P_x(A)$ a.s. P_x and since 1_A can only take on the values 0 and 1, it follows that $P_x(A) = 0$ or 1. □

Let L satisfy Assumption A on a domain $D \subseteq R^d$ and let $\{P_x, x \in \hat{D}\}$ denote the solution to the generalized martingale problem for L on D. Let $B \subset D$ be a domain. A point $a \in \partial B \cap D$ is called *regular* for B (with respect to L) if $P_a(\tau_{\bar{B} \cap D} = 0) = 1$, where $\tau_{\bar{B} \cap D} = \inf\{t \geq 0: X(t) \notin \bar{B} \cap D\}$. If there exists a sphere S in $R^d - B$ which satisfies $S \cap \bar{B} = \{a\}$, then B is said to satisfy an *exterior sphere condition* at $a \in \partial B \cap D$. If there exists a cone C in $R^d - B$ with base at $a \in \partial B \cap D$, then B is said to satisfy an *exterior cone condition* at $a \in \partial B \cap D$. (For later use, we note that B is said to satisfy an interior sphere (cone) condition at $a \in \partial B \cap D$ if $D - \bar{B}$ satisfies an exterior sphere (cone) condition at a.) A sufficient condition guaranteeing that B satisfy an exterior sphere (cone) condition at $a \in \partial B \cap D$ is that the boundary be C^2 (Lipschitz) at a. The next result shows that a sufficient condition for the regularity of a point $a \in \partial B \cap D$ for B with respect to any L satisfying Assumption A on D is that B satisfy an exterior cone condition at a.

Theorem 3.3. *Let L satisfy Assumption A on a domain $D \subseteq R^d$, $d \geq 2$, and let $\{P_x, x \in \hat{D}\}$ denote the solution to the generalized martingale problem for L on D. Let $B \subset D$ be a domain satisfying an exterior cone condition at $a \in \partial B \cap D$. Then*

(i) $\lim_{B \ni x \to a} P_x(\tau_B > t) = 0$, *for all* $t > 0$;
(ii) $\lim_{B \ni x \to a} E_x f(X(t \wedge \tau_B)) = f(a)$, *for all* $t > 0$ *and all bounded f which are continuous at a*;
(iii) *a is regular for B; that is,* $P_a(\tau_{\bar{B} \cap D} = 0) = 1$;
(iv) *Assume in addition that B satisfies an exterior cone condition at every point of* $\partial B \cap D$. *Then* $\tau_B = \tau_{\bar{B} \cap D}$ *a.s.* P_x *and* τ_B *and* $\tau_{\bar{B} \cap D}$ *are* P_x *− a.s. continuous extended real-valued functions on* $\hat{\Omega}_D$, *for each* $x \in D$.

Remark 1. By exercise 1.10, τ_U on $(\hat{\Omega}_D, \hat{\mathcal{F}}^D, \hat{\mathcal{F}}_t^D)$ is an $\hat{\mathcal{F}}_t^D$-stopping time if $U \subseteq D$ is open. Thus, from part (iv) of the theorem, if follows that $\tau_{\bar{B} \cap D}$ is a stopping time with respect to the P_x-completion of $\hat{\mathcal{F}}_t^D$. Since a (strong) Markov process with respect to $\hat{\mathcal{F}}_t^D$ is also a (strong) Markov process with respect to the completion of $\hat{\mathcal{F}}_t^D$ with the same transition probabilities, and since a martingale with respect to $\hat{\mathcal{F}}_t^D$ is also a martingale with respect to the completion of $\hat{\mathcal{F}}_t^D$, it follows that all of the results concerning martingales and stopping times and concerning Markov processes and stopping times may be applied to $\tau_{\bar{B} \cap D}$ if $B \subset D$ is a domain which satisfies an exterior cone condition at each point of $\partial B \cap D$.

Remark 2. In one dimension, every point is regular; see exercise 2.5.

Proof. We leave it to the reader to check that, in the light of Theorem 2.2 and the localization result of Theorem 1.13.2, we may assume that B is bounded, that $D = R^d$ and that the coefficients of L are bounded. We will first prove (i) ⇒ (ii) ⇒ (iii) ⇒ (iv); then we will prove (i).

(i) ⇒ (ii). This follows almost immediately in the light of Theorem 2.2.

(ii) ⇒ (iii). Since the coefficients of L are assumed bounded on R^d, $\{P_x, x \in R^d\}$ solves the martingale problem for L on R^d, in which case Theorem 3.2 may be used. We give a proof by contradiction. Assume that (iii) does not hold. Then, by Theorem 3.2, $P_a(\tau_{\bar{B}} = 0) = 0$ since $\{\tau_{\bar{B}} = 0\} \in \mathcal{F}_{0^+}$. Since, by Theorem 1.10.2, a density $p(t, x, y)$ exists for $P_x(X(t) \in dy)$, it follows from Fubini's theorem that $P_a(X(t) = a$ for some $t > 0) = 0$. Thus,

$$\lim_{r \to 0} P_a(X(\tau_{\bar{B}}) \in B_r(a)) = P_a(X(\tau_{\bar{B}}) = a) = 0.$$

Pick $r>0$ such that $P_a(X(\tau_{\bar{B}}) \in B_r(a)) \leq \frac{1}{4}$ and choose a decreasing sequence $\{\delta_n\}_{n=1}^{\infty}$ such that $0<\delta_n<r$ for all n and $\lim_{n\to\infty}\delta_n=0$. Letting

$$\tau_n = \inf\{t \geq 0: |X(t)-a| \geq \delta_n\},$$

we have $P_a(\lim_{n\to\infty}\tau_n=0)=1$ and thus $\lim_{n\to\infty}P_a(\tau_n<\tau_{\bar{B}})=1$. Note that on $\{\tau_n<\tau_{\bar{B}}\}$, we have $X(\tau_n)\in B$. For n large enough so that $P_a(\tau_n<\tau_{\bar{B}}) \geq \frac{1}{2}$, we have

$$\begin{aligned}\tfrac{1}{4} \geq P_a(X(\tau_{\bar{B}}) \in B_r(a)) &\geq P_a(X(\tau_{\bar{B}}) \in B_r(a),\, \tau_n<\tau_{\bar{B}})\\ &= E_a\{I_{\{\tau_n<\tau_{\bar{B}}\}}P_a(X(\tau_{\bar{B}}) \in B_r(a)|\mathcal{F}_{\tau_n})\}\\ &= \int_{\bar{B}\cap\partial B_{\delta_n}(a)} P_x(X(\tau_{\bar{B}}) \in B_r(a))P_a(\tau_n<\tau_{\bar{B}},\, X(\tau_n)\in \mathrm{d}x)\\ &\geq \tfrac{1}{2}\inf_{x\in\bar{B}\cap\partial B_{\delta_n}(a)} P_x(X(\tau_{\bar{B}}) \in B_r(a)).\end{aligned}$$

Therefore, for some $x_n \in \bar{B}\cap\partial B_{\delta_n}(a)$, $P_{x_n}(X(\tau_{\bar{B}}) \in B_r(a)) \leq \frac{1}{2}$. Now choose a bounded, continuous $f\colon R^d \to R$ such that $f=0$ outside $B_r(a)$, $0 \leq f \leq 1$ in $B_r(a)$, and $f(a)=1$. We have

$$\limsup_{n\to\infty} E_{x_n}f(X(t\wedge\tau_{\bar{B}})) \leq \limsup_{n\to\infty} P_{x_n}(X(\tau_{\bar{B}}) \in B_r(a)) \leq \tfrac{1}{2} < f(a),$$

which contradicts (ii).

(iii) $\Rightarrow$ (iv). For $x\in R^d-\bar{B}$, $\tau_B=\tau_{\bar{B}}=0$ a.s. P_x. Thus we may assume that $x\in\bar{B}$. Since B is bounded, by Theorem 1.1, $P_x(\tau_B<\infty)=1$. We have

$$\begin{aligned}P_x(\tau_{\bar{B}}>\tau_B) = E_xI_{\{\tau_{\bar{B}}>\tau_B\}} &= E_x(E_x(I_{\{\tau_{\bar{B}}>\tau_B\}}|\mathcal{F}_{\tau_B}))\\ &= E_xP_{X(\tau_B)}(\tau_{\bar{B}}>0)=0,\end{aligned}$$

where the final equality follows from (iii) and the fact that $X(\tau_B)\in\partial B$.

It is easy to check that τ_B is lower semicontinuous and that $\tau_{\bar{B}}$ is upper semicontinuous on Ω (exercise 2.1). Thus, since $\tau_{\bar{B}}=\tau_B$ a.s P_x, it follows that τ_B and $\tau_{\bar{B}}$ are P_x-a.s continuous on Ω.

Proof of (i). We will prove (i) under the stronger assumption that B satisfy an exterior sphere condition. To prove the result under an exterior cone condition, see exercise 2.4. Since B satisfies an exterior sphere condition at $a\in\partial B$, there exists a sphere $B_R(y)$ of radius R centered at $y\in B^c$ such that $\bar{B}\cap\overline{B_R(y)}=\{a\}$. Let $w(x)=c(R^{-\gamma}-|x-y|^{-\gamma})$, where $c,\gamma>0$. Using the boundedness of B, the boundedness of the coefficients of L on B and Assumption A(ii) and (iii), one can check that $Lw \leqslant -1$ on B, if c and γ are chosen suitably large. Note that $w > 0$ in $\bar{B} - \{a\}$ and $w(a) = 0$. The function w is called a *barrier* for L at $a \in \partial B$. Since $w(X(\tau_B \wedge t)) - \int_0^{t \wedge \tau_B} Lw(X(s))\,ds$ is a martingale, we have for $x \in B$,

$$0 \leqslant E_x w(X(\tau_B \wedge t)) = w(x) + E_x \int_0^{t \wedge \tau_B} Lw(X(s))\,ds$$
$$\leqslant w(x) - E_x \tau_B \wedge t.$$

Letting $t \to \infty$, we obtain

$$E_x \tau_B \leqslant w(x).$$

Thus, by Chebyshev's inequality,

$$P_x(\tau_B > t) \leqslant \frac{1}{t} E_x \tau_B \leqslant \frac{1}{t} w(x),$$

and (i) follows. □

The following extension of Theorem 3.3(i) will be required in Chapter 8.

***Corollary* 3.4.** *Let L satisfy Assumption A on a domain $D \subset R^d$ and let $\{P_x, x \in \hat{D}\}$ denote the solution to the generalized martingale problem for L on D. Let $a \in \partial D$ and assume that D satisfies an exterior cone condition at a. Assume in addition that the coefficients of L are bounded near a and that $\inf_{|v|=1} \sum_{i,j=1}^d a_{ij}(x) v_i v_j$ is uniformly positive over those $x \in D$ which are near a. Then $\lim_{x \to a} P_x(\tau_D > t) = 0$, for all $t > 0$.*

Proof. Choose domains U_1, $U_2 \subset R^d$ such that $a \in U_1 \subset\subset U_2$ and such that the coefficients of L are bounded on $U_2 \cap D$ and $\inf_{|v|=1} \sum_{i,j=1}^d a_{ij}(x) v_i v_j$ is uniformly positive on $U_2 \cap D$. Let $B \subset D$ be a bounded domain satisfying $\partial D \cap U_1 \subset \partial B$ and $\partial B \cap \partial D \subset U_2$. In the light of Theorem 2.2, it is enough to prove that

$$\lim_{x \to a} P_x(\tau_B > t) = 0, \text{ for all } t > 0. \tag{3.5}$$

Define w exactly as it was defined in the proof of Theorem 3.3(i). The only difference now between the proof of (3.5) and the proof of Theorem 3.3(i) is as follows. In the second and third to the last sentences of the proof of Theorem 3.3(i), replace B by B_n, where $\{B_n\}_{n=1}^\infty$ is a sequence of domains increasing to B such that $B_n \subset\subset D$. In the penultimate sentence, we now have '$E_x \tau_{B_n} \leqslant w(x)$'. Letting

$n \to \infty$, we obtain $E_x \tau_B \leqslant w(x)$, as in the penultimate line of the original. □

2.4 The Feynman–Kac formula

We begin with the following 'integration by parts' formula for martingales which is very useful.

Proposition 4.1. *Let $M(t)$ be an almost surely continuous martingale with respect to $(Y, \mathcal{M}, \mathcal{M}_t, P)$ and let $\eta(t) = \eta(t, q)$: $[0, \infty) \times Y \to R$ be progressively measurable with respect to $\mathcal{M}_t$ and such that the total variation $|\eta|(t, q)$ of $\eta(t, q)$ is finite, for all $t \geqslant 0$ and $q \in Y$. If for all $t \geqslant 0$,*

$$E|\eta|(t) \sup_{0 \leqslant s \leqslant t} |M(s)| < \infty, \tag{4.1}$$

then $M(t)\eta(t) - \int_0^t M(s)\,\mathrm{d}\eta(s)$ is a martingale with respect to $(Y, \mathcal{M}, \mathcal{M}_t, P)$.

Proof. It is not difficult to show that $\int_0^t M(s)\eta(\mathrm{d}s)$ can be defined as a progressively measurable function. By (4.1), $M(t)\eta(t) - \int_0^t M(t)\eta(\mathrm{d}s)$ is integrable. Now, for $0 \leqslant t_1 \leqslant t_2$ and $A \in \mathcal{M}_{t_1}$, we have

$$E\left(M(t_2)\eta(t_2) - M(t_1)\eta(t_1) - \int_{t_1}^{t_2} M(s)\eta(\mathrm{d}s); A\right) = E\left(\int_{t_1}^{t_2} (M(t_2) - M(s))\eta(\mathrm{d}s); A\right),$$

since $E((M(t_2) - M(t_1))\eta(t_1); A) = 0$. We have

$$\int_{t_1}^{t_2} (M(t_2) - M(s))\eta(\mathrm{d}s) = \lim_{n \to \infty} \sum_{k=1}^{n} \left(M(t_2) - M\left(t_1 + \frac{k}{n}(t_2 - t_1)\right)\right) \times \left(\eta\left(t_1 + \frac{k}{n}(t_2 - t_1)\right) - \eta\left(t_1 + \frac{k-1}{n}(t_2 - t_1)\right)\right) \text{ a.s.}$$

By (4.1) and the dominated convergence theorem, the above convergence is also in $L^1(Y, P)$. Since

$$E\left(\left(M(t_2) - M\left(t_1 + \frac{k}{n}(t_2 - t_1)\right)\right)\left(\eta\left(t_1 + \frac{k}{n}(t_2 - t_1)\right) - \eta\left(t_1 + \frac{k-1}{n}(t_2 - t_1)\right)\right); A\right) = 0,$$

for all $n \geq 0$ and $1 \leq k \leq n$, it follows that

$$E\left(\int_{t_1}^{t_2}(M(t_2) - M(s))\eta(ds);\, A\right) = 0.$$

This completes the proof. □

As an almost immediate consequence of Proposition 4.1, we obtain the following version of the *Feynman–Kac formula*.

Theorem 4.2. *Let L_0 satisfy Assumption A on a domain $D \subseteq R^d$ and let $\{P_x, x \in \hat{D}\}$ denote the solution to the generalized martingale problem for L_0 on D.*

(i) *If $U \subset\subset D$ and V is bounded and measurable on U, then*

$$\exp\left(\int_0^{t\wedge\tau_U} V(X(s))\,ds\right)u(X(t\wedge\tau_U)) - \int_0^{t\wedge\tau_U}(L_0 + V)u(X(s))\exp\left(\int_0^s V(X(r))\,dr\right)ds$$

is a martingale with respect to $\{\hat{\Omega}_D, \hat{\mathcal{F}}^D, \hat{\mathcal{F}}_t^D, P_x)$, for all $u \in C^2(\bar{U})$ and all $x \in U$.

(ii) *If $\{P_x, x \in U\}$ in fact solves the martingale problem for L_0 on D and V is bounded from above on D, then*

$$\exp\left(\int_0^t V(X(s))\,ds\right)u(X(t)) - \int_0^t (L_0 + V)u(X(s))\exp\left(\int_0^s V(X(r))\,dr\right)ds.$$

is a martingale with respect to $(\Omega, \mathcal{F}, \mathcal{F}_t, P_x)$, for all $u \in C_b^2(D)$ and all $x \in D$.

Remark. If $(L_0 + V)u = 0$, then taking expectations in (i) with $t = 0$ and with general t, and equating them, one obtains

$$u(x) = E_x \exp\left(\int_0^{t\wedge\tau_U} V(X(s))\,ds\right)u(X(t\wedge\tau_U)).$$

Letting $t \to \infty$, under appropriate conditions (see Theorem 3.6.6 in Chapter 3) one obtains

$$u(x) = E_x \exp\left(\int_0^{\tau_U} V(X(s))\,ds)\right)\phi(X(\tau_U)), \tag{4.2}$$

where ϕ is the restriction of u to ∂U. Similarly in (ii), if $(L_0 + V)u = 0$, then

$$u(x) = E_x \exp\left(\int_0^t V(X(s))\,\mathrm{d}s\right) u(X(t)). \tag{4.3}$$

In the literature, the Feynman–Kac formula usually refers to a formula such as (4.2) or (4.3); we have taken a broader definition.

Proof. The proof of part (i) follows by applying Proposition 4.1 to $M(t) = u(X(t \wedge \tau_U)) - \int_0^{t\wedge\tau_U} (L_0 u)(X(s))\,\mathrm{d}s$ and $\eta(t) = \exp\left(\int_0^{t\wedge\tau_U} V(X(s))\,\mathrm{d}s\right)$ and performing one integration by parts (of the standard kind). Similarly, for part (ii), apply Proposition 4.1 to

$$M(t) = u(X(t)) - \int_0^t (L_0 u)(X(s))\,\mathrm{d}s \text{ and } \eta(t) = \exp\left(\int_0^t V(X(s))\,\mathrm{d}s\right)$$

and perform one integration by parts. The calculation is left to the reader. □

Corollary 4.3. *Let L_0 satisfy Assumption A on a domain $D \subseteq R^d$ and let $\{P_x, x \in \hat{D}\}$ denote the solution to the generalized martingale problem for L_0 on D.*

(i) *If $U \subset\subset D$, then*

$$u(X(t \wedge \tau_U)) \exp\left(-\int_0^{t\wedge\tau_U} \frac{L_0 u}{u}(X(s))\,\mathrm{d}s\right)$$

is a martingale with respect to $(\hat{\Omega}_D, \hat{\mathcal{F}}^D, \hat{\mathcal{F}}_t^D, P_x)$, for all positive $u \in C^2(\bar{U})$ and all $x \in U$.

(ii) *If $\{P_x, x \in D\}$ in fact solves the martingale problem for L_0 on D, then*

$$u(X(t)) \exp\left(-\int_0^t \frac{L_0 u}{u}(X(s))\,\mathrm{d}s\right)$$

is a martingale with respect to $(\Omega, \mathcal{F}, \mathcal{F}_t, P_x)$, for all $u \in C_b^2(D)$ which satisfy $\lim_{x\in D} u > 0$, and all $x \in D$.

Proof. For part (i) (part (ii)) apply part (i) (part (ii)) of Theorem 4.2 with

$$V = -\frac{L_0 u}{u}.$$

□

2.5 Proof of Theorem 1.5.1

We can now give the proof of Theorem 1.5.1 for which we use Proposition 4.1, Corollary 4.3 and Theorem 2.2.

Part (a) (i) ⇒ (ii): A proof identical to that of Theorem 2.2(ii) shows that

$$P\left(\sup_{0\leq s\leq t}|\zeta(s)-\zeta(0)-\int_0^s b(u)\,\mathrm{d}u|\geq\lambda\right)\leq 2d\exp\left(\frac{-\lambda^2}{2A\,\mathrm{d}t}\right), \quad (5.1)$$

where

$$A=\sup_{\substack{0\leq s\leq t\\ q\in Y}}\sup_{|v|=1}\langle v, a(s,q)v\rangle.$$

From this it follows that for $0\leq t_1<t_2$ and $B\in\mathcal{F}_{t_1}$, both sides of the equality

$$E(X_\theta(t_2);\,B)=E(X_\theta(t_1);\,B)$$

are entire functions of θ. Thus the equality holds with $\mathrm{i}\theta$ in place of θ. This proves (ii).

(ii) ⇒ (iii). Using basic Fourier analysis, it is enough to prove (iii) for f of the form $f(x)=\exp(\mathrm{i}\langle\theta,x\rangle)$, for $\theta\in R^d$.

Let

$$\eta(t)=\exp\left(\mathrm{i}\left\langle\theta,\int_0^t b(s)\,\mathrm{d}s\right\rangle-\frac{1}{2}\int_0^t\langle\theta,a(s)\theta\rangle\,\mathrm{d}s\right).$$

By Proposition 4.1,

$$X_{\mathrm{i}\theta}(t)\eta(t)-\int_0^t X_{\mathrm{i}\theta}(s)\,\mathrm{d}\eta(s)$$

is a martingale with respect to $(Y,\mathcal{M},\mathcal{M}_t,P)$. But this expression in fact reduces to

$$\exp(-\mathrm{i}\langle\theta,\zeta(0)\rangle)\left(f(\zeta(t))-\int_0^t(L_sf)(\zeta(s))\,\mathrm{d}s\right),\text{ with}$$

$$f(x)=\exp(\mathrm{i}\langle\theta,x\rangle).$$

(iii) ⇒ (i). Let $f\in C_b^2(R^d)$ be uniformly positive. It follows from Corollary 4.3(ii) that

$$f(\zeta(t))\exp\left(-\int_0^t\left(\frac{L_sf}{f}\right)(\zeta(s))\,\mathrm{d}s\right)\text{ is a martingale with respect to }(Y,\mathcal{M},\mathcal{M}_t,P). \quad (5.2)$$

Now (i) would follow by choosing $f(x)=\exp(\langle\theta,x\rangle)$; however, f is unbounded so we cannot do this directly. For each $n\geq 1$, let $f_n\in C_b^2(R^d)$ satisfy $f_n(x)=\exp(\langle\theta,x\rangle)$, for $|x|\leq n$. Define

$\tau_n = \inf\{t \geq 0: |\zeta(t)| \geq n\}$. By Doob's optional sampling theorem (Theorem 1.2.1) and (5.2), it follows that

$$f_n(\zeta(t \wedge \tau_n)) \exp\left(-\int_0^{t\wedge\tau_n}\left(\frac{L_s f_n}{f_n}\right)(\zeta(s))\,ds\right) \text{ is a martingale,}$$

$$\text{for } n = 1, 2, \ldots. \quad (5.3)$$

Since $f_n = \exp(\langle\theta, x\rangle)$ on $|x| \leq n$, we conclude from (5.3) that $X_\theta(t \wedge \tau_n)$ is a martingale with respect to $(Y, \mathcal{M}, \mathcal{M}_t, P)$, for $n = 1, 2, \ldots$. Equivalently, for $0 \leq t_1 \leq t_2$ and $B \in \mathcal{M}_{t_1}$,

$$E(X_\theta(t_2 \wedge \tau_n); B) = E(X_\theta(t_1 \wedge \tau_n); B). \quad (5.4)$$

We will show that $\{X_\theta(t \wedge \tau_n)\}_{n=1}^\infty$ is uniformly integrable. From this and (5.4) it will follow that $E(X_\theta(t_2); B) = E(X_\theta(t_1); B)$, for $0 \leq t_1 < t_2$ and $B \in \mathcal{M}_{t_1}$ or, equivalently, that $X_\theta(t)$ is a martingale with respect to $(Y, \mathcal{M}, \mathcal{M}_t, P)$. We have

$$X_\theta^2(t \wedge \tau_n) = X_{2\theta}(t \wedge \tau_n)\exp\left(\int_0^{t\wedge\tau_n}\langle\theta, a(s)\theta\rangle\,ds\right) \leq cX_{2\theta}(t \wedge \tau_n),$$

for some constant $c > 0$, and $EX_{2\theta}(t \wedge \tau_n) = EX_{2\theta}(0) = 1$. Thus, $EX_\theta^2(t \wedge \tau_n) \leq c$; the uniform integrability follows from this.

Since (iii) is a special case of (iv), to complete the proof of part (a), it remains to show that

(iii) $\Rightarrow$ (iv). Assume that $f \in C_b^2([0, \infty) \times R^d)$. Then for $0 \leq t_1 < t_2$ and $A \in \mathcal{F}_{t_1}$, we have

$$\begin{aligned}
&E(f(t_2, \zeta(t_2)) - f(t_1, \zeta(t_1)); A) \\
&\quad = E(f(t_2, \zeta(t_2)) - f(t_1, \zeta(t_2)); A) \\
&\qquad + E(f(t_1, \zeta(t_2)) - f(t_1, \zeta(t_1)); A) \\
&\quad = E\left(\int_{t_1}^{t_2}\left(\frac{\partial f}{\partial s}\right)(s, \zeta(t_2))\,ds; A\right) + E\left(\int_{t_1}^{t} L_s f(t_1, \zeta(s))\,ds; A\right) \\
&\quad = E\left(\int_{t_1}^{t_2}\left(\frac{\partial f}{\partial s}\right)(s, \zeta(s))\,ds; A\right) \\
&\qquad + E\left(\int_{t_1}^{t_2}\left[\left(\frac{\partial f}{\partial s}\right)(s, \zeta(t_2)) - \left(\frac{\partial f}{\partial s}\right)(s, \zeta(s))\right]ds; A\right) \\
&\qquad + E\left(\int_{t_1}^{t}(L_s f)(s, \zeta(s))\,ds; A\right) \\
&\qquad + E\left(\int_{t_1}^{t_2}[(L_s f)(t_1, \zeta(s)) - (L_s f)(s, \zeta(s))]\,ds; A\right)
\end{aligned}$$

$$= E\left(\int_{t_1}^{t_2}\left(\frac{\partial f}{\partial s} + L_s f\right)(s, \zeta(s))\,\mathrm{d}s;\ A\right)$$
$$+ E\left(\int_{t_1}^{t_2}\mathrm{d}s\int_{s}^{t_2}\left(L_u\frac{\partial f}{\partial s}\right)(s, \zeta(u))\,\mathrm{d}u;\ A\right)$$
$$- E\left(\int_{t_1}^{t_2}\mathrm{d}u\int_{t_1}^{u}\left(\frac{\partial}{\partial s}L_u f\right)(s, \zeta(u))\,\mathrm{d}s;\ A\right)$$
$$= E\left(\int_{t_1}^{t_2}\left(\frac{\partial f}{\partial s} + L_s f\right)(s, \zeta(s))\,\mathrm{d}s;\ A\right),$$

where the third equality uses (iii) applied to $f(t_1, x)$ with t_1 fixed, the fifth equality uses (iii) applied to

$$\frac{\partial f}{\partial s}(s, x)$$

with s fixed and the final equality follows since

$$\frac{\partial}{\partial s}L_u f = L_u\frac{\partial f}{\partial s}$$

and the double integrals in which they appear as integrands are over the same region.

Part (b). Assume that $E\exp(\lambda|\zeta(0)| < \infty$, for all $\lambda \in R$. Then from (5.1), it follows that $E\exp(\lambda|\zeta(t)|) < \infty$, for all $\lambda \in R$ and $t > 0$. Now let $f \in C^2(R^d)$ and assume that f and all its partial derivatives up to order 2 are bounded in absolute value by $c_1\exp(c_2|x|)$.

Using an appropriate mollifier, it is possible to choose a sequence $\{f_n\}_{n=1}^{\infty} \in C_0^{\infty}(R^d)$ such that

$$\lim_{n\to\infty}\frac{\partial^\alpha f_n}{\partial x^\alpha}(x) = \frac{\partial^\alpha f}{\partial x^\alpha}(x), \quad \text{for all } x \in R^d \text{ and } 0 \leqslant |\alpha| \leqslant 2 \tag{5.5}$$

and

$$\left|\frac{\partial^\alpha f_n}{\partial x^\alpha}(x)\right| \leqslant c_3\exp(c_4 x), \quad \text{for all } x \in R^d \text{ and } 0 \leqslant |\alpha| \leqslant 2. \tag{5.6}$$

Let $Z_g(t) = g(\zeta(t)) - \int_0^t (Lg)(\zeta(s))\,\mathrm{d}s$, for $g \in C_b^2(R^d)$. Since $Z_{f_n}(t)$ is a martingale, we have $E(Z_{f_n}(t_2); B) = E(Z_{f_n}(t_1); B)$, for all $0 \leqslant t_1 < t_2$ and $B \in \mathcal{F}_{t_1}$. By (5.5), (5.6) and the fact that $E\exp(\lambda|\zeta(t)|) < \infty$, for all $\lambda \in R$, it follows from the dominated convergence theorem that $E(Z_f(t_2); B) = E(Z_f(t_1); B)$, for all $0 \leqslant t_1 < t_2$ and $B \in \mathcal{F}_{t_1}$. Thus

$$f(\zeta(t)) - \int_0^t Lf(\zeta(s))\,\mathrm{d}s$$

is a martingale with respect to $(Y, \mathcal{M}, \mathcal{M}_t, P)$.

It remains to prove (1.5.9) and (1.5.10). By choosing $f(x) = \langle \theta, x\rangle$, we obtain (1.5.9). In the special case $b \equiv 0$, (1.5.10) follows by choosing $f(x) = \langle \theta, x\rangle^2$. Since

$$\bar{\zeta}(t) \equiv \zeta(t) - \int_0^t b(s)\,\mathrm{d}s$$

satisfies part (a) of the theorem with the same matrix $a(t, q)$ and with $b(t, q)$ replaced by zero, it then follows that (1.5.10) holds for $\bar{\zeta}(t)$. That is, $\langle \theta, \bar{\zeta}(t)\rangle^2 - \int_0^t \langle \theta, a(s)\theta\rangle\,\mathrm{d}s$ is a martingale. Substituting $\bar{\zeta}(t) = \zeta(t) - \int_0^t b(s)\,\mathrm{d}s$ now gives (1.5.10). □

2.6 The Stroock–Varadhan support theorem

Theorem 6.1. *Let L satisfy Assumption A on R^d and assume that $\{P_x, x \in R^d)$ solves the martingale problem for L on R^d. Then for each $x \in R^d$,* supp $(P_x) = \Omega_x = \{\omega \in \Omega\colon X(0, \omega) = x\}$.

Remark. This theorem is known as the *Stroock–Varadhan support theorem*. There is a more involved version which identifies the support of P_x in the case where $a(x)$ is allowed to have degeneracies (see the notes at the end of the chapter).

To prove the theorem, we will need a couple of auxiliary results.

Lemma 6.2. *Let*

$$L = \sum_{i,j=1}^d a_{ij}(x)\frac{\partial^2}{\partial x_i \partial x_j},$$

where a satisfies $c_1|v|^2 \leq \sum_{i,j=1}^d a_{ij}(x)v_i v_j \leq c_2|v|^2$, for all $v, x \in R^d$, where $0 < c_1 \leq c_2 < \infty$. Assume that $\{P_x, x \in R^d\}$ solves the martingale problem for L on R^d. Fix $x \in R^d$ and let $\phi\colon [0, \infty) \to R^d$ be continuously differentiable with a bounded derivative ϕ' and such that $\phi(0) = x$. Then there exists a measure Q_x on Ω such that

$$\left.\frac{\mathrm{d}Q_x}{\mathrm{d}P_x}\right|_{\mathcal{F}_t} = N(t)$$

$$\equiv \exp\left(\int_0^t \langle a^{-1}(X(s))\phi'(s), \mathrm{d}X(s)\rangle - \tfrac{1}{2}\int_0^t \langle \phi'(s), a^{-1}(X(s))\phi'(s)\rangle\,\mathrm{d}s\right),$$

for all $t \geq 0$. The process $X(t)$ on $(\Omega, \mathcal{F}, \mathcal{F}_t, Q_x)$ is an $I_d(a(X(\cdot)), \phi'(\cdot))$-process or, equivalently, Q_x solves the martingale problem starting from x for the time-inhomogeneous operator

$$L^\phi = \frac{1}{2} \sum_{i,j=1}^{d} a_{ij}(x) \frac{\partial^2}{\partial x_i \partial x_j} + \sum_{i=1}^{d} \phi_i'(t) \frac{\partial}{\partial x_i}.$$

Proof. If we define

$$L_t = \frac{1}{2} \sum_{i,j=1}^{d} a_{ij}(t, X(t, \omega)) \frac{\partial^2}{\partial x_i \partial x_j} + \sum_{i=1}^{d} \phi_i'(t) \frac{\partial}{\partial x_i},$$

then L_t is of the form in (1.5.8) and Theorem 1.5.1 is in force. Thus, the equivalence noted at the end of the statement of the lemma holds. The proof of the lemma is exactly like the proof of Theorem 1.9.1. □

Theorem 6.3. *Let $(Y, \mathcal{M}, P)$ be a probability space with a filtration $\mathcal{M}_t$ and let $B(t)$ be a d-dimensional Brownian motion on $(Y, \mathcal{M}, \mathcal{M}_t, P)$. Let σ: $[0, \infty) \times Y \to R^d \otimes R^d$ satisfy $c_1|v|^2 \leq \sum_{i,j=1}^d a_{ij}(t, q) v_i v_j \leq c_2|v|^2$, for all $v \in R^d$, $t \in [0, \infty)$ and $q \in Y$, where $0 < c_1 \leq c_2 < \infty$ and $a = \sigma\sigma^{\mathrm{T}}$. Define $\zeta(t) = \int_0^t \sigma(s)\, \mathrm{d}B(s)$. Then, for each $\varepsilon > 0$ and $t > 0$,*

$$P(\sup_{0 \leq s \leq t} |\zeta(s)| \leq \varepsilon) > 0.$$

Remark. In the case $\zeta(t) = \int_0^t (\sigma(s))\, \mathrm{d}B(s)$, the result is a direct consequence of Theorem 3.6.1 in Chapter 3. Because of the bounds on the matrix a, it is intuitively clear that the result should hold in general. However, the proof requires several technical points that we have not introduced. The reader is referred to [Stroock (1971)].

Proof of Theorem 6.1 Using the localization result of Theorem 1.10.3, it is easy to reduce to the case where b is bounded and a satisfies

$$c_1|v|^2 \leq \sum_{i,j=1}^{d} a_{ij}(x) v_i v_j \leq c_2|v|^2, \quad \text{for all } x, v \in R^d, \tag{6.1}$$

where $0 < c_1 \leq c_2 < \infty$. By the Cameron–Martin–Girsanov formula, a solution exists to the martingale problem for

$$\frac{1}{2} \sum_{i,j=1}^{d} a_{ij}(x) \frac{\partial^2}{\partial x_i \partial x_j} + \sum_{i=1}^{d} b_i(x) \frac{\partial}{\partial x_i}$$

if and only if one exists for

$$\frac{1}{2}\sum_{i,j=1}^{d} a_{ij}(x)\frac{\partial^2}{\partial x_i \partial x_j},$$

and if these solutions do exist, then starting from a common $x \in R^d$, they are mutually absolutely continuous on $\mathcal{F}_t$, for all $t \geqslant 0$. Thus, it suffices to consider the case in which

$$L = \frac{1}{2}\sum_{i,j=1}^{d} a_{ij}(x)\frac{\partial^2}{\partial x_i \partial x_j},$$

where a satisfies (6.1).

Assume that $\{P_x, x \in R^d\}$ solves the martingale problem on R^d for such an L. Since the class of functions ϕ appearing in Lemma 6.2 is dense in Ω, to prove the theorem it is enough to show that for each such ϕ and every $t > 0$, $\varepsilon > 0$ and $x \in R^d$,

$$P_x(\sup_{0 \leqslant s \leqslant t} |X(s) - \phi(s)| < \varepsilon) > 0. \tag{6.2}$$

Fix $x \in R^d$ and a $\phi \in \Omega$ of the class described above and let Q_x be as in Lemma 6.2. Since Q_x and P_x are mutually absolutely continuous on $\mathcal{F}_t$, for any $t \geqslant 0$, in order to show that (6.2) holds it suffices to show that it holds with Q_x in place of P_x. By Lemma 6.2, $X(t)$ is an $I_d(a(X(\cdot)), \phi'(\cdot))$-process under Q_x and then, by the definition of an Ito process,

$$\bar{X}(t) = X(t) - \int_0^t \phi'(s)\,\mathrm{d}s = X(t) - \phi(t) + x$$

is an $I_d(a(X(\cdot)), 0)$-process. Letting $\sigma = a^{1/2}$, it follows from Theorem 1.6.2 that under Q_x, $\bar{B}(t) \equiv \int_0^t \sigma^{-1}(X(s))\,\mathrm{d}\bar{X}(s)$ is an $I_d(I, 0)$-process, that is, a Brownian motion. By Lemma 1.7.4.

$$\bar{X}(t) = X(0) + \int_0^t \sigma(X(s))\,\mathrm{d}\bar{B}(s).$$

But then, by Theorem 6.3, we have for all $\varepsilon > 0$ and $t > 0$,

$$Q_x(\sup_{0 \leqslant s \leqslant t} |\bar{X}(s) - X(0)| \leqslant \varepsilon) > 0.$$

This proves the Theorem since $\bar{X}(t) - X(0) = X(t) - \phi(t)$ a.s. Q_x. □

For $U \subset R^d$, define the *first entrance time* of U by $\sigma_U = \inf\{t \geqslant 0: X(t) \in U\}$. Note that $\sigma_U = \tau_{R^d - U}$. The following theorem is an immediate consequence of Theorem 6.1 and the localization result of Theorem 1.13.2.

Theorem 6.4. *Let L satisfy Assumption A on a domain $D \subseteq R^d$ and let $\{P_x, x \in \hat{D}\}$ denote the solution to the generalized martingale problem for L on D. Let $D_1, D_2 \subset D$ be open and satisfy $D_1 \cap D_2 = \varnothing$. If x is contained in the connected component of $D - \bar{D}_2$ which contains D_1, then*

$$P_x\{\sigma_{D_1} < \sigma_{D_2}) > 0.$$

2.7 Transience and recurrence for diffusions on R^d

In the literature, one may find several different definitions of transience and recurrence; however, they are all equivalent in the context of diffusion processes associated with elliptic operators satisfying Assumption A. Thus, we will choose convenient definitions for transience and recurrence and will then prove that no diffusion is both transient and recurrent, and that every diffusion is either transient or recurrent.

We will assume throughout this section that L satisfies Assumption A on R^d. Let $\{P_x, x \in \hat{R}^d\}$ denote the solution to the generalized martingale problem for L on R^d. Recall that $\sigma_U = \inf\{t \geq 0: X(t) \in U\}$ and that τ_{R^d} is the lifetime of the process defined in Section 1.15; $B_\varepsilon(y)$ denotes the ball of radius ε centered at y.

Definition. *The diffusion process on R^d corresponding to L is called recurrent if for all $x, y \in R^d$ and $\varepsilon > 0$, $P_x(\sigma_{B_\varepsilon(y)} < \infty) = 1$.*

Definition. *The diffusion process on R^d is called transient if for all $x \in R^d$, $P_x(\lim_{t \to \tau_{R^d}} |X(t)| = \infty) = 1$.*

We begin with the following lemma.

Lemma 7.1. *Let L satisfy Assumption A on R^d and let $\{P_x, x \in \hat{R}^d\}$ denote the solution to the generalized martingale problem for L on R^d. If the corresponding diffusion process is recurrent, then $P_x(\tau_{R^d} = \infty) = 1$, for all $x \in R^d$. In particular, $\{P_x, x \in R^d\}$ solves the martingale problem for L on R^d.*

Proof. Define $\eta_1 = \inf\{t \geq 0: |X(t)| \geq 1\}$ and, by induction, define $\sigma_n = \inf\{t \geq \eta_n: |X(t)| < \frac{1}{2}\}$ and $\eta_{n+1} = \inf\{t \geq \sigma_n: |X(t)| \geq n+1\}$, $n = 1, 2, \ldots$. We prove by induction that

$$P_x(\sigma_n < \infty) = P_x(\eta_n < \infty) = 1, \quad \text{for all } x \in R \text{ and } n = 1, 2, \ldots.$$

By Theorem 1.1 (or by the recurrence assumption), $P_x(\eta_1 < \infty) = 1$, for all $x \in R^d$. Assume now that $P_x(\eta_n < \infty) = 1$, for all $x \in R^d$. Then, by the strong Markov property,

$$P_x(\sigma_n < \infty) = P_x(\sigma_n < \infty, \eta_n < \infty) = E_x I_{\{\eta_n<\infty\}} I_{\{\sigma_n<\infty\}}$$
$$= E_x I_{\{\eta_n<\infty\}} E_x(I_{\{\sigma_n<\infty\}}|\mathcal{F}_{\eta_n}) = E_x I_{\{\eta_n<\infty\}} E_{X(\eta_n)} I_{\{\sigma_{B_{1/2}(0)}<\infty\}}$$
$$= E_x P_{X(\eta_n)}(\sigma_{B_{1/2}(0)} < \infty)$$
$$= \int_{R^d} P_z(\sigma_{B_{1/2}(0)} < \infty) P_x(X(\eta_n) \in dz) = 1$$

since, by recurrence, $P_z(\sigma_{B_{1/2}(0)} < \infty) = 1$, for all $z \in R^d$. A similar proof shows that if $P_x(\sigma_n < \infty) = 1$, for all $x \in R^d$, then $P_x(\eta_{n+1} < \infty) = 1$, for all $x \in R^d$.

Now if $\tau_{R^d}(\omega) < \infty$, for some $\omega \in \hat{\Omega}_{R^d}$, then, since ω is a continuous path on $\hat{R}^d$, $\lim_{t\to\tau_{R^d}}|X(t)| = \infty$ and thus, $\sigma_n(\omega) = \infty$, for n sufficiently large. Since we have just proved that $P_x(\sigma_n < \infty) = 1$, for all $x \in R^d$ and $n = 1, 2, \ldots$, we conclude that $P_x(\tau_{R^d} < \infty) = 0$. □

Recurrence is in fact equivalent to the following stronger formulation which states that a recurrent diffusion returns to $B_\varepsilon(y)$ infinitely often.

Proposition 7.2. *Let L satisfy Assumption A on R^d and let $\{P_x, x \in \hat{R}^d\}$ denote the solution to the generalized martingale problem for L on R^d. If the corresponding diffusion process is recurrent, then*

$$P_x(X(t) \in B_\varepsilon(y) \textit{ for arbitrarily large } t) = 1,$$

for all x, $y \in R^d$ and $\varepsilon > 0$.

Proof. We will show that for arbitrary $s > 0$,

$$P_x(X(t) \in B_\varepsilon(y) \text{ for some } t \geq s) = 1.$$

By Lemma 7.1, $\tau_{R^d} = \infty$ a.s. P_x. We have from the Markov property that

$$P_x(X(t) \in B_\varepsilon(y) \text{ for some } t \geq s) = E_x P_{X(s)}(\sigma_{B_\varepsilon(y)} < \infty) = 1,$$

since $P_x(\sigma_{B_\varepsilon(y)} < \infty) = 1$, for all $x \in R^d$. □

The behavior in Proposition 7.2 is clearly incompatible with transience. Thus, we have the following corollary.

Corollary 7.3 *Let L satisfy Assumption A on R^d. Then the corresponding diffusion process cannot be both recurrent and transient.*

The next theorem is the principal result of this section.

Theorem 7.4. *Let L satisfy Assumption A on R^d. Then the corresponding diffusion process is either transient or recurrent.*

For the proof of the theorem, we will need a lemma.

Lemma 7.5. *Let L satisfy Assumption A on R^d and let $\{P_x, x \in \hat{R}^d\}$ denote the solution to the generalized martingale problem for L on R^d. Let D, $U \subset R^d$ be domains satisfying an interior cone condition at each point of their boundaries. Then*

(i) *$P_x(\sigma_D < \infty)$ is continuous and positive for $x \in R^d$.*
(ii) *If $\bar{U} \cap \bar{D} = \phi$, then $P_x(\sigma_D < \sigma_U)$ is continuous for all $x \in R^d$ and positive for all x contained in the connected component of $R^d - \bar{U}$ which contains D.*

Remark. The proof of the lemma utilizes Theorem 3.3 whose proof in turn relied on the existence of a density $p(t, x, y)$. The proof of Theorem 7.4 does not require the continuity claim in the lemma but just the uniform positivity of $P_x(\sigma_D < \infty)$ for x in compacts, and the uniform positivity of $P_x(\sigma_D < \sigma_U)$ for x in compact subsets of the connected component of $R^d - \bar{U}$ which contains D. An alternative proof of this, which does not rely on the existence of a density, is suggested in exercise 4.5. To understand why we emphasize this, see the remark following Theorem 8.1 in the next section and see the introduction to Chapter 4.

Proof. The positivity claim in both parts of the lemma follows from Theorem 6.4. We now consider the continuity claim. For part (i), by Theorem 1.3.1, it is enough to show that $1_{\{\sigma_D<\infty\}}$ is P_x-a.s. continuous for all $x \in R^d$. If $x \in D$, then, clearly, $1_{\{\sigma_D<\infty\}} = 1$ a.s P_x. If $x \in R^d - D$, then $1_{\{\sigma_D<\infty\}}$ is P_x-a.s. continuous by part (iv) of Theorem 3.3. The proof of part (ii) is similar and is left to the reader. □

Proof of Theorem 7.4. Assume that $X(t)$ is not recurrent, that is, assume that for some $x_0, y_0 \in R^d$ and $\varepsilon_0 > 0$,

$$P_{x_0}(\sigma_{B_{\varepsilon_0}(y_0)} < \infty) < 1. \tag{7.1}$$

We must show that $X(t)$ is transient; that is, that

$$P_x(\lim_{t\to\tau_{R^d}} |X(t)| = \infty) = 1, \text{ for all } x \in R^d.$$

In fact, we will prove the following statement which is clearly equivalent to transience: for all $x \in R^d$ and $r > 0$,

$$P_x(\liminf_{t\to\tau_{R^d}} |X(t)| \leqslant r) = 0.$$

The proof will be given in three steps.

Step 1. $P_x(\sigma_{B_{\varepsilon_0}(y_0)} < \infty) < 1$, for all $x \in R^d - \overline{B_{\varepsilon_0}(y_0)}$.

Proof. Let $x \in R^d - \overline{B_{\varepsilon_0}(y_0)}$. By (7.1) and Lemma 7.5, there exists a $\delta > 0$ and a $\gamma < 1$ such that $\sup_{|z-x_0|\leqslant\delta} P_z(\sigma_{B_{\varepsilon_0}(y_0)} < \infty) = \gamma$. Of course, $\overline{B_\delta(x_0)} \cap \overline{B_{\varepsilon_0}(y_0)} = \phi$ and thus, by Theorem 6.4, $P_x(\sigma_{B_\delta(x_0)} < \sigma_{B_{\varepsilon_0}(y_0)}) > 0$.

By the strong Markov property,

$$\begin{aligned} P_x(\sigma_{B_{\varepsilon_0}(y_0)} = \infty) &\geqslant P_x(\sigma_{B_\delta(x_0)} < \sigma_{B_{\varepsilon_0}(y_0)}, \sigma_{B_{\varepsilon_0}(y_0)} = \infty) \\ &= \int_{|z-x_0|\geqslant\delta} P_z(\sigma_{B_{\varepsilon_0}(y_0)} = \infty) P_x(\sigma_{B_\delta(x_0)} < \sigma_{B_{\varepsilon_0}(y_0)}, X(\sigma_{B_\delta(x_0)}) \in dz) \\ &\geqslant (1-\gamma) P_x(\sigma_{B_\delta(x_0)} < \sigma_{B_{\varepsilon_0}(y_0)}) > 0. \end{aligned}$$

Step 2. $P_x(\sigma_{B_\varepsilon(y)} < \infty) < 1$, for all x, $y \in R^d$ and all $\varepsilon > 0$ such that $x \notin \overline{B_\varepsilon(y)}$.

Proof. Assume to the contrary that for some x_1, $y_1 \in R^d$ and some $\varepsilon_1 > 0$ such that $x_1 \notin \overline{B_{\varepsilon_1}(y_1)}$, $P_{x_1}(\sigma_{B_{\varepsilon_1}(y_1)} < \infty) = 1$. Then, by Step 1,

$$P_x(\sigma_{B_{\varepsilon_1}(y_1)} < \infty) = 1, \text{ for all } x \in R^d. \tag{7.2}$$

Let U be an open ball containing $\overline{B_{\varepsilon_1}(y_1)} \cup \overline{B_{\varepsilon_0}(y_0)}$ and define

$$\eta_1 = \inf\{t \geqslant 0\colon X(t) \in B_{\varepsilon_1}(y_1)\},\ \eta_{2n} = \inf\{t \geqslant \eta_{2n-1}\colon X(t) \notin U\}$$

and

$$\eta_{2n+1} = \inf\{t \geqslant \eta_{2n}\colon X(t) \in B_{\varepsilon_1}(y_1)\}, \text{ for } n = 1, 2, \ldots.$$

By Theorem 1.1 and (7.2), it follows that

$$P_x(\eta_n < \infty) = 1, \text{ for all } x \in R^d \text{ and } n = 1, 2, \ldots.$$

By Lemma 7.5,

$$\lambda \equiv \inf_{z\in\partial B_{\varepsilon_1}(y_1)} P_z(\sigma_{B_{\varepsilon_0}(y_0)} < \sigma_{U^c}) > 0.$$

Let $A_n = \{\sigma_{B_{\varepsilon_0}(y_0)} \notin (\eta_{2n-1}, \eta_{2n})\}$, $n = 1, 2, \ldots$. Then

$$\{\sigma_{B_{\varepsilon_0}(y_0)} = \infty\} \subset \bigcap_{n=1}^{\infty} A_n.$$

By the strong Markov property, for $m \geqslant 2$,

$$P_{x_0}\left(\bigcap_{n=1}^{m} A_n\right) = E_{x_0}\prod_{n=1}^{m} 1_{A_n} = E_{x_0}\left(\prod_{n=1}^{m-1} 1_{A_n} E_{x_0}(1_{A_m}|\mathcal{F}_{\eta_{2m-1}})\right)$$

$$= E_{x_0}\prod_{n=1}^{m-1} 1_{A_n} E_{X(\eta_{2m-1})} 1_{A_1}.$$

But $X(\eta_{2m-1}) \in \partial B_{\varepsilon_1}(y_1)$ and, for all $z \in \partial B_{\varepsilon_1}(y_1)$,

$$E_z 1_{A_1} = P_z(A_1) = P_z(\sigma_{U^C} < \sigma_{B_{\varepsilon_0}(y_0)}) \leqslant 1 - \lambda.$$

Thus

$$P_{x_0}\left(\bigcap_{n=1}^{m} A_n\right) \leqslant (1-\lambda) E_{x_0}\prod_{n=1}^{m-1} 1_{A_n} = (1-\lambda) P_{x_0}\left(\bigcap_{n=1}^{m-1} A_n\right).$$

By induction $P_{x_0}(\bigcap_{n=1}^{m} A_n) \leqslant (1-\lambda)^m$ and thus,

$$P_{x_0}(\sigma_{B_{\varepsilon_0}(y_0)} = \infty) \leqslant P_{x_0}\left(\bigcap_{n=1}^{\infty} A_n\right) = 0;$$

this contradicts (7.1).

Step 3. For all $x \in R^d$ and any $r > 0$,

$$P_x(\liminf_{t \to \tau_{R^d}} |X(t)| \leqslant r) = 0.$$

Proof. We will use the notation $B_\delta \equiv B_\delta(0)$. Let $0 < r < r_1$. By Step 2, $P_z(\sigma_{B_r} < \infty) < 1$, for all $z \in \partial B_{r_1}$ and thus, by Lemma 7.5,

$$\rho \equiv \sup_{z \in \partial B_{r_1}} P_z(\sigma_{B_r} < \infty) < 1. \tag{7.3}$$

Define $\nu_1 = \inf\{t \geqslant 0: |X(t)| = r_1\}$, $\nu_{2n} = \inf\{t \geqslant \nu_{2n-1}: |X(t)| = r\}$ and $\nu_{2n+1} = \inf\{t \geqslant \nu_{2n}: |X(t)| = r_1\}$, for $n = 1, 2, \ldots$. Let $G_n = \{\nu_n < \infty\}$. By Theorem 1.1,

$$P_x\{\nu_{2n+1} < \infty | \nu_{2n} < \infty) = 1, \quad \text{for all } x \in R^d. \tag{7.3}$$

Thus $\{\liminf_{t \to \tau_{R^d}} |X(t)| \leqslant r\} = \bigcap_{n=1}^{\infty} G_n$ a.s. P_x. By the strong Markov property, for any positive integer m and any $x \in R^d$,

$$P_x\left(\bigcap_{n=1}^{2m} G_n\right) = E_x\prod_{n=1}^{2m} 1_{G_n} = E_x\left(\bigcap_{n=1}^{2m-1} 1_{G_n} E_x(1_{G_{2m}}|\mathcal{F}_{\nu_{2m-1}})\right)$$

$$= E_x \prod_{n=1}^{2m-1} 1_{G_n} P_{X(\nu_{2m-1})}(\sigma_{B_r} < \infty).$$

But $|X(\nu_{2m-1})| = r_1$, so by (7.3), $P_{X(\nu_{2m-1})}(\sigma_{B_r} < \infty) \leq \rho < 1$. Thus we have $P_x(\bigcap_{n=1}^{2m} G_n) \leq \rho P_x(\bigcap_{n=1}^{2m-2} G_n)$ and, by induction $P_x(\bigcap_{n=1}^{2m} G_n) \leq \rho^m$. We conclude that

$$P_x(\liminf_{t \to \tau_{R^d}} |X(t)| \leq r) = P_x\left(\bigcap_{n=1}^{\infty} G_n\right) = 0, \text{ for all } x \in R^d. \qquad \square$$

The following alternative characterization of transience is often convenient.

Theorem 7.6. *Let L satisfy Assumption A on R^d and let $\{P_x, x \in \hat{R}^d\}$ denote the solution to the generalized martingale problem for L on R^d.*

(i) *If there exists an open set $U \subset R^d$ and a point $x_0 \in R^d$ such that $P_{x_0}(\sigma_U < \infty) < 1$, then the corresponding diffusion process is transient.*
(ii) *If the corresponding diffusion process is transient and U is bounded, then $P_x(\sigma_U < \infty) < 1$, for every x which is contained in the unbounded connected component of $R^d - \bar{U}$.*

Proof. (i) This follows immediately from Theorem 7.4.

(ii) Without loss of generality, assume that U is open and has a smooth boundary. By assumption, for all

$$x \in R^d,\ P_x(\lim_{t \to \infty} |X(t)| = \infty) = 1.$$

Therefore, letting $\tau_n = \inf(t \geq 0\colon\ |X(t)| = n)$, it follows that $P_x(X(t) \notin \bar{U}$ for all $t \geq \tau_n) > 0$, for some n. However, by the strong Markov property, $P_x(X(t) \notin \bar{U}$, for all $t \geq \tau_n) = E_x P_{X(\tau_n)}(\sigma_U = \infty)$. Thus it follows that $P_{z_0}(\sigma_U < \infty) < 1$, for some z_0 with $|z_0| = n$. As demonstrated in the proof of Lemma 7.5, $1_{\{\sigma_U < \infty\}}$ is P_x-a.s. continuous for $x \in R^d$. Thus there exists a ball B containing z_0 such that

$$\sup_{z \in \bar{B}} P_z(\sigma_U < \infty) < 1. \tag{7.5}$$

(For an alternative proof of this, see exercise 4.5.)

Of course, $\bar{B}$ is contained in the unbounded connected component of $R^d - \bar{U}$. Let x be any point in the unbounded connected component of $R^d - \bar{U}$. Using the strong Markov property and, in the final inequality, (7.5) and Lemma 7.5, we have

$$\begin{aligned} P_x(\sigma_U = \infty) &\geq P_x(\sigma_U = \infty,\ \sigma_{\bar{B}} < \sigma_U) \\ &\geq \inf_{z \in \bar{B}} P_z(\sigma_U = \infty) P_x(\sigma_{\bar{B}} < \sigma_U) > 0. \qquad \square \end{aligned}$$

2.8 Transience and recurrence for diffusions on $\mathbf{D} \subseteq R^d$

In this section, we extend the concept of transience and recurrence to diffusions on a domain $D \subseteq R^d$. Let $\{P_x, x \in \hat{D}\}$ solve the generalized martingale problem for L on D.

Definition. *The diffusion process on D corresponding to L is called* recurrent (on D) *if for all* $x, y \in D$ *and* $\varepsilon > 0$, $P_x(\sigma_{B_\varepsilon(y)} < \infty) = 1$.

Definition. *The diffusion process on D corresponding to L is called* transient (on D) *if for all* $x \in D$, $P_x(X(t)$ *is eventually in* $D_n) = 1$, *for* $n = 1, 2, \ldots,$ *where* $\{D_n\}_{n=1}^{\infty}$ *is a sequence of domains satisfying* $D_n \subset\subset D_{n+1}$ *and* $\bigcup_{n=1}^{\infty} D_n = D$.

The results of the previous section hold in the present context with essentially identical proofs. At the points in the proofs where exit times from the balls $B_n = \{|x| < n\}$, $n = 1, 2, \ldots$ were used, in the present context one should replace them by exit times from D_n, $n = 1, 2, \ldots$. The following theorem is the analog of Lemma 7.1, Proposition 7.2, Corollary 7.3 and Theorem 7.4.

Theorem 8.1. *Let L satisfy Assumption A on a domain* $D \subseteq R^d$ *and let* $\{P_x, x \in \hat{D}\}$ *denote the solution to the generalized martingale problem for L on D.*

(i) *The diffusion process cannot be both recurrent on D and transient on D.*
(ii) *The diffusion process is either recurrent on D or transient on D.*
(iii) *If the diffusion process is recurrent on D, then* $P_x(\tau_D = \infty) = 1$, *for all* $x \in D$. *In particular, then* $\{P_x, x \in D\}$ *solves the martingale problem for L on D.*
(iv) *If the diffusion process is recurrent on D, then for all* $x, y \in D$ *and all* $\varepsilon > 0$, $P_x(X(t) \in B_\varepsilon(y)$ *for arbitrarily large* $t) = 1$

Remark. Theorem 8.1 is of seminal importance in our approach to Chapter 4. Indeed, the proof of almost every result in that chapter depends on it.

The analog of Theorem 7.6 is as follows.

Theorem 8.2. *Let L satisfy Assumption A on a domain* $D \subseteq R^d$ *and let* $\{P_x, x \in \hat{D}\}$ *denote the solution to the generalized martingale problem for L on D.*

(i) *If there exists an open set* $U \subset D$ *and a point* $x_0 \in D$ *such that* $P_{x_0}(\sigma_U < \infty) < 1$, *then the diffusion process is transient on D.*

(ii) *If the diffusion process is transient on* D *and* $U \subset\subset D$, *then* $P_x(\sigma_U < \infty) < 1$, *for every* x *which is contained in a connected component,* A, *of* $D - \bar{U}$ *which satisfies* $P_y(X(t)$ *is eventually in* $A) > 0$, *for some (or equivalently, all)* $y \in D$.

Proof. The proof of (i) is immediate from Theorem 8.1. The proof of (ii) is left as exercise 2.3. □

Exercises

2.1 Show that $\tau_A = \inf\{t \geq 0 : X(t) \notin A\}$ defined on $(\Omega, \mathcal{F})$ is lower semicontinuous if $A \subset R^d$ is open and upper semicontinuous if $A \subset R^d$ is closed.

2.2 Recall the law of the iterated logarithm: if $\beta(t)$ is a standard Brownian motion, then

$$\limsup_{t \to 0} \frac{\beta(t)}{\left(2t \log\log \frac{1}{t}\right)^{1/2}} = 1 \text{ a.s.}$$

Use this to prove Theorem 3.3(iii) in the case $a = I$. (Hint: Prove it for $b \equiv 0$ and use the Cameron–Martin–Girsanov formula to extend it to the case of non-zero b.)

2.3 Prove part (ii) of Theorem 8.2, along the lines of the proof of Theorem 7.6(ii).

2.4 Theorem 3.3(i) was proved under the condition that B satisfy an exterior sphere condition at $a \in \partial B$. Prove the result in the case where B satisfies an exterior cone condition at a by replacing the barrier function $w(x) = c(R^{-\gamma} - |x - y|^{-\gamma})$ appearing in the proof by a barrier function of the form $w(x) = |x - a|^{\mu} f(\theta)$, where θ is the angle between $x - a$ and the axis of the exterior cone.

2.5 Let L satisfy Assumption A on R. Use the law of the interated logarithm (see exercise 2.2) and the time change formula (exercise 1.15) to show that every point is regular.

Notes

Results of the type given in Section 1 have their inception in the work of [Kakutani (1944b, 1945)], who gave the beautiful probabilistic representation to the Dirichlet problem. Section 1 follows [Friedman (1975)]. Theorem 2.1 is a standard theorem to workers in the field; however I failed to uncover a direct proof in the literature. The proof of Theorem 2.2 follows [Stroock and Varadhan (1979)]. The proof of Theorem 3.2 follows [Friedman (1975)] and the proof of Theorem 3.3 follows [Karatzas and Shreve (1988)]. The Feynman–Kac formula in its initial formulation goes back to [Kac (1949)]; the treatment in Section 4 follows [Stroock and Varadhan (1979)]. Other references for the Feynman–Kac formula are, for example, [Karatzas and Shreve (1988)] and [Freidlin (1985)]. Section 5 follows [Stroock and Varadhan (1979)]. The Stroock–Varadhan support theorem and its proof come from [Stroock and Varadhan (1970)] which actually proves a more general support theorem for diffusions which may be degenerate. Section 7 follows [Bhattacharya (1978)].

3

The spectral theory of elliptic operators on smooth bounded domains

In the first three sections of this chapter, we recall a number of standard results from functional analysis and elliptic PDE theory. The majority of these results are stated without proof; the reader is referred to the notes at the end of the chapter.

3.1 The spectral theory of compact operators

Let $\mathcal{B}$ be a Banach space and let A be a linear operator defined on a dense subspace $\mathcal{D} \subset \mathcal{B}$ and taking values in $\mathcal{B}$. The operator A is called *closed* if its graph $\{(x, Ax); x \in \mathcal{D}\}$ is a closed subset of $\mathcal{B} \times \mathcal{B}$. It is called *closable* if it can be extended to a closed operator. Clearly, A is closeable if and only if, whenever $x_n \to 0$ and Ax_n converges, then, in fact, $Ax_n \to 0$. It follows from the closed graph theorem that a closed operator defined on a Banach space $\mathcal{B}$ is bounded. Thus, if A is closed and unbounded, its domain of definition $\mathcal{D}$ must necessarily be a proper subspace of $\mathcal{B}$. The *resolvent* set $\rho(A)$ of a closed densely defined operator A is defined as the collection of complex numbers λ for which $\lambda - A$ is a bijection of $\mathcal{D}$ onto $\mathcal{B}$. If $\lambda \in \rho(A)$, then $(\lambda - A)^{-1}$ is a closed operator from $\mathcal{B}$ into $\mathcal{B}$ and it thus follows from the closed graph theorem that $(\lambda - A)^{-1}$ is in fact a bounded operator. The resolvent set can be shown to be an open subset of $\mathbb{C}$. The *spectrum* $\sigma(A)$ of A is defined as the complement of $\rho(A)$ in $\mathbb{C}$ and, as such, is a closed subset of $\mathbb{C}$. A point $\lambda \in \sigma(A)$ is called an *eigenvalue* if there exists an $x \in \mathcal{D}$ such that $Ax = \lambda x$. The subspace $\{x \in \mathcal{D}: Ax = \lambda x\}$ is the *eigenspace* associated with the eigenvalue λ. If this eigenspace is finite (infinite) dimensional, then λ is called an eigenvalue of finite (infinite) multiplicity.

A bounded linear operator $A: \mathcal{B} \to \mathcal{B}$ is called *compact* if, for every bounded sequence $\{x_n\}_{n=1}^{\infty} \subset \mathcal{B}$, the sequence $\{Ax_n\}_{n=1}^{\infty}$ has a convergent subsequence. This is equivalent to the condition that the image under A of any bounded set be precompact. The spectral theory of compact operators is particularly simple.

Theorem 1.1 (Riesz–Schauder). *Let $\mathcal{B}$ be an infinite-dimensional Banach space and let $A: \mathcal{B} \to \mathcal{B}$ be a compact operator. Then the spectrum $\sigma(A)$ of A is a compact subset of $\mathbb{C}$ containing 0 as its unique accumulation point. Every non-zero point in $\sigma(A)$ is an eigenvalue of finite multiplicity.*

An immediate consequence of this theorem is the following Fredholm alternative.

Corollary 1.2 (Fredholm alternative). *Let $\mathcal{B}$ be a Banach space and let $A: \mathcal{B} \to \mathcal{B}$ be a compact operator. Then either:*

(a) *the equation $x - Ax = y$ has a unique solution $x \in \mathcal{B}$ for each $y \in \mathcal{B}$; or*
(b) *the equation $x - Ax = 0$ has non-trivial solutions which form a finite-dimensional subspace of $\mathcal{B}$.*

Remark. The alternative (b) holds, of course, if and only if $1 \in \sigma(A)$.

The dual space to a Banach space $\mathcal{B}$ is the Banach space of all bounded linear functionals from $\mathcal{B}$ to R and is denoted by $\mathcal{B}^*$. If $A: \mathcal{B} \to \mathcal{B}$ is a bounded linear operator, its adjoint $A^*: \mathcal{B}^* \to \mathcal{B}^*$ is defined by $(A^*\nu)(v) = \nu(Av)$ for $v \in \mathcal{B}$ and $\nu \in \mathcal{B}^*$; A^* is a bounded linear operator from $\mathcal{B}^*$ to $\mathcal{B}^*$. If A is compact, then A^* is also compact.

If A is an unbounded closed linear operator on $\mathcal{B}$ with dense domain $\mathcal{D}$, its adjoint A^* is defined as follows. Let $\mathcal{D}^* = \{\nu \in \mathcal{B}^*; \exists\, \eta \in \mathcal{B}^*$ satisfying $\nu(Av) = \eta(v)$, for all $v \in \mathcal{D}\}$. Note that for any $\nu \in \mathcal{D}^*$, the corresponding η in the definition is clearly unique since $\mathcal{D}$ is dense in $\mathcal{B}$. The adjoint to A is defined by $A^*\nu = \eta$, if $\nu \in \mathcal{D}^*$ and $\nu(Av) = \eta(v)$ for all $\nu \in \mathcal{D}$. The adjoint A^* is an unbounded closed linear operator on $\mathcal{B}^*$ with domain $\mathcal{D}^*$.

A subset K of a Banach space $\mathcal{B}$ is called a *proper convex cone* if the following three conditions hold:

(i) $x \in K \Rightarrow tx \in K$ for all $t \geq 0$;
(ii) $x, y \in K \Rightarrow x + y \in K$;
(iii) $x \in K \Rightarrow -x \notin K$.

The set $K^* = \{\nu \in \mathcal{B}^*: \nu(v) \geq 0$ for all $v \in K\}$ can easily be seen to be a closed convex cone in $\mathcal{B}^*$ whose interior is proper; it is called the dual cone to K. It is easy to check that if A is a linear operator and $AK \subset K$, then $A^*K^* \subset K^*$.

The next theorem is the infinite-dimensional generalization of the Perron–Frobenius theorem for matrices.

Theorem 1.3 (Krein–Rutman). *Let $\mathcal{B}$ be a Banach space, let $K \subset \mathcal{B}$ be a closed proper convex cone, let $A: \mathcal{B} \to \mathcal{B}$ be a compact operator, and let $\lambda_0 = \sup \mathrm{Re}(\sigma(A))$ denote the supremum of the real part of the spectrum of A.*

(i) *Assume that $AK \subset K$ and that A has a point in its spectrum different from zero. Then $\lambda_0 > 0$, $\lambda_0 \in \sigma(A)$ and $|\lambda| \le \lambda_0$ for all $\lambda \in \sigma(A)$, and there exists an eigenvector $v \in K$ corresponding to λ_0. Furthermore, $\lambda_0 \in \sigma(A^*)$ and $|\lambda| \le \lambda_0$ for all $\lambda \in \sigma(A^*)$ and there exists an eigenvector $v^* \in K^*$ corresponding to λ_0.*

(ii) *Assume that $AK \subset K$ and that for each $0 \ne x \in \partial K$, there exists a positive integer n such that $A^n x \in \mathrm{int}(K)$. Then $\lambda_0 > 0$, $\lambda_0 \in \sigma(A)$ and $|\lambda| < \lambda_0$ if $\lambda_0 \ne \lambda \in \sigma(A)$. The eigenvalue λ_0 is of multiplicity one and its eigenspace is generated by an element of $\mathrm{int}(K)$. The eigenvectors corresponding to eigenvalues in $\sigma(A) - \{\lambda_0\}$ do not belong to K. Furthermore, $\lambda_0 \in \sigma(A^*)$ and $|\lambda| < \lambda_0$ if $\lambda_0 \ne \lambda \in \sigma(A^*)$. The eigenvalue λ_0 is of multiplicity one and its eigenspace is generated by an element of $\mathrm{int}(K^*)$. The eigenvectors corresponding to eigenvalues in $\sigma(A^*) - \{\lambda_0\}$ do not belong to K^*.*

A compact operator $A: \mathcal{B} \to \mathcal{B}$ satisfying the condition $AK \subset K$ as in part (i) of the Krein–Rutman theorem is called *positive* with respect to the cone K. If A satisfies the stronger condition given in part (ii) of the theorem, then A is called *strongly positive* with respect to K. In the case that $\mathcal{B}$ is a function space and K is the cone of non-negative functions, we will simply call A positive or strongly positive.

3.2 Maximum principles and the Schauder estimates

Let

$$L = \tfrac{1}{2}\sum_{i,j=1}^{d} a_{ij}(x)\frac{\partial^2}{\partial x_i \partial x_j} + \sum_{i=1}^{d} b_i(x)\frac{\partial}{\partial x_i} + V(x).$$

For the maximum principles that we present, the following uniform ellipticity assumption on the operator L will be assumed.

Assumption E.

$$L = \tfrac{1}{2}\sum_{i,j=1}^{d} a_{ij}(x)\frac{\partial^2}{\partial x_i \partial x_j} + \sum_{i=1}^{d} b_i(x)\frac{\partial}{\partial x_i} + V$$

defined on a domain $D \subseteq R^d$ satisfies

(i) $0 < \lambda|v|^2 \leqslant \sum_{i,j=1}^d a_{ij}(x)v_iv_j \leqslant \Lambda|v|^2 < \infty$, *for all* $x \in D$ *and* $v \in R^d$;
(ii) $\sup_D |b(x)|, \sup_D |V(x)| < \infty$.

However, it is easy to see that Theorems 2.1, 2.2 and 2.6 and Corollary 2.3 below hold under the weaker condition that Assumption E holds in every subdomain $D' \subset\subset D$ with constants λ and Λ which may depend on D'.

We begin with the standard weak maximum principle. Define $u^+ = \max(u, 0)$ and $u^- = \min(u, 0)$.

Theorem 2.1. *Let L satisfy Assumption E on a bounded domain $D \subset R^d$ and let $u \in C^2(D) \cap C(\bar{D})$ satisfy $Lu \leqslant 0$ ($Lu \geqslant 0$) in D.*

(i) *If $V \equiv 0$ in D, then* $\inf_D u = \inf_{\partial D} u$ ($\sup_D u = \sup_{\partial D} u$);
(ii) *If $V \leqslant 0$ in D, then* $\inf_D u \geqslant \inf_{\partial D} u^-$($\sup_D u \leqslant \sup_{\partial D} u^+$);

Proof. Clearly, it is enough to prove the theorem in the case $Lu \leqslant 0$.

(i) First assume that $Lu < 0$ in D. If u attains its minimum in D at $x_0 \in D$, then

$$\nabla u(x_0) = 0 \text{ and } \left\{\frac{\partial^2 u}{\partial x_i \partial x_j}\right\}(x_0)$$

is non-negative definite. Since $\{a_{ij}(x_0)\}$ is positive definite, it follows that

$$\tfrac{1}{2}\sum_{i,j=1}^{d} a_{ij}(x_0)\frac{\partial^2}{\partial x_i \partial x_j}(x_0) \geqslant 0.$$

Thus, $Lu(x_0) \geqslant 0$, which is a contradiction and it follows that $\inf_D u = \inf_{\partial D} u$.

Now assume only that $Lu \leqslant 0$ in D. For $\varepsilon > 0$, let $v_\varepsilon(x) = u(x) - \varepsilon \exp(\gamma x_1)$. Then

$$\begin{aligned} Lv_\varepsilon &= Lu - \varepsilon\left(\frac{\gamma^2}{2}a_{11}(x) + \gamma b_1(x)\right)\exp(\gamma x_1) \\ &\leqslant -\varepsilon\left(\frac{\gamma^2}{2}a_{11}(x) + \gamma b_1(x)\right)\exp(\gamma x_1). \end{aligned}$$

By Assumption E, $a_{11}(x) \geqslant \lambda$ and $b_1(x)$ is bounded. Thus, by choosing γ sufficiently large,

$$\frac{\gamma^2}{2}a_{11}(x) + \gamma b(x) \geqslant 1, \text{ for all } x \in D, \text{ and } Lv_\varepsilon < 0 \text{ in } D.$$

Applying the first part of the proof to v_ε, we obtain $\inf_D v_\varepsilon = \inf_{\partial D} v_\varepsilon$, and letting $\varepsilon \to 0$ gives $\inf_D u = \inf_{\partial D} u$.

(ii) Let $D^- = D \cap \{u < 0\}$ and assume first that D^- is not empty. Define $L_0 = L - V$. We have $L_0 u = Lu - Vu \leqslant 0$ in D^- since $Lu \leqslant 0$ and $V \leqslant 0$ in D and $u < 0$ in D^-. Applying part (i) to the operator L_0 in D^-, and using the fact that $u = 0$ on $\partial D^- - \partial D$, we obtain

$$\inf_D u = \inf_{D^-} u = \inf_{\partial D^-} u = \inf_{\partial D^- \cap \partial D} u = \inf_{\partial D} u = \inf_{\partial D} u^-.$$

If D^- is empty, then trivially $\inf_D u \geqslant \inf_{\partial D} u^-$. Thus, in any case $\inf_D u \geqslant \inf_{\partial D} u^-$. □

If the condition $V \leqslant 0$ fails, then in certain circumstances a *generalized maximum principle* still holds.

Theorem 2.2. *Let L satisfy Assumption E on a bounded domain $D \subset R^d$ and let $u \in C^2(D) \cap C(\bar{D})$. Assume that there exists a function $h \in C^2(D) \cap C(\bar{D})$ such that $Lh \leqslant 0$ in D and $h > 0$ in $\bar{D}$. If $Lu \leqslant 0$ in D and $u \geqslant 0$ on ∂D, then $u \geqslant 0$ in D.*

Proof. Define the operator L^h by

$$L^h u = \frac{1}{h} L(hu).$$

Letting $L_0 = L - V$, a calculation reveals that

$$L^h = L_0 + a\frac{\nabla h}{h} \cdot \nabla + \frac{Lh}{h}.$$

The zeroth-order term in L^h is

$$\frac{Lh}{h},$$

which is non-positive by assumption. Thus, we may apply Theorem 2.1 to the operator L^h. Since

$$L^h\left(\frac{u}{h}\right) = \frac{1}{h} Lu \leqslant 0,$$

we conclude that

$$\inf_D \frac{u}{h} \geqslant \inf_{\partial D}\left(\frac{u}{h}\right)^- \geqslant 0.$$

Thus, $u \geqslant 0$ in D. □

The generalized maximum principle is enough to ensure uniqueness for the Dirichlet problem.

Corollary 2.3. *Under the conditions of Theorem 2.2, the Dirichlet problem $Lu = f$ in D and $u = \phi$ on ∂D has at most one solution $u \in C^2(D) \cap C(\bar{D})$.*

Proof. Let u_1 and u_2 be two such solutions. Applying Theorem 2.2 to $u_1 - u_2$ shows that $u_1 \geqslant u_2$ in D and applying it to $u_2 - u_1$ shows that $u_2 \geqslant u_1$ in D. □

Using Theorem 2.1, we can derive the following *a priori* bound.

Theorem 2.4. *Let L satisfy Assumption E on a bounded domain $D \subset R^d$ and assume that $V \leqslant 0$ in D. Let $u \in C^2(D) \cap C(\bar{D})$ satisfy $Lu \geqslant f (Lu = f)$ in D. Then*

$$\sup_D u \leqslant \sup_{\partial D} u^+ + \frac{C}{\lambda} \sup_D |f^-| \left(\sup_D |u| \leqslant \sup_{\partial D} |u| + \frac{C}{\lambda} \sup_D |f| \right),$$

where C depends only on diam(D) *and on*

$$\beta \equiv \frac{\sup_D |b|}{\lambda}.$$

Proof. The case $Lu = f$ follows from the case $Lu \geqslant f$ by applying the latter case to u and $-u$. Assume that $Lu \geqslant f$ in D and assume without loss of generality that D is contained in the slab $0 < x_1 < \gamma$. Let $L_0 = L - V$. Then for $\alpha \geqslant \beta + (\beta^2 + 2)^{\frac{1}{2}}$, we have

$$\begin{aligned} L_0 \exp(\alpha x_1) &= \left(\frac{\alpha^2}{2} a_{11}(x) + \alpha b_1(x) \right) \exp(\alpha x_1) \\ &\geqslant \lambda \left(\frac{\alpha^2}{2} - \alpha\beta \right) \exp(\alpha x_1) \geqslant \lambda. \end{aligned} \tag{2.1}$$

Let

$$w = \sup_{\partial D} u^+ + (\exp(\alpha\gamma) - \exp(\alpha x_1)) \sup_D |f^-| / \lambda.$$

Since $w > 0$, we have from (2.1), $Lw = (L_0 + V) w \leqslant -\sup_D |f^-|$. Thus,

$$L(w - u) \leqslant -\sup_D |f^-| - f \leqslant 0 \text{ in } D$$

and $w - u \geq 0$ on ∂D. Thus, by Theorem 2.1, $u(x) \leq w(x)$ in D and in particular,

$$u(x) \leq \sup_{\partial D} u^+ + \frac{C}{\lambda} \sup_D |f^-|,$$

where $C = \exp(\alpha\gamma) - 1$ and $\alpha = \beta + (\beta^2 + 2)^{1/2}$. □

We will also need the following two versions of the strong maximum principle.

Theorem 2.5. *Let L satisfy Assumption E on a domain $D \subset R^d$ and let $u \in C^2(D)$ satisfy $Lu \geq 0$ in D. Let $x_0 \in \partial D$ be such that u is continuous at x_0, $u(x_0) > u(x)$, for all $x \in D$, and ∂D satisfies an interior sphere condition at x_0. Assume that one of the following three conditions holds:*

(i) $V \equiv 0$;
(ii) $V \leq 0$ *and* $u(x_0) \geq 0$;
(iii) $u(x_0) = 0$.

Then the outer normal derivative of u at x_0, if it exists, satisfies

$$\frac{\partial u}{\partial n}(x_0) > 0.$$

Proof. We begin by reducing case (iii) to case (ii). In case (iii), we have $u < 0$ in D. Therefore, $(L - V^+)u \geq 0$ in D and case (ii) may be applied to the operator $L - V^+$. Consequently, from here on, we assume that either case (i) or case (ii) holds.

Since D satisfies an interior sphere condition at x_0, there exists a ball $B = B_R(y) \subset D$ of radius R and centered at $y \in D$, with $x_0 \in \partial B$. Fix $\rho \in (0, R)$ and define $w(x) = \exp(-\alpha r^2) - \exp(-\alpha R^2)$, for $\rho \leq r \leq R$, where $r = |x - y|$ and α is a positive constant yet to be determined. We have

$$\begin{aligned} Lw(x) = \exp(-\alpha r^2)\Bigg[2\alpha^2 \sum_{i,j=1}^{d} a_{ij}(x)(x_i - y_i)(x_j - y_j) \\ - \alpha(\mathrm{Tr}\,(a(x)) + 2b(x)\cdot(x - y))\Bigg] \\ + V(x)w(x) \\ \geq \exp(-\alpha r^2)[2\alpha^2\lambda r^2 - \alpha(\mathrm{Tr}\,(a(x)) + 2|b(x)|r) + V(x)], \end{aligned}$$

where $\lambda > 0$ is as in Assumption E. Since a, b and V are bounded in D, we may pick α sufficiently large so that $Lw \geqslant 0$ in the annulus $A \equiv B_R(y) - B_\rho(y)$. Since $u - u(x_0) < 0$ on $\partial B_\rho(y)$, there is a constant $\varepsilon > 0$ such that $u - u(x_0) + \varepsilon w \leqslant 0$ on ∂B_ρ. Since $w = 0$ on ∂B_R, we also have $u - u(x_0) + \varepsilon w \leqslant 0$ on ∂B_R. Therefore, $L(u - u(x_0) + \varepsilon w) \geqslant -u(x_0)V \geqslant 0$ in A and $u - u(x_0) + \varepsilon w \leqslant 0$ on ∂A. From Theorem 2.1, it then follows that $u - u(x_0) + \varepsilon w \leqslant 0$ in A. Now, taking the outer normal derivative at x_0 gives

$$\frac{\partial u}{\partial n}(x_0) \geqslant -\varepsilon \frac{\partial w}{\partial n}(x_0) = -\varepsilon(-2\alpha R)\exp(-\alpha R^2) > 0. \qquad \square$$

Theorem 2.6. *Let L satisfy Assumption E on a domain $D \subseteq R^d$ and assume that $u \in C^2(D) \cap C(\bar{D})$ satisfies $Lu \geqslant 0$ $(Lu \leqslant 0)$ in D. If $V \equiv 0$ and u attains its maximum (minimum) in the interior of D, then u is a constant. If $V \leqslant 0$, and u attains a non-negative maximum (non-positive minimum) in the interior of D, then it is constant. In the case of arbitrary V, if u attains the value zero as its maximum (minimum) in the interior of D, then it is identically zero.*

Proof. Assume to the contrary that u is non-constant and attains its maximum M in the interior of D, where M satisfies the restrictions imposed upon it in the statement of the theorem. Let $D^- = \{x \in D: u(x) < M\}$. By assumption, $\partial D^- \cap D \neq \emptyset$. Pick a point $x_0 \in D^-$ which is closer to ∂D^- than to ∂D and let B denote the largest ball contained in D^-, having x_0 as its center. Then $u(y) = M$ for some $y \in \partial B$ but $u < M$ in B. By Theorem 2.5, it then follows that $\nabla u(y) \neq 0$ which is impossible since y is an interior maximum. $\square$

Remark. In the sequel, we will sometimes refer to Theorem 2.5 as the *Hopf maximum principle* and to Theorem 2.6 as the *strong maximum principle*.

We now turn to the Schauder estimates. Let $D \subset R^d$ be a domain and define the seminorm

$$[f]_{\alpha;D} = \sup_{\substack{x,y \in D \\ x \neq y}} \frac{|f(x) - f(y)|}{|x - y|^\alpha}, \text{ for } \alpha \in (0, 1].$$

If $[f]_{\alpha;D} < \infty$, f is called *uniformly Hölder continuous with exponent α in D*. If $[f]_{\alpha;A} < \infty$ for all $A \subset\subset D$, then f is called *locally Hölder continuous with exponent α in D*. In the sequel, it will sometimes be convenient to denote the spaces $C(D)$ and $C(\bar{D})$ by $C^0(D)$ and $C^0(\bar{D})$

respectively. Define the Hölder space $C^{k,\alpha}(\bar{D})(C^{k,\alpha}(D))$ as the subspace of $C^k(\bar{D})(C^k(D))$ consisting of functions whose kth-order partial derivatives are uniformly (locally) Hölder continuous with exponent α in D. For $k=0$ and $\alpha \in (0,1)$, set $C^{0,\alpha}(\bar{D}) = C^\alpha(\bar{D})$ and $C^{0,\alpha}(D) = C^\alpha(D)$. On $C^k(\bar{D})$, define the norm

$$\|u\|_{k;D} = \sum_{0 \leq |\beta| \leq k} \sup_D \left| \frac{\partial^\beta f}{\partial x^\beta} \right|.$$

On $C^{k,\alpha}(\bar{D})$, define the norm

$$\|u\|_{k,\alpha;D} = \|u\|_{k;D} + \sup_{|\beta|=k} \left[\frac{\partial^\beta f}{\partial x^\beta} \right]_{\alpha;D}.$$

The norm of a vector or matrix-valued function is defined as the sum of the norms of its components.

A domain $D \subset R^d$ is said to have a $C^{k,\alpha}$-boundary if for each point $x_0 \in \partial D$, there is a ball B centered at x_0 and a one-to-one mapping ψ of B onto $A \subset R^d$ such that

(i) $\psi(B \cap D) \subset \{x \in R^d: x_n > 0\}$,
(ii) $\psi(B \cap \partial D) \subset \{x \in R^d: x_n = 0\}$, and
(iii) $\psi \in C^{k,\alpha}(B)$, $\psi^{-1} \in C^{k,\alpha}(A)$.

For the Schauder estimates, and in fact, for the rest of this chapter, we will assume that L is uniformly elliptic with coefficients which satisfy a uniform Hölder condition.

Assumption H.

$$L = \tfrac{1}{2} \sum_{i,j=1}^{d} a_{ij}(x) \frac{\partial^2}{\partial x_i \partial x_j} + \sum_{i=1}^{d} b_i(x) \frac{\partial}{\partial x_i} + V,$$

defined on the closure $\bar{D}$ of a domain $D \subseteq R^d$, satisfies

(i) $\sum_{i,j=1}^d a_{ij}(x) v_i v_j \geq \lambda |v|^2$, for all $x \in \bar{D}$, $v \in R^d$ and some $\lambda > 0$;
(ii) $\|a_{ij}\|_{\alpha;D}, \|b_i\|_{\alpha;D}, \|V\|_{\alpha;D} \leq \Lambda < \infty$.

In the sequel, we will write that 'L satisfies Assumption H on a domain $D \subseteq R^d$', rather than the more precise but more awkward 'L satisfies Assumption H on the closure $\bar{D}$ of a domain $D \subseteq R^d$'.

We will use the notation $D^2 u$ to denote the Hessian matrix

$$\left\{ \frac{\partial^2 u}{\partial x_i \partial x_j} \right\}_{i,j=1}^{d}.$$

Theorem 2.7 (**Schauder interior estimate**). *Let L satisfy Assumption H on a bounded domain $D \subset R^d$ and assume that $u \in C^{2,\alpha}(D)$ and $f \in C^{\alpha}(\bar{D})$ satisfy $Lu = f$ in D. If $D' \subset D$ is a domain satisfying* $\operatorname{dist}(D', \partial D) \geq \rho$, *then*

$$\rho\|\nabla u\|_{0;D'} + \rho^2\|D^2 u\|_{0;D'} + \rho^{2+\alpha}[D^2 u]_{\alpha;D'} \leq c(\|u\|_{0,D} + \|f\|_{0,\alpha;D}),$$

where $c = c(d, \alpha, \lambda, \Lambda, D)$.

Theorem 2.8 (**Schauder global estimate**). *Let L satisfy Assumption H on a domain $D \subseteq R^d$ with a $C^{2,\alpha}$-boundary. Let $\phi \in C^{2,\alpha}(\bar{D})$ and $f \in C^{\alpha}(\bar{D})$ and assume that $u \in C^{2,\alpha}(\bar{D})$ satisfies $Lu = f$ in D and $u = \phi$ on ∂D. Then $\|u\|_{2,\alpha;D} \leq c(\|u\|_{0;D} + \|\phi\|_{2,\alpha;D} + \|f\|_{0,\alpha;D})$, where $c = c(d, \alpha, \lambda, \Lambda, D)$.*

3.3 Existence and uniqueness for the Dirichlet problem

We begin by stating an existence and uniqueness theorem for the Dirichlet problem in the case where $V \leq 0$.

Theorem 3.1. *Let L satisfy Assumption H on a bounded domain $D \subset R^d$ and assume that $V \leq 0$ and that D has a $C^{2,\alpha}$-boundary. Let $f \in C^{\alpha}(\bar{D})$ and $\phi \in C^{2,\alpha}(\bar{D})$. Then the Dirichlet problem $Lu = f$ in D and $u = \phi$ on ∂D has a unique solution $u \in C^{2,\alpha}(\bar{D})$.*

Remark. The uniqueness of course follows from Corollary 2.3.

Now consider the Banach space

$$\mathcal{B}_\alpha = \{u \in C^{\alpha}(\bar{D}) \colon u = 0 \text{ on } \partial D\},$$

with the norm $\|\cdot\|_{0,\alpha;D}$. If L satisfies the condition of Theorem 3.1, then that theorem shows that $L^{-1}\colon C^{\alpha}(\bar{D}) \to \mathcal{B}_\alpha$ exists and that

$$\operatorname{Ran}((L^{-1}, C^{\alpha}(\bar{D}))) = C^{2,\alpha}(\bar{D}) \cap \mathcal{B}_\alpha.$$

By Theorem 2.8, for any $f \in C^{\alpha}(\bar{D})$

$$\|L^{-1}f\|_{2,\alpha;D} \leq c(\|L^{-1}f\|_{0,D} + \|f\|_{0,\alpha;D}).$$

By Theorem 2.4, $\|L^{-1}f\|_{0;D} \leq c\|f\|_{0;D}$. (Here and in what follows, c will stand for a constant which may vary from line to line.) Since $\|f\|_{0;D} \leq \|f\|_{0,\alpha;D}$, we conclude that

$$\|L^{-1}f\|_{2,\alpha;D} \leqslant c\|f\|_{0,\alpha;D}. \tag{3.1}$$

From the Ascoli–Arzela theorem, it then follows that L^{-1} is a compact operator from $C^{\alpha}(\bar{D})$ to $\mathcal{B}_{\alpha}$. Using this fact, we may now assert a Fredholm alternative concerning the existence of a solution to the Dirichlet problem in the case where the condition $V \leqslant 0$ fails.

***Theorem* 3.2.** *Let L satisfy Assumption H on D, and let D, ϕ and f be as in Theorem* 3.1. *Then either*:

(a) *the homogeneous problem $Lu = 0$ in D and $u = 0$ on ∂D has only the trivial solution, in which case the inhomogeneous problem $Lu = f$ in D and $u = \phi$ on ∂D has a unique solution; or*
(b) *the homogeneous problem $Lu = 0$ in D and $u = 0$ on ∂D has non-trivial solutions which form a finite-dimensional subspace of $C^{2,\alpha}(\bar{D})$.*

Proof. Letting $w = u - \phi$, the inhomogeneous Dirichlet problem $Lu = f$ in D and $u = \phi$ on ∂D may be written as $Lw = f - L\phi$ in D and $w = 0$ on ∂D and, by the assumption on ϕ, $f - L\phi \in C^{\alpha}(\bar{D})$. Thus, the general case may always be reduced to the case $\phi \equiv 0$. Therefore, we consider the Dirichlet problem

$$\begin{aligned} Lu &= f \text{ in } D, \quad f \in C^{\alpha}(\bar{D}). \\ u &= 0 \text{ on } \partial D. \end{aligned} \tag{3.2}$$

Let $\gamma = \sup_D V$ and define $L_{\gamma} = L - \gamma$. The zeroth-order term of L_{γ} is non-positive and thus, by the discussion preceding the statement of the theorem, L_{γ}^{-1}: $C^{\alpha}(\bar{D}) \to \mathcal{B}_{\alpha}$ is a compact operator and

$$\operatorname{Ran}((L_{\gamma}^{-1}, C^{\alpha}(\bar{D}))) = C^{2,\alpha}(\bar{D}) \cap \mathcal{B}_{\alpha}.$$

Now consider the equation

$$u + \gamma L_{\gamma}^{-1}u = L_{\gamma}^{-1}f, \; f \in C^{\alpha}(\bar{D}), \; u \in \mathcal{B}_{\alpha}. \tag{3.3}$$

Since $-\gamma L_{\gamma}^{-1}$ is compact, the Fredholm alternative applies to (3.3) and we conclude that either (a) there exists a finite-dimensional subspace of non-trivial solutions to (3.3) in the case $L_{\gamma}^{-1}f \equiv 0$, that is, the case $f \equiv 0$, or (b) (3.3) possesses a unique solution for each $f \in C^{\alpha}(\bar{D})$. Since $\operatorname{Ran}((L_{\gamma}^{-1}, C^{\alpha}(\bar{D}))) = C^{2,\alpha}(\bar{D}) \cap \mathcal{B}_{\alpha}$, it follows that if u solves (3.3), then $u = L_{\gamma}^{-1}f - \gamma L_{\gamma}^{-1}u \in C^{2,\alpha}(\bar{D})$. Thus, we may apply L_{γ} to both sides of (3.3) and we obtain $Lu = L_{\gamma}u + \gamma u = f$ which is (3.2). Consequently, there is a one-to-one correspondence between solutions to (3.2) and solutions to (3.3); this completes the proof. □

3.4 The semigroup and its generator on $\mathcal{B}_0 = \{u \in C(\bar{D}): u = 0$ on $\partial D\}$

Let L satisfy Assumption H on a bounded domain $D \subset R^d$ and assume that D has a $C^{2,\alpha}$-boundary. Let $L_0 = L - V$ and let $\{P_x, x \in \hat{D}\}$ denote the solution to the generalized martingale problem for L_0 on D. Define the Banach space

$$\mathcal{B}_0 = \{u \in C(\bar{D}): u = 0 \text{ on } \partial D\},$$

with the norm $\|\cdot\|_{0;D}$. For $t \geq 0$, define $T_t: \mathcal{B}_0 \to \mathcal{B}_b(\bar{D})$ by

$$T_t f(x) = \begin{cases} E_x\left(\exp\left(\int_0^t V(X(s))\,\mathrm{ds}\right) f(X(t));\ \tau_D > t\right), & x \in D \\ 0, \ x \in \partial D. \end{cases}$$

We will show that, in fact, T_t maps $\mathcal{B}_0$ into itself.

Remark. Recall that, from the definition of the generalized martingale problem on D, $X(t) = \Delta$ for all $t \geq \tau_D$ a.s., where Δ is the cemetery state, that is, the point at infinity in the one-point compactification of D. Therefore, if we were to consider each $f \in \mathcal{B}_0$ as a function on $\hat{D} = D \cup \{\Delta\}$ and were to define $f(\Delta) = 0$, then we could write

$$T_t f(x) = E_x \exp\left(\int_0^t V(X(s))\,\mathrm{ds}\right) f(X(t)), \quad x \in D.$$

It will be convenient to extend the coefficients a, b and V on $\bar{D}$ to continuous and bounded coefficients $\bar{a}$, $\bar{b}$ and $\bar{V}$ on R^d such that $\bar{L}_0$, the extension to R^d of L_0, satisfies Assumption A which appears at the beginning of Chapter 2. Denote by $\{\bar{P}_x, x \in R^d\}$ the solution to the martingale problem for $\bar{L}_0$ on R^d. Note that by Theorem 1.13.2,

$$E_x\left(\exp\left(\int_0^t V(X(s))\,\mathrm{ds}\right) f(X(t));\ \tau_D > t\right)$$
$$= \bar{E}_x\left(\exp\left(\int_0^t V(X(s))\,\mathrm{ds}\right) f(X(t));\ \tau_D > t\right), \quad (4.1)$$

since the coefficients of L and $\bar{L}$ coincide on D. Since $X(t)$ possesses a density under $\bar{P}_x$, we have $\bar{P}_x(\tau_D = t) = 0$. From this and Theorem 2.3.3(iv), it follows that for $x \in R^d$, the functional $\omega \to \exp(\int_0^t V(X(s,\omega))\,\mathrm{ds}) f(X(t,\omega)) 1\,(\omega)_{\{\tau_D > t\}}$ is $\bar{P}_x$ almost surely continuous on Ω. Since $X(t)$ is a Feller process on $(\Omega, \mathcal{F}, \mathcal{F}_t, \bar{P}_x)$, it follows from Theorem 1.3.1 and (4.1) that $T_t f(x)$ is continuous for $x \in \bar{D}$. Since $T_t f(x) = 0$, for $x \in \partial D$, we conclude that $T_t: \mathcal{B}_0 \to \mathcal{B}_0$.

By the Markov property, it follows that T_t defines a *semigroup* on $\mathcal{B}_0$, that is, $T_{t+s} = T_t T_s$. In fact, T_t is a *strongly continuous* semigroup on B_0, that is, $T_t f(x)$ is continuous in t for each $x \in \bar{D}$ and $f \in \mathcal{B}_0$. This follows from the bounded convergence theorem. Define

$$\mathcal{D}_0 = \left\{ f \in \mathcal{B}_0 \colon \lim_{t\to 0} \frac{T_t f - f}{t} \text{ exists} \right\},$$

where the limit is taken with respect to the norm $\|\cdot\|_{0;D}$ on $\mathcal{B}_0$. The *infinitessimal generator* $\mathcal{L}\colon \mathcal{D}_0 \to \mathcal{B}_0$ of the semigroup T_t is defined by

$$\mathcal{L} f = \lim_{t\to 0} \frac{T_t f - f}{t}, \quad \text{for } f \in \mathcal{D}_0.$$

$\mathcal{D}_0$ is called the *domain* of $\mathcal{L}$. A basic fact in the theory of strongly continuous semigroups states that $\mathcal{D}_0$ is dense in $\mathcal{B}_0$ and that $(\mathcal{L}, \mathcal{D}_0)$ is a closed operator on B_0 (exercise 3.3).

We will show that $\{u \in \mathcal{B}_0 \colon u \in C^2(\bar{D}) \text{ and } Lu = 0 \text{ on } \partial D\} \subset \mathcal{D}_0$ and that, for such u, $\mathcal{L}u = Lu$. Consider such a u and extend it to $\bar{u} \in C_0^2(R^d)$. Applying Theorem 2.4.2(i) to $\bar{u}$, $\bar{L}_0$ and $\bar{V}$, and using the fact that they equal u, L_0 and V on D, gives

$$\bar{E}_x \exp\left(\int_0^{t\wedge\tau_D} V(X(s))\,\mathrm{d}s\right) u(X(t\wedge\tau_D)) =$$
$$u(x) + \bar{E}_x \int_0^{t\wedge\tau_D} Lu(X(s)) \exp\left(\int_0^s V(X(r))\,\mathrm{d}r\right)\mathrm{d}s, \ x \in D. \tag{4.2}$$

Using (4.1) and the fact that $u \in \mathcal{B}_0$, we have for $x \in D$,

$$T_t u(x) = E_x\left(\exp\left(\int_0^t V(X(s))\,\mathrm{d}s\right) u(X(t)); \ \tau_D > t\right)$$
$$= \bar{E}_x \exp\left(\int_0^{t\wedge\tau_D} V(X(s))\,\mathrm{d}s\right) u(X(t\wedge\tau_D)).$$

Substituting this in (4.2) gives

$$T_t u(x) - u(x) = \bar{E}_x \int_0^{t\wedge\tau_D} Lu(X(s)) \exp\left(\int_0^s V(X(r))\,\mathrm{d}r\right)\mathrm{d}s, \ x \in D. \tag{4.3}$$

By Theorem 2.2.2,

$$\lim_{t\to 0} \sup_{x\in R^d} \bar{P}_x\left(\sup_{0\le s\le t} |X(s) - x| \geq \varepsilon\right) = 0. \tag{4.4}$$

From (4.3) and (4.4), it is easy to see that

$$\frac{T_t u(x) - u(x)}{t}$$

converges as $t \to 0$ to $Lu(x)$, uniformly over x in any compact subset of D. However, we must show that the convergence is uniform over $x \in D$. Let $\varepsilon > 0$. We will find a $t_\varepsilon > 0$ such that

$$\left| \frac{T_t u(x) - u(x)}{t} - Lu(x) \right| < \varepsilon,$$

for all $x \in D$ and all $0 < t < t_\varepsilon$. Choose $D_\varepsilon \subset\subset D$ such that

$$|Lu(x)| \leq \frac{\varepsilon}{3}, \text{ for } x \in D - D_\varepsilon.$$

By the above-mentioned uniform convergence on compact subsets of D, it suffices to find a $t_\varepsilon > 0$ such that

$$\left| \frac{T_t u(x) - u(x)}{t} - Lu(x) \right| < \varepsilon, \text{ for all } x \in D - D_\varepsilon \text{ and } 0 < t < t_\varepsilon.$$

For $x \in D - D_\varepsilon$, we write

$$\begin{aligned} \frac{T_t u(x) - u(x)}{t} &- Lu(x) \\ &= \frac{1}{t}\bar{E}_x \int_0^{t \wedge \tau_D} (Lu(X(s)) - Lu(x)) \exp\left(\int_0^s V(X(r))\, dr \right) ds \\ &\quad + Lu(x) \frac{1}{t} \bar{E}_x \int_0^{t \wedge \tau_D} \left(\exp\left(\int_0^s V(X(r))\, dr \right) - 1 \right) ds \\ &\quad - Lu(x) \frac{1}{t} \bar{E}_x (t - t \wedge \tau_D). \end{aligned} \tag{4.5}$$

By (4.4) again, the first and second terms on the righthand side of (4.5) converge to zero uniformly over $x \in D - D_\varepsilon$; thus there exists a $t_\varepsilon > 0$ such that for $0 < t < t_\varepsilon$, the absolute value of each of these terms is smaller than $\varepsilon/3$, for all $x \in D - D_\varepsilon$. The absolute value of the third term on the righthand side of (4.5) is smaller than $\varepsilon/3$ for all $t > 0$ and all $x \in D - D_\varepsilon$, by the definition of D_ε. This completes the proof that $\{u \in \mathcal{B}_0 \colon u \in C^2(\bar{D}) \text{ and } Lu = 0 \text{ on } \partial D\} \subset \mathcal{D}_0$ and that, for such u, $\mathcal{L}u = Lu$.

A subdomain $\mathcal{D}_0' \subset \mathcal{D}_0$ is called a *core* for $\mathcal{D}_0$ if $(\mathcal{L}, \mathcal{D}_0)$ is the closure of $(\mathcal{L}, \mathcal{D}_0')$. We will show that $\mathcal{D}_0' \equiv \{u \in \mathcal{B}_0 \colon u \in C^2(\bar{D}) \text{ and } Lu = 0 \text{ on } \partial D\}$ is a core for $\mathcal{D}_0$. Let $(\mathcal{L}, \hat{\mathcal{D}}_0)$ denote the closure of $(\mathcal{L}, \mathcal{D}_0')$. One of the assertions of the well-known Hille–Yoshida–Phillips theorem [Reed and Simon, Vol. 2 (1975), Theorem 10.47b] is that the resolvent set $\rho((\mathcal{L}, \mathcal{D}_0))$ contains all large real λ; that is, for all such λ, $(\lambda - \mathcal{L})\mathcal{D}_0 = \mathcal{B}_0$ and $\lambda - \mathcal{L}$ is one to one. Therefore, to show that $\hat{\mathcal{D}}_0 = \mathcal{D}_0$, it is enough to show that $(\lambda - \mathcal{L})\hat{\mathcal{D}}_0 = \mathcal{B}_0$.

Let $\lambda > \sup_D V$. By Theorem 3.1, for each $f \in C^\alpha(\bar{D}) \cap \mathcal{B}_0$, there exists a unique $u \in C^{2,\alpha}(\bar{D})$ such that $(L - \lambda)u = f$ in D and $u = 0$ on ∂D. In particular, then, $u \in \mathcal{D}_0'$ and we can write $(\mathcal{L} - \lambda)u = f$. Now let $f \in \mathcal{B}_0$. Since $C^\alpha(\bar{D}) \cap \mathcal{B}_0$ is dense in $\mathcal{B}_0$, one can choose a sequence $\{f_n\}_{n=1}^\infty \subset C^\alpha(\bar{D}) \cap \mathcal{B}_0$ such that $f = \lim_{n\to\infty} f_n$ in $\mathcal{B}_0$. Let $u_n \in \mathcal{D}_0'$ denote the solution to $(\mathcal{L} - \lambda)u_n = (L - \lambda)u_n = f_n$ in D and $u_n = 0$ on ∂D. By Theorem 2.4, $\|u_n - u_m\|_{0;D} \leq C\|f_n - f_m\|_{0;D}$. Therefore $\{u_n\}_{n=1}^\infty$ is Cauchy in $\mathcal{B}_0$ and there exists a $u \in \mathcal{B}_0$ such that $u = \lim_{n\to\infty} u_n$ in $\mathcal{B}_0$. By definition, $u \in \hat{D}_0$ and $(\mathcal{L} - \lambda)\, u = f$; therefore $(\lambda - \mathcal{L})\hat{D}_0 = \mathcal{B}_0$.

We have now proved the following result.

Theorem 4.1. *Let L satisfy Assumption H on a bounded domain $D \subset R^d$ with a $C^{2,\alpha}$-boundary and let $\{P_x, x \in \hat{D}\}$ denote the solution to the generalized martingale problem for $L_0 \equiv L - V$ on D. Define $\mathcal{B}_0 = \{u \in C(\bar{D})\colon u = 0 \text{ on } \partial D\}$. For $f \in \mathcal{B}_0$ and $t \geq 0$, let*

$$T_t f(x) = \begin{cases} E_x\left(\exp\left(\int_0^t V(X(s))\,ds\right) f(X(t));\ \tau_D > t\right), & x \in D \\ 0, & x \in \partial D. \end{cases}$$

Then T_t maps $\mathcal{B}_0$ into itself and is a strongly continuous semigroup on $\mathcal{B}_0$. Its infinitessimal generator, $(\mathcal{L}, \mathcal{D}_0)$, possesses as a core $(\mathcal{L}, \mathcal{D}_0')$, where $\mathcal{D}_0' = \{u \in \mathcal{B}_0\colon u \in C^2(\bar{D}) \text{ and } Lu = 0 \text{ on } \partial D\}$.

3.5 The spectrum of $(L, \mathcal{D}_\alpha)$ on $\mathcal{B}_\alpha \equiv \{u \in C^\alpha(\bar{D})\colon u = 0 \text{ on } \partial D\}$

In this section, we maintain the assumptions and notation of the previous section. The discussion appearing between Theorems 3.1 and 3.2 showed that $L_\gamma^{-1}\colon C^\alpha(\bar{D}) \to \mathcal{B}_\alpha$ is a compact operator and that $\text{Ran}\,((L_\gamma^{-1}, C^\alpha(\bar{D}))) = C^{2,\alpha}(\bar{D}) \cap \mathcal{B}_\alpha$. Therefore, $L_\gamma^{-1}\colon \mathcal{B}_\alpha \to \mathcal{B}_\alpha$ is also a compact operator and $\text{Ran}\,((L_\gamma^{-1}, \mathcal{B}_\alpha)) \subset C^{2,\alpha}(\bar{D}) \cap \mathcal{B}_\alpha$. Define

$$\mathcal{D}_\alpha = \text{Ran}\,((L_\gamma^{-1}, \mathcal{B}_\alpha)).$$

The operator $(L, \mathcal{D}_\alpha)$ on $\mathcal{B}_\alpha$ is closed and densely defined. The closure property is clear. We will not prove that $\mathcal{D}_\alpha$ is dense in $\mathcal{B}_\alpha$, but will suffice with a brief elucidation. Note that $\mathcal{D}_\alpha =$

$\{u \in C^{2,\alpha}(\bar{D}) \cap \mathcal{B}_\alpha: Lu = 0 \text{ on } \partial D\}$. It is standard although a bit tedious to show that $C^{2,\alpha}(\bar{D}) \cap \mathcal{B}_\alpha$ is dense in $\mathcal{B}_\alpha$. (Approximations obtained through mollification must be modified so that they vanish on ∂D.) To complete the proof that $\mathcal{D}_\alpha$ is dense in $\mathcal{B}_\alpha$, one must show that $\mathcal{D}_\alpha$ is dense in $C^{2,\alpha}(\bar{D}) \cap \mathcal{B}_\alpha$ with respect to the norm on $\mathcal{B}_\alpha$. That is, one must show that given any $\psi \in C^\alpha(\bar{D})$ and any $\varepsilon > 0$, one can find $u \in C^{2,\alpha}(\bar{D}) \cap \mathcal{B}_\alpha$ satisfying $Lu = \psi$ on ∂D and $\|u\|_{0,\alpha;D} < \varepsilon$. This is true because one can construct functions whose C^2-norms are of order O(1) but whose C^1-norms are of order o(1). A rigorous proof, which is quite tedious, requires a localization and partition of unity argument, and uses the fact that ∂D is a $C^{2,\alpha}$-boundary.

Proposition 5.1. *$-L_\gamma^{-1}$ is a positive operator on $\mathcal{B}_\alpha$.*

Proof. Let $0 \leqslant f \in \mathcal{B}_\alpha$ and define $u = -L_\gamma^{-1} f \in \mathcal{D}_\alpha$. Since $L_\gamma u = -f \leqslant 0$ on D and $u = 0$ on ∂D, and since the zeroth-order part of the operator L_γ is non-positive, it follows from Theorem 2.1 that $u \geqslant 0$ in D. □

Since $-L_\gamma^{-1}$ is a positive compact operator on $\mathcal{B}_\alpha$, part (i) of the Krein–Rutman theorem (Theorem 1.3) applies. Recall that the stronger conclusions of part (ii) of that theorem require that $(-L_\gamma^{-1}, \mathcal{B}_\alpha)$ be strongly positive. However, to verify strong positivity directly for $(-L_\gamma^{-1}, \mathcal{B}_\alpha)$ seems very difficult. In order to surmount this obstacle, we consider the operator $-L_\gamma^{-1}$ restricted to the Banach space $\mathcal{B}_1 \equiv C^1(\bar{D}) \cap \{u = 0 \text{ on } \partial D\}$ with the norm $\|\cdot\|_{1;D}$. We will verify strong positivity for $(-L_\gamma^{-1}, \mathcal{B}_1)$ using the Hopf maximum principle and the strong maximum principle and then show that the spectral properties of $(-L_\gamma^{-1}, \mathcal{B}_\alpha)$ and $(-L_\gamma^{-1}, \mathcal{B}_1)$ coincide. This will then allow us to obtain the conclusions of part (ii) of the Krein–Rutman theorem for $(-L_\gamma^{-1}, \mathcal{B}_\alpha)$.

Define $\mathcal{D}_1 = \text{Ran}((L_\gamma^{-1}, \mathcal{B}_1))$. Since $\|\cdot\|_{0,\alpha;D} \leqslant \|\cdot\|_{1;D}$, it follows from (3.1) that L_γ^{-1} is a compact operator on $\mathcal{B}_1$. The method mentioned above for showing that $\mathcal{D}_\alpha$ is dense in $\mathcal{B}_\alpha$ can be used to show that $\mathcal{D}_1$ is dense in $\mathcal{B}_1$. Thus $(L - \lambda, \mathcal{D}_1)$ is a closed, densely defined operator on $\mathcal{B}_1$ for all $\lambda \in R$.

Proposition 5.2. *$-L_\gamma^{-1}$ is strongly positive on $\mathcal{B}_1$.*

Proof. Let $K = \{u \in \mathcal{B}_1: u \geqslant 0 \text{ on } \bar{D}\}$. Let $n(x)$ denote the outer unit normal to D at $x \in \partial D$. Note that

$$\operatorname{int}(K) = \{u \in K: u > 0 \text{ on } D \text{ and } \inf_{x \in \partial D} |(\nabla u \cdot n)(x)| > 0\}.$$

Let $f \in \mathcal{B}_1$ satisfy $f \not\equiv 0$ and $f \geqslant 0$ and define $u = -L_\gamma^{-1} f$. We will show that $u \in \operatorname{int}(K)$. We have $L_\gamma u = -f$ in D and $u = 0$ on ∂D. By the strong maximum principle (Theorem 2.6), either $u > 0$ in D or $u \equiv 0$. Since $f \not\equiv 0$, it follows that $u > 0$ in D. Similarly, by the Hopf maximum principle (Theorem 2.5), $\inf_{x \in D} |(\nabla u \cdot n)(x)| > 0$. Thus $u \in \operatorname{int}(K)$. □

Proposition 5.3. *The spectra of $(L_\gamma^{-1}, \mathcal{B}_\alpha)$ and $(L_\gamma^{-1}, \mathcal{B}_1)$ coincide.*

Proof. Since $(L_\gamma^{-1}, \mathcal{B}_\alpha)$ and $(L_\gamma^{-1}, \mathcal{B}_1)$ are both compact, their spectra consist only of eigenvalues. Let λ be an eigenvalue for $(L_\gamma^{-1}, \mathcal{B}_1)$ and let ϕ belong to the corresponding eigenspace. Since $\mathcal{B}_1 \subset \mathcal{B}_\alpha$, it follows that λ is also an eigenvalue for $(L_\gamma^{-1}, \mathcal{B}_\alpha)$ and that ϕ belongs to the corresponding eigenspace. Thus, $\sigma((L_\gamma^{-1}, \mathcal{B}_1)) \subset \sigma((L_\gamma^{-1}, \mathcal{B}_\alpha))$. Now let λ be an eigenvalue for $(L_\gamma^{-1}, \mathcal{B}_\alpha)$ and let ϕ belong to the corresponding eigenspace. Since $L_\gamma^{-1} \phi = \lambda \phi$, it follows that $\phi \in \operatorname{Ran}((L_\gamma^{-1}, \mathcal{B}_\alpha)) = \mathcal{D}^\alpha \subset C^{2,\alpha}(\bar{D}) \cap \mathcal{B}_\alpha$. Therefore, $\phi \in \mathcal{B}_1$ and λ is an eigenvalue for $(L_\gamma^{-1}, \mathcal{B}_1)$. Thus $\sigma((L_\gamma^{-1}, \mathcal{B}_\alpha)) \subset \sigma((L_\gamma^{-1}, \mathcal{B}_1))$. □

Proposition 5.4. *The spectrum of $(L, \mathcal{D}_\alpha)$ consists only of eigenvalues, and $\lambda \in \sigma((L, \mathcal{D}_\alpha))$ if and only if $-1/(\gamma - \lambda) \in \sigma((L_\gamma^{-1}, \mathcal{B}_\alpha))$.*

Proof. Consider the following two equations:

$$u + (\gamma - \lambda) L_\gamma^{-1} u = L_\gamma f,\ f \in \mathcal{B}_\alpha,\ u \in \mathcal{D}_\alpha. \tag{5.1}$$

$$(L - \lambda) u = f \text{ in } D,\ f \in \mathcal{B}_\alpha,\ u \in \mathcal{D}_\alpha. \tag{5.2}$$

$$u = 0 \text{ on } \partial D$$

One can show that (5.1) and (5.2) are equivalent in exactly the same way that (3.2) and (3.3) were shown to be equivalent in the proof of Theorem 3.2. Now $\lambda \in \rho((L, \mathcal{D}_\alpha))$ if and only if (5.2) has a unique solution $u \in \mathcal{D}_\alpha$ for every $f \in \mathcal{B}_\alpha$ or, equivalently, if and only if (5.1) has a unique solution $u \in \mathcal{D}_\alpha$ for every $f \in \mathcal{B}_\alpha$. Since $(L_\gamma^{-1}, \mathcal{B}_\alpha)$ is compact, it follows from the Fredholm alternative (Corollary 1.2) that (5.1) has a unique solution for each $f \in C^\alpha(\bar{D})$ if and only if

$$\frac{-1}{\gamma - \lambda} \in \rho((L_\gamma^{-1}, \mathcal{B}_\alpha)).$$

This proves the proposition. □

From Proposition 5.2, it follows that the conclusions of part (ii) of the Krein–Rutman Theorem hold for $(L_\gamma^{-1}, \mathcal{B}_1)$ and, by Proposition 5.3, it follows that these conclusions also hold for $(L_\gamma^{-1}, \mathcal{B}_\alpha)$. Proposition 5.4 allows us to deduce the spectral properties of $(L, \mathcal{D}_\alpha)$ from those of $(L_\gamma^{-1}, \mathcal{B}_\alpha)$. In particular then, we have proved the following theorem.

Theorem 5.5. *Let L satisfy Assumption H on a bounded domain $D \subset R^d$ with a $C^{2,\alpha}$-boundary. Then there exists a real number $\lambda_0 \in \sigma((L, \mathcal{D}_\alpha))$, the spectrum of $(L, \mathcal{D}_\alpha)$, which is of multiplicity one and whose corresponding eigenspace is generated by a function $\phi_0 \in \mathcal{D}_\alpha \subset C^{2,\alpha}(\bar{D}) \cap \mathcal{B}_\alpha$ which satisfies $\phi_0 > 0$ in D. If $\lambda \in \sigma((L, \mathcal{D}_\alpha))$ and $\lambda \neq \lambda_0$, and if ϕ is a corresponding eigenfunction, then ϕ changes sign in D.*

3.6 The existence and the behavior of the principal eigenvalue

The analysis in the previous section does not yield any useful information about the position that the eigenvalue λ_0 appearing in Theorem 5.5 occupies in $\sigma((L, \mathcal{D}_\alpha))$. From Proposition 5.4 and the Krein–Rutman theorem, all that can be said is that if $\lambda \in \sigma((L, \mathcal{D}_\alpha))$ and $\lambda \neq \lambda_0$, then

$$\left|\frac{1}{\gamma - \lambda}\right| < \left|\frac{1}{\gamma - \lambda_0}\right|.$$

In order to obtain more information about the eigenvalue λ_0, we consider the semigroup T_t introduced in Section 4. Define the operator norm

$$\|T_t\| \equiv \sup_{\substack{f \in \mathcal{B}_0 \\ \|f\|_{0;D} \leq 1}} \|T_t f\|_{0;D}$$

and define

$$\lambda_0^* = \lim_{t \to \infty} \frac{1}{t} \log \|T_t\|. \tag{6.1}$$

The limit in (6.1) exists by a well-known subadditivity argument (exercise 3.1). By taking an increasing sequence, $\{f_n\}_{n=1}^{\infty} \subset \mathcal{B}_0$, with $0 \leq f_n \leq 1$ and $\lim_{n\to\infty} f_n(x) = 1$ for all $x \in D$, it follows that

$$\|T_t\| = \sup_{x \in D} E_x\left(\exp\left(\int_0^t V(X(s))\, ds\right); \tau_D > t\right).$$

Thus, we can also write

$$\lambda_0^* = \lim_{t\to\infty} \frac{1}{t} \log \sup_{x\in D} E_x\left(\exp\left(\int_0^t V(X(s))\,\mathrm{d}s\right);\ \tau_D > t\right). \tag{6.2}$$

We will now prove the following important theorem.

Theorem 6.1. *Let L satisfy Assumption H on a bounded domain $D \subset R^d$ with a $C^{2,\alpha}$-boundary. Then the eigenvalue λ_0 in Theorem 5.5 satisfies:*

(i) $\lambda_0 = \sup \operatorname{Re}(\sigma((L, \mathcal{D}_\alpha)))$;

(ii) $\lambda_0 = \lambda_0^* \equiv \lim_{t\to\infty} \frac{1}{t} \log \sup_{y\in D} E_y\left(\exp\left(\int_0^t V(X(s))\,\mathrm{d}s\right);\ \tau_D > t\right)$

$$= \lim_{t\to\infty} \frac{1}{t} \log E_x\left(\exp\left(\int_0^t V(X(s))\,\mathrm{d}s\right);\ \tau_D > t\right), \textit{ for all } x \in D.$$

In light of Theorem 6.1, we will call λ_0 the *principal eigenvalue* for L on D. The following corollary is an immediate consequence of Theorem 6.1.

Corollary 6.2. *Under the conditions of Theorem 6.1, λ_0 is monotone non-decreasing in its dependence on the domain D.*

Proof of Theorem 6.1. The proof will require our comparing λ_0 and λ_0^* for different domains; therefore we will use the notation $\lambda_0(D)$ and $\lambda_0^*(D)$ to signify the dependence on D. We give a five-step proof of (ii). The first step of the proof is that $\sup \operatorname{Re}(\sigma((L, \mathcal{D}_\alpha))) \leq \lambda_0^*(D)$. This, in conjunction with (ii), gives a proof of (i).

Step 1. $\sup \operatorname{Re}(\sigma((L, \mathcal{D}_\alpha))) \leq \lambda_0^*(D)$

Proof. Let $\lambda > \lambda_0^*(D)$. We will show that $\lambda \in \rho((L, \mathcal{D}_\alpha))$. By the definition of $\lambda_0^*(D)$, it follows that $\int_0^\infty \exp(-\lambda t)\|T_t\|\,\mathrm{d}t$ is a convergent integral and also that $\lim_{t\to\infty} \exp(-\lambda t)\|T_t\| = 0$. Recall that $(\mathcal{L}, \mathcal{D}_0)$ denotes the infinitessimal generator of T_t on $\mathcal{B}_0$. Let $f \in \mathcal{D}_0$. We will show that $\int_0^\infty \exp(-\lambda t) T_t f\,\mathrm{d}t \in \mathcal{D}_0$. Since $\int_0^\infty \exp(-\lambda t)\|T_t\|\,\mathrm{d}t < \infty$, $\int_0^\infty \exp(-\lambda t) T_t f\,\mathrm{d}t$ is well defined and it follows from the dominated convergence theorem that $\int_0^\infty \exp(-\lambda t) T_t f\,\mathrm{d}t \in \mathcal{B}_0$. Similarly, it is easy to see that $T_h \int_0^\infty \exp(-\lambda t) T_t f\,\mathrm{d}t = \int_0^\infty \exp(-\lambda t) T_{t+h} f\,\mathrm{d}t$. Thus,

$$\frac{T_h - I}{h}\left(\int_0^\infty \exp(-\lambda t) T_t f\,\mathrm{d}t\right) = \int_0^\infty \exp(-\lambda t) T_t\left(\frac{T_h - I}{h}\right) f\,\mathrm{d}t.$$

Since

$$\lim_{h\to 0} \frac{T_h - I}{h} f = \mathcal{L}f \text{ in } \mathcal{B}_0,$$

it follows from the dominated convergence theorem that

$$\int_0^\infty \exp(-\lambda t) T_t f \, dt \in \mathcal{D}_0$$

and

$$\mathcal{L}\left(\int_0^\infty \exp(-\lambda t) T_t f \, dt\right) = \int_0^\infty \exp(-\lambda t) T_t \mathcal{L} f \, dt.$$

Since $T_t\mathcal{L} = \mathcal{L}T_t$ on $\mathcal{D}_0$, we have

$$\mathcal{L}\left(\int_0^\infty \exp(-\lambda t) T_t f \, dt\right) = \int_0^\infty \exp(-\lambda t) \mathcal{L} T_t f \, dt$$

$$= \int_0^\infty \exp(-\lambda t) \frac{d}{dt}(T_t f) \, dt.$$

Thus,

$$(\mathcal{L} - \lambda)\int_0^\infty \exp(-\lambda t) T_t f \, dt = \int_0^\infty \frac{d}{dt}(\exp(-\lambda t) T_t f) \, dt. \tag{6.3}$$

Since $\lim_{t\to\infty} \exp(-\lambda t)\|T_t\| = 0$, we conclude from (6.3) that

$$(\mathcal{L} - \lambda)\int_0^\infty \exp(-\lambda t) T_t f \, dt = -f, \text{ for } f \in \mathcal{D}_0. \tag{6.4}$$

Now let $f \in \mathcal{B}_0$ and let $\{f_n\}_{n=1}^\infty \subset \mathcal{D}_0$ satisfy $\lim_{n\to\infty} f_n = f$ in $\mathcal{B}_0$. Since $\lim_{n\to\infty} \int_0^\infty \exp(-\lambda t) T_t f_n \, dt = \int_0^\infty \exp(-\lambda t) T_t f \, dt$ in $\mathcal{B}_0$, and since $\mathcal{L} - \lambda$ is a closed operator, it follows that $\int_0^\infty \exp(-\lambda t) T_t f \, dt \in \mathcal{D}_0$ and $(\mathcal{L} - \lambda)\int_0^\infty \exp(-\lambda t) T_t f \, dt = -f$, for all $f \in \mathcal{B}_0$. This proves that $\lambda \in \rho((\mathcal{L}, \mathcal{D}_0))$ and thus $\lambda \in \rho((L, \mathcal{D}_\alpha))$ since, by Theorem 4.1, $(\mathcal{L}, \mathcal{D}_0)$ is an extension of $(L, \mathcal{D}_\alpha)$.

Step 2. $\lambda_0(D) \le \lambda_0^*(D)$.

Proof. This follows immediately from Step 1 since $\lambda_0(D) \in \sigma((L, \mathcal{D}_\alpha))$.

Step 3. Let D_1 be a domain with a $C^{2,\alpha}$-boundary. Assume that $D_1 \subset\subset D$. Then $\lambda_0^*(D_1) \le \lambda_0(D)$ and $\lambda_0(D_1) \le \lambda_0^*(D)$.

Proof. We will prove that $\lambda_0^*(D_1) \leqslant \lambda_0(D)$; the proof that $\lambda_0(D_1) \leqslant \lambda_0^*(D)$ is very similar. For $x \in D$, define $H(t) = T_t\phi_0(x)$, where $\phi_0 > 0$ is the eigenvalue corresponding to $\lambda_0(D)$ in Theorem 5.5. Since $\phi_0 \in \mathcal{D}_\alpha \subset \mathcal{D}_0$ and $\mathcal{L}\phi_0 = L\phi_0$, we have

$$H'(t) = \mathcal{L}T_t\phi_0 = T_t\mathcal{L}\phi_0 = T_tL\phi_0 = \lambda_0(D)T_t\phi_0 = \lambda_0(D)H(t).$$

Since $H(0) = \phi_0(x)$, we conclude that $T_t\phi_0(x) = H(t) = \exp(\lambda_0(D)t)\phi_0(x)$. Now, for $x \in D_1$, we have

$$\begin{aligned}
&E_x\left(\exp\left(\int_0^t V(X(s))\,ds\right); \tau_{D_1} > t\right)\\
&\qquad \leqslant \frac{1}{\inf\limits_{y\in D_1}\phi_0(y)} E_x\left(\exp\left(\int_0^t V(X(s))\,ds\right)\phi_0(X(t)); \tau_{D_1} > t\right)\\
&\qquad \leqslant \frac{1}{\inf\limits_{y\in D_1}\phi_0(y)} E_x\left(\exp\left(\int_0^t V(X(s))\,ds\right)\phi_0(X(t)); \tau_{D} > t\right)\\
&\qquad = \frac{1}{\inf\limits_{y\in D_1}\phi_0(y)} T_t\phi_0(x) = \frac{\phi_0(x)}{\inf\limits_{y\in D_1}\phi_0(y)}\exp(\lambda_0(D)t).
\end{aligned}$$

From this and (6.2), it follows that $\lambda_0^*(D_1) \leqslant \lambda_0(D)$.

Step 4. Let D_1 be a bounded domain with a $C^{2,\alpha}$-boundary which satisfies $\bar{D} \subset D_1$. Extend the coefficients of L to $\bar{D}_1$ in such a way that Assumption H still holds. Let $\{D_n\}_{n=1}^\infty$ be a sequence of $C^{2,\alpha}$-domains satisfying $D_n \supset\supset D_{n+1}$, for $n \geqslant 1$, and $\bigcap_{n=1}^\infty D_n = \bar{D}$. Then $\lambda_0(D_n)$ is non-increasing and $\lim_{n\to\infty}\lambda_0(D_n) = \lambda_0(D)$.

Proof. Using Step 2 in the first and third inequalities below and Step 3 in the second and fourth ones, we have

$$\lambda_0(D) \leqslant \lambda_0^*(D) \leqslant \lambda_0(D_{n+1}) \leqslant \lambda_0^*(D_{n+1}) \leqslant \lambda_0(D_n).$$

From this it follows that $\lambda_0(D_n)$ is non-increasing in n and that $\lambda_0(D) \leqslant \lim_{n\to\infty}\lambda_0(D_n)$. To complete the proof that $\lambda_0(D) = \lim_{n\to\infty}\lambda_0(D_n)$, we will assume that $l \equiv \lim_{n\to\infty}\lambda_0(D_n) > \lambda_0(D)$ and arrive at a contradiction. Choose $\lambda \in (\lambda_0(D), l)$ such that $\lambda \in \rho((L, \mathcal{D}_\alpha))$. It follows from Theorem 3.2 that there exists an $f \in C^{2,\alpha}(\bar{D})$ such that $(L-\lambda)f = -2$ on D and $f = 1$ on ∂D. We will complete the proof using the fact that $f \geqslant 0$ on $\bar{D}$. Then we will return to prove this.

Defining $g = f + \varepsilon$, for $\varepsilon > 0$ sufficiently small, we have $(L - \lambda)g \leq -1$ on D, $g \geq 1$ on ∂D and $g > 0$ on $\bar{D}$. We can now pick n_0 sufficiently large so that there exists a function $h \in C^{2,\alpha}(\bar{D}_{n_0})$ which satisfies $h = g$ on D, $(L - \lambda)h \leq -\frac{1}{2}$ on D_{n_0}, $h \geq \frac{1}{2}$ on ∂D_{n_0} and $h > 0$ on $\bar{D}_{n_0}$. Then we have $(L - \lambda_0(D_{n_0}))h = (L - \lambda + \lambda - \lambda_0(D_{n_0}))h \leq -\frac{1}{2}$. By Theorem 2.2, the maximum principle holds for $L - \lambda_0(D_{n_0})$ on D_{n_0}. This contradicts Theorem 5.5 which states the existence of an eigenvalue ϕ which satisfies $\phi > 0$ in D_{n_0}, $(L - \lambda_0(D_n))\phi = 0$ in D_{n_0} and $\phi = 0$ on ∂D_{n_0}.

It remains to show that $f \geq 0$ in $\bar{D}$. Assume to the contrary. Then there exists a domain $D_0 \subset\subset D$ such that $(L - \lambda)f = -2$ on D_0, $f = 0$ on ∂D_0 and $f < 0$ in D_0. By the Feynman–Kac formula (Theorem 2.4.2) applied to $L - \lambda$, we have

$$
\begin{aligned}
&E_x \exp\left(\int_0^{t\wedge\tau_{D_0}} (V - \lambda)(X(s))\,ds\right) f\left(X(t\wedge\tau_{D_0})\right) \\
&\quad = f(x) + E_x \int_0^{t\wedge\tau_{D_0}} (L - \lambda)f(X(s)) \exp\left(\int_0^s (V - \lambda)(X(u))\,du\right) ds \\
&\quad = f(x) - 2E_x \int_0^{t\wedge\tau_{D_0}} \exp\left(\int_0^s (V - \lambda)(X(u))\,du\right) ds, \text{ for } x \in D_0. \qquad (6.5)
\end{aligned}
$$

But

$$
\begin{aligned}
&E_x \exp\left(\int_0^{t\wedge\tau_{D_0}} (V - \lambda)(X(s))\,ds\right) f(X(t\wedge\tau_{D_0})) \\
&\quad = E_x\left(\exp\left(\int_0^t (V - \lambda)(X(s))\,ds\right) f(X(t));\ \tau_{D_0} > t\right) \\
&\quad = \exp(-\lambda t) E_x\left(\exp\left(\int_0^t V(X(s))\,ds\right) f(X(t));\ \tau_{D_0} > t\right). \qquad (6.6)
\end{aligned}
$$

By Step 3, $\lambda_0^*(D_0) \leq \lambda_0(D) < \lambda$; thus it follows from (6.2) that the righthand side of (6.6) converges to zero as $t \to \infty$. From this and (6.5) we conclude that $f(x) > 0$ for $x \in D_0$, which is a contradiction.

Step 5. (ii) holds.

Proof. Let $\{D_n\}_{n=1}^{\infty}$ be as in Step 4. Then, by Step 3, $\lambda_0^*(D) \leq \lambda_0(D_n)$ for all $n \geq 1$ and by Step 4, $\lambda_0^*(D) \leq \lim_{n\to\infty} \lambda_0(D_n) = \lambda_0(D)$. But, by Step 2, $\lambda_0(D) \leq \lambda_0^*(D)$; thus, $\lambda_0(D) = \lambda_0^*(D)$.

Now let $x \in D$. In the proof of Step 3, it was shown that $T_t\phi_0(x) = \exp(\lambda_0(D)t)\phi_0(x)$. Thus,

$$E_x\left(\exp\left(\int_0^t V(X(s))\,\mathrm{d}s\right);\ \tau_D > t\right)$$

$$\geq \frac{1}{\sup_{y\in D}\phi_0(y)} E_x\left(\exp\left(\int_0^t V(X(s))\,\mathrm{d}s\right)\phi_0(X(t));\ \tau_D > t\right)$$

$$= \frac{1}{\sup_{y\in D}\phi_0(y)} T_t\phi_0(x) = \frac{\phi_0(x)}{\sup_{y\in D}\phi_0(y)} \exp(\lambda_0(D)t).$$

Thus, for any $x \in D$,

$$\lambda_0(D) = \lambda_0^*(D) \equiv \lim_{t\to\infty} \frac{1}{t} \log \sup_{y\in D} E_y\left(\exp\left(\int_0^t V(X(s))\,\mathrm{d}s\right);\ \tau_D > t\right)$$

$$\geq \liminf_{t\to\infty} \frac{1}{t} \log E_x\left(\exp\left(\int_0^t V(X(s))\,\mathrm{d}s\right);\ \tau_D > t\right)$$

$$\geq \liminf_{t\to\infty} \frac{1}{t} \log \frac{\phi_0(x)}{\sup_{y\in D}\phi_0(y)} \exp(\lambda_0(D)t) = \lambda_0(D).$$

This completes the proof of (ii) and, thereby, the proof of Theorem 6.1. □

In Step 1 of the proof of Theorem 6.1, we actually showed that if $\lambda > \lambda_0^*$, then $\lambda \in \rho((\mathcal{L}, \mathcal{D}_0))$. Furthermore, since $(\mathcal{L}, \mathcal{D}_0)$ is an extension of $(L, \mathcal{D}_\alpha)$, λ_0 is an eigenvalue for $(\mathcal{L}, \mathcal{D}_0)$ as well as for $(L, \mathcal{D}_\alpha)$. Thus, from the equality $\lambda_0 = \lambda_0^*$, we obtain the following result.

Theorem 6.3. *Let L satisfy Assumption H on a bounded domain $D \subset R^d$ with a $C^{2,\alpha}$-boundary and let $(\mathcal{L}, \mathcal{D}_0)$ be the infinitessimal generator of the semigroup corresponding to L on D. Let λ_0 denote the eigenvalue appearing in Theorem 5.5. Then $\lambda_0 \in \sigma((\mathcal{L}, \mathcal{D}_0))$ and $\lambda_0 = \sup \mathrm{Re}\,(\sigma((\mathcal{L}, \mathcal{D}_0)))$.*

The following result, which was essentially proved in the first step of Theorem 6.1, will be needed in the next chapter.

Theorem 6.4. *Let L satisfy Assumption H on a bounded domain $D \subset R^d$ with a $C^{2,\alpha}$-boundary. Then for any $f \in \mathcal{B}_\alpha$ and any $\lambda > \lambda_0$, the function $u \equiv \int_0^\infty \exp(-\lambda t) T_t f\,\mathrm{d}t \in \mathcal{D}_\alpha$ and is the unique solution of $(L - \lambda)u = -f$ in D and $u = 0$ on ∂D.*

Proof. In the proof of Step 1 of Theorem 6.1, it was shown that if $f \in \mathcal{B}_0$ and $\lambda > \lambda_0$, then $u \equiv \int_0^\infty \exp(-\lambda t) T_t f\,\mathrm{d}t \in \mathcal{D}_0$ and $(\mathcal{L} - \lambda)u =$

$-f$ in D. Now $\mathcal{L} - \lambda: \mathcal{D}_0 \to \mathcal{B}_0$ is an extension of $L - \lambda: \mathcal{D}_\alpha \to \mathcal{B}_\alpha$. Since both of these operators are bijections, it follows that $u \in \mathcal{D}_\alpha$. □

The next theorem shows that if $\lambda_0 < 0$, then the Dirichlet problem is solvable and the generalized maximum principle holds.

Theorem 6.5. *Let L satisfy Assumption H on a bounded domain $D \subset R^d$ with a $C^{2,\alpha}$-boundary and assume that $\lambda_0 < 0$. Then for each $f \in C^\alpha(\bar{D})$ and each $\phi \in C^{2,\alpha}(\bar{D})$, there exists a unique solution $u \in C^{2,\alpha}(\bar{D})$ to the Dirichlet problem $Lu = -f$ in D and $u = \phi$ on ∂D. For any domain $D' \subseteq D$, if $u \in C^2(D') \cap C(\bar{D}')$ satisfies $Lu \leq 0$ in D' and $u \geq 0$ on $\partial D'$, then $u \geq 0$ in D'. If $u \not\equiv 0$, then $u > 0$ in D'.*

Proof. Since $\lambda_0 < 0$, by Theorem 6.1, $0 \notin \sigma((L, D_\alpha))$; thus, in Theorem 3.2, alternative (a) holds. This proves the existence and uniqueness of u. To prove the generalized maximum principle, note that since $\lambda_0 = \sup \operatorname{Re}(\sigma((L, \mathcal{D}_\alpha))) < 0$, it follows from Theorem 3.2 that there exists a unique solution $h \in C^{2,\alpha}(\bar{D})$ to $Lh = 0$ in D and $h = 1$ on ∂D. We will show that $h > 0$ on $\bar{D}$; the generalized maximum principle will then follow from Theorem 2.2.

Let $\{D_n\}_{n=1}^\infty$ be a sequence of domains satisfying $D_n \subset\subset D_{n+1}$ and $\bigcup_{n=1}^\infty D_n = D$. By the Feynman–Kac formula (Theorem 2.4.2),

$$h(x) = E_x \exp\left(\int_0^{t \wedge \tau_{D_n}} V(X(s))\, ds\right) h(X(t \wedge \tau_{D_n})), \quad n = 1, 2, \ldots. \tag{6.7}$$

Since $h \not\equiv 0$, it follows from (6.7) that if $h \geq 0$, then in fact $h > 0$ on $\bar{D}$.

Now assume that h changes sign in D. Then, since $h \in C^{2,\alpha}(\bar{D})$, there exists a domain $U \subset D$ with a $C^{2,\alpha}$-boundary such that $Lh = 0$ in U, $h < 0$ in U and $h = 0$ on ∂U. It then follows from Theorem 5.5 that $\lambda_0(U) = 0$. This contradicts the monotonicity of λ_0 as a function of its domain (Corollary 6.2).

Finally, the strong maximum principle follows from the generalized maximum principle and Theorem 2.6. □

The following theorem is also sometimes useful. It may be thought of as a particular version of the Feynman–Kac formula (see the remark following Theorem 2.4.2).

Theorem 6.6. *Let L and D satisfy the conditions of Theorem 6.1 and assume that $\lambda_0 < 0$.*

(i) *For any* $D' \subseteq D$,

$$\sup_{x \in D'} E_x \exp\left(\int_0^{\tau_{D'}} V(X(s))\,ds\right) < \infty.$$

(ii) *Let* $D' \subset\subset D$ *and let* $u \in C^2(D') \cap C(\bar{D}')$ *satisfy* $Lu = 0$ *in* D' *and* $u = \phi$ *on* $\partial D'$. *Then*

$$u(x) = E_x \exp\left(\int_0^{\tau_{D'}} V(X(s))\,ds\right)\phi(X(\tau_{D'})).$$

(iii) *Let* $u \in C^2(D) \cap C(\bar{D})$ *satisfy* $Lu = 0$ in D *and* $u = \phi$ *on* ∂D. *Let* $\bar{L}$ *be a smooth and bounded extension of* L *to all of* R^d, *and let* $\{\bar{P}_x, x \in R^d\}$ *denote the solution to the martingale problem for* $\bar{L}$ *on* R^d. *Then*

$$u(x) = \bar{E}_x \exp\left(\int_0^{\tau_D} V(X(s))\,ds\right)\phi(X(\tau_D)). \tag{6.8}$$

Remark. It is notationally incorrect to write (6.8) with $\bar{E}_x$ replaced by E_x. Indeed, in the framework of the generalized martingale problem, $X(\tau_D) = \Delta$ a.s. P_x, where Δ is the point at infinity in the one-point compactification $\hat{D}$ of D.

Proof. (i) Clearly, (6.7) holds with D_n replaced by $D_n \cap D'$. Consider (6.7) with $D_n \cap D'$ in place of D_n, let $n \to \infty$ and use Fatou's lemma and the strict positivity of h on $\bar{D}$.

(ii) As in (6.7), we have

$$u(x) = E_x \exp\left(\int_0^{t \wedge \tau_{D'}} V(X(s))\,ds\right)u(X(t \wedge \tau_{D'})). \tag{6.9}$$

By the monotone convergence theorem

$$\lim_{t\to\infty} E_x\left(\exp\left(\int_0^{\tau_{D'}} V(X(s))\,ds\right)u(X(\tau_{D'}));\ \tau_{D'} \le t\right) =$$
$$E_x \exp\left(\int_0^{\tau_{D'}} V(X(s))\,ds\right)u(X(\tau_{D'})), \tag{6.10}$$

and by Theorem 6.1(ii),

$$E_x\left(\exp\left(\int_0^t V(X(s))\,ds\right)u(X(t));\ \tau_{D'} > t\right) \le$$
$$\lim_{t\to\infty} E_x\left(\exp\left(\int_0^t V(X(s))\,ds\right)u(X(t));\ \tau_D > t\right) = 0. \tag{6.11}$$

Now (ii) follows from (6.9)–(6.11).

(iii) The proof is similar to the proof of (ii). □

3.7 A mini-max variational formula for the principal eigenvalue

In this section, we derive an explicit analytic representation for λ_0 in terms of the coefficients of L. We will need to guarantee that L and its formal adjoint $\tilde{L}$ both satisfy Assumption H. For the representation, it will be convenient to express L in divergence form: $L = \frac{1}{2}\nabla \cdot a\nabla + b \cdot \nabla + V$ and $\tilde{L} = \frac{1}{2}\nabla \cdot a\nabla - b \cdot \nabla - \nabla \cdot b + V$.

Assumption $\tilde{H}$. *$L = \frac{1}{2}\nabla \cdot a\nabla + b \cdot \nabla + V$ is defined on the closure $\bar{D}$ of a domain $D \subseteq R^d$ and satisfies a_{ij}, $b_i \in C^{1,\alpha}(\bar{D})$, $V \in C^{\alpha}(\bar{D})$ and $\sum_{i,j=1}^{d} a_{ij}(x) v_i v_j > 0$, for all $v \in R^d - \{0\}$ and all $x \in \bar{D}$.*

As with Assumption H, we will write that 'L satisfies Assumption $\tilde{\mathrm{H}}$ on a domain $D \subseteq R^d$' rather than the more precise 'L satisfies Assumption $\tilde{\mathrm{H}}$ on the closure $\bar{D}$ of a domain $D \subseteq R^d$'.

Denote by $\tilde{\mathcal{D}}_\alpha$ the domain of the operator $\tilde{L}$ on $\mathcal{B}_\alpha$, analogous to $\mathcal{D}_\alpha$ for L. Note that since the adjoint, $(L^*, \mathcal{D}_\alpha^*)$ on $\mathcal{B}_\alpha^*$, of $(L, \mathcal{D}_\alpha)$ is an extension of $(\tilde{L}, \tilde{\mathcal{D}}_\alpha)$, it follows by the Krein–Rutman theorem and by Theorem 6.1 applied to L and $\tilde{L}$, that $\lambda_0 = \sup \mathrm{Re}\,(\sigma((L, \mathcal{D}_\alpha))) = \sup \mathrm{Re}\,(\sigma((\tilde{L}, \tilde{\mathcal{D}}_\alpha)))$. Also, there exist eigenfunctions ϕ_0 and $\tilde{\phi}_0$ for L and $\tilde{L}$, respectively, which correspond to the eigenvalue λ_0 and which satisfy $\phi_0, \tilde{\phi}_0 \in C^{2,\alpha}(\bar{D})$ and $\phi_0, \tilde{\phi}_0 > 0$ in D.

Theorem 7.1. *Let $L = \frac{1}{2}\nabla \cdot a\nabla + b \cdot \nabla + V$ satisfy Assumption $\tilde{\mathrm{H}}$ on a bounded domain $D \subset R^d$ with a $C^{2,\alpha}$-boundary. Then*

$$\lambda_0 = \sup_{\mu} \inf_{\substack{u \in C^2(\bar{D}) \\ u>0 \text{ on } \bar{D}}} \int_D \frac{Lu}{u} \, d\mu$$

$$= \sup_{\substack{g \in C^1(\bar{D}) \\ g \geq 0 \text{ on } D \\ g=0 \text{ on } \partial D \\ \int_D g^2 dx = 1}} \inf_{h \in C^1(\bar{D})} \Bigg[\frac{1}{2} \int_D (\nabla h - a^{-1}b) a (\nabla h - a^{-1}b) g^2 \, dx$$

$$- \frac{1}{2} \int_D \left(\frac{\nabla g}{g} - a^{-1}b \right) a \left(\frac{\nabla g}{g} - a^{-1}b \right) g^2 \, dx + \int_D V g^2 \, dx \Bigg], \quad (7.1)$$

where $\sup_\mu$ is over all probability measures μ on D with densities f satisfying $g \equiv f^{1/2} \in C^1(\bar{D})$ and $g = 0$ on ∂D. Furthermore, the righthand mini-max in (7.1) is attained at $g = (\phi_0 \tilde{\phi}_0)^{1/2}$ and

$$h = \frac{1}{2} \log \frac{\tilde{\phi}_0}{\phi_0},$$

and the lefthand mini-max is attained at $\mu = \phi_0\tilde{\phi}_0\, dx$ *and* $u = \phi_0$, *where* ϕ_0 *and* $\tilde{\phi}_0$, *normalized by* $\int_D \phi_0\tilde{\phi}_0\, dx = 1$, *are the eigenfunctions associated with the eigenvalue* λ_0 *for* L *and* $\tilde{L}$ *respectively.*

Remark 1. It follows from Theorem 4.4.7 in Chapter 4 that the lefthand equality in (7.1) continues to hold if $\sup_\mu$ is over all probability measures on D.

Remark 2. In the special case that $a^{-1}b \equiv \nabla Q$ is a gradient, we can write $L = \frac{1}{2}\exp(-2Q)\nabla \cdot a\exp(2Q)\nabla + V$ and thus L is a symmetric operator on $L^2(D, \exp(2Q)\,dx)$. That is, $\int_D vLw\exp(2Q)\,dx = \int_D wLv\exp(2Q)\,dx$, for all $w,\ v \in C^2(\bar{D})$ which satisfy $w = v = 0$ on ∂D. Define $\langle w, v\rangle = \int_D wv\exp(2Q)\,dx$, for $w,\ v \in L^2(D, \exp(2Q)\,dx)$. Define the quadratic form $q(w, v) = \langle -Lw, v\rangle = \frac{1}{2}\int(\nabla wa\nabla v)\exp(2Q)\,dx - \int_D Vwv\exp(2Q)\,dx$, for $w,\ v \in C_0^\infty(D)$. Then q is bounded from below and the Friedrichs' extension theorem [Reed and Simon, Vol. 2 (1975), Theorem 10.23] states that q is closeable on $L^2(D, \exp(2Q)\,dx) \times L^2(D, \exp(2Q)\,dx)$ and that, if $\hat{q}$ defined on $\mathcal{D}_{\hat{q}} \times \mathcal{D}_{\hat{q}}$ denotes its closure, then there exists a unique self-adjoint operator $(-\hat{\mathcal{L}}, \hat{\mathcal{D}})$ on $L^2(D, \exp(2Q)\,dx)$, bounded from below, which satisfies $\hat{\mathcal{D}} \subset \mathcal{D}_{\hat{q}}$ and $\langle -\hat{\mathcal{L}}w, v\rangle = \hat{q}(w, v)$ for all w, $v \in \hat{\mathcal{D}}$. Because $(\hat{\mathcal{L}}, \hat{\mathcal{D}})$ has been obtained from the quadratic form q defined on $C_0^\infty(D)$, it is said to satisfy the Dirichlet boundary condition.

The Rayleigh–Ritz formula, which is the mini-max principle [Reed and Simon, Vol. 4 (1978), Theorems 13.1 and 13.2] applied to the principal eigenvalue, states that

$$\inf \operatorname{Re}(\sigma((-\mathcal{L}, \hat{\mathcal{D}}))) = \inf_{\substack{w \in \hat{\mathcal{D}} \\ \langle w,w\rangle = 1}} \langle -\hat{\mathcal{L}}w, w\rangle = \inf_{\substack{w \in \mathcal{D}_{\hat{q}} \\ \langle w,w\rangle = 1}} \hat{q}(w, w).$$

By the Beurling–Deny criterion [Reed and Simon, Vol. 4 (1978), Theorem 13.50 and problem 99 on p. 375],

$$\inf_{\substack{w \in \mathcal{D}_{\hat{q}} \\ \langle w,w\rangle = 1}} \hat{q}(w, w) = \inf_{\substack{w \in \mathcal{D}_{\hat{q}} \\ \langle w,w\rangle = 1 \\ w \geq 0}} \hat{q}(w, w).$$

Thus,

$$\inf \operatorname{Re}(\sigma((-\hat{\mathcal{L}}, \hat{\mathcal{D}}))) = \inf_{\substack{w \in \mathcal{D}_{\hat{q}} \\ \langle w,w\rangle = 1 \\ w \geq 0}} \hat{q}(w, w). \tag{7.2}$$

Using (7.1) and (7.2), we now show that λ_0, which by Theorems 6.1 and 6.3 is the supremum of the real part of the spectrum of both $(L, \mathcal{D}_\alpha)$ on $\mathcal{B}_\alpha$ and $(\mathcal{L}, \mathcal{D}_0)$ on $\mathcal{B}_0$, is also the supremum of the spectrum of $(\hat{\mathcal{L}}, \hat{\mathcal{D}})$ on $L^2(D, \exp(2Q)\,dx)$. (The spectrum of a self-adjoint operator is always real.) Since $a^{-1}b = \nabla Q$, by choosing $h = Q$ we obtain

$$\inf_{h \in C^2(\bar{D})} \int_D (\nabla h - a^{-1}b)a(\nabla h - a^{-1}b)g^2\,dx = 0.$$

Thus,

$$\lambda_0 = \sup_{\substack{g \in C^1(\bar{D}) \\ g \geq 0 \text{ on } D \\ g=0 \text{ on } \partial D \\ \int_D g^2\,dx = 1}} \left[\int_D Vg^2\,dx - \tfrac{1}{2}\int_D \left(\frac{\nabla g}{g} - \nabla Q\right)a\left(\frac{\nabla g}{g} - \nabla Q\right)g^2\,dx\right].$$

Now let $w = g\exp(-Q)$. By the assumption on a and b, it follows that $Q \in C^{2,\alpha}(\bar{D})$. Substituting into the expression for λ_0 above gives

$$\lambda_0 = \sup_{\substack{w \in C^2(\bar{D}) \\ w \geq 0 \text{ on } D \\ w=0 \text{ on } \partial D \\ \langle w,w \rangle = 1}} \left[\int_D Vw^2 \exp(2Q)\,dx - \tfrac{1}{2}\int_D (\nabla w a \nabla w)\exp(2Q)\,dx\right] =$$

$$\sup_{\substack{w \in C^2(\bar{D}) \\ w \geq 0 \text{ on } D \\ w=0 \text{ on } \partial D \\ \langle w,w \rangle = 1}} (-q(w, w)). \quad (7.3)$$

It is easy to show that $\mathcal{D}_{\hat{q}}$ contains $\{u \in C^2(\bar{D})\colon u = 0 \text{ on } \partial D\}$. Therefore, since $\{u \in C^2(\bar{D})\colon u = 0 \text{ on } \partial D\}$ contains $C_0^\infty(D)$ and since $(\hat{q}, \mathcal{D}_{\hat{q}})$ is the closure of $(q, C_0^\infty(D))$, it follows from (7.3) that

$$\lambda_0 = \sup_{\substack{w \in \mathcal{D}_{\hat{q}} \\ \langle w,w \rangle = 1 \\ w \geq 0}} (-\hat{q}(w, w)).$$

Comparing this to (7.2), we conclude that $\lambda_0 = \sup(\sigma((\hat{\mathcal{L}}, \hat{\mathcal{D}})))$.

Proof of Theorem 7.1. We began by proving the righthand equality in (7.1). Define

$$J(\mu) = \inf_{\substack{u \in C^2(\bar{D}) \\ u > 0 \text{ on } \bar{D}}} \int_D \frac{Lu}{u}\,d\mu, \text{ for probability measures } \mu \text{ on } D. \quad (7.4)$$

We will show that if μ is a probability measure on D, with density f satisfying $g \equiv f^{1/2} \in C^1(\bar{D})$ and $g = 0$ on ∂D, then

$$J(\mu) = \inf_{h \in C^1(\bar{D})} \left[\tfrac{1}{2}\int_D (\nabla h - a^{-1}b)a(\nabla h - a^{-1}b)g^2\,\mathrm{d}x \right.$$
$$\left. -\tfrac{1}{2}\int_D \left(\frac{\nabla g}{g} - a^{-1}b\right)a\left(\frac{\nabla g}{g} - a^{-1}b\right)g^2\,\mathrm{d}x + \int_D Vg^2\,\mathrm{d}x \right]. \quad (7.5)$$

Let $u = \exp(w) \in C^2(\bar{D})$. Then

$$\int_D \frac{Lu}{u}\,\mathrm{d}\mu = \int_D (\tfrac{1}{2}\nabla \cdot a\nabla w + b \cdot \nabla w + \tfrac{1}{2}\nabla w a \nabla w + V)\,\mathrm{d}\mu. \quad (7.6)$$

Extend the density f to all of R^d be defining $f(x) = 0$ for $x \notin \bar{D}$. Using the mollifier

$$G_\varepsilon(x) = (2\pi\varepsilon)^{-(d/2)} \exp\left(\frac{-|x|^2}{2\varepsilon}\right),$$

define $f_\varepsilon = G_\varepsilon^* f$. Then $f_\varepsilon \in C^\infty(\bar{D})$ and, as $\varepsilon \to 0$, $f_\varepsilon \to f$ and $\nabla f_\varepsilon \to \nabla f$, uniformly in D.

Let $\delta > 0$ and substitute $w = \frac{1}{2}\log(f_\varepsilon + \delta) - h$ into the righthand side of (7.6). Note that $h \in C^2(\bar{D})$. We obtain

$$\int_D \frac{Lu}{u} f\,\mathrm{d}x = \int_D \left(-\tfrac{1}{2}\nabla \cdot a\nabla h - b \cdot \nabla h + \tfrac{1}{2}\nabla h a \nabla h - \tfrac{1}{2}\frac{\nabla h a \nabla f_\varepsilon}{(f_\varepsilon + \delta)}\right) f\,\mathrm{d}x$$
$$+ \int_D \frac{\nabla f_\varepsilon a \nabla f_\varepsilon}{8(f_\varepsilon + \delta)^2} f\,\mathrm{d}x + \int_D \tfrac{1}{4} f \nabla \cdot \left(\frac{a\nabla f_\varepsilon}{f_\varepsilon + \delta}\right)\mathrm{d}x$$
$$+ \int_D \frac{b \cdot \nabla f_\varepsilon}{2(f_\varepsilon + \delta)} f\,\mathrm{d}x + \int_D Vf\,\mathrm{d}x.$$

Integrating by parts and using the fact that $f = 0$ on ∂D, we have

$$\int_D \tfrac{1}{2}\nabla \cdot (a\nabla h) f\,\mathrm{d}x = -\tfrac{1}{2}\int_D (\nabla f a \nabla h)\,\mathrm{d}x$$

and

$$\int_D \tfrac{1}{4} f \nabla \cdot \left(\frac{a\nabla f_\varepsilon}{f_\varepsilon + \delta}\right)\mathrm{d}x = -\int_D \frac{\nabla f a \nabla f_\varepsilon}{4(f_\varepsilon + \delta)}\,\mathrm{d}x.$$

Thus,

$$J(\mu) = \inf_{h \in C^2(\bar{D})} \left[\int_D \left(\tfrac{1}{2}(\nabla h a \nabla h) f - (b \cdot \nabla h) f - \tfrac{1}{2} \frac{\nabla h a \nabla f_\varepsilon}{(f_\varepsilon + \delta)} f + \tfrac{1}{2} \nabla f a \nabla h \right) \mathrm{d}x \right]$$
$$+ \int_D \frac{\nabla f_\varepsilon a \nabla f_\varepsilon}{8(f_\varepsilon + \delta)^2} f \,\mathrm{d}x - \int_D \frac{\nabla f a \nabla f_\varepsilon}{4(f_\varepsilon + \delta)} \mathrm{d}x + \int_D \frac{b \cdot \nabla f_\varepsilon}{2(f_\varepsilon + \delta)} f \,\mathrm{d}x$$
$$+ \int_D V f \,\mathrm{d}x. \tag{7.7}$$

At this point, we will prove (7.5) in the case where

$$\int_D \frac{|\nabla f|^2}{f} \mathrm{d}x \left(= 4 \int_D |\nabla g|^2 \,\mathrm{d}x \right) = \infty.$$

Picking $h \equiv 1$ in (7.7) and letting $\varepsilon \to 0$ gives

$$J(\mu) \leqslant - \int_D \frac{\nabla f a \nabla f}{8(f + \delta)^2} f \,\mathrm{d}x + \int_D \frac{b \cdot \nabla f}{2(f + \delta)} f \,\mathrm{d}x + \int_D V f \,\mathrm{d}x.$$

By assumption, $\nabla f a \nabla f \geqslant c |\nabla f|^2$ for some $c > 0$. Furthermore, using the inequality

$$|ab| \leqslant \frac{\gamma^2 a^2}{2} + \frac{b^2}{2\gamma^2},$$

for any $\gamma > 0$, we have

$$\left| \int_D \frac{b \cdot \nabla f}{2(f + \delta)} f \,\mathrm{d}x \right| \leqslant \gamma^2 \int_D \frac{|\nabla f|^2}{8 f^2} \mathrm{d}x + \frac{1}{2\gamma^2} \int_D \frac{|b|^2 f^3}{(f + \delta)^2} \mathrm{d}x.$$

Thus, letting $\delta \to 0$ gives $J(\mu) = -\infty$, if

$$\int_D \frac{|\nabla f|^2}{f} \mathrm{d}x = \infty.$$

Now, using the inequality

$$|ab| \leqslant \frac{\gamma^2 a^2}{2} + \frac{b^2}{2\gamma^2}$$

again, it follows that the righthand side of (7.5) also equals $-\infty$ if

$$\int_D \frac{|\nabla f|^2}{f} \mathrm{d}x \left(= 4 \int_D |\nabla g|^2 \,\mathrm{d}x \right) = \infty.$$

From here on, we will assume that

$$\int_D \frac{|\nabla f|^2}{f} \mathrm{d}x < \infty.$$

We want to let $\varepsilon \to 0$ in (7.7). We first show that one may interchange the order of $\inf_h$ and $\lim_{\varepsilon\to 0}$ on the righthand side of (7.7). To this end, denote the variational term on the righthand side of (7.7) by

$$\psi_{\varepsilon,\delta}(h) \equiv \int_D \left(\tfrac{1}{2}(\nabla h a \nabla h) f - (b \cdot \nabla h) f - \tfrac{1}{2}\frac{\nabla h a \nabla f_\varepsilon}{(f_\varepsilon + \delta)} f + \tfrac{1}{2}\nabla f a \nabla h\right) \mathrm{d}x,$$

for $\varepsilon \geq 0$ and $\delta > 0$ and also for $\varepsilon = \delta = 0$, where, for convenience, we define $f_0 \equiv f$.

Applying the inequality

$$|ab| \leq \frac{\gamma^2 a^2}{2} + \frac{b^2}{2\gamma^2}$$

to the last three terms in $\psi_{\varepsilon,\delta}(h)$, and using the fact that

$$\lim_{\delta\to 0}\lim_{\varepsilon\to 0}\int_D \frac{\nabla f_\varepsilon a \nabla f_\varepsilon}{(f_\varepsilon + \delta)^2} f \,\mathrm{d}x = \int_D \frac{\nabla f a \nabla f}{f}\,\mathrm{d}x < \infty,$$

and that $\nabla h a \nabla h \geq c|\nabla h|$, it follows that

$$\psi_{\varepsilon,\delta}(h) \geq c_1 \int_D |\nabla h|^2 f \,\mathrm{d}x - c_2,$$

for $\varepsilon \geq 0$ and $\delta > 0$ and also for $\varepsilon = \delta = 0$, where $c_1 > 0$ and c_1 and c_2 are independent of $0 \leq \varepsilon \leq 1$ and $0 \leq \delta \leq 1$. Thus, there exists an N, independent of $0 \leq \varepsilon \leq 1$ and $0 \leq \delta \leq 1$, such that

$$\inf_{h\in C^2(\bar{D})} \psi_{\varepsilon,\delta}(h) = \inf_{\substack{h\in C^2(\bar{D})\\ \int_D|\nabla h|^2 f\,\mathrm{d}x \leq N}} \psi_{\varepsilon,\delta}(h). \tag{7.8}$$

Using (7.8) we will show that

$$\lim_{\varepsilon\to 0}\inf_{h\in C^2(\bar{D})} \psi_{\varepsilon,\delta}(h) = \inf_{h\in C(\bar{D})} \psi_{0,\delta}(h), \text{ for } \delta > 0. \tag{7.9}$$

To prove (7.9), we must show that

$$\lim_{\varepsilon\to 0}\inf_{h\in C^2(\bar{D})} \psi_{\varepsilon,\delta}(h) = \lim_{h\in C^2(\bar{D})}\lim_{\varepsilon\to 0} \psi_{\varepsilon,\delta}(h).$$

In light of (7.8), it is in fact enough to show that

$$\lim_{\varepsilon\to 0}\inf_{\substack{h\in C^2(\bar{D})\\ \int_D|\nabla h|^2 f\,\mathrm{d}x \leq N}} \psi_{\varepsilon,\delta}(h) = \inf_{\substack{h\in C^2(\bar{D})\\ \int_D|\nabla h|^2 f\,\mathrm{d}x \leq N}} \lim_{\varepsilon\to 0} \psi_{\varepsilon,\delta}(h). \tag{7.10}$$

To prove (7.10) we will show that the one term in $\psi_{\varepsilon,\delta}(h)$ which depends on ε,

$$\int_D \frac{\nabla h a \nabla f_\varepsilon}{2(f_\varepsilon + \delta)} f\, \mathrm{d}x,$$

converges to

$$\int_D \frac{\nabla h a \nabla f}{2(f + \delta)} f\, \mathrm{d}x$$

uniformly over $\{h \in C^2(\bar{D}): \int_D |\nabla h|^2 f\, \mathrm{d}x \leq N\}$. Using the Schwarz inequality, we have

$$\left|\int_D \frac{\nabla h a \nabla f_\varepsilon}{(f_\varepsilon + \delta)} f\, \mathrm{d}x - \int_D \frac{\nabla h a \nabla f}{f + \delta} f\, \mathrm{d}x\right|^2 = \left|\int_D \nabla h a\left(\frac{f \nabla f_\varepsilon}{f_\varepsilon + \delta} - \frac{f \nabla f}{f + \delta}\right) \mathrm{d}x\right|^2$$

$$\leq \int_D (\nabla h a \nabla h) f\, \mathrm{d}x \cdot \int_D \left(\frac{f^{1/2} \nabla f_\varepsilon}{f_\varepsilon + \delta} - \frac{f^{1/2} \nabla f}{f + \delta}\right) a \left(\frac{f^{1/2} \nabla f_\varepsilon}{f_\varepsilon + \delta} - \frac{f^{1/2} \nabla f}{f + \delta}\right) \mathrm{d}x$$

$$\leq \|a\|^2 N \int_D \left|\frac{f^{1/2} \nabla f_\varepsilon}{f_\varepsilon + \delta} - \frac{f^{1/2} \nabla f}{f + \delta}\right|^2 \mathrm{d}x \to 0, \quad \text{as } \varepsilon \to 0.$$

We can now let $\varepsilon \to 0$ in (7.7). Using (7.9) we conclude that

$$J(\mu) = \inf_{h \in C^2(\bar{D})} \psi_{0,\delta}(h) + \int_D \frac{\nabla f a \nabla f}{8(f + \delta)^2} f\, \mathrm{d}x - \int_D \frac{\nabla f a \nabla f}{4(f + \delta)} \mathrm{d}x$$

$$+ \int_D \frac{b \cdot \nabla f}{2(f + \delta)} f\, \mathrm{d}x + \int_D V f\, \mathrm{d}x. \tag{7.11}$$

The same argument used to prove (7.9) shows that

$$\lim_{\delta \to 0} \inf_{h \in C^2(\bar{D})} \psi_{0,\delta}(h) = \inf_{h \in C^2(\bar{D})} \psi_{0,0}(h).$$

Also, clearly,

$$\inf_{h \in C^1(\bar{D})} \psi_{0,0}(h) = \inf_{h \in C^2(\bar{D})} \psi_{0,0}(h).$$

Thus, letting $\delta \to 0$ in (7.11) gives

$$J(\mu) = \inf_{h \in C^1(\bar{D})} \psi_{0,0}(h) - \int_D \frac{\nabla f a \nabla f}{8f} \mathrm{d}x + \int_D \tfrac{1}{2}(b \cdot \nabla f)\, \mathrm{d}x + \int_D V f\, \mathrm{d}x. \tag{7.12}$$

Adding and subtracting the term $\frac{1}{2}\int_D (b a^{-1} b) g^2\, \mathrm{d}x$ on the righthand side of (7.12) and then substituting g^2 for f gives (7.5).

We now prove the lefthand inequality in (7.1). Let $\mu_0(\mathrm{d}x) = \phi_0(x)\tilde{\phi}_0(x)\, \mathrm{d}x$. By choosing $u = \phi_0 + \varepsilon$ in (7.4) and then letting $\varepsilon \to 0$, we obtain

$$J(\mu_0) \leqslant \lim_{\varepsilon\to 0}\int_D \frac{L(\phi_0+\varepsilon)}{\phi_0+\varepsilon}\phi_0\tilde{\phi}_0\,\mathrm{d}x = \int_D \tilde{\phi}_0 L\phi_0\,\mathrm{d}x = \lambda_0\int_D \phi_0\tilde{\phi}_0\,\mathrm{d}x = \lambda_0.$$

We will now show that in fact $J(\mu_0) = \lambda_0$. Define

$$H(w;\mu_0) = \int \frac{L\exp(w)}{\exp(w)}\,\mathrm{d}\mu_0,$$

for w in the convex hull of $C^2(\bar{D}) \cup \{\log\phi_0\}$, and denote this convex hull by A. The above calculation showed that

$$\lim_{\varepsilon\to 0} H(\log(\phi_0+\varepsilon);\mu_0) = H(\log\phi_0;\mu_0) = \lambda_0.$$

Let $v_1, v_2 \in C^2(\bar{D})$. A direct calculation reveals that

$$H(tv_1+(1-t)v_2;\mu_0) = tH(v_1;\mu_0) + (1-t)H(v_2;\mu_0)$$

$$-\tfrac{1}{2}t(1-t)\int_D \exp(2(v_1-v_2))\nabla(\exp(v_2-v_1))a\nabla(\exp(v_2-v_1))\mu_0(\mathrm{d}x).$$

Thus, $H(v;\mu_0)$ is convex for $v \in C^2(\bar{D})$. The convexity of $H(v;\mu_0)$ extends to $v \in A$. Indeed, it is easy to see that the above calculation applied with $v_1 = \log\phi_0$ and $v_2 \in C^2(\bar{D})$ is valid as long as

$$\int_D \left|\frac{\nabla\phi_0}{\phi_0}\right|^2 \phi_0\tilde{\phi}_0\,\mathrm{d}x < \infty,$$

and the finiteness of this integral follows from the fact that, by Theorem 2.5,

$$\frac{\tilde{\phi}_0}{\phi_0}(x)$$

is bounded on D. In light of this convexity,

$$J(\mu_0) = \inf_{w\in C^2(\bar{D})} H(w;\mu_0) = \inf_{w\in A} H(w;\mu_0).$$

To complete the proof that $J(\mu_0) = \lambda_0$, it suffices to show that

$$\frac{\mathrm{d}}{\mathrm{d}\varepsilon}H(\log\phi_0+\varepsilon v;\mu_0)|_{\varepsilon=0} = 0, \quad \text{for all } v \in A.$$

For $v \in C^2(\bar{D})$, a direct calculation and integration by parts reveal that

$$\frac{\mathrm{d}}{\mathrm{d}\varepsilon}H(\log\phi_0+\varepsilon v;\mu_0)|_{\varepsilon=0} = \int_D \tilde{\phi}_0 L(\phi_0 v)\,\mathrm{d}x - \int_D v\tilde{\phi}_0 L\phi_0\,\mathrm{d}x$$

$$= \int_D \phi_0 v\tilde{L}\tilde{\phi}_0\,\mathrm{d}x - \int_D v\tilde{\phi}_0 L\phi_0\,\mathrm{d}x = \lambda\int_D \phi_0\tilde{\phi}_0 v\,\mathrm{d}x - \lambda\int_D \phi_0\tilde{\phi}_0 v\,\mathrm{d}x = 0.$$

The same calculation and integration by parts is also valid in the case $v = \log \phi_0$ since the terms $\phi_0 \log \phi_0$ and $\tilde{\phi}_0 \log \phi_0$ remain bounded in D. We conclude that $J(\mu_0) = \lambda_0$.

Call a probability measure μ on D admissible if, as in the statement of the theorem, it possesses a density f which satisfies $g \equiv f^{1/2} \in C^1(\bar{D})$ and $g = 0$ on ∂D. To conclude the proof of the lefthand inequality in (7.1), we must show that $J(\mu_0) = \sup_\mu J(\mu)$, where the supremum is over all admissible μ. Define $\psi(h; \mu) = \frac{1}{2}\int_D (\nabla h - a^{-1}b)a(\nabla h - a^{-1}b)\, d\mu$, for μ an admissible measure and $h \in W^{1,2}(D, d\mu)$, the Sobelev space of functions with one generalized $L^2(D, d\mu)$—derivative. Clearly, $\inf_{h \in C^1(\bar{D})} \psi(h; \mu) = \inf_{h \in W^{1,2}(D, d\mu)} \psi(h; \mu)$. We will now show that, up to an additive constant, there exists a unique $h \in W^{1,2}(D, \mu)$ at which the infimum above is attained. From the inequality,

$$|ab| \leq \frac{\gamma^2}{2}a^2 + \frac{b^2}{2\gamma^2}, \quad \text{for any } \gamma > 0,$$

it follows that $\psi(h; \mu) \geq c_1 \int_D |\nabla h|^2\, d\mu - c_2$, where $c_1 > 0$ and c_1 and c_2 are independent of h. Thus, if $\{h_n\}_{n=1}^{\infty}$ is a minimizing sequence in $W^{1,2}(D, d\mu)$ for $\psi(h; \mu)$, it follows that $\int_D |\nabla h_n|^2\, d\mu \leq M$, for $n = 1, 2, \ldots$ and some M. We may assume of course that $h_n(x_0) = 1$, for some fixed $x_0 \in D$ and $n = 1, 2, \ldots$. Thus $\{h_n\}_{n=1}^{\infty}$ is weakly relatively compact in $W^{1,2}(D, d\mu)$. Since the norm is lower semicontinuous with respect to weak convergence, this proves the existence of a minimizer h_μ.

To prove uniqueness, note that since h_μ is a minimizer,

$$0 = \frac{d}{d\varepsilon}\psi(h_\mu + \varepsilon q)|_{\varepsilon=0}, \quad \text{for } q \in W^{1,2}(D, d\mu). \tag{7.13}$$

Calculating (7.13) gives

$$\int_D \nabla q a(\nabla h_\mu - a^{-1}b)\, d\mu = 0, \quad \text{for } q \in W^{1,2}(D, d\mu). \tag{7.14}$$

Now if $\hat{h}_\mu$ is also a minimizer, then (7.14) also holds with $\hat{h}_\mu$ in place of h_μ. Subtracting then gives

$$\int_D \nabla q a(\nabla h_\mu - \nabla \hat{h}_\mu)\, d\mu = 0, \quad \text{for } q \in W^{1,2}(D, d\mu). \tag{7.15}$$

Choosing $q = h_\mu - \hat{h}_\mu$ in (7.15) gives

$$\int_D (\nabla h_\mu - \nabla \hat{h}_\mu)a(\nabla h_\mu - \nabla \hat{h}_\mu)\, d\mu = 0.$$

Thus $\nabla(h_\mu - \hat{h}_\mu) = 0$ a.s. and it follows that $h_\mu = \hat{h}_\mu + \text{const.}$ a.s.

In light of (7.5) and the above analysis, for an admissible μ, we have

$$J(\mu) = \psi(h_\mu; \mu) - \psi(\tfrac{1}{2}\log f; \mu) + \int_D V\,\mathrm{d}\mu,$$

where h_μ minimizes $\psi(h; \mu)$ and

$$\frac{\mathrm{d}\mu}{\mathrm{d}x} = f.$$

Let μ_1 be an arbitrary admissible measure and define $M(p) = J((1-p)\mu_0 + p\mu_1)$, for $0 \leq p \leq 1$. From the original definition of $J(\mu)$ in (7.4), it follows that $J(\mu)$ is concave; thus $M(p)$ is concave in p. Therefore, to complete the proof that $J(\mu_0) = \sup_\mu J(\mu)$, it suffices to show that

$$M'(0^+) = 0. \tag{7.16}$$

Define $\mu_p = (1-p)\mu_0 + p\mu_1$ and

$$f_p = \frac{\mathrm{d}\mu_p}{\mathrm{d}x},$$

and let $h_p \equiv h_{\mu_p}$, for $0 \leq p \leq 1$. In particular, $f_0 = \phi_0\tilde{\phi}_0$. We now show that

$$h_0 = \tfrac{1}{2}\log\frac{\tilde{\phi}_0}{\phi_0}.$$

From (7.14), upon integrating by parts, it follows that if h_0 is smooth, then it must satisfy the following equation for h:

$$\left.\begin{aligned} &\nabla\cdot f_0 a(\nabla h - a^{-1}b) = 0 \text{ in } D,\\ &\lim_{D\ni x\to\partial D} f_0(x)(\nabla han)(x) = 0, \end{aligned}\right\} \tag{7.17}$$

where n is the unit outward normal at ∂D. Conversely, if a smooth function h satisfies (7.17), then $h = h_0$. Using the fact that $L\phi_0 = \lambda_0\phi_0$ and $\tilde{L}\tilde{\phi}_0 = \lambda_0\tilde{\phi}_0$, one can check that

$$h = \tfrac{1}{2}\log\frac{\tilde{\phi}_0}{\phi_0}$$

satisfies (7.17); thus

$$h_0 = \tfrac{1}{2}\log\frac{\tilde{\phi}_0}{\phi_0}.$$

Let G_i, $i = 0, 1$, denote the space of all functions $z\colon R^d \to R^d$ of the form $z = \nabla q + ka^{-1}b$, for $q \in W^{1,2}(D, \mathrm{d}\mu_i)$ and $k \in R$, and define the norm $\|z\|_{G_i} = \int_D (zaz)\,\mathrm{d}\mu_i$. Then G_i is a Hilbert space. We will prove (7.16) under the assumption that

$$H \equiv \lim_{p\to 0^+} \frac{\nabla h_p - \nabla h_0}{p} \tag{7.18}$$

exists as a weak limit in $(G_0, \|\cdot\|_{G_0})$. Then we will return to prove (7.18).

We have

$$M(p) = \tfrac{1}{2}\int_D (\nabla h_p - a^{-1}b)a(\nabla h_p - a^{-1}b)f_p\,\mathrm{d}x - \tfrac{1}{2}\int_D \frac{(\frac{1}{2}\nabla f_p - a^{-1}bf_p)a(\frac{1}{2}\nabla f_p - a^{-1}bf_p)}{f_p}\,\mathrm{d}x + \int_D Vf_p\,\mathrm{d}x \tag{7.19}$$

and, formally,

$$M'(0^+) = \int_D (\nabla h_0 - a^{-1}b)aHf_0\,\mathrm{d}x + \tfrac{1}{2}\int_D (\nabla h_0 - a^{-1}b)a(\nabla h_0 - a^{-1}b)(f_1 - f_0)\,\mathrm{d}x$$
$$- \int_D \Bigg[\frac{(\frac{1}{2}\nabla(f_1 - f_0) - a^{-1}b(f_1 - f_0))a(\frac{1}{2}\nabla f_0 - a^{-1}bf_0)}{f_0} - \frac{(\frac{1}{2}\nabla f_0 - a^{-1}bf_0)a(\frac{1}{2}\nabla f_0 - a^{-1}bf_0)(f_1 - f_0)}{2f_0^2}\Bigg]\mathrm{d}x$$
$$+ \int_D V(f_1 - f_0)\,\mathrm{d}x. \tag{7.20}$$

To show that (7.20) holds rigorously, we must show that

$$\lim_{p\to 0} \frac{1}{p}\Bigg[\int_D (\nabla h_p - a^{-1}b)a(\nabla h_p - a^{-1}b)f_p\,\mathrm{d}x - \int_D (\nabla h_0 - a^{-1}b)a(\nabla h_0 - a^{-1}b)f_0\,\mathrm{d}x\Bigg]$$
$$= 2\int_D (\nabla h_0 - a^{-1}b)aHf_0\,\mathrm{d}x + \int_D (\nabla h_0 - a^{-1}b)a(\nabla h_0 - a^{-1}b)(f_1 - f_0)\,\mathrm{d}x. \tag{7.21}$$

We have

$$\frac{1}{p}\left[\int_D (\nabla h_p - a^{-1}b)a(\nabla h_p - a^{-1}b)f_p\,dx - \int_D (\nabla h_0 - a^{-1}b)a(\nabla h_0 - a^{-1}b)f_0\,dx\right]$$

$$= \frac{1}{p}\int_D \Big[(\nabla h_p - a^{-1}b)a(\nabla h_p - a^{-1}b) - (\nabla h_0 - a^{-1}b)a(\nabla h_0 - a^{-1}b)\Big]f_0\,dx$$

$$+ \int_D (\nabla h_p - a^{-1}b)a(\nabla h_p - a^{-1}b)(f_1 - f_0)\,dx$$

$$= \frac{1}{p}\int_D ((\nabla h_p - a^{-1}b) + (\nabla h_0 - a^{-1}b))a((\nabla h_p - a^{-1}b) - (\nabla h_0 - a^{-1}b))f_0\,dx$$

$$+ \int_D (\nabla h_p - a^{-1}b)a(\nabla h_p - a^{-1}b)(f_1 - f_0)\,dx$$

$$= \int_D ((\nabla h_p - a^{-1}b) + (\nabla h_0 - a^{-1}b))a\left(\frac{\nabla h_p - \nabla h_0}{p}\right)f_0\,dx$$

$$+ \int_D (\nabla h_p - a^{-1}b)a(\nabla h_p - a^{-1}b)(f_1 - f_0)\,dx. \qquad (7.22)$$

Since, by (7.18),

$$\frac{\nabla h_p - \nabla h_0}{p}$$

converges weakly in $(G_0, \|\cdot\|_{G_0})$, it is norm bounded, and thus, $\nabla h_p \to \nabla h_0$ strongly in $(G_0, \|\cdot\|_{G_0})$. By Theorem 2.5, it follows that ϕ_0 and $\tilde{\phi}_0$ each have first-order zeros at ∂D and thus that f_0 has a second-order zero at ∂D. By the admissibility of μ_1, it follows then that $f_1 \leqslant kf_0$ in D for some $k > 0$. Consequently, we also have $\nabla h_p \to \nabla h_0$ strongly in $(G_1\|\cdot\|_{G_1})$. The weak convergence of

$$\frac{\nabla h_p - \nabla h_0}{p}$$

to H in $(G_0, \|\cdot\|_{G_0})$ and the strong convergence of ∇h_p to ∇h_0 in $(G_0, \|\cdot\|_{G_0})$ and in $(G_1, \|\cdot\|_{G_1})$ show that as $p \to 0$, the righthand side of (7.22) converges to the righthand side of (7.21). This verifies (7.21).

From (7.14) and (7.18) it follows that $\int_D (\nabla h_0 - a^{-1}b)aHf_0\,dx = 0$. Thus (7.20) becomes

$$M'(0^+) = \tfrac{1}{2}\int_D (\nabla h - a^{-1}b)a(\nabla h_0 - a^{-1}b)(f_1 - f_0)\,\mathrm{d}x$$
$$- \int_D \left[\frac{(\tfrac{1}{2}\nabla(f_1 - f_0) - a^{-1}b(f_1 - f_0))a(\tfrac{1}{2}\nabla f_0 - a^{-1}bf_0)}{f_0}\right.$$
$$\left. - \frac{(\tfrac{1}{2}\nabla f_0 - a^{-1}bf_0)a(\tfrac{1}{2}\nabla f_0 - a^{-1}bf_0)(f_1 - f_0)}{2f_0^2}\right]\mathrm{d}x$$
$$+ \int_D V(f_1 - f_0)\,\mathrm{d}x. \tag{7.23}$$

From the fact that $\phi_0 = f_0^{1/2}\exp(-h_0)$ and $\tilde{\phi}_0 = f_0^{1/2}\exp(h_0)$, one can verify that f_0 satisfies

$$\tfrac{1}{2}\nabla\cdot a\nabla f_0 - \tfrac{1}{4}\frac{\nabla f_0 a\nabla f_0}{f_0} + f_0(\nabla h_0 a\nabla h_0 - \nabla\cdot b - 2\nabla h_0 b) + Vf_0 = 0. \tag{7.24}$$

If we set $q = f_1 - f_0$ in (7.20), and then integrate by parts and use (7.24) and the fact that $q = 0$ on ∂D, we obtain $M'(0^+) = 0$.

It remains to show (7.18). Since $\int_D(\nabla h_p - b)a(\nabla h_p - b)f_p\,\mathrm{d}x$ is bounded independently of p, it is clear that

$$\overline{\lim_{p\to 0}}\int_D |\nabla h_p|^2 f_0\,\mathrm{d}x < \infty. \tag{7.25}$$

By the argument following (7.22), it then follows that

$$\overline{\lim_{p\to 0}}\int_D |\nabla h_p|^2 f_1\,\mathrm{d}x < \infty. \tag{7.26}$$

The argument following (7.22) also shows that $W^{1,2}(D, \mathrm{d}\mu_0) = W^{1,2}(D, \mathrm{d}\mu_p)$, for $0 < p < 1$. Thus, from (7.14) we have $\int_D(\nabla h_p - a^{-1}b)a\nabla q f_p\,\mathrm{d}x = 0$, for $0 \leq p < 1$ and all $q \in W^{1,2}(D, \mathrm{d}\mu_0)$. Considering this equation for any $p \in (0, 1)$ and also for $p = 0$, and subtracting one from the other, gives

$$\int_D (\nabla h_p - \nabla h_0)a\nabla q f_0\,\mathrm{d}x + p\int_D (\nabla h_p - a^{-1}b)a\nabla q(f_1 - f_0)\,\mathrm{d}x = 0. \tag{7.27}$$

From (7.25), (7.26) and (7.27), we have

$$\lim_{p\to 0}\int_D (\nabla h_p - \nabla h_0)a\nabla q f_0\,\mathrm{d}x = 0, \quad \text{for all } q \in W^{1,2}(D, \mathrm{d}\mu_0). \tag{7.28}$$

From the argument following (7.22), it then follows that

$$\lim_{p\to 0}\int_D (\nabla h_p - \nabla h_0)a\nabla q f_1\,dx = 0, \quad \text{for all } q \in W^{1,2}(D, d\mu_0). \quad (7.29)$$

Using (7.28) and (7.29), we obtain from (7.27) that

$$\lim_{p\to 0}\int_D \left(\frac{\nabla h_p - \nabla h_0}{p}\right)a\nabla q f_0\,dx = -\int_D (\nabla h_0 - b)a\nabla q(f_1 - f_0)\,dx,$$

$$\text{for all } q \in W^{1,2}(D, d\mu_0). \quad (7.30)$$

Substituting $q = h_p - h_0$ in (7.14) and substituting $q = h_0$ in (7.28), and then subtracting one from the other gives

$$\int_D \left(\frac{\nabla h_p - \nabla h_0}{p}\right)bf_0\,dx = 0. \quad (7.31)$$

Now (7.18) follows from (7.30) and (7.31). (In fact we have proved that

$$H = (\nabla h_0 - b)\frac{f_0 - f_1}{f_0}.)$$

This completes the proof of the lefthand equality in (7.1).

The fact that the righthand mini-max in (7.1) is attained at $g = (\phi_0\tilde{\phi}_0)^{1/2}$ and

$$h = \tfrac{1}{2}\log\frac{\tilde{\phi}_0}{\phi_0}$$

and that the lefthand mini-max in (7.1) is attained at $\mu = \phi_0\tilde{\phi}_0\,dx$ and $u = \phi_0$, follows from the above proof of the rest of the theorem. □

We conclude this section with a result concerning the asymptotic behavior of the principal eigenvalue when the domain is a small ball.

Theorem 7.2. *Let L satisfy Assumption $\tilde{H}$ on a domain $D \subseteq R^d$. Let $r_0 > 0$. Then there exist constants $0 < c_1 < c_2$ depending only on*

$$r_0,\ d,\ \sup_D|V|,\ \sup_D|b|,\ \sup_D\sup_{|v|=1}\sum_{i,j=1}^d a_{ij}v_iv_j \text{ and } \inf_D,\ \inf_{|v|=1}\sum_{i,j=1}^d a_{ij}v_iv_j$$

such that

$$\frac{-c_2}{r^2} \leq \lambda_0(B_r(x)) \leq \frac{-c_1}{r^2},$$

for all $x \in D$ and $r \in (0, r_0)$ such that $B_r(x) \subset D$, where $B_r(x)$ denotes the ball of radius r centered at x.

Proof. From Theorem 6.1(ii), it is easy to see that it suffices to treat the case $V \equiv 0$. In this case,

$$\lambda_0(D) = \lim_{t\to\infty} \frac{1}{t} \log P_y(\tau_D > t), \quad \text{for any } y \in D. \tag{7.32}$$

First consider the case

$$L = \tfrac{1}{2}\frac{d^2}{dx^2}$$

on $D = (0, l)$. The principal eigenfunction is

$$\phi_0(x) = \sin \frac{\pi}{l} x$$

and thus

$$\lambda_0(D) = \frac{-\pi^2}{2l^2}.$$

From (7.32), it follows that

$$\lim_{t\to\infty} \log P_y(\tau_D > t) = \frac{-\pi^2}{2l^2}, \ y \in D.$$

It then follows by the independence of the one-dimensional Brownian motions which constitute the coordinates of a d-dimensional Brownian motion that in the case $L = \frac{1}{2}\Delta$ and

$$D = (0, l)^d, \ \lim_{t\to\infty} \frac{1}{t} \log P_y(\tau_D > t) = \frac{-d\pi^2}{2l^2}, \ y \in D.$$

One can inscribe in $B_r(x)$ a hypercube whose sides are of length

$$l = \frac{2}{\sqrt{d}} r$$

and one can circumscribe about $B_r(x)$ a hypercube whose sides are of length $l = 2r$. By comparison, it follows that for $x \in R^d$ and $r > 0$,

$$\frac{-d^2\pi^2}{8r^2} \leq \lim_{t\to\infty} \frac{1}{t} \log P_y(\tau_{B_r}(x) > t) \leq \frac{-d\pi^2}{8r^2}, \ y \in B_r(x).$$

From (7.32), it follows that in the case where $L = \frac{1}{2}\Delta$,

$$\frac{-d^2\pi^2}{8r^2} \leq \lambda_0(B_r(x)) \leq \frac{-d\pi^2}{8r^2}, \tag{7.33}$$

for all $r>0$ and $x \in R^d$. This proves the theorem in the case where $L=\frac{1}{2}\Delta$.

Now consider the case where $L=\frac{1}{2}\nabla \cdot a\nabla$. We will write $\lambda_0^{(a)}$ to denote the dependence of λ_0 on a. By (7.3), $\lambda_0^{(a)}(B_r(x)) = -\inf_w \frac{1}{2}\int_{B_r(x)}(\nabla w a \nabla w)(y)\,dy$, where the infimum is taken over all $w \in C^2(\overline{B_r(x)})$ which satisfy $w>0$ in $B_r(x)$, $w=0$ on $\partial B_r(x)$ and $\int_{B_r(x)} w^2(y)\,dy = 1$. Therefore,

$$-\gamma_2\lambda_0^{(I)}(B_r(x)) \leq \lambda_0^{(a)}(B_r(x)) \leq -\gamma_1\lambda_0^{(I)}(B_r(x)), \tag{7.34}$$

where

$$\gamma_1 = \inf_D \inf_{|v|=1} \sum_{i,j=1}^d a_{ij}v_iv_j \text{ and } \gamma_2 = \sup_D \sup_{|v|=1} \sum_{i,j=1}^d a_{ij}v_iv_j,$$

and I denotes the d-dimensional identity matrix. Since $\lambda_0(B_r(x))$ appearing in (7.33) is actually $\lambda_0^{(I)}(B_r(x))$, the theorem now follows from (7.33) and (7.34) in the case where $L=\frac{1}{2}\nabla \cdot a\nabla$.

We now treat the general case $L=\frac{1}{2}\nabla \cdot a\nabla + b\cdot\nabla$. Let $\bar{L}=\frac{1}{2}\nabla\cdot a\nabla$, let $\{P_x,\ x\in\hat{D}\}$ denote, as usual, the solution to the generalized martingale problem for L on D and let $\{Q_x,\ x\in\hat{D}\}$ denote the solution to the generalized martingale problem for $\bar{L}$ on D. Expectations corresponding to P_x will be denoted, as usual, by E_x; expectations corresponding to Q_x will be denoted by E_x^Q. Applying the Cameron–Martin–Girsanov formula (Theorem 1.9.1) with $c=a^{-1}b$, we have for $y\in D$,

$$P_y(\tau_{B_r(x)}>t) = E_y^Q\Bigg(\exp\bigg(\int_0^t \langle (a^{-1}b)(X(s)),\, d\bar{X}(s)\rangle - \frac{1}{2}\int_0^t \langle b, a^{-1}b\rangle(X(s))\,ds\bigg);\ \tau_{B_r(x)}>t\Bigg), \tag{7.35}$$

where $\bar{X}(t)=X(t)-\int_0^t(\nabla\cdot a)(X(s))\,ds$.

From the Cameron–Martin–Girsanov formula (or from the definition of an Îto-process), it follows that

$$E_y^Q\exp\bigg(\theta\int_0^t\langle (a^{-1}b)(X(s)),\, d\bar{X}(s)\rangle - \frac{\theta^2}{2}\int_0^t\langle b, a^{-1}b\rangle(X(s))\,ds\bigg) = 1, \tag{7.36}$$

for all $\theta\in R$.

Thus, applying the Schwarz inequality and using (7.36) with $\theta=2$, we obtain from (7.35),

$$P_y(\tau_{B_r(x)} > t) \leq (Q_y(\tau_{B_r(x)} > t))^{1/2}\left(E_y^Q \exp\left(2\int_0^t \langle (a^{-1}b)(X(s)), \mathrm{d}\bar{X}(s)\rangle - \int_0^t \langle b, a^{-1}b\rangle (X(s))\,\mathrm{d}s\right)\right)^{1/2}$$
$$\leq (Q_y(\tau_{B_r(x)} > t))^{1/2} \exp\left(\tfrac{1}{2}t \sup_{B_r(x)} \langle b, a^{-1}b\rangle\right).$$

Therefore,

$$\lim_{t\to\infty} \frac{1}{t} \log P_y(\tau_{B_r(x)} > t) \leq \tfrac{1}{2} \sup_{B_r(x)} \langle b, a^{-1}b\rangle + \tfrac{1}{2} \lim_{t\to\infty} \log Q_y(\tau_{B_r(x)} > t). \tag{7.37}$$

The upper bound follows from (7.37) by applying (7.32) to Q_y and using the fact that the theorem has already been proved for $\bar{L}$.

For the lower bound, we use the Cameron–Martin–Girsanov formula to obtain, analogous to (7.35),

$$Q_y(\tau_{B_r(x)} > t) = E_y\left(\exp\left(-\int_0^t \langle (a^{-1}b)(X(s)), \mathrm{d}\bar{X}(s)\rangle - \tfrac{1}{2}\int_0^t \langle b, a^{-1}b\rangle (X(s))\,\mathrm{d}s\right); \tau_{B_r(x)} > t\right), \tag{7.38}$$

where $\bar{X}(t) = X(t) - \int_0^t (\nabla \cdot a)(X(s))\,\mathrm{d}s - \int_0^t b(X(s))\,\mathrm{d}s$.

Analogous to (7.36), we have

$$E_y \exp\left(-\theta\int_0^t \langle (a^{-1}b)(X(s)), \mathrm{d}\bar{X}(s)\rangle - \frac{\theta^2}{2}\int_0^t \langle b, a^{-1}b\rangle (X(s))\,\mathrm{d}s\right) = 1, \quad \text{for all } \theta \in R. \tag{7.39}$$

Applying the Schwarz inequality and using (7.39) with $\theta = 2$, we obtain from (7.38),

$$Q_y(\tau_{B_r(x)} > t) \leq (P_y(\tau_{B_r(x)} > t))^{1/2}\left(E_y^Q \exp\left(-2\int_0^t (a^{-1}b)(X(s)), \mathrm{d}\bar{X}(s)\rangle - \int_0^t \langle b, a^{-1}b\rangle (X(s))\,\mathrm{d}s\right)\right)^{1/2}$$
$$\leq (P_y(\tau_{B_r(x)} > t))^{1/2} \exp\left(\tfrac{1}{2}t \sup_{B_r(x)} \langle b, a^{-1}b\rangle\right).$$

Thus

$$\lim_{t\to\infty} \frac{1}{t} \log Q_y(\tau_{B_r(x)} > t) \leq \tfrac{1}{2} \sup_{B_r(x)} \langle b, a^{-1}b\rangle + \tfrac{1}{2} \lim_{t\to\infty} \log P_y(\tau_{B_r(x)} > t). \tag{7.40}$$

The lower bound follows from (7.40) by applying (7.32) to Q_x and using the fact that the theorem has already been proved for $\bar{L}$. □

Exercises

3.1 Prove that

$$\lambda_0^* \equiv \lim_{t\to\infty} \frac{1}{t} \log \|T_t\|$$

exists. (Hint: Define $\lambda_t = \log \|T_t\|$ and show that the subadditivity property holds: $\lambda_{t+s} \leqslant \lambda_t + \lambda_s$, for $s,\ t \geqslant 0$. Fix $T > 0$ and for $t \geqslant T$, write $t = mT + r$, where m is a positive integer and $0 \leqslant r < T$. Use the subadditivity property to show that

$$\limsup_{t\to\infty} \frac{\lambda_t}{t} \leqslant \frac{\lambda_T}{T}.$$

Now prove that

$$\lambda_0^* \equiv \lim_{t\to\infty} \frac{1}{t} \log \|T_t\| = \inf_{t>0} \frac{1}{t} \log \|T_t\|.)$$

3.2 Under the conditions of Theorem 7.1, let $L_1 = \frac{1}{2}\nabla \cdot a\nabla + a\nabla Q \cdot \nabla + V$ and let $L_2 = L_1 + b \cdot \nabla$ in a bounded domain $D \subset R^d$. Denote the corresponding principal eigenvalues by $\lambda_0^{V,1}$ and $\lambda_0^{V,2}$ respectively. If $\nabla \cdot b + 2b \cdot \nabla Q = 0$ in D, and $b \not\equiv 0$, then $\lambda_0^{V,1} > \lambda_0^{V,2}$, for all V. (Hint: From the condition on b, it follows that $\int_D (b\nabla q) \exp(2Q)\,\mathrm{d}x = 0$, for all $q \in C^1(D) \cap C(\bar{D})$ which vanish on ∂D. In particular, then,

$$\tfrac{1}{2}\int_D b\nabla(g^2 \exp(-2Q)) \exp(2Q)\,\mathrm{d}x = \int_D b(\nabla g - g\nabla Q)g\,\mathrm{d}x = 0,$$

if $g \in C^1(\bar{D})$ and $g = 0$ on ∂D.)

3.3 Let T_t be a strongly continuous semigroup on a Banach space $\mathcal{B}_0$ and let $\mathcal{L}$ be its infinitesimal generator with domain $\mathcal{D}$.

(i) Show that T_t maps $\mathcal{D}_0$ into itself.
(ii) Show that $\mathcal{D}_0$ is dense in $\mathcal{B}_0$.

$$\left(\text{Hint: Let } f_\varepsilon(x) = \frac{1}{\varepsilon}\int_0^\varepsilon T_t f(x)\,\mathrm{d}t.\right)$$

(iii) Show that $T_t\mathcal{L}f = \mathcal{L}T_t f$, for $f \in \mathcal{D}_0$.
(iv) Show that $(\mathcal{L}, \mathcal{D}_0)$ is a closed operator on $\mathcal{B}_0$.

$$\text{(Hint: For } f \in \mathcal{D},\ \frac{\mathrm{d}}{\mathrm{d}t}(T_t f) = \mathcal{L}T_t f.$$

Now express

$$\frac{T_t f - f}{t}$$

as an integral.)

3.4 Let L satisfy Assumption $\tilde{\mathrm{H}}_{\mathrm{loc}}$ (see Section 4.0) with $V \equiv 0$ on a domain $D \subseteq R^d$ and let $\{P_x, x \in \hat{D}\}$ denote the solution to the generalized martingale problem for L on D. Let $U \subset\subset D$. Then $\inf_{x \in U} P_x(\tau_D > t) > 0$, for all $t > 0$. (Hint: One may assume without loss of generality that D is bounded and has a $C^{2,\alpha}$-boundary. As in Step 3 of the proof of Theorem 6.1, apply the semigroup T_t to ϕ_0, the positive eigenfunction corresponding to the principal eigenvalue λ_0 of L on D.)

Notes

The concepts introduced in Section 1 can be found in numerous texts on functional analysis and operator theory. A proof of the Riesz–Schauder theorem (Theorem 1.1) may be found in [Yosida (1965)] or in [Brown and Page (1970)]. For Theorem 1.3, see the original article [Krein and Rutman (1948)] or [Krasnosel'skii, Lifshits and Sobolev (1989)].

The treatment of the maximum principle given in Section 2 follows [Gilbarg and Trudinger (1983)]; see also [Protter and Weinberger (1984)]. [Gilbarg and Trudinger (1983)] may also be consulted for the Schauder estimates and for the existence and uniqueness of the solution to the Dirichlet problem. An alternative reference for this is [Miranda (1970)].

I was unable to locate a proof of Theorem 4.1 in the literature, although workers in the field would no doubt call it a 'well-known fact'. With regard to that theorem [Ethier and Kurtz (1986)] show that the closure of the operator L on the domain $\{u \in \mathcal{B}_0;\ u \in C^{2,\alpha}(\bar{D})$ and $Lu = 0$ on $\partial D\}$ is the infinitessimal generator of a semigroup. They note that the semigroup corresponds 'at least intuitively' to the solution of what the present work calls the generalized martingale problem for L on D; however, they do not prove this correspondence.

Theorem 5.5 and Theorem 6.1(i) are frequently grouped together as one theorem and there is a tendency among workers in the field to pass it off as a direct consequence of the Krein–Rutman theorem. This is incorrect; the proof of Theorem 5.5 is an indirect application of the Krein–Rutman theorem, while the proof of Theorem 6.1(i) does not use the Krein–Rutman theorem at all (see the first paragraph of Section 6). In Theorem 5.5, for the strict positivity of the eigenfunction ϕ_0 on D and for the simplicity of the corresponding eigenvalue, one needs a strongly positive operator. As was noted in the text, it does not seem an easy task to prove that $(-L_\gamma^{-1}, \mathcal{B}_\alpha)$ is strongly positive; hence the detour via $(-L_\gamma^{-1}, \mathcal{B}_1)$. We note in passing that $\mathcal{B}_0$ has an empty interior so the strong version of the Krein–Rutman theorem can never be applied directly to an operator on $\mathcal{B}_0$.

Theorem 6.1(i) goes back to [Protter and Weinberger (1966)], who gave a proof using maximum principle techniques. Modulo small technical differences, Theorem 6.1 and the lefthand equality in Theorem 7.1 appear as one theorem in [Donsker and Varadhan (1976)]. One key step of their proof uses the Sion mini-max principle. The proof given in the text is completely different; it is a 'bare hands' approach and avoids the Sion mini-max principle. The righthand equality in Theorem 7.1 is also due to [Donsker and Varadhan (1976)]; the proof in the text essentially follows their proof. The expression $g(x) \equiv E_x \exp(\int_0^{\tau_D} V(X(s))\,ds)$ appearing in Theorem 6.6 is called the *gauge* of L on D. Under appropriate conditions on L and D, it can be shown that $g(x) < \infty$ for some $x \in D$ if and only if $\sup_{x \in D} g(x) < \infty$ if and only if there exists a function u satisfying $Lu = 0$ and $0 < c_1 \leqslant u \leqslant c_2 < \infty$ in D. See, for example, [Chung and Rao (1981)], [Cranston, Fabes and Zhao (1988)], [Falkner (1983)], [Zhao (1983)] and [Zhao 1986]. The proof of Theorem 7.2 uses the Cameron–Martin–Girsanov transformation. A purely analytic proof may be found in [Berestycki, Nirenberg, and Varadhan (1994), Lemma 1.1 and Theorem 2.5]. Actually, Theorem 2.5 in that paper gives lower bounds for general domains.

4
Generalized spectral theory for elliptic operators on arbitrary domains

4.0 Introduction

In Chapter 3, we developed the spectral theory of elliptic operators L on smooth bounded domains D, concentrating in particular on the principal eigenvalue λ_0. This theory hinged on the fact that L possessed a compact resolvent, since only then could the Fredholm theory be applied. In the case of an arbitrary domain D, L does not possess a compact resolvent in general, and the above spectral theory breaks down. Indeed, it is no longer even clear on what space to define L or on what space to define the corresponding semigroup T_t. In this chapter, we develop a generalized spectral theory for elliptic operators L on arbitrary domains. Specifically, we will extend the definition of the principal eigenvalue λ_0 via the existence or non-existence of positive harmonic functions for $L - \lambda$ in D, that is, functions u satisfying $(L - \lambda)u = 0$ and $u > 0$ in D.

In this chapter, we will assume that L satisfies Assumption $\tilde{\text{H}}$ locally:

Assumption $\tilde{H}_{loc}$. $L = \frac{1}{2}\nabla \cdot a\nabla + b \cdot \nabla + V$ *is defined on a domain* $D \subseteq R^d$ *and satisfies Assumption* $\tilde{H}$ *(defined in Chapter 3, Section 7) on every subdomain* $D' \subset\subset D$.

(In Exercise 4.16, the reader is asked to check that all the theorems in this chapter which involve only L and not $\tilde{L}$ hold if L satisfies Assumption H (defined in Chapter 3, Section 2) locally.)

An indispensable tool in this chapter is *Harnack's inequality*, which we state here for operators in non-divergence form since the formulation is simpler in this case. See the notes at the end of the chapter for references.

Theorem 0.1 (Harnack's inequality. *Let*

$$L = \tfrac{1}{2}\sum_{i,j=1}^{d} a_{ij}\frac{\partial^2}{\partial x_i \partial x_j} + \sum_{i=1}^{d} b_i \frac{\partial}{\partial x_i} + V(x)$$

in a domain $D \subseteq R^d$. Assume that there exist positive constants λ, Λ and γ such that

(i) $\lambda|v|^2 \leqslant \sum_{i,j=1}^{d} a_{ij}(x)v_i v_j \leqslant \Lambda|v|^2$, *for all $v \in R^d$ and $x \in D$;*

(ii) $\dfrac{|b(x)|}{\lambda}, \dfrac{|V(x)|}{\lambda}, \dfrac{\Lambda}{\lambda} \leqslant \gamma$, *for all $x \in D$.*

Let $u \in C^2(D)$ satisfy $Lu = 0$ and $u > 0$ in D. Then for any subdomain $D' \subset\subset D$, there exists a constant $c = c(\gamma, d, D', D)$ such that

$$\sup_{x \in D'} u(x) \leqslant c \inf_{x \in D'} u(x).$$

Throughout this chapter, $\{P_x, x \in \hat{D}\}$ will denote the solution to the generalized martingale problem for $L_0 = L - V$ on D. For $t \geqslant 0$ and $x \in D$, define the *transition measure* $p(t, x, dy)$ for the operator L on D by

$$p(t, x, B) = E_x\left(\exp\left(\int_0^t V(X(s))\,ds\right); X(t) \in B\right),$$

for measurable $B \subseteq D$.

The transition measure for $\tilde{L}$ on D will be denoted by $\tilde{p}(t, x, dy)$.

Remark 1. It is important to recall that, by definition of the generalized martingale problem for L_0 on D, $P_x(X(t) = \Delta \text{ for all } t \geqslant \tau_D) = 1$, for all $x \in \hat{D}$. Thus, for example, if $V \equiv 0$, then $p(t, x, D) = P_x(\tau_D > t)$. Also, note that for a function $g \in C_0(D)$, we have

$$\int_D p(t, x, dy)g(y) = E_x\left(\exp\left(\int_0^t V(X(s))\,ds\right)g(X(t)); \tau_D > t\right).$$

However, if we extend the definition of V and g to $\hat{D}$ by defining $g(\Delta) = 0$ and defining $V(\Delta)$ arbitrarily, then we can write

$$\int_D p(t, x, dy)g(y) = E_x \exp\left(\int_0^t V(X(s))\,ds\right)g(X(t)).$$

Remark 2. If V is unbounded from above, then it is possible to have $p(t, x, B) = \infty$ for $B \subset\subset D$, in which case the transition measure does not exist. In the case where $L = \frac{1}{2}\Delta$ and $D = R^d$, sufficient conditions for the existence of the transition measure may be given in terms of Kato class functions [Simon (1982), section B]. As will become clear in the sequel, our primary interest will be in the case that $\int_0^\infty \exp(-\lambda t)p(t,x,B)\,dt < \infty$ for $B \subset\subset D$ and sufficiently large λ. The theorems in Section 5 give sufficient conditions and necessary conditions for this to occur.

Remark 3. In Section 1.13, the existence of a density $p(t,x,y)$ for $p(t,x,dy)$ was stated in the case where $V \equiv 0$ and, assuming the existence in the case where $V \equiv 0$, and assuming that $p(t,x,D) < \infty$, the existence in the general case is trivial. The proof of the existence of $p(t,x,y)$ is quite easy when $L = \frac{1}{2}\Delta + b \cdot \nabla$; see exercise 4.1; however, to extend to the case of a general diffusion matrix is not easy. (In parabolic PDE theory, the parametrix method is usually employed.) It seems reasonable to try to develop the theory of elliptic operators without having to resort to techniques from parabolic PDE theory. Therefore, with one exception, in this chapter we will work with $p(t,x,dy)$ without assuming the existence of a density. In fact, this will cause us to work harder only at one point in the exposition – in Theorem 2.5 in the proof of the existence of a density for $\int_0^\infty p(t,x,dy)\,dt$. The exception alluded to above is Theorem 9.9; the reason for this exception is that this result is a result from parabolic PDEs and there seems to be no way of avoiding using the density $p(t,x,y)$. On the other hand, the rest of the 'parabolic' results in Sections 8 and 9 are proved without relying on the existence of a density for $p(t,x,dy)$.

4.1 The *h*-transform

An operator of the form $L = L_0 + V$ is a much more tame object in the case $V \le 0$ than in the alternative case. The analyst's laconic explanation for this is that 'the maximum principle automatically holds'; the probabilist's is that 'integrability holds automatically in the Feynman–Kac formula'. These properties which hold 'automatically' in the case $V \le 0$ may in fact also hold in the alternative case. Indeed, in Theorem 3.2.2, a generalized maximum principle was given for certain operators which do not satisfy the condition $V \le 0$. In the proof of that theorem, the original operator was compared to a related operator for which the condition $V \le 0$ does hold. This simple technique is amazingly useful and will be employed throughout the rest of this book, particularly in this chapter and in Chapter 7. We define this technique formally.

Definition. *Let $h \in C^{2,\alpha}(D)$ satisfy $h > 0$ in D. The operator L^h defined by*

$$L^h f = \frac{1}{h} L(hf)$$

is called the h-transform of the operator L. Written out explicitly, one has

$$L^h = L_0 + a\frac{\nabla h}{h} \cdot \nabla + \frac{Lh}{h}.$$

Remark. Note that if L satisfies Assumption $\tilde{\text{H}}_{\text{loc}}$, then so does L^h.

From the point of view of probability theory, the h-transform is very important because it can be used to define conditioned diffusion processes; this theme is taken up in Chapter 7, Section 2. In the present chapter only the analytic properties of the h-transform will be used.

The virtue of the h-transform lies in its simplicity. As we now show, this simplicity is manifested in the behavior of the transition measure under such a transform. We introduce the following notation. Noting that the zeroth-order part of the operator

$$L^h - \frac{Lh}{h}$$

vanishes, let $\{P_x^h, x \in \hat{D}\}$ denote the solution to the generalized martingale problem for

$$L^h - \frac{Lh}{h} \text{ on } D,$$

and denote the expectation corresponding to P_x^h by E_x^h. Also, let $p^h(t, x, dy)$ denote the transition measure for L^h on D.

Theorem 1.1. *Let L satisfy Assumption $\tilde{\text{H}}_{\text{loc}}$ on a domain $D \subseteq R^d$ and let $h \in C^{2,\alpha}(D)$ satisfy $h > 0$ in D. Assume that the transition measure $p(t, x, dy)$ exists for L on D. Then $p^h(t, x, y)$, the transition measure for L^h on D, satisfies*

$$p^h(t, x, dy) = \frac{1}{h(x)} p(t, x, dy) h(y).$$

Proof. By definition, we have for $f \in C_0^2(D)$,

$$E_x\left(\exp\left(\int_0^t V(X(s))\,ds\right)f(X(t));\ \tau_D > t\right) = \int_D p(t,x,dy)f(y)\,dy$$

and

$$E_x^h\left(\exp\left(\int_0^t \frac{Lh}{h}(X(s))\,ds\right)f(X(t));\ \tau_D > t\right) = \int_D p^h(t,x,dy)f(y)\,dy.$$

Thus, to prove the theorem, it is enough to show that for all $f \in C_0(D)$,

$$\frac{1}{h(x)}E_x\left(\exp\left(\int_0^t V(X(s))\,ds\right)(hf)(X(t));\ \tau_D > t\right)$$
$$= E_x^h\left(\exp\left(\int_0^t \frac{Lh}{h}(X(s))\,ds\right)f(X(t));\ \tau_D > t\right). \quad (1.1)$$

Let $\{D_n\}_{n=1}^{\infty}$ be a sequence of bounded domains satisfying $D_n \subset\subset D_{n+1}$ and $\bigcup_{n=1}^{\infty} D_n = D$. Clearly, it is enough to show that (1.1) holds with τ_D replaced by τ_{D_n} and $f \in C_0(D_n)$.

Recalling that L is in divergence form, let $\hat{b} \equiv b + \nabla \cdot a$ denote the coefficient of the first-order term when L is written out in non-divergence form. Let

$$N(t) = \exp\left(\int_0^t\left\langle\frac{\nabla h}{h}(X(s)), d\bar{X}(s)\right\rangle - \tfrac{1}{2}\int_0^t\left\langle\frac{\nabla h}{h}(X(s)), a\frac{\nabla h}{h}(X(s))\right\rangle ds\right),$$

where $\bar{X}(t) = X(t) - \int_0^t \hat{b}(X(s))\,ds$. An application of the Cameron–Martin–Girsonov formula (Theorem 1.9.1) gives

$$\left.\frac{dP_x^h}{dP_x}\right|_{\mathcal{F}^D_{\tau_{D_n}\wedge t}} = N(\tau_{D_n} \wedge t) \text{ (see exercise 4.2).} \quad (1.2)$$

Since

$$\exp\left(\int_0^t \frac{Lh}{h}(X(s))\,ds\right)f(X(t))1_{\{\tau_{D_n}>t\}}$$

is $\mathcal{F}^D_{\tau_{D_n}\wedge t}$-measurable, we have

$$E_x^h\left(\exp\left(\int_0^t \frac{Lh}{h}(X(s))\,ds\right)f(X(t));\ \tau_{D_n} > t\right)$$
$$= E_x\left(N(\tau_{D_n}\wedge t)\exp\left(\int_0^t \frac{Lh}{h}(X(s))\,ds\right)f(X(t));\ \tau_{D_n} > t\right)$$
$$= E_x\left(N(\tau_{D_n}\wedge t)\exp\left(\int_0^{\tau_{D_n}\wedge t} \frac{Lh}{h}(X(s))\,ds\right)f(X(\tau_{D_n}\wedge t));\ \tau_{D_n} > t\right).$$
$$(1.3)$$

We can now use Ito's formula (Theorem 1.6.3). In the notation of that theorem we choose $f = \log h$ and replace t by $\tau_{D_n} \wedge t$ (see exercise 4.3). We obtain

$$\log h(X(\tau_{D_n} \wedge t)) = \log h(x) + \int_0^{\tau_{D_n}\wedge t}\left\langle \frac{\nabla h}{h}(X(s)), \mathrm{d}\bar{X}(s)\right\rangle + \int_0^{\tau_{D_n}\wedge t}(L_0(\log h))(X(s))\,\mathrm{d}s$$

$$= \log h(x) + \int_0^{\tau_{D_n}\wedge t}\left\langle \frac{\nabla h}{h}(X(s)), \mathrm{d}\bar{X}(s)\right\rangle + \int_0^{\tau_{D_n}\wedge t}\frac{L_0 h}{h}(X(s))\,\mathrm{d}s - \frac{1}{2}\int_0^{\tau_{D_n}\wedge t}\left\langle \frac{\nabla h}{h}(X(s)), a\frac{\nabla h}{h}(X(s))\right\rangle \mathrm{d}s, \text{ a.s. } [P_x]. \tag{1.4}$$

Substituting for

$$\int_0^{D_n}\left\langle \frac{\nabla h}{h}(X(s)), \mathrm{d}\bar{X}(s)\right\rangle$$

via (1.4), we have

$$N(\tau_{D_n} \wedge t) = \frac{h(X(\tau_{D_n} \wedge t))}{h(x)} \exp\left(-\int_0^{\tau_{D_n}\wedge t}\frac{L_0 h}{h}(X(s))\,\mathrm{d}s\right).$$

Using this in (1.3) gives

$$E_x^h\left(\exp\left(\int_0^t \frac{Lh}{h}(X(s))\,\mathrm{d}s\right)f(X(t));\ \tau_{D_n} > t\right)$$

$$= \frac{1}{h(x)}E_x\left(\exp\left(\int_0^{\tau_{D_n}\wedge t} V(X(s))\,\mathrm{d}s\right)(hf)(X(\tau_{D_n} \wedge t));\ \tau_{D_n} > t\right)$$

$$= \frac{1}{h(x)}E_x\left(\exp\left(\int_0^t V(X(s))\,\mathrm{d}s\right)(hf)(X(t));\ \tau_{D_n} > t\right),$$

which is (1.1) with τ_D replaced by τ_{D_n}. This completes the proof of the theorem. □

It is sometimes convenient to have the following extension of Theorem 1.1.

Corollary 1.2. *Let L satisfy Assumption $\tilde{\mathrm{H}}_{\mathrm{loc}}$ on a domain $D \subseteq R^d$ and let $h \in C^{2,\alpha}(D)$ satisfy $h > 0$ in D. Then for any $B \subset \hat{\mathcal{F}}_t^D$,*

$$E_x^h\left(\exp\left(\int_0^t \frac{Lh}{h}(X(s))\,ds\right);\, B,\, \tau_D > t\right) = \frac{1}{h(x)} E_x\left(\exp\left(\int_0^t V(X(s))\,ds\right) h(X(t));\, B,\, \tau_D > t\right).$$

Remark. Note that in the special case $B \in \sigma(X(t))$, Corollary 1.2 reduces to Theorem 1.1.

Proof. By standard measure-theoretic considerations, it is enough to prove the proposition in the case $B = \{X(s_1) \in A_1, \ldots, X(s_n) \in A_n\}$, where $A_i \in \mathcal{B}(D)$ and $0 \leq s_1 < s_2 < \cdots < s_n = t$. Using the Markov property, the definition of $p(t, x, dy)$ and Theorem 1.1, we have

$$E_x^h\left(\exp\left(\int_0^t \frac{Lh}{h}(X(s))\,ds\right);\, B,\, \tau_D > t\right)$$

$$= \int_{B_n} \cdots \int_{B_1} p^h(s_1, x, dy_1) p^h(s_2 - s_1, y_1, dy_2) \cdots p^h(t - s_{n-1}, y_{n-1}, dy_n)$$

$$= \int_{B_n} \cdots \int_{B_1} \frac{1}{h(x)} p(s_1, x, dy_1) \cdots p(t - s_{n-1}, y_{n-1}, dy_n) h(y_n)$$

$$= \frac{1}{h(x)} E_x\left(\exp\left(\int_0^t V(X(s))\right) h(X(t));\, B,\, \tau_D > t\right). \qquad \square$$

4.2 The Green's measure and the Green's function and criteria for their existence

Note that

$$\int_0^\infty p(t, x, B)\,dt = E_x \int_0^\infty \exp\left(\int_0^t V(X(s))\,ds\right) 1_B(X(t))\,dt$$

$$= E_x \int_0^{\tau_D} \exp\left(\int_0^t V(X(s))\,ds\right) 1_B(X(t))\,dt, \text{ for } B \subseteq \mathcal{B}(D).$$

We give the following definition.

Definition. *If*

$$\int_0^\infty p(t, x, B)\,dt = E_x \int_0^\infty \exp\left(\int_0^t V(X(s))\,ds\right) 1_B(X(t))\,dt < \infty,$$

for all $x \in D$ and all $B \subset\subset D$, then

$$G(x, dy) \equiv \int_0^\infty p(t, x, dy)\, dt = E_x \int_0^\infty \exp\left(\int_0^t V(X(s))\, ds\right) 1_{(dy)}(X(t))\, dt$$

is called the Green's measure *for L on D. If the above condition fails, then the Green's measure for L on D is said not to exist. The Green's measure for $\tilde{L}$ on D will be denoted by $\tilde{G}(x, dy)$.*

Remark 1. In Theorem 3.6, it is shown that if the Green's measure for L on D does not exist, then $\int_0^\infty p(t, x, B)\, dt = \infty$, for all $x \in D$ and all open $B \subseteq D$.

Remark 2. If D is bounded with a smooth boundary and the coefficients of L are smooth up to the boundary, then the Green's measure may be interpreted as the kernel of an inverse operator as follows. Let $G_\lambda(x, dy)$ denote the Green's measure for $L - \lambda$ on D, if it exists. Using the definition of the Green's measure, Theorem 3.6.4 may be reformulated as follows: 'Let L satisfy Assumption H on a bounded domain $D \subset R^d$ with a $C^{2,\alpha}$-boundary. Then, for any $\lambda > \lambda_0$, $G_\lambda(x, dy)$ exists and, for any $f \in \mathcal{B}_\alpha$, the function $u(x) = \int_D f(y) G(x, dy)$ belongs to $\mathcal{D}_\alpha$ and is the unique solution of $(L - \lambda)u = -f$ in D and $u = 0$ on ∂D.' Thus, $G_\lambda(x, dy)$ is the kernel of the inverse operator $(\lambda - L)^{-1}$ on $\mathcal{B}_\alpha$. In the general case, the Green's measure may be thought of as a kind of generalized inverse kernel; see Theorem 3.8 and the remark which follows it.

We begin with the following fundamental theorem.

Theorem 2.1. *Let L satisfy Assumption $\tilde{\mathrm{H}}_{\mathrm{loc}}$ on a domain $D \subseteq R^d$ and let $L_0 = L - V$.*

- (i) *If $V \equiv 0$ and the diffusion process on D corresponding to L_0 is recurrent, then L does not possess a Green's measure on D.*
- (ii) *If $V \equiv 0$ and the diffusion process on D corresponding to L_0 is transient, then L possesses a Green's measure on D.*
- (iii) *If $V \le 0$ and $V \not\equiv 0$, then L possesses a Green's measure on D.*

Proof. (i) We will show that $\int_0^\infty p(t, x, B)\, dt = \infty$ for every $x \in D$ and every ball $B \subset\subset D$. Fix $x \in D$ and a ball $B \subset\subset D$ and let B_1 be a ball satisfying $B_1 \subset\subset B$. Define $\tau_1 = \inf\{t \ge 0: X(t) \notin B_1\}$ and, by induction, define $\sigma_n = \inf\{t > \tau_n: X(t) \in \partial B\}$ and $\tau_{n+1} = \inf\{t > \sigma_n: X(t) \in \partial B_1\}$, $n = 1, 2, \ldots$. By the recurrence assumption, $P_x(\tau_n < \infty) = P_x(\sigma_n < \infty) = 1$, for all n. By Theorems 2.1.2 and 3.3.1, the function

$v(x) \equiv E_x \sigma_1$, $x \in B$, satisfies $L_0 v = -1$ in B and $v = 0$ on ∂B. Thus, by the strong maximum principle (Theorem 3.2.6), $\gamma \equiv \inf_{x \in \partial B_1} v(x) > 0$.

Using the strong Markov property, we have for $n \geq 2$,

$$E_x(\sigma_n - \tau_n) = E_x E_x(\sigma_n - \tau_n | \mathcal{F}_{\tau_n}) = E_x E_{X(\tau_n)} \sigma_1 \geq \gamma.$$

Thus,

$$\int_0^\infty p(t, x, B)\, dt = E_x \int_0^\infty 1_B(X(t))\, dt \geq E_x \sum_{n=2}^\infty (\sigma_n - \tau_n) \geq \sum_{n=2}^\infty \gamma = \infty.$$

(ii) Clearly, it is enough to show that $\int_0^\infty p(t, x, B)\, dt < \infty$ for each $x \in D$ and each ball $B \subset\subset D$. Let B_1 and B_2 be balls satisfying $B \subset\subset B_1 \subset\subset B_2 \subset\subset D$. Define $\sigma_1 = \inf\{t \geq 0: X(t) \in \bar{B}_1\}$ and, by induction, define $\tau_n = \inf\{t > \sigma_n: X(t) \in \partial B_2\}$ and $\sigma_{n+1} = \inf\{t > \tau_n: X(t) \in \partial B_1\}$, $n = 1, 2, \ldots$. By Theorem 2.1.1,

$$\gamma \equiv \sup_{z \in \bar{B}_1} E_z \tau_1 < \infty. \tag{2.1}$$

Since the diffusion process is assumed to be transient, it follows from Theorem 2.8.2 that $P_z(\sigma_1 < \infty) < 1$, for all $z \in \partial B_2$. In fact, then,

$$\rho \equiv \sup_{z \in \partial B_2} P_z(\sigma_1 < \infty) < 1. \tag{2.2}$$

The proof of this is left to the reader as exercise 4.5.

Using the convention $\tau_n - \sigma_n = 0$ if $\tau_n = \sigma_n = \infty$, we have

$$\int_0^\infty p(t, x, B)\, dt = E_x \int_0^\infty 1_B(X(t))\, dt \leq E_x \sum_{n=1}^\infty (\tau_n - \sigma_n). \tag{2.3}$$

By the strong Markov property and (2.1),

$$\begin{aligned} E_x(\tau_n - \sigma_n) &= E_x E_x(\tau_n - \sigma_n | \sigma_n < \infty) \\ &= E_x 1_{\{\sigma_n < \infty\}} E_{X(\sigma_n)} \tau_1 \leq \gamma P_x(\sigma_n < \infty). \end{aligned} \tag{2.4}$$

By the strong Markov property and Theorem 2.1.1, $P_x(\tau_n < \infty) = P_x(\sigma_n < \infty)$. Thus, another application of the strong Markov property along with (2.2) gives

$$P_x(\sigma_{n+1} < \infty) \leq \rho P_x(\tau_n < \infty) = \rho P_x(\sigma_n < \infty).$$

By induction,

$$P_x(\sigma_n < \infty) \leq \rho^{n-1}, \text{ for } x \in D. \tag{2.5}$$

From (2.3), (2.4) and (2.5), we conclude that

$$\int_0^\infty p(t, x, B)\,dt \leqslant \sum_{n=1}^{\infty} \gamma\rho^{n-1} < \infty.$$

(iii) Since for any $B \in \mathcal{B}(D)$, $\int_0^\infty p(t, x, B)\,dt$ is monotone non-decreasing in its dependence on V, it suffices to consider the case $V = -c1_A$, where $c > 0$ and A is a domain satisfying $A \subset\subset D$. If L_0 corresponds to a transient diffusion, then the existence of the Green's measure follows from part (ii) and the above-mentioned monotonicity. Thus we may assume that L_0 corresponds to a recurrent diffusion.

As in part (ii), it is enough to show that $\int_0^\infty p(t, x, B)\,dt < \infty$ for each $x \in D$ and each ball $B \subset\subset D$. Fix $x \in D$. Let D_1 and D_2 be domains satisfying $B \cup A \subset\subset D_1 \subset\subset D_2 \subset\subset D$. Define $\sigma_1 = \inf\{t \geqslant 0: X(t) \in \bar{D}_1\}$ and, by induction, define $\tau_n = \inf\{t > \sigma_n: X(t) \in \partial D_2\}$ and $\sigma_{n+1} = \inf\{t > \tau_n: X(t) \in \partial D_1\}$, $n = 1, 2, \ldots$. By the Stroock–Varadhan support theorem (Theorem 2.6.1), $E_z \exp(-c\int_0^{\tau_1} 1_A(X(s))\,ds) < 1$, for all $z \in \partial D_1$. In fact, then,

$$\rho \equiv \sup_{z \in \partial D_1} E_z \exp\left(-c\int_0^{\tau_1} 1_A(X(s))\,ds\right) < 1. \tag{2.6}$$

The proof of this is left as exercise 4.6.

Using the strong Markov property and (2.6), we have for $n \geqslant 1$,

$$\begin{aligned} E_z \exp\left(-c\int_0^{\sigma_{n+1}} 1_A(X(t))\,dt\right) &= E_z \exp\left(-c\int_0^{\tau_n} 1_A(X(t))\,dt\right) \\ &= E_z \exp\left(-c\int_0^{\sigma_n} 1_A(X(t))\,dt\right) E_{X(\sigma_n)} \exp\left(-c\int_0^{\tau_1} 1_A(X(t))\,dt\right) \\ &\leqslant \rho E_z \exp\left(-c\int_0^{\sigma_n} 1_A(X(t))\,dt\right). \end{aligned}$$

Since $E_x \exp(-c\int_0^{\sigma_1} 1_A(X(t))\,dt) \leqslant 1$, we obtain by induction that

$$E_x \exp\left(-c\int_0^{\sigma_n} 1_A(X(t))\,dt\right) \leqslant \rho^{n-1},\ n \geqslant 1. \tag{2.7}$$

Also, by Theorem 2.1.1,

$$\gamma \equiv \sup_{z \in \partial D_1} E_z \int_0^{\tau_1} 1_B(X(t))\,dt < \infty. \tag{2.8}$$

Now, using the strong Markov property along with (2.7) and (2.8), we have

$$\int_0^\infty p(t, x, B)\,\mathrm{d}t = E_x\int_0^\infty \exp\left(-c\int_0^t 1_A(X(s))\,\mathrm{d}s\right)1_B(X(t))\,\mathrm{d}t$$

$$= E_x\sum_{n=1}^\infty \int_{\sigma_n}^{\tau_n} \exp\left(-c\int_0^t 1_A(X(s))\,\mathrm{d}s\right)1_B(X(t))\,\mathrm{d}t$$

$$= E_x\sum_{n=1}^\infty \exp\left(-c\int_0^{\sigma_n} 1_A(X(s))\,\mathrm{d}s\right)\int_{\sigma_n}^{\tau_n} \exp\left(-c\int_{\sigma_n}^t 1_A(X(s))\,\mathrm{d}s\right)1_B(X(t))\,\mathrm{d}t$$

$$\leqslant E_x\sum_{n=1}^\infty \exp\left(-c\int_0^{\sigma_n} 1_A(X(s))\,\mathrm{d}s\right)\int_{\sigma_n}^{\tau_n} 1_B(X(t))\,\mathrm{d}t$$

$$= \sum_{n=1}^\infty E_x \exp\left(-c\int_0^{\sigma_n} 1_A(X(s))\,\mathrm{d}s\right)E_{X(\sigma_n)}\int_0^{\tau_1} 1_B(X(t))\,\mathrm{d}t$$

$$\leqslant \sum_{n=1}^\infty \gamma\rho^{n-1} < \infty.$$ □

If $h \in C^{2,\alpha}(D)$ and $h > 0$ on D, then, if it exists, denote the Green's measure for L^h on D, the h-transform of L on D, by $G^h(x, \mathrm{d}y)$.

Proposition 2.2. *Let L satisfy Assumption $\widetilde{\mathrm{H}}_{\mathrm{loc}}$ on a domain $D \subseteq R^d$ and let $h \in C^{2,\alpha}(D)$ satisfy $h > 0$ on D. Then the Green's measure $G(x, \mathrm{d}y)$ exists for L on D if and only if the Green's measure $G^h(x, \mathrm{d}y)$ exists for L^h on D. If these measures exist, then*

$$G^h(x, \mathrm{d}y) = \frac{1}{h(x)}G(x, \mathrm{d}y)h(y).$$

Proof. The proof is immediate from Theorem 1.1. □

Proposition 2.3. *Let L satisfy Assumption $\widetilde{\mathrm{H}}_{\mathrm{loc}}$ on a domain $D \subseteq R^d$. If there exists a function $h \in C^{2,\alpha}(D)$ such that $h > 0$, $Lh \leqslant 0$ and $Lh \not\equiv 0$, then the Green's measure exists for L on D.*

Proof. Consider L^h the h-transform of L. Then

$$\frac{Lh}{h},$$

the zeroth-order part of L^h, is non-positive and not identically zero. Therefore, by Theorem 2.1(iii), L^h possesses a Green's measure and, by Proposition 2.2, L also possesses one. □

Remark. The converse of Proposition 2.3 also holds; see Theorem 3.9.

In the case that L satisfies Assumption $\tilde{\text{H}}$ on D and D is bounded with a $C^{2,\alpha}$-boundary, we have the following theorem which will serve as a building block for the general case.

Theorem 2.4. *Let L satisfy Assumption $\tilde{H}$ on D, where $D \subset R^d$ is a bounded domain with a $C^{2,\alpha}$-boundary. Let $\lambda_0 = \sup \operatorname{Re}(\sigma((L, \mathcal{D}_\alpha)))$ denote the principal eigenvalue as in Theorem* 3.6.1. *Then the Green's measure for $L - \lambda$ on D exists if and only if $\lambda > \lambda_0$.*

Remark. Theorem 7.1 generalizes Theorem 2.4 to the case where D is an arbitrary bounded domain and L is uniformly elliptic with bounded coefficients on D. The proof of Theorem 7.1 requires quite a bit of work. It uses the perturbation theory developed in Section 6 and also relies on a key estimate on solutions to elliptic equations. Even under the smoothness conditions of Theorem 2.4, we were unable to produce an analytic proof which relied only on the aspects of the theory developed up to here; thus the proof below is probabilistic. An analytic proof can be given using Theorem 2.8 and the Hopf maximum principle, and assuming that the Green's function, if it exists, enjoys the properties enumerated in Theorem 2.5 (see exercise 4.19).

Proof. Let $\phi_0 > 0$ denote the eigenfunction corresponding to λ_0 as in Theorem 3.5.5. First, assume that $\lambda > \lambda_0$. Then $(L - \lambda)\phi_0 < 0$ and the theorem follows from Proposition 2.3. (Alternatively, the existence of the Green's measure is an immediate consequence of Theorem 3.6.1(ii).)

Now assume that $\lambda = \lambda_0$. Consider the h-transform of $L - \lambda_0$ via the function ϕ_0. We obtain

$$(L - \lambda_0)^{\phi_0} = L_0 + a\frac{\nabla \phi_0}{\varphi_0} \cdot \nabla.$$

We will show that the diffusion process corresponding to $(L - \lambda_0)^{\phi_0}$ is recurrent on D. By Theorem 2.1 and Proposition 2.2, this will prove that $L - \lambda_0$ does not possess a Green's measure on D.

Let $\{P_x^{\phi_0,\lambda_0}, x \in \hat{D}\}$ denote the solution to the generalized martingale problem for $(L - \lambda_0)^{\phi_0}$ on D and let $X(t)$ denote the canonical diffusion process. Let $\{D_n\}_{n=1}^{\infty}$ be a sequence of domains satisfying $D_n \subset\subset D_{n+1}$ and $\bigcup_{n=1}^{\infty} D_n = D$. We will apply the martingale formulation to the functions ϕ_0^{-1} and ϕ_0.

The function ϕ_0^{-1} satisfies

$$(L-\lambda_0)^{\phi_0}\phi_0^{-1} = \frac{1}{\varphi_0}(L-\lambda_0)(\phi_0\phi_0^{-1}) = (V-\lambda_0)\phi_0^{-1} \leq M\phi_0^{-1},$$

where $M = \sup(V-\lambda_0)$. Thus, we have for $x \in D$,

$$\begin{aligned}
E_x^{\phi_0,\lambda_0}\phi_0^{-1}(X(t\wedge\tau_{D_n})) &= \phi_0^{-1}(x) + E_x^{\phi_0,\lambda_0}\int_0^{t\wedge\tau_{D_n}}(L-\lambda_0)^{\phi_0}(\phi_0^{-1})(X(s))\,ds \\
&\leq \phi_0^{-1}(x) + ME_x^{\phi_0,\lambda_0}\int_0^{t\wedge D_n}\phi_0^{-1}(X(s))\,ds \\
&= \phi_0^{-1}(x) + ME_x^{\phi_0,\lambda_0}\int_0^{t\wedge\tau_{D_n}}\phi_0^{-1}(X(s\wedge\tau_{D_n}))\,ds \\
&\leq \phi_0^{-1}(x) + ME_x^{\phi_0,\lambda_0}\int_0^{t}\phi_0^{-1}(X(s\wedge\tau_{D_n}))\,ds \\
&= \phi_0^{-1}(x) + M\int_0^{t}E_x^{\phi_0,\lambda_0}\phi_0^{-1}(X(s\wedge\tau_{D_n}))\,ds.
\end{aligned}$$

By Gronwall's inequality, it then follows that

$$E_x^{\phi_0,\lambda_0}\phi_0^{-1}(X(t\wedge\tau_{D_n})) \leq \phi_0^{-1}(x)\exp(Mt);$$

letting $n\to\infty$ and using Fatou's lemma, we have

$$E_x^{\phi_0,\lambda_0}\phi_0^{-1}(X(t\wedge\tau_{D})) \leq \phi_0^{-1}(x)\exp(Mt).$$

Since $\phi_0 = 0$ on ∂D, we conclude from this that

$$P_x^{\phi_0,\lambda_0}(\tau_D<\infty) = 0, \quad \forall x \in D. \tag{2.9}$$

We now apply the martingale formulation to ϕ_0. We have

$$\begin{aligned}
(L-\lambda_0)^{\phi_0}(\phi_0) = \frac{1}{\varphi_0}(L-\lambda_0)(\phi_0^2) &= 2L_0\phi_0 + (V-\lambda_0)\phi_0 + \frac{\nabla\phi_0 a\nabla\phi_0}{\varphi_0} \\
&= (\lambda_0 - V)\phi_0 + \frac{\nabla\phi_0 a\nabla\phi_0}{\varphi_0}.
\end{aligned}$$

Thus,

$$\begin{aligned}
E_x^{\phi_0,\lambda_0}\phi_0(X(t\wedge\tau_{D_n})) = \phi_0(x) &+ E_x^{\phi_0,\lambda_0}\int_0^{t\wedge\tau_{D_n}}((\lambda_0-V)\phi_0)(X(s))\,ds \\
&+ E_x^{\phi_0,\lambda_0}\int_0^{t\wedge\tau_{D_n}}\frac{\nabla\phi_0 a\nabla\phi_0}{\phi_0}(X(s))\,ds.
\end{aligned}$$

Letting $n\to\infty$ and using (2.9), the boundedness of ϕ_0, the bounded convergence theorem and the monotone convergence theorem gives

$$E_x^{\phi_0,\lambda_0}\int_0^t \frac{\nabla\phi_0 a\nabla\phi_0}{\phi_0}(X(s))\,ds = E^{\phi_0,\lambda_0}\phi_0(X(t)) - \phi_0(x)$$
$$+ E_x^{\phi_0,\lambda_0}\int_0^t ((V-\lambda_0)\phi_0)(X(s))\,ds. \quad (2.10)$$

To complete the proof of the theorem, we will assume that $X(t)$ is transient and come to a contradiction. By (2.9), it follows that $\tau_D = \infty$ a.s. $P_x^{\phi_0,\lambda_0}$. Thus, by the boundedness of D and by the definition of transience on D,

$$\lim_{t\to\infty} \operatorname{dist}(X(t), \partial D) = 0 \text{ a.s. } P_x^{\phi_0,\lambda_0}. \quad (2.11)$$

Now by the Hopf maximum principle (Theorem 3.2.5), it follows that $\nabla\phi_0$ does not vanish on ∂D. Since $\phi_0 = 0$ on ∂D, we conclude that

$$\lim_{\operatorname{dist}(x,\partial D)\to 0} \frac{\nabla\phi_0 a\nabla\phi_0}{\phi_0}(x) = \infty. \quad (2.12)$$

From (2.11) and (2.12), it follows that for any $N > 0$, the lefthand side of (2.10) is larger than Nt for sufficiently large t. On the other hand, the righthand side of (2.10) is clearly bounded by $c_1 t + c_2$, for appropriate constants, c_1 and c_2. This contradiction proves that $X(t)$ is recurrent on D. □

The next theorem shows that the Green's measure $G(x, dy)$ possesses a density $G(x, y)$ which enjoys several nice properties.

Theorem 2.5. *Let L satisfy Assumption $\widetilde{H}_{loc}$ on a domain $D \subseteq R^d$ and assume that the Green's measure $G(x, dy)$ exists for L on D. Then the Green's measure $\widetilde{G}(x, dy)$ for $\widetilde{L}$ on D also exists. For each $x \in D$, the Green's measures $G(x, dy)$ and $\widetilde{G}(x, dy)$ are absolutely continuous and their densities $G(x, y)$ and $\widetilde{G}(x, y)$ satisfy the following properties:*

(i) $\widetilde{G}(x, y) = G(y, x)$;
(ii) *$G(x, y)$ is positive and jointly continuous on $U_1 \times U_2$, whenever U_1, $U_2 \subset\subset D$ and $\bar{U}_1 \cap \bar{U}_2 = \phi$;*
(iii) *For each $y \in D$, $G(\cdot, y) \in C^{2,\alpha}(D - \{y\})$ and $LG(\cdot, y) = 0$ on $D - \{y\}$.*
(iv) *For each $x \in D$, $G(x, \cdot) \in C^{2,\alpha}(D - \{x\})$ and $\widetilde{L}G(x, \cdot) = 0$ on $D - \{x\}$.*

Before proving the theorem, we give the following definition.

Definition. *If the Green's measure $G(x, dy)$ for L on D exists, then its density $G(x, y)$ is called the Green's function for L on D. It satisfies the* *properties enumerated in Theorem 2.5. If the Green's measure does not exist for L on D, then the Green's function for L on D is said not to exist.*

Proof of Theorem 2.5. In the proof, c will denote a positive constant that may change from line to line. We will first prove the theorem under the assumption that D is bounded with a $C^{2,\alpha}$-boundary. Then we will extend this to the general case.

By Theorem 2.4, since D is bounded with a $C^{2,\alpha}$-boundary, and since the Green's measure exists, we have $\lambda_0 < 0$. Since λ_0 is also the principal eigenvalue for $\tilde{L}$ on D, applying Theorem 2.4 to $\tilde{L}$ shows that the Green's measure also exists for $\tilde{L}$. Since $\lambda_0 < 0$, by Theorem 3.6.5, for each $f \in C^\alpha(\bar{D})$, there exists a $u \in C^{2,\alpha}(\bar{D})$ such that

$$\left.\begin{aligned} Lu &= -f \text{ in } D, \\ u &= 0 \text{ on } \partial D. \end{aligned}\right\} \tag{2.13}$$

Similarly, there exists a solution $\tilde{\psi} \in C^{2,\alpha}(\bar{D})$ to $\tilde{L}\tilde{\psi} = -1$ in D and $\tilde{\psi} = 0$ on ∂D. Integrating by parts gives

$$\int_D u\,\mathrm{d}x = -\int_D u\tilde{L}\tilde{\psi}\,\mathrm{d}x = \int_D \tilde{\psi} f\,\mathrm{d}x. \tag{2.14}$$

Using the definition of the Green's measure and using Theorem 3.6.4 with $\lambda = 0$, we have the representation $u(x) = \int_D G(x, \mathrm{d}y)f(y)$. Using this in (2.14) gives

$$\int_D \mathrm{d}x \int_D G(x, \mathrm{d}y)f(y) \leqslant c\int_D f(y)\,\mathrm{d}y, \quad \text{for } 0 \leqslant f \in C^\alpha(\bar{D}), \tag{2.15}$$

where c is independent of f.

Choose open sets A, B and U such that $B \subset\subset D$, $A \subset\subset U \subset\subset D$ and $\bar{B} \cap \bar{U} = \phi$. Let f satisfy $0 \leqslant f \in C^\alpha(D)$ and $\operatorname{supp}(f) \subset B$, and let u satisfy (2.13) with this choice of f. Then $u > 0$ in D by Theorem 3.6.5. Since $Lu = 0$ in U, by Harnack's inequality there exists a positive constant c_1, independent of f, such that

$$u(x) \leqslant c_1 u(z), \quad \text{for } x, z \in A. \tag{2.16}$$

From (2.15) and (2.16), we have for $x \in A$,

$$\begin{aligned} \int_B G(x, \mathrm{d}y)f(y) = u(x) &\leqslant \frac{c_1}{l(A)}\int_D u(z)\,\mathrm{d}z = \frac{c_1}{l(A)}\int_D \mathrm{d}z \int_D G(z, \mathrm{d}y)f(y) \\ &\leqslant \frac{cc_1}{l(A)}\int_D f(y)\,\mathrm{d}y = \frac{cc_1}{l(A)}\int_B f(y)\,\mathrm{d}y, \end{aligned}$$

where l denotes Lebesgue measure. From this it follows that for each $x \in A$, the measure $G(x, dy)$ on B possesses a density $G(x, y)$. In fact, since A and B are arbitrary open sets satisfying $\bar{A} \cap \bar{B} = \phi$, it follows that for each $x \in D$, the measure $G(x, dy)$ on $D - \{x\}$ possesses a density $G(x, y)$. We emphasize that, at this stage, $G(x, y)$ is only defined for $y \in D - \{x\}$ up to sets of measure zero. Applying the same type of argument to the Green's measure $\tilde{G}(x, dy)$ for $\tilde{L}$ shows that for each $x \in D$, $\tilde{G}(x, dy)$ possesses a density $\tilde{G}(x, y)$ defined for $y \in D - \{x\}$ up to sets of measure zero.

Let $g \in C^\alpha(\bar{D})$ and let v be the solution of

$$\tilde{L}v = -g \text{ in } D,$$

$$v = 0 \text{ on } \partial D.$$

By Theorem 3.6.4 again, $v(x) = \int_D \tilde{G}(x, y) g(y)\, dy$. Letting u satisfy (2.13) with $f \in C^\alpha(\bar{D})$, we have, upon integrating by parts, $\int_D gu\, dx = \int_D fv\, dx$. Writing this out in terms of the Green's functions, we obtain

$$\int_D \int_D G(x, y) f(y) g(x)\, dx\, dy = \int_D \int_D \tilde{G}(x, y) g(y) f(x)\, dx\, dy. \quad (2.17)$$

Since (2.17) also holds with $f(y)g(x)$ and $g(y)f(x)$ replaced by linear combinations of the form $\sum_{j=1}^n f_j(y) g_j(x)$ and $\sum_{j=1}^n g_j(y) f_j(x)$, where f_j, $g_j \in C^\alpha(\bar{D})$, and since such linear combinations are dense in $C(\bar{D})$, it follows that

$$G(x, y) = \tilde{G}(y, x) \quad \text{for almost all } (x, y) \in D \times D. \quad (2.18)$$

Fix A, B and U again as in the beginning of the proof. For $n = 1, 2, \ldots$, let $g_n \in C^\infty(R^d)$ be supported in $B_{1/n}(0)$, the ball of radius $1/n$ centered at the origin, and satisfy $g_n \geq 0$ and $\int_{R^d} g_n\, dx = 1$. Let $f_{n,y}(x) = g_n(x - y)$, for $y \in B$. Then for each $y \in B$, $\int_B f_{n,y}\, dx = 1$ for sufficiently large n. Let $u_{n,y}$ denote the solution to (2.13) with $f_{n,y}$ in place of f and define

$$v_{n,y}(x) = \frac{u_{n,y}(x)}{\int_D u_{n,y}(z)\, dz}, \quad x \in U.$$

For each $y \in B$ and sufficiently large n, we have for $x \in U$,

$$v_{n,y}(x) = \frac{\int_B G(x, w) f_{n,y}(w)\, dw}{\int_D dz \int_B G(z, w) f_{n,y}(w)\, dw}.$$

From Theorem 3.6.4, $\tilde{\psi}(x) = \int_D \tilde{G}(x, y)\, dy$. Thus from (2.18) and Fubini's theorem, $\tilde{\psi}(y) = \int_D G(z, y)\, dz$, for almost all $y \in D$. By

Theorem 3.6.5, $\tilde{\psi} > 0$ on D. From this and Lebesgue's integration theorem, we obtain

$$\text{for each } x \in A, \lim_{n\to\infty} v_{n,y}(x) = \frac{G(x, y)}{\tilde{\psi}(y)}, \quad \text{for almost all } y \in B. \tag{2.19}$$

From Fubini's theorem, it follows that

$$\text{for almost all } y \in B, \lim_{n\to\infty} v_{n,y}(x) = \frac{G(x, y)}{\tilde{\psi}(y)}, \quad \text{for almost all } x \in A. \tag{2.20}$$

For $y \in B$, we have $Lv_{n,y} = 0$ in U for large n. Also, since $\int_A v_{n,y}\, dx \leq 1$, there exists an $x_{n,y} \in A$ such that

$$v_{n,y}(x_{n,y}) \leq \frac{1}{l(A)}.$$

Therefore, by Harnack's inequality, it follows that

$$\sup_{\substack{n \\ y\in B}} \|v_{n,y}\|_{0;A} \leq c.$$

Using this in the interior Schauder estimate (Theorem 3.2.7) gives

$$\sup_{\substack{n \\ y\in B}} \|v_{n,y}\|_{2,\alpha;A} \leq c. \tag{2.21}$$

From (2.20), (2.21) and the proof of the Ascoli–Arzela theorem, it follows that there exists a set $N_B \subset B$ of measure zero such that for $y \in B - N_B$,

$$\lim_{n\to\infty} v_{n,y}(x) \text{ exists for all } x \in A. \tag{2.22}$$

From (2.21) and the Ascoli–Arzela theorem, it follows, that $\{v_{n,y}\}_{n=1}^{\infty}$ is precompact in the $\|\cdot\|_{2;A}$-norm. From this and (2.22) it follows that for $y \in B - N_B$, $u_y(x) \equiv \lim_{n\to\infty} v_{n,y}(x)$ exists for $x \in A$, $u_y \in C^2(\bar{A})$ and $Lu_y = 0$ in A. From (2.21), it is clear that

$$\sup_{y\in B-N_B} \|u_y\|_{2,\alpha;A} \leq c. \tag{2.23}$$

In particular, then, $u_y \in C^{2,\alpha}(\bar{A})$. From (2.19) and the fact that $G(x, y)$ is only defined in y up to sets of measure zero, we may set

$$G(x, y) = \tilde{\psi}(y)u_y(x), \text{ for } x \in A \text{ and } y \in B - N_B. \tag{2.24}$$

Applying the same type of argument to $\widetilde{G}(x, y)$ on $B \times A$ shows that there exists a set $N_A \subset A$ of measure zero such that we may set $\widetilde{G}(x, y) = \psi(y)v_y(x)$ for $x \in B$ and $y \in A - N_A$, where $\psi \in C^{2,\alpha}(\bar{D})$ solves $L\psi = -1$ in D and $\psi = 0$ on ∂D, and where for each $y \in A - N_A$, $v_y \in C^{2,\alpha}(\bar{B})$ and satisfies $\widetilde{L}v_y = 0$ in B. From this and (2.18), we obtain

$$G(x, y) = \psi(x)v_x(y), \quad \text{for almost all } (x, y) \in (A - N_A) \times B. \tag{2.25}$$

From (2.24) and (2.25), we have

$$u_y(x) = \frac{\psi(x)}{\widetilde{\psi}(y)}v_x(y), \quad \text{for almost all } (x, y) \in (A - N_A) \times (B - N_B). \tag{2.26}$$

Let $y \in B$. By Fubini's theorem, one can choose a sequence $\{x_m\}_{m=1}^{\infty} \subset A$ which is dense in A and a sequence $\{y_n\}_{n=1}^{\infty} \subset B$ satisfying $\lim_{n\to\infty} y_n = y$ such that (2.26) holds for (x_m, y_n), $m, n = 1, 2, \ldots$. Since the righthand side of (2.26) is continuous in y, it follows that $\lim_{n\to\infty} u_{y_n}(x_m)$ exists for $m = 1, 2, \ldots$. From this along with (2.23) and the proof of the Ascoli–Arzela theorem, it then follows that $\bar{u}_y(x) \equiv \lim_{n\to\infty} u_{y_n}(x)$ exists for every $x \in A$, that $\bar{u}_y \in C^{2,\alpha}(\bar{A})$, and that $L\bar{u}_y = 0$. It follows also that $\bar{u}_y$ does not depend on the choice of $\{y_n\}_{n=1}^{\infty}$. From (2.26), $\bar{u}_y(x) = u_y(x)$ for almost all $(x, y) \in A \times (B - N_B)$. Since $\bar{u}_y$ and u_y are both continuous in x, in fact $\bar{u}_y(x) = u_y(x)$ for all $x \in A$ and almost all $y \in B - N_B$. By construction, $\bar{u}_y(x)$ is jointly continuous for $(x, y) \in A \times B$. In light of this and (2.24), and recalling that $G(x, y)$ is only defined in y up to sets of measure zero, we may choose a version of $G(x, y)$ which is jointly continuous for $(x, y) \in A \times B$; namely,

$$G(x, y) = \widetilde{\psi}(y)\bar{u}_y(x). \tag{2.27}$$

The same type of construction shows that $v_x(y)$ can be extended to $\bar{v}_x(y)$ which is jointly continuous for $(x, y) \in A \times B$ and such that for each $x \in A$, $\bar{v}_x \in C^{2,\alpha}(\bar{B})$ and $\widetilde{L}\bar{v}_x = 0$. From (2.25) and (2.27) we have

$$G(x, y) = \widetilde{\psi}(y)\bar{u}_y(x) = \psi(x)\bar{v}_x(y), \text{ for } (x, y) \in A \times B.$$

We conclude that $G(x, y)$ is jointly continuous on $A \times B$, that for each $y \in B$, $G(\cdot, y) \in C^{2,\alpha}(A)$ and $LG(\cdot, y) = 0$, and that for each $x \in A$, $G(x, \cdot) \in C^{2,\alpha}(B)$ and $\widetilde{L}G(x, \cdot) = 0$. The theorem now follows since $A, B \subset D$ are arbitrary open sets satisfying $\bar{A} \cap \bar{B} = \phi$.

We now extend the theorem to the case of arbitrary D. Let $\{D_n\}_{n=1}^{\infty}$ be a sequence of domains with $C^{2,\alpha}$-boundaries satisfying $D_n \subset\subset D_{n+1}$ and $\bigcup_{n=1}^{\infty} D_n = D$, let $G_n(x, dy)$ and $\tilde{G}_n(x, dy)$ denote the Green's measures for L and $\tilde{L}$ on D_n, and let $G_n(x, y)$ and $\tilde{G}_n(x, y) = G_n(y, x)$ denote the corresponding densities. From the definition of the Green's function, it is clear that $G_n(x, dy) \nearrow G(x, dy)$ and $\tilde{G}_n(x, dy) \nearrow \tilde{G}(x, dy)$; thus it follows that densities $G(x, y)$ and $\tilde{G}(x, y)$ exist for the measures $G(x, dy)$ and $\tilde{G}(x, dy)$, that

$$G(x, y) = \lim_{n\to\infty} G_n(x, y) \tag{2.28}$$

and that $\tilde{G}(x, y) = G(y, x)$.

Now fix $y \in D$ and define $U_n = D_n - \overline{B_{\varepsilon_n}(y)}$, where $\lim_{n\to\infty} \varepsilon_n = 0$. Fix $x_0 \in U_1$ and define

$$u_n(x) = \frac{G_n(x, y)}{G_n(x_0, y)} \text{ on } D_n - \{y\}.$$

We have $u_n \in C^{2,\alpha}(\bar{U}_m)$ for every positive integer $m < n$, $Lu_n = 0$ in $D_n - \{y\}$ and $u_n(x_0) = 1$. Thus, by Harnack's inequality, for each positive integer m, there exist constants $c_{m1}, c_{m2} > 0$ such that

$$c_{m1} \leq \inf_{n>m} \inf_{x\in U_m} u_n(x) \leq \sup_{n>m} \sup_{x\in \bar{U}_m} u_n(x) \leq c_{m2}. \tag{2.29}$$

Using this in the interior Schauder estimate (Theorem 3.2.7) gives $\sup_{n>m} \|u_n\|_{2,\alpha;U_m} \leq c_m$, for some $c_m > 0$. Thus, as above, we conclude that $\{u_n\}_{n=1}^{\infty}$ is precompact in the $\|\cdot\|_{2;U_m}$-norm and that any accumulation point $\bar{u}$ satisfies $L\bar{u} = 0$ in U_m and $\bar{u} \in C^{2,\alpha}(\bar{U}_m)$. However, by (2.28),

$$\frac{G(x, y)}{G(x_0, y)}$$

is the unique accumulation point of $\{u_n\}_{n=1}^{\infty}$. Thus $G(\cdot, y) \in C^{2,\alpha}(\bar{U}_m)$ and $LG(\cdot, y) = 0$ in U_m. Also, from (2.29), $G(x, y) > 0$ in U_m. Since m and $y \in D$ are arbitrary, we obtain $G(\cdot, y) \in C^{2,\alpha}(D - \{y\})$, $LG(\cdot, y) = 0$ in $D - \{y\}$ and $G(x, y) > 0$ for $x, y \in D$ and $x \neq y$. A similar argument shows that $G(x, \cdot) \in C^{2,\alpha}(D - \{x\})$ and that $\tilde{L}G(x, \cdot) = 0$ in $D - \{x\}$. By the Schauder estimates obtained in the construction of $G(x, y)$, it follows that if $U_1, U_2 \subset\subset D$ and $\bar{U}_1 \cap \bar{U}_2 = \phi$, then $\{G(x, \cdot), x \in U_1\}$ is equicontinuous on $\bar{U}_2$ and $\{G(\cdot, y), y \in U_2\}$ is equicontinuous on $\bar{U}_1$. The joint continuity of $G(x, y)$ on $U_1 \times U_2$ is an easy consequence of this. □

From the proof of Theorem 2.5, we obtain the following corollary.

Corollary 2.6. *Let L satisfy Assumption $\widetilde{\mathrm{H}}_{\mathrm{loc}}$ on a domain $D \subseteq R^d$. Let $\tilde{p}(t, x, \mathrm{d}y)$ denote the transition measure for $\tilde{L}$ on D. Assume that for sufficiently large λ, the Green's function exists for $L - \lambda$ on D. Then for $t > 0$ and for all f, $g \in C_0(D)$,*

$$\int_D \int_D f(x)g(y)p(t, x, \mathrm{d}y)\,\mathrm{d}x = \int_D \int_D g(x)f(y)\tilde{p}(t, x, \mathrm{d}y)\,\mathrm{d}x.$$

The equality above continues to hold (possibly in the extended sense) for non-negative f, $g \in C(D)$.

Remark. Under the assumption that a density $p(t, x, y)$ exists for $p(t, x, \mathrm{d}y)$, the corollary may be stated as $p(t, x, y) = \tilde{p}(t, y, x)$.

Proof. Let λ be sufficiently large so that G_λ and $\widetilde{G}_\lambda$, the Green's functions for $L - \lambda$ and $\tilde{L} - \lambda$ on D, exist. From the definition of the Green's measure, $G_\lambda(x, \mathrm{d}y) = \int_0^\infty \exp(-\lambda t)p(t, x, \mathrm{d}y)\,\mathrm{d}t$ and $\widetilde{G}_\lambda(x, \mathrm{d}y) = \int_0^\infty \exp(-\lambda t)\tilde{p}(t, x, \mathrm{d}y)\,\mathrm{d}t$. Let f, $g \in C_0(D)$ be non-negative. Applying Theorem 2.5(i) to G_λ and $\widetilde{G}_\lambda$, we conclude that

$$\int_0^\infty \mathrm{d}t \exp(-\lambda t) \int_D \int_D f(x)g(y)p(t, x, \mathrm{d}y)\,\mathrm{d}x =$$
$$\int_0^\infty \mathrm{d}t \exp(-\lambda t) \int_D \int_D g(x)f(y)\tilde{p}(t, x, \mathrm{d}y)\,\mathrm{d}x.$$

Since this equality holds for all large λ, and since

$$h(t) \equiv \int_D \int_D f(x)g(y)p(t, x, \mathrm{d}y)\,\mathrm{d}x \text{ and}$$

$$\tilde{h}(t) \equiv \int_D \int_D g(x)f(y)\tilde{p}(t, x, \mathrm{d}y)\,\mathrm{d}x$$

are non-negative functions of t, it follows by the uniqueness theorem for Laplace transforms that

$$\int_D \int_D f(x)g(y)p(t, x, \mathrm{d}y)\,\mathrm{d}x = \int_D \int_D g(x)f(y)\tilde{p}(t, x, \mathrm{d}y)\,\mathrm{d}x.$$

The result now follows easily. □

Example 2.7. Let $L = \frac{1}{2}\Delta$ in R^d, $d \geq 1$. Then the corresponding diffusion process is Brownian motion and the transition probability density is given by the Gaussian distribution

$$p(t, x, y) = (2\pi t)^{-d/2} \exp\left(\frac{-|x - y|^2}{2t}\right).$$

It is clear that for $x \neq y$, $\int_0^\infty p(t, x, y)\,dt$ is finite if $d \geq 3$ and infinite if $d \leq 2$. Thus, from Theorem 2.1, Brownian motion is recurrent if $d \leq 2$ and transient if $d \geq 3$. In the case where $d \geq 3$, making the change of variables

$$s = \frac{t}{|x|^2},$$

one obtains

$$G(x, y) = \int_0^\infty p(t, x, y)\,dt = c_d|x - y|^{2-d},$$

where

$$c_d = \int_0^\infty \frac{\exp\left(-\frac{1}{2s}\right)}{(2\pi s)^{d/2}}\,ds.$$

Some other calculations involving Green's functions are carried out in the examples at the end of Chapter 7, Section 3.

Our approach to the Green's function does not give any information on the behavior of $G(x, y)$ as $x \to y$. Example 2.7 shows that $G(x, y)$ has a singularity on the order of $|x - y|^{2-d}$ if $L = \frac{1}{2}\Delta$ and $D = R^d$, $d \geq 3$. A similar integration shows that if $d = 2$, then $\int_0^1 p(t, x, y)\,dt = O(-\log|x - y|)$ as $|x - y| \to 0$. In fact, as is well known, this behavior holds in general for Green's functions. We state the following result without proof; for references, see the notes at the end of the chapter.

Theorem 2.8. *Let L satisfy Assumption $\widetilde{H}_{\text{loc}}$ on a domain $D \subseteq R^d$ and assume that the Green's function exists. Then for each $x \in D$, there exist positive constants c_1 and c_2 and a number $r_0 \in (0, 1)$ such that*

$$c_1|x - y|^{2-d} \leq G(x, y) \leq c_2|x - y|^{2-d}, \text{ for } 0 < |y - x| < r_0, \quad \text{if } d \geq 3,$$

$$-c_1 \log|x - y| \leq G(x, y) \leq -c_2 \log|x - y|, \quad \text{for } 0 < |y - x| < r_0, \text{ if } d = 2,$$

and

$$c_1 \leq G(x, y) \leq c_2, \text{ for } 0 < |y - x| < r_0, \text{ if } d = 1.$$

Remark. In the sequel, Theorem 2.8 will be used only in the following three instances: (i) Theorem 11.1, which extends the criticality theory of this chapter to compact manifolds, uses Theorem 2.9 below which in turn relies on the fact that $\lim_{x\to y} G(x, y) = \infty$ if $d \geq 2$; (ii) Theorem 7.3.9 in Chapter 7 uses the fact that $\lim_{x\to y} G(x, y) = \infty$ if $d \geq 2$; (iii) the proof of Theorem 7.6.2 in Chapter 7 uses the upper bound in Theorem 2.8 with $d \geq 2$.

The following result is a corollary of Theorem 2.8.

Theorem 2.9. *Let L satisfy Assumption $\widetilde{\mathrm{H}}_{\mathrm{loc}}$ on a domain $D \subseteq R^d$, $d \geq 2$, and assume that $V \equiv 0$. Let $\{P_x, x \in \hat{D}\}$ denote the solution to the generalized martingale problem for L on D. For each $y \in D$, let $\sigma_y = \inf\{t \geq 0\colon X(t) = y\}$. Then $P_x(\sigma_y < \infty) = 0$, for $x \in D - \{y\}$.*

Proof. Fix $y \in D$ and let B_s denote the ball of radius s centered at y. Choose $r > 0$ such that $B_r \subset\subset D$. Let $\tau_r = \inf\{t \geq 0\colon X(t) \notin B_r\}$ and let $\sigma_\varepsilon = \inf\{t \geq 0\colon X(t) \in \bar{B}_\varepsilon\}$. For $\varepsilon \in (0, r)$, let $u_\varepsilon(x) = P_x(\sigma_\varepsilon < \tau_r\}$, $x \in \bar{B}_r - B_\varepsilon$. We will show below that

$$\lim_{\varepsilon\to 0} u_\varepsilon(x) = 0, \text{ for } x \in B_r - \{0\}. \tag{2.30}$$

The theorem follows from (2.30) and the strong Markov property.

By Theorems 3.6.5 and 3.6.6, $u_\varepsilon \in C^{2,\alpha}(\bar{B}_r - B_\varepsilon)$, $Lu_\varepsilon = 0$ in $B_r - \bar{B}_\varepsilon$, $u_\varepsilon = 1$ on ∂B_ε and $u_\varepsilon = 0$ on ∂B_r. Thus, it follows by Theorem 2.5 and the maximum principle (Theorem 3.2.1) that

$$u_\varepsilon(x) \leq \frac{G(x, y)}{\inf\limits_{z\in\partial B_\varepsilon} G(z, y)}, \ x \in \bar{B}_r - B_\varepsilon. \tag{2.31}$$

By Theorem 2.8, $\lim_{\varepsilon\to 0} \inf_{z\in\partial B_\varepsilon} G(z, y) = \infty$; thus letting $\varepsilon \to 0$ in (2.31) gives (2.30). □

4.3 Criticality and the generalized principal eigenvalue

A function $u \in C^2(D)$ satisfying $Lu = 0$ and $u > 0$ in D is called a *positive harmonic function* for L on D or, when there is no ambiguity, simply, a positive harmonic function. Using the Schauder estimates and the fact that L satisfies Assumption $\widetilde{\mathrm{H}}_{\mathrm{loc}}$ on D, it is not difficult to show that if u is a positive harmonic function, then $u \in C^{2,\alpha}(D)$ (exercise 4.4). The cone of all positive harmonic functions will be denoted as follows:

$$C_L = C_L(D) = \{u \in C^{2,\alpha}(D)\colon Lu = 0 \text{ and } u > 0 \text{ in } D\}\,.$$

Theorem 3.1. *Let L satisfy Assumption $\widetilde{H}_{\text{loc}}$ on a domain $D \subseteq R^d$. If the Green's function $G(x, y)$ exists for L on D, then $C_L(D)$, the cone of positive harmonic functions for L on D, is non-empty.*

Proof. Let $\{D_n\}_{n=1}^{\infty}$ be a sequence of domains satisfying $D_n \subset\subset D_{n+1}$ and $\bigcup_{n=1}^{\infty} D_n = D$. Let $\{y_n\}_{n=1}^{\infty}$ be a sequence of points such that $y_n \in D_{n+1} - \bar{D}_n$ and fix a point $x_0 \in D_1$. Define

$$u_n(x) = \frac{G(x, y_n)}{G(x_0, y_n)}.$$

Since $Lu_n = 0$ in D_n and $u_n(x_0) = 1$, it follows from Harnack's inequality that for each positive integer m, there exist constants $c_{m1}, c_{m2} > 0$ such that

$$c_{m1} \leqslant \inf_{\substack{x \in D_m \\ n > m}} u_n \leqslant \sup_{\substack{x \in D_m \\ n > m}} u_n \leqslant c_{m2}. \tag{3.1}$$

Using this in the interior Schauder estimate (Theorem 3.2.7) shows that for each positive integer m, there exists a constant $c_m > 0$ such that $\sup_{n>m} \|u_n\|_{2,\alpha;D_m} \leqslant c_m$. Therefore $\{u_n\}$ is precompact in the $\|\cdot\|_{2;D_m}$-norm for each m. By diagonalization, we can extract a subsequence $\{u_{n_k}\}_{k=1}^{\infty}$ which is convergent in the $\|\cdot\|_{2;D_m}$-norm for all m. Clearly, the function $u \equiv \lim_{k\to\infty} u_{n_k}$ satisfies $u \in C^2(D)$, $Lu = 0$ in D and, by (3.1), $u > 0$ in D. By the fact that for each m, $\sup_{n>m} \|u_n\|_{2,\alpha;D_m} \leqslant c_m$ or, alternatively, by exercise 4.4, it follows that $u \in C^{2,\alpha}(D)$. □

Remark. The converse of Theorem 3.1 is, of course, false. Indeed, if D is bounded with a $C^{2,\alpha}$-boundary and λ_0 denotes the principal eigenvalue for L on D, then by Theorem 2.4, the Green's function does not exist for $L - \lambda_0$ on D; however, the positive eigenfunction ϕ_0 corresponding to λ_0 belongs to $C_L(D)$.

The theorem and remark above motivate the following classification of operators.

Definition.

(i) L on D is called *subcritical* if it possesses a Green's function;
(ii) L on D is called *critical* if it is not subcritical, but $C_L(D)$ is not empty;
(iii) L on D is called *supercritical* if it is neither critical nor subcritical.

Remark. In the present context, because of Theorem 3.1, (iii) is of course equivalent to the following definition: (iii)′ *L on D is called supercritical if $C_L(D)$ is empty*. However, in Section 11, where we extend the criticality theory to operators on compact manifolds without boundaries, we will need to use the definition in (iii) because Theorem 3.1 will no longer be valid.

We have the following fundamental theorem concerning the above classification.

Theorem 3.2. *Let L satisfy Assumption $\tilde{H}_{\text{loc}}$ on a domain $D \subseteq R^d$. Then either*

(i) *$L - \lambda$ on D is supercritical for all $\lambda \in (-\infty, \infty)$*, or
(ii) *there exists a $\lambda_c = \lambda_c(D) \in (-\infty, \infty)$ such that $L - \lambda$ is subcritical for $\lambda > \lambda_c$, supercritical for $\lambda < \lambda_c$ and either critical or subcritical for $\lambda = \lambda_c$.*

If V is bounded from above, then (ii) *holds. If D is bounded with a $C^{2,\alpha}$-boundary, and L satisfies Assumption $\tilde{H}$ on D, then $\lambda_c(D) = \lambda_0(D)$, where $\lambda_0(D)$ is the principal eigenvalue of L on D, and $L - \lambda_c(D)$ is critical.*

Proof. We will prove that either (i) or (ii) must hold by proving the following four statements:

(a) If $L - \lambda$ is critical or subcritical for some λ_1, then $L - \lambda$ is subcritical for all $\lambda > \lambda_1$.
(b) If $L - \lambda_1$ is critical or supercritical for some λ_1, then $L - \lambda$ is supercritical for all $\lambda < \lambda_1$.
(c) The set $\{\lambda: L - \lambda \text{ is subcritical or critical}\}$ is closed.
(d) The set $\{\lambda: L - \lambda \text{ is supercritical}\}$ is not empty.

Proof of (a). Let $\phi \in C_{L-\lambda_1}$. Then for $\lambda > \lambda_1$, $(L - \lambda)\phi < 0$ and thus, by Proposition 2.3, $L - \lambda$ is subcritical.

Proof of (b). Assume to the contrary that $L - \lambda$ is not supercritical for some $\lambda < \lambda_1$, and let $\phi \in C_{L-\lambda}(D)$. Then $(L - \lambda_1)\phi < 0$ and thus, by Proposition 2.3, $L - \lambda_1$ is subcritical, a contradiction.

Proof of (c). We use a compactness argument similar to the ones employed in the proofs of Theorem 3.1 and Theorem 2.5. Assume that $\lambda = \lim_{n\to\infty} \lambda_n$ and that $C_{L-\lambda_n}(D)$ is not empty. We will show that $C_{L-\lambda}(D)$ is also not empty. Let $\{D_n\}_{n=1}^{\infty}$ be a sequence of domains satisfying $D_n \subset\subset D_{n+1}$ and $\bigcup_{n=1}^{\infty} D_n = D$. Pick $u_n \in C_{L-\lambda_n}(D)$, fix $x_0 \in D_1$ and define

$$v_n(x) = \frac{u_n(x)}{u_n(x_0)}.$$

Since $v_n(x_0) = 1$, it follows from Harnack's inequality that for each m, there exist constants $c_{m1}, c_{m2} > 0$ such that

$$c_{m1} \leqslant \inf_{\substack{x \in D_m \\ n \geqslant 1}} v_n(x) \leqslant \sup_{\substack{x \in D_m \\ n \geqslant 1}} v_n(x) \leqslant c_{m2}. \tag{3.2}$$

Using the upper bound in (3.2) in the Schauder interior estimate (Theorem 3.2.7), it follows that for each m there exists a constant $c_m > 0$ such that

$$\sup_{n \geqslant 1} \|v_n\|_{2,\alpha;D_m} \leqslant c_m.$$

Thus, for each m, $\{v_n\}_{n=1}^{\infty}$ is precompact in the $\|\cdot\|_{2;D_m}$-norm. By diagonalization, it follows that there exists a $v \in C^2(D)$ satisfying $Lv = 0$ in D. The lower bound in (3.2) guarantees that $v > 0$ in D. Since $\sup_{n \geqslant 1} \|v_n\|_{2,\alpha;D_m} \leqslant c_m$, it follows that $v \in C^{2,\alpha}(D)$.

Proof of (d). Let $B \subset\subset D$ be a ball and let $\lambda_0(B)$ denote the principal eigenvalue for L on B. By Theorem 2.4, $L - \lambda_0(B)$ is critical on B and thus, by statement (b), $L - \lambda$ is supercritical on B for $\lambda < \lambda_0(B)$. Clearly, supercritically on B implies supercritically on D.

If V is bounded from above and $\lambda > \sup V$, then it is clear from its definition that the Green's function exists for $L - \lambda$; thus (ii) must hold if V is bounded from above.

The final statement of the theorem follows from (ii) and Theorem 2.4. □

Remark. In the sequel, any compactness argument such as the ones used in the proofs of Theorem 2.5, Theorem 3.1 and part (c) of Theorem 3.2 will be referred to as a 'standard compactness argument'.

Theorem 3.2 motivates the following definition.

Definition. $\lambda_c = \lambda_c(D)$ appearing in Theorem 3.2 is called the *generalized principal eigenvalue* for L on D. If (i) holds in Theorem 3.2, then define $\lambda_c = \lambda_c(D) = \infty$.

We have seen that if D is bounded with a $C^{2,\alpha}$-boundary, then $\lambda_c = \lambda_0$ and $L - \lambda_c$ is critical. In Section 7, we will show that if D is bounded and the coefficients of L are uniformly elliptic and bounded, then $L - \lambda_c$ is critical. In the general case, $L - \lambda_c$ may be critical or may be subcritical. The simplest example is $L = \frac{1}{2}\Delta$ in R^d. From the definition of the Green's function, it is clear that $L + \varepsilon$ is not subcritical for $\varepsilon > 0$. Since $1 \in C_L$, it then follows that $\lambda_c = 0$. However, by Example 2.7 and Theorem 2.1, $L - \lambda_c = \frac{1}{2}\Delta$ is critical if $d \leqslant 2$ and subcritical if $d \geqslant 3$.

The next theorem reformulates some of the results of the previous section in terms of the classification introduced in this section.

Theorem 3.3. *Let L satisfy Assumption $\tilde{\mathrm{H}}_{\mathrm{loc}}$ on a domain $D \subseteq R^d$ and let $L_0 = L - V$.*

(i) *If $V \equiv 0$ and the diffusion process on D corresponding to L_0 is recurrent, then L is critical on D.*
(ii) *If $V \equiv 0$ and the diffusion process on D corresponding to L_0 is transient, then L is subcritical on D.*
(iii) *If $V \leqslant 0$ and $V \not\equiv 0$, then L is subcritical on D.*
(iv) *If L is subcritical (critical, supercritical) on D and $h \in C^{2,\alpha}(D)$ satisfies $h > 0$, then L^h, the h-transform of L, is also subcritical (critical, supercritical) on D. Therefore, $\lambda_c^h = \lambda_c$, where λ_c^h is the generalized principal eigenvalue for L^h.*
(v) *L is subcritical (critical, supercritical) on D if and only if its formal adjoint $\tilde{L}$ is subcritical (critical, supercritical) on D. Therefore, $\tilde{\lambda}_c = \lambda_c$, where $\tilde{\lambda}_c$ is the generalized principal eigenvalue for $\tilde{L}$.*

Proof. (i) follows from Theorem 2.1(i) and the fact that the constant function $1 \in C_L$; (ii) and (iii) follow from Theorem 2.1(ii) and (iii) respectively; (iv) follows from Proposition 2.2 and the fact that if $\phi \in C_L(D)$, then

$$\frac{\phi}{h} \in C_{L^h}(D).$$

(v) follows from Theorem 2.5 and Theorem 3.2. □

In the case where L is critical, we have the following theorem.

Theorem 3.4. *Let L satisfy Assumption $\tilde{H}_{\mathrm{loc}}$ on a domain $D \subseteq R^d$. If L is critical on D, then $C_L(D)$ is one-dimensional.*

The following corollary is immediate.

Corollary 3.5. *If $V \equiv 0$ and L corresponds to a recurrent diffusion process on D, then $C_L(D)$ contains only the positive constants.*

Before proving the theorem, we make the following definition.

Definition. If $L - \lambda_c$ is critical on D, the unique function (up to a constant multiple) in $C_{L-\lambda_c}(D)$ is called the *ground state* of $L - \lambda_c$. It will be denoted by ϕ_c. Of course, if D is bounded with a $C^{2,\alpha}$-boundary, then $\phi_c = \phi_0$, where ϕ_0 is the positive eigenfunction corresponding to the principal eigenvalue $\lambda_0 = \lambda_c$ of L on D. The ground state for $\tilde{L} - \lambda_c$ will be denoted by $\tilde{\phi}_c$.

Remark 1. The appellation *ground state* has its origins in the mathematical physics literature. The ground state of a critical operator enjoys an important 'minimal growth' property that is not shared by any $u \in C_L(D)$ in the case where L is subcritical on D. This property is developed in Chapter 7, Section 3. See, in particular, Theorems 7.3.8 and 7.3.9, and Examples 3.10–3.12.

Remark 2. If L is subcritical, then C_L may still be one-dimensional; however, frequently it is larger. For example, if $D \subseteq R^1$ and L is subcritical, then $C_L(D)$ is always two-dimensional. (See Proposition 5.1.3 in Chapter 5.) The structure of C_L is treated in Chapters 7 and 8.

Proof of Theorem 3.4. First we show how to reduce to the case where $V \equiv 0$ and L corresponds to a recurrent diffusion process. By Theorem 3.3(iv) and its proof, if $h \in C_L$, then L^h is also critical and there is a one to one correspondence between $C_L(D)$ and $C_{L^h}(D)$. Thus, it suffices to prove the theorem for L^h. Since the zeroth-order part of L^h vanishes, it follows from Theorem 3.3(i) and (ii) that L^h generates a recurrent diffusion process.

From this reduction, we may assume without the loss of generality that $V \equiv 0$ and that the diffusion process corresponding to L is recurrent. Obviously, the constant function 1 belongs to $C_L(D)$. We will prove that there are no non-constant functions in $C_L(D)$.

Let $\phi \in C_L(D)$ and let x, $y \in D$. Choose an open set U such that $y \in U \subset\subset D$ and let $\{D_n\}_{n=1}^{\infty}$ be a sequence of domains such that $D_n \subset\subset D_{n+1}$ and $\bigcup_{n=1}^{\infty} D_n = D$. Define $\sigma_{\bar{U}} = \inf\{t \geq 0\colon\ X(t) \in \bar{U})$ and $\tau_{D_n} = \inf\{t \geq 0\colon\ X(t) \notin D_n)$. Since $\phi \in C_L(D)$, $\phi(X(t \wedge \tau_{D_n} \wedge \sigma_{\bar{U}}))$ is a P_x-martingale; thus $E_x\phi(t \wedge \tau_{D_n} \wedge \sigma_{\bar{U}}) = \phi(x)$. By recurrence, $\sigma_U < \infty$ a.s. P_x and, by Theorem 2.8.1, $\tau_D \equiv \lim_{n\to\infty} \tau_{D_n} = \infty$ a.s. P_x. Thus, letting $n \to \infty$ and then letting $t \to \infty$, and using Fatou's lemma each time, we obtain $E_x\phi(X(\sigma_{\bar{U}})) \leq \phi(x)$. Since U may be chosen to be an arbitrarily small neighbourhood of y, and since $X(\sigma_{\bar{U}}) \in \bar{U}$, it follows that $\phi(y) \leq \phi(x)$. As x, $y \in D$ are arbitrary, it follows that ϕ is constant. □

We now state and prove a number of results regarding criticality, which will be useful in the sequel.

Theorem 3.6. *Let L satisfy Assumption $\tilde{H}_{\text{loc}}$ on a domain $D \subseteq R^d$. If L is not subcritical on D, then*

$$\int_0^\infty p(t, x, B)\, dt = \infty, \quad \textit{for all } x \in D \textit{ and all open } B \subseteq D.$$

Proof. By Theorem 1.1, it is enough to prove the theorem for any h-transform L^h of L. First consider the case where L is critical. Let $h \in C_L(D)$. By Theorem 3.3(i), (ii) and (iv), L^h corresponds to a recurrent diffusion process. Letting $p^h(t, x, dy)$ denote the transition measure for L^h, it follows from the proof of Theorem 2.1(i) that $\int_0^\infty p^h(t, x, B)\, dt = \infty$ for all $x \in D$ and all open $B \subseteq D$.

Now consider the supercritical case, that is, the case in which $\lambda_c(D) > 0$. Let $\{D_n\}_{n=1}^\infty$ be an increasing sequence of domains with $C^{2,\alpha}$-boundaries such that $D_n \subset\subset D_{n+1}$ and $\bigcup_{n=1}^\infty D_n = D$. By Theorem 4.1 in Section 4, $\lambda_c(D_n) > 0$ for sufficiently large n. Fix such an n. By Theorem 3.2, $L - \lambda_c(D_n)$ is critical on D_n. Let $h \in C_{L-\lambda_c(D_n)}(D_n)$ and let $p_n^{h,\lambda_c(D_n)}(t, x, dy)$ denote the transition measure for $(L - \lambda_c(D_n))^h$ on D_n. By Theorem 3.3(i), (ii) and (iv), $(L - \lambda_c(D_n))^h$ corresponds to a recurrent diffusion process on D_n; therefore, from the proof of Theorem 2.1(i) again,

$$\int_0^\infty p_n^{h,\lambda_c(D_n)}(t, x, B)\, dt = \infty, \quad \text{for all } x \in D \text{ and all open } B \subseteq D_n. \tag{3.3}$$

But $p_n^{h,\lambda_c(D_n)}(t, x, dy) = \exp(-\lambda_c(D_n)t) p_n^h(t, x, dy)$, where $p_n^h(t, x, dy)$ is the transition measure for L^h on D_n and, by Theorem 1.1,

$$p_n^h(t, x, dy) = \frac{1}{h(x)} p_n(t, x, dy) h(y),$$

where $p_n(t, x, dy)$ is the transition measure for L on D_n. Clearly $p_n(t, x, dy) \leq p(t, x, dy)$, for $x \in D_n$. Thus

$$p_n^{h,\lambda_c(D_n)}(t, x, dy) \leq \frac{1}{h(x)} p(t, x, dy) h(y) \exp(-\lambda_c(D_n))t, \text{ for } x \in D_n.$$

The theorem follows from this and (3.3). □

In a similar vein, we have the following result.

Theorem 3.7. *Let L satisfy Assumption $\widetilde{H}_{\text{loc}}$ on a domain $D \subseteq R^d$. Let $\{D_n\}_{n=1}^{\infty}$ be an increasing sequence of domains such that $D_n \subset\subset D_{n+1}$ and $\bigcup_{n=1}^{\infty} D_n = D$. Assume that for each n, L is subcritical on D_n and let $G_n(x, y)$ denote the corresponding Green's function. The sequence $\{G_n(x, y)\}_{n=1}^{\infty}$ is monotone increasing in x and in y.*

(i) *If L is subcritical on D, then the Green's function $G(x, y)$ satisfies $G(x, y) = \lim_{n\to\infty} G_n(x, y)$ for $x, y \in D$ with $x \neq y$.*
(ii) *If L is critical on D, then $\lim_{n\to\infty} G_n(x, y) = \infty$ for $x, y \in D$ with $x \neq y$.*

Proof. From the definition of the Green's function, it follows that $\{G_n(x, y)\}_{n=1}^{\infty}$ is monotone non-decreasing in x and y. The strict monotonicity follows easily from the Stroock–Varadhan support theorem (Theorem 2.6.1). Part (i) was proved in the proof of Theorem 2.5. Part (ii) is left as exercise 4.9. □

The next theorem characterizes subcriticality in terms of the solvability of certain inhomogeneous Dirichlet problems.

Theorem 3.8. Let L satisfy Assumption $\widetilde{H}_{\text{loc}}$ on a domain $D \subseteq R^d$.

(i) *If L is subcritical on D, then for each non-negative $f \in C_0^{\alpha}(D)$ which is not identically zero, there exist positive solutions $u \in C^{2,\alpha}(D)$ of $Lu = -f$ in D. The smallest such solution is given by $u_0(x) \equiv \int_D G(x, y)f(y)\,dy$. Every other such solution is of the form $u = u_0 + \phi$, where $\phi \in C_L(D)$.*
(ii) *If L is not subcritical on D, then for each non-negative $f \in C_0^{\alpha}(D)$ which is not identically zero, there are no positive solutions of $Lu = -f$ in D.*

Remark. (i) should be compared to Remark 2 following the definition of the Green's measure at the beginning of Section 2.

Proof. (i) Let $\{D_n\}_{n=1}^{\infty}$ be a sequence of domains with $C^{2,\alpha}$-boundaries such that $D_n \subset\subset D_{n+1}$ and $\bigcup_{n=1}^{\infty} D_n = D$. Let $G_n(x, y)$ denote the Green's function for L on D_n. By Theorem 3.6.4 and the definition of the Green's function, $u_n(x) \equiv \int_{D_n} G_n(x, y)f(y)\,dy$ is the unique solution to $Lu_n = -f$ in D_n and, $u_n = 0$ on ∂D_n. By Theorem 3.7, $u_0(x) \equiv \lim_{n\to\infty} u_n(x) = \int_D G(x, y)f(y)\,dy$. By a standard compactness argument, $u_0 \in C^{2,\alpha}(D)$ and $Lu_0 = -f$ in D. By Theorem 4.1(iii) in Section 4, $\lambda_c(D_n) < 0$. Thus, if u is any positive solution of $Lu = -f$ in D, then by the maximum principle in Theorem 3.6.5, $u \geqslant u_n$ on D_n. Thus, $u \geqslant u_0$ on D and it follows that u_0 is in fact the smallest positive solution. Clearly, $\phi \equiv u - u_0 \in C_L(D)$.

(ii) Let $\{D_n\}_{n=1}^{\infty}$ be as in (i). Assume that u is a positive solution of $Lu = -f$ in D. Then by the Feynman–Kac formula (Theorem 2.4.2),

$$E_x \exp\left(\int_0^{t\wedge\tau_{D_n}} V(X(s))\,ds\right) u(X(t\wedge\tau_{D_n}))$$
$$= u(x) - E_x \int_0^{t\wedge\tau_{D_n}} f(X(s)) \exp\left(\int_0^s V(X(r))\,dr\right) ds.$$

Letting $t\to\infty$ and then $n\to\infty$, we obtain

$$u(x) \geqslant E_x \int_0^{\tau_D} f(X(t)) \exp\left(\int_0^t V(X(s))\,ds\right) dt$$
$$= \int_D dy \int_0^{\infty} dt p(t, x, dy) f(y) = \infty,$$

by Theorem 3.6. □

We can now give the following very useful characterization of subcriticality.

Theorem 3.9. *Let L satisfy Assumption $\widetilde{H}_{\text{loc}}$ on a domain $D \subseteq R^d$. Then L is subcritical if and only if there exists a $\phi \in C^{2,\alpha}(D)$ satisfying $\phi > 0$, $L\phi \leqslant 0$ and $L\phi \neq 0$.*

Proof. The proof follows immediately from Theorem 3.8. □

Corollary 3.10. *Let L satisfy Assumption $\widetilde{H}_{\text{loc}}$ on a domain $D \subseteq R^d$ and let $\phi \in C^{1,\alpha}(D)$ satisfy $\phi > 0$ in D. Then ϕL also satisfies Assumption $\widetilde{H}_{\text{loc}}$ on D and ϕL is subcritical (critical, supercritical) on D if and only if L is subcritical (critical, supercritical) on D.*

Proof. ϕL satisfies Assumption $\widetilde{H}_{\text{loc}}$ since it can be written in the form $\phi L = \frac{1}{2}\nabla\cdot\phi a\nabla - \frac{1}{2}a\nabla\phi\cdot\nabla + \phi b\cdot\nabla + \phi V$. Since $C_{\phi L}(D) = C_L(D)$, ϕL is supercritical on D if and only if L is supercritical on D. By Theorem 3.9, ϕL is subcritical on D if and only if L is subcritical on D. □

Remark. An alternative probabilistic proof of Corollary 3.10 is suggested in exercise 4.11.

The following result is known as the *resolvent equation*.

Theorem 3.11. *Let L satisfy Assumption $\widetilde{H}_{\text{loc}}$ on a domain $D \subseteq R^d$. Let $W \in C^{\alpha}(D)$ satisfy either $W \geqslant 0$ or $W \in C_0^{\alpha}(D)$. Assume that L and*

$L+W$ are subcritical on D and denote their respective Green's functions by $G(x,y)$ and $G_W(x,y)$. Then

$$G_W(x,y) = G(x,y) + \int_D G(x,z)W(z)G_W(z,y)\,dz.$$

Proof. Let $f \in C_0^\alpha(D)$ and satisfy $f \geq 0$. Let $\{D_n\}_{n=1}^\infty$ be a sequence of domains with $C^{2,\alpha}$-boundaries such that $D_n \subset\subset D_{n+1}$ and $\bigcup_{n=1}^\infty D_n = D$, and denote the Green's functions for L on D_n and $L+W$ on D_n by $G_n(x,y)$ and $G_{W,n}(x,y)$. Let u_n denote the unique solution to $(L+W)u_n = -f$ in D_n and $u_n = 0$ on ∂D_n. As in the proof of Theorem 3.8, we have

$$u_n(x) = \int_{D_n} G_{W,n}(x,y)f(y)\,dy. \tag{3.4}$$

On the other hand, writing $Lu_n = -f - Wu_n$, it follows from the same considerations that

$$u_n(x) = \int_{D_n} G_n(x,y)(f(y) + W(y)u_n(y))\,dy. \tag{3.5}$$

Combining (3.4) and (3.5), we obtain

$$\int_{D_n} G_{W,n}(x,y)f(y)\,dy = \int_{D_n} [G_n(x,y) + \int_{D_n} G_n(x,z)W(z)G_{W,n}(z,y)\,dz]f(y)\,dy.$$

Letting $n \to \infty$ and using Theorem 3.7 along with either the monotone convergence theorem or the dominated convergence theorem, depending on whether $W \geq 0$ or $W \in C_0^\alpha(D)$, we conclude that

$$\int_D G_W(x,y)f(y)\,dy = \int_D [G(x,y) + \int_D G(x,z)W(z)G_W(z,y)\,dz]f(y)\,dy,$$

for $x \in D$ and non-negative $f \in C_0^\alpha(D)$; the theorem follows from this. □

We close this section with a classical example.

Example 3.12. Let

$$L = \tfrac{1}{2}\Delta + \frac{k}{|x|^2} \text{ on } D = \begin{cases} R^d - \{0\}, & \text{if } d \geq 2 \\ (0,\infty), & \text{if } d = 1. \end{cases}$$

We will show that L is subcritical on D if

$$k < \frac{(d-2)^2}{8},$$

critical on D if

$$k = \frac{(d-2)^2}{8}$$

and supercritical on D if

$$k > \frac{(d-2)^2}{8}.$$

First consider the case where $d = 2$. Two-dimensional Brownian motion is recurrent on D since it is recurrent on R^d and since, starting away from the origin, the probability of it ever reaching the origin is zero (Example 5.1.2 in Chapter 5). Thus, by Theorem 3.3(i), L is critical if $k = 0$. If $k < 0$, then $L1 < 0$ on D and the subcriticality follows from Theorem 3.9. Now let $k > 0$ and assume that L is not supercritical. Let $\phi \in C_L(D)$. Then $\frac{1}{2}\Delta\phi < 0$ in D and, by Theorem 3.9, $\frac{1}{2}\Delta$ is subcritical on R^2, a contradiction. Thus, L is supercritical for $k > 0$.

Now consider the case $d \neq 2$. Note that

$$L|x|^\gamma = \left(\frac{\gamma^2}{2} + \frac{(d-2)\gamma}{2} + k\right)|x|^{\gamma-2}.$$

In particular, if we define $h(x) = |x|^{(2-d)/2}$, then

$$Lh(x) = \left(k - \frac{(d-2)^2}{8}\right)|x|^{-2}h(x).$$

If

$$k < \frac{(d-2)^2}{8},$$

then $Lh < 0$ in D and, by Theorem 3.9, L is subcritical on D. If

$$k = \frac{(d-2)^2}{8},$$

then $Lh = 0$ in D and

$$L^h = \tfrac{1}{2}\Delta + \frac{\nabla h}{h}\cdot\nabla.$$

In polar coordinates, $r = |x|$ and

$$\phi = \frac{x}{|x|} \in S^{d-1},$$

we have

$$L^h = \begin{cases} \frac{1}{2}\left(\frac{\partial^2}{\partial r^2} + \frac{1}{r}\frac{\partial}{\partial r}\right) + \frac{1}{2r^2}\Delta_{S^{d-1}}, & d \geqslant 3 \\ \frac{1}{2}\left(\frac{d^2}{dr^2} + \frac{1}{r}\frac{d}{dr}\right) & d = 1, \end{cases}$$

where $\Delta_{S^{d-1}}$ is the angular part of the Laplacian, known as the *Laplace–Beltrami* operator on S^{d-1}. Since the coefficients of the terms involving differentiation in r depend only on r, the process $r(t) = |X(t)|$ is also a diffusion; it corresponds to the operator

$$\frac{1}{2}\left(\frac{d^2}{dr^2} + \frac{1}{r}\frac{d}{dr}\right)$$

(see exercise 1.13). Now it is easy to see from Theorem 2.8.1 that $X(t)$ is recurrent on $R^d - \{0\}$ if and only if $r(t) = |X(t)|$ is recurrent on $(0, \infty)$. We recognize

$$\frac{1}{2}\left(\frac{d^2}{dr^2} + \frac{1}{r}\frac{d}{dr}\right)$$

as the radial part of the two-dimensional Laplacian; therefore $r(t) = |X(t)|$ is equal in distribution to the radial part of a two-dimensional Brownian motion. As noted above, two-dimensional Brownian motion is recurrent and, starting away from the origin, the probability of it ever reaching the origin is zero; thus $r(t)$ is recurrent on $(0, \infty)$. We conclude that L^h corresponds to a recurrent diffusion and, by Theorem 3.3(i) and (iv), L is critical for

$$k = \frac{(d-2)^2}{8}.$$

The same argument used for supercriticality in the case $d = 2$ shows that L is supercritical for

$$k > \frac{(d-2)^2}{8}.$$

Example 3.12 shows that it is possible to add to the Laplacian a positive potential decaying at infinity on the order $|x|^{-2}$ and still obtain a subcritical operator. In fact, the order $|x|^{-2}$ is maximal.

Proposition 3.13. *Let $V \in C^{\alpha}(R^d)$, $d \geq 2$, satisfy $\lim_{|x|\to\infty} |x|^2 V(x) = \infty$. Then $\frac{1}{2}\Delta + V$ on R^d is supercritical. Similarly, if $V \in C((0, \infty))$ satisfies*

$$\lim_{x\to\infty} x^2 V(x) = \infty, \text{ then } \tfrac{1}{2}\frac{d^2}{dx^2} + V \text{ on } (0, \infty)$$

is supercritical.

Remark. In fact, the result holds whenever there exists an open set $A \subset S^{d-1}$ such that

$$\lim_{\substack{|x|\to\infty \\ x/|x| \in A}} |x|^2 V(x) = \infty.$$

See exercise 4.14.

Proof. Let λ denote the principal eigenvalue of $\frac{1}{2}\Delta$ in the annulus (interval, if $d = 1$) $\{1 < |x| < 2\}$. An easy scaling argument shows that the principal eigenvalue of $\frac{1}{2}\Delta$ in $\{R < |x| < 2R\}$ is given by

$$\frac{\lambda}{R^2}.$$

If R is chosen sufficiently large, then

$$V(x) \geq \frac{5\lambda}{|x|^2} \geq \frac{5\lambda}{4R^2}, \text{ on } \{R < |x| < 2R\}.$$

Since

$$\tfrac{1}{2}\Delta + \frac{\lambda}{R^2} + \varepsilon \text{ on } \{R < |x| < 2R\}$$

is supercritical for any $\varepsilon > 0$, it follows by comparison that $\frac{1}{2}\Delta + V(x)$ is supercritical on $\{R < |x| < 2R\}$ and, *a fortiori* on $R^d((0, \infty)$, if $d = 1)$. □

4.4 The generalized principal eigenvalue: continuity properties and mini-max principles

The following theorem concerns the behavior of the generalized principal eigenvalue $\lambda_c(D)$ as a function of the domain D.

Theorem 4.1. *Let L satisfy Assumption $\widetilde{H}_{\mathrm{loc}}$ on a domain $D \subseteq R^d$.*

(i) *If $\{D_n\}_{n=1}^{\infty}$ is an increasing sequence of domains such that $\bigcup_{n=1}^{\infty} D_n = D$, then $\lim_{n\to\infty} \lambda_c(D_n) = \lambda_c(D)$.*
(ii) *If $D' \subset\subset D$ is a domain with a $C^{2,\alpha}$-boundary and $\{D_n\}_{n=1}^{\infty}$ is a decreasing sequence of domains such that $D' \subseteq \bigcap_{n=1}^{\infty} D_n \subseteq \bar{D}'$, then $\lim_{n\to\infty} \lambda_c(D_n) = \lambda_c(D')$.*
(iii) *Let D_1, $D_2 \subset D$ be domains satisfying $D_1 \subset D_2$, $D_2 - \bar{D}_1$ is non-empty, and $\lambda_c(D_1) < \infty$. Assume either that $L - \lambda_c(D_1)$ is critical on D_1, or that D_1 is bounded and* dist $(D_1, \partial D_2) > 0$. *Then $\lambda_c(D_1) < \lambda_c(D_2)$.*

Remark 1. A. Sznitman has outlined for me a proof that (ii) holds in general without either the smoothness assumption on $\partial D'$ or the boundedness assumption on D'. The proof uses probabilistic and potential theoretic techniques.

Remark 2. The boundedness assumption on D_1 in (iii) is necessary; for example, if $L = \frac{1}{2}\Delta$ and $D_1 = \{x : x_1 > 1\}$ and $D_2 = \{x : x_1 > 0\}$, then it is clear that $\lambda_c(D_1) = \lambda_c(D_2)$.

Remark 3. The condition in (iii) that $L - \lambda_c(D_1)$ be critical in fact holds in some generality – see Theorem 4.7.1.

Remark 4. The condition in (iii) that $D_2 - \bar{D}_1$ be non-empty is needed to exclude, for example, the case in which $D_1 = D_2 - \{x_0\}$, where $x_0 \in D_2$.

Proof of Theorem 4.1. (i) It is clear that $\lambda_c(D)$ is a monotone non-decreasing function of D; thus

$$\lambda_c(D) \geqslant \lim_{n\to\infty} \lambda_c(D_n). \tag{4.1}$$

If $\lim_{n\to\infty} \lambda_c(D_n) = \infty$, there is nothing more to prove; therefore, assume that $l \equiv \lim_{n\to\infty} \lambda_c(D_n) < \infty$. Let $\phi_n \in C_{L-\lambda(D_n)}(D_n)$. A standard compactness argument shows that there exists a $\phi \in C_{L-l}(D)$. Thus $\lambda_c(D) \leqslant l$ and (i) follows from this and (4.1).

(ii) Since $\lambda_c(D)$ is monotone non-decreasing in D, it suffices to consider the case in which D_n possesses a $C^{2,\alpha}$-boundary and $D_{n+1} \subset\subset D_n$. But this case was proved as Step 4 in the proof of Theorem 3.6.1.

(iii) We will assume that $\lambda_c(D_2) < \infty$ since otherwise there is nothing to prove. First assume that D_1 is bounded and dist $(D_1, \partial D_2) > 0$. Then we can pick a bounded domain D' with a $C^{2,\alpha}$-boundary satisfying $D_1 \subset D' \subset\subset D_2$. By monotonicity it is enough to show that $\lambda_c(D') \neq \lambda_c(D_2)$. Let $u \in C_{L-\lambda(D_2)}(D_2)$. Then u is strictly positive on D' and by Corollary 3.2.3 it follows that the only solution ϕ of $(L - \lambda_c(D_2))\phi = 0$ in D' and $\phi = 0$ on $\partial D'$ is $\phi \equiv 0$. Since, by Theorem 3.5.5, the eigenfunction ϕ_0 corresponding to the principal eigenvalue $\lambda_0(D') = \lambda_c(D')$ for L on D' satisfies $(L - \lambda_c(D'))\phi_0 = 0$ in D' and $\phi_0 = 0$ on $\partial D'$, it follows that $\lambda_c(D') \neq \lambda_c(D_2)$.

To prove (iii) under the assumption that $L - \lambda_c(D_1)$ is critical, we will show that if $\lambda_c(D_1) = \lambda_c(D_2)$, then $L - \lambda_c(D_1)$ on D_1 must be subcritical. Assume that $\lambda_c(D_1) = \lambda_c(D_2) \equiv \gamma$ and let $h \in C_{L-\gamma}(D_2)$. By assumption, $D_2 - \bar{D}_1$ is a non-empty open set. Thus there exists an open set $U \subset\subset D_2$ such that $U \cap D_1 \neq \phi$, $U \cap (D_2 - \bar{D}_1) \neq \phi$ and such that every continuous curve in U from a point $x_1 \in D_1$ to a point $x_2 \in D_2 - \bar{D}_1$ must intercept ∂D_1. Note that $h \in C^{2,\alpha}(\bar{U})$ and $\inf_U h > 0$. Consider the h-transform

$$(L - \gamma)^h = L_0 + a\frac{\nabla h}{h} \cdot \nabla$$

of $L - \gamma$ on D_1 and let $\{P_x^{h,\gamma}, x \in \hat{D}_1\}$ denote the solution of the generalized martingale problem for $(L - \gamma)^h$ on D_1. Since the coefficients of $(L - \gamma)^h$ are bounded on U, it follows from the Stroock–Varadhan support theorem (Theorem 2.6.1) and the construction of U that $P_x^{h,\gamma}(\tau_{D_1} < \infty) > 0$ for $x \in D_1 \cap U$. Thus, by Theorem 2.8.1(iii), $(L - \gamma)^h$ corresponds to a transient diffusion process on D_1 and by Theorem 3.3(ii) and (iv), $L - \gamma$ on D_1 is subcritical. □

Remark. A much simpler and more satisfying proof of Theorem 4.1(iii) in the case where $L - \lambda_c(D_1)$ is assumed critical can be given using Theorem 6.3 (see exercise 4.20).

Corollary 4.2. *Let L satisfy Assumption $\tilde{H}_{\text{loc}}$ on a domain $D \subseteq R^d$. Let $D_1 \subset D$ and assume that $D - \bar{D}_1$ is non-empty. If L is critical on D_1, then L is supercritical on D; if L is critical on D, then L is subcritical on D_1.*

Proof. If L is critical on D_1, then $\lambda_c(D_1) = 0$ and, by Theorem 4.1(iii), $\lambda_c(D) > 0$; thus L is supercritical on D. Now assume that L is critical on D. If $\lambda_c(D_1) < 0$, then, clearly, L is subcritical on D_1. If $\lambda_c(D_1) = \lambda_c(D) = 0$, then $L = L - \lambda_c(D_1)$ is subcritical by Theorem 4.1(iii). □

We now consider the dependence of λ_c on the zeroth-order term V.

Theorem 4.3. *Let $L = L_0 + V$ satisfy Assumption $\widetilde{H}_{\text{loc}}$ on a domain $D \subseteq R^d$. In its dependence on V, λ_c is convex and is Lipschitz continuous with Lipschitz constant one (using the sup-norm on V).*

Remark. See Theorem 7.7 for the dependence of λ_c on the drift b and for more on the dependence of λ_c on V in the case where D is bounded and L is uniformly elliptic with bounded coefficients.

Proof. The convexity is an immediate consequence of Theorem 6.1 in Section 6; the Lipschitz continuity with Lipschitz constant one follows, for example, from the definition of λ_c and Theorem 3.9, or from Theorem 3.6.1(ii) and Theorem 4.1(i). □

The next theorem gives a probabilistic characterization of $\lambda_c(D)$.

Theorem 4.4. *Let L satisfy Assumption $\widetilde{H}_{\text{loc}}$ on a domain $D \subseteq R^d$. Then*

$$\lambda_c(D) = \sup_{A \subset\subset D} \lim_{t\to\infty} \frac{1}{t} \log \sup_{y \in A} E_y\left(\exp\left(\int_0^t V(X(s))\,ds\right);\ \tau_A > t\right)$$

$$= \sup_{\substack{A \subset\subset D \\ \partial A \text{ a} \\ C^{2,\alpha}\text{-boundary}}} \lim_{t\to\infty} \frac{1}{t} \log E_x\left(\exp\left(\int_0^t V(X(s))\,ds\right);\ \tau_A > t\right), \text{ for any } x \in D.$$

Also, if $\lambda_c(D) \leq 0$ so that $C_L(D) \neq \phi$, then

$$\lambda_c(D) = \sup_{A \subset\subset D} \lim_{t\to\infty} \frac{1}{t} \log \sup_{y \in A} P_y^h(\tau_A > t)$$

$$= \sup_{\substack{A \subset\subset D \\ \partial A \text{ a} \\ C^{2,\alpha}\text{-boundary}}} \lim_{t\to\infty} \frac{1}{t} \log P_x^h(\tau_A > t), \tag{4.2}$$

for any $x \in D$ and $h \in C_L(D)$, where $\{P_x^h, x \in \hat{D}\}$ is the solution to the generalized martingale problem for L^h on D. In particular, if $V \equiv 0$, then

$$\lambda_c(D) = \sup_{A \subset\subset D} \lim_{t\to\infty} \frac{1}{t} \log \sup_{y \in A} P_y(\tau_A > t)$$

$$= \sup_{\substack{A \subset\subset D \\ \partial A \text{ a} \\ C^{2,\alpha}\text{-boundary}}} \lim_{t\to\infty} \frac{1}{t} \log P_x(\tau_A > t),$$

for any $x \in D$.

Remark. From (4.2), it follows that if $\lambda_c(D) \le 0$, then $\lambda_c(D)$ may be characterised as the *exponential rate of escape from compact sets* of the diffusion process corresponding to L^h, for any $h \in C_L(D)$. In Chapter 8, Section 2, we develop the connection between h-transforms and diffusions conditioned to exit D in a specified manner. In this context, (4.2) states that the exponential rate of escape from compact sets for a diffusion conditioned to exit D in a specified manner does not depend on the particular conditioning.

Proof. Note that if $A \subset\subset D$ is a domain with a $C^{2,\alpha}$-boundary, then

$$\lim_{t\to\infty} \frac{1}{t} \log E_x\left(\exp\left(\int_0^t V(X(s))\,ds\right);\ \tau_A > t\right)$$

exists for all $x \in D$ by Theorem 3.6.1. For arbitrary $A \subset\subset D$, the technique of exercise 3.1 shows that

$$\lim_{t\to\infty} \frac{1}{t} \log \sup_{y\in A} E_y\left(\exp\left(\int_0^t V(X(s))\,ds\right);\ \tau_A > t\right),$$

exists. The first part of the theorem now follows from Theorem 4.1(i), Theorem 3.6.1 and the monotonicity of $\sup_{y\in A} E_y(\exp(\int_0^t V(X(s))\,ds);$ $\tau_A > t)$ as a function of A. The second part of the theorem follows by applying the first part to L^h and using Theorem 3.3(iv). □

We now give three different mini-max representations for λ_c.

Theorem 4.5. *Let L satisfy Assumption $\widetilde{H}_{\mathrm{loc}}$ on a domain $D \subseteq R^d$. Then*

$$\lambda_c = \inf_{\substack{u\in C^{2,\alpha}(D)\\ u>0 \text{ on } D}} \sup_{x\in D} \frac{Lu}{u}(x). \tag{4.3}$$

Theorem 4.6. *Let L satisfy Assumption $\widetilde{H}_{\mathrm{loc}}$ on a domain $D \subseteq R^d$ and assume that L is subcritical on D. Then*

$$(-\lambda_c)^{-1} = \inf_{\phi\in C^\alpha(D)} \sup_{x\in D} \frac{1}{\phi(x)} \int_D G(x,y)\phi(y)\,dy. \tag{4.4}$$

Remark. Theorem 4.6 can be applied even when L is not subcritical as long as one can identify a $\bar{\lambda}$ such that $L - \bar{\lambda}$ is subcritical. Then (4.4) holds with λ_c replaced by $\lambda_c - \bar{\lambda}$ and $G(x,y)$ replaced by $G_{\bar{\lambda}}(x,y)$, the Green's function for $L - \bar{\lambda}$.

***Theorem* 4.7.** *Let L satisfy Assumption $\widetilde{H}_{\mathrm{loc}}$ on a domain $D \subseteq R^d$. Then*

$$\lambda_c = \sup_{\mathcal{P}(D)} \inf_{\substack{u \in C^2(D) \\ u>0}} \int_D \frac{Lu}{u}(x)\mu(dx), \tag{4.5}$$

where $\mathcal{P}(D)$ is the space of probability measures on D.

Remark. In the case where D is bounded with a $C^{2,\alpha}$-boundary and L satisfies Assumption $\widetilde{H}$ on D, Theorem 4.7 is similar to the lefthand equality in equation (3.7.1). The difference is that in (3.7.1), the supremum is over probability measures μ with densities f satisfying $g \equiv f^{1/2} \in C^1(\bar{D})$ and $g = 0$ on ∂D, whereas in Theorem 4.7 the supremum is over all probability measures μ on D.

Proof of Theorem 4.5. Denote the righthand side of (4.3) by γ. Let $\varepsilon > 0$ and let $u \in C^{2,\alpha}(D)$ satisfy $u > 0$ and

$$\sup_{x \in D} \frac{Lu}{u}(x) \leqslant \gamma + \varepsilon.$$

Then $(L - (\gamma + \varepsilon))u \leqslant 0$ in D, and it follows from Theorem 3.9 and the definition of λ_c that $\lambda_c \leqslant \gamma + \varepsilon$. Since $\varepsilon > 0$ is arbitrary, we obtain $\lambda_c \leqslant \gamma$. To obtain the reverse inequality, let $u_0 \in C^{2,\alpha}(D)$ satisfy $u_0 > 0$ and $(L - \lambda_c)u_0 \leqslant 0$ in D; such a u_0 exists by the definition of λ_c. Then

$$\sup_{x \in D} \frac{Lu_0}{u_0}(x) \leqslant \lambda_c$$

and it follows that $\gamma \leqslant \lambda_c$. □

Proof of Theorem 4.6. Since L is subcritical, $\lambda_c \leqslant 0$. Let $\phi \in C_{L-\lambda_c}(D)$. Let $\{D_n\}_{n=1}^{\infty}$ be a sequence of domains with $C^{2,\alpha}$-boundaries satisfying $D_n \subset\subset D_{n+1}$ and $\bigcup_{n=1}^{\infty} D_n = D$, and let $G_n(x, y)$ denote the Green's function for L on D_n. By Theorem 3.6.4 and the definition of the Green's function, $u_n(x) = -\lambda_c \int_{D_n} G_n(x, y)\phi(y)\,dy$ is the unique solution to $Lu_n = \lambda_c \phi$ in D_n and $u_n = 0$ on ∂D_n. Since $L\phi = \lambda_c \phi$ in D_n and $\phi > 0$ on ∂D_n, applying the maximum principle of Theorem 3.6.5 to L on D_n, it follows that $u_n \leqslant \phi$ on D_n; that is,

$$(-\lambda_c)^{-1} \geqslant (\phi(x))^{-1} \int_{D_n} G_n(x, y)\phi(y)\,dy, \text{ for all } x \in D_n.$$

Letting $n \to \infty$ and using Theorem 3.7, we have

$$(-\lambda_c)^{-1} \geq \sup_{x \in D} (\phi(x))^{-1} \int_D G(x, y)\phi(y)\, dy.$$

This proves that the lefthand side of (4.4) is larger or equal to the righthand side of (4.4). To prove the reverse inequality, assume on the contrary that the lefthand side of (4.4) is strictly larger than the righthand side. Then there exists an $\varepsilon > 0$ and a $0 < \phi_\varepsilon \in C^\alpha(D)$ such that

$$\sup_{x \in D} (\phi_\varepsilon(x))^{-1} \int_D G(x, y)\phi_\varepsilon(y)\, dy = (-\lambda_c + \varepsilon)^{-1}. \qquad (4.6)$$

Let $v_\varepsilon(x) = \int_D G(x, y)\phi_\varepsilon(y)\, dy$. Then $Lv_\varepsilon = -\phi_\varepsilon$ in D. (To prove this, choose $\phi_\varepsilon^{(n)} \in C_0^\alpha(D)$ with $0 \leq \phi_\varepsilon^{(n)} \leq \phi_\varepsilon$ and $\phi_\varepsilon^{(n)} \nearrow \phi_\varepsilon$. Then, by Theorem 3.8, $v_\varepsilon^{(n)}(x) \equiv \int_D G(x, y)\phi_\varepsilon^{(n)}(y)\, dy$ satisfies $Lv_\varepsilon^{(n)} = -\phi_\varepsilon^{(n)}$ in D. Clearly, $v_\varepsilon(x) = \lim_{n\to\infty} v_\varepsilon^{(n)}(x)$, and a standard compactness argument then shows that $Lv_\varepsilon = -\phi_\varepsilon$.) Thus, from (4.6) and the definition of v_ε, we have $(L - \lambda_c + \varepsilon)v_\varepsilon = -\phi_\varepsilon + (-\lambda_c + \varepsilon)v_\varepsilon \leq 0$. It then follows, from Theorem 3.9 and the definition of λ_c, that $L - \lambda_c + \varepsilon$ is either subcritical or critical; this contradicts the definition of λ_c. □

Proof of Theorem 4.7. Denote the righthand side of (4.5) by γ. Since

$$\sup_{x \in D} \frac{Lu}{u}(x) \geq \int_D \frac{Lu}{u}(x)\mu(dx),$$

for any probability measure μ on D, it follows from Theorem 4.5 that $\lambda_c \geq \gamma$. To prove the reverse inequality, let $\{D_n\}_{n=1}^\infty$ be a sequence of domains with $C^{2,\alpha}$-boundaries satisfying $D_n \subset\subset D_{n+1}$ and $\bigcup_{n=1}^\infty = D$. Define

$$A_n = \left\{\mu \in \mathcal{P}(D) \colon \operatorname{supp}\mu \subseteq D_n,\, f \equiv \frac{d\mu}{dx} \text{ exists, } g \equiv f^{1/2} \in C^1(D_n) \right.$$
$$\left. \text{and } g = 0 \text{ on } \partial D_n \right\}.$$

Let λ_n denote the principal eigenvalue for L on D_n. Then, by Theorem 3.7.1,

$$\gamma \geq \sup_{\mu \in A_n} \inf_{\substack{u \in C^2(D) \\ u>0 \text{ on } D}} \int_{D_n} \frac{Lu}{u}(x)\mu(dx) = \lambda_n.$$

Letting $n \to \infty$ and using Theorem 4.1(i) gives $\gamma \geq \lambda_c$. □

4.5 Criteria for the finiteness or infiniteness of the generalized principal eigenvalue

If V is bounded on D, then by Theorem 3.2, $\lambda_c(D) < \infty$. In the unbounded case, it is possible to have $\lambda_c(D) = \infty$.

Theorem 5.1. *Let $L = L_0 + V$ on a bounded domain $D \subset R^d$ with a $C^{2,\alpha}$-boundary. Assume that L_0 satisfies Assumption $\tilde{H}$ on D and that $V \in C^\alpha(D)$ satisfies $\lim_{D \ni x \to \partial D} (\text{dist}(x, \partial D))^2 V^+(x) = 0$, where $V^+ = \max(0, V)$. Then $\lambda_c(D) < \infty$.*

Theorem 5.2. *Let $L = L_0 + V$ satisfy Assumption $\tilde{\mathrm{H}}_{\mathrm{loc}}$ on a domain $D \subset R^d$. Assume that there exists an open set $U \subset\subset R^d$ such that:* (i) *$U \cap \partial D \neq \phi$;* (ii) *D satisfies an interior sphere condition at some point in $\partial D \cap U$;* (iii) *L_0 satisfies Assumption $\tilde{H}$ on $D \cap U$; and* (iv) *$\lim_{D \cap U \ni x \to \partial D} (\text{dist}(x, \partial D))^2 V(x) = \infty$. Then $\lambda_c(D) = \infty$.*

Theorem 5.3. *Let $D \subset R^d$ be a domain of the form $D = U - \{x_0\}$, where $U \subseteq R^d$ is a domain and $x_0 \in U$, and let $L = L_0 + V$ satisfy Assumption $\tilde{\mathrm{H}}_{\mathrm{loc}}$ on D. Assume that L_0 satisfies Assumption E (defined in Section* 3.2) *near x_0.*

(i) *Assume that V is bounded from above away from x_0 and that $\lim_{x \to x_0} |x - x_0|^2 V^+(x) = 0$. Then $\lambda_c(D) < \infty$.*
(ii) *Assume that $\lim_{x \to x_0} |x - x_0|^2 V(x) = \infty$. Then $\lambda_c(D) = \infty$.*

Proof of Theorem 5.1. By the smoothness and boundedness assumptions on D, there exists an $\varepsilon > 0$ such that to every $x \in D$ satisfying $\text{dist}(x, \partial D) < \varepsilon$, there exists a unique $y \in \partial D$ for which $\text{dist}(x, y) = \text{dist}(x, \partial D)$. Furthermore, letting $A_\varepsilon = \{x \in D: \text{dist}(x, \partial D) < \varepsilon\}$, then the map T: $A_\varepsilon \to (0, \infty) \times \partial D$ defined by $Tx = (r, y)$, where $r = \text{dist}(x, \partial D)$ and $y \in \partial D$ satisfies $y = \text{dist}(x, \partial D)$, is a $C^{2,\alpha}$-map. In this coordinate system, L is restricted to A_ε and has the form

$$L = a(r, y)\frac{\partial^2}{\partial r^2} + b(r, y)\frac{\partial}{\partial r} + \mathcal{A} + V,$$

where $\inf_{A_\varepsilon} a > 0$ and b is bounded on A_ε, and where $\mathcal{A}f = 0$ whenever f is a function only of r. Define u on A_ε by $u(r, y) = r^{1/2}$. Then $(Lu)(r, y) = -\frac{1}{4}a(r, y)r^{-3/2} + \frac{1}{2}b(r, y)r^{-1/2} + r^{1/2}V(r, y)$. By the assumption on V and by choosing ε even smaller if necessary, we have $Lu \leqslant -1$ on A_ε. Now continue u to all of D in such a manner that $u \in C^{2,\alpha}(D)$ and $\inf_{D-A_\varepsilon} u > 0$. Clearly, then, $(L - \lambda)u \leqslant -1$ in D for sufficiently large λ. Thus, by Theorem 3.9, $L - \lambda$ is subcritical; hence $\lambda_c(D) < \infty$. □

Proof of Theorem 5.2. It follows by the interior sphere assumption that for some $r_0 > 0$, and each $r \in (0, r_0)$, there exists a ball B_r whose boundary intersects ∂D in one point and which satisfies $B_r \subset U \cap D$. Since L_0 satisfies Assumption $\tilde{\text{H}}$ on $U \cap D$, it follows by Theorem 3.7.2 that there exists a constant $c > 0$ independent of $r \in (0, r_0)$ such that the principal eigenvalue of L_0 on B_r is larger than

$$\frac{-c}{r^2}.$$

Thus, for

$$r \in (0, r_0),\ L_0 + \frac{c}{r^2}$$

is supercritical on B_r. By the assumption on V,

$$V \geqslant \frac{c_r}{r^2} \text{ in } B_r,$$

where $\lim_{r \to 0} c_r = \infty$. Therefore, given any λ, there exists an $r > 0$ such that

$$V - \lambda \geqslant \frac{c}{r^2} \text{ in } B_r.$$

By comparison, it follows that $L - \lambda = L_0 + V - \lambda$ is supercritical on B_r and, *a fortiori* on D. Since λ is arbitrary, $L - \lambda$ is supercritical on D for all λ and, thus, $\lambda_c(D) = \infty$. □

Proof of Theorem 5.3. The proof is left to the reader in exercise 4.18. □

4.6 Perturbation theory for critical and subcritical operators

In this section, we consider perturbations of the operator L on D of the form $L + tW$, where W is a function with compact support in D. We will investigate how such perturbations affect criticality. With the exception of Theorem 6.1, the results will be stated and proved for compactly supported perturbations. In fact, many of the results hold more generally for sufficiently small non-compactly supported perturbations; see the notes at the end of the chapter. In the proofs below, the reader will detect a dash of probabilistic flavor here and there; the proofs for non-compactly supported perturbations are a little more difficult and lose the probabilistic flavor. Some explicit calculations regarding the perturbation theory of this section are carried out in Chapter 5, Section 2 for operators of the form $\frac{1}{2}\Delta + tV(|x|)$.

In this section, the notation $\lambda_c = \lambda_c(D)$ will always refer to the operator L. For the operator $L + tW$, the notation $\lambda_c^{(t)} = \lambda_c^{(t)}(D)$ will be used. We begin with a result which will be used to prove several of the theorems in this section and which is also of independent interest.

Theorem 6.1. *Let L satisfy Assumption $\widetilde{H}_{\text{loc}}$ on a domain $D \subseteq R^d$ and let $V_0, V_1 \in C^\alpha(D)$. For $t \in (0, 1)$, define $L_t = L + (1-t)V_0 + tV_1$. Assume that $V_0 \not\equiv V_1$ and that $C_{L+V_0}(D) \neq \phi$ and $C_{L+V_1}(D) \neq \phi$. Then L_t is subcritical for $0 < t < 1$. Also, if the generalized principal eigenvalue of $L + V_0$ or $L + V_1$ is negative, then the generalized principal eigenvalue of L_t is negative for all $t \in (0, 1)$.*

Proof. Let $\phi_0 \in C_{L+V_0}(D)$ and $\phi_1 \in C_{L+V_1}(D)$ and define $\phi_t = \phi_0^{1-t}\phi_1^t$ for $0 < t < 1$. A straightforward calculation reveals that

$$L_t\phi_t = (1-t)\phi_t\phi_0^{-1}(L + V_0)\phi_0 + t\phi_t\phi_1^{-1}(L + V_1)\phi_1$$

$$-\tfrac{1}{2}t(1-t)\phi_t\left(\frac{\phi_1}{\phi_0}\right)^2 \sum_{i,j=1}^d a_{ij} \frac{\partial\left(\dfrac{\phi_0}{\phi_1}\right)}{\partial x_i} \frac{\partial\left(\dfrac{\phi_0}{\phi_1}\right)}{\partial x_j}.$$

Now $\phi_0/\phi_1 \not\equiv$ constant since $V_1 \not\equiv V_2$. Furthermore, $(L + V_i)\phi_i = 0$, $i = 0, 1$. Therefore, for $0 < t < 1$, $L_t\phi_t \le 0$ and $L_t\phi_t \not\equiv 0$ and, by Theorem 3.9, L is subcritical.

Now assume that $\lambda_c^{V_0}$, the generalized principal eigenvalue of $L + V_0$, is negative. Applying the first part of the theorem with V_0 replaced by $V_0 - \lambda_c^{V_0}$ shows that $L_t - (1-t)\lambda_c^{V_0}$ is subcritical; thus the generalized principal eigenvalue of L_t is negative for all $t \in (0, 1)$. □

The following result is very useful.

Theorem 6.2. *Let L satisfy Assumption $\widetilde{H}_{\text{loc}}$ on a domain $D \subseteq R^d$ and assume that L is subcritical on D. Then for any non-negative $W \in C_0^\alpha(D)$, there exists an $\varepsilon_0 > 0$ such that $L + \varepsilon W$ is subcritical on D for all $\varepsilon < \varepsilon_0$.*

Proof. If $h \in C_L(D)$, then the zeroth-order part of the L^h vanishes. Therefore, by Theorem 3.3(iv), we may assume that $V \equiv 0$. Extend W to $\hat{D}$ be defining $W(\Delta) = 0$ and exclude the trivial case, $W \equiv 0$. By Theorem 3.8, $u(x) \equiv \int_D G(x, y) W(y)\, dy = E_x \int_0^\infty W(X(s))\, ds$ is the smallest positive solution of $Lu = -W$ in D. Let $A = \text{supp}(W) \subset\subset D$. By the strong Markov property,

$$u(x) = E_x \int_0^\infty W(X(s))\, ds = E_x\left(E_{X(\sigma_A)} \int_0^\infty W(X(s))\, ds;\, \sigma_A < \infty \right)$$
$$= E_x(u(X(\sigma_A));\, \sigma_A < \infty), \tag{6.1}$$

where $\sigma_A = \inf\{t \geq 0\colon X(t) \in A\}$. Since u is bounded on A, it follows from (6.1) that u is in fact bounded on D. We have $(L + \varepsilon W)u = W(\varepsilon u - 1)$ on D. Since u is bounded, for sufficiently small $\varepsilon > 0$, $(L + \varepsilon W)u \leq 0$ and $(L + \varepsilon W)u \not\equiv 0$ on D; thus by Theorem 3.9, $L + \varepsilon W$ is subcritical on D. □

Remark. Note that the reduction via an h-transform in Theorem 6.2 was critical for our method of proof. Indeed, the above technique using the strong Markov property to show that the smallest solution u to $Lu = -W$ is bounded in D does not work if V does not have compact support in D.

The next theorem is the analog of Theorem 6.2 for critical and supercritical operators.

Theorem 6.3. *Let L satisfy Assumption $\widetilde{H}_{\text{loc}}$ on a domain $D \subseteq R^d$ and let $W \in C_0^\alpha(D)$ satisfy $W \geq 0$ and $W \not\equiv 0$.*

(i) *If L is critical on D, then $L + \varepsilon W$ is subcritical on D for $\varepsilon < 0$ and supercritical on D for $\varepsilon > 0$.*
(ii) *If L is supercritical on D, then there exists an $\varepsilon_0 > 0$ such that $L - \varepsilon W$ is supercritical on D for $\varepsilon < \varepsilon_0$.*

Proof. (i) Let $\phi_c \in C_L(D)$ be the ground state. Then for $\varepsilon < 0$, $(L + \varepsilon W)\phi_c \leq 0$ and $(L + \varepsilon W)\phi_c \not\equiv 0$. Thus by Theorem 3.9, $L + \varepsilon W$ is subcritical. Now let $\varepsilon > 0$ and assume that $L + \varepsilon W$ is not supercritical. Let $\phi \in C_{L+\varepsilon W}(D)$. Then $L\phi \leq 0$ and $L\phi \not\equiv 0$. By Theorem 3.9, it follows that L is subcritical, a contradiction. Thus, in fact, $L + \varepsilon W$ is supercritical for all $\varepsilon > 0$.

(ii) Assume to the contrary that, for all $\varepsilon > 0$, $L - \varepsilon W$ is not supercritical. For each such ε, there exists a $\phi_\varepsilon \in C_{L-\varepsilon W}(D)$. But then, a standard compactness argument shows that $C_L(D)$ is non-empty, a contradiction. □

For $W \in C_0^\alpha(D)$, define

$$S_+ = \{t \in R\colon L + tW \text{ is subcritical}\},$$

$$S_0 = \{t \in R\colon L + tW \text{ is critical}\},$$

and

$$S_- = \{t \in R\colon L + tW \text{ is supercritical}\}.$$

In the case where $W \geqslant 0$, we have the following theorem.

Theorem 6.4. *Let L satisfy Assumption $\widetilde{H}_{\text{loc}}$ on a domain $D \subseteq R^d$ and assume that L is subcritical on D. Let $W \in C_0^\alpha(D)$ satisfy $W \geqslant 0$ and $W \not\equiv 0$. Then there exists a $t_0 > 0$ such that $S_+ = (-\infty, t_0)$, $S_0 = \{t_0\}$ and $S_- = (t_0, \infty)$. Also, if the generalized principal eigenvalue of $L + tW$ is negative for some $t \in S_+$, then it is negative for all $t \in S_+$.*

Proof. By Theorem 6.2, S_+ is an open set; thus either $S_+ = (-\infty, \infty)$ or $S_+ = (-\infty, t_0)$, where $t_0 > 0$. We show now that $S_+ \neq (-\infty, \infty)$. We can find a domain $A \subset\subset D$ with a $C^{2,\alpha}$-boundary and an $\varepsilon > 0$ such that $W \geqslant \varepsilon$ on A. Now $L - \lambda$ is supercritical on A if $\lambda < \lambda_0(A) = \lambda_c(A)$. Since for sufficiently large t, $tW > -\lambda_c(A)$ on A, it follows by comparison that $L + tW$ is supercritical on A and thus, *a fortiori* on D. We conclude that $S_+ = (-\infty, t_0)$. A standard compactness argument shows that $t_0 \in S_0$, and applying Theorem 6.3(i) to $L + t_0 W$ shows that $S_- = (t_0, \infty)$. The final statement of the theorem follows easily from Theorem 6.1. □

We now prove the analog of Theorem 6.4 for the case in which W changes sign in D.

Theorem 6.5. *Let L satisfy Assumption $\widetilde{H}_{\text{loc}}$ on a domain $D \subseteq R^d$ and assume that L is subcritical on D. Let $W \in C_0^\alpha(D)$ satisfy $\inf_D W < 0$ and $\sup_D W > 0$. Then there exist t_-, $t_+ \in R$ satisfying $t_- < 0 < t_+$ such that $S_+ = (t_-, t_+)$, $S_0 = \{t_-, t_+\}$ and $S_- = (-\infty, t_-) \cup (t_+, \infty)$. Also, if the generalized principal eigenvalue for $L + tW$ is negative for some $t \in S_+$, then it is negative for all $t \in S_+$.*

Proof. We first show that $S_+ \cup S_0$ is bounded. We proceed as we did in the proof of Theorem 6.4. Pick a domain $A \subset\subset D$ with a $C^{2,\alpha}$-boundary and an $\varepsilon > 0$ such that $W \geqslant \varepsilon$ on A. Since $L - \lambda$ is supercritical on A for $\lambda < \lambda_c(A)$, and since $tW > -\lambda_c(A)$ on A for sufficiently large t, it follows by comparison that $L + tW$ is supercritical on A for sufficiently large t and, *a fortiori*, on D. By choosing an $\varepsilon > 0$ and a domain $B \subset\subset D$ with a $C^{2,\alpha}$-boundary such that $W < -\varepsilon$ on B and repeating the above argument, it follows that $L + tW$ is supercritical for all $t < 0$ for which $|t|$ is sufficiently large. This proves that $S_+ \cup S_0$ is bounded.

We now show that S_+ is open. Assume that $t \in S_+$. By Theorem 6.2, for $\varepsilon > 0$ sufficiently small, $L + tW + \varepsilon W^+$ is subcritical and thus, by comparison, $L + (t + \varepsilon)W$ is also subcritical. An identical argument using W^- in place of W^+ shows that $L + (t - \varepsilon)W$ is also subcritical for sufficiently small $\varepsilon > 0$.

A standard compactness argument shows that $S_+ \cup S_0$ is closed. By assumption, $0 \in S_+$. Thus, to complete the characterization of S_+, S_0 and S_-, it suffices to show that if $t_0 < t_1$ and $t_0, t_1 \in S_0$, then $t \in S_+$ for all $t \in (t_0, t_1)$. But this follows directly from Theorem 6.1 by letting $V_0 = t_0 W$ and $V_1 = t_1 W$. The final statement of the theorem also follows easily from Theorem 6.1. □

Theorems 6.4 and 6.5 treated perturbations of subcritical operators in the case where $W \geq 0$ and in the case where W changes sign, respectively. Theorem 6.3(i) treated perturbations of critical operators in the case where $W \geq 0$. The most delicate situation is the case in which the operator is critical and W changes sign.

Theorem 6.6. *Let L satisfy Assumption $\tilde{H}_{\text{loc}}$ on a domain $D \subseteq R^d$. Assume that L and, consequently, $\tilde{L}$ are critical on D and let ϕ_c and $\tilde{\phi}_c$ denote their ground states. Let $0 \not\equiv W \in C_0^\alpha(D)$.*

(i) *If $\int_D W \phi_c \tilde{\phi}_c \, dx < 0$, then there exists a $t_0 > 0$ such that $L + tW$ is subcritical for all $t \in (0, t_0)$.*
(ii) *If $\int_D W \phi_c \tilde{\phi}_c \, dx \geq 0$, then $L + tW$ is supercritical for all $t > 0$.*

Remark. Let S_+, S_0 and S_- be as defined above Theorem 6.4. From Theorem 6.6 and the theory developed in the preceding theorems, it is easy to deduce the following result. Let L and W be as in Theorem 6.6. If $\int_D W \phi_c \tilde{\phi}_c \, dx = 0$, then $S_+ = \phi$, $S_0 = \{0\}$, and $S_- = R - \{0\}$. If $\int_D W \phi_c \tilde{\phi}_c \, dx > 0$, then $S_+ = (t_0, 0)$, where $t_0 < 0$, $S_0 = \{t_0, 0\}$, and $S_- = (-\infty, t_0) \cup (0, \infty)$. If $\int_D w \phi_c \tilde{\phi}_c \, dx < 0$, then $S_+ = (0, t_0)$, where $t_0 > 0$, $S_0 = \{0, t_0\}$, and $S_- = (-\infty, 0) \cup (t_0, \infty)$.

Proof. We may assume that W changes sign in D since otherwise the theorem follows from Theorem 6.3. We will prove the theorem in three steps.

Step 1. Assume that D is bounded with a $C^{2,\alpha}$-boundary and that L satisfies Assumption $\tilde{\mathrm{H}}$ on D. If $\int_D W\phi_c\tilde{\phi}_c\,dx > 0$, then $L + tW$ is supercritical for all $t > 0$.

Proof of Step 1. Let $\lambda_c^{(t)}(D)$ denote the generalized principal eigenvalue for $L + tW$ on D. We will show that $\lambda_c^{(t)}(D) > 0$ for $t > 0$. By the assumptions on L and D, $\lambda_c^{(t)}(D)$ and $\lambda_c(D)$ are given by the mini-max variational formulas of Theorem 3.7.1. In particular,

$$\lambda_c(D) = \sup_{\mu} \inf_{\substack{u \in C^2(D) \\ u>0 \text{ on } \bar{D}}} \int_D \frac{Lu}{u}\,d\mu,$$

where $\sup_\mu$ is over all probability measures μ on D with densities f satisfying $f^{1/2} \in C^1(D)$ and $f = 0$ on ∂D. Furthermore, the mini-max is attained at $\mu = \phi_c\tilde{\phi}_c\,dx$ and $u = \phi_c$, where ϕ_c and $\tilde{\phi}_c$ have been normalized by $\int_D \phi_c\tilde{\phi}_c\,dx = 1$. Therefore, since $\lambda_c(D) = 0$, we have

$$\inf_{\substack{u \in C^2(D) \\ u>0 \text{ on } \bar{D}}} \int_D \frac{Lu}{u}\phi_c\tilde{\phi}_c\,dx = 0. \tag{6.2}$$

Applying the mini-max formula to $\lambda_c^{(t)}(D)$ and using (6.2) for the second equality below gives for $t > 0$,

$$\begin{aligned}\lambda_c^{(t)}(D) &= \sup_{\mu} \inf_{\substack{u \in C^2(D) \\ u>0 \text{ on } \bar{D}}} \int_D \frac{(L + tW)u}{u}\,d\mu \\ &\geqslant \inf_{\substack{u \in C^2(D) \\ u>0 \text{ on } \bar{D}}} \int_D \frac{(L + tW)u}{u}\phi_c\tilde{\phi}_c\,dx \\ &= t\int_D W\phi_c\tilde{\phi}_c\,dx > 0.\end{aligned}$$

Step 2. Part (i) holds.

Proof of Step 2. If $L + t_1W$ is subcritical for some $t_1 > 0$, then by Theorem 6.1, $L + tW$ is subcritical for all $t \in (0, t_1)$. Thus to prove Step 2, we will assume that

$$L + tW \text{ is not subcritical}, \quad \text{for each } t > 0, \tag{6.3}$$

and show that $\int_D W\phi_c\tilde{\phi}_c\,dx \geqslant 0$. Let $\{D_n\}_{n=1}^{\infty}$ be a sequence of domains with $C^{2,\alpha}$-boundaries satisfying $D_n \subset\subset D_{n+1}$ and $\bigcup_{n=1}^{\infty} D_n = D$.

By Theorem 4.1 or Corollary 4.2, L is subcritical on D_n and by Theorem 6.5, it then follows that there exists a $t_n > 0$ such that $L + t_n W$ is critical on D_n and $L + (t_n - t)W$ is subcritical on D_n for $0 < t < t_n$. From Step 1 applied to the critical operator $L + t_n W$ on D_n and the perturbation $-W$, we conclude that

$$\int_{D_n} W \phi_n \tilde{\phi}_n \, dx \geq 0, \tag{6.4}$$

where ϕ_n and $\tilde{\phi}_n$ are the ground states for $L + t_n W$ and $\tilde{L} + t_n W$ on D_n, normalized by $\phi_n(x_0) = \tilde{\phi}_n(x_0) = 1$, for some $x_0 \in D_1$. We have

$$\lim_{n \to \infty} t_n = 0. \tag{6.5}$$

To see this, assume on the contrary that $\limsup_{n \to \infty} t_n = t_0 > 0$. By a standard compactness argument, $C_{L+t_0W}(D) \neq \phi$ and then, by Theorem 6.1, $L + tW$ is subcritical for $0 < t < t_0$; this contradicts (6.3). From (6.5), a standard compactness argument, and Theorem 3.4, it follows that $\lim_{n \to \infty} \phi_n = \phi_c$ and $\lim_{n \to \infty} \tilde{\phi}_n = \tilde{\phi}_c$, uniformly on compact subsets of D, where ϕ_c and $\tilde{\phi}_c$ have been normalized by $\phi_c(x_0) = \tilde{\phi}_c(x_0) = 1$. Since W has compact support in D, it follows from this and (6.4) that $\int_D W \phi_c \tilde{\phi}_c \, dx \geq 0$.

Step 3. Part (ii) holds.

Proof of Step 3. First consider the case $\int_D W \phi_c \tilde{\phi}_c \, dx > 0$. Then applying part (i) to $-W$ shows that $L - t_1 W$ is subcritical for some $t_1 > 0$. Now if for some $t_2 > 0$, $L + t_2 W$ were not supercritical, then it would follow from Theorem 6.1 that L is subcritical, which is a contradiction. This proves part (ii) in the case $\int_D W \phi_c \tilde{\phi}_c \, dx > 0$. Now consider the case $\int_D W \phi_c \tilde{\phi}_c \, dx = 0$. Assume that $L + t_1 W$ is subcritical for some $t_1 > 0$. Let $V_0 \in C_0^\infty(D)$ satisfy $V \geq 0$ and $V \not\equiv 0$. By Theorem 6.2, there exists an $\varepsilon > 0$ such that

$$L + t_1 W + \varepsilon V_0 = L + t_1 \left(W + \frac{\varepsilon}{t_1} V_0 \right)$$

is subcritical on D. Since

$$\int_D \left(W + \frac{\varepsilon}{t_1} V_0 \right) \phi_c \tilde{\phi}_c \, dx > 0,$$

this contradicts the case we have already proved. □

The following result is sometimes useful; it is used for the applications in Chapter 5, Section 2.

Theorem 6.7. *Let L satisfy Assumption $\widetilde{H}_{\text{loc}}$ on a domain $D \subseteq R^d$ and let $W \in C_0^\alpha(D)$. Assume that L is either critical or subcritical on D and that $L + W$ is supercritical on D, and let $\lambda_c^W > 0$ denote the generalized principal eigenvalue for $L + W$ on D. Then $L + W - \lambda_c^W$ is critical on D.*

Proof. Assume contrary to the statement of the theorem that $L + W - \lambda_c^W$ is subcritical. Then, by Theorem 6.4 or Theorem 6.5, there exists an $\varepsilon > 0$ such that $L + (1 + \varepsilon)W - \lambda_c^W$ is also subcritical. Since L is either critical or subcritical, and $\lambda_c^W > 0$, the generalized principal eigenvalue of $L - \lambda_c^W$ is negative. Applying Theorem 6.1 with L replaced by $L - \lambda_c^W$ and with $V_0 = 0$ and $V_1 = (1 + \varepsilon)W$, it follows that the generalized principal eigenvalue of $L + W - \lambda_c^W$ is negative. But this contradicts the definition of λ_c^W. □

We conclude this section with an interesting result which holds in the case where $L_0 = \frac{1}{2}\Delta$. In the mathematical physics literature, this result is known as the 'localization of binding'.

Theorem 6.8. *Let $W_1, W_2 \in C_0^\alpha(R^d)$ and assume that $\frac{1}{2}\Delta + W_1$ and $\frac{1}{2}\Delta + W_2$ are subcritical operators on R^d, $d \geq 3$. Let $\hat{e}$ be a unit vector in R^d and define $V_R(x) = W_1(x) + W_2(x - R\hat{e})$. Then for sufficiently large R, $\frac{1}{2}\Delta + V_R$ is subcritical on R^d.*

Before proving Theorem 6.8, we present two lemmas which are of independent usefulness. The first one is used to prove the second one, and the second one is used in the proof of Theorem 6.8.

Lemma 6.9. *Let $L = L_0 + V$ satisfy Assumption $\widetilde{H}_{\text{loc}}$ on a domain $D \subseteq R^d$. Assume that L is subcritical on D and that $V \in C_0^\alpha(D)$. Then there exists a positive function $\phi \in C^{2,\alpha}(D)$ which is bounded and bounded away from zero and satisfies $L\phi \leq 0$.*

Proof. Let $V^+ = \max(0, V)$. Extend V^+ to $\hat{D}$ by $V^+(\Delta) = 0$. By Theorem 3.8 and the definition of the Green's function,

$$u(x) \equiv \int_D G(x, y)V^+(y)\,dy = E_x \int_0^\infty V^+(X(t)) \exp\left(\int_0^t V(X(s))\,ds\right) dt$$

satisfies $Lu = -V^+$ in D and $u \in C^{2,\alpha}(D)$. Let $A = \text{supp}(V)$ and let $\sigma_A = \inf\{t \geq 0\colon X(t) \in A\}$. By the strong Markov property and the fact that V is supported in A, we have for $x \in D$,

$$u(x) = E_x\left(E_{X(\sigma_A)}\int_0^\infty V^+(X(t))\exp\left(\int_0^t V(X(s))\,ds\right)dt;\ \sigma_A < \infty\right)$$
$$= E_x(u(X(\sigma_A));\ \sigma_A < \infty).$$

Since u is bounded on A, it follows that in fact u is bounded on D. Let $\phi = 1 + u$. Then $\phi \in C^{2,\alpha}(D)$, ϕ is bounded and bounded away from zero, and $L\phi \leq 0$. □

Lemma 6.10. *Let $L = L_0 + V$ satisfy Assumption $\widetilde{H}_{\mathrm{loc}}$ on a domain $D \subseteq R^d$. Assume that L is subcritical on D and that $V \in C_0^\infty(D)$. Then there exists a constant c such that $E_x(\exp(\int_0^\sigma V(X(s))\,ds);\ \tau_D > \sigma) \leq c$, for all $x \in D$ and all stopping times σ for $X(\cdot)$.*

Proof. By Lemma 6.9, there exists a $\phi \in C^{2,\alpha}(D)$ which is bounded and bounded away from zero and satisfies $L\phi \leq 0$. Let $\{D_n\}_{n=1}^\infty$ be a sequence of domains satisfying $D_n \subset\subset D_{n+1}$ and $\bigcup_{n=1}^\infty D_n = D$. Since $L\phi \leq 0$, by the Feynman–Kac formula (Theorem 2.4.2), $\exp(\int_0^{t\wedge\tau_{D_n}} V(X(s))\,ds)\phi(X(t \wedge \tau_{D_n}))$ is a supermartingale. In fact, the same proof shows that $\exp(\int_0^{t\wedge\sigma\wedge\tau_{D_n}} V(X(s))\,ds)\phi(X(t \wedge \sigma \wedge \tau_{D_n}))$ is a supermartingale, for any stopping time σ. Therefore,

$$E_x \exp\left(\int_0^{t\wedge\sigma\wedge\tau_{D_n}} V(X(s))\,ds\right)$$
$$\leq \frac{1}{\inf_D \phi} E_x \exp\left(\int_0^{t\wedge\sigma\wedge\tau_{D_n}} V(X(s))\,ds\right)\phi(X(t \wedge \sigma \wedge \tau_{D_n}))$$
$$\leq \frac{\phi(x)}{\inf_D \phi} \leq \frac{\sup_D \phi}{\inf_D \phi}.$$

Letting $t \to \infty$ and using Fatou's lemma gives

$$E_x \exp\left(\int_0^{\sigma\wedge\tau_{D_n}} V(X(s))\,ds\right) \leq \frac{\sup_D \phi}{\inf_D \phi},\ \text{for } n = 1, 2, \ldots$$

and any stopping time σ. Thus,

$$E_x\left(\exp\left(\int_0^\sigma V(X(s))\,ds\right);\ \tau_D > \sigma\right)$$
$$= \lim_{n\to\infty} E_x\left(\exp\left(\int_0^\sigma V(X(s))\,ds\right);\ \tau_{D_n} > \sigma\right)$$

$$= \lim_{n\to\infty} E_x\left(\exp\left(\int_0^{\sigma\wedge\tau_{D_n}} V(X(s))\,ds\right);\ \tau_{D_n} > \sigma\right)$$

$$\leqslant \limsup_{n\to\infty} E_x \exp\left(\int_0^{\sigma\wedge\tau_{D_n}} V(X(s))\,ds\right) \leqslant \frac{\sup_D \phi}{\inf_D \phi}. \qquad \square$$

Proof of Theorem 6.8. Choose r so that W_1 and W_2 are supported inside the sphere of radius r centered at the origin. Let $R > 2r$ and define

$$\Gamma_1 = \left\{x \in R^d\colon |x| = \frac{R}{2} \text{ or } |x - R\hat{e}| = \frac{R}{2}\right\}$$

and $\Gamma_2 = \{x \in R^d\colon |x| = r \text{ or } |x - R\hat{e}| = r\}$. Define $\sigma_1 = \inf\{t \geqslant 0\colon X(t) \in \Gamma_1\}$ and, by induction, define $\tau_n = \inf\{t > \sigma_n\colon X(t) \in \Gamma_2\}$ and $\sigma_{n+1} = \inf\{t > \tau_n\colon X(t) \in \Gamma_1\}$, $n = 1, 2, \ldots.$ Let $A_n = \{X(\cdot)\colon \tau_n = \infty, \tau_{n-1} < \infty\}$, $n \geqslant 2$, and let $A_1 = \{X(\cdot)\colon \tau_1 = \infty\}$. By the transience of d-dimensional Brownian motion for $d \geqslant 3$,

$$P_x\left(\bigcup_{n=1}^{\infty} A_n\right) = 1. \tag{6.6}$$

By Theorem 6.4 or 6.5, we can choose an $\varepsilon > 0$ such that $\frac{1}{2}\Delta + (1 + \varepsilon)W_i$ is subcritical for $i = 1, 2$. By Lemma 6.10 and the translation invariance of the Laplacian, there exists a constant $c_\varepsilon \geqslant 1$ such that for $x \in R^d$ and any stopping time σ,

$$\left.\begin{aligned} E_x \exp\left(\int_0^\sigma (1 + \varepsilon)W_1(X(t))\,dt\right) &\leqslant c_\varepsilon, \\ E_x \exp\left(\int_0^\sigma (1 + \varepsilon)W_2(X(t) - R\hat{e})\,dt\right) &\leqslant c_\varepsilon. \end{aligned}\right\} \tag{6.7}$$

By Example 5.1.2 in Chapter 5, the probability that d-dimensional Brownian motion starting from modulus r_2 ever attains modulus $r_1 < r_2$ is

$$\left(\frac{r_1}{r_2}\right)^{d-2}.$$

Thus for $x \in \Gamma_1$,

$$P_x(\tau_1 < \infty) \leqslant 2\left(\frac{r}{(\frac{R}{2})}\right)^{d-2} = 2\left(\frac{2r}{R}\right)^{d-2}. \tag{6.8}$$

Note that for any $t \geq 0$, if $|x| < R/2$, then

$$E_x \exp\left(\int_0^{\sigma_1 \wedge t} (1+\varepsilon) V_R(X(s))\, ds\right) = E_x \exp\left(\int_0^{\sigma_1 \wedge t} (1+\varepsilon) W_1(X(s))\, ds\right);$$

if $|x - R\hat{e}| < \dfrac{R}{2}$, then

$$E_x \exp\left(\int_0^{\sigma_1 \wedge t} (1+\varepsilon) V_R(X(s))\, ds\right) = E_x \exp\left(\int_0^{\sigma_1 \wedge t} (1+\varepsilon) W_2(X(s) - R\hat{e})\, ds\right);$$

and if

$$|x| \geq \frac{R}{2} \text{ and } |x - R\hat{e}| \geq \frac{R}{2},$$

then

$$E_x \exp\left(\int_0^{\sigma_1 \wedge t} (1+\varepsilon) V_R(X(s))\, ds\right) = 1.$$

Thus, it follows for (6.7) that

$$E_x \exp\left(\int_0^{\sigma_1 \wedge t} (1+\varepsilon) V_R(X(s))\, ds\right) \leq c_\varepsilon, \text{ for } x \in R^d \text{ and } t \geq 0. \quad (6.9)$$

From the definition of A_n and the fact that $A_n \subset \{\sigma_n < \infty\}$, we have

$$\begin{aligned} E_x&\left(\exp\left(\int_0^t (1+\varepsilon) V_R(X(s))\, ds\right); A_n\right) \\ &= E_x\left(\exp\left(\int_0^{t \wedge \sigma_n} (1+\varepsilon) V_R(X(s))\, ds\right); A_n\right) \\ &\leq E_x\left(\exp\left(\int_0^{t \wedge \sigma_n} (1+\varepsilon) V_R(X(s))\, ds\right); \sigma_n < \infty\right). \quad (6.10) \end{aligned}$$

In order to avoid ambiguities in the calculation below involving conditional expectations, the variable with respect to the outer integration E_x will be denoted, as usual, by $X(\cdot)$, while the variable with respect to the inner integration $E_{X(\tau_{n-1})}$ will be denoted by $X'(\cdot)$. Furthermore, inside the inner integration, we will write $\sigma_1(X'(\cdot))$ and $\tau_{n-1}(X(\cdot))$ to indicate that the former term is a variable with respect to the $E_{X(\tau_{n-1})}$ integration, but that the latter term does not depend on $X'(\cdot)$ and is, thus, a constant with respect to the $E_{X(\tau_{n-1})}$ integration. Using the strong Markov property, (6.8), (6.9) and the fact that, up to sets of P_x-measure zero, $\{\sigma_n < \infty\} = \{\tau_{n-1} < \infty\}$, we have

$$
\begin{aligned}
&E_x\left(\exp\left(\int_0^{t\wedge\sigma_n}(1+\varepsilon)V_R(X(s))\,\mathrm{d}s\right);\ \sigma_n<\infty\right)\\
&\quad= E_x\exp\left(\int_0^{t\wedge\tau_{n-1}}(1+\varepsilon)V_R(X(s))\,\mathrm{d}s\right)1_{\{\tau_{n-1}<\infty\}}\\
&\quad\cdot E_{X(\tau_{n-1})}\exp\left(\int_0^{(t-t\wedge\tau_{n-1}(X(\cdot)))\wedge\sigma_1(X'(\cdot))}(1+\varepsilon)V_R(X'(s))\,\mathrm{d}s\right)\\
&\quad\leqslant c_\varepsilon E_x\left(\exp\left(\int_0^{t\wedge\tau_{n-1}}(1+\varepsilon)V_R(X(s))\,\mathrm{d}s\right);\ \tau_{n-1}<\infty\right)\\
&\quad= c_\varepsilon E_x\left(\exp\left(\int_0^{t\wedge\sigma_{n-1}}(1+\varepsilon)V_R(X(s))\,\mathrm{d}s\right);\ \tau_{n-1}<\infty\right)\\
&\quad= c_\varepsilon E_x\exp\left(\int_0^{t\wedge\sigma_{n-1}}(1+\varepsilon)V_R(X(s))\,\mathrm{d}s\right)1_{\{\sigma_{n-1}<\infty\}}P_{X(\sigma_{n-1})}(\tau_1<\infty)\\
&\quad\leqslant 2\left(\frac{2r}{R}\right)^{d-2}c_\varepsilon E_x\left(\exp\left(\int_0^{t\wedge\sigma_{n-1}}(1+\varepsilon)V_R(X(s))\,\mathrm{d}s\right);\ \sigma_{n-1}<\infty\right).
\end{aligned}
$$

Thus, by induction and (6.9),

$$
E_x\left(\exp\left(\int_0^{t\wedge\sigma_n}(1+\varepsilon)V_R(X(s))\,\mathrm{d}s\right);\ \sigma_n<\infty\right)\leqslant c_\varepsilon\left[2\left(\frac{2r}{R}\right)^{d-2}c_\varepsilon\right]^{n-1}. \tag{6.11}
$$

From (6.6), (6.10) and (6.11), we obtain

$$
E_x\exp\left(\int_0^t(1+\varepsilon)V_R(X(s))\,\mathrm{d}s\right)\leqslant c_\varepsilon\sum_{n=0}^{\infty}\left[2\left(\frac{2r}{R}\right)^{d-2}c_\varepsilon\right]^n<\infty,
$$

$$
\text{if } R>2r(2c_\varepsilon)^{1/(d-2)}. \tag{6.12}
$$

We will now show that $\frac{1}{2}\Delta+V_R$ is subcritical on R^d for $R>2r(2c_\varepsilon)^{1/(d-2)}$. Assume on the contrary that $\frac{1}{2}\Delta+V_R$ is not subcritical. Then applying Theorem 6.4 or Theorem 6.5 (depending on whether or not V_R changes sign) with $L=\frac{1}{2}\Delta$ and $W=V_R$, it follows that $\frac{1}{2}\Delta+(1+\varepsilon)V_R$ is supercritical on R^d, where ε is as above. Thus $\lambda_c=\lambda_c(R^d)>0$. Thus, from Theorem 4.1(i), the principal eigenvalue for $\frac{1}{2}\Delta+(1+\varepsilon)V_R$ in $B_n=\{|x|<n\}$ is also positive if n is sufficiently large. Fix such an n. Then it follows from Theorem 3.6.1(ii) that

$$
\lim_{t\to\infty}E_x\left(\exp\left(\int_0^t(1+\varepsilon)V_R(X(s))\,\mathrm{d}s\right);\ \tau_{B_n}>t\right)=\infty.
$$

Thus, *a fortiori*,

$$\lim_{t\to\infty} E_x \exp\left(\int_0^t (1+\varepsilon)V_R(X(s))\,\mathrm{d}s\right) = \infty.$$

But this contradicts (6.12). □

4.7 $L - \lambda_c$ is critical if D is bounded and L is uniformly elliptic with bounded coefficients

In this section, we will prove the following generalization of Theorem 2.4.

Theorem 7.1. *Let L satisfy Assumption $\tilde{H}_{\mathrm{loc}}$ on a bounded domain $D \subset R^d$. Assume in addition that L satisfies Assumption E on D (Assumption E is defined in Chapter 3, Section 2). Then $L - \lambda_c$ is critical on D.*

In order to prove Theorem 7.1, we will need to establish several auxiliary results which are of independent interest.

Definition. *Let $\lambda_{c,\infty} = \lambda_{c,\infty}(D) = \inf\{\lambda_c(D'):\ D'$ a domain satisfying $D - D' \subset\subset D\}$.*

Remark. In the case where L is symmetric, $\lambda_{c,\infty}$ is the supremum of the essential spectrum of the self-adjoint realization of L (see notes at the end of the chapter).

Theorem 7.2. *Let L satisfy Assumption $\tilde{H}_{\mathrm{loc}}$ on a domain $D \subseteq R^d$. Assume that $\lambda_{c,\infty} < \lambda_c < \infty$. Then $L - \lambda_c$ is critical on D.*

For the proof of Theorem 7.2, we need the following lemma.

Lemma 7.3. *Let L satisfy Assumption $\tilde{H}_{\mathrm{loc}}$ on a domain $D \subseteq R^d$. Assume that $\lambda_{c,\infty} < \lambda_c$ and let $\lambda \in (\lambda_{c,\infty}, \lambda_c)$. Let $U \subset\subset D$ be a simply connected domain such that $\lambda_c(D - \bar{U}) \leq \lambda$. Then for any non-negative $W \in C^\alpha(D)$ satisfying $W > 0$ on $\bar{U}$, there exists a $t_0 > 0$ such that $L - \lambda - t_0 W$ is subcritical on D.*

Proof. Let $u \in C_{L-\lambda}(D - \bar{U})$ and let U_1 be a domain satisfying $U \subset\subset U_1 \subset\subset D$ and such that $W > 0$ on $\bar{U}_1$. Let $\zeta \in C^\infty(D)$ satisfy $\operatorname{supp}\zeta \subset D - U$, $0 \leq \zeta \leq 1$, and $\zeta = 1$ on $D - U_1$. Define $v(x) = \zeta(x)u(x) + (1 - \zeta(x))$ and

$$q = \frac{Lv}{v}.$$

By construction, $v > 0$ on D, $v \in C^{2,\alpha}(D)$, $\operatorname{supp} q \subseteq \bar{U}_1$ and $q \in C^{\alpha}(D)$. Since $W > 0$ on $\bar{U}_1$, it follows that there exists a $t_0 > 0$ such that $q \leqslant t_0 W$ on D and $q < t_0 W$ on U_1. Since $(L - q)v \equiv 0$, it follows that $(L - t_0 W)v \leqslant 0$ and $(L - t_0 W)v \not\equiv 0$. Thus, by Theorem 3.9, $L - t_0 W$ is subcritical on D. □

Proof of Theorem 7.2. Let $\lambda_1 \in (\lambda_{c,\infty}, \lambda_c)$. By Lemma 7.3, we can choose a $W \in C_0^{\alpha}(D)$ such that $W \geqslant 0$ and $C_{L-\lambda_1-W}(D) \neq \emptyset$. Define $\lambda_t = (1 - t)\lambda_c + t\lambda_1$, for all $t \in [0, 1)$ and define

$$\nu(t) = \inf\{s \in R: C_{L-\lambda_t-sW}(D) \neq \emptyset\}, \qquad t \in [0, 1].$$

Since $W \not\equiv 0$ and since $C_{L-\lambda_1-W}(D) \neq \emptyset$, $\nu(t)$ is finite for $t \in [0, 1]$. It then follows from Theorem 6.1 that ν is convex; thus, in particular, $\nu(t)$ is continuous on $[0, 1]$. Since $\lambda_t < \lambda_c$, for $t \in (0, 1]$, we have $\nu(t) > 0$ for $t \in (0, 1]$. On the other hand, since $\lambda_0 = \lambda_c$, we have $\nu(0) \leqslant 0$. By continuity, it follows that $\nu(0) = 0$. By Theorem 6.2, if $\nu(0) = 0$, then $L - \lambda_c$ is critical on D. □

The following theorem, which we state without proof, is due to Alexandroff and may be found (in a more refined form) in [Gilbarg and Trudinger (1983), Theorem 9.1]. The proof, though by no means trivial, uses only advanced calculus and geometric considerations.

Theorem 7.4. *Let L and D satisfy the assumptions of Theorem 7.1 and assume in addition that $V \leqslant 0$. Let $u \in C^{2,\alpha}(D) \cap C(\bar{D})$ satisfy $Lu \geqslant f$ in D. Then*

$$\sup_D u \leqslant \sup_{\partial D} u^+ + c\|f\|_{L^d(D)},$$

where c depends only on d, on the diameter of D, and on

$$\inf_{\substack{x\in D\\|v|=1}} \sum_{i,j=1}^{d} a_{ij}(x)v_i v_j, \ \sup_{\substack{x\in D\\|v|=1}} \sum_{i,j=1}^{d} a_{ij}(x)v_i v_j, \quad \text{and} \ \sup_{x\in D} |b|.$$

Corollary 7.5. *Let L satisfy the Assumptions of Theorem 7.4 and let $G(x, y)$ be the Green's function for L on D. Then*

$$\sup_{x\in D} \int_D G(x, y)\,dy \leqslant c|D|^{1/d},$$

where c is as in Theorem 7.4.

Proof. Let $\{D_n\}_{n=1}^{\infty}$ be a sequence of domains with $C^{2,\alpha}$-boundaries satisfying $D_n \subset\subset D_{n+1}$ and $\bigcup_{n=1}^{\infty} D_n = D$, and let $G_n(x, y)$ denote the Green's function for L on D_n. For each n, let $\{f_{n,m}\}_{m=1}^{\infty}$ be an increasing sequence of non-negative functions satisfying $f_{n,m} \in C_0^{\alpha}(D_n)$ and $\lim_{m\to\infty} f_{n,m}(x) = 1$, for $x \in D$, and let

$$u_{n,m}(x) = \int_{D_n} G_n(x, y) f_{n,m}(y)\,dy, \text{ for } x \in D_n.$$

By Theorem 3.8, $u_{n,m} \in C^{2,\alpha}(D_n)$ and $Lu_{n,m} = -f_{n,m}$ on D_n. By Theorem 7.3.2 in Chapter 7, it follows that $u_{n,m} \in C(\bar{D}_n)$ and $u_{n,m} = 0$ on ∂D_n. Applying Theorem 7.4 to $u_{n,m}$, and using the fact that $|f_{n,m}| \leqslant 1$, we have

$$\sup_{D_n} u_{n,m} \leqslant c|D_n|^{1/d}, \tag{7.1}$$

where c is as in Theorem 7.4.

Letting $m \to \infty$ and then letting $n \to \infty$ and using Theorem 3.7, it follows from (7.1) that

$$\sup_{x\in D} \int_D G(x, y)\,dy \leqslant c|D|^{1/d}.$$ □

Theorem 7.6. *Under the conditions of Theorem* 7.1, $\lambda_{c,\infty} = -\infty$.

Proof. By assumption, V is bounded on D; thus it is enough to prove the theorem in the case where $V \leqslant 0$. In particular, then, L is subcritical on D. Let $\{U_n\}_{n=1}^{\infty}$ be a sequence of simply connected domains satisfying $U_n \subset\subset U_{n+1}$ and $\bigcup_{n=1}^{\infty} U_n = D$. Let $D_n = D - \bar{U}_n$, and let $G_n(x, y)$ denote the Green's function for L on D_n. Applying Theorem 4.6 to L on D_n, and using the test function $\phi \equiv 1$, we obtain

$$(-\lambda_c(D_n))^{-1} \leqslant \sup_{x\in D_n} \int_{D_n} G_n(x, y)\,dy. \tag{7.2}$$

By (7.2), Corollary 7.5, and the assumptions on L, there exists a constant c, independent of n, such that $(-\lambda_c(D_n))^{-1} \leqslant c|D_n|^{1/d}$. The theorem follows from this since $\lim_{n\to\infty} |D_n| = 0$. □

Proof of Theorem 7.1. Theorem 7.1 follows immediately from Theorem 7.2 and Theorem 7.6. □

Under the assumptions of Theorem 7.1, we have the following theorem concerning the dependence of the generalized principal eigenvalue on the coefficients of L.

Theorem 7.7. *Let L and D satisfy the conditions of Theorem* 7.1.

(i) *In its dependence on V, λ_c is strictly convex.*
(ii) *In its dependence on b, λ_c is locally Lipschitz continuous.*

Proof. (i) follows immediately from Theorem 6.1 and Theorem 7.1. (ii) is left as exercise 4.12. □

We end this section by noting that under the conditions of Theorem 7.1, a generalized maximum principle holds for L [Berestycki, Nirenberg, and Varadhan (1994)]. We state this result without proof.

Theorem 7.8. *Let L and D satisfy the assumptions of Theorem* 7.1. *Assume that L is subcritical on D; that is, $\lambda_c < 0$.*

(i) *There exists a function $u \in C^{2,\alpha}(D)$ and constants c_1, $c_2 > 0$ such that $Lu \leqslant 0$ and $c_1 < u < c_2$ in D.*
(ii) *If $v \in C^{2,\alpha}(D) \cap C(\bar{D})$ satisfies $Lv \leqslant 0$ in D and $v \geqslant 0$ on ∂D, then $v \geqslant 0$ in D.*

Remark. This theorem generalizes Theorem 3.6.5. Note that (ii) follows from (i) and Theorem 3.2.2. (i) is equivalent to

$$\sup_{x \in D} E_x \exp\left(\int_0^{\tau_D} V(X(s))\,\mathrm{d}s\right) < \infty.$$

4.8 Invariant measures and invariant functions for transition measures

Definition. *A σ-finite measure $v \not\equiv 0$ on D is called an* invariant measure *for the transition measure $p(t,x,\mathrm{d}y)$ if $\int_D p(t,x,\mathrm{d}y)\nu(\mathrm{d}x) = \nu(\mathrm{d}y)$ for all $t \geqslant 0$. If ν is absolutely continuous and $\nu(\mathrm{d}x) = \phi(x)\,\mathrm{d}x$, then ϕ is called an* invariant density *for $p(t,x,\mathrm{d}y)$. In the case $V \equiv 0$, $\nu(\phi)$ is also called an invariant measure (density) for the diffusion process corresponding to L on D and if, in addition, $\nu(D) = 1$, then $\nu(\phi)$ is also called an* invariant probability measure (density) *for the diffusion process corresponding to L on D.*

Definition. *A measureable positive function ϕ on D is called an* invariant positive function *for the transition measure $p(t,x,\mathrm{d}y)$ if $\int_D p(t,x,\mathrm{d}y)\phi(y) = \phi(x)$, for all $x \in D$.*

We begin by showing that invariant measures and positive functions do not exist in the supercritical case.

Theorem 8.1. *Let L satisfy Assumption $\widetilde{H}_{\text{loc}}$ on a domain $D \subseteq R^d$ and assume that L is supercritical on D. Then there are no invariant measures and no invariant positive functions for the transition measure $p(t, x, \mathrm{d}y)$.*

Proof. Assume that ν is an invariant measure. Then for $B \subset\subset D$, $\int_D p(t, x, B)\nu(\mathrm{d}x) = \nu(B) < \infty$. Therefore, for $\lambda > 0$,

$$\int_0^\infty \mathrm{d}t \int_D \exp(-\lambda t) p(t, x, B)\nu(\mathrm{d}x) = \nu(B)\int_0^\infty \exp(-\lambda t)\,\mathrm{d}t < \infty.$$

Thus, for some $x \in D$, $\int_0^t \exp(-\lambda t)p(t, x, B)\,\mathrm{d}t < \infty$ and, by Theorem 3.6, it follows that $L - \lambda$ is subcritical for all $\lambda < 0$. Consequently, $\lambda_c \leq 0$ and L is not supercritical on D, a contradiction. A similar proof works for invariant positive functions. □

The following two propositions, which are immediate consequences of results we have already proved, will be useful in the sequel. Recall that $\tilde{p}(t, x, \mathrm{d}y)$ denotes the transition measure for $\widetilde{L}$.

Proposition 8.2. *Let L satisfy Assumption $\widetilde{H}_{\text{loc}}$ on a domain $D \subseteq R^d$. Then*

(i) *ϕ is an invariant density for $p(t, x, \mathrm{d}y)$ if and only if it is an invariant positive function for $\tilde{p}(t, x, \mathrm{d}y)$;*
(ii) *ϕ is an invariant positive function for $p(t, x, \mathrm{d}y)$ if and only if it is an invariant density for $\tilde{p}(t, x, \mathrm{d}y)$.*

Proof. The proof follows from Corollary 2.6 and the fact that an invariant density must be positive on D since

$$\phi(y)\,\mathrm{d}y = \int_D p(t, x, \mathrm{d}y)\phi(x)\,\mathrm{d}x. \qquad \square$$

Proposition 8.3. *Let L satisfy Assumption $\widetilde{H}_{\text{loc}}$ on a domain $D \subseteq R^d$ and let $h \in C^{2,\alpha}(D)$ satisfy $h > 0$ in D. Denote by $p^h(t, x, \mathrm{d}y)$ the transition measure for L^h, the h-transform of L. Then*

(i) *ϕ is an invariant density for $p(t, x, \mathrm{d}y)$ if and only if ϕh is an invariant density for $p^h(t, x, \mathrm{d}y)$;*
(ii) *ϕ is an invariant positive function for $p(t, x, \mathrm{d}y)$ if and only if ϕ/h is an invariant positive function for $p^h(t, x, \mathrm{d}y)$.*

Proof. The proposition follows from Theorem 1.1. □

The following theorem gives the connection between positive harmonic functions and invariant positive functions and densities.

Theorem 8.4. *Let L satisfy Assumption $\widetilde{H}_{\mathrm{loc}}$ on a domain $D \subseteq R^d$.*

(i) *If $\phi \in C^{2,\alpha}(D)$ is an invariant density for $p(t, x, dy)$ on D, then $\widetilde{L}\phi = 0$ and $\phi > 0$ on D; that is, $\phi \in C_{\widetilde{L}}(D)$.*
(ii) *If $\phi \in C^{2,\alpha}(D)$ is an invariant positive function for $p(t, x, dy)$ on D, then $L\phi = 0$ on D; that is, $\phi \in C_L(D)$.*

Proof. We will prove (i), the proof of (ii) being similar. By Theorem 8.1, we may assume that $C_L(D)$ is not empty. By Proposition 8.3 it follows that if $h \in C_L(D)$, then ϕ is an invariant density for $p(t, x, dy)$ if and only if $h\phi$ is an invariant density for $p^h(t, x, dy)$, the transition measure corresponding to L^h on D. It is easy to check that the formal adjoint of L^h is equal to $(\widetilde{L})^{1/h}$. Furthermore, $\phi \in C_{\widetilde{L}}(D)$ if and only if $h\phi \in C_{(\widetilde{L})^{1/h}}(D)$. Since the zeroth-order part of L^h vanishes, this argument shows that we may assume without loss of generality that $V \equiv 0$.

Assume that $V \equiv 0$ and let $g \in C_0^2(D)$. Extend g and Lg to $\hat{D}$ by defining $g(\Delta) = Lg(\Delta) = 0$. Then from the definition of the generalized martingale problem,

$$E_x g(X(t)) = g(x) + E_x\int_0^t Lg(X(s))\,ds = g(x) + \int_0^t E_x Lg(X(s))\,ds.$$

Multiplying by $\phi(x)$ and integrating and using the invariance of ϕ gives

$$0 = \int_0^t\left(\int_D E_x Lg(X(s))\phi(x)\,dx\right)ds = t\int_D Lg(x)\phi(x)\,dx.$$

Integrating by parts gives $\int_D g(x)\widetilde{L}\phi(x)\,dx = 0$ for all $g \in C_0^2(D)$; hence $\widetilde{L}\phi = 0$ on D. Since $\phi \geqslant 0$ on D and $\phi \not\equiv 0$, it follows from Theorem 3.2.6 that $\phi > 0$ on D. □

The proof of Theorem 8.4(i) in fact shows that if ν is an invariant measure, then $\int_D Lg(x)\nu(dx) = 0$ for all $g \in C_0^2(D)$. Similarly, the proof of part (ii), which was left to the reader, shows that if ϕ is any invariant positive function, then $\int_D \widetilde{L}g(x)\phi(x)\,dx = 0$ for all $g \in C_0^2(D)$. In the language of distribution theory, this means that $L\phi \equiv 0$ and $\widetilde{L}\nu \equiv 0$ in the sense of distributions. Then, by elliptic regularity theory, it follows that in fact $\phi \in C^{2,\alpha}(D)$, and that the Radon–Nikodym derivative

$$\frac{d\nu}{dx} \in C^{2,\alpha}(D).$$

An alternative method of arriving at this conclusion is via parabolic PDE theory, which guarantees that $p(t, x, dy)$ possesses a density $p(t, x, y)$ which is smooth in x and y and whose derivatives satisfy certain bounds. Then one writes $\nu(dy) = (\int_D p(t, x, y)\nu(dx))\,dy$ and $\phi(x) = \int_D p(t, x, y)\phi(y)\,dy$. (See notes at end of chapter.)

The above discussion in conjunction with Theorem 8.4 shows that an invariant measure must possess a density which belongs to $C_{\tilde{L}}(D)$ and that an invariant positive function must belong to $C_L(D)$.

The following theorem gives a probabilistic criterion for determining whether a particular element of $C_L(D)$ or $C_{\tilde{L}}(D)$ is an invariant positive function or an invariant density.

Theorem 8.5. *Let L satisfy Assumption $\tilde{H}_{\text{loc}}$ on a domain $D \subseteq R^d$. Then*

(i) *$\phi \in C_{\tilde{L}}(D)$ is an invariant density for $p(t, x, dy)$ if and only if the diffusion process corresponding to the solution $\{\tilde{P}_x^\phi, x \in \hat{D}\}$ of the generalized martingale problem for $\tilde{L}^\phi$ on D does not explode; that is, if and only if $\tilde{P}_x^\phi(\tau_D = \infty) = 1$, for all $x \in D$.*

(ii) *$\phi \in C_L(D)$ is an invariant positive function for $p(t, x, dy)$ if and only if the diffusion process corresponding to the solution $\{P_x^\phi, x \in \hat{D}\}$ of the generalized martingale problem for L^ϕ on D does not explode; that is, if and only if $P_x^\phi(\tau_D = \infty) = 1$, for all $x \in D$.*

Proof. We will prove (i); the proof of (ii) is similar. Let $\phi \in C_{\tilde{L}}(D)$. By Proposition 8.2, it suffices to show that ϕ is an invariant positive function for $\tilde{p}(t, x, dy)$. By Proposition 8.3, ϕ is an invariant positive function for $\tilde{p}(t, x, dy)$ if and only if 1 is an invariant positive function for $\tilde{p}^\phi(t, x, dy)$, the transition measure for $\tilde{L}^\phi$. But

$$\int_D \tilde{p}^\phi(t, x, dy) = \tilde{p}^\phi(t, x, D) = \tilde{P}_x^\phi(X(t) \in D) = \tilde{P}_x^\phi(\tau_D > t).$$

Thus, 1 is invariant if and only if $\tilde{P}_x^\phi(\tau_D > t) = 1$, for all $t > 0$ and $x \in D$. □

As a first application of Theorem 8.5, consider the case in which L is critical on D. Recall that if L is critical, then by Theorem 3.4, $C_L(D)$ and $C_{\tilde{L}}(D)$ are one-dimensional, and their unique elements (up to positive multiples) are denoted by ϕ_c and $\tilde{\phi}_c$ and are called ground states. The following result completely solves the question of invariant densities and positive functions in the critical case.

Theorem 8.6. *Let L satisfy Assumption $\tilde{H}_{\text{loc}}$ on a domain $D \subseteq R^d$ and assume that L is critical on D. Then the ground state ϕ_c of L on D is* *(up to positive multiples) the unique invariant positive function for the transition measure $p(t, x, dy)$ and the ground state $\tilde{\phi}_c$ of $\tilde{L}$ on D is (up to positive multiples) the unique invariant density for $p(t, x, dy)$.*

Proof. The uniqueness follows from Theorem 8.4 and the discussion that follows it and the fact that $C_L(D)$ and $C_{\tilde{L}}(D)$ are one-dimensional. By Theorem 8.5, to show that $\tilde{\phi}_c$ (ϕ_c) is an invariant density (positive function), one must show that $\tilde{P}_x^{\tilde{\phi}_c}(\tau_D = \infty) = 1$ ($P_x^{\phi_c}(\tau_D = \infty) = 1$), for all $x \in D$. By Theorem 3.3(i), (ii), (iv) and (v) and the criticality assumption, it follows that the diffusion processes corresponding to L^{ϕ_c} on D and $\tilde{L}^{\tilde{\phi}_c}$ on D are recurrent on D. The Theorem now follows from Theorem 2.8.1. □

In the case where L is subcritical on D, the connection between positive harmonic functions and invariant positive functions and measures is not clear cut. For example, consider the following three possibilities:

1. $C_L(D)$ ($C_{\tilde{L}}(D)$) contains more than one linearly independent element and all of its elements are invariant positive functions (densities).
2. $C_L(D)$ ($C_{\tilde{L}}(D)$) contains more than one linearly independent element, some of which are invariant positive functions (densities) and some of which are not.
3. $C_L(D)$ ($C_{\tilde{L}}(D)$) contains more than one linearly independent element and none of its elements are invariant positive functions (densities).

In Example 5.1.9 in Chapter 5, we present one-dimensional examples which show that if $\lambda_c < 0$, then each of the three possibilities above can occur and that if $\lambda_c = 0$, then possibilities 1 and 2 above can occur. However, we believe that the third possibility above cannot occur if $\lambda_c = 0$. More generally, regardless of how many linearly independent elements exist in $C_L(D)$ or $C_{\tilde{L}}(D)$, we conjecture that there is always at least one invariant one. This conjecture was proposed in [Stroock (1982)].

4.9 The product L^1 property and its connection to the asymptotic behavior of the transition measure; positive and null recurrence

We begin with the following results for subcritical operators.

Theorem 9.1. *Let L satisfy Assumption $\tilde{H}_{\text{loc}}$ on a domain $D \subseteq R^d$ and assume that L is subcritical on D. Then $\lim_{t\to\infty} p(t, x, B) = 0$, for all $x \in D$ and all $B \subset\subset D$.*

***Theorem* 9.2.** *Let L satisfy Assumption $\widetilde{H}_{\text{loc}}$ on a domain $D \subseteq R^d$ and assume that L is subcritical on D. Let $\phi \in C_L(D)$ and $\tilde{\phi} \in C_{\tilde{L}}(D)$ and assume either that ϕ is an invariant positive function or that $\tilde{\phi}$ is an invariant density for $p(t, x, dy)$. Then $\int_D \phi\tilde{\phi}\,dx = \infty$, that is, $\phi\tilde{\phi} \notin L^1(D)$.*

As a consequence of Theorem 9.2, we obtain the following corollaries.

***Corollary* 9.3.** *Let L satisfy Assumption $\widetilde{H}_{\text{loc}}$ on a domain $D \subseteq R^d$ and assume that $V \equiv 0$ and that L corresponds to a transient diffusion process. Then the diffusion process corresponding to L on D does not possess an invariant probability density.*

Proof. The proof follows from Theorem 9.2 since $\phi \equiv 1$. □

***Corollary* 9.4.** *Let L satisfy Assumption $\widetilde{H}_{\text{loc}}$ on a domain $D \subseteq R^d$ and assume that $V \equiv 0$ and that L corresponds to a transient diffusion process. If $\tilde{\phi} \in C_{\tilde{L}}(D)$ and $\int_D \tilde{\phi}\,dx < \infty$, then the diffusion process explodes.*

Proof. Explosion is equivalent to $\phi \equiv 1$ not being an invariant positive function. By assumption $\int \phi\tilde{\phi}\,dx < \infty$; thus, by Theorem 9.2, $\phi \equiv 1$ is not invariant. □

Proof of Theorem 9.1. Let $\phi \in C_L(D)$. By Theorem 1.1, it suffices to prove the theorem for $p^\phi(t, x, B)$, the transition measure for L^ϕ. Let $\{P_x^\phi, x \in \hat{D}\}$ denote the solution to the generalized martingale problem for L^ϕ on D. Then, $p^\phi(t, x, B) = P_x^\phi(X(t) \in B)$. Since L is subcritical, by Theorem 3.3(i), (ii) and (iv), L^ϕ corresponds to a transient diffusion and thus, by Theorem 2.8.1, $\lim_{t\to\infty} P_x^\phi(X(t) \in B) = 0$, for $B \subset\subset D$. □

Proof of Theorem 9.2. First assume that $\tilde{\phi}$ is an invariant density for $p(t, x, dy)$. Then, by Proposition 8.3, $\phi\tilde{\phi}$ is an invariant density for $p^\phi(t, x, dy)$, the transition measure for L^ϕ. Therefore, for $B \subset\subset D$,

$$\int_D p^\phi(t, x, B)\phi(x)\tilde{\phi}(x)\,dx = \int_B \phi(x)\tilde{\phi}(x)\,dx. \tag{9.1}$$

We have $0 \leq p^\phi(t, x, B) \leq 1$ and, by Theorem 9.1, $\lim_{t\to\infty} p^\phi(t, x, B) = 0$ for $B \subset\subset D$. Now if $\phi\tilde{\phi}$ belonged to $L^1(D)$, then it would follow from the dominated convergence theorem that the lefthand side of (9.1) would approach zero as $t \to \infty$, for all $B \subset\subset D$. This would yield $\int_B \phi(x)\tilde{\phi}(x)\,dx = 0$, for $B \subset\subset D$, which is a contradiction.

Now assume that ϕ is an invariant positive function for $p(t, x, dy)$. Then, by Propositions 8.2 and 8.3, $\phi\tilde{\phi}$ is an invariant density for $\tilde{p}^{\phi}(t, x, dy)$, the transition measure for $\tilde{L}^{\phi}$. The rest of the proof is now the same as in the first part above. □

We now turn to the critical case. Let ϕ_c and $\tilde{\phi}_c$ denote the ground states for L and $\tilde{L}$ on D; that is, ϕ_c and $\tilde{\phi}_c$ are (up to positive multiples) the unique elements of $C_L(D)$ and $C_{\tilde{L}}(D)$ respectively. It turns out that $\phi_c\tilde{\phi}_c$ sometimes does belong to $L^1(D)$ and sometimes does not. Furthermore, the asymptotic behavior of $p(t, x, dy)$ will depend dramatically on whether or not $\phi_c\tilde{\phi}_c$ belongs to $L^1(D)$.

Definition. *A critical operator L on D is called* product L^1 critical *if* $\int_D \phi_c\tilde{\phi}_c\,dx < \infty$.

If L is critical and $V \equiv 0$, then $\phi_c \equiv 1$; thus the product L^1 criticality condition becomes $\int_D \tilde{\phi}_c\,dx < \infty$. From this, Theorem 8.6 and Corollary 9.3, it follows that, without any assumptions on the criticality of L, if $V \equiv 0$, then L is product L^1 critical if and only if there exists an invariant probability density for the corresponding diffusion process.

Definition. *If $V \equiv 0$ and L is critical, then the recurrent diffusion process corresponding to L is called* positive recurrent *if* $\int_D \tilde{\phi}_c\,dx < \infty$ *and* null recurrent *if* $\int_D \tilde{\phi}_c\,dx = \infty$.

See Theorem 9.6 below and the remark which follows it for another characterization of positive and null recurrence in terms of the expectation of the exit time of the diffusion from an exterior domain.

Theorem 9.5. *Let L satisfy Assumption $\tilde{H}_{\text{loc}}$ on a domain $D \subseteq R^d$ and assume that L is critical on D.*

(i) *If $\int_D \phi_c\tilde{\phi}_c\,dx = \infty$, then*

$$\lim_{t\to\infty} \frac{1}{t}\int_0^t p(s, x, B)\,ds = 0,$$

for all $x \in D$ and $B \subset\subset D$. Furthermore, in the case where $V \equiv 0$,

$$\lim_{t\to\infty} \frac{1}{t}\int_0^t 1_B(X(s))\,ds = 0$$

a.s. P_x, for all $x \in D$ and all $B \subset\subset D$, where $\{P_x, x \in D\}$ denotes the solution to the martingale problem for L on D.

(ii) *If* $\int_D \phi_c \tilde{\phi}_c \, dx < \infty$, *then*

$$\lim_{t\to\infty} \frac{1}{t}\int_0^t \left(\int_D p(s, x, dy) f(y)\right) ds = \frac{\phi_c(x)\int_D f(y)\tilde{\phi}_c(y)\,dy}{\int_D \phi_c(y)\tilde{\phi}_c(y)\,dy}, \text{ for all } x \in D$$

and all bounded measurable f with compact support in D. Furthermore, in the case where $V \equiv 0$ *or, equivalently,* $\phi_c \equiv 1$, *the above limit holds for all bounded measurable f on D and*

$$\lim_{t\to\infty} \frac{1}{t}\int_0^t f(X(s))\,ds = \frac{\int_D f(y)\tilde{\phi}_c(y)\,dy}{\int_D \tilde{\phi}_c(y)\,dy} \text{ a.s. } P_x, \text{ for all } x \in D$$

and all bounded measurable f on D, where $\{P_x, x \in D\}$ *denotes the solution to the martingale problem for L on D.*

For the proof of part (i) of Theorem 9.5, we need part (i) of the following theorem which is also of independent interest.

Theorem 9.6. *Let L satisfy Assumption* $\tilde{H}_{\text{loc}}$ *on a domain* $D \subseteq R^d$, $d \geq 2$. *Assume that* $V \equiv 0$ *and that L corresponds to a recurrent diffusion so that* $\phi_c \equiv 1$. *Let* $B \subset\subset D$ *be a simply connected domain so that* $D - B$ *is connected, and define* $\sigma_{\bar{B}} = \inf\{t \geq 0: X(t) \in \bar{B}\}$.

(i) *Assume that* $\int_D \tilde{\phi}_c \, dx = \infty$. *Then* $E_x \sigma_{\bar{B}} = \infty$, *for* $x \in D - \bar{B}$. *Equivalently, there are no positive solutions* $u \in C^2(D - \bar{B})$ *of*

$$Lu = -1 \text{ on } D - \bar{B}. \tag{9.2}$$

(ii) *Assume that* $\int_D \tilde{\phi}_c \, dx < \infty$. *Then* $E_x \sigma_{\bar{B}} < \infty$, *for* $x \in D - \bar{B}$. *Equivalently, there exists a positive solution to* (9.2).

Remark 1. Theorem 9.6 states that a recurrent diffusion is positive recurrent if the expected hitting time of, say, a closed ball in D, starting from outside the closed ball, is finite, and null recurrent if the expected hitting time is infinite.

Remark 2. The analog of Theorem 9.6 in the one-dimensional case is as follows. Let $D = (\alpha, \beta)$ and let $B = (c_1, c_2)$, where $-\infty \leq \alpha < c_1 < c_2 < \beta \leq \infty$. Assume that $V \equiv 0$ and that L corresponds to a recurrent diffusion so that $\phi_c \equiv 1$. Let $\sigma_x = \inf\{t \geq 0: X(t) = x\}$. Then the following dichotomy holds:

(i) Assume that $\int_D \tilde{\phi}_c \, dx = \infty$. Then at least one of the following two conditions holds:
 (a) $E_x \sigma_{c_2} = \infty$, for $x \in (c_2, \beta)$,
 (b) $E_x \sigma_{c_1} = \infty$, for $x \in (\alpha, c_1)$.
 (a) is equivalent to the non-existence of positive solutions $u \in C^2((c_2, \beta))$

of $Lu = -1$ in (c_2, β) and (b) is equivalent to the non-existence of positive solutions $u \in C^2((\alpha, c_1))$ of $Lu = -1$ in (α, c_1).

(ii) Assume that $\int_D \tilde{\phi}_c \, dx < \infty$. Then $E_x \sigma_{c_2} < \infty$, for $x \in (c_2, \beta)$ and $E_x \sigma_{c_1} < \infty$, for $x \in (\alpha, c_1)$.

With the exception of the change in notation, the proof is identical to the proof of Theorem 9.6.

Remark 3. Since the proof of part (ii) of Theorem 9.6 uses Theorem 9.5, we emphasize that the proof of Theorem 9.5 relies only on part (i) of Theorem 9.6.

Proof. We begin by proving the equivalence noted in each part of the theorem. Clearly, we may assume that $B \subset\subset D$ possesses a $C^{2,\alpha}$-boundary. Let $\{D_n\}_{n=1}^{\infty}$ be a sequence of domains with $C^{2,\alpha}$-boundaries satisfying $B \subset\subset D_1$, $D_n \subset\subset D_{n+1}$ and $\bigcup_{n=1}^{\infty} D_n = D$. By Theorem 4.1(iii) and the recurrence assumption, $\lambda_c(D_n - \bar{B}) < \lambda_c(D) = 0$; therefore, by Theorem 3.6.5, there exists a unique solution u_n to

$$\left.\begin{aligned} Lu_n &= -1 \text{ in } D_n - \bar{B}, \\ u_n &= 0 \text{ on } \partial D_n \cup \partial B. \end{aligned}\right\} \tag{9.3}$$

Also by Theorem 3.6.5, $u_n > 0$ on $D_n - \bar{B}$; using this and applying Theorem 3.6.5 to $u_{n+1} - u_n$ shows that u_n is non-decreasing in n. Applying Theorem 3.6.5 to $u - u_n$, where u is any positive solution of (9.2), we conclude that

$$\bar{u} \equiv \lim_{n\to\infty} u_n \leqslant u, \quad \text{for any } u > 0 \text{ which solves (9.2).} \tag{9.4}$$

By Theorem 2.1.2, $u_n(x) = E_x \sigma_B \wedge \tau_{D_n}$ and, by the recurrence assumption, $\bar{u}(x) = \lim_{n\to\infty} u_n(x) = E_x \sigma_B$. By exercise 4.8, either $\bar{u}(x) = \infty$ for all $x \in D - \bar{B}$ or $\bar{u}$ is locally bounded on $D - \bar{B}$. In the latter case, a standard compactness argument shows that $\bar{u}$ satisfies (9.2). Thus, either $\bar{u}(x) = E_x \sigma_B = \infty$, for $x \in D - \bar{B}$, and, from (9.4), there are no positive solutions to (9.2), or $\bar{u}(x) = E_x \sigma_B < \infty$, for all $x \in D - \bar{B}$, and $\bar{u}$ solves (9.2). This completes the proof of the equivalence.

We now prove part (i). Let B and $\{D_n\}_{n=1}^{\infty}$ be as above and fix $x_0 \in B$. By Theorem 3.2, $L - \lambda_c(D_n)$ is critical on D_n. By Theorem 3.5.5, the ground state $\tilde{\phi}_n$ of $\tilde{L} - \lambda_c(D_n)$ satisfies $\tilde{\phi}_n \in C^{2,\alpha}(\bar{D}_n)$ and $\tilde{\phi}_n = 0$ on ∂D_n. Normalize $\tilde{\phi}_n$ by $\tilde{\phi}_n(x_0) = 1$. Let u_n denote the solution to (9.3) and let A be a domain satisfying $B \subset\subset A \subset\subset D_1$. Integrating by parts and using the boundary conditions, we have

$$\int_{D_n - A} \tilde{\phi}_n \, \mathrm{d}x = -\int_{D_n - A} \tilde{\phi}_n L u_n \, \mathrm{d}x$$

$$= \int_{\partial A} ((\tfrac{1}{2}\tilde{\phi}_n a \nabla u_n - \tfrac{1}{2} u_n a \nabla \tilde{\phi}_n + \tilde{\phi}_n u_n b) \cdot v) \sigma(\mathrm{d}x)$$

$$- \int_{D_n - A} \lambda_c(D_n) \tilde{\phi}_n u_n \, \mathrm{d}x, \tag{9.5}$$

where v denotes the unit outward normal to A on ∂A and $\sigma(\mathrm{d}x)$ denotes Lebesgue surface measure on ∂A. It follows from Harnack's inequality applied to the operators $\tilde{L} - \lambda_c(D_n)$ that $\tilde{\phi}_n$ is bounded on ∂A, independent of n. Using this in the interior Schauder estimate (Theorem 3.2.7), it follows that $|\nabla \tilde{\phi}_n|$ is also bounded on ∂A independent of n.

To prove part (i), we will assume that there is a positive solution to (9.2) and come to the conclusion that $\int_D \tilde{\phi}_c \, \mathrm{d}x < \infty$. By the proof above of the equivalence claim, it follows that $\bar{u} = \lim_{n\to\infty} u_n < \infty$. Thus u_n is bounded on ∂A independent of n and, again by the Schauder interior estimate, $|\nabla u_n|$ is bounded on ∂A, independent of n. By Theorem 4.1(i), $\lim_{n\to\infty} \lambda_c(D_n) = \lambda_c(D) = 0$. Using this in (9.5), we conclude that there exists a constant c, independent of n, such that

$$\int_{D_n - A} \tilde{\phi}_n \, \mathrm{d}x \leqslant c. \tag{9.6}$$

By a standard compactness argument along with the uniqueness of the ground state $\tilde{\phi}_c$, it follows that $\tilde{\phi}_c = \lim_{n\to\infty} \tilde{\phi}_n$, where $\tilde{\phi}_c$ has been normalized by $\tilde{\phi}_c(x_0) = 1$. By Fatou's lemma and (9.6), it follows that $\int_{D-A} \tilde{\phi}_c \, \mathrm{d}x \leqslant c$ and, thus, $\int_D \tilde{\phi}_c \, \mathrm{d}x < \infty$.

We now turn to the proof of part (ii). We will show that if $E_x \sigma_B = \infty$, then $\int_D \tilde{\phi}_c \, \mathrm{d}x = \infty$. The proof below of part (i) of Theorem 9.5 shows that if $E_x \sigma_B = \infty$, then

$$\lim_{t\to\infty} \frac{1}{t} \int_0^t p(s, x, B) \, \mathrm{d}s = 0.$$

Now if it were true that $\int_D \tilde{\phi}_c \, \mathrm{d}x < \infty$, then since $p(t, x, \mathrm{d}y)$ is a probability measure for all $t > 0$, it would follow from the dominated convergence theorem that

$$\lim_{t\to\infty} \int_D \left(\frac{1}{t} \int_0^t p(s, x, B) \, \mathrm{d}s \right) \tilde{\phi}_c(x) \, \mathrm{d}x = 0. \tag{9.7}$$

However, by Theorem 8.6, $\tilde{\phi}_c$ is invariant; consequently,

$$\int_D \left(\frac{1}{t} \int_0^t p(s, x, B)\, ds \right) \tilde{\phi}_c(x)\, dx = \int_B \tilde{\phi}_c(x)\, dx.$$

This contradicts (9.7). □

Proof of Theorem 9.5. We will assume that $d \geq 2$ since the proof in the case where $d = 1$, while conceptually identical, requires different notation. Since

$$p^{\phi_c}(t, x, dy) = \frac{1}{\phi_c(x)} p(t, x, dy) \phi_c(y),$$

and since the ground state of L^{ϕ_c} is 1 and the ground state of $\tilde{L}^{\phi_c}$ is $\phi_c \tilde{\phi}_c$, it is enough to prove the theorem for L^{ϕ_c}, whose zeroth-order term vanishes, in place of L. Thus, we may assume without loss of generality that $V \equiv 0$, in which case $\phi_c \equiv 1$ and the diffusion process corresponding to L is recurrent. However, we should note that it is in this reduction that the additional requirement that f have compact support if $V \not\equiv 0$ enters in.

Part (i). Clearly, it suffices to assume that B is a ball. Let B_1 and B_2 be balls satisfying $B \subset\subset B_1 \subset\subset B_2 \subset\subset D$. Define $\sigma_0 = 0$, define $\tau_1 = \inf\{t \geq 0: X(t) \in \partial B_2\}$ and, by induction, define

$$\sigma_n = \inf\{t \geq \tau_n: X(t) \in \partial B_1\} \text{ and } \tau_{n+1} = \inf\{t \geq \sigma_n: X(t) \in \partial B_2\},$$

$n = 1, 2, \ldots$. By the recurrence assumption, $\tau_n < \infty$ and $\sigma_n < \infty$ for all n, a.s. P_x, for all $x \in D$. For $x \in D$, define $\pi_1(x, dy) = P_x(X(\tau_1) \in dy)$. For $x \in \partial B_2$, define $\pi_2(x, dy) = P_x(X(\sigma_1) \in dy)$. By exercise 4.7, there exists a positive constant c such that

$$\pi_2(z, dy) \leq c\pi_2(w, dy) \text{ for all } z, w \in \partial B_2 \tag{9.8}$$

and

$$\pi_1(z, dy) \leq c\pi_1(w, dy), \text{ for } z, w \in \partial B_1. \tag{9.9}$$

Now $\pi_1(x, dy)$, for $x \in \partial B_1$, is a transition probability measure from ∂B_1 to ∂B_2 and $\pi_2(x, dy)$, for $x \in \partial B_2$, is a transition probability measure from ∂B_2 to ∂B_1. By (9.8) and (9.9), the Doeblin condition holds [Doob (1953), p. 192] and, consequently, there exist unique probability measures μ_1 and μ_2 on ∂B_1 and ∂B_2 such that

$$\int_{\partial B_1} \pi_1(x, dy)\mu_1(dx) = \mu_2(dy) \text{ and } \int_{\partial B_2} \pi_2(x, dy)\mu_2(dx) = \mu_1(dy). \tag{9.10}$$

Let $N(t) = \max\{n: \tau_n \leqslant t\}$. Then

$$\frac{1}{t}\int_0^t 1_B(X(s))\,ds \leqslant \frac{\sum_{j=1}^{N(t)+1}(\tau_j - \sigma_{j-1})}{\sum_{j=1}^{N(t)}(\tau_j - \sigma_{j-1}) + \sum_{j=1}^{N(t)-1}(\sigma_j - \tau_j)}. \tag{9.11}$$

Now from (9.10) and the strong Markov property, it follows that $\{\tau_j - \sigma_{j-1}\}_{j=1}^{\infty}$ and $\{\sigma_j - \tau_j\}_{j=1}^{\infty}$ are sequences of independent identically distributed random variables under $P_{\mu_1} \equiv \int_{\partial B_1} P_x \mu_1(dx)$. By Theorem 2.1.1, $E_{\mu_1}(\tau_j - \sigma_{j-1}) < \infty$, and by Theorem 9.6 and the assumption that $\int \tilde{\phi}_c\,dx = \infty$, $E_{\mu_1}(\sigma_j - \tau_j) = \infty$. Clearly, $\lim_{t\to\infty} N(t) = \infty$ a.s. P_{μ_1}; therefore, by (9.11) and the strong law of large numbers [Chung (1974) or Durrett (1991)],

$$\lim_{t\to\infty}\frac{1}{t}\int_0^t 1_B(X(s))\,ds = 0 \text{ a.s. } P_{\mu_1}. \tag{9.12}$$

Let $\mathcal{F}^{\sigma_1} \equiv \sigma(X(t), t \geqslant \sigma_1)$. From the definition of σ_1 along with (9.8) and (9.9) and the recurrence assumption, $\{P_x, x \in D\}$ is a collection of mutually absolutely continuous probability measures on $\mathcal{F}^{\sigma_1}$; therefore it follows from (9.12) that

$$\lim_{t\to\infty}\frac{1}{t}\int_0^t 1_B(X(s))\,ds = 0 \text{ a.s. } P_x,$$

for all $x \in D$. Also, since $E_x\int_0^t 1_B(X(s))\,ds = \int_0^t p(s, x, B)\,ds$, it follows from the bounded convergence theorem that

$$\lim_{t\to\infty}\frac{1}{t}\int_0^t p(s, x, B)\,ds = 0.$$

Part (ii). By Theorem 8.6, $X(t)$ is a stationary Markov process under $P_{\tilde{\phi}_c} \equiv \int_D P_x\tilde{\phi}_c(x)\,dx$ with marginal density

$$\frac{\tilde{\phi}_c(x)}{\int_D \tilde{\phi}_c(y)\,dy}.$$

Also, of course, $X(t)$ is irreducible under $P_{\tilde{\phi}_c}$; that is, $P_{\tilde{\phi}_c}(X(t) \in B$ for some $t > 0) > 0$ for all open $B \subseteq D$. A stationary, irreducible Markov process is always ergodic [Breiman (1968) or (1992)] and so by the pointwise ergodic theorem [Breiman (1968) or (1992) or Durrett (1991)],

$$\lim_{t\to\infty}\frac{1}{t}\int_0^t 1_B(X(s))\,ds = \frac{\int_B \tilde{\phi}_c(y)\,dy}{\int_D \tilde{\phi}_c(y)\,dy} \text{ a.s. } P_{\tilde{\phi}_c},$$

for all measurable $B \subseteq D$. From the definition of $P_{\tilde{\phi}_c}$, it follows that for each measurable $B \subseteq D$,

$$\lim_{t\to\infty}\frac{1}{t}\int_0^t 1_B(X(s))\,ds = \frac{\int_B \tilde{\phi}_c(y)\,dy}{\int_D \tilde{\phi}_c(y)\,dy} \text{ a.s. } P_x,$$

for almost all $x \in D$. The extension to all $x \in D$ is left as exercise 4.10. (We note that if we were to assume the existence of a density $p(t, x, y)$, then the extension to all x would follow from the fact that for $A \in \sigma(X(t), t \geq 1)$, $P_x(A) = \int_D P_y(X(\cdot) \in \theta_1 A)p(1, x, y)\,dy$, where θ_1 is the shift operator defined in Chapter 2, Section 3.) Taking expectations gives

$$\lim_{t\to\infty}\frac{1}{t}\int_0^t p(s, x, B)\,ds = \frac{\int_B \tilde{\phi}_c(y)\,dy}{\int_D \tilde{\phi}_c(y)\,dy},$$

for all $x \in D$ and all measurable $B \subseteq D$. This proves part (ii) in the case that $f(x) = 1_B(x)$; the extension to bounded measurable f is via a standard technique. □

Using the following theorem, we can prove a much stronger version of Theorem 9.5(ii).

***Theorem* 9.7.** *Let L satisfy Assumption $\widetilde{H}_{loc}$ on a domain $D \subseteq R^d$ and assume that $V \equiv 0$ and that L corresponds to a recurrent diffusion process on D. Then for $t > 0$, $\{\int_D p(t, \cdot, dy)f(y)$; f bounded and measurable, $\sup_D |f| \leq 1\}$ is an equicontinuous family of functions on any compact subset of D.*

Remark. This theorem follows, for example, from [Stroock and Varadhan (1979), Theorem 7.2.4]. The proof in the general case is long and highly technical. However, the case $L = \frac{1}{2}\Delta$ on $D = R^d$ may be verified easily and the extension from this case to the case where $L = \frac{1}{2}\Delta + b \cdot \nabla$ on R^d can be achieved via the Cameron–Martin–Girsanov formula [Varadhan (1980), pp. 251–8].

We will actually need the following corollary of Theorem 9.7.

***Corollary* 9.8.** *Under the conditions of Theorem 9.7, for any $\varepsilon > 0$, $\{\int_D p(t, \cdot, dy)f(y)$; f bounded and measurable, $\sup_D |f| \leq 1$, $t \geq \varepsilon\}$ is an equicontinuous family of functions on any compact subset of D.*

Proof. For $t > \varepsilon$, $\int_D p(t, x, dy)f(y) = \int_D p(\varepsilon, x, dy)h(y)$, where $h(y) = \int_D p(t - \varepsilon, y, dz)f(z)$. Since $p(t, x, D) = 1$, $\sup_D |h| \leq \sup_D |f|$. □

We can now prove the following theorem.

Theorem 9.9. *Let L satisfy Assumption* $\widetilde{H}_{\mathrm{loc}}$ *on a domain* $D \subseteq R^d$ *and assume that L is critical on D and that* $\int_D \phi_c \tilde{\phi}_c \, dx < \infty$. *Normalize* ϕ_c *and* $\tilde{\phi}_c$ *by* $\int_D \phi_c \tilde{\phi}_c \, dx = 1$. *Then for any bounded measurable f with compact support in D, and any compact* $K \subset D$,

$$\limsup_{t\to\infty} \sup_{x \in K} \left| \int_D p(t, x, dy) f(y) - \phi_c(x) \int_D \tilde{\phi}_c(y) f(y)\, dy \right| = 0. \quad (9.13)$$

If $V \equiv 0$, *or, equivalently,* $\phi_c \equiv 1$, *then* (9.13) *holds for any bounded measurable f and, in fact,*

$$\limsup_{t\to\infty} \sup_{x \in K} \sup_{|f| \leq 1} \left| \int_D p(t, x, dy) f(y) - \int_D \tilde{\phi}_c(y) f(y)\, dy \right| = 0. \quad (9.14)$$

Proof. As in the proof of Theorem 9.5, it is easy to use the h-transform to reduce to the case $V \equiv 0$ and $\phi_c \equiv 1$. However, in this reduction, the stronger result (9.14) must be replaced by the weaker result (9.13). We will prove (9.14) under the assumption that $V \equiv 0$. Consider the domain $D \times D \subseteq R^d \times R^d$ and denote points $x \in D \times D$ by $x = (x_1, x_2)$ with $x_1, x_2 \in D$. Let L_{x_i}, $i = 1, 2$, denote the operator L in the variable x_i and let $\bar{L} = L_{x_1} + L_{x_2}$ on $D \times D$. Let $\bar{P}_x = P_{x_1} \times P_{x_2}$. By exercise 1.14, $\{\bar{P}_x, x \in D \times D\}$ is the solution to the martingale problem for $\bar{L}$ on $D \times D$. Therefore $\bar{p}(t, x, dy)$, the transition probability measure for $\bar{L}$ on $D \times D$, is given by $\bar{p}(t, x, dy) = p(t, x_1, dy_1) p(t, x_2, dy_2)$. Since $\tilde{\phi}_c$ is the invariant probability density for $p(t, x, dy)$, it follows that $\tilde{\phi}_c(x_1)\tilde{\phi}_c(x_2)$ is the invariant probability density for $\bar{p}(t, x, dy)$ and, by Corollary 9.3, the diffusion on $D \times D$ corresponding to $\bar{P}_x$ is recurrent.

Let $X(t) = (X_1(t), X_2X(t))$ denote the canonical diffusion process on $D \times D$ corresponding to $\bar{P}_x$. Fix $x_0 \times x_0 \in D \times D$ and let $B_\delta((x_0, x_0)) \subset R^d \times R^d$ denote the ball of radius $\delta > 0$ centered at (x_0, x_0). Let $\sigma_\delta = \inf(t \geq 0: X(t) \in B_\delta(x_0, x_0))$. By recurrence, $\lim_{t\to\infty} \bar{P}_x(\sigma_\delta \leq t) = 1$. In fact, we will show that for any compact $K \subset D$,

$$\lim_{t\to\infty} \inf_{x_1, x_2 \in K} \bar{P}_{x_1,x_2}(\sigma_\delta \leq t) = 1. \quad (9.15)$$

To prove (9.15), we will use the fact that $\bar{p}(t, x, dy)$ possesses a density. Since $\bar{P}_{x_1,x_2}(\sigma_\delta = t) = 0$, for $t > 0$, it follows from Theorem 2.3.3(iv), Theorem 1.3.1 and the Feller property that $\bar{P}_{x_1,x_2}(\sigma_\delta \leq t)$ is continuous in (x_1, x_2) for $t > 0$; (9.15) then follows from Dini's theorem.

Let $U(x, t) = E_x f(X(t)) = \int_D p(t, x, dy) f(y)$, for f bounded and measurable on D and satisfying $\sup_D |f| \le 1$. Then for $t \ge 1$, and $\delta > 0$ such that $B_\delta((x_0, x_0)) \subset\subset D \times D$, we have

$$\begin{aligned}
U(x_1, t) - U(x_2, t) &= \bar{E}_{x_1,x_2}(f(X_1(t)) - f(X_2(t))) \\
&= \bar{E}_{x_1,x_2}(f(X_1(t)) - f(X_2(t)); \sigma_\delta \le t - 1) \\
&\quad + \bar{E}_{x_1,x_2}(f(X_1(t)) - f(X_2(t)); \sigma_\delta > t - 1) \\
&= \bar{E}_{x_1,x_2}(U(X_1(\sigma_\delta), t - \sigma_\delta) - U(X_2(\sigma_\delta), t - \sigma_\delta); \sigma_\delta \le t - 1) \\
&\quad + \bar{E}_{x_1,x_2}(f(X_1(t)) - f(X_2(t)); \sigma_\delta > t - 1). \qquad (9.16)
\end{aligned}$$

Clearly, $|X_1(\sigma_\delta) - X_2(\sigma_\delta)| \le 2\delta$. Therefore, by Corollary 9.8,

$$\lim_{\delta \to 0} \sup_{t \ge 1} \sup_{x_1, x_2 \in D} |\bar{E}_{x_1,x_2}(U(X_1(\sigma_\delta), t - \sigma_\delta) - U(X_2(\sigma_\delta), t - \sigma_\delta); \sigma_\delta \le t - 1)| = 0, \qquad (9.17)$$

and the convergence is uniform over $\{f: \sup_D |f| \le 1\}$. From (9.15)–(9.17), we conclude that for any compact $K \subset D$,

$$\lim_{t\to\infty} \sup_{|f| \le 1} \sup_{x_1 x_2 \in K} \left| \int_D p(t, x_1, dy) f(y) - \int_D p(t, x_2, dy) f(y) \right| = 0. \qquad (9.18)$$

Using (9.18) and the fact that $\int_D p(t, x, dy) \tilde{\phi}_c(x)\, dx = \tilde{\phi}_c(y)\, dy$ and that $\int_D \tilde{\phi}_c(y)\, dy = 1$, we can now prove (9.14). We have

$$\begin{aligned}
&\left| \int_D p(t, x, dy) f(y) - \int_D \tilde{\phi}_c(y) f(y)\, dy \right| \\
&= \left| \int_D p(t, x, dy) f(y) - \int_D \left(\int_D p(t, z, dy) \tilde{\phi}_c(z)\, dz \right) f(y) \right| \\
&= \left| \int_D \left(\int_D p(t, x, dy) f(y) - \int_D p(t, z, dy) f(y) \right) \tilde{\phi}_c(z)\, dz \right| \\
&\le \int_D \left| \int_D p(t, x, dy) f(y) - \int_D p(t, z, dy) f(y) \right| \tilde{\phi}_c(z)\, dz.
\end{aligned}$$

Since $|\int_D p(t, x, dy) f(y) - \int_D p(t, z, dy) f(y)| \le 2 \sup_D |f|$, (9.14) now follows from (9.18) and the dominated convergence theorem. □

4.10 The product L^1 property and the existence of eigenfunctions in the symmetric case

In the case where $b = a\nabla Q$, where $Q \in C^{2,\alpha}(D)$, $L = \frac{1}{2}\nabla \cdot a\nabla + b \cdot \nabla + V$ can be written in the form $L = \frac{1}{2}\exp(-2Q)\nabla \cdot$

$a \exp(2Q)\nabla + V$, and it is easy to check that $\phi \in C_L(D)$ if and only if $\phi \exp(2Q) \in C_{\tilde{L}}(D)$. In particular, if L is critical on D, then $\tilde{\phi}_c = \phi_c \exp(2Q)$ and the product L^1 property, $\phi_c\tilde{\phi}_c \in L^1(D)$, becomes $\phi_c \in L^2(D, \exp(2Q)\, dx)$. The operator L with domain $C_0^\infty(D)$ is symmetric on the Hilbert space $L^2(D, \exp(2Q)\, dx)$. Denote the inner product on $L^2(D, \exp(2Q)\, dx)$ by $(\cdot, \cdot)$; that is $(f, g) = \int_D fg \exp(2Q)\, dx$, for f, $g \in L^2(D, \exp(2Q)\, dx)$. As noted in the second remark following Theorem 3.7.1, if one defines the quadratic form $q(f, g) = \frac{1}{2}\int_D(\nabla f a \nabla g) \exp(2Q)\, dx - \int_D Vfg \exp(2Q)\, dx = -(Lf, g)$, for f, $g \in C_0^\infty(D)$, then the Friedrichs' extension theorem [Reed and Simon, Vol. 2 (1975), Theorem 10.23] states that q can be extended to a closed quadratic form $\hat{q}$ on $L^2(D, \exp(2Q)\, dx)$ with domain $\hat{\mathcal{D}}_{\hat{q}}$ and that $(\hat{q}, \hat{\mathcal{D}}_{\hat{q}})$ is the quadratic form of a self-adjoint operator $(-\hat{\mathcal{L}}, \hat{\mathcal{D}})$ which is bounded from below and which is the unique self-adjoint extension of $(-L, C_0^\infty(D))$ satisfying $\hat{\mathcal{D}} \subset \hat{\mathcal{D}}_{\hat{q}}$.

The spectrum of a self-adjoint operator is real. Let $\hat{\lambda}_0 \equiv \hat{\lambda}_0(D) = \sup \sigma((\hat{\mathcal{L}}, \hat{\mathcal{D}}))$, where $\sigma((\hat{\mathcal{L}}, \hat{\mathcal{D}}))$ denotes the spectrum of $(\hat{\mathcal{L}}, \hat{\mathcal{D}})$. In Theorem 3.2, it was shown that the generalized principal eigenvalue coincides with the classical principal eigenvalue in the case where the domain is bounded with a smooth boundary and the coefficients of the operator are smooth. Under the same conditions and assuming that the operator is symmetric, it was shown in Remark 2 following Theorem 3.7.1 that the principal eigenvalue coincides with the supremum of the spectrum of the self-adjoint realization of the operator. The proposition below shows that in general, if the operator is symmetric, then the generalized principal eigenvalue coincides with the supremum of the spectrum of the self-adjoint realization of the operator.

Proposition 10.1. *Let $L = \frac{1}{2}\nabla \cdot a\nabla + b \cdot \nabla + V$ satisfy Assumption $\tilde{H}_{\text{loc}}$ on a domain $D \subseteq R^d$ and assume that b is of the form $b = a\nabla Q$ so that L is symmetric with respect to the density* $\exp(2Q)$. *Let $(\hat{\mathcal{L}}, \hat{\mathcal{D}})$ denote the corresponding self-adjoint operator obtained via the Friedrichs' extension, and let $\hat{\lambda}_0(D)$ denote the supremum of the spectrum of $(\hat{\mathcal{L}}, \mathcal{D})$. Then*

$$\hat{\lambda}_0(D) = \lambda_c(D).$$

Proof. The Rayleigh-Ritz formula, which is the mini-max principle [Reed and Simon, Vol. 4 (1978), Theorems 13.1 and 13.2] applied to the principal eigenvalue, asserts that

$$\hat{\lambda}_0(D) = \sup \sigma((\hat{\mathcal{L}}, \hat{\mathcal{D}})) = \sup_{v \in \hat{\mathcal{D}}} (\hat{\mathcal{L}}v, v) = \sup_{v \in \hat{\mathcal{D}}_{\hat{q}}} (-\hat{q}(v, v))$$

$$= \sup_{v \in C_0^\infty(D)} (-q(v, v)). \tag{10.1}$$

Let $\{D_n\}_{n=1}^\infty$ be a sequence of domains with $C^{2,\alpha}$-boundaries satisfying $D_n \subset\subset D_{n+1}$ and $\bigcup_{n=1}^\infty D_n = D$ and recall that in Remark 2 following Theorem 3.7.1, it was shown that $\hat{\lambda}_0(D_n) = \lambda_c(D_n)$. By Theorem 4.1(i), $\lim_{n\to\infty} \lambda_c(D_n) = \lambda_c(D)$. Clearly, $\sup_{v \in C_0^\infty(D)} (-q(v, v)) = \lim_{n\to\infty} \sup_{v \in C_0^\infty(D_n)} (-q(v, v))$. Thus, by (10.1),

$$\hat{\lambda}_0(D) = \sup_{v \in C_0^\infty(D)} (-q(v, v)) = \lim_{n\to\infty} \sup_{v \in C_0^\infty(D_n)} (-q(v, v))$$

$$= \lim_{n\to\infty} \hat{\lambda}_0(D_n) = \lim_{n\to\infty} \lambda_c(D_n) = \lambda_c(D). \quad \square$$

If $\hat{\lambda}_0 = \lambda_c$ is an eigenvalue for $(\hat{\mathcal{L}}, \hat{\mathcal{D}})$, then it is a simple eigenvalue and the corresponding eigenfunction $\phi_0 \in \hat{\mathcal{D}}$ belongs to $C_L(D)$. To see this, note that the transition measure $p(t, x, dy)$ for L on D is *positivity improving*; that is, if $f \in C(D)$, $f \geq 0$ and $f \not\equiv 0$, then $\int_D p(t, x, dy) f(y) > 0$, for all $x \in D$. In such a case, if $\hat{\lambda}_0$ is an eigenvalue, then it is a simple eigenvalue and the corresponding eigenfunction ϕ_0 is positive on D [Reed and Simon, Vol. 4 (1978), Theorem 13.44]. By the elliptic regularity theory alluded to in the remark following Theorem 8.4, $\phi_0 \in C^{2,\alpha}(D)$; thus $\phi_0 \in C_L(D)$. Conversely, if $\phi \in C_{L-\lambda_c}(D) \cap \hat{\mathcal{D}}$, then ϕ is, of course, an eigenfunction. Thus $C_{L-\lambda_c}(D) \cap \hat{\mathcal{D}}$ is either empty or one-dimensional.

The following theorem determines when $\hat{\lambda}_0 = \lambda_c$ is an eigenvalue or, equivalently, when $C_{L-\lambda_c}(D) \cap \hat{\mathcal{D}} \neq \phi$.

Theorem 10.2. *Let $L = \frac{1}{2}\nabla \cdot a\nabla + a\nabla Q \cdot \nabla + V$ satisfy Assumption $\widetilde{H}_{\text{loc}}$ on a domain $D \subseteq R^d$ and let $(-\hat{\mathcal{L}}, \hat{\mathcal{D}})$ denote the self-adjoint operator which is the Friedrichs' extension corresponding to the quadratic form*

$$q(f, g) = \tfrac{1}{2}\int_D (\nabla f a \nabla g) \exp(2Q)\, dx - \int_D Vfg \exp(2Q)\, dx \text{ on } C_0^\infty(D).$$

(i) *If $L - \lambda_c$ is subcritical on D, then no element of $C_{L-\lambda_c}(D) \cap L^2(D, \exp(2Q)\, dx)$ belongs to $\hat{\mathcal{D}}$. Thus $\hat{\lambda}_0 = \lambda_c$ is not an eigenvalue for $(\hat{\mathcal{L}}, \hat{\mathcal{D}})$.*

(ii) *If $L - \lambda_c$ is critical on D, but not product L^1 critical, then the ground state $\phi_c \notin L^2(D, \exp(2Q)\, dx)$. Thus $\hat{\lambda}_0 = \lambda_c$ is not an eigenvalue for $(\hat{\mathcal{L}}, \hat{\mathcal{D}})$.*

(iii) *If $L - \lambda_c$ is product L^1 critical on D, then the ground state $\phi_c \in \hat{\mathcal{D}}$, and thus, ϕ_c is an eigenfunction for $(\hat{\mathcal{L}}, \hat{\mathcal{D}})$ corresponding to the eigenvalue $\hat{\lambda}_0 = \lambda_c$.*

Proof. Part (ii) follows by definition. (Recall that in the symmetric case, the condition $\phi_c\tilde{\phi}_c \in L^1(D)$ becomes $\phi_c \in L^2(D, \exp(2Q)\,dx)$.)

For the proof of parts (i) and (iii), let $\phi \in C_{L-\lambda_c}(D) \cap L^2(D, \exp(2Q)\,dx)$. We must show that $\phi \in \hat{\mathcal{D}}$ if $L - \lambda_c$ is critical on D and that $\phi \notin \hat{\mathcal{D}}$ if $L - \lambda_c$ is subcritical on D. It is easy to check that L^ϕ with domain $C_0^\infty(D)$ is symmetric on $L^2(D, \phi^2 \exp(2Q)\,dx)$ and that if $(\hat{\mathcal{L}}^\phi, \hat{\mathcal{D}}^\phi)$ denotes the Friedrichs' extension corresponding to L^ϕ on D, then

$$\hat{\mathcal{L}}^\phi f = \frac{1}{\phi}\hat{\mathcal{L}}(\phi f) \text{ and } \hat{\mathcal{D}}^\phi = \{f\colon \phi f \in \hat{\mathcal{D}}\}.$$

Also, clearly, the Friedrichs' extension of $L^\phi - \lambda_c = (L - \lambda_c)^\phi$ is $(\hat{\mathcal{L}}^\phi - \lambda_c, \hat{\mathcal{D}}^\phi)$. From this and the fact that $1 \in C_{(L-\lambda_c)^{\phi_c}}(D) \cap L^2(D, \phi^2 \exp(2Q)\,dx)$ if and only if $\phi \in C_{L-\lambda_c}(D) \cap L^2(D, \exp(2Q)\,dx)$, it follows that we may assume that $V \equiv 0$, that $\lambda_c = 0$, and that $\phi \equiv 1 \in L^2(D, \exp(2Q)\,dx)$.

Let $\{D_n\}_{n=1}^\infty$ be a sequence of domains with $C^{2,\alpha}$-boundaries satisfying $D_n \subset\subset D_{n+1}$ and $\bigcup_{n=1}^\infty D_n = D$. Define

$$\lambda_n = \inf_{\substack{u \in C_0^\infty(D) \\ u=1 \text{ on } D_1 \\ u=0 \text{ on } D_n^c}} q(u, u),\ n = 1, 2, \ldots.$$

By Theorem 6.4.1 in Chapter 6, if L is subcritical (that is, transient), then $\lim_{n\to\infty} \lambda_n > 0$, while if L is critical (that is, recurrent), then $\lim_{n\to\infty} \lambda_n = 0$. From this and the definition of $\hat{\mathcal{D}}_{\hat{q}}$, it follows that in the subcritical case, $1 \notin \hat{\mathcal{D}}_{\hat{q}}$ and, *a fortiori*, $1 \notin \hat{\mathcal{D}}$. This proves (i). It also follows that in the critical case, $1 \in \hat{\mathcal{D}}_{\hat{q}}$. We must show that, in fact, $1 \in \hat{\mathcal{D}}$. It is easy to check that L is symmetric on $C_0^\infty(D) \cup \{\text{constant functions}\}$. Then $\bar{q}$ on $C_0^\infty(D) \cup \{\text{constant functions}\}$ is an extension of q on $C_0^\infty(D)$ and since $1 \in \hat{\mathcal{D}}_{\hat{q}}$, it follows that $\hat{q}$ on $\hat{\mathcal{D}}_{\hat{q}}$ is the closure of each of these two forms. By the uniqueness part of Friedrichs' extension, it follows that $(\hat{\mathcal{L}}, \hat{\mathcal{D}})$ is an extension of L on $C_0^\infty(D) \cup \{\text{constants}\}$. Thus $1 \in \hat{D}$. □

4.11 Extension to the case where *D* is a non-compact open manifold or a compact manifold without boundary

Both the martingale theory and the criticality theory extend readily to the case where L is a second-order elliptic operator on D where D is an open non-compact Riemannian manifold. Assume that L satisfies the obvious analog of Assumption $\tilde{H}_{\text{loc}}$ on D. The adjoint $\tilde{L}$ of L is defined with respect to the natural Riemannian volume element on the manifold. One can solve the generalized martingale problem locally on coordinate patches and then use the strong Markov property to patch the solutions together and obtain a solution to the generalized martingale problem on the entire domain. The criticality theory extends readily since the Schauder estimates still hold and since, if $V \leq 0$, the Dirichlet problem can still be solved for smooth bounded subdomains using, for example, the Perron method of subharmonic functions.

Now consider the case in which L is defined on a compact Riemannian manifold D without boundary. Assume that L satisfies the obvious analog of Assumption $\widetilde{H}$ on D. One can define a solution to the generalized martingale problem by the method noted above. In fact, since there is no boundary, $\tau_D \equiv \infty$ and thus the solution actually solves the martingale problem for L on D. Also, since there is no boundary, the diffusion cannot satisfy the condition for transience. We will show below that the diffusion is positive recurrent. As above, the adjoint operator $\widetilde{L}$ is defined with respect to the natural Riemannian volume element. The Green's function is defined to be the Radon–Nikodym derivative of the Green's measure with respect to the Riemannian volume element. The criticality theory can be developed, but there will be one difference – $C_L(D)$ will be empty in the subcritical case; indeed, letting $G(x, y)$ denote the Green's function, note that the singularity at y in

$$k(\cdot\,; y) \equiv \frac{G(\cdot\,, y)}{G(x_0, y)}$$

can no longer be 'pushed' out to the boundary to obtain a positive harmonic function on all of D as was done in Theorem 3.1. We sum up the criticality theory in a theorem and sketch its proof.

Theorem 11.1. *Let D be a compact Riemannian manifold without boundary of class $C^{2,\alpha}$ and let $l(\mathrm{d}x)$ denote its natural Riemannian volume element. Assume that $L = L_0 + V$ satisfies the analog of Assumption $\widetilde{H}$ on D and let $\widetilde{L}$ denote the adjoint of L with respect to $l(\mathrm{d}x)$.*

(i) *There exists a unique solution $\{P_x, x \in D\}$ to the martingale problem for L_0 on D and the corresponding diffusion process is positive recurrent.*
(ii) *Let $\lambda_0 = \inf\{\lambda$: the Green's function exists for $L - \lambda$ on $D\}$. Then $\lambda_0 \leq \sup V$. If V is constant, then $\lambda_0 = V$.*
(iii) *$C_{L-\lambda}(D)$ is empty if $\lambda \neq \lambda_0$.*

(iv) *For $\lambda > \lambda_0$, let $G_\lambda(x, y)$ denote the Green's function for $L - \lambda$ on D. Then for each $f \in C^\alpha(D)$, there exists a unique solution $u \in C^{2,\alpha}(D)$ to $(L - \lambda)u = -f$. The solution is given by*

$$u(x) = \int_D G_\lambda(x, y) f(y) l(dy).$$

(v) *For $\lambda > \lambda_0$, $(\lambda - L)^{-1}$: $C^\alpha(D) \to C^{2,\alpha}(D)$ is well defined and there exists a constant C_λ such that $\|(\lambda - L)f\|_{2,\alpha;D} \leq c_\lambda \|f\|_{0,\alpha;D}$. Thus L possesses a compact resolvent.*

(vi) *λ_0 is an eigenvalue of multiplicity one for the operator $(L, C^{2,\alpha}(D))$ on $C^\alpha(D)$. The corresponding eigenspace is generated by a function $\phi_0 \in C^{2,\alpha}(D)$ which satisfies $\phi_0 > 0$ on D.*

(vii) *λ_0 is also an eigenvalue of multiplicity one for $(\widetilde{L}, C^{2,\alpha}(D))$ and the corresponding eigenspace is generated by a function $\tilde{\phi}_0 \in C^{2,\alpha}(D)$ which satisfies $\tilde{\phi}_0 > 0$ on D.*

(viii)

$$\sup \operatorname{Re}(\sigma((L, C^{2,\alpha}(D)))) = \lambda_0 = \lim_{t\to\infty} \frac{1}{t} \log \sup_{y \in D} E_y \exp\left(\int_0^t V(X(s))\, ds\right)$$

$$= \lim_{t\to\infty} \frac{1}{t} \log E_x \exp\left(\int_0^t V(X(s))\, ds\right),$$

for all $x \in D$. In particular, λ_0 is the principal eigenvalue for $(L, C^{2,\alpha}(D))$.

(ix) *$L - \lambda_0$ is product L^1 critical.*

Sketch of proof. (i) We noted above how to solve the martingale problem for L_0 on D. Since D is compact, it can be covered by a finite number of coordinate patches. Applying the Stroock–Varadhan support theorem (Theorem 2.6.1) in each one, it is easy to show that for each $x \in D$, $E_x f(X(1)) > 0$, whenever $f \in C(D)$, $f \geq 0$ and $f \not\equiv 0$. By the Feller property, $E_x f(X(1))$ is continuous in x; thus $\inf_{x\in D} E_x f(X(1)) > 0$. Since this holds for all such f, we have $\inf_{x\in D} P_x(X(1) \in U) > 0$ for any open $U \subset D$. Iterating using the Markov property now shows that $P_x(X(t) \in U$ for some $t \geq n) = 1$, for all $x \in D$ and all positive integers n. This proves recurrence. The positive recurrence follows as a particular case of (ix).

(ii) Since L_0 corresponds to a recurrent diffusion, the Green's function does not exist for L_0. On the other hand, from its definition, it is clear that the Green's function does exist for $L_0 + W$ if $\sup W < 0$. (ii) follows from this.

(iii) Assume that $h \in C_{L-\lambda}(D)$. Let $(L - \lambda)^h$ denote the h-transform of $L - \lambda$. Since its zeroth-order term vanishes, it corresponds to a diffusion process and, by (i), the diffusion process is recurrent. It then follows from Theorem 3.3 that $L - \lambda$ is critical.

(iv) Fix $x_0 \in D$ and let $G_\lambda^{(x_0)}(x, y)$ denote the Green's function for $L - \lambda$ on $D_{x_0} \equiv D - \{x_0\}$. Since D_{x_0} is an open manifold without boundary, it follows from Theorem 3.8 (with Lebesgue measure replaced by $l(dx)$) that for each $f \in C_0^\alpha(D_{x_0})$, the function $u_{x_0}(x) \equiv \int_{D_{x_0}} G_\lambda^{(x_0)}(x, y) f(y) l(dy) \in C^{2,\alpha}(D_{x_0})$ and $(L - \lambda)u_{x_0} = -f$. By Theorem 2.9, $P_x(\tau_{x_0} < \infty) = 0$, for $x \neq x_0$, where $\tau_{x_0} = \inf\{t \geq 0: X(t) = x_0\}$. Therefore, it follows from the definition of the Green's function that $G_\lambda^{(x_0)}(x, y) = G_\lambda(x, y)$, for $x \neq x_0$ and $y \neq x$. Thus, the function $u(x) \equiv \int_D G_\lambda(x, y) f(y) l(dy)$, $x \in D$, coincides with u_{x_0} on D_{x_0}. Since x_0 is arbitrary, we conclude that $u \in C^{2,\alpha}(D)$ and $Lu = -f$ in D. This proves the result if f vanishes on an open set of D. The extension to general f follows by linearity since any f can be written as the sum of two functions each of which vanishes on an open set of D. Uniqueness is left as exercise 4.17.

(v) Let $f \in C^\alpha(D)$ and let $u \in C^{2,\alpha}(D)$ be the unique solution to $Lu = -f$ in D. Since D is compact, we can choose proper bounded subdomains U_1 and U_2 of D such that $U_1 \cup U_2 = D$. Applying the Schauder interior estimate (Theorem 3.2.7) gives

$$\|u\|_{2,\alpha;U_i} \leq c[\|u\|_{0;D} + \|f\|_{0,\alpha;D}],\ i = 1, 2. \tag{11.1}$$

By (iv),

$$\|u\|_{0;D} \leq \sup_{x \in D} \int_D G(x, y) l(dy)) \|f\|_{0;D}. \tag{11.2}$$

Clearly,

$$\|f\|_{0;D} \leq \|f\|_{0,\alpha;D}. \tag{11.3}$$

The result now follows from (11.1)–(11.3).

(vi) This is the analog of Theorem 3.5.5; the proof is essentially the same.

(vii) This follows from the Krein–Rutman theorem (see the paragraph preceding Theorem 3.7.1).

(viii) This is the analog of Theorem 3.6.1; however, as we now show, its proof is simpler since ϕ_0 is uniformly positive in the present case. Step 1 of the proof of Theorem 3.6.1 goes through in the same way; thus

$$\sup \operatorname{Re}(\sigma((L, C^{2,\alpha}(D)))) \leq \lim_{t \to \infty} \frac{1}{t} \log \sup_{y \in D} E_y \exp\left(\int_0^t V(X(s))\, ds\right). \tag{11.4}$$

From the beginning of Step 3 (which relies on Step 1), we obtain

$$E_x \exp\left(\int_0^t V(X(s))\,\mathrm{d}s\right)\phi_0(X(t)) = \exp(\lambda_0 t)\phi_0(x), \text{ for } x \in D.$$

Thus,

$$\frac{\inf \phi_0}{\sup \phi_0} \exp(\lambda_0 t) \leq E_x \exp\left(\int_0^t V(X(s))\,\mathrm{d}s\right) \leq \frac{\sup \phi_0}{\inf \phi_0} \exp(\lambda_0 t),$$
$$\text{for } x \in D. \quad (11.5)$$

The second and third equalities in (viii) follow from (11.5) and the first equality in (viii) follows from (11.4), (11.5) and the fact that $\lambda_0 \in \sigma((L, C^{2,\alpha}(D)))$.
(ix) Since $\phi_0, \tilde{\phi}_0 \in C^{2,\alpha}(D)$, $\int_D \phi_0 \tilde{\phi}_0 l(\mathrm{d}x) < \infty$. □

Exercises

In the exercises below, $\{D_n\}_{n=1}^{\infty}$ denotes a sequence of domains with $C^{2,\alpha}$-boundaries satisfying $D_n \subset\subset D_{n+1}$ and $\bigcup_{n=1}^{\infty} D_n = D$.

4.1 Let $D \subseteq R^d$ and let $L = \frac{1}{2}\Delta + b \cdot \nabla$, where b is locally bounded and measurable. Show that $p(t, x, \mathrm{d}y)$, the transition subprobability measure on D, is absolutely continuous and, thus, possesses a density on D. (*Hint*: For $b \equiv 0$ and $D = R^d$ it is obvious. It then holds *a fortiori* for $b \equiv 0$ and arbitrary D. Use the Cameron–Martin–Girsanov formula to obtain the result for bounded b. For locally bounded b, use the fact that for $B \subset\subset D$,

$$\begin{aligned} p(t, x, B) &= P_x(X(t) \in B) = P_x(X(t) \in B;\, \tau_D > t) \\ &= \lim_{n\to\infty} P_x(X(t) \in B, \tau_{D_n} > t) \\ &= \lim_{n\to\infty} p^{(n)}(t, x, B), \text{ where } p^{(n)}(t, x, \mathrm{d}y) \end{aligned}$$

corresponds to L on D_n.)

4.2 Show that (1.2) holds. (*Hint*: Let a_n, b_n and h_n be nice bounded coefficients on R^d with $\inf h_n > 0$ and such that $a = a_n$, $b = b_n$ and $h = h_n$ on D_n.)

4.3 Use the technique of exercise 4.2 to show that Ito's formula holds with $f = \log h$ and with t replaced by $\tau_{D_n} \wedge t$ as in (1.4).

4.4 Let L satisfy Assumption $\widetilde{\mathrm{H}}_{\mathrm{loc}}$ on a domain $D \subseteq R^d$. If $u \in C^2(D)$ satisfies $Lu = 0$ and $u > 0$ in D, then $u \in C^{2,\alpha}(D)$. (*Hint*: For any $D' \subset\subset D$ with $C^{2,\alpha}$-boundary, $\lambda_c(D') = \lambda_0(D') < 0$. There exists a unique solution $v \in C^{2,\alpha}(D')$ to $Lv = 0$ in D' and $v = u$ on $\partial D'$; this follows essentially from [Gilbarg and Trudinger, Theorem 6.13].

4.5 Let

$$L_0 = \frac{1}{2}\sum_{i,j=1}^{d} a_{ij}\frac{\partial^2}{\partial x_i \partial x_j} + \sum_{i=1}^{d} b_i \frac{\partial}{\partial x_i}$$

on a domain $D \subseteq R^d$. Assume that Harnack's inequality holds and that there exists a unique solution to the generalized martingale problem for L_0 on D.

(i) Assume that the corresponding diffusion process is transient on D. Let B be a ball satisfying $B \subset\subset D$ and define $\sigma_{\bar{B}} = \inf\{t \geq 0: X(t) \in \bar{B}\}$. By Theorem 2.8.2, $P_x(\sigma_{\bar{B}} < \infty) < 1$, for $x \in D - \bar{B}$. Prove that $\sup_{x \in U} P_x(\sigma_{\bar{B}} < \infty) < 1$, for $U \subset D$ satisfying $\bar{U} \cap \bar{B} = \phi$. (*Hint*: By the strong Markov property, it is enough to show it for $U = \partial B_1$, where B_1 is a ball satisfying $B \subset\subset B_1 \subset\subset D$. Choose the sequence $\{D_n\}_{n=1}^{\infty}$ so that $B_1 \subset\subset D_1$. Let $u_n(x) = P_x(\sigma_{\bar{B}} > \tau_{D_n})$, where $\tau_{D_n} = \inf\{t \geq 0: X(t) \notin D_n\}$. By Theorem 2.1.2 and Theorem 3.3.1, $u_n \in C^{2,\alpha}(D_n - \bar{B})$ and $Lu_n = 0$ in $D_n - \bar{B}$. Also $\lim_{n\to\infty} u_n(x) = P_x(\sigma_{\bar{B}}) = \infty)$. Now use Harnack's inequality.)

(ii) Let U_1, $U_2 \subset D$ be domains satisfying $\bar{U}_1 \cap \bar{U}_2 = \phi$ and let $\sigma_{\bar{U}_i} = \inf\{t \geq 0: X(t) \in \bar{U}_i\}$. By the Stroock–Varadhan support theorem, $P_x(\sigma_{\bar{U}_1} < \sigma_{\bar{U}_2}) > 0$, for all x in the connected component of $D - \bar{U}_2$ which contains U_1. Prove that $\inf_{x\in K} P_x(\sigma_{\bar{U}_1} < \sigma_{\bar{U}_2}) > 0$, for any compact K contained in the connected component of $D - \bar{U}_2$ which contains U_1. (*Hint*: Produce an argument similar to that in part (i).)

4.6 Prove (2.6). (*Hint*: Let $A_1 \subset\subset A$. By exercise 4.5,

$$\inf_{z\in\partial D_1} P_z(\sigma_{\bar{A}_1} < \tau_1) > 0,$$

where $\sigma_{\bar{A}_1} = \inf\{t \geq 0: X(t) \in \bar{A}_1\}$. Use Theorem 2.2.2 to show that there exist $\varepsilon > 0$ and $\delta > 0$ such that $P_x(\tau_A \geq \varepsilon) \geq \delta$ for all $x \in A_1$, where $\tau_A = \inf\{t \geq 0: X(t) \notin A\}$).

4.7 Let L_0 satisfy the assumption of exercise 4.5 on a domain $D \subseteq R^d$. Let B_1 and B_2 be domains satisfying $B_1 \subset\subset B_2 \subset\subset D$ and let $\sigma_{\bar{B}_1} = \inf\{t \geq 0: X(t) \in \bar{B}_1\}$ and $\tau_{B_2} = \inf\{t \geq 0: X(t) \notin B_2\}$. Show that there exists a constant c such that $E_x\,\phi(X(\tau_{B_2})) \leq cE_y\phi(X(\tau_{B_2}))$, for all x, $y \in \bar{B}_1$ and all non-negative, bounded measurable ϕ on ∂B_2. Show also that there exists a constant c such that

$$E_x(\phi(X(\sigma_{\bar{B}_1})); \sigma_{\bar{B}_1} < \tau_D) \leq cE_y(\phi(X(\sigma_{\bar{B}_1})); \sigma_{\bar{B}_1} < \tau_D),$$

for x, $y \in \partial B_2$ and all non-negative, bounded measurable ϕ on ∂B_1. Furthermore, if the diffusion process corresponding to L_0 is recurrent, so that $P_x(\sigma_{\bar{B}_1} < \tau_D) = 1$, for all $x \in D$, then $E_x\phi(X(\sigma_{\bar{B}_1})) \leq cE_y\phi(X(\sigma_{\bar{B}_1}))$, for all x, $y \in D - B_2$. Therefore, the measures $P_x(X(\tau_{B_2}) \in \cdot)$ and $P_y(X(\tau_{B_2}) \in \cdot)$ with x, $y \in \bar{B}_1$ and the measures $P_x(X(\sigma_{\bar{B}_1}) \in \cdot, \sigma_{\bar{B}_1} < \tau_D)$ and $P_y(X(\sigma_{\bar{B}_1}) \in \cdot, \sigma_{\bar{B}_1} < \tau_D)$ with $x, y \in \partial B_2$ (or, in the recurrent case, $P_x(X(\sigma_{\bar{B}_1}) \in \cdot)$ and $P_y(X(\sigma_{\bar{B}_1}) \in \cdot)$ with x, $y \in D - B_2$) are mutually absolutely continuous with Radon–Nikodym derivatives which are bounded and bounded away from zero. (*Hint*: By approximation, it suffices to prove it for $\phi \in C^{2,\alpha}(\partial B_2)$. Use Harnack's inequality, Theorem 2.1.2 and Theorem 3.3.1.)

4.8 Let L_0 be as in exercise 4.5 on a domain $D \subseteq R^d$ and assume that the corresponding diffusion process is recurrent on D. Let $B \subset\subset D$ be such that $D - B$ is connected and choose $\{D_n\}_{n=1}^{\infty}$ so that $B \subset\subset D_1$. Let $\sigma_{\bar{B}} = \inf\{t \geq 0;\ X(t) \in \bar{B}\}$ and let $\tau_{D_n} = \inf\{t \geq 0:\ X(t) \notin D_n\}$. Define $u_n(x) = E_x\sigma_{\bar{B}} \wedge \tau_{D_n}$, for $x \in D_n - \bar{B}$, and let $u(x) \equiv E_x\sigma_{\bar{B}} = \lim_{n\to\infty} u_n(x)$, for $x \in D - \bar{B}$. Prove that either $u(x) = \infty$, for all $x \in D - \bar{B}$, or $u(x)$ is locally bounded on $D - \bar{B}$. (*Hint*: Choose A_1, A_2 and A_3 with $C^{2,\alpha}$-boundaries such that $B \subset\subset A_1 \subset\subset A_2 \subset\subset A_3 \subset\subset D$. Let $\sigma = \inf\{t \geq 0:\ X(t) \notin A_3 - \bar{A}_1\}$. By Theorem 2.1.1, $\sup_{x\in\partial A_2} E_x\sigma < \infty$. By exercise 4.7, there exists a constant c such that $P_x(X(\sigma) \in B) \leq cP_y(X(\sigma) \in B)$, for all x, $y \in \partial A_2$ and all $B \subset \partial A_1 \cup \partial A_3$. Use these facts and the strong Markov property to conclude that if $u(x) = \infty$ for some $x \in \partial A_2$, then $u(x) = \infty$ for all $x \in \partial A_2$. Now use the Stroock–Varadhan support theorem and the connectedness of $D - B$ along with the strong Markov property to show that if $u(x) = \infty$ for all $x \in \partial A_2$, then $u(x) = \infty$ for all $x \in D - \bar{B}$.)

4.9 Prove part (ii) of Theorem 3.7. (*Hint*: Use Harnack's inequality.)

4.10 Let L_0 satisfy the assumptions of exercise 4.5 and assume that there exists a number γ and an $x \in D$ such that

$$\lim_{t\to\infty} \frac{1}{t}\int_0^t 1_B(X(s))\,ds = \gamma \text{ a.s. } P_x.$$

Without assuming the existence of a density $p(t, x, y)$, prove that

$$\lim_{t\to\infty} \frac{1}{t}\int_0^t 1_B(X(s))\,ds = \gamma \text{ a.s. } P_y,$$

for any $y \in D$. (*Hint*: Fix x, $y \in D$ and choose a domain $U \subset\subset D$ such that x, $y \in U$. Let $\tau_U = \inf\{t \geq 0: X(t) \notin U\}$. Write

$$\frac{1}{t}\int_0^t 1_B(X(s))\,ds = \frac{1}{t}\left[\int_0^{\tau_U\wedge t} 1_B(X(s))\,ds + \int_{\tau_U\wedge t}^t 1_B(X(s))\,ds\right]$$

and use the strong Markov property and exercise 4.7.)

4.11 Prove Corollary 3.10 via exercise 1.15. (*Hint*: Reduce to the case where $V \equiv 0$ and show that the property of transience or recurrence is invariant under time changes of the type in exercise 1.15.)

4.12 Prove Theorem 7.7(ii). (*Hint*: Let $L_1 = L + b_1 \cdot \nabla$ with $|b_1| < \delta$. Let ϕ_c denote the ground state of $L - \lambda_c$ and let $h = (\phi_c)^\alpha$, for $\alpha \in (0, 1)$. Using the fact that $(L - \lambda_c)\phi_c = 0$, show that for an appropriate choice of α, $(L_1 - \lambda_c)h \leq \gamma\delta h$, for some $\gamma > 0$. Now use Theorem 3.9.)

4.13 Let

$$L = \tfrac{1}{2}\Delta + \frac{k}{|x|^2} \text{ on } D = \begin{cases} R^d - \{0\} \text{ if } d \geq 2 \\ (0, \infty) \text{ if d} = 1 \end{cases},$$

where

$$-\infty < k < \frac{(d-2)^2}{8}.$$

(See Example 3.12.) Show that $\lambda_c = 0$.

4.14 Prove that $L = \frac{1}{2}\Delta + V(x)$ on R^d, $d \geq 2$, is supercritical if there exists an open set $A \subset S^{d-1}$ such that

$$\lim_{\substack{|x|\to\infty \\ \frac{x}{|x|}\in A}} |x|^2 V(x) = \infty.$$

4.15 Prove Theorem 1.15.1. (*Hint*: Use exercise 4.7.)

4.16 Consider the following assumption:

Assumption H_{loc}. $L = \frac{1}{2}\sum_{i,j=1}^{d} a_{ij}(x)\frac{\partial^2}{\partial x_i \partial x_j} + \sum_{i=1}^{d} b_i(x)\frac{\partial}{\partial x} + V$ *is defined on a domain* $D \subseteq R^d$ *and satisfies Assumption H (defined in Chapter* 3, *Section* 2) *on every subdomain* $D' \subset\subset D$.

Check that all the theorems in this chapter which involve only L and not $\tilde{L}$ hold if L satisfies Assumption H_{loc} on $D \subseteq R^d$. (*Hint*: The spectral theory of Chapter 3, Sections 3–6, holds for L on any subdomain $D' \subset\subset D$.)

4.17 Prove the uniqueness part of Theorem 11.1(iv). (*Hint*: Use the Feynman–Kac formula (Theorem 2.4.2), Theorem 11.1(viii) and the definition of the Green's function.)

4.18 Prove Theorem 5.3. (*Hint*: The proof is similar to the proofs of Theorems 5.1 and 5.2.)

4.19 Assuming that a Green's function, if it exists, enjoys the properties enumerated in Theorem 2.5, use Theorem 2.8 and the Hopf maximum principle (Theorem 3.2.5) to prove Theorem 2.4. (*Hint*: Assume that the Green's function $G_{\lambda_0}(x, y)$ exists for $L - \lambda_0$, and let ϕ_0 denote the positive eigenfunction corresponding to λ_0. Fix $y \in D$. Since $(L - \lambda_0)G_{\lambda_0}(\cdot, y) = 0$, in $D - \{y\}$, and $(L - \lambda_0)\phi_0 = 0$ in D, it follows from the Hopf maximum principle and the compactness of ∂D that there exists a $c > 0$ such that

$$\frac{1}{c}\phi_0(x) \leq G(x, y) \leq c\phi_0(x),$$

for x close to ∂D. By Theorem 2.8, there exists an $\varepsilon > 0$ such that $G(x, y) \geq 2c\phi_0(x)$, for $x \in \partial B_\varepsilon(y)$. Now let $\gamma_0 = \sup\{\gamma\colon G(x, y) - \gamma\phi_0(x) > 0$, for all $x \in D - B_\varepsilon(y)\}$. Let $v(x) = G(x, y) - \gamma_0\phi_0(x)$. Show that $\gamma_0 > 0$ and that $v > 0$ on $\partial B_\varepsilon(y)$. But then, applying the same considerations to v and ϕ_0, show that $\gamma_1 = \sup\{\gamma\colon v(x) - \gamma\phi_0(x) > 0$, for all $x \in D - B_\varepsilon(y)\} > 0$. This contradicts the maximality of γ_0.)

4.20 Use Theorem 6.3 to prove Theorem 4.1(iii) in the case where $L - \lambda_c(D_1)$ is assumed critical on D_1. (*Hint*: Assume on the contrary that $\lambda_c(D_2) = \lambda_c(D_1)$. Then $L - \lambda_c(D_1)$ is critical on D_2. Choose a non-negative and not identically zero function W which is compactly supported in $D_2 - \bar{D}_1$. Apply Theorem 6.3 to $L - \lambda_c(D_1) - W$ on D_2, and then consider its restriction to D_1.)

Notes

The Harnack inequality for elliptic operators goes back to [Serrin, (1955–56)]. However, in that paper, the regularity requirements on the coefficients were stronger than those appearing in Theorem 0.1. The state of the art Harnack inequality which requires much less regularity than in Theorem 0.1 follows from the work of [Krylov and Safanov (1979), (1980)] and [Safanov 1980]; one can find a treatment of this in [Gilbarg and Trudinger (1983)].

The h-transform theory goes back to [Doob (1957)]. Although, the results of Section 1 are standard to those working in the field, I was unable to locate rigorous proofs in the literature.

The material on Green's functions which appears in textbooks on PDEs is surprisingly sparse. The one standard source for the construction of the Green's function is [Miranda (1970)]. Miranda defines the Green's function as a fundamental solution which satisfies an appropriate boundary condition. The construction of the Green's function involves the theory of integral equations and, specifically, the theory of single and double layer potentials. I have chosen a completely different tack which seems to me to be more in harmony with the spirit of this book. The Green's measure is defined as the time integral of the transition measure; that is, $G(x, dy) = \int_0^\infty p(t, x, dy)\, dt$. Using the parametrix method [Friedman (1964)] or using the methods in [Stroock and Varadhan (1979)], one can show that $p(t, x, dy)$ possesses a density, from which it follows that $G(x, dy)$ also possesses a density, the Green's function. However, I have refrained from using these highly technical methods from parabolic PDE theory. Instead, three fundamental results from elliptic PDE theory are used: the existence theorem for the Dirichlet problem, the basic Schauder estimate and Harnack's inequality. With these basic tools, I have given a 'bare-handed' proof that the Green's function exists and enjoys certain fundamental properties (Theorem 2.5). The one fundamental property that is not proved concerns the behavior of the Green's function $G(x, y)$ near the diagonal $x = y$ (Theorem 2.8). For this, the reader may consult [Miranda (1970)].

The classification of an operator as *subcritical*, *critical* or *supercritical* was initiated by [Simon (1981)], who defined these terms for Schrodinger operators. This was then developed by [Murata (1984), (1986)]. However, the definitions do not generalize to the case of non-self-adjoint operators. The definition given in the text for general second-order elliptic operators is due to [Pinchover (1988)]. Theorem 3.4 is due to [Agmon (1982)]; see also [Murata (1986)] for the symmetric case. The simple, probabilistic proof given in the text is completely different from the treatment in the two articles mentioned above. Many of the other results of Section 3 are standard to those working in linear elliptic PDE theory; however, I could not find references in the literature. The material in this section has been worked out with a probabilistic flavor.

Theorem 4.1(ii) and (iii) seems to be new. Corollary 4.2 appears in [Murata (1986)] for symmetric operators and in [Pinchover (1988)] for general operators. In the case where $V = 0$, the characterization of λ_c given in Theorem 4.4 and the remark which follows it was noted in [Stroock (1982)]. Theorem 4.5 is due essentially to [Protter and Weinberger (1966)]; note that after the theory of Section 3, its proof becomes trivial. Theorems 4.6 and 4.7 are due to [Nussbaum and Pinchover (1992)]. Theorems 5.1 and 5.2 are pretty basic; however, I was unable to find any references in the literature. Theorem 5.3 is even easier.

Theorem 6.1 can be found in [Pinchover (1989)]. Theorems 6.2–6.5 are due to [Murata (1986)] in the symmetric case and to [Pinchover (1988), (1989)] in the general case. Theorem 6.6 is due to [Pinchover (1990a)]. In the case where $L = \Delta$, Theorem 6.6 appears as the last result in [Reed and Simon, Vol. 4, (1978)]. Pinchover's techniques are simpler that those of Reed and Simon. The proof in the text follows in the spirit of Pinchover's proof but contains several technical differences. Theorem 6.7 is due to [Pinchover (1990b)]. The 'localization of binding' (Theorem 6.8) is due to [Simon (1980)]. The proof in the text is self-contained; it follows Simon's proof until that proof unleashes some spectral theoretic machinery.

Theorem 7.1 is due to [Berestycki, Nirenberg, and Varadhan (1994)]. The proof given in the text is completely different; it was suggested by Y. Pinchover. For a class of symmetric operators, [Persson (1960)] showed that $\lambda_{c,\infty}$ is the supremum of the essential spectrum; see also [Agmon (1982)]. Theorem 7.2 is due to [Pinchover (1990b)]. Theorems 7.7 and 7.8 are due to [Berestycki, Nirenberg and Varadhan (1994)].

For results in the spirit of Section 8, see [Stroock (1982)] and [Sullivan (1986)]. The elliptic regularity theory discussed after the proof of Theorem 8.4 can be found in [Gilbarg and Trudinger (1983)].

Theorem 9.5(i) has been proved in [Hasminskii (1980)] using probabilistic techniques and in [Pinchover (1992)] using PDE techniques. The proof in the text differs from either of them. In Theorem 9.6, the equivalence between the finiteness of $E_x\sigma_{\bar{B}}$ and the existence of a positive solution to $Lu = -1$ on $D - \bar{B}$ was noted in [Hasminskii (1980)]. In [Hasminskii, (1980)] it is also shown that if $E_x\sigma_{\bar{B}} < \infty$, for $x \in D - \bar{B}$, then an invariant probability density exists; this is essentially Theorem 9.6(i). The proof of Theorem 9.9 in the text follows [Varadhan (1980)], which treated the case $V = 0$. In fact, the stronger result that the density $p(t, x, y)$ satisfies $\lim_{t\to\infty} p(t, x, y) = \phi_c(x)\tilde{\phi}_c(y)$ has been proved by [Hasminskii (1980)] using probabilistic techniques and by [Pinchover (1992)] using PDE techniques. This stronger version of Theorem 9.9, along with the analogous stronger versions of Theorem 9.5(i), namely $\lim_{t\to\infty} p(t, x, y) = 0$, have been proved in the symmetric case by [Chavel and Karp (1991)] and by [Simon (1993)].

Proposition 10.1 and Theorem 10.2 have been proved in [Sullivan (1987)]; the approach taken in the text to Theorem 10.2 is new.

The extension to manifolds is fairly straightforward; for the construction of diffusions on a manifold, see, for example, [Ikeda and Watanabe (1981)].

5
Applications to the one-dimensional case and the radially symmetric multi-dimensional case

5.0 Introduction

In this chapter, we apply the results of Chapter 4 to one-dimensional operators of the form

$$L = \tfrac{1}{2}a(x)\frac{d^2}{dx^2} + b(x)\frac{d}{dx} + V(x)$$

on a domain $D = (\alpha, \beta)$, where $-\infty \leqslant \alpha < \beta \leqslant \infty$. It is easy to see that if $\hat{L}$ is a radially symmetric operator on an annulus in R^d, then the criticality properties of $\hat{L}$ can be reduced to those of a one-dimensional operator L on D as above (see exercise 5.5).

In the first section, we obtain integral criteria to answer questions related to criticality. In the case where $V \equiv 0$, criteria for transience and recurrence, for explosion and non-explosion, and for positive and null recurrence are given in terms of a and b. In the case where $V \neq 0$, the analogous criteria related to criticality cannot be given in terms of a, b and V but, rather, in terms of a, b and $\phi \in C_L$. In the second section, we study certain criticality properties of the Laplacian in R^d perturbed by a radially symmetric function of compact support.

When $V \equiv 0$, we will usually use Assumption A, which appears at the beginning of Chapter 2, while when $V \not\equiv 0$, we will use Assumption $\tilde{\mathrm{H}}_{\mathrm{loc}}$, which appears at the beginning of Chapter 4. However, even when Assumption $\tilde{\mathrm{H}}_{\mathrm{loc}}$ is in force, we will persist in writing operators in non-divergence form:

$$L = \tfrac{1}{2}a\frac{d^2}{dx^2} + b\frac{d}{dx} + V.$$

We note that in the one-dimensional case, if L satisfies Assumption $\tilde{\mathrm{H}}_{\mathrm{loc}}$, then it is automatically symmetric, as can be seen by writing

$$L = \tfrac{1}{2}a\frac{\mathrm{d}^2}{\mathrm{d}x^2} + b\frac{\mathrm{d}}{\mathrm{d}x} + V = \tfrac{1}{2}\frac{\mathrm{d}}{\mathrm{d}x}a\frac{\mathrm{d}}{\mathrm{d}x} + aQ'\frac{\mathrm{d}}{\mathrm{d}x} + V,$$

where

$$Q(x) = \int_{x_0}^{x} \frac{b}{a}(y)\,\mathrm{d}y - \tfrac{1}{2}\log a(x),$$

for some $x_0 \in (\alpha, \beta)$. Thus, the L^2-spectral theory for self-adjoint operators is in effect (see Chapter 4, Section 10).

5.1 Criticality properties

We begin with necessary and sufficient conditions for transience and recurrence in the case where $V \equiv 0$.

Theorem 1.1. *Let*

$$L = \tfrac{1}{2}a\frac{\mathrm{d}^2}{\mathrm{d}x^2} + b\frac{\mathrm{d}}{\mathrm{d}x}$$

satisfy Assumption A on $D = (\alpha, \beta)$, where $-\infty \leqslant \alpha < \beta \leqslant \infty$, and let $\{P_x, x \in \hat{D}\}$ denote the solution to the generalized martingale problem for L on D. Let $x_0 \in (\alpha, \beta)$.

(i) *If*

$$\int_{\alpha}^{x_0} \exp\left(-\int_{x_0}^{x} \frac{2b}{a}(s)\,\mathrm{d}s\right)\mathrm{d}x = \infty \text{ and } \int_{x_0}^{\beta} \exp\left(-\int_{x_0}^{x} \frac{2b}{a}(s)\,\mathrm{d}s\right)\mathrm{d}x = \infty,$$

then the diffusion corresponding to L is recurrent.

(ii) *If*

$$\int_{\alpha}^{x_0} \exp\left(-\int_{x_0}^{x} \frac{2b}{a}(s)\,\mathrm{d}s\right)\mathrm{d}x = \infty(<\infty) \text{ and}$$

$$\int_{x_0}^{\beta} \exp\left(-\int_{x_0}^{x} \frac{2b}{a}(s)\,\mathrm{d}s\right)\mathrm{d}x < \infty(=\infty),$$

then the diffusion corresponding to L is transient and, in particular, $P_x(\lim_{t\to\tau_D} X(t) = \beta) = 1(P_x(\lim_{t\to\tau_D} X(t) = \alpha) = 1)$, for all $x \in (\alpha, \beta)$.

(iii) *If*

$$\int_{\alpha}^{x_0} \exp\left(-\int_{x_0}^{x} \frac{2b}{a}(s)\,\mathrm{d}s\right)\mathrm{d}x < \infty \text{ and } \int_{x_0}^{\beta} \exp\left(-\int_{x_0}^{x} \frac{2b}{a}(s)\,\mathrm{d}s\right)\mathrm{d}x < \infty,$$

then the diffusion corresponding to L is transient and, in particular,

$$P_x(\lim_{t\to\tau_D} X(t) = \alpha) = \frac{\int_{x}^{\beta} \exp\left(-\int_{x_0}^{y} \frac{2b}{a}(s)\,\mathrm{d}s\right)\mathrm{d}y}{\int_{\alpha}^{\beta} \exp\left(-\int_{x_0}^{y} \frac{2b}{a}(s)\,\mathrm{d}s\right)\mathrm{d}y}$$

and

$$P_x(\lim_{t\to\tau_D} X(t)=\beta)=\frac{\displaystyle\int_\alpha^x \exp\left(-\int_{x_0}^y \frac{2b}{a}(s)\,\mathrm{d}s\right)\mathrm{d}y}{\displaystyle\int_\alpha^\beta \exp\left(-\int_{x_0}^y \frac{2b}{a}(s)\,\mathrm{d}s\right)\mathrm{d}y}.$$

Proof. We will assume that $b\in C((\alpha,\beta))$. The extension to the case that b is measurable and locally bounded is left to the reader as exercise 5.1. We have

$$u(x)\equiv\int_{x_0}^x \exp\left(-\int_{y_0}^y \frac{2b}{a}(s)\,\mathrm{d}s\right)\mathrm{d}y\in C^2((\alpha,\beta)) \text{ and } Lu=0 \text{ in } (\alpha,\beta).$$

Let $\tau_x=\inf\{t\geq 0: X(t)=x\}$. Then, by Theorem 2.1.2,

$$u(x)=u(x_1)P_x(\tau_{x_1}<\tau_{x_2})+u(x_2)P_x(\tau_{x_2}<\tau_{x_1}), \text{ if } x_1<x<x_2.$$

Since $P_x(\tau_{x_1}<\tau_{x_2})=1-P_x(\tau_{x_2}<\tau_{x_1})$, we obtain

$$P_x(\tau_{x_1}<\tau_{x_2})=\frac{u(x_2)-u(x)}{u(x_2)-u(x_1)}. \tag{5.1}$$

Similarly,

$$P_x(\tau_{x_2}<\tau_{x_1})=\frac{u(x)-u(x_1)}{u(x_2)-u(x_1)}. \tag{5.2}$$

Assume now that the conditions in (i) hold. Then from (5.1) and (5.2) we obtain

$$P_x(\tau_{x_2}<\infty)=\lim_{x_1\to\alpha}P_x(\tau_{x_2}<\tau_{x_1})=1 \quad\text{and}\quad P_x(\tau_{x_1}<\infty)$$
$$=\lim_{x_2\to\beta}P_x(\tau_{x_1}<\tau_{x_2})=1.$$

Since x, x_1 and x_2 are arbitrary, this proves recurrence.

Assume now that the conditions in (ii) hold. For concreteness, assume that

$$\int_\alpha^{x_0}\exp\left(-\int_{x_0}^x \frac{2b}{a}(s)\,\mathrm{d}s\right)\mathrm{d}x=\infty$$

and that

$$\int_{x_0}^{\beta}\exp\left(-\int_{x_0}^x \frac{2b}{a}(s)\,\mathrm{d}s\right)\mathrm{d}x<\infty.$$

Then from (5.1) we have

$$P_x(\tau_{x_1} < \infty) = \lim_{x_2 \to \beta} P_x(\tau_{x_1} < \tau_{x_1}) = \frac{u(\beta) - u(x)}{u(\beta) - u(x_1)} < 1,$$

which proves transience. Furthermore,

$$\lim_{x_1 \to \alpha} P_x(\tau_{x_1} < \infty) = \lim_{x_1 \to \alpha} \frac{u(\beta) - u(x)}{u(\beta) - u(x_1)} = 0,$$

which proves that $P_x(\lim_{t \to \tau_D} X(t) = \alpha) = 0$. Since, by the definition of transience, $P_x(\lim_{t \to \tau_D} X(t) = \alpha \text{ or } \beta) = 1$, it follows that

$$P_x(\lim_{t \to \tau_D} X(t) = \beta) = 1.$$

Under the condition in (iii), the process is transient by the same argument used above in the proof of (ii). Furthermore,

$$\lim_{x_1 \to \alpha} P_x(\tau_{x_1} < \infty) = \frac{u(\beta) - u(x)}{u(\beta) - u(\alpha)} \quad \text{and} \quad \lim_{x_2 \to \beta} P_x(\tau_{x_2} < \infty)$$

$$= \frac{u(x) - u(\alpha)}{u(\beta) - u(\alpha)}.$$

The rest of (iii) now follows. □

Example 1.2. The radial part of $\frac{1}{2}\Delta$ in R^d, $d \geq 2$, is given by

$$\frac{1}{2}\frac{\mathrm{d}^2}{\mathrm{d}r^2} + \frac{d-1}{2r}\frac{\mathrm{d}}{\mathrm{d}r} \text{ on } (0, \infty).$$

Applying Theorem 1.1 shows that Brownian motion is recurrent if $d = 2$ and transient if $d \geq 3$, as was shown by a different method in Example 4.2.7. Also, letting $\tau_r = \inf\{t \geq 0: |X(t)| = r\}$, it follows that for Brownian motion,

$$P_x(\tau_r < \infty) = \left(\frac{r}{|x|}\right)^{d-2} \text{ and } P_x(\tau_0 < \infty) = 0,$$

$$\text{for } x \in R^d - \{0\} \text{ and } 0 < r < |x|.$$

In the case where $V \equiv 0$, it is easy to check that 1 and

$$\int_{x_0}^{x} \exp\left(-\int_{x_0}^{y} \frac{2b}{a}(s)\, \mathrm{d}s\right) \mathrm{d}y$$

form a complete set of linearly independent solutions to $Lu = 0$ in (α, β). Therefore, under (i) of Theorem 1.1, $C_L = \{c \in R: c > 0)$; under (ii) of Theorem 1.1, C_L is generated by

$u_1(x) = 1$ and $u_2(x)$

$$= \int_x^\beta \exp\left(-\int_{x_0}^y \frac{2b}{a}(s)\,ds\right) dy \left(u_2(x) = \int_\alpha^x \exp\left(-\int_{x_0}^y \frac{2b}{a}(s)\,ds\right) dy\right);$$

and under (iii) of Theorem 1.1, C_L is generated by

$$u_1(x) = \int_\alpha^x \exp\left(-\int_{x_0}^y \frac{2b}{a}(s)\,ds\right) dy$$

and

$$u_2(x) = \int_x^\beta \exp\left(-\int_{x_0}^y \frac{2b}{a}(s)\,ds\right) dy.$$

In particular, C_L is one-dimensional in the recurrent case (as follows also from Theorem 4.3.4) and two-dimensional in the transient case.

Now consider the case in which V is not necessarily zero. Assume that L satisfies Assumption $\tilde{H}_{loc}$ at the beginning of Chapter 4. If L is critical, then by Theorem 4.3.4, C_L is one-dimensional. Now, assume that L is subcritical. Let $\phi \in C_L$. Then

$$L^\phi = L_0 + a\frac{\phi'}{\phi}\frac{d}{dx}$$

and, by Theorem 4.3.3, L^ϕ corresponds to a transient diffusion. Thus, by the paragraph above, C_{L^ϕ} is two-dimensional. Since there is a one-to-one correspondence between C_L and C_{L^ϕ}, it follows that C_L is also two-dimensional. We have proved the following result.

Proposition 1.3. *Assume that L satisfies Assumption $\tilde{H}_{loc}$ on $D = (\alpha, \beta)$, where $-\infty \leqslant \alpha < \beta \leqslant \infty$. If L is subcritical, then C_L is two-dimensional.*

Remark. Unlike the state of affairs in one dimension, there are many examples of subcritical operators L on domains $D \subseteq R^d$, $d \geqslant 2$, for which $C_L(D)$ is only one-dimensional. (See Chapters 7 and 8.)

In contrast to Theorem 1.1 for the case where $V \equiv 0$, in the case where $V \not\equiv 0$ there is no explicit condition on a, b and V which determines whether L is supercritical, critical or subcritical. However, if $\phi \in C_L$, a condition for distinguishing between the critical and the subcritical cases can be given in terms of a, b and ϕ.

Theorem 1.4. *Let*

$$L = \tfrac{1}{2}a\frac{d^2}{dx^2} + b\frac{d}{dx} + V$$

satisfy Assumption $\widetilde{\mathrm{H}}_{\mathrm{loc}}$ *on* $D = (\alpha, \beta)$, *where* $-\infty \leqslant \alpha < \beta \leqslant \infty$. *Assume that L is either critical or subcritical and let* $\phi \in C_L$. *Let* $x_0 \in (\alpha, \beta)$. *Then L is critical if and only if*

$$\int_\alpha^{x_0} \mathrm{d}x \frac{1}{\phi^2(x)} \exp\left(-\int_{x_0}^x \frac{2b}{a}(s)\,\mathrm{d}s\right) = \int_{x_0}^\beta \mathrm{d}x \frac{1}{\phi^2(x)} \exp\left(-\int_{x_0}^x \frac{2b}{a}(s)\,\mathrm{d}s\right) = \infty.$$

Proof. The theorem follows by applying Theorem 1.1 to

$$L^\phi = L_0 + a\frac{\phi'}{\phi}\frac{\mathrm{d}}{\mathrm{d}x}$$

and using Theorem 4.3.3. □

We now investigate the existence of invariant densities and invariant positive functions. Theorem 4.8.5 showed that $\phi \in C_L(D)(\phi \in C_{\tilde{L}}(D))$ is an invariant positive function (invariant density) if and only if the diffusion process on D corresponding to $L^\phi(\tilde{L}^\phi)$ does not explode. The following theorem, known as *Feller's test for explosion*, gives necessary and sufficient conditions for explosion.

Theorem 1.5. *Let*

$$L = \tfrac{1}{2}a(x)\frac{\mathrm{d}^2}{\mathrm{d}x^2} + b(x)\frac{\mathrm{d}}{\mathrm{d}x}$$

satisfy Assumption A on $D = (\alpha, \beta)$, *where* $-\infty \leqslant \alpha < \beta \leqslant \infty$, *and let* $x_0 \in (\alpha, \beta)$. *Define* $\tau_x = \inf\{t \geqslant 0: X(t) = x\}$, *for* $x \in (\alpha, \beta)$, *and define* $\tau_\alpha = \lim_{x\to\alpha} \tau_x$ *and* $\tau_\beta = \lim_{x\to\beta} \tau_x$. *Then the diffusion corresponding to L explodes if and only if at least one of the following two conditions holds:*

(i) $$\int_\alpha^{x_0} \mathrm{d}x \exp\left(-\int_{x_0}^x \frac{2b}{a}(s)\,\mathrm{d}s\right)\int_x^{x_0} \mathrm{d}y \frac{1}{a(y)} \exp\left(\int_{x_0}^y \frac{2b}{a}(s)\,\mathrm{d}s\right) < \infty;$$

(ii) $$\int_{x_0}^\beta \mathrm{d}x \exp\left(-\int_{x_0}^x \frac{2b}{a}(s)\,\mathrm{d}s\right)\int_{x_0}^x \mathrm{d}y \frac{1}{a(y)} \exp\left(\int_{x_0}^y \frac{2b}{a}(s)\,\mathrm{d}s\right) < \infty.$$

If (i) *holds, then* $E_x\tau_{x_0} \wedge \tau_\alpha < \infty$, *for* $x \in (\alpha, x_0)$, *and*

$$\lim_{x\to\alpha} E_x\tau_{x_0} \wedge \tau_\alpha = 0.$$

If (ii) *holds, then* $E_x\tau_{x_0} \wedge \tau_\beta < \infty$, *for* $x \in (x_0, \beta)$, *and*

$$\lim_{x\to\beta} E_x\tau_{x_0} \wedge \tau_\beta = 0.$$

Proof. We will assume that $b \in C((\alpha, \beta))$. The extension to the case where b is measurable and locally bounded is left to the reader as exercise 5.1. By Theorem 1.1, if

$$\int_{\alpha}^{x_0} \exp\left(-\int_{x_0}^{x} \frac{2b}{a}(s)\,ds\right) dx = \int_{x_0}^{\beta} \exp\left(-\int_{x_0}^{x} \frac{2b}{a}(s)\,ds\right) dx = \infty,$$

then the diffusion is recurrent and, thus, it does not explode. Since the integrals in (i) and (ii) are clearly infinite in this case, we may assume that at least one of the two integrals above is finite. Assume, for example, that

$$\int_{x}^{\beta} \exp\left(-\int_{x_0}^{x} \frac{2b}{a}(s)\,ds\right) dx < \infty.$$

For large n, let

$$\beta_n = \beta - \frac{1}{n}$$

and let u_n be the solution of $(L-1)u_n = 0$ in (x_0, β_n), $u_n'(x_0) = 0$ and $u_n(\beta_n) = 1$. From the differential equation, we obtain

$$\begin{aligned} u_n(x) = {} & c_n \int_{x_0}^{x} \exp\left(-\int_{x_0}^{y} \frac{2b}{a}(s)\,ds\right) dy \\ & + \int_{x_0}^{x} dy \exp\left(-\int_{x_0}^{y} \frac{2b}{a}(s)\,ds\right) \int_{x_0}^{y} dz \frac{2u_n(z)}{a(z)} \exp\left(\int_{x_0}^{z} \frac{2b}{a}(s)\,ds\right), \end{aligned}$$

where

$$c_n = \frac{1 - \displaystyle\int_{x_0}^{\beta_n} dy \exp\left(-\int_{x_0}^{y} \frac{2b}{a}(s)\,ds\right) \int_{x_0}^{y} dz \frac{2u_n(z)}{a(z)} \exp\left(\int_{x_0}^{z} \frac{2b}{a}(s)\,ds\right)}{\displaystyle\int_{x_0}^{\beta_n} \exp\left(-\int_{x_0}^{y} \frac{2b}{a}(s)\,ds\right) dy}.$$

Let $\tau_x = \inf\{t \geq 0: X(t) = x\}$. By the Feynman–Kac formula (Theorem 2.4.2), $u_n(x) = E(\exp(-\tau_{\beta_n}); \tau_{\beta_n} < \tau_{x_0})$. Define

$$u(x) = \lim_{n\to\infty} u_n(x) = E_x(\exp(-\tau_\beta); \tau_\beta < \tau_{x_0}).$$

Note that $u(x)$ is bounded away from zero for x near β if and only if $P_x(\tau_\beta < \infty) > 0$, for x near β and, thus, in fact for all $x \in (\alpha, \beta)$. If u is bound away from zero for x near β, then the integral in (ii) is finite, since otherwise we would obtain $\lim_{n\to\infty} c_n = -\infty$, which is an impossibility since $0 \leq u_n \leq 1$. We have thus shown that if $P_x(\tau_\beta < \infty) > 0$, then the integral in (ii) is finite.

Now assume that the integral in (ii) is finite. Let v_n solve $Lv_n = -1$ in (x_0, β_n) and $v_n(x_0) = v_n(\beta_n) = 0$. Then

$$v_n(x) = c_n \int_{x_0}^{x} \exp\left(-\int_{x_0}^{y} \frac{2b}{a}(s)\,\mathrm{d}s\right)\mathrm{d}y - \int_{x_0}^{x} \mathrm{d}y \exp\left(-\int_{x_0}^{y}\frac{2b}{a}(s)\,\mathrm{d}s\right)\int_{x_0}^{y}\mathrm{d}z\frac{2}{a(z)}\exp\left(\int_{x_0}^{z}\frac{2b}{a}(s)\,\mathrm{d}s\right),$$

where

$$c_n = \frac{\int_{x_0}^{\beta_n}\mathrm{d}y \exp\left(-\int_{x_0}^{y}\frac{2b}{a}(s)\,\mathrm{d}s\right)\int_{x_0}^{y}\mathrm{d}z\frac{2}{a(z)}\exp\left(\int_{x_0}^{z}\frac{2b}{a}(s)\,\mathrm{d}s\right)}{\int_{x_0}^{\beta_n}\exp\left(-\int_{x_0}^{y}\frac{2b}{a}(s)\,\mathrm{d}s\right)\mathrm{d}y}.$$

By Theorem 2.1.2, $v_n(x) = E_x\tau_{x_0} \wedge \tau_{\beta_n}$. Let $v(x) = \lim_{n\to\infty} v_n(x) = E_x\tau_{x_0} \wedge \tau_\beta$. Since the integral in (ii) is finite, it follows that $\lim_{n\to\infty} c_n < \infty$ and we obtain $E_x\tau_{x_0} \wedge \tau_\beta < \infty$, for $x \in (x_0, \beta)$ and $\lim_{x\to\beta} E_x\tau_{x_0} \wedge \tau_\beta = 0$. By Theorem 1.1, the finiteness of

$$\int_{x_0}^{\beta}\exp\left(-\int_{x_0}^{x}\frac{2b}{a}(s)\,\mathrm{d}s\right)\mathrm{d}x$$

ensures that $P_x(\tau_{x_0} = \infty) > 0$, for $x \in (x_0, \beta)$. Therefore, we conclude that $P_x(\tau_\beta < \infty) > 0$, for $x \in (x_0, \beta)$, and thus in fact, for $x \in (\alpha, \beta)$. A similar proof holds at the lefthand endpoint α. □

The following two examples are applications of Theorem 1.5.

Example 1.6. Let

$$L = \tfrac{1}{2}a\frac{\mathrm{d}^2}{\mathrm{d}x^2} + b\frac{\mathrm{d}}{\mathrm{d}x}$$

on $D = (\alpha, \beta)$. It is easy to see that if α is finite, a is bounded away from zero near α, and b is bounded from above near α, or if β is finite, a is bounded away from zero near β, and b is bounded below near β, then the diffusion explodes.

Example 1.7. Let

$$L = \tfrac{1}{2}\frac{\mathrm{d}^2}{\mathrm{d}x^2} + cx^\gamma\frac{\mathrm{d}}{\mathrm{d}x} \text{ on } (0, \infty),$$

where $c > 0$. Then $P_x(\tau_D < \infty, \lim_{t\to\tau_D} X(t) = \infty) > 0$, if $\gamma > 1$ and

$P_x(\tau_D < \infty, \lim_{t\to\tau_D} X(t) = \infty) = 0$, if $\gamma \leqslant 1$. To see this, let $\beta = \infty$ in Theorem 1.5. Then the integral in part (ii) of the theorem is

$$\int_{x_0}^{\infty} \exp\left(-\frac{2c}{1+\gamma}x^{\gamma+1}\right)\int_{x_0}^{x} \mathrm{d}y \exp\left(\frac{2c}{1+\gamma}y^{\gamma+1}\right).$$

Applying l'Hôpital's rule to

$$\frac{\int_{x_0}^{x} \mathrm{d}y \exp\left(\frac{2c}{1+\gamma}y^{\gamma+1}\right)}{x^{-\gamma}\exp\left(\frac{2c}{1+\gamma}x^{\gamma+1}\right)}$$

shows that the above integrand is of the order $x^{-\gamma}$ as $x \to \infty$.

We now give conditions for invariant densities and positive functions.

Theorem 1.8. *Let*

$$L = \tfrac{1}{2}a\frac{\mathrm{d}^2}{\mathrm{d}x^2} + b\frac{\mathrm{d}}{\mathrm{d}x} + V$$

satisfy Assumption $\widetilde{\mathrm{H}}_{\mathrm{loc}}$ *on* $D = (\alpha, \beta)$, *where* $-\infty \leqslant \alpha < \beta \leqslant \infty$. *Let* $x_0 \in (\alpha, \beta)$.

(a) $\phi \in C_L$ *is an invariant positive function if and only if the following two conditions hold:*

(i) $$\int_{\alpha}^{x_0} \mathrm{d}x \frac{1}{\phi^2(x)} \exp\left(-\int_{x_0}^{x}\frac{2b}{a}(s)\,\mathrm{d}s\right)\int_{x}^{x_0}\mathrm{d}y\frac{\phi^2(y)}{a(y)}\exp\left(\int_{x_0}^{y}\frac{2b}{a}(s)\,\mathrm{d}s\right) = \infty;$$

(ii) $$\int_{x_0}^{\beta} \mathrm{d}x \frac{1}{\phi^2(x)} \exp\left(-\int_{x_0}^{x}\frac{2b}{a}(s)\,\mathrm{d}s\right)\int_{x_0}^{x}\mathrm{d}y\frac{\phi^2(y)}{a(y)}\exp\left(\int_{x_0}^{y}\frac{2b}{a}(s)\,\mathrm{d}s\right) = \infty.$$

(b) $\phi \in C_{\tilde{L}}$ *is an invariant density if and only if the following two conditions hold:*

(i) $$\int_{\alpha}^{x_0} \mathrm{d}x \frac{1}{a^2(x)\phi^2(x)} \exp\left(\int_{x_0}^{x}\frac{2b}{a}(s)\,\mathrm{d}s\right)\int_{x}^{x_0}\mathrm{d}y a(y)\phi^2(y) \times \exp\left(-\int_{x_0}^{y}\frac{2b}{a}(s)\,\mathrm{d}s\right) = \infty;$$

(ii) $$\int_{x_0}^{\beta} \mathrm{d}x \frac{1}{a^2(x)\phi^2(x)} \exp\left(\int_{x_0}^{x}\frac{2b}{a}(s)\,\mathrm{d}s\right)\int_{x_0}^{x}\mathrm{d}y a^2(y)\phi^2(y) \times \exp\left(-\int_{x_0}^{y}\frac{2b}{a}(s)\,\mathrm{d}s\right) = \infty.$$

Proof. The theorem follows by applying Theorem 1.5 to the operators

$$L^{\phi} = \tfrac{1}{2}a\frac{d^2}{dx^2} + b\frac{d}{dx} + a\frac{\phi'}{\phi}\frac{d}{dx}$$

and

$$\widetilde{L}^{\phi} = \tfrac{1}{2}a\frac{d^2}{dx^2} + a'\frac{d}{dx} - b\frac{d}{dx} + a\frac{\phi' d}{\phi dx},$$

and then invoking Theorem 4.8.5. □

We now supply the example promised in the discussion at the end of Chapter 4, Section 8.

Example 1.9. In this example, whenever we state that ϕ_1 and ϕ_2 generate C_L, we mean that any function $u \in C_L$ can be written in the form $u = c_1\phi_1 + c_2\phi_2$, where $c_1, c_2 \geqslant 0$. In the language of Chapter 7, ϕ_1 and ϕ_2 are thus minimal positive harmonic functions. From the definition of an invariant function, it follows that if both ϕ_1 and ϕ_2 are invariant, then every $\phi \in C_L$ is invariant, while if neither ϕ_1 nor ϕ_2 is invariant, then no $\phi \in C_L$ is invariant.

(a) Let

$$L = \tfrac{1}{2}\frac{d^2}{dx^2} + b\frac{d}{dx} \text{ on } (-\infty, \infty),$$

where $b \neq 0$ is a constant. Then $\phi_1(x) = 1$ and $\phi_2(x) = \exp(-2bx)$ generate C_L. By Example 1.7 and exercise 5.3, neither $L^{\phi_1} = L$ nor

$$L^{\phi_2} = \tfrac{1}{2}\frac{d^2}{dx^2} - b\frac{d}{dx}$$

corresponds to an explosive diffusion. Therefore, every $\phi \in C_L$ is invariant. It is easy to show (exercise 8.19 in Chapter 8) that

$$\lambda_c = \frac{-b^2}{2}.$$

(b) Let

$$L = \tfrac{1}{2}\frac{d^2}{dx^2} + b\frac{d}{dx} \text{ on } (0, \infty),$$

where $b > 0$ is a constant. Then

$$\phi_1(x) = \exp(-2bx) \text{ and } \phi_2(x) = \frac{1 - \exp(-2bx)}{2b}$$

generate C_L. By Example 1.6, the diffusion corresponding to

$$L^{\phi_1} = \tfrac{1}{2}\frac{d^2}{dx^2} - b\frac{d}{dx}$$

explodes; thus ϕ_1 is not invariant. However, the diffusion corresponding to

$$L^{\phi_2} = \tfrac{1}{2}\frac{d^2}{dx^2} + \frac{2b\exp(-2bx)}{1-\exp(-2bx)}\frac{d}{dx}$$

does not explode; thus ϕ_2 is invariant. To see this, apply Theorem 1.5 to L^ϕ and use the fact that

$$\frac{2b\exp(-2bx)}{1-\exp(-2bx)} = \frac{1}{x} + O(1), \text{ as } x \to 0.$$

We note that, again,

$$\lambda_c = \frac{-b^2}{2}.$$

(c) Let

$$L = \tfrac{1}{2}\frac{d^2}{dx^2} \text{ on } (0, 1).$$

Then $\phi_1(x) = 1 - x$ and $\phi_2(x) = x$ generate C_L. By Example 1.6,

$$L^{\phi_1} = \tfrac{1}{2}\frac{d^2}{dx^2} - \frac{1}{1-x}\frac{d}{dx} \text{ and } L^{\phi_2} = \tfrac{1}{2}\frac{d^2}{dx^2} + \frac{1}{x}\frac{d}{dx}$$

correspond to explosive diffusions. Thus, there are no invariant positive functions. Note that

$$\lambda_c = \frac{-\pi^2}{2}.$$

(d) Let

$$L = \tfrac{1}{2}\frac{d^2}{dx^2} + \frac{1}{1+|x|^\gamma}\frac{d}{dx} \text{ on } (-\infty, \infty), \text{ where } 0 < \gamma < 1.$$

Then

$$\phi_1(x) = 1 \text{ and } \phi_2(x) = \int_x^\infty \exp\left(-\int_0^y \frac{2}{1+|z|^\gamma}\,dz\right)dy$$

generate C_L. From Example 1.7 and exercise 5.3, it is easy to see that the diffusion corresponding to $L^{\phi_1} = L$ does not explode. Using exercises 5.2 and 5.3, it is easy to show that the diffusion corresponding to L^{ϕ_2} does not explode. Thus, every $\phi \in C_L$ is invariant. By exercise 5.4, $\lambda_c = 0$.

(e) Let

$$L = \tfrac{1}{2}\frac{d^2}{dx^2} \text{ on } (0, \infty).$$

Then $\phi_1(x) = 1$ and $\phi_2(x) = x$ generate C_L. By Example 1.6, the diffusion corresponding to $L^{\phi_1} = L$ explodes; thus ϕ_1 is not invariant. However, the diffusion corresponding to

$$L^{\phi_2} = \tfrac{1}{2}\frac{d^2}{dx^2} + \frac{1}{x}\frac{d}{dx}$$

does not explode (apply Theorem 1.5 or, more simply, note that the diffusion in question is the radial part of a three-dimensional Brownian motion); thus ϕ_2 is invariant. It is easy to show that $\lambda_c = 0$.

We end this section by considering when a critical operator

$$L = \tfrac{1}{2}a\frac{d^2}{dx^2} + b\frac{d}{dx} + V$$

is product L^1 critical or, in the case where $V = 0$, when the corresponding diffusion is positive recurrent. We can write L in divergence form as follows:

$$L = \tfrac{1}{2}\frac{d}{dx}a\frac{d}{dx} + \left(b - \frac{a'}{2}\right)\frac{d}{dx} + V = \tfrac{1}{2}\frac{d}{dx}a\frac{d}{dx} + aQ'\frac{d}{dx} + V,$$

where

$$Q(x) = \int_{x_0}^{x}\frac{b}{a}(y)\,dy - \tfrac{1}{2}\log a(x).$$

Thus, by the discussion at the beginning of Chapter 4, Section 10, L is symmetric with respect to the density $\exp(2Q)$ and, if ϕ_c denotes the ground state of L, then the ground state $\tilde{\phi}_c$ of $\tilde{L}$ satisfies $\tilde{\phi}_c = \phi_c \exp(2Q)$. Thus, L is product L^1 critical if and only if

$$\int_\alpha^\beta \frac{\phi_c^2(x)}{a(x)}\exp\left(\int_{x_0}^{x}\frac{2b}{a}(y)\,dy\right)dx < \infty.$$

We record this as follows.

Theorem 1.10. *Let*

$$L = \tfrac{1}{2}a\frac{d^2}{dx^2} + b\frac{d}{dx} + V$$

satisfy Assumption $\tilde{\text{H}}_{\text{loc}}$ on $D = (\alpha, \beta)$, where $-\infty \leqslant \alpha < \beta \leqslant \infty$. Assume that L is critical and let ϕ_c denote the ground state. Let $x_0 \in (\alpha, \beta)$. Then the ground state $\tilde{\phi}_c$ of $\tilde{L}$ is given by

$$\tilde{\phi}_c(x) = \frac{\phi_c(x)}{a(x)}\exp\left(\int_{x_0}^{x}\frac{2b}{a}(y)\,dy\right)$$

and L is product L^1 critical if and only if

$$\int_\alpha^\beta \frac{\phi_c^2(x)}{a(x)}\exp\left(\int_{x_0}^{x}\frac{2b}{a}(y)\,dy\right)dx < \infty.$$

Corollary 1.11. *Let*

$$L = \tfrac{1}{2}a\frac{\mathrm{d}^2}{\mathrm{d}x^2} + b\frac{\mathrm{d}}{\mathrm{d}x}$$

satisfy Assumption $\widetilde{\mathrm{H}}_{\mathrm{loc}}$ on $D = (\alpha, \beta)$, where $-\infty \leqslant \alpha < \beta \leqslant \infty$. Let $x_0 \in (\alpha, \beta)$. Then the diffusion corresponding to L is positive recurrent if and only if the following two conditions hold:

(i) $$\int_\alpha^{x_0} \exp\left(-\int_{x_0}^x \frac{2b}{a}(s)\,\mathrm{d}s\right)\mathrm{d}x = \int_{x_0}^\beta \exp\left(-\int_{x_0}^x \frac{2b}{a}(s)\,\mathrm{d}s\right)\mathrm{d}x = \infty;$$

(ii) $$\int_\alpha^\beta \frac{1}{a(x)}\exp\left(\int_{x_0}^x \frac{2b}{a}(y)\,\mathrm{d}y\right)\mathrm{d}x < \infty.$$

Proof. By Theorem 1.1, condition (i) is necessary and sufficient for the diffusion to be recurrent; that is, for L to be critical. Since $V \equiv 0$, we have $\phi_c \equiv 1$ and, thus, condition (ii) is just the condition of Theorem 1.10. □

5.2 The Laplacian perturbed by radially symmetric and compactly supported functions

The operator $\frac{1}{2}\Delta$ on R^d is critical if $d \leqslant 2$ and subcritical if $d \geqslant 3$. In the critical case, $\phi_c = \tilde{\phi}_c = 1$ and, thus, it follows from Theorem 4.6.6 that $\frac{1}{2}\Delta + \varepsilon V$ is supercritical for all $\varepsilon > 0$, if $V \in C_0^\alpha(R^d)$ satisfies $\int_{\mathbb{R}^d} V\,\mathrm{d}x \geqslant 0$. In this case, the generalized principal eigenvalue $\lambda_c^{(\varepsilon)}$ of $\frac{1}{2}\Delta + \varepsilon V$ on R^d is positive. We will study the behavior of $\lambda_c^{(\varepsilon)}$ as $\varepsilon \to 0$ in the case where V is radially symmetric.

In the subcritical case, it follows from Theorem 4.6.4 that if $V \in C_0^\alpha(R^d)$ satisfies $V \geqslant 0$ and $V \not\equiv 0$, then there exists a $t_0 > 0$ such that $\frac{1}{2}\Delta + tV$ on R^d is subcritical for $t < t_0$ and critical for $t = t_0$. We will derive inequalities for t_0 in terms of integrals involving V. In one special case, we will calculate t_0 explicitly.

We begin with the critical case.

Theorem 2.1. *Let $V \in C_0^\alpha(R)$ satisfy $\int_R V\,\mathrm{d}x \geqslant 0$ and $V \not\equiv 0$. Then $\lambda_c^{(\varepsilon)}$, the generalized principal eigenvalue for*

$$\tfrac{1}{2}\frac{\mathrm{d}^2}{\mathrm{d}x^2} + \varepsilon V \quad \text{on } R,$$

satisfies

$$(\lambda_c^{(\varepsilon)})^{1/2} = \frac{\sqrt{2}\varepsilon}{2}\int_R V(x)\,dx - \frac{\sqrt{2}}{2}\varepsilon^2\int_R\int_R V(x)|x-y|V(y)\,dx\,dy + O(\varepsilon^3),$$

as $\varepsilon \to 0$.

If $\int_R V(x)\,dx = 0$, *then* $\int_R\int_R V(x)|x-y|V(y)\,dx\,dy < 0$.

Theorem 2.2. *Let* $V \in C_0^\alpha(R^2)$ *be radially symmetric and satisfy* $\int_{R^2} V\,dx \geq 0$ *and* $V \not\equiv 0$. *Let* $\lambda_c^{(\varepsilon)}$ *denote the generalized principal eigenvalue for* $\frac{1}{2}\Delta + \varepsilon V$ *on* R^2.

(i) *If*

$$\int_{R^2} V\,dx > 0, \text{ then } \lambda_c^{(\varepsilon)} \approx \exp\left(-\frac{2\pi}{\varepsilon\int_{R^2} V\,dx}\right), \text{ as } \varepsilon \to 0.$$

(ii) *If*

$$\int_{R^2} V\,dx = 0, \text{ then } \lambda_c^{(\varepsilon)} \approx \exp\left(-\frac{2\pi^2}{\varepsilon^2\gamma}\right), \text{ as } \varepsilon \to 0,$$

where

$$\gamma = \int_0^\infty \frac{1}{r}\,dr\left(\int_{|x|<r} V(x)\,dx\right)^2.$$

$$\lambda_c^{(\varepsilon)} \approx \exp\left(\frac{-c}{\varepsilon}\right)$$

means that for any

$$\delta > 0, \exp\left(\frac{-c(1+\delta)}{\varepsilon}\right) \leq \lambda_c^{(\varepsilon)} \leq \exp\left(\frac{-c(1-\delta)}{\varepsilon}\right),$$

for small ε;

$$\lambda_c^{(\varepsilon)} \approx \exp\left(\frac{-c}{\varepsilon^2}\right)$$

is defined analogously.)

Remark 1. These theorems actually hold under more general conditions on V. See the notes at the end of the chapter.

Remark 2. The proofs of Theorems 2.1 and 2.2 will reveal that the dramatically different asymptotics in the two theorems are a result of the following difference between one-dimensional and two-dimensional Brownian motion. For $a > 0$, let $\tau_a = \inf\{t \geq 0: |X(t)| = a\}$. Let $w_\lambda(x) = E_x^{(1)} \exp(-\lambda\tau_a)$, for $x \geq a$ and $\lambda > 0$, where $E_x^{(1)}$ corresponds to one-dimensional Brownian motion, and let $v_\lambda(x) = E_x^{(2)} \exp(-\lambda\tau_a)$, for $x \in R^2$ with $|x| \geq a$ and $\lambda > 0$, where $E_x^{(2)}$ corresponds to two-dimensional Brownian motion. Then $w'_\lambda(a) = -\sqrt{2\lambda}$ and

$$\nabla v_\lambda(x) = \left(\frac{2}{a}\frac{1}{\log \lambda} + o\left(\frac{1}{\log \lambda}\right)\right)\frac{x}{|x|}, \quad \text{as } \lambda \to 0, \quad \text{for } |x| = a.$$

Thus, for small λ, v_λ is decreasing at $|x| = a$ much more rapidly than w_λ is decreasing at $x = a$. The very rapid decrease of v_λ is due to the fact that two-dimensional Brownian motion is 'borderline' recurrent.

Proof of Theorem 2.1. By Theorem 4.6.7,

$$\frac{1}{2}\frac{\mathrm{d}^2}{\mathrm{d}x^2} + \varepsilon V - \lambda_c^{(\varepsilon)} \text{ on } R$$

is critical; denote its ground state by ϕ_ε, and normalize by $\phi_\varepsilon(0) = 1$. Choose $a > 0$ such that $\operatorname{supp}(V) \subset [-a, a]$. Since ϕ_ε solves $\frac{1}{2}\phi''_\varepsilon = \lambda_c^{(\varepsilon)}\phi_\varepsilon$ on $\{|x| > a\}$, it follows that

$$\phi_\varepsilon(x) = c_1 \exp(-\sqrt{2\lambda_c^{(\varepsilon)}}x) + c_2 \exp(\sqrt{2\lambda_c^{(\varepsilon)}}x),$$

for $x \geq a$, and

$$\phi_\varepsilon(x) = d_1 \exp(\sqrt{2\lambda_c^{(\varepsilon)}}x) + d_2 \exp(-\sqrt{2\lambda_c^{(\varepsilon)}}x), \text{ for } x \leq -a.$$

Since ϕ_ε is the ground state of a critical operator, it follows from Theorem 1.4 that $c_2 = d_2 = 0$. Thus,

$$\phi_\varepsilon(x) = c_1 \exp(-\sqrt{2\lambda_c^{(\varepsilon)}}x), \quad \text{for } x \geq a, \quad \text{and } \phi_\varepsilon(x) = d_1 \exp(\sqrt{2\lambda_c^{(\varepsilon)}}x), \quad \text{for } x \leq -a. \tag{2.1}$$

It follows from (2.1) that $\phi'_\varepsilon(x_0) = 0$, for some $x_0 \in (-a, a)$. Since the interval $[-a, a]$ is an arbitrary interval containing the support of V, it is clear that we may assume without loss of generality that $x_0 = 0$, that is, $\phi'_\varepsilon(0) = 0$. Thus,

$$\phi'_\varepsilon(x) = 2\int_0^x (\lambda_c^{(\varepsilon)} - \varepsilon V(y))\phi_\varepsilon(y)\,\mathrm{d}y, \quad x \in [-a, a], \tag{2.2}$$

and

$$\phi_\varepsilon(x) = 1 + 2\int_0^x \mathrm{d}y \int_0^y \mathrm{d}z (\lambda_c^{(\varepsilon)} - \varepsilon V(z))\phi_\varepsilon(z), \quad x \in [-a, a]. \tag{2.3}$$

From (2.1) and (2.2), we have

$$\frac{\phi'_\varepsilon(a)}{\phi_\varepsilon(a)} = -(2\lambda_c^{(\varepsilon)})^{1/2} = 2\int_0^a (\lambda_c^{(\varepsilon)} - \varepsilon V(x))\frac{\phi_\varepsilon(x)}{\phi_\varepsilon(a)}\,\mathrm{d}x \tag{2.4}$$

and

$$\frac{\phi_\varepsilon'(-a)}{\phi_\varepsilon(-a)} = (2\lambda_c^{(\varepsilon)})^{1/2} = 2\int_0^{-a} (\lambda_c^{(\varepsilon)} - \varepsilon V(x))\frac{\phi_\varepsilon(x)}{\phi_\varepsilon(-a)}\,dx. \qquad (2.5)$$

From (2.4) and (2.5), it is easy to see that $\lambda_c^{(\varepsilon)}$ is not larger than order ε^2. From this and (2.3), we have

$$\frac{\phi_\varepsilon(x)}{\phi_\varepsilon(a)} = 1 - 2\int_x^a dy\int_0^y dz(\lambda_c^{(\varepsilon)} - \varepsilon V(z)) + O(\varepsilon^2), \quad \text{as } \varepsilon \to 0,$$

$$\text{for } x \in [0, a]$$

and

$$\frac{\phi_\varepsilon(x)}{\phi_\varepsilon(-a)} = 1 - 2\int_{-a}^x dy\int_y^0 dz(\lambda_c^{(\varepsilon)} - \varepsilon V(z)) + O(\varepsilon^2), \quad \text{as } \varepsilon \to 0,$$

$$\text{for } x \in [-a, 0].$$

Substituting this in (2.4) and (2.5), subtracting one equation from the other, and using the fact that $\lambda_c^{(\varepsilon)}$ is not larger than order ε^2, we obtain

$$2(2\lambda_c^{(\varepsilon)})^{1/2} = 2\varepsilon\int_{-a}^a V(x)\,dx - 4a\lambda_c^{(\varepsilon)} + 4\varepsilon^2\left(\int_{-a}^0 dx V(x)\int_{-a}^x dy\int_y^0 dz V(z)\right.$$
$$\left. + \int_0^a dx V(x)\int_x^a dy\int_0^y dz V(z)\right) + O(\varepsilon^3), \quad \text{as } \varepsilon \to 0. \qquad (2.6)$$

Let $b = \int_{-a}^a V(x)\,dx$ and let

$$c = \int_{-a}^0 dx V(x)\int_{-a}^x dy\int_y^0 dz V(z) + \int_0^a dx V(x)\int_x^a dy\int_0^y dz V(z).$$

Solving (2.6) as a quadratic equation in the variable $(\lambda_c^{(\varepsilon)})^{1/2}$ and choosing the appropriate root gives

$$(\lambda_c^{(\varepsilon)})^{1/2} = \frac{\sqrt{2}}{2}b\varepsilon + \left(\sqrt{2}c - \frac{\sqrt{2}}{2}ab^2\right)\varepsilon^2 + O(\varepsilon^3), \quad \text{as } \varepsilon \to 0, \quad (2.7)$$

After some integrations by parts and some algebra, one finds that

$$c = \int_{-a}^a \left(\int_0^x V(y)\,dy\right)^2 dx \quad \text{and that} \quad \int_{-a}^a\int_{-a}^a V(x)|x - y|V(y)\,dx\,dy$$
$$= -2c + ab^2.$$

(The latter calculation is a bit tedious; it is considerably simpler if $V(x) = V(-x)$.) In particular, if $b = 0$, then $\int_{-a}^a\int_{-a}^a V(x)|x - y| V(y)\,dx\,dy < 0$. Substituting this in (2.7) proves the theorem. □

For the proof of Theorem 2.2, we need the following lemma.

Lemma 2.3. *Let $a > 0$. For each $\lambda > 0$, there exists a unique positive solution u_λ to $\frac{1}{2}u''_\lambda + \frac{1}{2x}u'_\lambda - \lambda u_\lambda = 0$ in (a, ∞), $u_\lambda(a) = 1$ and*

$$\lim_{x\to\infty} \frac{u_\lambda(x)}{x^m} = 0, \text{ for some } m > 0.$$

The solution is given by $u_\lambda(x) = E_x \exp(-\lambda\tau_a)$, where E_x is the measure corresponding to the radial part of a two-dimensional Brownian motion (that is, E_x corresponds to the solution of the martingale problem for

$$\tfrac{1}{2}\frac{\mathrm{d}^2}{\mathrm{d}x^2} + \frac{1}{2x}\frac{\mathrm{d}}{\mathrm{d}x} \text{ on } (0, \infty), \text{ and } \tau_a = \inf\{t \geqslant 0 : X(t) = a\}.$$

Furthermore,

$$u'_\lambda(a) = \frac{2}{a}\frac{1}{\log\lambda} + \mathrm{o}\left(\frac{1}{\log\lambda}\right), \quad \text{as } \lambda \to 0.$$

Proof. Fix $\lambda > 0$. Let v_n be the solution to

$$\tfrac{1}{2}v''_n + \frac{1}{2x}v'_n - \lambda v_n = 0 \text{ in } (a, n),\ v_n(a) = 1 \text{ and } v_n(n) = 0.$$

By the Feynman–Kac formula,

$$v_n(x) = E_x(\exp(-\lambda\tau_a);\ \tau_a < \tau_n).$$

A standard compactness argument and the monotone convergence theorem now show that $u_\lambda(x) \equiv E_x \exp(-\lambda\tau_a) = \lim_{n\to\infty} v_n(x)$ and that u_λ satisfies

$$\tfrac{1}{2}u''_\lambda + \frac{1}{2x}u'_\lambda - \lambda u_\lambda = 0 \quad \text{in } (a, \infty).$$

Let v be any positive solution of

$$\tfrac{1}{2}v'' + \frac{1}{2x}v' - \lambda v = 0$$

in (a, ∞) and $v(a) = 1$. Assume first that $\liminf_{x\to\infty} v(x) < \infty$. Let $\{x_n\}_{n=1}^{\infty}$ be a sequence increasing to infinity such that $\lim_{n\to\infty} v(x_n) < \infty$. An application of the Feynman–Kac formula gives

$$v(x) = E_x(\exp(-\lambda\tau_a);\ \tau_a < \tau_{x_n}) + v(x_n)E_x(\exp(-\lambda\tau_{x_n});\ \tau_{x_n} < \tau_a),$$

for $a \leqslant x \leqslant x_n$. Letting $n \to \infty$ shows that $v = u_\lambda$. Assume now that $\lim_{x\to\infty} v(x) = \infty$. An integration gives

$$v'(x) = \frac{a}{x}v'(a) + \frac{2\lambda}{x}\int_a^x r v(r)\,\mathrm{d}r,$$

from which it follows easily that

$$\lim_{x\to\infty} \frac{v(x)}{x^m} = \infty,$$

for all $m > 0$. It remains to prove the final statement of the lemma concerning $u'_\lambda(a)$.

The equality

$$P_x(\tau_a > \tau_r) = \frac{\log x - \log a}{\log r - \log a}, \quad \text{for } a < x < r,$$

follows from Theorem 1.1. For x, γ and t such that $a < x < t^\gamma$, we have

$$P_x(\tau_a > t) \leq P_x(\tau_a > \tau_{t^\gamma}) + P_x(\tau_{t^\gamma} > t) = \frac{\log x - \log a}{\gamma \log t - \log a} + P_x(\tau_{t^\gamma} > t) \tag{2.8}$$

and

$$P_x(\tau_a > t) \geq P_x(\tau_a > \tau_{t^\gamma}) - P_x(\tau_{t^\gamma} < t) = \frac{\log x - \log a}{\gamma \log t - \log a} - P_x(\tau_{t^\gamma} < t). \tag{2.9}$$

The well-known Brownian scaling property gives

$$P_x(\tau_{t^\gamma} > t) = P_{x/t^\gamma}(\tau_1 > t^{1-2\gamma}) \quad \text{and} \quad P_x(\tau_{t^\gamma} < t) = P_{x/t^\gamma}(\tau_1 < t^{1-2\gamma}).$$

By Theorem 3.6.1, for example, it follows that if $\gamma < \frac{1}{2}$, then

$$P_{x/t^\gamma}(\tau_1 > t^{1-2\gamma}) \leq \exp(-ct^{1-2\gamma}),$$

for $x \in [a, a+1]$, large t and some $c > 0$. By Theorem 2.2.2, it follows that if $\gamma > \frac{1}{2}$, then $P_{x/t^\gamma}(\tau_1 < t^{1-2\gamma}) \leq \exp(-ct^{2\gamma-1})$, for $x \in [a, a+1]$, large t and some $c > 0$. Using this along with (2.8) and (2.9), it follows that

$$P_x(\tau_a > t) = \frac{2\log\dfrac{x}{a}}{\log t} + o\left(\frac{1}{\log t}\right), \quad \text{as } t \to \infty,$$

$$\text{uniformly for } x \in [a, a+1]. \tag{2.10}$$

An integration by parts gives

$$u_\lambda(x) = E_x \exp(-\lambda \tau_a) = 1 - \int_0^\infty \lambda \exp(-\lambda t) P_x(\tau_a > t)\, dt. \tag{2.11}$$

Using (2.10) and the fact that $\lim_{\lambda\to 0}\log\lambda\int_0^{\lambda^{-1/2}}\lambda\exp(-\lambda t)\,dt = 0$, we have

$$\lim_{\lambda\to 0}(\log\lambda)\int_0^\infty \lambda\exp(-\lambda t)P_x(\tau_a > t)\,dt$$

$$= 2\log\frac{x}{a}\lim_{\lambda\to 0}\int_{\lambda^{-1/2}}^\infty \frac{\lambda\exp(-\lambda t)\log\lambda}{\log t}\,dt$$

$$= 2\log\frac{x}{a}\lim_{\lambda\to 0}\int_{\lambda^{1/2}}^\infty \frac{e^{-s}\log\lambda}{\log s - \log\lambda}\,ds = -2\log\frac{x}{a},$$

uniformly for $x \in [a, a+1]$,

where the last step follows from the dominated convergence theorem. Using this with (2.11) gives

$$\lim_{\lambda\to 0}(\log\lambda)\frac{u_\lambda(x) - u_\lambda(a)}{x - a}$$

$$= -\frac{1}{x-a}\lim_{\lambda\to 0}(\log\lambda)\int_0^\infty \lambda\exp(-\lambda t)P_x(\tau_a > t)\,dt$$

$$= \frac{2\log\frac{x}{a}}{x-a}, \quad \text{uniformly for } x \in [a, a+1]. \tag{2.12}$$

Letting $x \to a$ in (2.12), we obtain

$$\lim_{\lambda\to 0}(\log\lambda)u_\lambda'(a) = \frac{2}{a}. \qquad \square$$

Proof of Theorem 2.2. The proof is similar to that of Theorem 2.1. Let $W(x)$, $x \geq 0$, satisfy $W(x) = V(y)$, for $|y| = x$. By Theorem 4.6.7, $\frac{1}{2}\Delta + \varepsilon V(x) - \lambda_c^{(\varepsilon)}$ on R^2 is critical and, thus, by exercise 5.5,

$$\frac{1}{2}\frac{d^2}{dx^2} + \frac{1}{2x}\frac{d}{dx} + \varepsilon W(x) - \lambda_c^{(\varepsilon)}$$

is critical on $(0, \infty)$. Denoting the respective ground states by $\hat{\phi}_\varepsilon(y)$, $y \in R^2$, and $\phi_\varepsilon(x)$, $x \in (0, \infty)$, it follows easily that (properly normalized) $\hat{\phi}_\varepsilon(y) = \phi_\varepsilon(x)$, if $|y| = x > 0$, and that ϕ_ε extends smoothly to $[0, \infty)$ by defining $\phi_\varepsilon(0) = \hat{\phi}_\varepsilon(0)$. Normalize ϕ_ε by $\phi_\varepsilon(0) = 1$. Choose $a > 0$ such that $\text{supp}(W) \subset [0, a]$. Since ϕ_ε is the ground state of the critical operator

$$\frac{1}{2}\frac{d^2}{dx^2} + \frac{1}{2x}\frac{d}{dx} + \varepsilon W(x) - \lambda_c^{(\varepsilon)},$$

it follows from Theorem 1.4 and Lemma 2.3 that $\phi_\varepsilon(x) = ku_{\lambda_c^{(\varepsilon)}}(x)$, for $x \geqslant a$, where $k > 0$ and $u_{\lambda_c^{(\varepsilon)}}$ is as in Lemma 2.3. Thus, by Lemma 2.3,

$$\frac{\phi_\varepsilon'(a)}{\phi_\varepsilon(a)} = \frac{2}{a}\frac{1}{\log \lambda_c^{(\varepsilon)}} + o\left(\frac{1}{\log \lambda_c^{(\varepsilon)}}\right), \quad \text{as } \varepsilon \to 0. \tag{2.13}$$

Since $(x\phi_\varepsilon'(x))' = 2x(\lambda_c^{(\varepsilon)} - \varepsilon W(x))\phi_\varepsilon(x)$, we have

$$x\phi_\varepsilon'(x) = 2\int_0^x (\lambda_c^{(\varepsilon)} - \varepsilon W(y))y\phi_\varepsilon(y)\,\mathrm{d}y, \quad \text{for } 0 \leqslant x \leqslant a, \tag{2.14}$$

and

$$\phi_\varepsilon(x) = 1 + 2\int_0^x \mathrm{d}y\frac{1}{y}\int_0^y \mathrm{d}z(\lambda_c^{(\varepsilon)} - \varepsilon W(z))z\phi_\varepsilon(z), \quad \text{for } 0 \leqslant x \leqslant a. \tag{2.15}$$

From (2.13) and (2.14), we have

$$\frac{2}{a}\frac{1}{\log \lambda_c^{(\varepsilon)}} + o\left(\frac{1}{\log \lambda_c^{(\varepsilon)}}\right) = \frac{2}{a}\int_0^a ((\lambda_c^{(\varepsilon)} - \varepsilon W(x))x\frac{\phi_\varepsilon(x)}{\phi_\varepsilon(a)}\,\mathrm{d}x, \quad \text{as } \varepsilon \to 0. \tag{2.16}$$

It is easy to see from (2.16) that $\lambda_c^{(\varepsilon)}$ decays to zero as $\varepsilon \to 0$ faster than any power of ε. Thus, from (2.15) we have

$$\frac{\phi_\varepsilon(x)}{\phi_\varepsilon(a)} = 1 - 2\int_x^a \mathrm{d}y\frac{1}{y}\int_0^y \mathrm{d}z(\lambda_c^{(\varepsilon)} - \varepsilon W(z))z\phi_\varepsilon(z) + O(\varepsilon^2),$$
$$\text{as } \varepsilon \to 0, \text{ for } x \in [0, a].$$

Substituting this in (2.16) and using the fact that $\lambda_c^{(\varepsilon)}$ decays faster than any power of ε gives

$$\frac{2}{a}\frac{1}{\log \lambda_c^{(\varepsilon)}} + o\left(\frac{1}{\log \lambda_c^{(\varepsilon)}}\right) = -\frac{2}{a}\varepsilon\int_0^a xW(x)\,\mathrm{d}x$$
$$-\frac{4}{a}\varepsilon^2\int_0^a \mathrm{d}x\, xW(x)\int_x^a \mathrm{d}y\frac{1}{y}\int_0^y \mathrm{d}z\, zW(z) + O(\varepsilon^3), \quad \text{as } \varepsilon \to 0. \tag{2.17}$$

Note that $\int_{R^2} V(z)\,\mathrm{d}z = 2\pi\int_0^a xW(x)\,\mathrm{d}x$. Also, an integration by parts gives

$$\int_0^a \mathrm{d}x\, xW(x)\int_x^a \mathrm{d}y\frac{1}{y}\int_0^y \mathrm{d}z\, zW(z) = \int_0^a \mathrm{d}x\frac{1}{x}\left(\int_0^x yW(y)\,\mathrm{d}y\right)^2$$
$$= \frac{1}{4\pi^2}\int_0^a \mathrm{d}r\frac{1}{r}\left(\int_{|z|<r} V(z)\,\mathrm{d}z\right)^2.$$

The result follows from this and (2.17). □

We now turn to the subcritical case.

Theorem 2.4. *Let $d \geq 3$ and let $V \in C_0^\alpha([0, \infty))$ satisfy $V \geq 0$ and $V \not\equiv 0$. Define $t_0 = \sup\{t: \frac{1}{2}\Delta + tV(|x|)$ on R^d is subcritical$\}$. Let $a = \sup\{x: V(x) > 0\}$. Then*

$$\frac{2}{(d-2)a^{d-2}}\int_0^a x^{d-1}V(x)\,dx < t_0^{-1} < \frac{2}{(d-2)a^{d-2}}\int_0^a x^{d-1}V(x)\,dx + 2\int_0^a dx\, x^{1-d}\int_0^x dy\, y^{d-1}V(y). \quad (2.18)$$

In the case where $d = 3$ and $V = 1_{[0, a)}$, we can calculate t_0 explicitly. Of course this V is not a C^α-function; however the criticality theory is still valid, as will be explained in the proof of the following proposition.

Proposition 2.5. *Let $B_a(0)$ denote the ball in R^3 of radius a centered at 0. Then*

$$t_0 \equiv \sup\{t: \tfrac{1}{2}\Delta + t1_{B_a(0)} \text{ on } R^3 \text{ is subcritical}\} = \frac{\pi^2}{8a^2}.$$

Remark 1. The principal eigenvalue of $\frac{1}{2}\Delta$ on $B_a(0) \subset R^3$ is

$$\frac{\pi^2}{2a^2}\left(\text{with corresponding eigenfunction } \frac{\sin\frac{\pi}{a}|x|}{|x|}\right).$$

Thus, $\sup\{t: \frac{1}{2}\Delta + t1_{B_a(0)}$ on $B_a(0)$ is subcritical$\} = 4\sup\{t: \frac{1}{2}\Delta + t1_{B_a(0)}$ on R^3 is subcritical$\}$.

Remark 2. The proofs of Theorem 2.4 and Proposition 2.5 reveal that Theorem 2.4 holds for $V = 1_{B_a(0)}$ as in Proposition 2.5. When the bound in (2.18) is applied to the case in Proposition 2.5, we obtain

$$\frac{1}{a^2} < t_0 < \frac{3}{2a^2};$$

the actual value,

$$t_0 = \frac{\pi^2}{8a^2},$$

lies close to the center of this interval.

Proof of Theorem 2.4. By Theorem 4.6.4, $0 < t_0 < \infty$ and $\frac{1}{2}\Delta + t_0 V(|x|)$ on R^d is critical. Thus, by exercise 5.5,

$$\frac{1}{2}\frac{d^2}{dx^2} + \frac{d-1}{2x}\frac{d}{dx} + t_0 V(x)$$

on $(0, \infty)$ is also critical. Denoting the respective ground states by $\hat{\phi}_c(y)$, $y \in R^2$, and $\phi_c(x)$, $x \in (0, \infty)$, it follows easily that (properly normalized) $\hat{\phi}_c(y) = \phi_c(x)$, if $|y| = x > 0$, and that ϕ_c extends smoothly to $[0, \infty)$ by defining $\phi_c(0) = \hat{\phi}_c(0) = 0$.

Since ϕ_c solves

$$\tfrac{1}{2}\phi_c'' + \frac{d-1}{2x}\phi_c' = 0$$

on $x \geqslant a$, it follows that $\phi_c(x) = c_1 x^{2-d} + c_2$, for $x \geqslant a$. Since ϕ_c is the ground state of the critical operator

$$\frac{1}{2}\frac{d^2}{dx^2} + \frac{d-1}{2x}\frac{d}{dx} + t_0 V(x),$$

it follows from Theorem 1.4 that $c_2 = 0$. Thus,

$$\frac{\phi_c'(a)}{\phi_c(a)} = \frac{2-d}{a}. \tag{2.19}$$

Integrating the equation $(x^{d-1}\phi_c'(x))' = -2x^{d-1}t_0 V(x)\phi_c(x)$ gives

$$\phi_c'(a) = -t_0 a^{1-d}\int_0^a 2x^{d-1}V(x)\phi_c(x)\,dx, \tag{2.20}$$

and

$$\phi_c(x) = 1 - 2t_0\int_0^x dy\, y^{1-d}\int_0^y dz\, z^{d-1}V(z)\phi_c(z). \tag{2.21}$$

From (2.19) and (2.20), we have

$$d - 2 = 2t_0 a^{2-d}\int_0^a x^{d-1}V(x)\frac{\phi_c(x)}{\phi_c(a)}\,dx. \tag{2.22}$$

Since

$$\frac{\phi_c(x)}{\phi_c(a)} > 1, \quad \text{for } 0 \leqslant x < a,$$

we conclude from (2.22) that

$$t_0^{-1} > \frac{2}{(d-2)a^{d-2}}\int_0^a x^{d-1}V(x)\,dx;$$

this is the lower bound for t_0^{-1} in (2.18).

From (2.22) we have

$$\phi_c(a) = \frac{2a^{2-d}}{d-2} t_0 \int_0^a x^{d-1} V(x)\phi_c(x)\,dx.$$

Setting $x = a$ in (2.21) gives an alternative formula for $\phi_c(a)$. Equating these two formulas for $\phi_c(a)$, and solving the resulting equation for t_0 gives

$$t_0^{-1} = \frac{2}{(d-2)a^{d-2}} \int_0^a x^{d-1} V(x)\phi_c(x)\,dx + 2\int_0^a dy\, y^{1-d} \int_0^y dz V(z)\phi_c(z).$$

The upper bound for t_0^{-1} now follows since $\phi_c(x) \leqslant 1$. □

Proof of Proposition 2.5. Even though $I_{B_a(0)}$ is not a C^α-function, there is clearly no problem in defining the Green's function. It is also clear that the Green's function exists for $\frac{1}{2}\Delta + t 1_{B_a(0)}$ on R^3 if and only if it exists for

$$L_t \equiv \frac{1}{2}\frac{d^2}{dx^2} + \frac{1}{x}\frac{d}{dx} + t 1_{[0,a)}$$

on $(0, \infty)$. By monotonicity, if the Green's function exists for L_t, then it also exists for L_s if $s < t$. Since $1_{[0,a)}$ is not a C^α-function, we cannot expect to find $C^{2,\alpha}$-solutions h of $L_t h = 0$ in $(0, \infty)$. However, assume for the moment that we find an $h > 0$ which is $C^{1,\alpha}$ on $(0, \infty)$ and $C^{2,\alpha}$ on $(0, \infty) - \{a\}$, and satisfies $L_t h = 0$ on $(0, \infty) - \{a\}$. For this function, define the h-transform formally by

$$L_t^h \equiv \frac{1}{2}\frac{d^2}{dx^2} + \frac{1}{x}\frac{d}{dx} + \frac{h'}{h}\frac{d}{dx}.$$

Since L_t^h satisfies Assumption H_{loc} on $(0, \infty)$, the criticality theory holds for L_t^h (exercise 4.16). We claim that Proposition 4.2.2, is still in effect; that is,

$$G_t^h(x, y) = \frac{h(y)}{h(x)} G_t(x, y),$$

where G_t and G_t^h are the Green's functions for L_t, and L_t^h and, in particular, L_t is subcritical if and only if L_t^h is subcritical. Indeed, Proposition 4.2.2 is an immediate consequence of Theorem 4.1.1, and a check of Theorem 4.1.1 reveals that the only possible problem with applying that proof to the present h is in Itô's formula. However, for a C^1-function on $(0, \infty)$ which is C^2 except at isolated points, and whose second derivative is bounded, Itô's formula holds as before [Chung and Williams (1990), section 9.2].

Thus, we will prove the proposition as follows. For $t \in (0, \pi^2/8a^2)$, we will find an h as above such that

$$L_t^h = \frac{1}{2}\frac{d^2}{dx^2} + \frac{1}{x}\frac{d}{dx} + \frac{h'}{h}\frac{d}{dx}$$

on $(0, \infty)$ is subcritical, and for

$$t = \frac{\pi^2}{8a^2},$$

we will find an h as above such that L_t^h on $(0, \infty)$ is critical.

Assume first that $t \in (0, \pi^2/8a^2)$. We look for a solution h which satisfies

$$h(x) = \frac{c_1}{x}\sin\sqrt{2t}x,$$

for $x \in (0, a)$, and

$$h(x) = 1 + \frac{c_2}{x},$$

for $x \geqslant a$, where $c_1 > 0$ and $c_2 > -a$. In order that this h satisfy the appropriate smoothness conditions enumerated above, it suffices that $h(a^-) = h(a^+)$ and $h'(a^-) = h'(a^+)$. A little algebra shows that the above matching conditions can be met if and only if there exists a $c_2 \in (-a, \infty)$ satisfying

$$\sqrt{2t}\,a\cot\sqrt{2t}\,a = \frac{a}{a + c_2}.$$

By the assumption on t,

$$\sqrt{2t}\,a \in \left(0, \frac{\pi}{2}\right)$$

and, therefore, $\sqrt{2t}\,a\cot\sqrt{2t}\,a \in (0, \infty)$. Thus, there exists an appropriate choice of $c_2 \in (-a, \infty)$. Since

$$\frac{h'}{h}(x) = \frac{-c_2}{x^2 + c_2 x} \approx \frac{-c_2}{x^2}$$

for large x, it follows from Theorem 1.1 that the diffusion corresponding to L_t^h is transient; that is, L_t^h is subcritical.

Now let

$$t = \frac{\pi^2}{8a^2}$$

and let

$$h(x) = \frac{1}{x}\sin\frac{\pi}{2a}x, \quad \text{for } x \in (0, a),$$

$$\text{and } h(x) = \frac{1}{x}, \quad \text{for } x \geq a.$$

Then h is a $C^{1,1}$-function on $(0, \infty)$. Since

$$\frac{h'}{h}(x) = \frac{-1}{x}, \quad \text{for } x \geq a,$$

and

$$\frac{h'}{h}(x)$$

is bounded for x near zero, it follows from Theorem 1.1 that the diffusion corresponding to L_t^h is recurrent; that is, L_t^h is critical. □

Exercises

5.1 Prove Theorems 1.1 and 1.5 in the case where b is assumed to be only measurable and locally bounded. (*Hint:* Fix a subinterval $(x_1, x_2) \subset\subset (\alpha, \beta)$ and let b_ε^+ and b_ε^- be continuous on (x_1, x_2) and satisfy $0 \leq b_\varepsilon^+ - b < \varepsilon$ and $0 \leq b - b_\varepsilon^- < \varepsilon$. Let $x_0 \in (x_1, x_2)$. For Theorem 1.1, define

$$u_\varepsilon^\pm(x) = \int_{x_1}^x \exp\left(-\int_{x_0}^y \frac{2b_\varepsilon^\pm}{a}(z)\,dz\right)dy.$$

Then

$$Lu_\varepsilon^\pm(x) = (-b_\varepsilon^\pm + b)(x)\exp\left(-\int_{x_0}^x \frac{2b_\varepsilon^\pm}{a}(z)\,dz\right).$$

Make a similar construction for Theorem 1.5.)

5.2 Let

$$L = \tfrac{1}{2}a\frac{d^2}{dx^2} + b\frac{d}{dx}$$

on (α, β) be subcritical and assume that

$$\phi(x) \equiv \int_x^\beta \exp\left(-\int_{x_0}^y \frac{2b}{a}(z)\,dz\right)dt \in C_L((\alpha, \beta)).$$

Find appropriate conditions on b/a so that

$$\frac{\phi'(x)}{\phi(x)} \sim \frac{-2b}{a}(x), \quad \text{as } x \to \beta.$$

Thus, under appropriate conditions, the drift term in

$$L^{\phi} = L + a\frac{\phi'}{\phi}\frac{\mathrm{d}}{\mathrm{d}x}$$

behaves like $-b$ for x near β. In the language of Chapter 7, ϕ is a minimal positive harmonic function corresponding to the point α and thus, by Theorem 7.2.1, the transient diffusion on $D = (\alpha, \beta)$ corresponding to L^{ϕ} may be interpreted as the transient diffusion on (α, β) corresponding to L conditioned on the event $\{\lim_{t\to\tau_D} \operatorname{dist}(X(t), \alpha) = 0\}$. Thus, under appropriate conditions on b/a, the drift of the diffusion conditioned to exit (α, β) at α behaves near β like the negative of the drift of the original diffusion.

5.3 Let

$$L_i = \tfrac{1}{2}a\frac{\mathrm{d}^2}{\mathrm{d}x^2} + b_i\frac{\mathrm{d}}{\mathrm{d}x} \quad \text{on } (\alpha, \beta), \quad \text{for } i = 1, 2,$$

and assume that $b_1(x) \leq b_2(x)$, for x near β. Let $P_x^{(i)}$, $i = 1,2$, denote probabilities corresponding to L_i. If $P_x^{(1)}(\tau_D < \infty, \lim_{t\to\tau_D} X(t) = \beta) > 0$, then $P_x^{(2)}(\tau_D < \infty, \lim_{t\to\tau_D} X(t) = \beta) > 0$. (*Hint:* Show that the integral in Theorem 1.5(ii) is a decreasing function of b.)

5.4 Let

$$L = \tfrac{1}{2}\frac{\mathrm{d}^2}{\mathrm{d}x^2} + b\frac{\mathrm{d}}{\mathrm{d}x} \quad \text{on } (0, \infty).$$

If $\lim_{x\to\infty} (b^2(x) + b'(x)) = 0$, then $\lambda_c = 0$. (*Hint:* Make an h-transform to eliminate the drift term. Then make a comparison argument on the domain (n, ∞) with n large.)

5.5 Let

$$\hat{L} = \tfrac{1}{2}a(|y|)\Delta + b(|y|)\frac{y}{|y|} \cdot \nabla + V(|y|)$$

satisfy Assumption $\widetilde{\mathrm{H}}_{\mathrm{loc}}$ on $D = \{y \in R^d : \alpha < |y| < \beta\}$, where $0 \leq \alpha < \beta \leq \infty$, and let

$$L = \tfrac{1}{2}a(x)\frac{\mathrm{d}^2}{\mathrm{d}x^2} + \frac{d-1}{2x}a(x)\frac{\mathrm{d}}{\mathrm{d}x} + b(x)\frac{\mathrm{d}}{\mathrm{d}x} + V(x) \text{ on } (\alpha, \beta).$$

Show that the criticality properties of $\hat{L}$ on D and L on (α, β) coincide.

5.6 Let $L = \tfrac{1}{2}\Delta + V$ on R^d, $d \geq 3$, where

$$V(x) = \begin{cases} a|x|^{-2} & b \leq |x| \leq 1 \\ 0 & \text{otherwise} \end{cases},$$

$0 < b < 1$, and $-\infty < a < \infty$. If

$$a > \frac{(d-2)^2}{8},$$

let $c_d = \exp[-2(\pi - 2\theta_d)(8a - (d-2)^2)^{-1/2}$ and $\theta_d = \tan^{-1}[(8a - (d - 2)^2)^{1/2}/(d-2)]$, where

$$\theta_d \in \left(0, \frac{\pi}{2}\right).$$

Prove the following results.

(i) If

$$a \leqslant \frac{(d-2)^2}{8}$$

or if

$$a > \frac{(d-2)^2}{8}$$

and $b > c_d$, then L is subcritical on R^d.

(ii) If

$$a > \frac{(d-2)^2}{8}$$

and $b = c_d$, then L is critical on R^d.

(iii) If

$$a > \frac{(d-2)^2}{8}$$

and $b < c_d$, then L is supercritical on R^d. (*Hint:* Use the method of Section 2.) This result is due to [Murata (1986)] by a different method.

Notes

The material in Section 1 is quite standard. Theorems 1.1 and 1.5 go back to [Feller (1952)] and have appeared in a number of texts. The transience of d-dimensional Brownian motion, $d \geqslant 3$, was first proved by [Kakutani (1944a)], and the recurrence of two-dimensional Brownian motion was first proved by [Kakutani (1944b)].

In Section 2, Theorems 2.1 and 2.2 are due to [Simon (1976)]. The proof of Theorem 2.1 given in the text requires that V be compactly supported, while Simon's only requires that V satisfy $\int_R |V(x)|(1 + |x|^2)\,dx < \infty$. On the other hand, Simon's proof makes liberal use of fairly sophisticated machinery from functional analysis and spectral theory, while the proof in the text is just basic analysis. In [Klaus (1977)], the result was extended to the case where V satisfies $\int_R |V(x)|(1 + |x|)\,dx < \infty$. In that paper it was also shown that a weaker form of the result holds under the condition $\int_R |V(x)|\,dx < \infty$. The simple-minded approach taken in the text loses out to Simon's approach in a much bigger way in the two-dimensional case; in Theorem 2.2 in the text, V is required to be compactly supported and radially symmetric whereas Simon allows any V which satisfies $\int_{R^2} |V(x)|^{1+\delta}\,dx < \infty$ and $\int_{R^2} |V(x)|(1 + |x|^\delta)\,dx < \infty$, for some $\delta > 0$. Simon's proof of Theorem 2.2 is considerably more difficult than his proof of Theorem 2.1. The proof given in the text for the limited version of the theorem is just basic analysis along with Lemma 2.3. Lemma 2.3 is of some independent interest (see Remark 2 following Theorem 2.2). The proof of Lemma 2.3 uses probabilistic techniques; I resorted to it after failing to prove it via purely analytic techniques. I was unable to locate any reference in the literature. In [Englander and Pinsky (1994)], the method of proof in Theorems 2.1 and 2.2 is used to obtain the leading order asymptotics of the generalized principle eigenvalue for compactly supported perturbations of critical operators L of the form $L = (d^2/dx^2) + W$ on R, where $W(x) = l_1/x^2$, for $x \gg 1$, and $W(x) = l_2/x^2$, for $x \ll -1$, and l_1, l_2 are constants. To the best of my knowledge, Theorem 2.4 and Proposition 2.5 are new.

6
Criteria for transience or recurrence and explosion or non-explosion of diffusion processes

6.0 Introduction

In this chapter, we will assume that $V \equiv 0$. In Chapter 4, it was shown that in this case, criticality and subcriticality are equivalent to recurrence and transience, respectively, and that, in the critical (recurrent) case, the product L^1 property is equivalent to positive recurrence. It was also shown that the function $\phi \equiv 1$ is not invariant for the transition measure $p(t, x, dy)$ if and only if the diffusion explodes. In Chapter 5, integral tests were given which allow one to distinguish between transience and recurrence, between null recurrence and positive recurrence, and between explosion and non-explosion in the one-dimensional case. The multidimensional case is much more involved and does not yield a simple solution. In Section 1–3, we develop the Liapunov technique for considerations of transience and recurrence. For diffusions with appropriate coefficients, this technique is used to give sufficiency conditions for transience, for recurrence, for null recurrence and for positive recurrence in terms of integral tests. In Section 4, we present a variational approach to transience and recurrence for diffusions corresponding to symmetric operators. By exploiting the classical Dirichlet principle, a necessary and sufficient condition for transience or recurrence is given in terms of a variational formula. From this, one can extract sufficiency conditions for transience and for recurrence in terms of integral tests. The variational method also reveals an important comparison principle concerning transience and recurrence for such diffusions. In Section 5, we develop a generalized Dirichlet principle for general operators which is then used in Section 6 to give a necessary and sufficient condition for transience or recurrence in terms of a mini-max variational formula. Finally, in Section 7, we use the Liapunov method to give sufficiency conditions for explosion and for non-explosion in terms of integral tests.

In Sections 1, 2 and 7, we will assume that L satisfies Assumption A, which appears at the beginning of Chapter 2 and which is assumed throughout that chapter, while in Section 4–6, we will assume that L satisfies Assumption $\widetilde{\mathrm{H}}_{\mathrm{loc}}$, which appears at the beginning of Chapter 4 and is assumed throughout that chapter. A mixture of these two assumptions is used in Section 3. In particular, this means that in Section 1, 2 and 7, the coefficients a and b correspond to an operator in non-divergence form, while in Section 4–6, they correspond to an operator in divergence form.

6.1 The Liapunov method for transience, recurrence and positive recurrence

If u satisfies $Lu \leqslant 0$ in D, then it follows from the martingale formulation that $E_x u(X(t \wedge \tau_A)) \leqslant u(x)$, for $x \in D$ and $A \subset\subset D$, where $\tau_A = \inf\{t \geqslant 0: X(t) \notin A\}$. If, in addition, u satisfies certain monotonicity conditions, then it turns out that the above inequality may be exploited to prove results concerning transience and recurrence. The Liapunov method consists of using such test functions and exploiting the above equation. In this section, the Liapunov method will be used to prove the following three theorems.

Theorem 1.1. *Let L satisfy Assumption A on a domain $D \subseteq R^d$, $d \geqslant 2$, which contains $B \equiv R^d - \{|x| < r_0\}$, for some $r_0 > 0$, and let $\{P_x, x \in \hat{D}\}$ denote the solution to the generalized martingale problem for L on D. Assume that there exists an $x_0 \in R^d$ with $|x_0| > r_0$ and a bounded function $u \in C^2(B)$ such that $Lu(x) \leqslant 0$, for $x \in B$ and $u(x_0) < \inf_{|x|=r_0} u(x)$. Then $P_x(\tau_r < \infty) < 1$, for $r_0 \leqslant r < |x|$, where $\tau_r = \inf\{t \geqslant 0: |X(t)| = r\}$. If $\hat{L}$ satisfies Assumption A on R^d and $L = \hat{L}$ on B, then the diffusion on R^d corresponding to $\hat{L}$ is transient.*

Theorem 1.2. *Let L satisfy Assumption A on a domain $D \subseteq R^d$, $d \geqslant 2$, which contains $B \equiv R^d - \{|x| < r_0\}$, for some $r_0 > 0$, and let $\{P_x, x \in \hat{D}\}$ denote the solution to the generalized martingale problem for L on D. Assume that there exists a function $u \in C^2(B)$ such that $Lu(x) \leqslant 0$, for $x \in B$, and $\lim_{|x|\to\infty} u(x) = \infty$. Then $P_x(\tau_{r_0} < \infty) = 1$, for $|x| \geqslant r_0$, where $\tau_{r_0} = \inf\{t \geqslant 0: |X(t)| = r_0\}$. If $\hat{L}$ satisfies Assumption A on R^d and $L = \hat{L}$ on B, then the diffusion on R^d corresponding to $\hat{L}$ is recurrent.*

Theorem 1.3. *Let L satisfy Assumption A on a domain $D \subseteq R^d$, $d \geqslant 2$, which contains $B \equiv R^d - \{|x| < r_0\}$, for some $r_0 > 0$, and let*

$\{P_x, x \in \hat{D}\}$ denote the solution to the generalized martingale problem for L on D. Assume that there exists an $\varepsilon > 0$ and a function $u \in C^2(B)$ such that $Lu(x) \leq -\varepsilon$, for $x \in B$, and $\inf_{x \in B} u(x) > -\infty$. Then $E_x \tau_{r_0} < \infty$, for $|x| \geq r_0$, where $\tau_{r_0} = \inf\{t \geq 0: |X(t)| = r_0\}$. If $\hat{L}$ satisfies Assumption A on R^d and $L = \hat{L}$ on B, then the diffusion on R^d corresponding to $\hat{L}$ is positive recurrent.

Remark. The formulation in Theorems 1.1–1.3 can be simplified if it is assumed that L is defined and satisfies Assumption A on all of R^d. The present formulation has been employed because in the sequel it will sometimes be natural to consider operators defined on $R^d - \{0\}$. We note also that, clearly, Theorems 1.1–1.3 may be formulated and proved in a similar fashion in the case where R^d is replaced by a domain $\Omega \subset R^d$, $B \equiv R^d - \{|x| > r_0\}$ is replaced by $\Omega - A$, where $A \subset\subset \Omega$ is a simply connected subdomain, and D is now a domain satisfying $\Omega - A \subset D \subseteq \Omega$.

Proof of Theorem 1.1. For $m > |x_0|$,

$$u(X(t \wedge \tau_{r_0} \wedge \tau_m)) - \int_0^{t \wedge \tau_{r_0} \wedge \tau_m} Lu(X(s))\, ds$$

is a P_{x_0}-martingale; thus

$$E_{x_0} u(X(t \wedge \tau_{r_0} \wedge \tau_m)) - E_{x_0} \int_0^{t \wedge \tau_{r_0} \wedge \tau_m} Lu(X(s))\, ds = u(x_0).$$

Using the assumption on u, we obtain

$$E_{x_0} u(X(t \wedge \tau_{r_0} \wedge \tau_m)) \leq u(x_0). \tag{1.1}$$

Now assume, contrary to the statement of the theorem, that $P_{x_1}(\tau_{r_0} < \infty) = 1$, for some x_1 satisfying $|x_1| > r_0$. By the proof of Theorem 2.7.4, it then follows that $P_x(\tau_{r_0} < \infty) = 1$, for all x satisfying $|x| \geq r_0$. Thus, in particular,

$$\lim_{t \to \infty} \lim_{m \to \infty} t \wedge \tau_{r_0} \wedge \tau_m = \tau_{r_0} \text{ a.s. } P_{x_0}. \tag{1.2}$$

Letting $m \to \infty$ and then letting $t \to \infty$ in (1.1), and using (1.2) and the bounded convergence theorem, gives $E_{x_0} u(X(\tau_{r_0})) \leq u(x_0)$. But since $|X(\tau_{r_0})| = r_0$, this contradicts the assumption that $u(x_0) < \inf_{|x| = r_0} u(x)$. Thus, $P_x(\tau_{r_0} < \infty) < 1$, for $|x| > r_0$. To show that $P_x(\tau_r < \infty) < 1$, for $r_0 \leq r < |x|$, is left as exercise 6.2. Applying the localization result of Theorem 1.13.2 to L and $\hat{L}$, and then applying Theorem 2.7.6 to $\hat{L}$, it follows that the diffusion corresponding to $\hat{L}$ is transient. □

Proof of Theorem 1.2. The conditions on u imply that u is bounded from below. Since we may always add a constant to u, we will assume without loss of generality that $u \geqslant 0$. As in the proof of Theorem 1.1, we have for $r_0 \leqslant |x| \leqslant m$, $E_x u(X(t \wedge \tau_{r_0} \wedge \tau_m)) \leqslant u(x)$. Letting $t \to \infty$, it follows that

$$E_x u(X(\tau_{r_0} \wedge \tau_m)) \leqslant u(x). \tag{1.3}$$

Assume now, contrary to the statement of the theorem, that $P_x(\tau_{r_0} < \infty) < 1$, for some x with $|x| > r_0$. Since $P_x(\tau_{r_0} < \infty) = \lim_{m\to\infty} P_x(\tau_{r_0} < \tau_m)$, there exists a $\delta > 0$ such that

$$P_x(\tau_m < \tau_{r_0}) \geqslant \delta, \text{ for } m > |x|. \tag{1.4}$$

From (1.3), (1.4) and the non-negativity of u, we obtain

$$\delta \inf_{|y|=m} u(y) \leqslant u(x), \text{ for } m > |x|. \tag{1.5}$$

But, by assumption, $\lim_{m\to\infty} \inf_{|y|=m} u(y) = \infty$, which contradicts (1.5). Thus $P_x(\tau_{r_0} < \infty) = 1$, for all x with $|x| \geqslant r_0$.

The proof that the diffusion corresponding to $\hat{L}$ is recurrent follows from Theorem 2.7.6 and the localization result, Theorem 1.13.2. □

Proof of Theorem 1.3. For $r_0 \leqslant |x| \leqslant m$, we have, as in the proof of Theorem 1.1,

$$E_x u(X(t \wedge \tau_{r_0} \wedge \tau_m)) - E_x \int_0^{t\wedge\tau_{r_0}\wedge\tau_m} Lu(X(s))\, ds = u(x).$$

Using the assumption on u and letting $t \to \infty$ gives

$$E_x u(X(\tau_{r_0} \wedge \tau_m)) + \varepsilon E_x \tau_{r_0} \wedge \tau_m \leqslant u(x). \tag{1.6}$$

Since u satisfies the condition of Theorem 1.2, $P_x(\tau_{r_0} < \infty) = 1$. Therefore, $\lim_{m\to\infty} \tau_{r_0} \wedge \tau_m = \tau_{r_0}$ a.s. P_x. Thus, since u is bounded from below, letting $m \to \infty$ in (1.6) gives $E_x \tau_{r_0} < \infty$, for $0 < r_0 \leqslant |x|$.

The proof that the diffusion corresponding to $\hat{L}$ is positive recurrent follows from Theorem 4.9.6(i) and the localization result, Theorem 1.13.2. □

6.2 Application of the Liapunov method – Case I: uniformity in the non-radial variables

In this section, we apply the Liapunov method to obtain transience and recurrence criteria for operators whose coefficients satisfy certain conditions uniformly in the non-radial variables. We begin with some notation. Define

$$A(x) = \frac{1}{|x|^2}\sum_{i,j=1}^{d} a_{ij}(x)x_i x_j,$$

$$B(x) = \mathrm{Tr}\,(a(x)) = \sum_{i=1}^{d} a_{ii}(x),$$

and

$$C(x) = 2\sum_{i=1}^{d} x_i b_i(x).$$

Now define

$$\bar{\alpha}(r) = \sup_{|x|=r} A(x),\ \underline{\alpha}(r) = \inf_{|x|=r} A(x),$$

$$\bar{\beta}(r) = \sup_{|x|=r} \frac{B(x) - A(x) + C(x)}{A(x)} \text{ and } \underline{\beta}(r) = \inf_{|x|=r} \frac{B(x) - A(x) + C(x)}{A(x)}.$$

Finally, for $r \geqslant 1$, let

$$\bar{I}(r) = \int_1^r \frac{\bar{\beta}(s)}{s}\,\mathrm{d}s \text{ and } \underline{I}(r) = \int_1^r \frac{\underline{\beta}(s)}{s}\,\mathrm{d}s.$$

Theorem 2.1. *Let L satisfy Assumption A on R^d, $d \geqslant 2$.*

(i) *If $\int_1^\infty \exp(-\bar{I}(r))\,\mathrm{d}r = \infty$, then the diffusion corresponding to L is recurrent.*
(ii) *If $\int_1^\infty \exp(-\underline{I}(r))\,\mathrm{d}r < \infty$, then the diffusion corresponding to L is transient.*

Remark. The intuition for Theorem 2.1 is as follows. By Corollary 4.3.10, L is transient (recurrent) if and only if

$$\frac{1}{A(x)}L$$

is transient (recurrent). When

$$\frac{1}{A(x)}L$$

is expressed in spherical coordinates, the part of the operator involving differentiation in r is given by

$$\frac{1}{2}\frac{\partial^2}{\partial r^2}+\frac{B(x)-A(x)+C(x)}{2rA(x)}\frac{\partial}{\partial r}.$$

Now

$$\frac{\underline{\beta}(r)}{r}\leqslant\frac{B(x)-A(x)+C(x)}{rA(x)}\leqslant\frac{\bar{\beta}(r)}{r}.$$

By comparison with Theorem 5.1.1, one sees that Theorem 2.1 states that if the one-dimensional diffusion corresponding to

$$\frac{1}{2}\frac{d^2}{dr^2}+\frac{\underline{\beta}(r)}{r}\frac{d}{dr}\left(\frac{1}{2}\frac{d^2}{dr^2}+\frac{\bar{\beta}(r)}{r}\frac{d}{dr}\right)$$

and starting from $r>1$ reaches $r=1$ with probability less than one (equal to one), then the original diffusion is transient (recurrent).

***Theorem* 2.2.** *Let L satisfy Assumption A on R^d, $d\geqslant 2$, and assume that $\int_1^\infty \exp(-\bar{I}(r))\,dr=\infty$.*

(i) *If*

$$\int_1^\infty \frac{1}{\underline{\alpha}(r)}\exp(\bar{I}(r))\,dr<\infty,$$

then the diffusion corresponding to L is positive recurrent.

(ii) *If*

$$\lim_{n\to\infty}\frac{\int_1^n ds\exp(-\bar{I}(s))\int_1^s dt\frac{1}{\bar{\alpha}(t)}\exp(\bar{I}(t))}{\int_1^n \exp(-\underline{I}(s))\,ds}=\infty,$$

then the diffusion corresponding to L is null recurrent.

Remark. Compare this theorem to Corollary 5.1.11.

Proof of Theorem 2.1. We will prove the theorem under the additional condition that the drift b be continuous. The extension to the case where b is locally bounded and measurable is left to exercise 6.1.

(i) Let $F(r)=\int_1^r \exp(-\bar{I}(s))\,ds$, for $r\geqslant 1$, and note that, by assumption, $F\in C^2([1,\infty))$. Let $u\in C^2(R^d)$ satisfy $u(x)=F(|x|)$ for $|x|\geqslant 1$. A straightforward calculation reveals that u satisfies the conditions of Theorem 1.2; thus the diffusion is recurrent.

(ii) Let $G(r)=\int_r^\infty \exp(-\underline{I}(s))\,ds$, for $r\geqslant 1$. Again, by assumption, $G\in C^2([1,\infty))$. This time let $u\in C^2(R^d)$ satisfy $u(x)=G(|x|)$, for $|x|\geqslant 1$. A straightforward calculation reveals that u satisfies the conditions of Theorem 1.1; thus the diffusion is transient. □

Proof of Theorem 2.2. As in the proof of Theorem 2.1, we will assume that the drift b is continuous. The extension to the case that b is locally bounded and measurable is left to exercise 6.1. By assumption, $\int_1^\infty \exp(-\bar{I}(r))\,\mathrm{d}r = \infty$; thus, by Theorem 2.1, the diffusion is recurrent.

(i) Let

$$F(r) = \int_1^r \mathrm{d}s \exp(-\bar{I}(s)) \int_s^\infty \mathrm{d}t \frac{1}{\underline{\alpha}(t)} \exp(\bar{I}(t)), \text{ for } r \geq 1.$$

By assumption, $F \in C^2([1, \infty))$. Let $u \in C^2(R^d)$ satisfy $u(x) = F(|x|)$, for $|x| \geq 1$. A straightforward calculation reveals that u satisfies the conditions of Theorem 1.3; thus, the diffusion is positive recurrent.

(ii) Let

$$H(r) = \int_1^r \mathrm{d}s \exp(-\bar{I}(s)) \int_1^s \mathrm{d}t \frac{1}{\bar{\alpha}(t)} \exp(\bar{I}(t)), \text{ for } r \geq 1.$$

Again, by assumption, $H \in C^2([1, \infty))$. Let $v \in C^2(R^d)$ satisfy $v(x) = H(|x|)$, for $|x| \geq 1$. A straightforward calculation reveals that $Lv \leq \frac{1}{2}$. Let $1 < |x| < n$. Since $v(X(t \wedge \tau_1 \wedge \tau_n)) - \int_0^{t \wedge \tau_1 \wedge \tau_n} Lv(X(s))\,\mathrm{d}s$ is a P_x-martingale, we obtain

$$E_x v(X(t \wedge \tau_1 \wedge \tau_n)) - \tfrac{1}{2} E_x t \wedge \tau_1 \wedge \tau_n \leq v(x).$$

Letting $t \to \infty$ gives

$$E_x \tau_1 \wedge \tau_n \geq 2E_x v(X(\tau_1 \wedge \tau_n)) - 2v(x) \geq 2H(n) P_x(\tau_n < \tau_1) - 2v(x). \tag{2.1}$$

We will show that

$$P_x(\tau_n < \tau_1) \geq \frac{\int_1^{|x|} \exp(-\underline{I}(s))\,\mathrm{d}s}{\int_1^n \exp(-\underline{I}(s))\,\mathrm{d}s}. \tag{2.2}$$

From (2.1), (2.2) and the assumption of the theorem, we obtain

$$E_x \tau_1 = \lim_{n \to \infty} E_x \tau_1 \wedge \tau_n = \infty.$$

The null recurrence now follows from Theorem 4.9.6.

It remains to prove (2.2).

Let $J(r) = \int_1^r \exp(-\underline{I}(s))\,\mathrm{d}s$, for $r \geq 1$, and let $u \in C^2(R^d)$ satisfy $u(x) = J(|x|)$, for $|x| \geq 1$. Then since $Lu \geq 0$, for $|x| \geq 1$, and $u(X(t \wedge \tau_1 \wedge \tau_n)) - \int_0^{t \wedge \tau_1 \wedge \tau_n} Lu(X(s))\,\mathrm{d}s$ is a P_x-martingale, it follows upon letting $t \to \infty$ that $E_x u(X(\tau_1 \wedge \tau_n)) \geq u(x)$. Writing this in terms of J gives $J(n) P_x(\tau_n < \tau_1) \geq J(|x|)$, which is (2.2). □

6.3 Application of the Liapunov method – Case II: non-uniformity in the nonradial variables

The method of the previous section required a certain uniform behavior of the coefficients in the non-radial variables. Thus, for example, if $S^{d-1} = \overline{A \cup B}$ and L 'looks' transient on $[0, \infty) \times A$ and 'looks' recurrent on $[0, \infty) \times B$, then, clearly, the method of the previous section is inoperable. Although this situation is much more delicate than the previous one, it can sometimes be treated by the Liapunov method. The problems considered below are tractable either because a certain stochastic averaging takes place, or because the process eventually remains in the subdomain $(0, \infty) \times A$ where the coefficients do possess uniform behavior in the non-radial variables. The stochastic averaging is captured analytically by use of a Fredholm alternative.

Let

$$r = |x| \text{ and } \theta = \frac{x}{|x|} \in S^{d-1}$$

denote spherical coordinates in R^d. First we consider operators of the form

$$L = c_1(\theta)\frac{\partial^2}{\partial r^2} + \frac{c_2(\theta)}{r}\frac{\partial}{\partial r} + \frac{1}{r}\frac{\partial}{\partial r}D + \frac{1}{r^2}\mathcal{A}, \quad \text{on } R^d - \{0\},\ d \geq 2, \tag{3.1}$$

where $c_1 > 0$, $c_i \in C^\alpha(S^{d-1})$, $i = 1, 2$, D is a first-order operator on S^{d-1}, $\mathcal{A}$ is a second-order operator on S^{d-1}, $\mathcal{A}1 = D1 = 0$ (that is, the zeroth-order terms in $\mathcal{A}$ and D vanish), and $\mathcal{A} + \mathcal{D}$ satisfies the obvious analog of Assumption $\tilde{\text{H}}$ on S^{d-1}. By Theorem 4.11.1(ix), $\mathcal{A}$ is product L^1 critical on S^{d-1}; that is, the diffusion on S^{d-1} corresponding to $\mathcal{A}$ is positive recurrent. Let $\tilde{\mathcal{A}}$ denote the formal adjoint of $\mathcal{A}$ with respect to Lebesgue measure on S^{d-1}, and denote the ground state of $\tilde{\mathcal{A}}$ by $\tilde{\phi}_c^{\mathcal{A}}$; it is the invariant density for the diffusion on S^{d-1} corresponding to $\mathcal{A}$. Define

$$\rho = \int_{S^{d-1}} (c_2 - c_1)\tilde{\phi}_c^{\mathcal{A}}\, d\theta. \tag{3.2}$$

Also, let $\lambda_0(S^{d-1}; \mathcal{A} + 2D + 2c_1 + 2c_2)$ denote the principal eigenvalue of $\mathcal{A} + 2D + 2c_1 + 2c_2$ on S^{d-1}. (Note that $\lambda_0(S^{d-1}; \mathcal{A} + 2D) = 0$ by Theorem 4.11.1(ii).) We will prove the following theorem.

Theorem 3.1. *Let* L *be as in* (3.1), *let* ρ *be as in* (3.2) *and let* $\lambda_0(S^{d-1}; \mathcal{A} + 2D + 2c_1 + 2c_2)$ *denote the principal eigenvalue of* $\mathcal{A} + 2D + 2c_1 + 2c_2$ *on* S^{d-1}. *Define* $\tau_r = \inf\{t \geq 0\colon |X(t)| = r\}$, *for* $r > 0$, *and* $\tau_0 = \lim_{r\to 0} \tau_r$.

(i) *If* $\rho > 0$, *then* $P_x(\tau_r < \infty) < 1$, *for* $0 < r < |x|$, $P_x(\tau_0 < \infty) = 0$, *for* $x \in R^d - \{0\}$, *and* $P_x(\lim_{t\to\infty} |X(t)| = 0) = 0$, *for* $x \in R^d - \{0\}$.
(ii) *If* $\rho = 0$, *then* $P_x(\tau_r < \infty) = 1$, *for* $0 < r \leq |x|$, $P_x(\tau_0 < \infty) = 0$, *for* $x \in R^d - \{0\}$, *and* $P_x(\lim_{t\to\infty} |X(t)| = 0) = 0$, *for* $x \in R^d - \{0\}$.
(iii) *If* $\rho < 0$, *then* $P_x(\tau_0 < \infty) = 1$, *for* $x \in R^d - \{0\}$.
(iv) *If* $\lambda_0(S^{d-1}; \mathcal{A} + 2D + 2c_1 + 2c_2) < 0$, *then* $E_x\tau_r < \infty$, *for* $0 < r \leq |x|$.

Remark 1. Let $\hat{L}$ be an operator on all of R^d satisfying Assumption A and such that $L = \hat{L}$ on $\{|x| > 1\}$. From Theorem 2.7.6 and the localization result, Theorem 1.13.2, it follows that the diffusion on R^d corresponding to $\hat{L}$ is transient if $\rho > 0$ and recurrent if $\rho \leq 0$. From Theorem 4.9.6(i) and Theorem 1.13.2, it follows that if $\lambda_0(S^{d-1}; \mathcal{A} + 2D + 2c_1 + 2c_2) < 0$, then the diffusion on R^d corresponding to $\hat{L}$ is positive recurrent. In the case where c_1 and c_2 are constant, $\lambda_0(S^{d-1}; \mathcal{A} + 2D + 2c_1 + 2c_2) = 2c_1 + 2c_2$ and, in fact, the condition $c_1 + c_2 < 0$ is both necessary and sufficient for the positive recurrence of the diffusion corresponding to $\hat{L}$ on R^d. To see this, let

$$\bar{L} = c_1 \frac{d^2}{dx^2} + \frac{c_2}{x}\frac{d}{dx} \text{ on } (0, \infty),$$

and let $\bar{E}$ and $\hat{E}$ denote expectations for the diffusions corresponding to $\bar{L}$ and $\hat{L}$. By Remark 2 following Theorem 4.9.6 and by Corollary 5.1.11, $\bar{E}_x\tau_1 < \infty$, for $x > 1$, if and only if $c_1 + c_2 < 0$. It is easy to see then that $E_x\tau_1 < \infty$, for $|x| > 1$, if and only if $c_1 + c_2 < 0$ (use exercise 1.13, for example). Thus $\hat{E}_x\tau_1 < \infty$, for $|x| > 1$, if and only if $c_1 + c_2 < 0$. By Theorem 4.9.6 again, it follows that the diffusion corresponding to $\hat{L}$ is positive recurrent if and only if $c_1 + c_2 < 0$.

In the case where c_1 and c_2 are not constant, we do not know whether $\lambda_0(S^{d-1}; \mathcal{A} + 2D + 2c_1 + 2c_2) < 0$ is a necessary condition for positive recurrence.

Remark 2. An easy argument using Theorem 2.8.2 shows that the diffusion on $R^d - \{0\}$ correponding to L is recurrent if $\rho = 0$ and transient if $\rho \neq 0$.

Remark 3. The class of operators of the form

$$L = \sum_{i,j=1}^{d} a_{ij}\left(\frac{x}{|x|}\right)\frac{\partial^2}{\partial x_i \partial x_j} \text{ on } R^d - \{0\}$$

satisfies the conditions of the theorem.

Remark 4. In general, $\tilde{\phi}_c^{\mathcal{A}}$, and thus also ρ, cannot be written explicitly in terms of the coefficients of L. However, in the case where $d = 2$, the operator $\mathcal{A}$ is of the form

$$\mathcal{A} = c_3(\theta)\frac{\partial^2}{\partial\theta^2} + c_4(\theta)\frac{\partial}{\partial\theta}$$

and

$$\begin{aligned}\tilde{\phi}_c^{\mathcal{A}}(\theta) = \frac{1}{c_3(\theta)}\exp\left(\int_0^{\theta}\frac{c_4}{c_3}(s)\,\mathrm{d}s\right)\Bigg[\int_0^{\theta}\exp\left(-\int_0^t\frac{c_4}{c_3}(s)\,\mathrm{d}s\right)\mathrm{d}t \\ + \exp\left(\int_0^{2\pi}\frac{c_4}{c_3}(s)\,\mathrm{d}s\right)\int_{\theta}^{2\pi}\exp\left(-\int_0^t\frac{c_4}{c_3}(s)\,\mathrm{d}s\right)\mathrm{d}t\Bigg].\end{aligned}$$

Remark 5. The stochastic averaging in Theorem 3.1 is expressed through the number ρ.

The operator L in (3.1) is homogeneous in r of degree -2. We now investigate the case in which the degree of homogeneity for the terms involving differentiation in r differs from that of the terms involving differentiation in θ. Consider L of the form

$$L = c_1(\theta)\frac{\partial^2}{\partial r^2} + \frac{c_2(\theta)}{r}\frac{\partial}{\partial r} + \frac{1}{r^k}\frac{\partial}{\partial r}D + \frac{1}{r^{k+1}}\mathcal{A} \text{ on } R^d - \{0\},\ d \geq 2, \tag{3.3}$$

with the same assumptions on the coefficients as before. Theorem 3.1 treated the case where $k = 1$. If $k < 1$ ($k > 1$), then for large r the motion in the θ-variable takes place more rapidly (slowly) than in the case where $k = 1$, while for small r it takes place more slowly (rapidly). It turns out that $k = 1$ is the critical rate; if the motion in the θ-variable is below this rate, averaging will not occur. Thus, conditions for $P_x(\tau_r < \infty) = 1$ or $P_x(\tau_r < \infty) < 1$, with $r > 0$, will be dramatically different in the case where $k > 1$ as compared with the case where $k \leq 1$, and conditions for $P_x(\tau_0 < \infty) = 1$ or $P_x(\tau_0 < \infty) < 1$ will be dramatically different in the case where $k < 1$ as compared with the case where $k \geq 1$. Even in the case where averaging does occur with $k \neq 1$, the criteria are a little different from those in the case where $k = 1$. In the case where $k = 1$, the mixed derivative term

$$\frac{1}{r}\frac{\partial}{\partial r}D$$

plays no role in determining whether $P_x(\tau_r < \infty) = 1$ or $P_x(\tau_r < \infty) < 1$, since $\rho = \int_{S^{d-1}}(c_2 - c_1)\tilde{\phi}_c^{\mathcal{A}}\,d\theta$. When averaging occurs in the case where $k \neq 1$, we must work with

$$\rho_k = \int_{S^{d-1}} (c_2 - c_1)\tilde{\phi}_c^{\mathcal{A}+(k-1)D}\,d\theta, \tag{3.4}$$

where $\tilde{\phi}_c^{\mathcal{A}+(k-1)D}$ is the ground state of the formal adjoint of $\mathcal{A} + (k-1)D$ on S^{d-1}. Similarly, the sufficiency condition for $E_x\tau_r < \infty$ which was given in terms of $\lambda_0(S^{d-1}; \mathcal{A} + 2D + 2c_1 + 2c_2)$ will now be given in terms of

$$\lambda_0(S^{d-1}; \mathcal{A} + (k-1)D + 2D + 2c_1 + 2c_2) = \lambda_0(S^{d-1}; \mathcal{A} + (k+1)D + 2c_1 + 2c_2).$$

Theorem 3.2. *Let L be as in (3.3), let ρ_k be as in (3.4), and let $\lambda_0(S^{d-1}; \mathcal{A} + (k+1)D + 2c_1 + 2c_2)$ denote the principal eigenvalue of $\mathcal{A} + (k+1)D + 2c_1 + 2c_2$ on S^{d-1}. Define $\tau_r = \inf\{t \geq 0: |X(t)| = r\}$, for $r > 0$, and $\tau_0 = \lim_{r\to 0} \tau_r$.*

(a) *Let $k < 1$.*
   (i) *If $\rho_k > 0$, then $P_x(\tau_r < \infty) < 1$, for $0 < r < |x|$.*
   (ii) *If $\rho_k < 0$, then $P_x(\tau_r < \infty) = 1$, for $0 < r \leq |x|$.*
   (iii) *If $\lambda_0(S^{d-1}; \mathcal{A} + (k+1)D + 2c_1 + 2c_2) < 0$, then $E_x\tau_r < \infty$, for $0 < r \leq |x|$.*
   (iv) *If $\{\theta \in S^{d-1}: c_2(\theta) - c_1(\theta) < 0\} \neq \phi$, then $P_x(\tau_0 < \infty$ and*

$$\frac{X(t)}{|X(t)|} \in \{\theta \in S^{d-1}: c_2(\theta) - c_1(\theta) \leq 0\}$$

   for all t sufficiently close to $\tau_0) > 0$, for $x \in R^d - \{0\}$. If, in addition, $\rho_k < 0$, then

$$P_x(\tau_0 < \infty \text{ and } \frac{X(t)}{|X(t)|} \in \{\theta \in S^{d-1}: c_2(\theta) - c_1(\theta) \leq 0\}$$

   for all t sufficiently close to $\tau_0) = 1$, for $x \in R^d - \{0\}$.
   (v) *If $\{\theta \in S^{d-1}: c_2(\theta) - c_1(\theta) < 0\} = \phi$, then $P_x(\tau_0 < \infty) = 0$ and $P_x(\lim_{t\to\infty} |X(t)| = 0) = 0$, for $x \in R^d - \{0\}$.*

(b) *Let $k > 1$.*
   (i) *If $\{\theta \in S^{d-1}: c_2(\theta) - c_1(\theta) > 0\} \neq \phi$, then $P_x(\lim_{t\to\infty} |X(t)| = \infty$ and*

$$\frac{X(t)}{|X(t)|} \in \{\theta \in S^{d-1}: c_2(\theta) - c_1(\theta) \geq 0\}$$

   for all large $t) > 0$, for $x \in R^d - \{0\}$. In particular, $P_x(\tau_r < \infty) < 1$, for $0 < r < |x|$. If, in addition, $\rho_k > 0$, then $P_x(\lim_{t\to\infty} |X(t)| = \infty$ and

$$\frac{X(t)}{|X(t)|} \in \{\theta \in S^{d-1}: c_2(\theta) - c_1(\theta) \geq 0\}$$

for all large t) $= 1$, *for* $x \in R^d - \{0\}$.

(ii) *If* $\{\theta \in S^{d-1}: c_2(\theta) - c_1(\theta) > 0\} = \phi$, *then* $P_x(\tau_r < \infty) = 1$, *for* $0 < r \leq |x|$.

(iii) *If* $\sup_{\theta \in S^{d-1}} (c_1 + c_2)(\theta) < 0$, *then* $E_x\tau_r < \infty$, *for* $0 < r \leq |x|$.

(iv) *If* $\rho_k < 0$, *then* $P_x(\tau_0 < \infty) > 0$, *for* $x \in R^d - \{0\}$. *If, in addition,* $\{\theta \in S^{d-1}: c_2(\theta) - c_1(\theta) > 0\} = \phi$, *then* $P_x(\tau_0 < \infty) = 1$, *for* $x \in R^d - \{0\}$.

(v) *If* $\rho_k > 0$, *then* $P_x(\tau_0 < \infty) = 0$ *and* $P_x(\lim_{t\to\infty} |X(t)| = 0) = 0$, *for* $x \in R^d - \{0\}$.

Remark. Note that in contrast to Theorem 3.1, Theorem 3.2 does not handle the borderline case where $\rho_k = 0$. We suspect that if $\rho_k = 0$, then one obtains the result obtained in Theorem 3.1 in the case where $\rho = 0$. We known how to prove this in the special case where $D \equiv 0$. This is left as exercise 6.3.

For the classes of operators considered above, both the diffusion and the drift terms of $r(t) \equiv |X(t)|$ and

$$\theta(t) \equiv \frac{X(t)}{|X(t)|}$$

played a role in the stochastic averaging since ρ (ρ_k) depended on the entire operator $\mathcal{A}$ ($\mathcal{A} + (k-1)D$) through $\tilde{\phi}_c^{\mathcal{A}}$ ($\tilde{\phi}_c^{\mathcal{A}+(k-1)D}$) and on $c_1(\theta)$ and $c_2(\theta)$. Our final example of this section consists of a class of operators for which the drift alone determines the behavior of $P_x(\tau_r < \infty)$. For simplicity, we consider the case where $d = 2$. Define

$$L = c_1(\theta)\frac{\partial^2}{\partial r^2} + \frac{c_2(\theta)}{r}\frac{\partial}{\partial r} + \frac{1}{r}\frac{\partial}{\partial r}D + \frac{1}{r^2}\mathcal{A} + \frac{\gamma_1(\theta)}{r^\delta}\frac{\partial}{\partial r} + \frac{\gamma_2(\theta)}{r^{k+1}}\frac{\partial}{\partial \theta}, \text{ on } R^2 - \{0\}. \tag{3.5}$$

Assume that $c_1 > 0$, $c_1 \in C(S^1)$, c_2 is bounded and measurable, $\gamma_2 > 0$, $\gamma_i \in C^1(S^1)$, $i = 1, 2$, $\delta < 1$ and $k \in (-\infty, \infty)$. Let $\mathcal{A}$ and D on S^1 be as in Theorems 3.1 and 3.2. Define

$$\gamma = \int_0^{2\pi} \frac{\gamma_1}{\gamma_2}(\theta)\, d\theta. \tag{3.6}$$

Theorem 3.3. *Let L be as in* (3.5) *and let γ be as in* (3.6). *Define* $\tau_r = \inf\{t \geq 0: |X(t)| = r\}$, *for* $r > 0$, *and* $\tau_0 = \lim_{r\to 0} \tau_r$.

(a) *Let* $k \leqslant \delta$.
  (i) *If* $\gamma > 0$, *then* $P_x(\tau_r < \infty) < 1$, *for* $0 < r < |x|$.
  (ii) *If* $\gamma < 0$, *then* $E_x\tau_r < \infty$, *for* $0 < r < |x|$.
(b) *Let* $k > \delta$.
  (i) *If* $\{\theta \in S^1: \gamma_1(\theta) > 0\} \neq \phi$, *then* $P_x(\tau_r < \infty) < 1$, *for* $0 < r < |x|$, *and* $P_x(\lim_{t\to\infty} |X(t)| = \infty$ *and*

$$\frac{X(t)}{|X(t)|} \in \{\theta \in S^1: \gamma_1(\theta) > 0\} \text{ for all large } t) > 0.$$

  (ii) *If* $\{\theta \in S^1: \gamma_1(\theta) \geqslant 0\} = \phi$, *then* $E_x\tau_r < \infty$, *for* $0 < r < |x|$.

Remark. If $k \leqslant \delta$ and $\gamma = 0$, then the behavior of $P_x(\tau_r < \infty)$ will depend in general on c_1, c_2, $\mathcal{A}$ and D. We do not know how to handle the general case; particular cases are treated in exercise 6.4.

For the proofs of Theorem 3.1–3.3, we will need an extension of the Fredholm alternative of Corollary 3.1.2. If $\mathcal{L}$ is an operator satisfying the analog of Assumption $\tilde{\text{H}}$ on S^{d-1}, and its principal eigenvalue $\lambda_0(S^{d-1}; \mathcal{L}) = 0$, then, by Theorem 4.11.1, $\mathcal{L}$ is critical on S^{d-1}. Denote the ground states of $\mathcal{L}$ and $\tilde{\mathcal{L}}$ by $\phi_c^{\mathcal{L}}$ and $\phi_c^{\tilde{\mathcal{L}}}$. By Theorem 4.11.1, $\mathcal{L}$ has a compact resolvent. Thus, since $\mathcal{L}\phi_c^{\mathcal{L}} = 0$, it follows from Corollary 3.1.2 that the equation $\mathcal{L}u = h$ does not have a solution $u \in C^{2,\alpha}(S^{d-1})$ for certain choices of $h \in C^\alpha(S^{d-1})$. The following extension of the Fredholm alternative determines for which h the equation $\mathcal{L}u = h$ is solvable.

Fredholm alternative. Let $\mathcal{L}$ satisfy the analog of Assumption $\tilde{\text{H}}$ on S^{d-1} and assume that $\lambda_0(S^{d-1}; \mathcal{L}) = 0$. Let $\phi_c^{\tilde{\mathcal{L}}}$ denote the ground state of $\tilde{\mathcal{L}}$. Then for $h \in C^\alpha(S^{d-1})$, the equation $\mathcal{L}u = h$ possesses solutions if and only if $\int_{S^{d-1}} h\phi_c^{\tilde{\mathcal{L}}}\,d\theta = 0$.

Proof of Theorem 3.1. (i) Let $u(x) = u(r, \theta) = r^{-m}g(\theta)$, with $m > 0$ and g as yet undetermined. Then

$$Lu = r^{-m-2}[m(m+1)c_1g - mc_2g - mDg + \mathcal{A}g].$$

Writing g in the form $g = 1 + mg_1$, using the fact that $D1 = \mathcal{A}1 = 0$, and substituting above gives

$$Lu = mr^{-m-2}[c_1 - c_2 + \mathcal{A}g_1 + \mathrm{O}(m)], \text{ as } m \to 0. \tag{3.7}$$

By the Fredholm alternative, we can choose a solution g_1 to $\mathcal{A}g_1 = c_2 - c_1 - \rho$. With such a choice of g_1, and with m sufficiently small, it follows that $u > 0$ and, from (3.7), that $Lu < 0$. Also, for any

$r_0 > 0$, there exists an x_0 with $|x_0| > r_0$ such that $u(x_0) < \inf_{|x|=r_0} u(x)$. Thus, by Theorem 1.1, $P_x(\tau_r < \infty) < 1$, for $0 < r < |x|$. This proves the first claim in part (i).

Since $Lu \leqslant 0$, we have for $0 < \varepsilon < |x| < n$, $E_x u(X(\tau_\varepsilon \wedge \tau_n)) \leqslant u(x)$; letting $n \to \infty$ and using the fact that $u > 0$ gives

$$E_x(u(X(\tau_\varepsilon)); \tau_\varepsilon < \infty) \leqslant u(x). \tag{3.8}$$

Now, if it were true that $P_x(\tau_0 < \infty) > 0$ or that $P_x(\lim_{t\to\infty} |X(t)| = 0) > 0$, then as $\varepsilon \to 0$, the lefthand side of (3.8) would tend to infinity, which contradicts (3.8). This completes the proof of part (i).

(ii) Let $u(x) = u(r, \theta) = \log r + g(\theta)$, with g as yet undetermined. Then

$$Lu = \frac{1}{r^2}[-c_1 + c_2 + \mathcal{A}g].$$

Since $\rho = 0$, by the Fredholm alternative, we can choose a solution g of $\mathcal{A}g = c_1 - c_2$. With this choice of g, we have $Lu = 0$ and, by Theorem 1.2, $P_x(\tau_r < \infty) = 1$, for $0 < r < |x|$. This proves the first claim in part (ii).

If $P_x(\tau_0 < \infty)$ or $P_x(\lim_{t\to\infty} |X(t)| = 0)$ were positive for some $x \in R^d - \{0\}$, then it would follow that, for sufficiently large n, $\lim_{\varepsilon\to 0} E_x u(X(\tau_\varepsilon \wedge \tau_n)) = -\infty$. But this contradicts the equality $E_x u(X(\tau_\varepsilon \wedge \tau_n)) = u(x)$; thus $P_x(\tau_0 < \infty) = P_x(\lim_{t\to\infty} |X(t)| = 0) = 0$.

(iii) Let $u(x) = u(r, \theta) = r^m g(\theta)$, with $m > 0$ and g as yet undetermined. Then

$$Lu = r^{m-2}[m(m-1)c_1 g + mc_2 g + mDg + \mathcal{A}g].$$

Writing g in the form $g = 1 + mg_1$, using the fact that $D1 = \mathcal{A}1 = 0$, and substituting above gives

$$Lu = mr^{m-2}[-c_1 + c_2 + \mathcal{A}g_1 + \mathrm{O}(m)], \quad \text{as } m \to 0. \tag{3.9}$$

By the Fredholm alternative, we can choose a solution g_1 of $\mathcal{A}g_1 = c_1 - c_2 + \rho$. With such a choice of g_1, and with m sufficiently small, it follows that $u > 0$, that $\lim_{|x|\to\infty} u(x) = \infty$ and, from (3.9), that $Lu < 0$. In particular, for each $n > 0$, there exists a $\delta_n > 0$ such that $Lu(x) < -\delta_n$, if $0 < |x| < n$. It follows, then, that for $0 < \varepsilon < |x| < n$,

$$\delta_n E_x \tau_\varepsilon \wedge \tau_n \leqslant u(x) - E_x u(X(\tau_\varepsilon \wedge \tau_n)). \tag{3.10}$$

Letting $\varepsilon \to 0$ in (3.10), we obtain

$$E_x \tau_0 \wedge \tau_n < \infty, \quad \text{for } 0 < |x| < n \tag{3.11}$$

and

$$(\inf_{|x|=n} u(x)) P_x(\tau_n < \tau_0) \leqslant u(x), \quad \text{for } 0 < |x| < n. \tag{3.12}$$

Letting $n \to \infty$ in (3.12) shows that $\lim_{n\to\infty} P_x(\tau_0 < \tau_n) = 1 - \lim_{n\to\infty} P_x(\tau_n < \tau_0) = 1$. This in conjunction with (3.11) proves that $P_x(\tau_0 < \infty) = 1$.

(iv) Let $u(x) = u(r, \theta) = r^2 g(\theta)$, with g as yet undetermined. Then

$$Lu = 2c_1 g + 2c_2 g + 2Dg + \mathcal{A}g. \tag{3.13}$$

By Assumption, $\lambda_0(S^{d-1}; \mathcal{A} + 2D + 2c_1 + 2c_2) < 0$. By Theorem 4.11.1, $\mathcal{A} + 2D + 2c_1 + 2c_2 - \lambda_0(S^{d-1};\ \mathcal{A} + 2D + 2c_1 + 2c_2)$ is critical; choose g to be its ground state. Again by Theorem 4.11.1, $\inf_{S^{d-1}} g > 0$. Thus, from (3.13) we have

$$Lu = \lambda_0(S^{d-1}; \mathcal{A} + 2D + 2c_1 + 2c_2) g \leqslant -\varepsilon, \quad \text{for some } \varepsilon > 0.$$

By Theorem 1.3, it follows that $E_x \tau_r < \infty$, for $0 < r \leqslant |x|$. □

Proof of Theorem 3.2. (a(i)) Let $u(r, \theta) = r^{-m} \exp(r^{k-1} g(\theta))$ and let $g = 1 + m g_1$, where g_1 solves $(\mathcal{A} + (k-1)D) g_1 = c_2 - c_1 - \rho_k$; the existence of such a g_1 follows from the Fredholm alternative. If $m > 0$ is sufficiently small, then u satisfies the conditions of Theorem 1.1. Thus, there exists an $r_0 > 0$ such that $P_x(\tau_r < \infty) < 1$, for $r_0 \leqslant r < |x|$. An argument like that used in Step 1 and Step 2 of the proof of Theorem 2.7.4 shows that $P_x(\tau_r < \infty) < 1$, for $0 < r < |x|$.

(a(ii)) Let $u(r, \theta) = r^m \exp(r^{k-1} g(\theta))$ and let $g = 1 + m g_1$, where g_1 solves $(\mathcal{A} + (k-1)D) g_1 = c_1 - c_2 + \rho_k$; the existence of such a g_1 follows from the Fredholm alternative. If $m > 0$ is sufficiently small, then u satisfies the conditions of Theorem 1.2. Thus there exists an $r_0 > 0$ such that $P_x(\tau_{r_0} < \infty) = 1$, for $|x| \geqslant r_0$. An argument like that used in Step 1 and Step 2 of the proof of Theorem 2.7.4 shows that $P_x(\tau_r < \infty) = 1$, for $0 < r \leqslant |x|$.

(a(iii)) Let $u(r, \theta) = r^2 \exp(r^{k-1} g(\theta))$, where g, which satisfies $\inf_{S^{d-1}} g > 0$, is the ground state of

$$\mathcal{A} + (k+1)D + 2c_1 + 2c_2 - \lambda_0(S^{d-1}; \mathcal{A} + (k+1)D + 2c_1 + 2c_2)$$

on S^{d-1}. Then u satisfies the conditions of Theorem 1.3. Thus, there exists an $r_0 > 0$ such that $E_x \tau_{r_0} < \infty$, for $|x| \geqslant r_0$. To show from this that $E_x \tau_r < \infty$, for $0 < r < |x|$, is left as exercise 6.2.

(a(iv)) Let $A \subset S^{d-1}$ be a domain with a $C^{2,\alpha}$-boundary satisfying $A \subset\subset \{\theta \in S^{d-1}: c_2(\theta) - c_1(\theta) < 0\}$. We can choose an $m > 0$, a $\delta > 0$, and an $r_0 > 0$ such that the function $h(x) = h(r, \theta) = r^m$ satisfies $Lh(r, \theta) \leq -\delta$ for $(r, \theta) \in (0, r_0) \times A$. Let

$$\tau_A = \inf\left\{t \geq 0: \frac{X(t)}{|X(t)|} \notin A\right\}.$$

Then for $(r, \theta) \in (0, r_0) \times A$,

$$E_{r,\theta} h(X(\tau_0 \wedge \tau_{r_0} \wedge \tau_A)) \leq h(r, \theta) - \delta E_{r,\theta}(\tau_0 \wedge \tau_{r_0} \wedge \tau_A).$$

Thus,

$$E_{r,\theta}(\tau_0 \wedge \tau_{r_0} \wedge \tau_A) < \infty, \quad \text{for } (r, \theta) \in (0, r_0) \times A. \tag{3.14}$$

Let $\lambda_0(A; \mathcal{A})$ denote the principal eigenvalue of $\mathcal{A}$ on A and let $g = g(\theta)$ denote the ground state of $\mathcal{A} - \lambda_0(A; \mathcal{A})$ on A. Define $u(r, \theta) = (1 - r^\varepsilon) g^2(\theta)$, on $(0, \infty) \times \bar{A}$, where $\varepsilon > 0$ is yet to be determined. Note that $\mathcal{A} g^2 = 2g\mathcal{A}g + 2(\nabla_\theta g a \nabla_\theta g)$, where ∇_θ is the gradient on S^{d-1} and a is a positive definite $(d-1) \times (d-1)$ matrix-valued function on S^{d-1}. Then

$$Lu = \varepsilon r^{\varepsilon - 2}\Big[(1-\varepsilon)c_1 g^2 - c_2 g^2 - 2r^{1-k} g D g + \frac{2}{\varepsilon}(1 - r^\varepsilon) r^{1-k-\varepsilon} g \mathcal{A} g + \frac{2}{\varepsilon}(1 - r^\varepsilon) r^{1-k-\varepsilon}(\nabla_\theta g a \nabla_\theta g)\Big]. \tag{3.15}$$

By the Hopf maximum principle, $\nabla_\theta g \neq 0$ on ∂A; thus $\nabla_\theta g \nabla_\theta g$ is strictly positive on ∂A. Therefore, if $\varepsilon > 0$ is chosen so that $1 - k - \varepsilon > 0$ and so that $(1 - \varepsilon)c_1 - c_2 > 0$ on $\bar{A}$, then it follows from (3.15) that there exists an $r_0 > 0$ such that $Lu(r, \theta) \geq 0$, for $0 < r < r_0$ and $\theta \in A$. Thus,

$$E_{r,\theta} u(X(\tau_{r_0} \wedge \tau_\delta \wedge \tau_A)) \geq u(r, \theta), \text{ for } 0 < \delta < r < r_0 \text{ and } \theta \in A. \tag{3.16}$$

Let $\bar{\theta} \in A$ be a point where $\sup_A g$ is attained. Since

$$\begin{aligned} &E_{r,\bar{\theta}} u(X(\tau_{r_0} \wedge \tau_\delta \wedge \tau_A)) \\ &\quad \leq g^2(\bar{\theta})[(1 - r_0^\varepsilon) P_{r,\bar{\theta}}(\tau_{r_0} < \tau_\delta \wedge \tau_A) + (1 - \delta^\varepsilon) P_{r,\bar{\theta}}(\tau_\delta < \tau_{r_0} \wedge \tau_A)] \\ &\quad \leq g^2(\bar{\theta})[(1 - r_0^\varepsilon) + (1 - \delta^\varepsilon) P_{r,\bar{\theta}}(\tau_\delta < \tau_{r_0} \wedge \tau_A)], \end{aligned}$$

it follows from (3.16) that

$$P_{r,\bar{\theta}}(\tau_\delta < \tau_{r_0} \wedge \tau_A) \geqslant \frac{r_0^\varepsilon - r^\varepsilon}{1 - \delta^\varepsilon}, \quad \text{for } 0 < \delta < r < r_0. \tag{3.17}$$

From (3.17) and (3.14), it follows that

$$P_{r,\bar{\theta}}\left(\tau_0 < \infty \text{ and } \frac{X(t)}{|X(t)|} \in \{\theta \in S^{d-1}: c_2(\theta) - c_1(\theta) \leqslant 0\},\right.$$
$$\left.\text{for all } t \text{ sufficiently close to } \tau_0\right) > 0,$$

for $0 < r < r_0$. By Theorem 9.1.2 in Chapter 9

$$v(x) = v(r, \theta) \equiv P_{r,\bar{\theta}}\left(\tau_0 < \infty \text{ and } \frac{X(t)}{|X(t)|} \in\right.$$
$$\left.\{\theta \in S^{d-1}: c_2(\theta) - c_1(\theta) \leqslant 0\} \quad \text{for all } t \text{ sufficiently close to } \tau_0\right)$$

satisfies $v \in C^{2,\alpha}(R^d - \{0\})$ and $Lv = 0$. Since $v(r, \bar{\theta}) > 0$, for $0 < r < r_0$, it follows from the strong maximum principle (Theorem 3.2.6) that $v(x) > 0$, for all $x \in R^d - \{0\}$. Demonstration that the above probability actually equals one if $\rho_k < 0$ is left to the reader as exercise 6.7.

(a(v)) Let $u(x) = u(r, \theta) = \log r$. Then $Lu(x) \geqslant 0$, for $x \in R^d - \{0\}$ and thus $E_x u(X(\tau_\varepsilon \wedge \tau_n)) \geqslant u(x)$, for $0 < \varepsilon < |x| < n < \infty$. It follows then that $\lim_{\varepsilon \to 0} P_x(\tau_\varepsilon < \tau_n) = 0$, for $0 < x < n < \infty$.

(b(i)) Let $A \subset S^{d-1}$ be a domain with a $C^{2,\alpha}$-boundary satisfying $A \subset\subset \{\theta \in S^{d-1}: c_2(\theta) - c_1(\theta) > 0\}$. Let $g = g(\theta)$ be as in the proof of (a(iv)). Define $u(r, \theta) = (1 - r^{-\varepsilon})g^2(\theta)$, on $(0, \infty) \times \bar{A}$, where $\varepsilon > 0$ is yet to be determined. As noted in the proof of (a(iv)), $\mathcal{A}g^2 = 2g\mathcal{A}g + \nabla_\theta g \alpha \nabla_\theta g$ and $\nabla_\theta g \alpha \nabla_\theta g$ is strictly positive on ∂A. We have

$$Lu(r, \theta) = \varepsilon r^{-\varepsilon-2}\Big[-(\varepsilon + 1)c_1 g^2 + c_2 g^2 + 2r^{-(k-1)} gDg$$
$$+ \frac{2}{\varepsilon}(1 - r^{-\varepsilon})r^{\varepsilon-(k-1)} g\mathcal{A}g + \frac{2}{\varepsilon}(1 - r^{-\varepsilon})r^{\varepsilon-(k-1)}\nabla_\theta g \alpha \nabla_\theta g\Big]. \tag{3.18}$$

Therefore, if $\varepsilon > 0$ is chosen so that $\varepsilon - (k - 1) < 0$ and so that $c_2 - (1 + \varepsilon)c_1 > 0$ on $\bar{A}$, then it follows from (3.18) that there exists an $r_0 > 0$ such that $Lu(r, \theta) \geqslant 0$, for $r > r_0$ and $\theta \in A$. Thus, letting

$$\tau_A = \inf\left\{t \geqslant 0: \frac{X(t)}{|X(t)|} \notin A\right\},$$

we have for $r_0 < r < n$ and $\theta \in A$,

$$E_{r,\theta}u(X(\tau_{r_0} \wedge \tau_n \wedge \tau_A)) \geqslant u(r,\theta). \tag{3.19}$$

Let $\bar{\theta} \in A$ be a point where $\sup_A g$ is attained. Since

$$\begin{aligned}&E_{r,\bar{\theta}}u(X(\tau_{r_0} \wedge \tau_n \wedge \tau_A))\\ &\leqslant g^2(\bar{\theta})[(1-r_0^{-\varepsilon})P_{r,\bar{\theta}}(\tau_{r_0} < \tau_n \wedge \tau_A) + (1-n^{-\varepsilon})P_{r,\bar{\theta}}(\tau_n < \tau_{r_0} \wedge \tau_A)]\\ &\leqslant g^2(\bar{\theta})[(1-r_0^{-\varepsilon}) + (1-n^{-\varepsilon})P_{r,\bar{\theta}}(\tau_n < \tau_{r_0} \wedge \tau_A)],\end{aligned}$$

it follows from (3.19) that

$$P_{r,\bar{\theta}}(\tau_n < \tau_{r_0} \wedge \tau_A) \geqslant \frac{r_0^{-\varepsilon} - r^{-\varepsilon}}{1-n^{-\varepsilon}}, \quad \text{for } r_0 \leqslant r \leqslant n. \tag{3.20}$$

From (3.20), it follows that

$$P_{r,\bar{\theta}}\left(\lim_{t\to\infty} |X(t)| = \infty \text{ and } \frac{X(t)}{|X(t)|} \in \{\theta \in S^{d-1}\colon c_2(\theta) - c_1(\theta) \geqslant 0\}, \right.$$
$$\left. \text{for all large } t\right) > 0,$$

for $r > r_0$. By Theorem 9.1.2 in Chapter 9,

$$v(x) = v(r,\theta) \equiv P_{r,\bar{\theta}}\left(\lim_{r\to\infty} |X(t)| = \infty \right.$$
$$\left. \text{and } \frac{X(t)}{|X(t)|} \in \{\theta \in S^{d-1}\colon c_2(\theta) - c_1(\theta) \geqslant 0\}, \quad \text{for all large } t\right)$$

satisfies $v \in C^{2,\alpha}(R^d - \{0\})$ and $Lv = 0$. Since $v(r,\bar{\theta}) > 0$, for $r > r_0$, it follows from the strong maximum principle (Theorem 3.2.6) that $v(x) > 0$, for all $x \in R^d - \{0\}$. Demonstration that the above probability equals one if $\rho_k > 0$ is left to the reader as exercise 6.7.

(b(ii)) Let $u(x) = u(r,\theta) = \log r$. Then $Lu(x) \leqslant 0$, for $x \in R^d - \{0\}$, and the result follows from Theorem 1.2.

(b(iii)) Let $u(x) = u(r,\theta) = r^2$. Then $Lu(x) \leqslant -\varepsilon$, for some $\varepsilon > 0$ and all $x \in R^d - \{0\}$, and the result follows from Theorem 1.3.

(b(iv)) Let $u(x) = u(r,\theta) = r^m \exp(r^{k-1}g(\theta))$ and let $g(\theta) = 1 + mg_1(\theta)$, where g_1 solves $(\mathcal{A} + (k-1)D)g_1 = c_1 - c_2 + \rho_k$; the existence of such a g_1 follows from the Fredholm alternative. If $m > 0$ is chosen sufficiently small, then there exists an $R > 0$ and a $\delta > 0$ such that $Lu(x) \leqslant -\delta$, for $0 < |x| < R$. Thus, for $0 < \varepsilon < |x| < R$,

$$\delta E_x\tau_\varepsilon \wedge \tau_R \leqslant u(x) - E_x u(X(\tau_\varepsilon \wedge \tau_R)). \tag{3.21}$$

Letting $\varepsilon \to 0$ in (3.21), we obtain

$$E_x\tau_0 \wedge \tau_R < \infty, \quad \text{for } 0 < |x| < R \tag{3.22}$$

and

$$(\inf_{|x|=R} u(x))P_x(\tau_R < \tau_0) \leq u(x), \quad \text{for } 0 < |x| < R. \tag{3.23}$$

From (3.22) and (3.23), it follows that there exists a $c > 0$ and an $r_0 \in (0, R)$ such that $P_x(\tau_0 < \infty) \geq c$, for $0 < |x| < r_0$. An application of the strong Markov property then shows that $P_x(\tau_0 < \infty) > 0$, for all $x \in R^d - \{0\}$. Showing that $P_x(\tau_0 < \infty) = 1$, if $\{\theta \in S^{d-1}: c_2(\theta) - c_1(\theta) > 0\} = \phi$, is an easy argument using (b(ii)) and the strong Markov property.

(b(v)) Let $u(x) = u(r, \theta) = r^{-m} \exp(r^{k-1}g(\theta))$ and let $g(\theta) = 1 + mg_1(\theta)$, where g_1 solves $(\mathcal{A} + (k-1)D)g_1 = c_2 - c_1 - \rho_k$; the existence of such a g_1 follows from the Fredholm alternative. If $m > 0$ is sufficiently small, then there exists an $r_0 > 0$ such that $Lu(x) \leq 0$, for $0 < |x| < r_0$. Thus, for $0 < \varepsilon < r < r_0$,

$$E_x u(X(\tau_\varepsilon \wedge \tau_{r_0})) \leq u(x). \tag{3.24}$$

Now if $P_x(\tau_0 < \tau_{r_0})$ were positive or if $P_x(\lim_{t\to\infty} |X(t)| = 0$ and $\tau_{r_0} = \infty)$ were positive, then the lefthand side of (3.24) would approach infinity as $\varepsilon \to 0$, which contradicts (3.24). Thus $P_x(\tau_0 < \tau_{r_0}) = P_x(\lim_{t\to\infty} |X(t)| = 0$ and $\tau_{r_0} = \infty) = 0$, for $0 < |x| < r_0$. A simple application of the strong Markov property now shows that $P_x(\tau_0 < \infty) = P_x(\lim_{t\to\infty} |X(t)| = 0) = 0$, for all $x \in R^d - \{0\}$.

Proof of Theorem 3.3. (a(i)) Let $u(x) = u(r, \theta) = r^{-1} \exp(r^{k-\delta}g(\theta))$, where

$$g(\theta) = \int_0^\theta \frac{\gamma_1}{\gamma_2}(s)\, ds - \frac{\gamma}{2\pi}\theta.$$

Then u satisfies the conditions of Theorem 1.1. Thus, there exists an $r_0 > 0$ such that $P_x(\tau_r < \infty) < 1$, for $r_0 \leq r < |x|$. An argument like that used in Step 1 and Step 2 of the proof of Theorem 2.7.4 shows that $P_x(\tau_r < \infty) < 1$, for $0 < r < |x|$.

(a(ii)) Let $u(x) = u(r, \theta) = r^{1+\delta} \exp(-(1+\delta)r^{k-\delta}g(\theta))$, where g is as in (a(i)). Then u satisfies Theorem 1.3. Thus, there exists an $r_0 > 0$ such that $E_x\tau_{r_0} < \infty$, for $|x| \geq r_0$. To show from this that $E_x\tau_r < \infty$, for $0 < r \leq |x|$ is left as exercise 6.2.

(b(i)) By assumption, there exist $\varepsilon > 0$, $\delta > 0$ and $\bar{\theta} \in S^1$ such that $\gamma_1(\theta) \geq \varepsilon$, for $|\theta - \bar{\theta}| \leq \delta$. Let $u(x) = u(r, \theta) = (1 - r^{-q})g(\theta)$, where

$0 < q < \min(1-\delta, k-\delta)$ and $g(\theta) = 2\delta^2 - (\theta - \bar{\theta})^2$. Then there exists an $r_0 > 0$ such that

$$Lu(r, \theta) \geq 0, \quad \text{for } r \leq r_0 \text{ and } |\theta - \bar{\theta}| \leq \delta. \tag{3.25}$$

Choose r_0 sufficiently large so that

$$1 - r_0^{-q} > \tfrac{1}{2}. \tag{3.26}$$

Let $A = \{(r, \theta): r > r_0 \text{ and } |\theta - \bar{\theta}| < \delta\}$ and define $\tau_A = \inf\{t \geq 0: X(t) \notin A\}$. From (3.25), it follows that

$$E_{r,\bar{\theta}} u(X(t \wedge \tau_A)) \geq u(r, \bar{\theta}) = 2\delta^2(1 - r^{-q}), \quad \text{for } r > r_0. \tag{3.27}$$

From (3.26) and (3.27), we obtain

$$(1 - r_0^{-q}) 2\delta^2 P_{r,\bar{\theta}}(\tau_A \leq t) + 2\delta^2 P_{r,\bar{\theta}}(\tau_A > t) \geq 2\delta^2(1 - r^{-q}),$$
$$\text{for } r > r_0. \tag{3.28}$$

Letting $t \to \infty$ in (3.28) gives

$$P_{r,\bar{\theta}}(\tau_A = \infty) \geq \frac{r_0^{-q} - r^{-q}}{r_0^{-q}}, \quad \text{for } r > r_0. \tag{3.29}$$

In particular, the diffusion cannot be recurrent; thus, it is transient. Using the definition of transience along with (3.29), we have

$$P_{r,\bar{\theta}}\left(\lim_{t\to\infty} |X(t)| = \infty \text{ and } \frac{X(t)}{|X(t)|} \in \{\theta \in S^1: \gamma_1(\theta) > 0\}\right.$$
$$\left.\text{for all large } t\right) > 0, \text{ for } r > r_0.$$

By Theorem 9.1.2 in Chapter 9,

$$v(x) = v(r, \theta)$$
$$\equiv P_{r,\theta}\left(\lim_{t\to\infty} |X(t)| = \infty \text{ and } \frac{X(t)}{|X(t)|} \in \{\theta \in S^{d-1}: \gamma_1(\theta) > 0\},\right.$$
$$\left.\text{for all large } t\right)$$

satisfies $v \in C^{2,\alpha}(R^d - \{0\})$ and $Lv = 0$. Since $v(r, \bar{\theta}) > 0$ for $r > r_0$, it follows from the strong maximum principle (Theorem 3.2.6) that $v(x) > 0$, for all $x \in R^d - \{0\}$.

(b(ii)) Let $u(x) = u(r, \theta) = r^{1+\delta}$. Then u satisfies Theorem 1.3. Thus, there exists an $r_0 > 0$ such that $E_x \tau_{r_0} < \infty$, for $|x| \geq r_0$. Demonstration that $E_x \tau_r < \infty$, for $0 < r \leq |x|$, is left as exercise 6.2. □

6.4 Transience and recurrence via the Dirichlet principle for diffusions corresponding to symmetric operators: a variational approach

In this section, we will assume that L is of the form

$$L = \tfrac{1}{2}\nabla \cdot a\nabla + a\nabla Q \cdot \nabla$$

so that L is symmetric (see Chapter 4, Section 10). We recall the classical Dirichlet principle.

Classical Dirichlet principle. Let $L = \frac{1}{2}\nabla \cdot a\nabla + a\nabla Q \cdot \nabla$ satisfy Assumption $\widetilde{\mathrm{H}}$ on a bounded domain $D \subset R^d$ with a $C^{2,\alpha}$-boundary and let $\phi \in C^{2,\alpha}(\bar{D})$. Then

$$\inf_{\substack{u \in W^{1,2}(D) \\ u = \phi \text{ on } \partial D}} \tfrac{1}{2}\int_D (\nabla u a \nabla u) \exp(2Q)\,\mathrm{d}x$$

is attained uniquely at $u = u_0$, where $u_0 \in C^{2,\alpha}(\bar{D})$ is the unique solution of

$$Lu_0 = 0 \text{ in } D,$$
$$u_0 = \phi \text{ on } \partial D.$$

(A function $u \in W^{1,2}(D)$ is not necessarily continuous if $d \geq 2$. Thus, when we write $u \in W^{1,2}(D)$ and $u = \phi$ on ∂D, what we mean is that the trace of u on ∂D is equal to ϕ; that is,

$$\int_D f\frac{\mathrm{d}u}{\mathrm{d}x_j}\,\mathrm{d}x = -\int_D u\frac{\mathrm{d}f}{\mathrm{d}x_j}\,\mathrm{d}x + \int_{\partial D} f\phi e_j v\,\mathrm{d}x,$$
$$\text{for all } f \in C^1(\bar{D}) \text{ and } j = 1, 2, \ldots, n,$$

where $v(x)$ denotes the outward unit normal to D at $x \in \partial D$.

Assume now that $L = \frac{1}{2}\nabla \cdot a\nabla + a\nabla Q \cdot \nabla$ satisfies Assumption $\widetilde{\mathrm{H}}_{\mathrm{loc}}$ on a domain $D \subseteq R^d$, $d \geq 2$. Let $\{D_n\}_{n=1}^{\infty}$ be a sequence of domains with $C^{2,\alpha}$-boundaries satisfying $D_n \subset\subset D_{n+1}$ and $\bigcup_{n=1}^{\infty} D_n = D$. Also, let D_1 be simply connected so that $D_n - D_1$ is a domain. Define

$$\lambda_n = \inf_{\substack{u \in W^{1,2}(D_n - D_1) \\ u = 1 \text{ on } \partial D_1 \\ u = 0 \text{ on } \partial D_n}} \tfrac{1}{2}\int_{D_n - D_1} (\nabla u a \nabla u) \exp(2Q)\,\mathrm{d}x. \tag{4.1}$$

By the Dirichlet principle, the infimum on the righthand side of (4.1) is attained uniquely at $u = u_n$, where u_n is the unique solution of

$$Lu_n = 0 \text{ in } D_n - \bar{D}_1,$$

$$u_n = 1 \text{ on } \partial D_1,$$
$$u_n = 0 \text{ on } \partial D_n.$$

By Theorem 2.1.2, in fact

$$u_n(x) = P_x(\sigma_{\bar{D}_1} < \tau_{D_n}), \tag{4.2}$$

where $\tau_{D_n} = \inf\{t \geqslant 0: X(t) \notin D_n\}$ and $\sigma_{\bar{D}_1} = \inf\{t \geqslant 0: X(t) \in \bar{D}_1\}$. It is well known that if $u \in W^{1,2}(D_n - D_1)$, and $m > n$, then the function

$$v(x) = \begin{cases} u(x), & x \in D_n - D_1 \\ 0, & x \in D_m - D_n \end{cases}$$

belongs to $W^{1,2}(D_m - D_1)$. From this, we conclude that λ_n is non-increasing in n; thus, since $\lambda_n \geqslant 0$, it follows that $\lim_{n\to\infty} \lambda_n$ exists. We will prove the following theorem.

Theorem 4.1. *Let $L = \frac{1}{2}\nabla \cdot a\nabla + a\nabla Q \cdot \nabla$ satisfy Assumption $\widetilde{\mathrm{H}}_{\mathrm{loc}}$ on a domain $D \subseteq R^d$, $d \geqslant 2$. Let λ_n be as in (4.1). Then the diffusion corresponding to* L *is recurrent if $\lim_{n\to\infty} \lambda_n = 0$ and transient if $\lim_{n\to\infty} \lambda_n > 0$.*

An important comparison principle is an immediate consequence of Theorem 4.1.

Corollary 4.2. *Let $L = \frac{1}{2}\nabla \cdot a\nabla + a\nabla Q \cdot \nabla$ and $\hat{L} = \frac{1}{2}\nabla \cdot \hat{a}\nabla + \hat{a}\nabla\hat{Q} \cdot \nabla$ satisfy Assumption $\widetilde{\mathrm{H}}_{\mathrm{loc}}$ on a domain $D \subseteq R^d$, $d \geqslant 2$. Assume that $\hat{a}\exp(2\hat{Q}) \geqslant a\exp(2Q)$ (that is, $\hat{a}\exp(2\hat{Q}) - a\exp(2Q)$ is non-negative definite).*

(i) *If the diffusion corresponding to $\hat{L}$ is recurrent, then the diffusion corresponding to* L *is also recurrent.*
(ii) *If the diffusion corresponding to* L *is transient, then the diffusion corresponding to $\hat{L}$ is also transient.*

Example 4.3. If $L = \frac{1}{2}\nabla \cdot a\nabla + a\nabla Q \cdot \nabla$ satisfies Assumption $\widetilde{\mathrm{H}}_{\mathrm{loc}}$ on R^d, $d \geqslant 2$, and there exist constants $c_1, c_2 > 0$ such that $c_1 I \leqslant a\exp(2Q) \leqslant c_2 I$, then the diffusion corresponding to L is recurrent if $d = 2$ and transient if $d \geqslant 3$. This follows from Corollary 4.2 by comparison with

$$L_1 = \frac{c_1}{2}\Delta \text{ and } L_2 = \frac{c_2}{2}\Delta.$$

We emphasize that a result such as Example 4.3 does not hold in the non-symmetric case. In exercise 6.5, an example is given of two operators of the form $L_1 = \frac{1}{2}\nabla \cdot a_1\nabla + b \cdot \nabla$ and $L_2 = \frac{1}{2}\nabla \cdot a_2\nabla + b \cdot \nabla$

with $a_1 > a_2$ but such that the diffusion corresponding to L_1 is recurrent and the diffusion corresponding to L_2 is transient. On a different tack, let

$$L = \tfrac{1}{2} \sum_{i,j=1}^{d} a_{ij} \frac{\partial^2}{\partial x_i \partial x_j} \text{ in } R^d,\ d \geq 2,$$

where $c_1 I \leq a \leq c_2 I$, for constants $c_1, c_2 > 0$. Then one can obtain both transient and recurrent diffusions. To see this, let a be of the form

$$a(x) = a\left(\frac{x}{|x|}\right),$$

on $\{|x| > 1\}$ and refer to Theorem 3.1 and Remark 3 that follows it.

Proof of Theorem 4.1. By varying the Dirichlet form

$$\tfrac{1}{2}\int_{D_n - D_1} (\nabla u a \nabla u) \exp(2Q)\, dx$$

about the minimizer $u = u_n$, defined in (4.2), we obtain the variational equation

$$\int_{D_n - D_1} (\nabla u_n a \nabla q) \exp(2Q)\, dx = 0, \tag{4.3}$$

for all $q \in W^{1,2}(D_n - D_1)$ which satisfy $q = 0$ on $\partial D_n \cup \partial D_1$. For each k, extend the function u_k to $D - D_1$ by defining $u_k(x) = 0$ for $x \in D - D_k$. Then $u_k \in W^{1,2}_{\text{loc}}(D - D_1)$ and, for $k < n$, the function $q = u_k - u_n$ is admissible in (4.3). We obtain

$$2\lambda_n = \int_{D_n - D_1} (\nabla u_n a \nabla u_n) \exp(2Q)\, dx = \int_{D_n - D_1} (\nabla u_n a \nabla u_k) \exp(2Q)\, dx, \quad \text{for } k < n. \tag{4.4}$$

Using (4.4), we obtain for $k < n$,

$$\tfrac{1}{2}\int_{D_n - D_1} (\nabla u_n - \nabla u_k) a (\nabla u_n - \nabla u_k) \exp(2Q)\, dx = \lambda_k - \lambda_n. \tag{4.5}$$

Now construct the Hilbert space H corresponding to the norm

$$\|u\|_H = \int_D (\nabla u a \nabla u) \exp(2Q)\, dx + \int_D u^2 m\, dx,$$

where $m > 0$ and $\int_D m\, dx < \infty$. Define $u_\infty(x) = P_x(\sigma_{\bar{D}_1} < \infty)$. By (4.2), $u_\infty(x) = \lim_{n\to\infty} u_n(x)$. From this, (4.5), and the fact that $\lim_{n\to\infty} \lambda_n$

exists, it follows that $\{u_n\}_{n=1}^{\infty}$ is a Cauchy sequence in H converging to u_∞. Thus, in particular,

$$\lim_{n\to\infty}\lambda_n = \lim_{n\to\infty}\tfrac{1}{2}\int_{D-D_1}(\nabla u_n a \nabla u_n)\exp(2Q)\,\mathrm{d}x$$

$$= \tfrac{1}{2}\int_{D-D_1}(\nabla u_\infty a \nabla u_\infty)\exp(2Q)\,\mathrm{d}x. \tag{4.6}$$

A standard compactness argument shows that $u_\infty \in C^{2,\alpha}(D - \bar{D}_1)$. From Theorem 2.3.3 and the probabilistic definition of u_∞, we conclude that $u_\infty \in C^{2,\alpha}(D - \bar{D}_1) \cap C(D - D_1)$. Now if $\lim_{n\to\infty}\lambda_n = 0$, then from (4.6), u_∞ is constant on $D - D_1$; since $u_\infty = 1$ on ∂D_1, it follows that $u_\infty(x) = P_x(\sigma_{\bar{D}_1} < \infty) = 1$, for $x \in D - D_1$. By Theorem 2.8.2(ii), the diffusion is recurrent. If $\lim_{n\to\infty}\lambda_n > 0$, then, from (4.6), u_∞ is not constant on $D - D_1$. Thus $u_\infty(x) = P_x(\sigma_{\bar{D}_1} < \infty) < 1$ for some (and thus all) $x \in D - \bar{D}_1$ and, consequently, the diffusion is transient. □

Theorem 4.1 can be used to provide sufficiency conditions for recurrence and for transience in terms of integral tests. Let $L = \frac{1}{2}\nabla\cdot a\nabla + a\nabla Q\cdot\nabla$ on R^d, $d \geqslant 2$. Define

$$E_1(x) = \frac{(a(x)x, x)}{|x|^2}\exp(2Q(x)),$$

$$E_2(x) = \frac{|x|^2\exp(2Q(x))}{(a^{-1}(x)x, x)}$$

and

$$\bar{E}_1(r) = \int_{S^{d-1}} E_1(r\theta)\,\mathrm{d}\theta,$$

where $\mathrm{d}\theta$ is Lebesgue measure on S^{d-1}.

Theorem 4.4. *Let $L = \frac{1}{2}\nabla\cdot a\nabla + a\nabla Q\cdot\nabla$ satisfy Assumption $\widetilde{H}_{\mathrm{loc}}$ on R^d, $d \geqslant 2$.*

(i) *If $\int_1^\infty r^{1-d}\bar{E}_1^{-1}(r)\,\mathrm{d}r = \infty$, then the diffusion corresponding to L is recurrent.*
(ii) *If $\int_1^\infty r^{1-d}E_2^{-1}(r\theta)\,\mathrm{d}r < \infty$ on a subset of S^{d-1} with positive Lebesgue measure, then the diffusion corresponding to L is transient.*

Proof. Let $D_n = \{x \in R^d\colon |x| < n\}$.

(i) Let

$$\psi_n(x) = \frac{\int_{|x|}^{n} r^{1-d}\bar{E}_1^{-1}(r)\,\mathrm{d}r}{\int_1^n r^{1-d}\bar{E}_1^{-1}(r)\,\mathrm{d}r}, \text{ for } 1 \leqslant |x| \leqslant n.$$

Then

$$\nabla\psi_n(x) = \frac{-|x|^{1-d}\bar{E}_1^{-1}(|x|)\cdot\dfrac{x}{|x|}}{\int_1^n r^{1-d}\bar{E}_1^{-1}(r)\,\mathrm{d}r}$$

and $\int_{D_n-D_1}(\nabla\psi_n a\nabla\psi_n)\exp(2Q)\,\mathrm{d}x = (\int_1^n r^{1-d}\bar{E}_1^{-1}(r)\,\mathrm{d}r)^{-1}$. Since ψ_n is an admissible test function in (4.1), we have $\lambda_n \leqslant (2\int_1^n r^{1-d}\bar{E}_1^{-1}(r)\,\mathrm{d}r)^{-1}$. Thus, by assumption, $\lim_{n\to\infty}\lambda_n = 0$ and, by Theorem 4.1, the diffusion is recurrent.

(ii) We will show that for any $u \in W^{1,2}(D_n - D_1)$ satisfying $u = 1$ on ∂D_1 and $u = 0$ on ∂D_n, one has the inequality

$$\int_{D_n-D_1}(\nabla u a\nabla u)\exp(2Q)\,\mathrm{d}x \geqslant \int_{S_1^{d-1}} \frac{\mathrm{d}\theta}{\int_1^n r^{1-d}E_2^{-1}(r\theta)\,\mathrm{d}r}. \tag{4.7}$$

From this, it follows that for all n,

$$\begin{aligned}\lambda_n &= \inf_{\substack{u\in W^{1,2}(D_n-D_1)\\ u=1 \text{ on } \partial D_1\\ u=0 \text{ on } \partial D_n}} \tfrac{1}{2}\int_{D_n-D_1}(\nabla u a\nabla u)\exp(2Q)\,\mathrm{d}x \\ &\geqslant \tfrac{1}{2}\int_{S^{d-1}} \frac{\mathrm{d}\theta}{\int_1^n r^{1-d}E_2^{-1}(r\theta)\,\mathrm{d}r}.\end{aligned}$$

By assumption, then, $\lim_{n\to\infty}\lambda_n > 0$ and, by Theorem 4.1, the diffusion is transient. It remains to prove (4.7).

By the Schwarz inequality, $(x, v)^2 \leqslant (a^{-1}(x)x, x)\cdot(a(x)v, v)$, for x, $v \in R^d$. Thus

$$\begin{aligned}\tfrac{1}{2}\int_{D_n-D_1}(\nabla u a\nabla u)\exp(2Q)\,\mathrm{d}x &\geqslant \tfrac{1}{2}\int_{D_n-D_1}\frac{(x,\nabla u)^2}{(a^{-1}(x)x, x)}\exp(2Q)\,\mathrm{d}x \\ = \int_{D_n-D_1}E_2(x)\left(\frac{\partial u}{\partial r}(x)\right)^2\mathrm{d}x &= \int_{S^{d-1}}\mathrm{d}\theta\int_1^n r^{d-1}E_2(r\theta)\left(\frac{\partial u}{\partial r}(r\theta)\right)^2\mathrm{d}r.\end{aligned} \tag{4.8}$$

Applying the Schwarz inequality again, we obtain for each $\theta \in S^{d-1}$,

$$\left[\int_1^n r^{d-1} E_2(r\theta)\left(\frac{\partial u}{\partial r}(r\theta)\right)^2 \mathrm{d}r\right] \cdot \left[\int_1^n r^{1-d} E_2^{-1}(r\theta)\,\mathrm{d}r\right]$$

$$\geqslant \left[\int_1^n \frac{\partial u}{\partial r}(r\theta)\,\mathrm{d}r\right]^2 = (u(n\theta) - u(\theta))^2 = 1, \quad (4.9)$$

by the Assumption on u. Now (4.7) follows from (4.8) and (4.9). □

Remark. It can be shown that Theorem 2.1(i), when restricted to the case of symmetric operators, is contained in Theorem 4.4(i). (See notes at end of chapter.) Below we give two examples which are covered by Theorem 4.4 but not by Theorem 2.1.

Example 4.5. Let

$$L = (3 + \sin x_2)\frac{\partial^2}{\partial x_1^2} + \frac{\partial^2}{\partial x_2^2} \text{ on } R^2.$$

Then $L = \frac{1}{2}\nabla \cdot a\nabla$, where

$$a = \begin{pmatrix} 6 + 2\sin x_2 & 0 \\ 0 & 2 \end{pmatrix}$$

and recurrence follows from Example 4.3. (Alternatively, one can check that $\int_1^\infty r^{1-d}\bar{E}_1^{-1}(r)\,\mathrm{d}r = \infty$; thus by Theorem 4.4(i), the diffusion is recurrent.) On the other hand, one can check that $\int_1^\infty \exp(-\bar{I}(r))\,\mathrm{d}r < \infty$, so Theorem 2.1(i) fails.

Example 4.6. Let

$$L = \tfrac{1}{2}a_1(x_2)\frac{\partial^2}{\partial x_1^2} + \tfrac{1}{2}a_2(x_1)\frac{\partial^2}{\partial x_2^2} \text{ on } R^2,$$

where $a_1(x_2) \geqslant (\log|x_2|)^\alpha$, for large $|x_2|$ and $a_2(x_1) \geqslant (\log|x_1|)^\alpha$, for large $|x_1|$, and $\alpha > 1$. Then for $|x_1|$ and $|x_2|$ large,

$$E_2^{-1}(x) \leqslant \left(\frac{x_1^2}{(\log|x_2|)^\alpha} + \frac{x_2^2}{(\log|x_1|)^\alpha}\right)\frac{1}{|x|^2}.$$

It is easy to see that if

$$\theta \notin \left\{0, \frac{\pi}{2}, \pi, \frac{3\pi}{2}\right\},$$

then $\int_1^\infty r^{1-d}E_2^{-1}(r\theta)\,dr < \infty$; thus by Theorem 4.4(ii), the diffusion is transient. On the other hand, one can check that $\int_1^\infty \exp(-\underline{I}(r))\,dr = \infty$, so Theorem 2.1(ii) fails.

6.5 A generalized Dirichlet principle for non-symmetric operators

In the previous section, we exploited the classical Dirichlet principle to give a variational characterization of transience and recurrence for diffusions corresponding to symmetric operators. In this section, we develop a generalized Dirichlet principle for general second-order elliptic operators. We will use this in the next section to give a variational characterization of transience and recurrence in the general case.

Let $L = \frac{1}{2}\nabla \cdot a\nabla + b \cdot \nabla$ satisfy Assumption $\tilde{\mathrm{H}}$ on a bounded domain $D \subset R^d$ with a $C^{2,\alpha}$-boundary. Before stating the generalized Dirichlet principle, we give a bit of motivation. If L is symmetric, that is, $b = a\nabla Q$, then the Rayliegh–Ritz formula states that the principal eigenvalue λ_0 of L on D satisfies

$$-\lambda_0 = \inf_{\substack{\phi \in W^{1,2}(D) \\ \phi = 0 \text{ on } \partial D \\ \int_D \phi^2 \exp(2Q)\,dx = 1}} \frac{1}{2}\int_D (\nabla\phi a \nabla\phi) \exp(2Q)\,dx.$$

(See the second remark following Theorem 3.7.1.) Comparing the Rayliegh–Ritz formula with the classical Dirichlet principle which appears at the beginning of Section 4, one sees that the same functional is varied in both cases. The difference in the two variational principles appears solely in the particular subdomains of $W^{1,2}(D)$ over which the variations are taken. Now Theorem 3.7.1 constituted a mini-max variational formula for the principal eigenvalue which generalized the Rayleigh–Ritz formula to non-symmetric operators. In the light of the above discussion, in order to obtain a generalized Dirichlet principle, it is natural to try taking the functional appearing in the generalized Rayliegh–Ritz formula and searching for an appropriate subdomain of $W^{1,2}(D)$ over which to vary it.

Theorem 5.1. *Let $L = \frac{1}{2}\nabla \cdot a\nabla + b \cdot \nabla$ satisfy Assumption $\tilde{H}$ on a bounded domain $D \subset R^d$, $d \geq 2$, with a $C^{2,\alpha}$-boundary. Let $\phi \in C^{2,\alpha}(\bar{D})$ satisfy $\phi \geq 0$ on ∂D and $\phi \not\equiv 0$ on ∂D, and let $u_0 \in C^{2,\alpha}(\bar{D})$ denote the unique solution to $Lu_0 = 0$ in D and $u_0 = \phi$ on ∂D. Choose an arbitrary function $f \in C^{2,\alpha}(\bar{D})$ and let $\tilde{u}_f \in C^{2,\alpha}(\bar{D})$*

denote the unique solution to $\tilde{L}\tilde{u}_f = 0$ in D and $\tilde{u}_f = \phi \exp(2f)$ on ∂D. Then the mini-max variational formula

$$\mu^{(f)} = \inf_{\substack{g=\phi\exp(f) \text{ on } \partial D \\ g\in W^{1,2}(D) \\ (\operatorname{dist}(x,\partial D\cap\{\phi=0\}))^{-1}g(x)\in L^\infty(D)}} \sup_{\substack{h=f \text{ on } \partial D\cap\{\phi>0\} \\ h\in W^{1,2}(D,g^2\,dx)}} \left[\frac{1}{2}\int_D\left(\frac{\nabla g}{g} - a^{-1}b\right)a\left(\frac{\nabla g}{g} - a^{-1}b\right)g^2\,dx \right.$$

$$\left. - \frac{1}{2}\int_D(\nabla h - a^{-1}b)a(\nabla h - a^{-1}b)g^2\,dx\right] \tag{5.1}$$

is attained uniquely at

$$g = (u_0\tilde{u}_f)^{1/2} \text{ and } h = \tfrac{1}{2}\log\frac{\tilde{u}_f}{u_0}.$$

Thus

$$\mu^{(f)} = \frac{1}{2}\int_D\left(\frac{1}{2}\frac{\nabla\tilde{u}_f}{\tilde{u}_f} + \frac{1}{2}\frac{\nabla u_0}{u_0} - a^{-1}b\right)a\left(\frac{1}{2}\frac{\nabla\tilde{u}_f}{\tilde{u}_f} + \frac{1}{2}\frac{\nabla u_0}{u_0} - a^{-1}b\right)u_0\tilde{u}_f\,dx$$

$$- \frac{1}{2}\int_D\left(\frac{1}{2}\frac{\nabla\tilde{u}_f}{\tilde{u}_f} - \frac{1}{2}\frac{\nabla u_0}{u_0} - a^{-1}b\right)a\left(\frac{1}{2}\frac{\nabla\tilde{u}_f}{\tilde{u}_f} - \frac{1}{2}\frac{\nabla u_0}{u_0} - a^{-1}b\right)u_0\tilde{u}_f\,dx. \tag{5.2}$$

In the special case $\{\phi = 0 \text{ or } \phi = 1\} \cap \partial D = \partial D$, then

$$\mu^{(f)} = \frac{1}{2}\int_D \frac{(\nabla u_0 a \nabla u_0)}{u_0^2} u_0\tilde{u}_f\,dx. \tag{5.3}$$

(5.3) *also holds if $b = a\nabla Q$ where $Q = f$ on $\partial D \cap \{\phi > 0\}$.*

Remark 1. The positivity of u_0 and $\tilde{u}_f$ follows from the generalized maximum principle in Theorem 3.6.5 and the fact that, since $V \equiv 0$, $\lambda_0 < 0$. The generalized Dirichlet principle (as well as the classical one) can be extended easily to operators $L + V$, assuming that the principal eigenvalue remains negative. If L is replaced by $L + V$, then one simply adds the term $-\int_D V^2 g^2\,dx$ to the expression appearing on the righthand side of (5.1).

Remark 2. Theorem 5.1 may be thought of as a family, indexed by f, of generalized Dirichlet principles for the solution u_0. In the symmetric case, that is, the case where $b = a\nabla Q$, for some $Q \in C^{2,\alpha}(\bar{D})$, the generalized Dirichlet principle will reduce to the classical Dirichlet principle if and only if $Q = f$ on $\partial D \cap \{\phi > 0\}$. It doesn't seem possible to give one generalized Dirichlet principle which reduces to the classical one simultaneously for every symmetric operator. To prove this reduction, note that the term

$$\sup_{\substack{h=f \text{ on } \partial D\cap\{\phi>0\}\\ h\in W^{1,2}(D,g^2\mathrm{d}x)}}\left[-\tfrac{1}{2}\int_D(\nabla h-a^{-1}b)a(\nabla h-a^{-1}b)g^2\,\mathrm{d}x\right]$$

appearing in (5.1) vanishes if and only if $b=a\nabla Q$ and $Q=f$ on $\partial D\cap\{\phi>0\}$. Thus, if $b=a\nabla Q$ and $Q=f$ on $\partial D\cap\{\phi>0\}$, then we obtain

$$\mu^{(f)}=\inf_{\substack{g=\phi\exp(f)\text{ on }\partial D\\ g\in W^{1,2}(D)\\ (\mathrm{dist}(x,\partial D\cap\{\phi=0\}))^{-1}g(x)\in L^\infty(D)}}\tfrac{1}{2}\int_D\left(\frac{\nabla g}{g}-\nabla Q\right)a\left(\frac{\nabla g}{g}-\nabla Q\right)g^2\,\mathrm{d}x.$$

Substituting $u=g\exp(-Q)$ gives

$$\mu^{(f)}=\inf_{\substack{u=\phi\text{ on }\partial D\\ u\in W^{1,2}(D,\exp(2Q))\\ (\mathrm{dist}(x,\partial D\cap\{\phi=0\}))^{-1}u(x)\in L^\infty(D)}}\tfrac{1}{2}\int_D(\nabla u a\nabla u)\exp(2Q)\,\mathrm{d}x. \quad (5.4)$$

The classical Dirichlet principle states that

$$\inf_{\substack{u=\phi\text{ on }\partial D\\ u\in W^{1,2}(D,\exp(2Q)\,\mathrm{d}x)}}\tfrac{1}{2}\int_D(\nabla u a\nabla u)\exp(2Q)\,\mathrm{d}x$$

is attained uniquely at the solution u_0 of $Lu_0=0$ in D and $u_0=\phi$ on ∂D. By the Hopf maximum principle (Theorem 3.2.5),

$$(\mathrm{dist}\,(x,\partial D\cap\{\phi=0\}))^{-1}u_0(x)\in L^\infty(D).$$

Thus, the condition $(\mathrm{dist}\,(x,\partial D\cap\{\phi=0\}))^{-1}u(x)\in L^\infty(D)$ may be deleted from the righthand side of (5.4).

Remark 3. Since ∂D may not be connected (for example, if D is an annulus), the special case $\{\phi=0 \text{ or } \phi=1\}\cap\partial D=\partial D$ does not only include the trivial case in which $u_0\equiv 0$ or $u_0\equiv 1$. Indeed, analogous to the application of the classical Dirichlet principle in the previous section, in the next section we will apply the generalized Dirichlet principle to annuli and the special form (5.3) will be essential. As an aside, we note that the non-negativity of $\mu^{(f)}$ in the case where $\{\phi=0$ or $\phi=1\}\cap\partial D=\partial D$, which follows from (5.3), can be seen by inspection from (5.1) since, in this case (and only in this case), the function $h=\log g$ is admissible.

Remark 4. The classical Dirichlet principle allows the boundary value ϕ to change sign. For the generalized Dirichlet principle, we require that ϕ be non-negative and not identically zero on ∂D. Indeed, if ϕ were to change sign on ∂D, then u_0 and $\tilde{u}_f$ would change sign in D and $g = (u_0\tilde{u}_f)^{1/2}$ and

$$h = \tfrac{1}{2}\log\frac{\tilde{u}_f}{u_0}$$

would no longer be defined.

Remark 5. The condition $(\operatorname{dist}(x, \partial D \cap \{\phi = 0\}))^{-1}g(x) \in L^\infty(D)$ is probably unnecessary; however, our proof requires this technical condition.

Proof of Theorem 5.1. For $g \in W^{1,2}(D)$ satisfying $g = \phi\exp(f)$ on ∂D, let

$$H_g(h) = \int_D (\nabla h - a^{-1}b)a(\nabla h - a^{-1}b)g^2\,\mathrm{d}x.$$

First we show that

$$\inf_{\substack{h=f \text{ on } \partial D\cap\{\phi>0\}\\ h\in W^{1,2}(D,g^2\,\mathrm{d}x)}} H_g(h)$$

is attained at some $h = h_g \in W^{1,2}(D, g^2\,\mathrm{d}x)$ and that, if $g > 0$ in D, then h_g is unique. By the Schwarz inequality, it is easy to obtain $H_g(h) \geq c_1\int_D|\nabla h|^2g^2\,\mathrm{d}x - c_2$, for positive constants c_1 and c_2. Thus, if $\{h_n\}_{n=1}^\infty$ is a minimizing sequence, then $\int_D|\nabla h_n|^2g^2\,\mathrm{d}x$ is bounded independent of n. This, coupled with the fact that $h_n = f$ on $\partial D \cap \{\phi > 0\}$, shows that $\{h_n\}_{n=1}^\infty$ is norm-bounded in $W^{1,2}(D, g^2\,\mathrm{d}x)$; thus $\{h_n\}_{n=1}^\infty$ is weakly relatively compact in $W^{1,2}(D, g^2\,\mathrm{d}x)$. Now $H_g(h)$ is lower semicontinuous under weak convergence since the norm is lower semicontinuous under weak convergence. Thus, in fact, any limit $\hat{h}$ of a subsequence of $\{h_n\}_{n=1}^\infty$ must be a minimizer. Varying $H_g(h)$ at $\hat{h}$ gives

$$\int_D (\nabla\hat{h} - a^{-1}b)a\nabla q g^2\,\mathrm{d}x = 0, \tag{5.5}$$

for all $q \in W^{1,2}(D, g^2\,\mathrm{d}x)$ satisfying $q = 0$ on $\partial D \cap \{\phi > 0\}$. Now if $\tilde{h}$ is also a minimizer, then (5.5) also holds with $\tilde{h}$ in place of $\hat{h}$. Subtracting gives $\int_D(\nabla\hat{h} - \nabla\tilde{h})a\nabla q g^2\,\mathrm{d}x = 0$. Substituting $q = \hat{h} - \tilde{h}$ shows that $\nabla\hat{h} - \nabla\tilde{h} = 0$ a.e. $[g^2\,\mathrm{d}x]$. Since $\hat{h} = \tilde{h} = f$ on $\partial D \cap \{\phi > 0\}$, it is clear that if $g > 0$ on D, then $\hat{h} = \tilde{h}$ a.e. $[g^2\,\mathrm{d}x]$.

Let $g_0 \equiv (u_0\tilde{u}_f)^{1/2}$. By the Hopf maximum principle (Theorem 3.2.5), $g_0 \in W^{1,2}(D)$ and $(\operatorname{dist}(x, \partial D \cap \{\phi = 0\}))^{-1}g_0(x) \in L^\infty(D)$. Let $h_0 \equiv \frac{1}{2}\log(\tilde{u}_f/u_0)$. We claim that $h_0 = h_{g_0}$. Note that h_0 satisfies the appropriate boundary conditions since $u_0 = \phi$ and $\tilde{u}_f = \phi\exp(2f)$ on ∂D. It is also easy to check that $h_0 \in W^{1,2}(D, g_0^2\,\mathrm{d}x)$. We must check that (5.5) is satisfied with $\hat{h}$ replaced by h_0 and g replaced by g_0. From the fact that $Lu_0 = 0$ and $\tilde{L}\tilde{u}_f = 0$, one can check that h_0 satisfies

$$\nabla\cdot a\nabla h_0 + 2a\frac{\nabla g_0}{g_0}\nabla h_0 = \nabla\cdot b + 2\frac{\nabla g_0}{g_0}b. \tag{5.6}$$

Integrating the lefthand side of (5.5) by parts and using (5.6), one sees that (5.5) is indeed satisfied with $\hat{h} = h_0$ and $g = g_0$.

For the rest of the proof, it will be convenient to change the notation a bit. For $g \in W^{1,2}(D)$, let $\gamma = g^2$ and define for such γ and for $h \in W^{1,2}(D, \gamma\,\mathrm{d}x)$,

$$\psi(h, \gamma) = \tfrac{1}{2}\int_D (\nabla h - a^{-1}b)a(\nabla h - a^{-1}b)\gamma\,\mathrm{d}x.$$

Also define

$$J(\gamma) = \psi(\tfrac{1}{2}\log\gamma, \gamma) - \inf_{\substack{h=f \text{ on } \partial D\cap\{\phi>0\}\\ h\in W^{1,2}(D,\gamma\,\mathrm{d}x)}} \psi(h, \gamma).$$

Then the righthand side of (5.1) may be written as

$$\inf_{\substack{\gamma=\phi^2\exp(2f) \text{ on } \partial D\\ \gamma^{1/2}\in W^{1,2}(D)\\ (\operatorname{dist}(x,\partial D\cap\{\phi=0\}))^{-1}\gamma^{1/2}(x)\in L^\infty(D)}} J(\gamma).$$

To show that (a) the mini-max in (5.1) is attained at some pair (g, h_g); (b) the pair is unique; and (c) the pair is given by (g_0, h_0), it now suffices to show that $J(\gamma_0) < J(\gamma_1)$, for all $\gamma_1 \neq \gamma_0 \equiv g_0^2$ that satisfy $\gamma_1 = \phi^2\exp(2f)$ on ∂D, $\gamma_1^{1/2} \in W^{1,2}(D)$ and $(\operatorname{dist}(x, \partial D \cap \{\phi = 0\}))^{-1}\gamma_1^{1/2}(x) \in L^\infty(D)$. To do this, we will show that $M(p) \equiv J(\gamma_p)$ is strictly convex on $[0, 1]$ and satisfies $M'(0) = 0$, where $\gamma_p = p\gamma_1 + (1 - p)\gamma_0$.

Now

$$M(p) = J(\gamma_p) = \psi(\tfrac{1}{2}\log\gamma_p, \gamma_p) + \Big(-\inf_{\substack{h=f \text{ on } \partial D\cap\{\phi>0\}\\ h\in W^{1,2}(D,\gamma_p\,\mathrm{d}x)}} \psi(h, \gamma_p)\Big),$$

and it is easy to see that the second term on the righthand side is convex in p (the p appearing in the domain over which the infimum is calculated causes no problem). Thus, to show strict convexity, we need only show that $(d^2/dp^2)(\psi(\frac{1}{2}\log\gamma_p, \gamma_p)) > 0$, for $p \in [0, 1]$. We have

$$\psi(\tfrac{1}{2}\log\gamma_p, \gamma_p) = \tfrac{1}{2}\int_D \frac{(\frac{1}{2}\nabla\gamma_p - a^{-1}b\gamma_p)a(\frac{1}{2}\nabla\gamma_p - a^{-1}b\gamma_p)}{\gamma_p}\,dx$$

and

$$\frac{d^2}{dp^2}(\psi(\tfrac{1}{2}\log\gamma_p, \gamma_p))$$
$$= \int_D \left(\frac{\frac{1}{2}\nabla(\gamma_1 - \gamma_0) - a^{-1}b(\gamma_1 - \gamma_0)}{\gamma_p^{1/2}} - \frac{(\frac{1}{2}\nabla\gamma_p - a^{-1}b\gamma_p)(\gamma_1 - \gamma_0)}{\gamma_p^{3/2}}\right)$$
$$\cdot a\left(\frac{\frac{1}{2}\nabla(\gamma_1 - \gamma_0) - a^{-1}b(\gamma_1 - \gamma_0)}{\gamma_p^{1/2}} - \frac{(\frac{1}{2}\nabla\gamma_p - a^{-1}b\gamma_p)(\gamma_1 - \gamma_0)}{\gamma_p^{3/2}}\right)dx$$
$$\geqslant 0.$$

Equality occurs if and only if

$$\frac{\frac{1}{2}\nabla(\gamma_1 - \gamma_0) - a^{-1}b(\gamma_1 - \gamma_0)}{\gamma_p^{1/2}} = \frac{(\frac{1}{2}\nabla\gamma_p - a^{-1}b\gamma_p)(\gamma_1 - \gamma_0)}{\gamma_p^{3/2}},$$

that is, if and only if $(\nabla(\gamma_1 - \gamma_0))/(\gamma_1 - \gamma_0) = \nabla\gamma_p/\gamma_p$. For this to occur for even one fixed p requires that

$$\frac{\nabla\gamma_1}{\gamma_1} = \frac{\nabla\gamma_0}{\gamma_0},$$

that is, that $\nabla(\log\gamma_1) = \nabla(\log\gamma_0)$. Since $\gamma_1 = \gamma_0 = \phi^2\exp(2f)$ on ∂D, it would then follow that $\gamma_1 = \gamma_0$, which contradicts the assumption that $\gamma_1 \neq \gamma_0$. This proves the strict convexity.

We now prove that $M'(0) = 0$. Let $h_p \in W^{1,2}(D, \gamma_p\,dx)$ be a function at which

$$\inf_{\substack{h=f \text{ on } \partial D\cap\{\phi>0\} \\ h\in W^{1,2}(D,\gamma_p)}} \int_D (\nabla h - a^{-1}b)a(\nabla h - a^{-1}b)\gamma_p\,dx$$

is attained. (By what we have proved, h_p is unique if $0 \leqslant p < 1$.) We will show below that

$$M'(0) = \int_D \left[\frac{(\frac{1}{2}\nabla(\gamma_1 - \gamma_0) - a^{-1}b(\gamma_1 - \gamma_0))a(\frac{1}{2}\nabla\gamma_0 - a^{-1}b\gamma_0)}{\gamma_0}\right.$$

$$- \frac{(\frac{1}{2}\nabla\gamma_0 - a^{-1}b\gamma_0)a(\frac{1}{2}\nabla\gamma_0 - a^{-1}b\gamma_0)(\gamma_1 - \gamma_0)}{2\gamma_0^2}\Bigg]\,\mathrm{d}x$$

$$-\tfrac{1}{2}\int_D (\nabla h_0 - a^{-1}b)a(\nabla h_0 - a^{-1}b)(\gamma_1 - \gamma_0)\,\mathrm{d}x. \quad (5.7)$$

From the fact that $u_0 = g_0 \exp(-h_0)$ and $\tilde{u}_f = g_0 \exp(h_0)$, one can verify that $\gamma_0 = g_0^2$ satisfies

$$\tfrac{1}{2}\nabla\cdot a\nabla\gamma_0 - \tfrac{1}{4}\frac{\nabla\gamma_0 a\nabla\gamma_0}{\gamma_0} + \gamma_0(\nabla h_0 a\nabla h_0 - \nabla\cdot b - 2\nabla h_0 b) = 0. \quad (5.8)$$

Now (5.7) arises as the variation of $M(p)$ at $p = 0$ in the direction $\gamma_1 - \gamma_0$. If we set $q = \gamma_1 - \gamma_0$ in (5.7), and then integrate by parts and use (5.8) and the fact that $q = 0$ on ∂D, we obtain $M'(0) = 0$.

It remains to show (5.7). Recalling that $\gamma_0 = g_0^2 = u_0\tilde{u}_f$, applying the Hopf maximum principle (Theorem 3.2.5) to u_0 and $\tilde{u}_f$, and using the assumption that $(\operatorname{dist}(x, \partial D \cap \{\phi = 0\}))^{-1}\gamma_1^{1/2}(x) \in L^\infty(D)$, it follows that there exists a constant $c > 0$ such that

$$\gamma_1 \leqslant c\gamma_0. \quad (5.9)$$

Since $\gamma_p = (1-p)\gamma_0 + p\gamma_1$, and since

$$\int_D (\nabla h_p - a^{-1}b)a(\nabla h_p - a^{-1}b)\gamma_p\,\mathrm{d}x =$$

$$\inf_{\substack{h=f \text{ on } \partial D\cap\{\phi>0\}\\ h\in W^{1,2}(D,\gamma_p\,\mathrm{d}x)}} \int_D (\nabla h - a^{-1}b)a(\nabla h - a^{-1}b)\gamma_p\,\mathrm{d}x,$$

an application of the Schwarz inequality along with (5.9) shows that

$$\sup_{p\in[0,1]}\int_D |\nabla h_p|^2\gamma_i\,\mathrm{d}x < \infty, \quad \text{for } i = 0, 1. \quad (5.10)$$

It also follows from (5.9) that $W^{1,2}(D, \gamma_p\,\mathrm{d}x)$ does not depend on p for $p \in [0, 1)$. From this and (5.5) we have

$$\int_D (\nabla h_p - a^{-1}b)a\nabla q\gamma_p\,\mathrm{d}x = 0, \text{ for } 0 \leqslant p < 1$$

$$\text{and all } q \in W^{1,2}(D, \gamma_0\,\mathrm{d}x)$$

$$\text{satisfying } q = 0 \text{ on } \partial D \cap \{\phi > 0\}. \quad (5.11)$$

Considering (5.11) with $p > 0$ and with $p = 0$, and subtracting, we obtain

$$\int_D (\nabla h_p - \nabla h_0) a \nabla q \gamma_0 \, dx + p \int_D (\nabla h_p - a^{-1}b) a \nabla q (\gamma_1 - \gamma_0) \, dx = 0. \tag{5.12}$$

Substituting $q = h_p - h_0$ in (5.12), and using (5.9) and (5.10), we have

$$\lim_{p \to 0} \int_D |\nabla h_p - \nabla h_0|^2 \gamma_i \, dx = 0, \quad i = 1, 2. \tag{5.13}$$

Note that

$$M(p) = \tfrac{1}{2} \int_D \frac{(\tfrac{1}{2}\nabla \gamma_p - a^{-1}b\gamma_p) a (\tfrac{1}{2}\nabla \gamma_p - a^{-1}b\gamma_p)}{\gamma_p} \, dx$$
$$- \tfrac{1}{2} \int_D (\nabla h_p - a^{-1}b) a (\nabla h_p - a^{-1}b) \gamma_p \, dx.$$

Thus, to show (5.7), it suffices to show that

$$\lim_{p \to 0} \frac{1}{p} \left[\int_D (\nabla h_p - a^{-1}b) a (\nabla h_p - a^{-1}b) \gamma_p \, dx \right.$$
$$\left. - \int_D (\nabla h_0 - a^{-1}b) a (\nabla h_0 - a^{-1}b) \gamma_0 \, dx \right]$$
$$= \int_D (\nabla h_0 - a^{-1}b) a (\nabla h_0 - a^{-1}b)(\gamma_1 - \gamma_0) \, dx. \tag{5.14}$$

We have

$$\frac{1}{p} \left[\int_D (\nabla h_p - a^{-1}b) a (\nabla h_p - a^{-1}b) \gamma_p \, dx \right.$$
$$\left. - \int_D (\nabla h_0 - a^{-1}b) a (\nabla h_0 - a^{-1}b) \gamma_0 \, dx \right]$$
$$= \frac{1}{p} \int_D [(\nabla h_p - a^{-1}b) a (\nabla h_p - a^{-1}b)$$
$$- (\nabla h_0 - a^{-1}b) a (\nabla h_0 - a^{-1}b)] \gamma_0 \, dx$$
$$+ \int_D (\nabla h_p - a^{-1}b) a (\nabla h_p - a^{-1}b)(\gamma_1 - \gamma_0) \, dx$$
$$= \frac{1}{p} \int_D ((\nabla h_p - a^{-1}b) + (\nabla h_0 - a^{-1}b))$$
$$\cdot \, a((\nabla h_p - a^{-1}b) - (\nabla h_0 - a^{-1}b)) \gamma_0 \, dx$$
$$+ \int_D (\nabla h_p - a^{-1}b) a (\nabla h_p - a^{-1}b)(\gamma_1 - \gamma_0) \, dx$$

$$= \int_D ((\nabla h_p - a^{-1}b) + (\nabla h_0 - a^{-1}b))a\left(\frac{\nabla h_p - \nabla h_0}{p}\right)\gamma_0 \, dx$$
$$+ \int_D (\nabla h_p - a^{-1}b)a(\nabla h_p - a^{-1}b)(\gamma_1 - \gamma_0) \, dx. \qquad (5.15)$$

Using (5.11) with

$$q = \frac{h_p - h_0}{p},$$

we have

$$\int_D ((\nabla h_p - a^{-1}b) + (\nabla h_0 - a^{-1}b))a\left(\frac{\nabla h_p - \nabla h_0}{p}\right)\gamma_0 \, dx$$
$$= \int_D (\nabla h_p - a^{-1}b)a\left(\frac{\nabla h_p - \nabla h_0}{p}\right)\gamma_0 \, dx$$
$$= \int_D (\nabla h_p - a^{-1}b)a\left(\frac{\nabla h_p - \nabla h_0}{p}\right)(\gamma_0 - \gamma_p) \, dx$$
$$= \int_D (\nabla h_p - a^{-1}b)a(\nabla h_p - \nabla h_0)(\gamma_0 - \gamma_1) \, dx. \qquad (5.16)$$

By (5.10), (5.13) and the Schwarz inequality, it follows that

$$\lim_{p \to 0} \int_D (\nabla h_p - a^{-1}b)a(\nabla h_p - h_0)(\gamma_0 - \gamma_1) \, dx = 0. \qquad (5.17)$$

Now (5.14) follows from (5.15)–(5.17).

Now we derive (5.2) and (5.3). We obtain (5.2) simply by substituting $g_0 = (u_0\tilde{u}_f)^{1/2}$ and $h_0 = \frac{1}{2}\log(\tilde{u}_f/u_0)$ in (5.1). For the case $a^{-1}b = \nabla Q$ and $Q = f$ on $\partial D \cap \{\phi > 0\}$, (5.3) follows from (5.2) and the fact that $\tilde{u}_f = u_0 \exp(2Q)$ or, alternatively, from Remark 1 following the theorem. We now show that (5.3) holds in the case $\{\phi = 0$ or $\phi = 1\} \cap \partial D = \partial D$. Recalling that $\gamma_0 = g_0^2$, we have from (5.11) that

$$\int_D (\nabla h_0 - a^{-1}b)a\nabla q g_0^2 \, dx = 0,$$

for all $q \in W^{1,2}(D, g_0^2 \, dx)$ satisfying $q = 0$ on $\partial D \cap \{\phi > 0\}$. Because of the condition on ϕ, we may choose $q = \log g_0 - h_0$ to obtain

$$\int (\nabla h_0 - a^{-1}b)a\left(\frac{\nabla g_0}{g_0} - \nabla h_0\right)g_0^2 \, dx = 0. \qquad (5.18)$$

From (5.18) we have

$$\begin{aligned}\tfrac{1}{2}\int_D (\nabla h_0 - a^{-1}b)a(\nabla h_0 - a^{-1}b)g_0^2\,dx = {} & \int_D g_0 \nabla g_0 a \nabla h_0\,dx \\ & - \int_D g_0 b \nabla g_0\,dx \\ & - \tfrac{1}{2}\int_D (\nabla h_0 a \nabla h_0) g_0^2\,dx \\ & + \tfrac{1}{2}\int_D (b a^{-1} b) g_0^2\,dx.\end{aligned}$$

Substituting this in (5.1), we obtain

$$\begin{aligned}&\tfrac{1}{2}\int_D \left(\frac{\nabla g_0}{g_0} - a^{-1}b\right)a\left(\frac{\nabla g_0}{g_0} - a^{-1}b\right)g_0^2\,dx - \int_D g_0 \nabla g_0 a \nabla h_0\,dx \\ &\quad + \int_D g_0 b \nabla g_0\,dx + \tfrac{1}{2}\int_D (\nabla h_0 a \nabla h_0) g_0^2\,dx - \tfrac{1}{2}\int_D (b a^{-1} b) g_0^2\,dx \\ &= \tfrac{1}{2}\int_D (\nabla g_0 - g_0 \nabla h_0) a (\nabla g_0 - g_0 \nabla h_0)\,dx = \tfrac{1}{2}\int_D \left(\frac{(\nabla u_0 a \nabla u_0)}{u_0^2}\right) u_0 \tilde{u}_f\,dx,\end{aligned}$$

since $g_0 = (u_0 \tilde{u}_f)^{1/2}$ and $h_0 = \frac{1}{2}\log(\tilde{u}_f/u_0)$. □

6.6 Transience and recurrence for diffusions via the generalized Dirichlet principle: a mini-max approach

In this section we utilize the generalized Dirichlet principle of the previous section to prove the analog of Theorem 4.1 for diffusions corresponding to general second-order elliptic operators. Let $L = \frac{1}{2}\nabla \cdot a \nabla + b \cdot \nabla$ satisfy Assumption $\widetilde{H}_{\text{loc}}$ on a domain $D \subseteq R^d$ and let $\{D_n\}_{n=1}^{\infty}$ be a sequence of domains with $C^{2,\alpha}$-boundaries satisfying $D_n \subset\subset D_{n-1}$ and $\bigcup_{n=1}^{\infty} D_n = D$. Also, let D_1 be simply connected so that $D_n - \bar{D}_1$ is a domain. Choose an arbitrary function $f \in C^{2,\alpha}(D)$ and define, analogous to λ_n in (4.2),

$$\begin{aligned}\mu_n^{(f)} = \inf_{\substack{g \in W^{1,2}(D_n - D_1) \\ g = e^f \text{ on } \partial D_1, g = 0 \text{ on } \partial D_n \\ (\text{dist}(x, \partial D_n))^{-1} g(x) \in L^\infty(D_n - D_1)}} \sup_{\substack{h \in W^{1,2}(D_n - D_1, g^2 dx) \\ h = f \text{ on } \partial D_1}} & \left[\tfrac{1}{2}\int_{D_n - D_1}\left(\frac{\nabla g}{g} - a^{-1}b\right)a\left(\frac{\nabla g}{g} - a^{-1}b\right)g^2\,dx\right. \\ & \left. - \tfrac{1}{2}\int_{D_n - D_1} (\nabla h - a^{-1}b) a (\nabla h - a^{-1}b) g^2\,dx\right]. \qquad (6.1)\end{aligned}$$

This is the generalized Dirichlet principle of Section 5 with $D = D_n - \bar{D}_1$ and with $\phi = 1$ on ∂D_1 and $\phi = 0$ on ∂D_n. By (5.3),

$$\mu_n^{(f)} = \tfrac{1}{2}\int_{D_n - D_1} (\nabla u_n a \nabla u_n) \frac{\tilde{u}_n}{u_n} \, dx, \tag{6.2}$$

where u_n solves

$$\left.\begin{aligned} Lu_n &= 0 \text{ in } D_n - \bar{D}_1, \\ u_n &= 1 \text{ on } \partial D_1,\ u_n = 0 \text{ on } \partial D_n, \end{aligned}\right\} \tag{6.3}$$

and $\tilde{u}_n$ solves

$$\left.\begin{aligned} \tilde{L}\tilde{u}_n &= 0 \text{ in } D_n - \bar{D}_1, \\ \tilde{u}_n &= \exp(2f) \text{ on } \partial D_1,\ \tilde{u}_n = 0 \text{ on } \partial D_n. \end{aligned}\right\} \tag{6.4}$$

Of course, $u_n(x) = P_x(\sigma_{\bar{D}_1} < \tau_{D_n})$, where $\tau_{D_n} = \inf\{t \geq 0: X(t) \notin D_n\}$ and $\sigma_{\bar{D}_1} = \inf\{t \geq 0: X(t) \in \bar{D}_1\}$. Now $\mu_n^{(f)}$ is monotone non-increasing in n for the same reason that λ_n in (4.2) is, and, by (6.2), $\mu_n^{(f)} \geq 0$; therefore, $\lim_{n\to\infty} \mu_n^{(f)}$ exists. We will prove the following theorem.

Theorem 6.1. *Let* $L = \frac{1}{2}\nabla \cdot a\nabla + b \cdot \nabla$ *satisfy Assumption* $\tilde{H}_{\text{loc}}$ *on a domain* $D \subseteq R^d$, $d \geq 2$. *Define* $\mu_n^{(f)}$ *as above, where* $f \in C^{2,\alpha}(D)$ *is an arbitrary function. Then the diffusion corresponding to L is recurrent if* $\lim_{n\to\infty} \mu_n^{(f)} = 0$ *and transient if* $\lim_{n\to\infty} \mu_n^{(f)} > 0$.

Before proving the theorem, we state and prove the following application.

Theorem 6.2. *Let* $L = \frac{1}{2}\nabla \cdot a\nabla + a\nabla Q \cdot \nabla + b \cdot \nabla$ *and* $L_0 = \frac{1}{2}\nabla \cdot a\nabla + a\nabla Q \cdot \nabla$ *satisfy Assumption* $\tilde{H}_{\text{loc}}$ *on a domain* $D \subseteq R^d$, $d \geq 2$.

(i) *If* $\nabla \cdot (\exp(2Q)b) \geq 0$ *and* $\nabla \cdot (\exp(2Q)b) \not\equiv 0$, *then the diffusion corresponding to L is transient.*

(ii) *If* $\nabla \cdot (\exp(2Q)b) = 0$ *and the diffusion corresponding to* L_0 *is transient, then the diffusion corresponding to L is also transient.*

Remark. Theorem 8.2.7 in Chapter 8 is an application of Theorem 6.2.

Choosing $a \equiv I$, $Q \equiv 0$ and $D = R^d$ in Theorem 6.2, we obtain the following corollary.

Corollary 6.3. *Let* $L = \frac{1}{2}\Delta + b \cdot \nabla$, *where* $b \in C^{1,\alpha}(R^d)$.

(i) *If* $d = 2$, $\nabla \cdot b \geq 0$ *and* $\nabla \cdot b \not\equiv 0$, *then the diffusion corresponding to L is transient.*

(ii) *If* $d \geq 3$ *and* $\nabla \cdot b = 0$, *then the diffusion corresponding to L is transient.*

Remark 1. Part (i) of the corollary actually follows easily without reference to Theorem 6.2. Indeed, in this case, the zeroth-order term of $\widetilde{L}$ is non-positive and not identically zero (it equals $-\nabla \cdot b$). Thus, by Theorem 4.3.3(iii), $\widetilde{L}$ is subcritical, and then, by Theorem 4.3.3(v), L is also subcritical; equivalently, the diffusion corresponding to L is transient.

Remark 2. If $d = 2$ and $\nabla \cdot b = 0$, then the diffusion corresponding to $L = \frac{1}{2}\Delta + b \cdot \nabla$ may be recurrent and may be transient. For recurrence, take $b = 0$; for transience take b to be a non-zero constant vector, or take $b(x, y) = (x, -y)$.

Proof of Theorem 6.2. Let $f = Q$ and consider $\mu_n^{(Q)}$ given by (6.1). Note that the b which appears there must be replaced by $\tilde{b} \equiv a\nabla Q + b$. Picking $h = Q$, we obtain

$$\begin{aligned}\mu_n^{(Q)} \geqslant \inf\Bigg[&\frac{1}{2}\int_{D_n - D_1}\left(\frac{\nabla g}{g} - \nabla Q\right)a\left(\frac{\nabla g}{g} - \nabla Q\right)g^2\,\mathrm{d}x \\ &+ \int_{D_n - D_1}(g^2\nabla Q b - g\nabla g b)\,\mathrm{d}x\Bigg] \\ = \inf\Bigg[&\frac{1}{2}\int_{D_n - D_1}\left(\frac{\nabla g}{g} - \nabla Q\right)a\left(\frac{\nabla g}{g} - \nabla Q\right)g^2\,\mathrm{d}x \\ &+ \frac{1}{2}\int_{D_n - D_1}(\nabla \cdot b + 2\nabla Q b)g^2\,\mathrm{d}x\Bigg] \\ &- \frac{1}{2}\int_{\partial D_1}(b \cdot \nu)\exp(2Q)\,\mathrm{d}\sigma, \end{aligned} \tag{6.5}$$

where σ is Lebesgue measure on ∂D_1, ν is the outward unit normal to $D_n - D_1$, and where the infimum is calculated over $g \in W^{1,2}(D_n - D_1)$ which satisfy $g = \exp(Q)$ on ∂D_1 and $g = 0$ on ∂D_n. (The condition $(\operatorname{dist}(x, \partial D_n))^{-1}g(x) \in L^\infty(D_n - D_1)$ can be dropped once we have fixed an h to use for all g.) The last term on the righthand side of (6.5) satisfies $-\frac{1}{2}\int_{\partial D_1}(b \cdot \nu)\exp(2Q)\,\mathrm{d}\sigma = \frac{1}{2}\int_{D_1}\nabla \cdot (\exp(2Q)b)\,\mathrm{d}x$. Also, $(\nabla \cdot b + 2\nabla Q \cdot b) = \exp(-2Q)\nabla \cdot (\exp(2Q)b)$. Thus, from (6.5) we have

$$\begin{aligned}\mu_n^{(Q)} \geqslant \inf\Bigg[&\frac{1}{2}\int_{D_n - D_1}\left(\frac{\nabla g}{g} - \nabla Q\right)a\left(\frac{\nabla g}{g} - \nabla Q\right)g^2\,\mathrm{d}x \\ &+ \frac{1}{2}\int_{D_n - D_1}g^2\exp(-2Q)\nabla \cdot (\exp(2Q)b)\,\mathrm{d}x\Bigg] \\ &+ \frac{1}{2}\int_{D_1}\nabla \cdot (\exp(2Q)b)\,\mathrm{d}x. \end{aligned} \tag{6.6}$$

We first prove (ii). It follows from (6.6) and the assumption of the theorem that

$$\mu_n^{(Q)} \geqslant \inf\left[\frac{1}{2}\int_{D_n-D_1}\left(\frac{\nabla g}{g}-\nabla Q\right)a\left(\frac{\nabla g}{g}-\nabla Q\right)g^2\,dx\right]. \tag{6.7}$$

Substituting $u = g\exp(-Q)$, the righthand side of (6.7) becomes λ_n in (4.2) for the operator L_0. By assumption, the diffusion corresponding to L_0 is transient; thus, by Theorem 4.1, $\lambda_\infty \equiv \lim_{n\to\infty}\lambda_n > 0$. From (6.7) it follows that $\lim_{n\to\infty}\mu_n^{(Q)} \geqslant \lambda_\infty > 0$, and, by Theorem 6.1, the diffusion is transient.

We now prove (i). By assumption, $\nabla\cdot(\exp(2Q)b) \geqslant 0$ and $\nabla\cdot(\exp(2Q)b) \not\equiv 0$. Therefore, for some m, the term $\int_{D_m}\nabla\cdot(\exp(2Q)b)\,dx$ is positive. Now there was no reason in particular to use ∂D_1 as the inner boundary in the preceding theory; we could just as well have defined $\mu_n^{(f)}$, $n > m$, in (6.1) with D_1 replaced by D_m. With this new definition, D_1 is replaced by D_m in (6.6). Thus, for $n > m$, it follows that $\mu_n^{(Q)} > \int_{D_m}\nabla\cdot(\exp(2Q)b)\,dx > 0$ and, consequently, $\lim_{n\to\infty}\mu_n^{(Q)} > 0$. By Theorem 6.1, the diffusion is transient.

□

Proof of Theorem 6.1. Since the zeroth-order term in L vanishes, the common principal eigenvalue of L and of $\tilde{L}$ on $D_n - \bar{D}_1$ is negative. Thus, from the maximum principle in Theorem 3.6.5, it follows that the solutions u_n and $\tilde{u}_n$ of (6.3) and (6.4) satisfy $u_n > 0$ and $\tilde{u}_n > 0$ on $D_n - D_1$ and, for $m > n$, $u_m \geqslant u_n$ and $\tilde{u}_m \geqslant \tilde{u}_n$ on $D_n - D_1$. From this monotonicity property, one direction of the proof is easy. Assume that $\lim_{n\to\infty}\mu_n^{(f)} = 0$. Then, by (6.2),

$$\lim_{n\to\infty}\int_{D_n-D_1}(\nabla u_n a\nabla u_n)\frac{\tilde{u}_n}{u_n}\,dx = 0. \tag{6.8}$$

Fix an integer $N > 1$. By the monotonicity of $\tilde{u}_n(x)$, it follows that $\sup_{n\geqslant N+1}\inf_{x\in D_N-D_1}\tilde{u}_n(x) > 0$. This, together with (6.8) and the fact that $0 \leqslant u_n \leqslant 1$, gives

$$\lim_{n\to\infty}\int_{D_N-D_1}|\nabla u_n|^2\,dx = 0. \tag{6.9}$$

Let $u_\infty(x) = P_x(\sigma_{\bar{D}_1} < \infty)$. Then $u_\infty(x) = \lim_{n\to\infty}u_n(x)$, and a standard compactness argument shows that $u_\infty \in C^{2,\alpha}(D - \bar{D}_1)$. From Theorem 2.3.3, we conclude that $u_\infty \in C^{2,\alpha}(D - \bar{D}_1) \cap C(D - D_1)$. The bounded convergence theorem gives

$$\lim_{n\to\infty}\int_{D_N-D_1}(u_\infty - u_n)^2\,dx = 0. \tag{6.10}$$

From (6.9) and (6.10), it follows that $u_n \to u_\infty$ strongly in $W^{1,2}(D_N - D_1)$ and that $\nabla u_\infty = 0$ a.e. on $D_N - D_1$. Since $u_\infty \in C^{2,\alpha}(D - \bar{D}_1) \cap C(D - D_1)$ and $u_\infty = 1$ on ∂D_1, and since N is arbitrary, we conclude that $u_\infty(x) = 1$, for $x \in D - D_1$. Thus, $P_x(\sigma_{\bar{D}} < \infty) = 1$, for $x \in D - D_1$, and, by Theorem 2.8.2(ii), the diffusion is recurrent.

The proof of the other direction is considerably more involved. Assume that the diffusion corresponding to L on D is recurrent. Then $P_x(\sigma_{\bar{D}} < \infty) = u_\infty(x) = \lim_{n\to\infty} u_n(x) = 1$, for $x \in D - D_1$. Using the representation in (6.2), what we must prove is that

$$\lim_{n\to\infty}\mu_n^{(f)} = \lim_{n\to\infty}\tfrac{1}{2}\int_{D_n-D_1}(\nabla u_n a \nabla u_n)\frac{\tilde{u}_n}{u_n}\,dx = 0.$$

Since the diffusion corresponding to L on D is recurrent, it follows from Theorem 4.3.3 that $\tilde{L}$ is critical on D. Let $\tilde{\phi}_c$ denote the ground state of $\tilde{L}$ on D. As noted above, the maximum principle of Theorem 3.6.5 holds for $\tilde{L}$ on $D_n - D_1$. Thus, since $\tilde{u}_n = \exp(2f)$ on ∂D_1 and $\tilde{u}_n = 0$ on ∂D_n, it follows that

$$\tilde{u}_n(x) \leq k\tilde{\phi}_c(x), \text{ for } x \in D_n - D_1,\ n = 1, 2, \ldots, \tag{6.11}$$

where

$$k = \sup_{x\in\partial D_1}\frac{\exp(2f(x))}{\tilde{\phi}_c(x)}.$$

Define

$$L_n = L + a\frac{\nabla u_n}{u_n}\cdot\nabla$$

and its formal adjoint

$$\tilde{L}_n = \tilde{L} - a\frac{\nabla u_n}{u_n}\cdot\nabla - \nabla\cdot a\frac{\nabla u_n}{u_n}.$$

One may verify that $L_n(\log u_n) = \frac{1}{2}(\nabla u_n a \nabla u_n)/u_n^2$ and that $\tilde{L}_n(u_n\tilde{u}_n) = 0$. Now let $\psi \in C^\infty(D)$ satisfy $0 \leq \psi \leq 1$, $\psi = 0$ on D_1 and $\psi = 1$ on $D - D_2$. Let $\gamma_n = \tilde{L}_n(\psi u_n \tilde{u}_n)$. Note that $\gamma_n(x) = 0$, for $x \in D - D_2$. We now come to the key identity:

$$\tfrac{1}{2}\int_{D_n-D_1}(\nabla u_n a \nabla u_n)\frac{\tilde{u}_n}{u_n}\psi\,dx = \tfrac{1}{2}\int_{D_n-D_1}\gamma_n \log u_n\,dx$$

$$= \tfrac{1}{2}\int_{D_2-D_1} \gamma_n \log u_n \, dx. \qquad (6.12)$$

This is verified as follows. Since

$$L_n(\log u_n) = \tfrac{1}{2}\frac{\nabla u_n a \nabla u_n}{u_n^2},$$

we have

$$\begin{aligned}
\tfrac{1}{2}\int_{D_n-D_1} (\nabla u_n a \nabla u_n)\frac{\tilde{u}_n}{u_n}\psi \, dx &= \int_{D_n-D_1} u_n \tilde{u}_n \psi L_n(\log u_n)\, dx \\
&= \int_{D_n-D_1} (\log u_n)\tilde{L}_n(u_n\tilde{u}_n\psi) \\
&\quad + \text{boundary terms} \\
&= \int_{D_n-D_1} \gamma_n \log u_n \, dx + \text{boundary terms}.
\end{aligned}$$

All but one of the boundary terms that arise in the preceding integration by parts clearly vanish. The remaining boundary integral may be written as

$$\lim_{\varepsilon\to 0}\int_{\partial D_{n-\varepsilon}} \psi\tilde{u}_n(\log u_n)\nu_\varepsilon a \nabla u_n \, d\sigma,$$

where ν_ε is the outward unit normal to $D_{n-\varepsilon}$ on $\partial D_{n-\varepsilon}$. This boundary term also vanishes, as follows from the Hopf maximum principle (Theorem 3.2.5) which insures that

$$\frac{\tilde{u}_n}{u_n}$$

is bounded near ∂D_n.

Using the fact that $\tilde{L}_n(u_n\tilde{u}_n) = 0$, we have

$$\begin{aligned}
\gamma_n = \tilde{L}_n(\psi u_n\tilde{u}_n) &= u_n\tilde{u}_n\tilde{L}_n\psi + \psi\tilde{L}_n(u_n\tilde{u}_n) \\
&\quad + \left(\nabla\cdot b + \nabla\cdot a\frac{\nabla u_n}{u_n}\right)\psi u_n\tilde{u}_n + \nabla\psi a \nabla(u_n\tilde{u}_n) \\
&= u_n\tilde{u}_n\left(\tilde{L}\psi - \frac{\nabla\psi a \nabla u_n}{u_n} - \psi\nabla\cdot a\frac{\nabla u_n}{u_n}\right) \\
&\quad + \left(\nabla\cdot b + \nabla\cdot a\frac{\nabla u_n}{u_n}\right)\psi u_n\tilde{u}_n + (\nabla\psi a\nabla u_n)\tilde{u}_n + (\nabla\psi a \nabla\tilde{u}_n)u_n \\
&= u_n\tilde{u}_n\hat{L}\psi + (\nabla\psi a\nabla\tilde{u}_n)u_n,
\end{aligned}$$

where $\hat{L} = \frac{1}{2}\nabla \cdot a\nabla - b \cdot \nabla$. Thus we may rewrite the righthand side of (6.12) as

$$\int_{D_2-D_1} \gamma_n \log u_n \, dx = \int_{D_2-D_1} u_n \tilde{u}_n (\log u_n) \hat{L} \psi \, dx + \int_{D_2-D_1} (\nabla \psi a \nabla \tilde{u}_n) u_n \log u_n \, dx$$

$$= \int_{D_2-D_1} u_n \tilde{u}_n (\log u_n) \hat{L} \psi \, dx - \int_{D_2-D_1} u_n \tilde{u}_n (\log u_n) \nabla \cdot a \nabla \psi \, dx$$

$$- \int_{D_2-D_1} (\nabla \psi a \nabla u_n) \tilde{u}_n \log u_n \, dx - \int_{D_2-D_1} (\nabla \psi a \nabla u_n) \tilde{u}_n \, dx, \quad (6.13)$$

where we have used the fact that $u_n = 1$ on ∂D_1 and $\nabla \psi = 0$ on ∂D_2. By assumption, $\lim_{n\to\infty} \log u_n(x) = 0$, for $x \in D - D_1$. By (6.11), $\tilde{u}_n$ is bounded independent of n on $D_2 - D_1$. It is also clear that

$$\inf_{n \geq n_0} \inf_{x \in D_2 - D_1} u_n(x) = \inf_{x \in D_2 - D_1} u_{n_0}(x) > 0, \text{ for } n_0 > 2. \quad (6.14)$$

From these facts and from (6.13), we obtain

$$\lim_{n\to\infty} \int_{D_2-D_1} \gamma_n \log u_n \, dx = - \lim_{n\to\infty} \int_{D_2-D_1} (\nabla \psi a \nabla u_n)(\tilde{u}_n \log u_n + \tilde{u}_n) \, dx. \quad (6.15)$$

Now from (6.12), we obtain the inequality

$$\mu_n^{(f)} = \frac{1}{2} \int_{D_n-D_1} (\nabla u_n a \nabla u_n) \frac{\tilde{u}_n}{u_n} \, dx$$

$$\leq \frac{1}{2} \int_{D_2-D_1} (\nabla u_n a \nabla u_n) \frac{\tilde{u}_n}{u_n} \, dx + \int_{D_2-D_1} \gamma_n \log u_n \, dx. \quad (6.16)$$

In order to show that $\lim_{n\to\infty} \mu_n^{(f)} = 0$, it suffices from (6.11) and (6.14)–(6.16) to show that $\lim_{n\to\infty} \int_{D_2-D_1} |\nabla u_n|^2 \, dx = 0$, or equivalently, that

$$\lim_{n\to\infty} \int_{D_2-D_1} |\nabla v_n|^2 \, dx = 0, \quad (6.17)$$

where $v_n = 1 - u_n$. Since v_n is monotone and, by assumption, $\lim_{n\to\infty} v_n(x) = 0$, for $x \in D - D_1$, it follows from Dini's theorem that $\lim_{n\to\infty} v_n(x) = 0$, uniformly over $x \in \bar{D}_2 - D_1$. Thus, by the Schauder interior estimate (Theorem 3.2.7), $\lim_{n\to\infty} |\nabla v_n|(x) = 0$, for $x \in D_2 - \bar{D}_1$. Furthermore, by the Schauder global estimate (Theorem 3.2.8), $|\nabla v_n|$ is bounded on $\bar{D}_2 - D_1$, independent of n. Thus (6.17) now follows from the bounded convergence theorem. □

6.7 The Liapunov method and its application for explosion and non-explosion criteria

Recall that a diffusion on a domain $D \subseteq R^d$ is said to explode if $P_x(\tau_D < \infty) > 0$, for some (and thus all) $x \in D$.

Theorem 7.1. *Let L satisfy Assumption A on a domain $D \subseteq R^d$, $d \geq 2$.*

(i) *Assume that there exists a simply connected subdomain $\Omega \subset\subset D$, a $\lambda > 0$, and a function $u \in C^2(D)$ such that $Lu \leq \lambda u$ on $D - \Omega$ and $\lim_{x \to \partial D} u(x) = \infty$. Then the diffusion corresponding to L does not explode.*

(ii) *Assume that there exists a simply connected subdomain $\Omega \subset\subset D$, a $\lambda > 0$, a bounded function $u \in C^2(D)$ and an $x_0 \in D - \bar{\Omega}$ such that $Lu \geq \lambda u$ on $D - \Omega$ and $\sup_{\partial\Omega} u \leq u(x_0)$. Then the diffusion corresponding to L explodes.*

Proof. Let $\{D_n\}_{n=1}^{\infty}$ be a sequence of domains satisfying $\Omega \subset\subset D_1$, $D_n \subset\subset D_{n+1}$ and $\bigcup_{n=1}^{\infty} D_n = D$. Let $\tau_{D_n} = \inf\{t \geq 0: X(t) \notin D_n\}$ and $\sigma_{\bar{\Omega}} = \inf\{t \geq 0: X(t) \in \bar{\Omega}\}$.

(i) Since

$$\left(\frac{\partial}{\partial t} + L\right)(\exp(-\lambda t)u(x)) \leq 0, \quad \text{for } x \in D - \Omega,$$

it follows by Theorem 1.5.1(iv) that

$$E_x \exp(-\lambda(t \wedge \sigma_{\bar{\Omega}} \wedge \tau_{D_n}))u(X(t \wedge \sigma_{\bar{\Omega}} \wedge \tau_{D_n})) \leq u(x), \quad \text{for } x \in D_n - \Omega. \tag{7.1}$$

Fix $x_1 \in D - \bar{\Omega}$ and assume, contrary to the statement of the theorem, that the diffusion explodes. Then, clearly, there exists a $t_0 \in (0, \infty)$ such that

$$P_{x_1}(\tau_D < t_0 \wedge \sigma_{\bar{\Omega}}) > 0. \tag{7.2}$$

Letting $n \to \infty$ in (7.1) with $x = x_1$ and $t = t_0$, and using (7.2) and the fact that $\lim_{x \to \partial D} u(x) = \infty$, it follows that the lefthand side of (7.1) approaches infinity, which contradicts (7.1).

(ii) Since

$$\left(\frac{\partial}{\partial t} + L\right)(\exp(-\lambda t)u(x)) \geq 0, \text{ for } x \in D - \Omega,$$

it follows by Theorem 1.5.1(iv) that

$$E_{x_0} \exp(-\lambda(t \wedge \sigma_{\bar{\Omega}} \wedge \tau_{D_n}))u(X(t \wedge \sigma_{\bar{\Omega}} \wedge \tau_{D_n})) \geq u(x_0). \tag{7.3}$$

Assume now, contrary to the statement of the theorem, that the diffusion does not explode. Then $P_{x_0}(\tau_D < \infty) = 0$ and, letting $t \to \infty$ and $n \to \infty$ in (7.3), and using the fact that u is bounded, gives

$$E_{x_0}(\exp(-\lambda\sigma_{\bar{\Omega}})u(X(\sigma_{\bar{\Omega}})); \sigma_{\bar{\Omega}} < \infty) \geq u(x_0).$$

This is a contradiction since $\sup_{x \in \partial\Omega} u(x) \leq u(x_0)$. □

We use Theorem 7.1 to prove the following explosion and non-explosion criteria for diffusions on R^d. Let $A(x)$, $B(x)$, $C(x)$, $\bar{\alpha}(r)$, $\underline{\alpha}(r)$, $\bar{\beta}(r)$, $\underline{\beta}(r)$, $\bar{I}(r)$ and $\underline{I}(r)$ be as in Section 2.

Theorem 7.2. *Let L satisfy Assumption A on R^d, $d \geq 2$.*

(i) *If*

$$\int_1^\infty dr \exp(-\bar{I}(r)) \int_1^r ds \frac{1}{\bar{\alpha}(s)} \exp(\bar{I}(s)) = \infty,$$

then the diffusion corresponding to L does not explode.

(ii) *If*

$$\int_1^\infty dr \exp(-\underline{I}(r)) \int_1^r ds \frac{1}{\underline{\alpha}(s)} \exp(\underline{I}(s)) < \infty,$$

then the diffusion corresponding to L explodes.

Remark. The intuition for Theorem 7.2 in terms of Theorem 5.1.5 is similar to that for Theorem 2.1 in terms of 5.1.1, which was discussed in the remark following Theorem 2.1.

Proof. (i) We will prove the theorem under the additional condition that the drift b is continuous. The extension to the case where b is locally bounded and measurable is left to exercise 6.1. Define $v_0(r) = 1$, for $r \geq 1$ and, by induction, define

$$v_n(r) = 2\int_1^r ds \exp(-\bar{I}(s)) \int_1^s dt \frac{v_{n-1}(t)}{\bar{\alpha}(t)} \exp(\bar{I}(t)), \quad n \geq 1, \quad r \geq 1.$$

By induction, it follows that $v_n \geq 0$ and $v_n' \geq 0$. Thus,

$$v_n(r) \leq 2\int_1^r ds \exp(-\bar{I}(s)) v_{n-1}(s) \int_1^s dt \frac{1}{\bar{\alpha}(t)} \exp(\bar{I}(t)).$$

From this inequality and induction, it follows that

$$v_n(r) \leq \frac{2^n}{n!}\left(\int_1^r ds \exp(-\bar{I}(s)) \int_1^s dt \frac{1}{\bar{\alpha}(t)} \exp(\bar{I}(t))\right)^n.$$

We also have

$$v_n'' = (2/\bar{\alpha})v_{n-1} - \bar{\beta}v_n'. \tag{7.4}$$

From this it follows that $v(r) \equiv \sum_{n=0}^{\infty} v_n(r)$, $r \geq 1$, is a function that can be differentiated twice term by term. Let $u(x) = v(|x|)$, for $|x| \geq 1$. A straightforward calculation using (7.4) reveals that $Lu(x) \leq u(x)$, for $|x| \geq 1$. Also, $\lim_{|x|\to\infty} u(x) = \infty$, since $u(x) \geq v_1(|x|)$ and, by assumption, $\lim_{r\to\infty} v_1(r) = \infty$. Thus, from Theorem 7.1(i), the diffusion does not explode.

(ii) Define $v_0(r) = 1$, for $r \geq 1$, and by induction, define

$$v_n(r) = 2\int_1^r \mathrm{d}s \exp(-\underline{I}(s)) \int_1^s \mathrm{d}t \frac{v_{n-1}(t)}{\underline{\alpha}(t)} \exp(\underline{I}(t)), \quad n \geq 1, r \geq 1.$$

Then, as in part (i), we have $v_n \geq 0$, $v_n' \geq 0$,

$$v_n(r) \leq \frac{2^n}{n!}\left(\int_1^r \mathrm{d}s \exp(-\underline{I}(s)) \int_1^s \mathrm{d}t \frac{1}{\underline{\alpha}(t)} \exp(\underline{I}(t))\right)^n, \tag{7.5}$$

and

$$v_n'' = (2/\underline{\alpha})v_{n-1} - \underline{\beta}v_n'. \tag{7.6}$$

Let $v(r) = \sum_{n=0}^{\infty} v_n(r)$ and define $u(x) = v(|x|)$, for $|x| \geq 1$. A straightforward calculation using (7.6) reveals that $Lu(x) \geq u(x)$, for $|x| \geq 1$. By (7.5) and the integrability assumption, u is bounded and, by construction, $u(x) > u(y)$ if $|x| > |y|$. Thus, by Theorem 7.1(ii), the diffusion explodes. □

Exercises

6.1 Prove Theorems 2.1, 2.2 and 7.2 in the case where the drift is only assumed locally bounded and measurable. (Hint: For example, in Theorem 2.1, let b_ε be a smooth approximation of b and define $\bar{I}_\varepsilon$ and $\underline{I}_\varepsilon$ with b_ε replacing b. Applying the martingale formulation for $\{P_x, x \in \hat{D}\}$ to

$$F_\varepsilon(r) = \int_1^r \exp(-\bar{I}_\varepsilon(s))\,\mathrm{d}s \text{ and } G_\varepsilon(r) = \int_1^r \exp(-\underline{I}_\varepsilon(s))\,\mathrm{d}s,$$

obtain expressions for $P_x(\tau_n < \tau_1)$. Now let $\varepsilon \to 0$.)

6.2 Let L satisfy Assumption A on R^d. Let $\tau_r = \inf\{t \geq 0: |X(t)| = r\}$. Prove that if $r_0 > 0$ and $E_x\tau_{r_0} < \infty$, for $|x| > r_0$, then $E_x\tau_r < \infty$, for $0 < r < |x|$.

6.3 Let L and ρ_k be as in Theorem 3.2. Assume that $D \equiv 0$. Extend the theorem to the borderline case $\rho_k = 0$ as follows.
(i) If $k < 1$ and $\rho_k = 0$, then $P_x(\tau_r < \infty) = 1$, for $0 < r < |x|$.
(ii) If $k > 1$ and $\rho_k = 0$, then $P_x(\tau_0 < \infty) = 0$, and $P_x(\lim_{t\to\infty} |X(t)| = 0) = 0$, for $x \in R^d - \{0\}$.
(Hint: Look for a Liapunov function of the form $u(r, \theta) = \log\log r + f(r)g(\theta)$.)

6.4 Let L be as in Theorem 3.3. In the case $k = \delta < 1$, prove the following partial extension to the borderline case $\gamma = 0$: Let

$$f(\theta) = \int_0^\theta \frac{\gamma_1}{\gamma_2}(s)\,\mathrm{d}s.$$

(i) If

$$\int_0^{2\pi} \exp(-(1-\delta)f(\theta))\left(\frac{\mathcal{A}f + c_2 - c_1}{\gamma_2}\right)(\theta)\,\mathrm{d}\theta < 0,$$

then $P_x(\tau_r < \infty) = 1$, for $0 < r \leqslant |x|$.

(ii) If

$$\int_0^{2\pi} \exp(-(1-\delta)f(\theta))\left(\frac{\mathcal{A}f - Df - c_2 + c_1}{\gamma_2}\right)(\theta)\,\mathrm{d}\theta < 0,$$

then $P_x(\tau_r < \infty) < 1$, for $0 < r < |x|$.

In particular, if $\gamma_1 \equiv 0$, then $f \equiv 0$ and one obtains the following conditions:

(i)′ If

$$\int_0^{2\pi} \frac{c_2 - c_1}{\gamma_2}(\theta)\,\mathrm{d}\theta < 0,$$

then $P_x(\tau_r < \infty) = 1$, for $0 < r \leqslant |x|$.

(ii)′ If

$$\int_0^{2\pi} \frac{c_2 - c_1}{\gamma_2}(\theta)\,\mathrm{d}\theta > 0,$$

then $P_x(\tau_r < \infty) < 1$, for $0 < r < |x|$.

Note that (i)′ and (ii)′ may be thought of as an extension of Theorem 3.2 to a particular case in which the operator $\mathcal{A}$ in (3.3) is degenerate. Recall that ρ_k, defined in (3.4), is the average of $(c_2 - c_1)$ with respect to the invariant density of the diffusion on S^{d-1} corresponding to $\mathcal{A} + (k-1)D$. In the present case, $d = 2$, $D \equiv 0$ and

$$\mathcal{A} = \gamma_2 \frac{\partial}{\partial\theta}$$

is degenerate. The degenerate 'diffusion' on S^1 corresponding to

$$\mathcal{A} + (k-1)D = \gamma_2 \frac{\partial}{\partial \theta}$$

is just a deterministic dynamical system whose invariant density is

$$\frac{1}{\gamma_2}.$$

Therefore, the natural way to extend ρ_k to the present case is by defining

$$\rho_k = \int_0^{2\pi} \frac{c_2 - c_1}{\gamma_2}(\theta)\, d\theta.$$

Thus, (i)′ and (ii)′ give an extension of Theorem 3.2 to a particular degenerate case. However, unlike in Theorem 3.2, the case $\rho_k = 0$ is not solved here. (*Hint*: For (i), look for a Liapunov function in the form $u(r, \theta) = \log r + g(\theta) + r^{\delta - 1}h(\theta)$. For (ii), look for a Liapunov function in the form $u(r, \theta) = r^{-l}g(\theta) + r^{-m}h(\theta)$. Use this to obtain a sufficiency condition for $P_x(\tau_r < \infty) < 1$, and then let $l \to 0$.)

6.5 Let

$$L_1 = \tfrac{1}{2}\Delta + \frac{\gamma(\theta)}{r}\frac{\partial}{\partial \theta}$$

and

$$L_2 = \tfrac{1}{4}\frac{\partial^2}{\partial x^2} + \tfrac{1}{2}\frac{\partial^2}{\partial y^2} + \frac{\gamma(\theta)}{r}\frac{\partial}{\partial \theta} = \tfrac{1}{2}\nabla \cdot \begin{pmatrix} \frac{1}{2} & 0 \\ 0 & 1 \end{pmatrix} \nabla + \frac{\gamma(\theta)}{r}\frac{\partial}{\partial \theta}$$

on $R^2 - \{0\}$, and let $P_x^{(i)}$, $i = 1, 2$, denote the probabilities for the corresponding diffusions.

(i) Show that $P_x^{(1)}(\tau_r < \infty) = 1$, for $0 < r \leqslant |x|$. (*Hint*: Since $\frac{1}{2}\Delta$ does not depend on θ, the angular drift

$$\frac{\gamma(\theta)}{r}\frac{\partial}{\partial \theta}$$

plays no role in transience or recurrence considerations.)

(ii) Use (ii)′ of exercise (6.4) to show that γ can be chosen so that $P_x^{(2)}(\tau_r < \infty) < 1$, for $0 < r < |x|$.

This shows that a comparison principle analogous to Corollary 4.2 does not hold in the non-symmetric case. This example is due to [R. Pinsky 1988c]

6.6 Let L be as in Theorem 3.3. Prove the following result:

(i) If $k \leqslant \delta$ and $\delta < -1$, then the diffusion explodes if $\gamma > 0$.

(ii) If $k > \delta$ and $\delta < -1$, then the diffusion explodes if $\{\theta \in S^1: \gamma_1(\theta) > 0\} \neq \phi$.

6.7 Prove the second statement in Theorem 3.2(a(iv)) and in Theorem 3.2(b(i)).

6.8 Let $L = \frac{1}{2}\Delta + b \cdot \nabla$ on R^d, where $|b(x)| \leqslant c(1 + |x|)$. Prove that the diffusion corresponding to L does not explode.

Notes

Techniques in the spirit of the results in Section 1 have been around for a long time; I do not know who deserves credit for their discovery. Theorem 2.1 is due to [Friedman (1975)] and Theorem 2.2 is due to [Bhattacharya (1978)]. Results similar to those in Theorems 2.1 and 2.2 were announced but not proved in [Hasminskii (1960)]. Theorems 3.1 and 3.3 extend results in [R. Pinsky (1987)]; Theorem 3.2 is new. Theorem 4.1 was proved by [Brown (1971)] and was extended by [Ichihara (1978)] to operators with measurable coefficients. In both of these papers, $Q \equiv 0$; however, the extension to non-zero Q is trivial. Theorem 4.4 and Examples 4.5 and 4.6 are due to [Ichihara (1978)]. He also showed that Theorem 2.1(i), when restricted to symmetric operators, is contained in Theorem 4.4(i). Section 5 is due to [R. Pinsky (1988a)] and Section 6 is due to [R. Pinsky (1988b)]. Section 7 is due to [Stroock and Varadhan (1979)].

7
Positive harmonic functions and the Martin boundary: general theory

1.0 Introduction

We begin with a presentation of the classical Martin boundary theory and its connection to the behavior of diffusions as $t \to \tau_D$. We then develop an alternative probabilistic approach in terms of positive solutions of minimal growth at ∂D and diffusions conditioned to hit a compact set. This latter approach turns out to be more useful for calculating Martin boundaries explicitly.

7.1 The Martin boundary

Let $L = L_0 + V$ satisfy Assumption $\tilde{\text{H}}_{\text{loc}}$ on a domain $D \subseteq R^d$. Assume that L is subcritical on D and let $G(x, y)$ denote the Green's function. Fix a point $x_0 \in D$ and define the *Martin kernel* by

$$k(x; y) = \begin{cases} \dfrac{G(x, y)}{G(x_0, y)}, & y \neq x, y \neq x_0 \\ 0, & y = x_0, x \neq x_0 \\ 1, & y = x_0 = x. \end{cases}$$

Let $\{y_n\}_{n=1}^{\infty} \subset D$ be a sequence with no accumulation points in D and consider the sequence $\{k(x; y_n)\}_{n=1}^{\infty}$. Recall that, by Theorem 4.2.5, for $y \in D$, $LG(\cdot, y) = 0$ on $D - \{y\}$. Using this, the Schauder estimates and Harnack's inequality, the proof of Theorem 4.3.1 showed that there exists a subsequence $\{y_{n_j}\}_{j=1}^{\infty}$ such that $\lim_{j\to\infty} k(x; y_{n_j})$ exists for all $x \in D$ and the limiting function belongs to $C_L(D)$. A sequence $\{y_n\}_{n=1}^{\infty} \subset D$ for which $\lim_{n\to\infty} k(x; y_n) \in C_L(D)$ is called a *Martin sequence*. If $\{y_n\}_{n=1}^{\infty}$ and $\{y'_n\}_{n=1}^{\infty}$ are Martin sequences and $\lim_{n\to\infty} k(x; y_n) = \lim_{n\to\infty} k(x; y'_n)$, then the two Martin sequences are called *equivalent*. The collection of such equivalence classes is called the *Martin boundary* for L on D and will be denoted by Λ. A point on the Martin boundary will frequently be denoted by ξ or ζ. An element of $C_L(D)$ corresponding to ξ will be denoted by $k(x; \xi)$; that is, up to positive multiples, $k(x; \xi) = \lim_{n\to\infty} k(x; y_n)$, where $\{y_n\}_{n=1}^{\infty}$ is any representative of the equivalence class ξ.

Fix a domain $u \subset\subset D$ and define ρ: $(D \cup \Lambda) \times (D \cup \Lambda) \to [0, \infty)$ by

$$\rho(z_1, z_2) = \int_U \frac{|k(x; z_1) - k(x; z_2)|}{1 + |k(x; z_1) - k(x; z_2)|} \, dx. \tag{1.1}$$

Proposition 1.1. *ρ is a metric on $D \cup \Lambda$. The relative topology on D induced by ρ is equivalent to the Euclidean topology. Under ρ, D is open, Λ is the boundary of D and $D \cup \Lambda$ is compact.*

Proof. It is clear that ρ is non-negative and symmetric. To complete the proof that ρ is a metric, it remains to show that ρ satisfies the triangle inequality and that $\rho(z_1, z_2) > 0$ if $z_1 \neq z_2$; this is left as exercise 7.1.

From the continuity of $k(x; y)$ as a function of $y \in D - \{x\}$ and the bounded convergence theorem, it follows that the relative topology of D induced by ρ is equivalent to the Euclidean topology on D. It also follows from this that under ρ, D is open in $D \cup \Lambda$. From the definition of the Martin boundary, it follows that under ρ, Λ is the boundary of D. The standard compactness argument for positive harmonic functions using the Schauder estimates and Harnack's inequality shows that $D \cup \Lambda$ is sequentially compact under ρ. Thus, to prove that $D \cup \Lambda$ is compact, it suffices to show that Λ is separable. Let $\{x_j\}_{j=1}^{\infty} \subset D$ be a countable dense subset of D. For each pair (n, m) of positive integers, select (if possible) a point $\xi_{n,m} \in \Lambda$ such that

$$\rho(x_n, \xi_{n,m}) \leq \frac{1}{m}.$$

If no such point exists, choose $\xi_{n,m} \in \Lambda$ arbitrarily. It is easy to see that $\{\xi_{n,m}\}_{n,m=1}^{\infty}$ is dense in Λ under ρ. □

The topology on $D \cup \Lambda$ induced by ρ is called the *Martin topology*. We will use the notation 'm-lim' to denote convergence in the Martin topology and to distinguish it from convergence in the Euclidean topology which, of course, is denoted by 'lim'.

A function $u \in C_L(D)$ is called *minimal* if whenever $v \in C_L(D)$ and $v \leq u$, then in fact $v = cu$ for some constant $c \in (0, 1]$. A point $\xi \in \Lambda$ is called a *minimal Martin boundary point* if $k(x;\xi)$ is minimal. The collection of all minimal Martin boundary points is called the *minimal Martin boundary* and will be denoted by Λ_0. It is a Borel set under the Martin topology [Helms (1969)].

The following theorem is known as the *Martin representation theorem*.

Theorem 1.2. *Let L satisfy Assumption $\tilde{H}_{\text{loc}}$ on a domain $D \subseteq R^d$ and assume that L is subcritical. Then for each $u \in C_L(D)$, there exists a unique finite measure μ_u supported on the minimal Martin boundary Λ_0 such that*

$$u(x) = \int_{\Lambda_0} k(x;\xi)\mu_u(\mathrm{d}\xi).$$

Conversely, for each finite measure μ supported on the minimal Martin boundary Λ_0,

$$u(x) \equiv \int_{\Lambda_0} k(x;\xi)\mu(\mathrm{d}\xi) \in C_L(D).$$

Remark. From this theorem, it follows that the Martin boundary is independent of the choice of fixed point $x_0 \in D$. On the other hand, $k(x;\xi)$ and μ_0 do of course depend on x_0. In particular, note that since $k(x_0;\xi) = 1$, for all $\xi \in \Lambda$, it follows that $\mu_u(\Lambda_0) = u(x_0)$.

The proof of the Martin representation theorem requires certain potential theoretic developments which go beyond the scope of this book. References appear in the notes at the end of the chapter.

Corollary 1.3. *$u \in C_L(D)$ is minimal if and only if $u(x) = k(x;\xi)$ for some $\xi \in \Lambda_0$.*

Proof. One direction follows by definition. For the other direction, we use the uniqueness part of the Martin representation theorem. It suffices to show that if the support of μ_u contains more than one point, then u is not minimal. Assume that the support of μ_u contains more than one point and let $A \subset \Lambda_0$ satisfy $0 < \mu_u(A) < 1$. Define $v(x) = \int_A k(x;\xi)\mu_u(\mathrm{d}\xi) = \int_{\Lambda_0} k(x;\xi)1_A(\xi)\mu_u(\mathrm{d}\xi)$. Clearly, $v \leq u$ and it is easy to show that $v \in C_L(D)$; thus $\mu_v(\mathrm{d}\xi) = 1_A(\xi)\mu_u(\mathrm{d}\xi)$ and, by the uniqueness of μ_u, it follows that $v \neq cu$. □

The following proposition shows that the Martin boundary is invariant under h-transforms.

Proposition 1.4. *Let L satisfy Assumption $\widetilde{H}_{\text{loc}}$ on a domain $D \subseteq R^d$. Assume that L is subcritical and let $h \in C_L(D)$. Then the Martin boundaries of L and L^h coincide. Furthermore, $k^h(x;\xi)$, the element of $C_{L^h}(D)$ corresponding to $\xi \in \Lambda$ satisfies*

$$k^h(x;\xi) = \frac{h(x_0)}{h(x)} k(x;\xi).$$

Proof. The proof follows from Proposition 4.2.2, which states that $G^h(x,y)$, the Green's function for L^h, satisfies

$$G^h(x,y) = \frac{1}{h(x)} G(x,y)h(y).$$

□

Similarly, in the symmetric case, the Martin boundaries of L and $\widetilde{L}$ coincide.

Proposition 1.5. *Let L satisfy Assumption $\widetilde{H}_{\text{loc}}$ on a domain $D \subseteq R^d$ and assume that L is subcritical. If L is symmetric, that is, $b = a\nabla Q$ for some $Q \in C^{2,\alpha}(D)$, then the Martin boundaries of L and $\widetilde{L}$ coincide.*

Proof. If L is symmetric with $b = a\nabla Q$, then it is easy to show that $\exp(2Q(x))G(x,y) = \exp(2Q(y))G(y,x)$ (exercise 7.2). Since $\widetilde{G}(x,y) = G(y,x)$, it follows that the Martin kernels $k(x,y)$ and $\widetilde{k}(x,y)$ for L and $\widetilde{L}$ satisfy

$$\widetilde{k}(x,y) = \frac{\widetilde{G}(x,y)}{\widetilde{G}(x_0,y)} = \frac{\exp(2Q(x))}{\exp(2Q(x_0))}\frac{G(x,y)}{G(x_0,y)}$$
$$= \frac{\exp(2Q(x))}{\exp(2Q(x_0))} k(x,y).$$

□

7.2 Conditioned diffusions and *h*-transforms and the behavior of a transient diffusion as it exits its domain

In this section, Theorem 2.1 will be presented without proof as its proof requires potential theoretic arguments beyond the scope of this book; the other results, Theorem 2.2 and Corollary 2.3, will follow readily from Theorem 2.1. Although these results will occasionally be used in the sequel as motivation, with two exceptions they will not be used to prove additional results. (The exceptions are Theorem 8.2.13 and parts (d) and (e) of Theorem 8.6.1.) Exercises 7.3–7.5 outline a complete proof, without relying on Theorem 2.1 of a result in the spirit of Theorem 2.2 and Corollary 2.3 below.

In Chapter 4, we exploited the purely analytic properties of the h-transform in our development of a generalized spectral theory. In this section, we discuss the beautiful probabilistic interpretation of h-transforms in terms of conditioned diffusions.

Theorem 2.1. *Let L satisfy Assumption $\tilde{H}_{\text{loc}}$ on a domain $D \subseteq R^d$ and assume that L is subcritical on D. For $\xi \in \Lambda_0$, let $k(x; \xi)$ denote the corresponding minimal positive harmonic function. Then for each $\xi \in \Lambda_0$,*

$$P_x^{k(\cdot;\xi)}(m - \lim_{t \to \tau_D} X(t) = \xi) = 1, \quad \text{for } x \in D.$$

As noted above, the proof of Theorem 2.1 requires certain potential theoretic developments beyond the scope of this book; the reader is referred to the notes at the end of the chapter. We now consider the connection between conditioned diffusions and h-transforms in the case where $h \in C(D)$ is arbitrary.

Theorem 2.2. *Let L satisfy Assumption $\tilde{H}_{\text{loc}}$ on a domain $D \subseteq R^d$ and assume that L is subcritical on D. Let $h \in C_L(D)$. Then for $x \in D$,*

$$P_x^h(m - \lim_{t \to \tau_D} X(t) \text{ exists}) = 1,$$

and for any $B \in \hat{\mathcal{F}}^D$ and measurable $A \subset \Lambda_0$,

$$P_x^h(B, m - \lim_{t \to \tau_D} X(t) \in A) = \frac{1}{h(x)} \int_A P_x^{k(\cdot;\xi)}(B) k(x; \xi) \mu_h(d\xi), \tag{2.1}$$

where μ_h is the measure in the Martin representation theorem for h. Equivalently,

$$P_x^h(m - \lim_{t \to \tau_D} X(t) \in A) = \frac{1}{h(x)} \int_A k(x; \xi) \mu_h(d\xi) \tag{2.2}$$

and

$$P_x^h(\cdot \mid m - \lim_{t \to \tau_D} X(t) = \xi) = P_x^{k(\cdot;\xi)}(\cdot), \text{ for } \xi \in \Lambda_0. \tag{2.3}$$

We note the following particular case of Theorem 2.2.

Corollary 2.3. *Let the conditions of Theorem 2.2 hold and let $V \equiv 0$ so that $1 \in C_L(D)$. Then for any measurable $A \subseteq \Lambda_0$,*

$$P_x(m-\lim_{t\to\tau_D} X(t)\in A)=\int_A k(x;\xi)\mu_1(d\xi)$$

and

$$P_x(\cdot\,|\,m-\lim_{t\to\tau_D} X(t)=\xi)=P_x^{k(\cdot\,;\xi)}(\cdot),\quad \textit{for } \xi\in\Lambda_0.$$

Proof of Theorem 2.2. (2.2) follows from (2.1) by setting $B=\hat{\Omega}_D$, and then (2.3) follows from (2.1), (2.2) and the definition of conditional expectation. It remains to prove (2.1).

As usual, $p(t,x,dy)$ denotes the transition measure corresponding to L. By the Martin representation theorem, $h(y)=\int_{\Lambda_0}k(y;\xi)\mu_h(d\xi)$, for $y\in D$; therefore, by Theorem 4.1.1, $p^h(t,x,dy)$, the transition measure corresponding to L^h, may be written in the form

$$\begin{aligned}p^h(t,x,dy)&=\frac{1}{h(x)}p(t,x,dy)h(y)\\&=\frac{1}{h(x)}\int_{\Lambda_0}\left(\frac{1}{k(x;\xi)}p(t,x,dy)k(y;\xi)\right)k(x;\xi)\mu_h(d\xi)\\&=\frac{1}{h(x)}\int_{\Lambda_0}p^{k(\cdot\,;\xi)}(t,x,dy)k(x;\xi)\mu_h(d\xi),\end{aligned}$$

where $p^{k(\cdot\,;\xi)}(t,x,dy)$ is the transition measure corresponding to $L^{k(\cdot\,;\xi)}$. From this, it follows that

$$P_x^h(X(t)\in U)=\frac{1}{h(x)}\int_{\Lambda_0}P_x^{k(\cdot\,;\xi)}(X(t)\in U)k(x;\xi)\mu_h(d\xi)$$

for $t\geq 0$ and measurable $U\subseteq\hat{D}$. The same proof used to extend Theorem 4.1.1 to Corollary 4.1.2 shows that

$$P_x^h(B)=\frac{1}{h(x)}\int_{\Lambda_0}P_x^{k(\cdot\,;\xi)}(B)k(x;\xi)\mu_h(d\xi),\tag{2.4}$$

for all measurable rectangles, $B\in\hat{\mathcal{F}}_t^D$, $t>0$, and a monotone class argument extends (2.4) to all $B\in\hat{\mathcal{F}}^D$. Now (2.1) follows from (2.4) and Theorem 2.1. □

If $V\equiv 0$ and $A\subset\Lambda_0$, then by Corollary 2.3, $P_x(m-\lim_{t\to\tau_D}X(t)\in A)>0$ if and only if $\mu_1(A)>0$. The next corollary considers the form of the process conditioned on the event $\{m-\lim_{t\to\tau_D}X(t)\in A\}$ in the case where $\mu_1(A)>0$.

Corollary 2.4. *Let the conditions of Theorem 2.2 hold and let $V \equiv 0$ so that $1 \in C_L(D)$. Let $A \subset \Lambda_0$ be measurable and satisfy $\mu_1(A) > 0$, and define $h(x) = P_x(m - \lim_{t\to\tau_D} X(t) \in A)$. Then*

$$P_x(\cdot \,|m - \lim_{t\to\tau_D} X(t) \in A) = P_x^h(\cdot).$$

Proof. By Corollary 2.3, $h(x) = \int_A k(x;\xi)\mu_1(d\xi)$ and thus, by the uniqueness of the measure appearing in the Martin representation theorem (Theorem 1.2), it follows that $\mu_h(d\xi) = 1_A(\xi)\mu_1(d\xi)$. Applying (2.1) with the h appearing there set to 1, it follows that

$$P_x(B, m - \lim_{t\to\tau_D} X(t) \in A) = \int_A P_x^{k(\cdot;\xi)}(B)k(x;\xi)\mu_1(d\xi),$$

for any $B \in \hat{\mathcal{F}}^D$. Thus,

$$\begin{aligned}
P_x(B|m - \lim_{t\to\tau_D} X(t) \in A) &= \frac{P_x(B, m - \lim_{t\to\tau_D} X(t) \in A)}{P_x(m - \lim_{t\to\tau_D} X(t) \in A)}\\
&= \frac{1}{h(x)}\int_A P_x^{k(\cdot;\xi)}(B)k(x;\xi)\mu_1(d\xi)\\
&= \frac{1}{h(x)}\int_{\Lambda_0} P_x^{k(\cdot;\xi)}(B)k(x;\xi)\mu_h(d\xi)\\
&= P_x^h(B, m - \lim_{t\to\tau_D} X(t) \in \Lambda_0)\\
&= P_x^h(B),
\end{aligned}$$

where the next to the last inequality follows from (2.1). □

Example 2.5 (Brownian motion conditioned to hit a ball). Let $L = \frac{1}{2}\Delta$ in $D = R^d - \bar{B}_{r_0}(0)$, $d \geq 3$, where $B_{r_0}(0) \subset R^d$ denotes the r_0-ball centered at $0 \in R^d$. The corresponding process $X(t)$ is, of course, Brownian motion in $R^d - \bar{B}_{r_0}(0)$. We now describe the process obtained by conditioning the Brownian motion to hit $\bar{B}_{r_0}(0)$ or, more precisely, to exit D at ∂B_{r_0} rather than at infinity.

Let $\tau_{r_0} = \inf\{t \geq 0: |X(t)| = r_0\}$. By Theorem 8.1.4 in Chapter 8, the Euclidean boundary $\partial B_{r_0}(0)$ may be identified with the part of the Martin boundary corresponding to sequences $\{x_n\}_{n=1}^{\infty}$ satisfying $\sup_{n\geq 1}|x_n| < \infty$. We leave it to the reader to check that $\{\lim_{t\to\tau_D}|X(t)| = r_0\} = \{\tau_{r_0} < \infty\}$ a.s. P_x, for $|x| > r_0$. Thus, setting $A = \partial B_{r_0}(0)$ in Corollary 2.4, it follows from that corollary that the Brownian motion in $D = R^d - \bar{B}_{r_0}(0)$, conditioned to hit $\bar{B}_{r_0}(0)$, is described by $\{P_x^h,\ x \in \hat{D}\}$, where $h(x) = P_x(\tau_{r_0} < \infty)$. By Example 5.1.2,

$$h(x) = \left(\frac{r_0}{|x|}\right)^{d-2}.$$

Thus, $\{P_x^h,\ x \in \hat{D}\}$ is the solution to the generalized martingale problem for

$$L^h = \tfrac{1}{2}\Delta + \frac{\nabla h}{h}\cdot\nabla = \tfrac{1}{2}\Delta + \frac{2-d}{|x|}\frac{x}{|x|}\cdot\nabla \text{ in } R^d - \bar{B}_{r_0}(0).$$

In spherical coordinates,

$$r = |x| \text{ and } \theta = \frac{x}{|x|} \in S^{d-1},$$

we have

$$\tfrac{1}{2}\Delta = \tfrac{1}{2}\frac{\partial^2}{\partial r^2} + \frac{d-1}{2r}\frac{\partial}{\partial r} + \frac{1}{2r^2}\Delta_{S^{d-1}},$$

where $\Delta_{S^{d-1}}$ is the Laplace–Beltrami operator on S^{d-1}, and

$$\frac{2-d}{|x|}\frac{x}{|x|}\cdot\nabla = \frac{2-d}{r}\frac{\partial}{\partial r}.$$

Thus, in spherical coordinates, the operator corresponding to Brownian motion in $R^d - \bar{B}_0(r_0)$, conditioned to hit $\bar{B}_0(r_0)$, is given by

$$\tfrac{1}{2}\frac{\partial^2}{\partial r^2} + \frac{3-d}{2r}\frac{\partial}{\partial r} + \frac{1}{2r^2}\Delta_{S^{d-1}}.$$

Note that in the three-dimensional case, the radial part of the conditioned Brownian motion behaves like a standard one-dimensional Brownian motion.

Example 2.6 (A diffusion conditioned to approach a fixed point). Let $L = L_0 = \frac{1}{2}\nabla\cdot a\nabla + b\cdot\nabla$ satisfy Assumption $\tilde{\text{H}}_{\text{loc}}$ on a domain $D \subseteq R^d$, $d \geq 2$, and as usual, let $\{P_x,\ x \in \hat{D}\}$ denote the solution to the generalized martingale problem for L on D. Assume that L is subcritical on D; that is, assume that the diffusion corresponding to L on D is transient. Fix $y_0 \in D$ and let $D_{y_0} = D - \{y_0\}$. By Theorem 4.2.9, $P_x(\sigma_{y_0} < \infty) = 0$, for $x \neq y_0$, where $\sigma_{y_0} = \inf\{t \geq 0\colon X(t) = y_0\}$. Thus, if we identify $\hat{\Omega}_{D_{y_0}} \cap \{\sigma_{y_0} = \infty\}$ with a subset of $\hat{\Omega}_D$ in the natural way, it follows that the solution to the generalized martingale problem for L on D_{y_0} is given by $\{P_x,\ x \in \hat{D}_{y_0}\}$.

The Euclidean point y_0 may be identified with a unique Martin boundary point for the operator L on D_{y_0}; that is, all sequences $\{y_n\}_{n=1}^{\infty} \subset D_{y_0}$ satisfying $\lim_{n\to\infty} y_n = y_0$ are equivalent Martin sequences. To see this, let $G(x, y)$ denote the Green's function for L on D and let $G^{(y_0)}(x, y)$ denote the Green's function for L on D_{y_0}. By Theorem 4.2.9 and the definition of the Green's function, it follows that $G^{(y_0)}(x, y) = G(x, y)$, for $x, y \in D_{y_0}$. Thus, if $\{y_n\}_{n=1}^{\infty} \subset D_{y_0}$ and $\lim_{n\to\infty} y_n = y_0$, it follows that

$$\lim_{n\to\infty} \frac{G^{(y_0)}(x, y_n)}{G^{(y_0)}(x_0, y_n)} = \frac{G(x, y_0)}{G(x_0, y_0)}.$$

The above argument also shows that, up to positive multiples, $G(\cdot, y_0)$ is the element of $C_L(D)$ corresponding to the Martin boundary point y_0 for L on D_{y_0}. Using the theory of positive solutions of minimal growth, which is developed in Section 3, it is easy to show that $G(\cdot, y_0)$ is a minimal positive harmonic function; see exercise 7.9. Let $\{P_x^{G(\cdot, y_0)}, x \in \hat{D}_{y_0}\}$ denote the solution to the generalized martingale problem for the h-transformed operator $L^{G(\cdot, y_0)}$ on D_{y_0}. It then follows from Corollary 2.3 that $P_x(\cdot \mid m - \lim_{t\to\tau_{D_{y_0}}} X(t) = y_0) = P_x^{G(\cdot, y_0)}(\cdot)$. The diffusion corresponding to $P_x^{G(\cdot, y_0)}$ may be thought of as the diffusion corresponding to L on D and 'conditioned to approach y_0'.

As a specific example, consider $L = \frac{1}{2}\Delta$ on R^d, $d \geq 3$, and $y_0 = 0$. Then $G(x, 0) = c|x|^{2-d}$ and thus

$$\begin{aligned} \tfrac{1}{2}\Delta^{G(\cdot, 0)} &= \tfrac{1}{2}\Delta + \frac{\nabla G(\cdot, 0)}{G(\cdot, 0)} \cdot \nabla = \tfrac{1}{2}\Delta + \frac{2-d}{|x|}\frac{x}{|x|} \cdot \nabla \\ &= \frac{1}{2}\frac{\partial^2}{\partial r^2} + \frac{3-d}{2r}\frac{\partial}{\partial r} + \frac{1}{2r^2}\Delta_{S^{d-1}}. \end{aligned}$$

Note that this is the same operator as the one obtained in Example 2.5.

The probabilistic connection to the Martin boundary via Theorem 2.2 and Corollary 2.3, though very pretty, is not very useful for calculating Martin boundaries explicitly. A discussion of this appears in Chapter 8, Section 8. In the sections that follow, a useful alternative probabilistic approach to the Martin boundary is developed.

7.3 Positive solutions of minimal growth and applications to Green's functions and ground states

Let $L = L_0 + V$ on a domain $D \subseteq R^d$ and let $\Omega \subset\subset D$ be a subdomain. Note that $D - \bar{\Omega}$ is not necessarily a domain; that is, it is not necessarily connected. ($D - \bar{\Omega}$ will be a domain if and only if $d \geq 2$ and Ω is simply connected.) Since $\Omega \subset\subset D$, it is easy to see that $D - \bar{\Omega}$ has a finite number of components. These components will be denoted by $(D - \bar{\Omega})^{(1)}, \ldots, (D - \bar{\Omega})^{(k)}$. In this section and the two that follow, we will study the properties of certain positive solutions u of $Lu = 0$ in $D - \bar{\Omega}$. These results will be used in Section 6 to give an alternative characterization of the Martin boundary for subcritical operators L on D. However, it should be emphasized that the concepts developed in these sections are important in their own right. Some applications of the theory developed in this section appear in Examples 3.10–3.12 at the end of the section. The general theory developed in these sections applies equally well to critical and subcritical operators. The reason for this is as follows. Let $\Omega' \subset\subset \Omega \subset\subset D$. By Proposition 3.1 below, if L is critical on D, then it is subcritical on each connected component of $D - \bar{\Omega}$. Clearly, the study of positive solutions for L on $D - \bar{\Omega}$ does not depend on whether L is defined on $D - \bar{\Omega}'$ where it is subcritical, or on D, where it is critical.

Proposition 3.1. *Let L satisfy Assumption $\tilde{H}_{\text{loc}}$ on a domain $D \subseteq R^d$ and assume that L is critical on D. Then L is subcritical on each connected component of $D - \bar{\Omega}$ for any subdomain $\Omega \subset\subset D$.*

Proof. The proof follows from Corollary 4.4.2. □

Definition. *Let $D \subset R^d$. A function h defined on the complement $D - \Omega$ of some subdomain $\Omega \subset\subset D$ is called a* positive solution of minimal growth *at ∂D for L on D if $h \in C^2(D - \bar{\Omega}) \cap C(D - \Omega)$ and satisfies the following conditions*:

(i) *$Lh = 0$ and $h > 0$ on $D - \bar{\Omega}$;*
(ii) *If Ω' satisfies $\Omega \subseteq \Omega' \subset\subset D$ and if $u \in C^2(D - \bar{\Omega}') \cap C(D - \Omega')$ satisfies $Lu = 0$ and $u > 0$ in $D - \bar{\Omega}'$, and $u = h$ on $\partial\Omega'$, then $h(x) \leq u(x)$, for all $x \in D - \Omega'$.*

An existence theorem and a representation theorem for positive solutions of minimal growth at ∂D will be proved below. However, we need to estabish several preliminary results. Let $\sigma_{\bar{\Omega}} = \inf\{t \geq 0: X(t) \in \bar{\Omega}\}$.

Theorem 3.2. *Let $L = L_0 + V$ satisfy Assumption $\tilde{H}$ on a bounded domain $D \subset R^d$ with a $C^{2,\alpha}$-boundary. Assume that L is subcritical on D and let $G(x, y)$ denote the Green's function for L on D. Then for each $y \in D$,*

$$\lim_{D \ni x \to \partial D} G(x, y) = 0. \tag{3.1}$$

Also, if $\Omega \subset\subset D$ is a subdomain, then

$$G(x, y) = E_x\left(\exp\left(\int_0^{\sigma_\Omega} V(X(s))\mathrm{d}s\right) G(X(\sigma_{\bar\Omega}), y); \sigma_{\bar\Omega} < \tau_D\right),$$

$$\text{for } x \in D - \Omega \text{ and } y \in \Omega. \tag{3.2}$$

Proof. To prove (3.1), it will be convenient to have $V \equiv 0$. Thus, we first show that such an assumption can be made without loss of generality. By Theorem 4.3.2, the principal eigenvalue for L on D is negative. Thus, by Theorem 3.6.5, there exists an $h \in C^{2,\alpha}(\bar D)$ which satisfies $Lh = 0$ and $h > 0$ in $\bar D$. Then L^h, whose zeroth-order term vanishes identically, also satisfies Assumption $\tilde{\mathrm{H}}$ on D and, by Proposition 4.2.2, the Green's function $G^h(x, y)$ for L^h satisfies

$$G^h(x, y) = \frac{1}{h(x)} G(x, y)h(y).$$

Thus, we may assume without loss of generality that $V \equiv 0$.

Fix $y \in D$ and let U and Ω be domains satisfying $y \in U \subset\subset \Omega \subset\subset D$. Since $V \equiv 0$, we have from the strong Markov property,

$$G(x, U) = \int_U G(x, z)\mathrm{d}z = E_x \int_0^\infty 1_U(X(s))\,\mathrm{d}s = E_x G(X(\sigma_{\bar U}), U). \tag{3.3}$$

Thus, $M \equiv \sup_{x \in D} G(x, U) = \sup_{x \in \bar U} G(x, U) < \infty$. Applying the Markov property to the second expression from the right in (3.3), we obtain the inequality $G(x, U) \leq t + MP_x(\tau_D > t)$, for any $t > 0$. By Theorem 2.3.3(i), for any $x_0 \in \partial D$, $\lim_{D \ni x \to x_0} P_x(\tau_D > t) = 0$. Thus, since $t > 0$ is arbitrary, we conclude that

$$\lim_{D \ni x \to \partial D} \int_U G(x, z)\,\mathrm{d}z = 0. \tag{3.4}$$

By Theorem 4.2.5, for each $x \in D - \Omega$, $G(x, \cdot)$ satisfies $\tilde L G(x, \cdot) = 0$ in Ω. Thus, by Harnack's inequality, there exist positive constants c_1 and c_2, independent of $x \in D - \Omega$, such that

$$c_1 G(x, y) \leq G(x, z) \leq c_2 G(x, y), \quad \text{for } z \in U. \tag{3.5}$$

From (3.4) and (3.5), it follows that $\lim_{D \ni x \to \partial D} G(x, y) = 0$.

We now prove (3.2). We no longer assume that $V \equiv 0$. Since V is bounded on D, there exists an $\varepsilon > 0$ such that the principal eigenvalue of $L + \varepsilon V$ is still negative. By Theorem 3.6.5, there exists an $h \in C^{2,\alpha}(\bar D)$ such that $(L + \varepsilon V)h = 0$ and $h > 0$ on $\bar D$. Let $\Omega \subset\subset D$

be a subdomain and let $\{D_n\}_{n=1}^{\infty}$ be a sequence of domains satisfying $\Omega \subset\subset D_1$, $D_n \subset\subset D_{n+1}$ and $\bigcup_{n=1}^{\infty} D_n = D$. Then by the Feynman–Kac formula (Theorem 2.4.2),

$$h(x) = E_x \exp\left((1+\varepsilon)\int_0^{t\wedge\sigma_{\bar{\Omega}}\wedge\tau_{D_n}} V(X(s))\,ds\right) h(X(t\wedge\sigma_{\bar{\Omega}}\wedge\tau_D)),$$
$$\text{for } x \in D_n - \Omega,\ t \geqslant 0 \text{ and } n = 1, 2, \ldots.$$

Thus, there exists a constant c such that

$$E_x \exp\left((1+\varepsilon)\int_0^{t\wedge\sigma_{\bar{\Omega}}\wedge\tau_{D_n}} V(X(s))\,ds\right) \leqslant c,$$
$$\text{for } x \in D_n - \Omega, \quad t \geqslant 0 \text{ and } n = 1, 2, \ldots.$$

By Theorem 4.2.5, for $y \in \Omega$, $LG(\cdot\,; y) = 0$ on $D - \Omega$. Thus, by the Feynman–Kac formula,

$$G(x, y) = E_x \exp\left(\int_0^{t\wedge\sigma_{\bar{\Omega}}\wedge\tau_{D_n}} V(X(s))\,ds\right) G(X(t\wedge\sigma_{\bar{\Omega}}\wedge\tau_{D_n}), y),$$
$$\text{for } x \in D_n - \Omega,\ y \in \Omega \text{ and } t \geqslant 0. \quad (3.7)$$

By (3.6), $\exp\left(\int_0^{t\wedge\sigma_{\bar{\Omega}}\wedge\tau_{D_n}} V(X(s))\,ds\right)$ is uniformly integrable in t and n. Thus, (3.2) follows from (3.7) by letting $n \to \infty$ and then letting $t \to \infty$ and using (3.1). □

The next theorem shows that the representation (3.2) continues to hold in the case that $D \subseteq R^d$ is an arbitrary domain.

Theorem 3.3. *Let $L = L_0 + V$ satisfy Assumption $\widetilde{H}_{\text{loc}}$ on a domain $D \subseteq R^d$. Assume that L is subcritical on D and let $G(x, y)$ denote the Green's function for L on D. Let $\Omega \subset\subset D$ be a subdomain. Then*

$$G(x, y) = E_x\left(\exp\left(\int_0^{\sigma_{\bar{\Omega}}} V(X(s))\,ds\right) G(X(\sigma_{\bar{\Omega}}), y);\ \sigma_{\Omega} < \tau_D\right),$$
$$\text{for } x \in D - \Omega \text{ and } y \in \Omega. \quad (3.8)$$

Proof. Let $\{D_n\}_{n=1}^{\infty}$ be a sequence of domains with $C^{2,\alpha}$-boundaries satisfying $\Omega \subset\subset D_1$, $D_n \subset\subset D_{n+1}$ and $\bigcup_{n=1}^{\infty} D_n = D$, and let $G_n(x, y)$ denote the Green's function for L on D_n. By Theorem 4.3.7, $G(x, y) = \lim_{n\to\infty} G_n(x, y)$, for $x, y \in D$ and $x \neq y$. By Theorem 3.2,

$$G_n(x, y) = E_x\left(\exp\left(\int_0^{\sigma_{\bar{\Omega}}} V(X(s))\,ds\right) G_n(X(\sigma_{\bar{\Omega}}), y);\ \sigma_{\bar{\Omega}} < \tau_{D_n}\right),$$
$$\text{for } x \in D_n - \Omega \text{ and } y \in \Omega.$$

The theorem now follows from the monotone convergence theorem. □

The following lemma will be very useful in the sequel.

Lemma 3.4. *Let $L = L_0 + V$ satisfy Assumption $\widetilde{H}_{\text{loc}}$ on a domain $D \subseteq R^d$ and assume that L is either critical or subcritical on D. Let $\Omega \subset\subset D$ be a subdomain and let $\phi \in C(D - \Omega)$.*

(i) *$E_x(\exp(\int_0^{\sigma_{\bar{\Omega}}} V(X(s))\,ds);\ \sigma_{\bar{\Omega}} < \tau_D)$ is locally bounded on $D - \Omega$.*

(ii) *If $\{\Omega_n\}_{n=1}^{\infty}$ is a sequence of domains satisfying $\Omega_1 \subset\subset D$, $\Omega_{n+1} \subset\subset \Omega_n$ and $\bigcap_{n=1}^{\infty} \Omega_n = \bar{\Omega}$, then*

$$\lim_{n\to\infty} E_x\left(\exp\left(\int_0^{\sigma_{\bar{\Omega}_n}} V(X(s))\,ds\right)\phi(X(\sigma_{\bar{\Omega}_n}));\ \sigma_{\bar{\Omega}_n} < \tau_D\right)$$
$$= E_x\left(\exp\left(\int_0^{\sigma_{\bar{\Omega}}} V(X(s))ds\right)\phi(X(\sigma_{\bar{\Omega}}));\ \sigma_{\bar{\Omega}} < \tau_D\right), \quad \textit{for } x \in D - \Omega.$$

(iii) *$E_x(\exp(\int_0^{\sigma_{\bar{\Omega}}} V(X(s))\,ds)\phi(X(\sigma_{\bar{\Omega}}));\ \sigma_{\bar{\Omega}} < \tau_D)$ is a $C^{2,\alpha}$-function of $x \in D - \bar{\Omega}$.*

Remark. The proof of part (iii) of the lemma utilizes Theorem 3.5 below. The reason the lemma is positioned here is that parts (i) and (ii) are used for the proof of Theorem 3.5. Part (i) is, of course, a consequence of part (iii).

Proof. We will first prove the three parts of the lemma under the assumption that L is subcritical on D. Then we will note the minor modification that must be made to accommodate the critical case. Let $G(x, y)$ denote the Green's function for L on D.

(i) By (3.8),

$$E_x\left(\exp\left(\int_0^{\sigma_{\bar{\Omega}}} V(X(s))\,ds\right);\ \sigma_{\bar{\Omega}} < \tau_D\right) \leqslant \frac{G(x, y)}{\inf\limits_{z \in \partial\Omega} G(z, y)},$$

for $x \in D - \Omega$ and any $y \in \Omega$.

(ii) Fix $y \in \Omega$. Define

$$f_n(X(\cdot)) = \exp\left(\int_0^{\sigma_{\bar{\Omega}_n}} V(X(s))\,ds\right) 1_{\{\sigma_{\bar{\Omega}_n} < \tau_D\}} \phi(X(\sigma_{\bar{\Omega}_n}))$$

and

$$g_n(X(\cdot)) = \exp\left(\int_0^{\sigma_{\bar{\Omega}_n}} V(X(s))\,ds\right) 1_{\{\sigma_{\bar{\Omega}_n} < \tau_D\}} G(X(\sigma_{\bar{\Omega}_n}, y)).$$

Define $f(X(\cdot))$ and $g(X(\cdot))$ to be the functions obtained by replacing Ω_n by Ω in the definition of f_n and g_n. We need to show that $E_x f = \lim_{n\to\infty} E_x f_n$. If a path $X(\cdot) \in \hat{\Omega}_D$ with $X(0) \in D - \bar{\Omega}$ satisfies $f(X(\cdot)) \neq \lim_{n\to\infty} f_n(X(\cdot))$, then $\sigma_{\bar{\Omega}_n}(X(\cdot)) < \infty$ for $n = 1, 2, \ldots,$

and $\sigma_{\bar{\Omega}}(X(\cdot)) = \infty$. This necessitates $X(t) \in D - \bar{\Omega}$ for all $t \geq 0$ and $X(t) \in \Omega_1 - \bar{\Omega}$ for arbitrarily large t. But such behavior is compatible with neither transience nor recurrence. Since, by Theorem 2.8.1, $\{P_x, x \in \hat{D}\}$ must correspond to either a transient or a recurrent diffusion, it follows that $f(X(\cdot)) = \lim_{n\to\infty} f_n(X(\cdot))$, a.s. P_x, for $x \in D - \bar{\Omega}$. By (3.8), $E_x g = E_x g_n$ for $x \in D - \Omega_n$. Also, there exists a constant c such that $|f_n| \leq c g_n$, for all n. Therefore, by a slightly extended version of the dominated convergence theorem, $E_x f = \lim_{n\to\infty} E_x f_n$.

(iii) Let $\{\Omega_n\}_{n=1}^{\infty}$ be a sequence of domains which satisfy an interior cone condition at each point of their boundaries and which satisfy $\Omega_1 \subset\subset D$, $\Omega_{n+1} \subset\subset \Omega_n$ and $\bigcap_{n=1}^{\infty} \Omega_n = \bar{\Omega}$. Let

$$h_n(x) \equiv E_x\left(\exp\left(\int_0^{\sigma_{\bar{\Omega}_n}} V(X(s))\,ds\right)\phi(X(\sigma_{\bar{\Omega}_n}));\quad \sigma_{\bar{\Omega}_n} < \tau_D\right),\qquad x \in D - \Omega_n.$$

By part (ii),

$$h(x) \equiv \lim_{n\to\infty} h_n(x) = E_x\left(\exp\left(\int_0^{\sigma_{\bar{\Omega}}} V(X(s))\,ds\right)\phi(X(\sigma_{\bar{\Omega}}));\quad \sigma_{\bar{\Omega}} < \tau_D\right),\qquad x \in D - \bar{\Omega}.$$

By Theorem 3.5 below, $h_n \in C^{2,\alpha}(D - \bar{\Omega}_n)$. Thus, a standard compactness argument shows that $h \in C^{2,\alpha}(D - \bar{\Omega})$.

We now turn to the critical case. Let $\Omega' \subset\subset \Omega$ be a subdomain. By Proposition 3.1, L is subcritical on each connected component of $D - \bar{\Omega}'$. Using the Green's functions for L on these domains in the above proofs will prove the lemma in the critical case. □

We can now prove an existence theorem and a representation theorem for positive solutions of minimal growth at ∂D.

Theorem 3.5. *Let $L = L_0 + V$ satisfy Assumption $\tilde{H}_{\mathrm{loc}}$ on a domain $D \subseteq R^d$ and assume that L is either critical or subcritical on D. Let $\Omega \subset\subset D$ be a subdomain which satisfies an interior cone condition at each point of its boundary. Then for each non-negative and not identically zero $\phi \in C(\partial\Omega)$, there exists a positive solution $h \in C^{2,\alpha}(D - \bar{\Omega}) \cap C(D - \Omega)$ of minimal growth at ∂D for L on D which satisfies $h = \phi$ on $\partial\Omega$. The solution is given by*

$$h(x) = E_x\left(\exp\left(\int_0^{\sigma_{\bar{\Omega}}} V(X(s))\,ds\right)\phi(X(\sigma_{\bar{\Omega}}));\quad \sigma_{\bar{\Omega}} < \tau_D\right),\qquad x \in D - \Omega. \tag{3.9}$$

Theorem 3.6. *Let $L = L_0 + V$ satisfy Assumption $\widetilde{H}_{\text{loc}}$ on a domain $D \subseteq R^d$ and assume that L is either critical or subcritical on D. Let $\Omega \subset\subset D$ be a subdomain and let $h \in C^{2,\alpha}(D - \bar{\Omega}) \cap C(D - \Omega)$ be a non-negative and not identically zero. Then h is a positive solution of minimal growth at ∂D for L on D if and only if*

$$h(x) = E_x\left(\exp\left(\int_0^{\sigma_{\bar{\Omega}}} V(X(s))\,\mathrm{d}s\right)h(X(\sigma_{\bar{\Omega}}));\quad \sigma_{\bar{\Omega}} < \tau_D\right),\quad x \in D - \Omega. \tag{3.10}$$

In particular, positive solutions of minimal growth at ∂D satisfy the maximum principle in the following sense: if $h_1, h_2 \in C^{2,\alpha}(D - \bar{\Omega}) \cap C(D - \Omega)$ are positive solutions of minimal growth at ∂D and $h_1 \leqslant h_2$ on $\partial\Omega$, then $h_1 \leqslant h_2$ on $D - \Omega$.

From (3.8) and Theorem 3.5 or 3.6, the following theorem is immediate.

Theorem 3.7. *Let $L = L_0 + V$ satisfy Assumption $\widetilde{H}_{\text{loc}}$ on a domain $D \subseteq R^d$. Assume that L is subcritical on D and let $G(x, y)$ denote the Green's function for L on D. Then for each $y \in D$, $G(\cdot, y)$ is a positive solution of minimal growth at ∂D for L on D.*

Proof of Theorem 3.5. First assume that $\partial\Omega$ is a $C^{2,\alpha}$-boundary and that $\phi \in C^{2,\alpha}(D - \Omega)$. Let $\{D_n\}_{n=1}^{\infty}$ be an increasing sequence of domains with $C^{2,\alpha}$-boundaries such that $\Omega \subset\subset D_1$, $D_n \subset\subset D_{n+1}$ and $\bigcup_{n=1}^{\infty} D_{n+1} = D$. Since $\lambda_c(D) \leqslant 0$, it follows from Theorem 4.4.1(iii) that the principal eigenvalue of each connected component of $D_n - \bar{\Omega}$ is negative. Thus, by Theorem 3.6.5 applied to each connected component of $D - \bar{\Omega}$, there exists a unique solution $h_n \in C^2(\bar{D}_n - \Omega)$ to $Lh_n = 0$ in $D_n - \bar{\Omega}$, $h_n = \phi$ on $\partial\Omega$ and $h_n = 0$ on ∂D_n. By Theorem 3.6.6,

$$h_n(x) = E_x\left(\exp\left(\int_0^{\sigma_{\bar{\Omega}}} V(X(s))\,\mathrm{d}s\right)\phi(X(\sigma_{\Omega}));\, \sigma_{\bar{\Omega}} < \tau_{D_n}\right). \tag{3.11}$$

Thus $h_n \geqslant 0$, $h_{n+1} \geqslant h_n$ and, by Lemma 3.4(i), $\limsup_{n\to\infty} h_n(x) < \infty$, for $x \in D - \Omega$. Thus $h(x) \equiv \lim_{n\to\infty} h_n(x)$ exists for $x \in D - \Omega$ and, by a standard compactness argument, $h \in C^{2,\alpha}(D - \bar{\Omega})$ and $Lh = 0$ in $D - \bar{\Omega}$. By construction $h = \phi$ on $\partial\Omega$. We now show that $h \in C(D - \Omega)$.

By Theorem 3.6.6, for $x \in D_1 - \Omega$ and $n = 1, 2, \ldots,$

$$h_n(x) = E_x \exp\left(\int_0^{\sigma_{\bar{\Omega}} \wedge \tau_{D_1}} V(X(s))\,\mathrm{d}s\right)h_n(X(\sigma_{\bar{\Omega}} \wedge \tau_{D_1})). \tag{3.12}$$

Using Theorems 2.3.3(i) and 2.2.2(i), the bounded convergence theorem, the boundedness of V on D_1 and of $\{h_n\}_{n=1}^{\infty}$ on ∂D_1, and the fact that $h_n = \phi \in C(\partial\Omega)$ on $\partial\Omega$, it follows that for $x_0 \in \partial\Omega$,

$$\lim_{x\to x_0} \sup_n \left| E_x\left(\exp\left(\int_0^{\sigma_{\bar\Omega}\wedge\tau_{D_1}} V(X(s))\,ds\right) h_n(X(\sigma_{\bar\Omega}\wedge\tau_{D_1}));\right.\right.$$

$$\left.\left.\sigma_{\bar\Omega}\wedge\tau_{D_1} \leqslant 1\right) - \phi(x_0)\right| = 0. \quad (3.13)$$

Since V is bounded on $D_1 - \Omega$, and since $\lambda_0(D_1 - \bar\Omega) < 0$, for sufficiently small $\varepsilon > 0$, the principal eigenvalue of $L + \varepsilon V$ on $D_1 - \bar\Omega$ is negative. Thus, from Theorem 3.6.6,

$$\sup_{x\in D_1-\Omega} E_x \exp\left((1+\varepsilon)\int_0^{\sigma_{\bar\Omega}\wedge\tau_{D_1}} V(X(s))\,ds\right) < \infty. \quad (3.14)$$

By Hölder's inequality and the boundedness of $\{h_n\}_{n=1}^{\infty}$ on ∂D_1, there exists a $c > 0$ such that

$$E_x\left(\exp\left(\int_0^{\sigma_{\bar\Omega}\wedge\tau_{D_1}} V(X(s))\,ds\right) h_n(X(\sigma_{\bar\Omega}\wedge\tau_{D_1}));\quad \sigma_{\bar\Omega}\wedge\tau_{D_1} > 1\right) \quad (3.15)$$
$$\leqslant c\left(E_x \exp\left((1+\varepsilon)\int_0^{\sigma_{\bar\Omega}\wedge\tau_{D_1}} V(X(s))\,ds\right)\right)^{\frac{1}{1+\varepsilon}} \left(P_x(\sigma_{\bar\Omega}\wedge\tau_{D_1} > 1)\right)^{\frac{\varepsilon}{1+\varepsilon}}.$$

By Theorem 2.3.3(i) and 2.2.2(i),

$$\lim_{x\to x_0} P_x(\sigma_{\bar\Omega}\wedge\tau_{D_1} > 1) = 0. \quad (3.16)$$

From (3.12)–(3.16), it follows that $\lim_{x\to x_0}\sup_n |h_n(x) - \phi(x_0)| = 0$; thus $\lim_{x\to x_0} |h(x) - \phi(x_0)| = 0$, and $h \in C(D - \Omega)$.

To show that h is a positive solution of minimal growth at ∂D, suppose that Ω' is a domain satisfying $\Omega \subseteq \Omega' \subset\subset D$ and that $u \in C^{2,\alpha}(D - \bar\Omega') \cap C(D - \Omega')$ satisfies $Lu = 0$ and $u > 0$ in $D - \bar\Omega'$ and $u = h$ on $\partial\Omega'$. For sufficiently large n, $\Omega' \subset\subset D_n$, and for such n, $u \geqslant h_n$ on $\partial D_n \cup \partial\Omega'$; therefore, by the maximum principle in Theorem 3.6.5, $u \geqslant h_n$ on $D_n - \Omega'$, and we conclude that $u \geqslant h$ on $D - \Omega'$. Thus, h is a positive solution of minimal growth at ∂D.

Finally, (3.9) follows from (3.11) and the fact that $h = \lim_{n\to\infty} h_n$.

We now extend the proof to the case where Ω satisfies an interior cone condition at each point of its boundary and $\phi \in C(\partial\Omega)$. Extend ϕ to $D - \Omega$ so that $\phi \in C^{2,\alpha}(D - \bar\Omega) \cap C(D - \Omega)$. Let $\{\Omega_n\}_{n=1}^{\infty}$ be a sequence of domains with $C^{2,\alpha}$-boundaries satisfying $\Omega_1 \subset\subset D$, $\Omega_{n+1} \subset\subset \Omega_n$ and $\bigcap_{n=1}^{\infty}\Omega_n = \bar\Omega$. By what we have already proved,

$$h_n(x) \equiv E_x\left(\exp\left(\int_0^{\sigma_{\bar\Omega_n}} V(X(s))\,ds\right)\phi(X(\sigma_{\bar\Omega_n}));\, \sigma_{\bar\Omega_n} < \tau_D\right),\, x \in D - \Omega_n,$$

is a positive solution of minimal growth at ∂D. By Lemma 3.4(ii),

$$h(x) \equiv \lim_{n\to\infty} h_n(x) = E_x \exp\left(\left(\int_0^{\sigma_{\bar{\Omega}}} V(X(s))\,ds\right)\phi(X(\sigma_{\bar{\Omega}}));\ \sigma_{\bar{\Omega}} < \tau_D\right).$$

By a standard compactness argument, $h \in C^{2,\alpha}(D - \bar{\Omega})$ and $Lh = 0$; in $D - \bar{\Omega}$. The proof that $h \in C(D - \Omega)$ is the same as the proof above under the additional assumptions, since Theorem 2.3.3 holds for boundaries satisfying appropriate cone conditions. (Theorem 2.3.3 is stated in terms of an exterior cone condition, but when applied to the present setup, the condition required on Ω is an interior cone condition.) The proof that it is a positive solution of minimal growth at ∂D is similar to the proof above under the additional assumptions. □

Proof of Theorem 3.6. First assume that h is a positive solution of minimal growth at ∂D. Let $\{\Omega_n\}_{n=1}^{\infty}$ be a sequence of domains satisfying an interior cone condition at each point of their boundaries and such that $\Omega_1 \subset\subset D$, $\Omega_{n+1} \subset\subset \Omega_n$ and $\bigcap_{n=1}^{\infty} \Omega_n = \bar{\Omega}$. Then by Theorem 3.5,

$$h(x) = E_x\left(\exp\left(\int_0^{\sigma_{\bar{\Omega}_n}} V(X(s))\,ds\right)h(X(\sigma_{\bar{\Omega}_n}));\quad \sigma_{\bar{\Omega}_n} < \tau_D\right),$$

$$\text{for } x \in D - \Omega_n.$$

Letting $n \to \infty$ and using Lemma 3.4(ii), it follows that h satisfies (3.10).

Now assume that h satisfies (3.10). Let Ω' be a domain satisfying $\Omega \subseteq \Omega' \subset\subset D$ and let $u \in C^{2,\alpha}(D - \bar{\Omega}') \cap C(D - \Omega')$ satisfy $Lu = 0$ and $u > 0$ in $D - \bar{\Omega}'$ and $u = h$ on $\partial\Omega'$. We need to show that $u \geqslant h$ on $D - \Omega'$. An application of the strong Markov property shows that

$$h(x) = E_x\left(\exp\left(\int_0^{\sigma_{\bar{\Omega}'}} V(X(s))\,ds\right)h(X(\sigma_{\bar{\Omega}'}));\right.$$

$$\left.\sigma_{\bar{\Omega}'} < \tau_D\right),\quad x \in D - \Omega'. \quad (3.17)$$

Let $\{\Omega'_n\}_{n=1}^{\infty}$ be a sequence of domains satisfying an interior cone condition at each point of their boundaries and such that $\Omega'_1 \subset\subset D$, $\Omega'_{n+1} \subset\subset \Omega'_n$ and $\bigcap_{n=1}^{\infty} \Omega'_n = \bar{\Omega}'$. By Theorem 3.5,

$$u(x) \geqslant E_x\left(\exp\left(\int_0^{\sigma_{\bar{\Omega}'_n}} V(X(s))\,ds\right)u(X(\sigma_{\bar{\Omega}'_n}));\quad \sigma_{\bar{\Omega}'_n} < \tau_D\right),$$

$$x \in D - \Omega'_n.$$

Letting $n \to \infty$ and using Lemma 3.4(ii) and (3.17) gives $u(x) \geqslant h(x)$.

Finally, since (3.17) holds for arbitrary Ω', it follows from Theorem 3.5 and (3.17) that $Lh = 0$ in $D - \bar{\Omega}$. □

The next two theorems show, in particular, that in the critical case the ground state is a positive solution of minimal growth at ∂D, and that in the subcritical case no element of $C_L(D)$ is a positive solution of minimal growth at ∂D.

Theorem 3.8. *Let L satisfy Assumption $\tilde{H}_{\text{loc}}$ on a domain $D \subseteq R^d$. Assume that L is critical on D and let ϕ_c denote the ground state.*

(i) *Let $W \in C_0^\alpha(D)$ and assume that $L - W$ is subcritical on D. (In particular, by Theorem 4.6.3, this will be true if $W \geq 0$ and $W \not\equiv 0$.) Let $G_W(x, y)$ denote the Green's function for $L - W$ on D. Then ϕ_c satisfies*

$$\phi_c(x) = \int_D G_W(x, y) W(y) \phi_c(y)\, dy. \tag{3.18}$$

(ii) *ϕ_c is a positive solution of minimal growth at ∂D for L on D.*

(iii) *Let W be as in part (i). For any domain $\Omega \subset\subset D$ and any $y_0 \in \Omega$, there exist positive constants c_1 and c_2 (depending on Ω and y_0) such that*

$$c_1 G_W(x, y_0) \leq \phi_c(x) \leq c_2 G_W(x, y_0), \quad \text{for } x \in D - \Omega. \tag{3.19}$$

Proof. (i) For $\alpha > 0$, let G_α denote the Green's function for $L - \alpha W$ on D; thus $G_1 \equiv G_W$. By the resolvent equation (Theorem 4.3.11),

$$G_\alpha(x, y) = G_1(x, y) + (1 - \alpha) \int_D G_1(x, z) W(z) G_\alpha(z, y)\, dz. \tag{3.20}$$

Fix $x_0 \in \Omega$ and let $\{\alpha_k\}_{k=1}^\infty$ be a decreasing sequence satisfying $\alpha_1 = 1$ and $\lim_{k\to\infty} \alpha_k = 0$. Note that

$$0 < \frac{G_1(x_0, y)}{G_{\alpha_k}(x_0, y)} \leq 1, \text{ for } k = 1, 2, \ldots \quad \text{and } y \in D - \{x_0\}.$$

Thus, by diagonalization, we can choose a sequence $\{y_n\}_{n=1}^\infty \subset D$ such that for all $k = 1, 2, \ldots, \{y_n\}_{n=1}^\infty$ is a Martin sequence for

$$L - \alpha_k W \text{ and } \gamma_k \equiv \lim_{n\to\infty} \frac{G_1(x_0, y_n)}{G_{\alpha_k}(x_0, y_n)} \text{ exists.}$$

Let u_k denote the element in $C_{L-\alpha_k W}(D)$ corresponding to the Martin sequence $\{y_n\}_{n=1}^\infty$. Dividing each side of (3.20) by $G_\alpha(x_0, y)$, substituting α_k for α and y_n for y, and then letting $n \to \infty$, we obtain

$$u_k(x) = \gamma_k u_1(x) + (1 - \alpha_k) \int_D G_1(x, z) W(z) u_k(z)\, dz. \tag{3.21}$$

By a standard compactness argument and Theorem 4.3.4,

$$\lim_{k\to\infty} u_k(x) = \phi_c(x), \tag{3.22}$$

where ϕ_c has been normalized by $\phi_c(x_0) = 1$. We now show that

$$\lim_{k\to\infty} \gamma_k = 0. \tag{3.23}$$

Let $\Omega = \operatorname{supp} W$. By Theorem 4.3.7, $\lim_{k\to\infty} G_{\alpha_k}(x_0, y) = \infty$, for all $y \in D - \{x_0\}$. Thus, by Harnack's inequality, there exists a sequence $\{\varepsilon_k\}$ with $\lim_{k\to\infty} \varepsilon_k = 0$ such that $G_1(x_0, y) \leqslant \varepsilon_k G_{\alpha_k}(x_0, y)$, for $y \in \partial\Omega$. Now, using Theorem 3.7 to represent $G_1(x_0, y)$ and $G_{\alpha_k}(x_0, y)$ for $y \in D - \Omega$, and using the fact that $W = 0$ on $D - \Omega$, it follows that $G_1(x_0, y) \leqslant \varepsilon_k G_{\alpha_k}(x_0, y)$, for $y \in D - \Omega$; hence (3.23) holds. Letting $k \to \infty$ in (3.21), and using (3.22) and (3.23), we obtain (3.18).

(ii) Let $\Omega \subset\subset D$ be a domain with a smooth boundary and fix $y_0 \in \Omega$. Choose W supported in Ω and satisfying $W \geqslant 0$ and $W \not\equiv 0$. Denote the Green's function for $L - W$ by $G_W(x, y)$. By (3.18) and Harnack's inequality, there exist constants $c_1, c_2 > 0$ such that

$$c_1 G_W(x, y_0) \leqslant \phi_c(x) \leqslant c_2 G_W(x, y_0), \quad \text{for } x \in D - \Omega. \tag{3.24}$$

Define

$$u(x) = E_x\left(\exp\left(\int_0^{\sigma_{\bar\Omega}} V(X(s))\,ds\right)\phi_c(X(\sigma_{\bar\Omega}));\ \sigma_{\bar\Omega} < \tau_D\right);\ x \in D - \Omega.$$

By Theorem 3.5, u is a positive solution of minimal growth at ∂D for L on D which satisfies $u = \phi_c$ on $\partial\Omega$. By the definition of minimality,

$$\phi_c \geqslant u \text{ on } D - \Omega. \tag{3.25}$$

By (3.24), Theorem 3.3, and the fact that W is supported in Ω, there exist constants $c_3, c_4 > 0$ such that

$$c_3\phi_c(x) \leqslant u(x) \leqslant c_4\phi_c(x), \quad \text{for } x \in D - \Omega. \tag{3.26}$$

Let $\gamma = \sup\{t \geqslant 0\colon u - t\phi_c > 0 \text{ in } D - \bar\Omega\}$. By (3.25) and (3.26), we have $\gamma \in (0, 1]$. To complete the proof, we must show that $\gamma = 1$. Assume on the contrary that $\gamma < 1$ and let $w = u - \gamma\phi_c$. Then $w > 0$ on $\partial\Omega$ and $Lw = 0$ in $D - \bar\Omega$. By the strong maximum principle (Theorem 3.2.6), $w > 0$ in $D - \bar\Omega$. Let v denote the positive solution of minimal growth at ∂D which satisfies $v = w$ on $\partial\Omega$. By minimality, $v \leqslant w$. By Theorem 3.3, Theorem 3.5, and (3.24), there exists a constant $c_5 > 0$ such that $v \geqslant c_5\phi_c$ on $D - \Omega$; thus $w \geqslant c_5\phi_c$ on $D - \Omega$. From this it follows that $\sup\{t > 0\colon w - t\phi_c > 0 \text{ on } D - \bar{\Omega}\} > 0$, which contradicts the definition of γ.

(iii) Let $\Omega_0 \subset\subset D$ contain the support of W. By (ii), ϕ_c is a positive solution of minimal growth and, by Theorem 3.7, $G_W(x, y)$ is also one. Since G_W and ϕ_c are strictly positive on $\partial\Omega_0$, it then follows from Theorem 3.6 that (3.19) holds with $\Omega = \Omega_0$. The extension to arbitrary $\Omega \subset\subset D$ is immediate since G_W and ϕ_c are positive functions. □

Theorem 3.9. *Let L satisfy Assumption $\tilde{H}_{\mathrm{loc}}$ on a domain $D \subseteq R^d$ and assume that L is subcritical on D. Fix $y_0 \in D$ and fix a domain $\Omega \subset\subset D$ satisfying $y_0 \in \Omega$.*

(i) *If $u \in C_L(D)$, then*

$$G(x, y_0) \leqslant cu(x), \quad \text{for } x \in D - \Omega,$$

where

$$c = \sup_{x \in \partial\Omega} \frac{G(x, y_0)}{u(x)}.$$

(ii) *There is no $u \in C_L(D)$ which satisfies*

$$u(x) \leqslant cG(x, y_0), \text{ for } x \in D - \Omega \text{ and some } c > 0.$$

In particular then, there is no $u \in C_L(D)$ which is a positive solution of minimal growth at ∂D for L on D.

Proof. Let $\{\Omega_m\}_{m=1}^{\infty}$ and $\{D_n\}_{n=1}^{\infty}$ be sequences of domains satisfying $\Omega_1 \subset\subset \Omega$, $\Omega_{m+1} \subset\subset \Omega_m$, $\bigcap_{m=1}^{\infty}\Omega_m = \{y_0\}$, $\Omega \subset\subset D_1$, $D_n \subset\subset D_{n+1}$ and $\bigcup_{n=1}^{\infty} D_n = D$.

(i) Let $G_n(x, y)$ denote the Green's function for L on D_n. Then by (3.1) and the maximum principle, it follows that

$$G_n(x, y_0) \leqslant c_n u(x), \quad \text{for } x \in D_n - \Omega,$$

where

$$c_n = \sup_{x \in \partial\Omega} \frac{G_n(x, y_0)}{u(x)}.$$

Now let $n \to \infty$ and invoke Theorem 4.3.7(i).

(ii) Assume on the contrary that there exists a $u \in C_L(D)$ and a $c > 0$ such that $u(x) \leqslant cG(x, y_0)$, for $x \in D - \Omega$. By Theorem 4.4.1, the generalized principal eigenvalue of L on $D_n - \bar{\Omega}_m$ is negative; thus by Theorem 3.6.6,

$$G(x, y_0) = E_x\left(\exp\left(\int_0^{\sigma_{\bar{\Omega}_m}} V(X(s))\,ds\right)G(X(\sigma_{\bar{\Omega}_m}), y_0);\ \sigma_{\bar{\Omega}_m} < \tau_{D_n}\right)$$
$$+ E_x\left(\exp\left(\int_0^{\tau_{D_n}} V(X(s))\,ds\right)G(X(\tau_{D_n}), y_0);\ \tau_{D_n} < \sigma_{\bar{\Omega}_m}\right),$$
$$x \in D_n - \Omega_m,\ m, n = 1, 2, \ldots. \quad (3.27)$$

On the other hand, by Theorem 3.3,

$$G(x, y_0) = E_x\left(\exp\left(\int_0^{\sigma_{\bar{\Omega}_m}} V(X(s))\,ds\right)G(X(\sigma_{\bar{\Omega}_m}), y_0);\ \sigma_{\bar{\Omega}_m} < \tau_D\right),$$
$$x \in D - \Omega_m,\ m = 1, 2, \ldots. \quad (3.28)$$

Letting $n \to \infty$ in (3.27), and using (3.28) we have

$$\lim_{n\to\infty} E_x\left(\exp\left(\int_0^{\tau_{D_n}} V(X(s))\,ds\right)G(X(\tau_{D_n}), y_0);\ \tau_{D_n} < \sigma_{\bar{\Omega}_m}\right) = 0,$$
$$x \in D - \Omega_m,\ m = 1, 2, \ldots. \quad (3.29)$$

Applying Theorem 3.6.6 to u on $D_n - \bar{\Omega}_m$ gives

$$u(x) = E_x\left(\exp\left(\int_0^{\sigma_{\bar{\Omega}_m}} V(X(s))\,ds\right)u(X(\sigma_{\bar{\Omega}_m}));\ \sigma_{\bar{\Omega}_m} < \tau_{D_n}\right)$$
$$+ E_x\left(\exp\left(\int_0^{\tau_{D_n}} V(X(s))\,ds\right)u(X(\tau_{D_n}));\ \tau_{D_n} < \sigma_{\bar{\Omega}_m}\right),$$
$$x \in D_n - \Omega_m,\ m, n = 1, 2, \ldots. \quad (3.30)$$

Since $u(x) \le cG(x, y_0)$, for $x \in D - \Omega$, it follows from (3.29) and (3.30), upon letting $n \to \infty$, that

$$u(x) = E_x\left(\exp\left(\int_0^{\sigma_{\bar{\Omega}_m}} V(X(s))\,ds\right)u(X(\sigma_{\bar{\Omega}_m}));\ \sigma_{\bar{\Omega}_m} < \tau_D\right),$$
$$x \in D - \Omega_m,\ m = 1, 2, \ldots. \quad (3.31)$$

By Theorem 4.2.9, $\lim_{x\to y_0} G(x, y_0) = \infty$. Using this with (3.28) and (3.31), it follows that there exists a sequence $\{\varepsilon_m\}_{m=1}^{\infty}$ satisfying $\lim_{m\to\infty} \varepsilon_m = 0$ such that $u(x) \le \varepsilon_m G(x, y_0)$, for $x \in D - \Omega_m$. Thus, $u \equiv 0$, which contradicts the assumption that $u \in C_L(D)$.

For the second statement in part (ii), note that if $u \in C_L(D)$ is a positive solution of minimal growth at ∂D, then, by Theorem 3.3 and Theorem 3.6, $u(x) \le cG(x, y_0)$, for $x \in D - \Omega$ and some $c > 0$. □

We close with a sampling of interesting examples which can be proved quite simply using the results developed in this section.

Example 3.10. Consider the operators $\frac{1}{2}\Delta + tW$ and $\frac{1}{2}\Delta + tV$ on R^d, $d \geq 3$, where $V, W \in C_0^\alpha(R^d)$, $V, W \not\equiv 0$, $W \geq 0$ and V changes sign. By Theorem 4.6.4, the operator $\frac{1}{2}\Delta + tW$ on R^d is critical for a unique value $t = t_0 > 0$; denote the ground state of $\frac{1}{2}\Delta + t_0 W$ on R^d by ϕ_c. By Theorem 4.6.5, there exist exactly two values of t, $t = t^+ > 0$ and $t = t^- < 0$, for which $\frac{1}{2}\Delta + tV$ on R^d is critical; denote the ground state of $\frac{1}{2}\Delta + t^\pm V$ on R^d by $\phi_c^\pm$. The operators $\frac{1}{2}\Delta + t_0 W$, $\frac{1}{2}\Delta + t^+ W$ and $\frac{1}{2}\Delta + t^- W$ are symmetric with respect to Lebesgue measure. Thus, as discussed in Chapter 4, Section 10, each of these operators, defined initially on $C_0^\infty(R^d)$, can be extended by the Friedrichs' extension to a self-adjoint operator on $L^2(R^d)$. Since the generalized principal eigenvalue is, of course, equal to zero for the critical operators $\frac{1}{2}\Delta + t_0 W$, $\frac{1}{2}\Delta + t^+ V$ and $\frac{1}{2}\Delta + t^- V$ on R^d, it follows from Proposition 4.10.1 that the supremum of the spectrum for each of the associated self-adjoint operators is also equal to zero.

We now investigate whether the point zero is in fact an eigenvalue for the above self-adjoint operators. By Theorem 4.10.2, zero will be an eigenvalue for one of these operators if and only if its corresponding ground state belongs to $L^2(R^d)$. We utilize Theorem 3.8(iii). Let $\frac{1}{2}\Delta + t_0 W$, $\frac{1}{2}\Delta + t^+ V$ or $\frac{1}{2}\Delta + t^- V$ play the role of the critical operator L in that theorem, let $t_0 W$, $t^+ V$ or $t^- V$ play the role of W, and let $\frac{1}{2}\Delta$ play the role of the subcritical operator $L - W$. By Example 4.2.7, the Green's function for $\frac{1}{2}\Delta$ on R^d is given by $c|x - y|^{2-d}$. Thus, by Theorem 3.8(iii), there exist positive constants c_1 and c_2 such that $c_1|x|^{2-d} \leq \phi_c, \phi_c^+, \phi_c^- \leq c_2|x|^{2-d}$, for $|x| > 1$. Since

$$\int_{|x|>1} (|x|^{2-d})^2 \, dx = \omega_d \int_1^\infty r^{4-2d} r^{d-1} \, dr = \omega_d \int_1^\infty r^{3-d} \, dr,$$

where ω_d is the surface area of the unit sphere in R^d, it follows that ϕ_c, ϕ_c^+, $\phi_c^- \in L^2(R^d)$ if and only if $d \geq 5$. Thus, the point zero, which is the supremum of the spectrum of the Friedrichs' extensions of $\frac{1}{2}\Delta + t_0 W$, $\frac{1}{2}\Delta + t^+ V$ and $\frac{1}{2}\Delta + t^- V$ on $C_0^\infty(R^d)$, is an eigenvalue if $d \geq 5$ and is not an eigenvalue if $d = 3$ or 4.

The operator $\frac{1}{2}\Delta$ on R^2 is critical; however, it is subcritical on the domain $D_{\theta_0} \equiv \{x = (r, \phi) \in R^2\colon\ r > 0 \text{ and } 0 < \theta < \theta_0\}$, where $\theta_0 \in (0, 2\pi]$. Thus, the above theory may be applied to $\frac{1}{2}\Delta$ on $D_{\theta_0} \subset R^2$. Define V, W, t_0, t^+ and t^- analogously. In exercise 7.8 the reader is asked to show that zero is an eigenvalue for the Friedrichs' extensions of $\frac{1}{2}\Delta + t_0 W$, $\frac{1}{2}\Delta + t^+ V$ and $\frac{1}{2}\Delta + t^- V$ on $C_0^\infty(D_{\theta_0})$ if $\theta_0 \in (0, 2\pi)$, but is not an eigenvalue if $\theta_0 = 2\pi$.

Remark 1. It was noted at the beginning of Chapter 4, Section 6, that most of the perturbation results of that section which were stated and proved for compactly supported functions, continue to hold for sufficiently small non-compactly supported functions. Similarly, Theorem 3.8 holds for sufficiently small non-compactly supported functions. In particular, if $L = \frac{1}{2}\Delta$, then it turns out that functions which are bounded by

$$\frac{c}{(1+|x|)^{2+\varepsilon}},$$

for some $\varepsilon > 0$, are sufficiently small. (See the notes at the end of Chapter 4). Using this, it can be shown that the results of Example 3.10 continue to hold if $|V|$ and W are bounded by

$$\frac{c}{(1+|x|)^{2+\varepsilon}},$$

for some $\varepsilon > 0$.

Remark 2. Consider $\frac{1}{2}\Delta + t/x^2$ on R^d, $d \geqslant 3$. By Example 4.3.12,

$$\tfrac{1}{2}\Delta + \frac{t}{|x|^2}$$

on R^d, $d \geqslant 3$, is critical if

$$t = \frac{(d-2)^2}{8},$$

and the corresponding ground state is $\phi_c = |x|^{(2-d)/2}$. Thus, $\int_{R^d} \phi_c^2 \, dx = \infty$, for all $d \geqslant 3$.

Example 3.11. Consider the operator $L = \frac{1}{2}\Delta - \gamma$ on R^d, $d \geqslant 1$, where $\gamma > 0$. Then

$$G(x, y) = \int_0^\infty \exp(-\gamma t) \frac{1}{(2\pi t)^{d/2}} \exp\left(-\frac{|x-y|^2}{2t}\right) dt.$$

Define

$$G(r) = \int_0^\infty \exp(-\gamma t) \frac{1}{(2\pi t)^{d/2}} \exp\left(\frac{-r^2}{2t}\right) dt$$

so that $G(r) = G(x, y)$, if $|x - y| = r$. Since the radial part of $\frac{1}{2}\Delta$ is given by

$$\frac{1}{2}\frac{\partial^2}{\partial r^2} + \frac{d-1}{2r}\frac{\partial}{\partial r},$$

it follows from Theorem 4.2.5 that

$$\tfrac{1}{2}G'' + \frac{d-1}{2r}G' - \gamma G = 0,$$

for $r > 0$. Letting $v(r) = r^{(d-1)/2}G(r)$, it follows that v satisfies

$$\tfrac{1}{2}v'' + \left(\frac{(3-d)(d-1)}{8r^2} - \gamma\right)v = 0, \quad \text{for } r > 0.$$

If $d = 1$ or 3, then by inspection, one sees that $v = \exp(-\sqrt{2\gamma}\, r)$ and $v = 1$ form two linearly independent solutions to the above differential equation. By Theorem 3.7, $G(x, y)$ is a positive solution of minimal growth; therefore, at least for large r, $v(r) = c\exp(-\sqrt{2\gamma}\, r)$. Now, if $v(r) = c\exp(-\sqrt{2\gamma}\, r)$ for $r \geqslant r_0$ and $v(r) = c_1\exp(-\sqrt{2\gamma}\, r) + c_2$, for $r \leqslant r_0$ then $v(r)$ is differentiable at $r = r_0$ if and only if $c_1 = c$ and $c_2 = 0$. From this we conclude that, in fact, $v(r) = c\exp(-\sqrt{2\gamma}\, r)$, for all $r > 0$. Thus, it follows that $G(x, y) = c\exp(-\sqrt{2\gamma}\,|x - y|)$, if $d = 1$, and

$$G(x, y) = \frac{c}{|x - y|}\exp(-\sqrt{2\gamma}\,|x - y|), \quad \text{if } d = 3.$$

Note that $G(x, y)$ has the appropriate behavior at $x = y$ specified in Theorem 4.2.8. If $d \neq 1$ or 3, then the solutions to the differential equation above for v are given in terms of modified Bessel functions. As before, the appropriate solution must have minimal growth at infinity. It also must have the appropriate behavior at $r = 0$. Using Theorem 4.2.8 and the fact that $G(r)\ f = r^{(1-d)/2}v(r)$, it follows that $v(r)$ must behave on the order $r^{(3-d)/2}$ near $r = 0$ if $d \geqslant 4$ and on the order $-r^{1/2}\log r$ near $r = 0$ if $d = 2$. Standard results from the theory of Bessel functions show that $v(r)\exp(\sqrt{2\gamma}\, r)$ tends to a positive constant as $r \to \infty$. Thus, in any dimension d,

$$G(x, y) = c|x - y|^{(1-d)/2}\exp(-\sqrt{2\gamma}\,|x - y|) + \text{lower order terms},$$

as $|x - y| \to \infty$.

The construction in Example 3.10 can be applied to $L = \frac{1}{2}\Delta - \gamma$ on R^d, $d \geqslant 1$. Let W, $V \in C_0^\alpha(R^d)$ and assume that V, $W \neq 0$, that $W \geqslant 0$, and that V changes sign. As in Example 3.10, let $t_0 > 0$ be such that $\frac{1}{2}\Delta + t_0 W$ on R^d is critical, and let $t^+ > 0$ and $t^- < 0$ be such that $\frac{1}{2}\Delta + t^{\pm}V$ on R^d is critical. Let ϕ_c denote the ground state for $\frac{1}{2}\Delta + t_0 W$ on R^d and let $\phi_c^{\pm}$ denote the ground state for $\frac{1}{2}\Delta + t^{\pm}V$ on R^d. By Theorem 3.8(iii) and the Green's function estimates above, there exist positive constants c_1 and c_2 such that

$$c_1|x|^{(1-d)/2}\exp(-\sqrt{2\gamma}\,|x|) \leq \phi_c(x),\ \phi_c^{+}(x),\ \phi_c^{-}(x)$$
$$\leq c_2|x|^{(1-d)/2}\exp(-\sqrt{2\gamma}\,|x|), \quad \text{for } |x| > 1.$$

In particular then, using the reasoning of Example 3.10, it follows that zero is an eigenvalue for the Friedrichs' extensions of $\frac{1}{2}\Delta - \gamma + t_0 W$, $\frac{1}{2}\Delta - \gamma + t^{+}V$, and $\frac{1}{2}\Delta - \gamma + t^{-}V$ on $C_0^{\infty}(R^d)$.

Example 3.12. Let $W \in C_0^{\alpha}(R^2)$ satisfy $\int_{R^d} W\,dx < 0$. Then, by Theorem 4.6.6, $\frac{1}{2}\Delta + tW$ on R^2 is subcritical for sufficiently small $t > 0$. Replacing W by tW is necessary, we will assume that $L = \frac{1}{2}\Delta + W$ on R^2 is subcritical. Denote the Green's function by $G_W(x, y)$. Choose $R > 0$ so that the ball $B_R(0) \subset R^2$ of radius R and centered at 0 contains the support of W. By Theorem 3.7, $G_W(x, 0)$ on $R^2 - B_R(0)$ is a positive solution of minimal growth for $\frac{1}{2}\Delta + W$ on R^2. Since $W \equiv 0$ on $R^2 - B_R(0)$, it follows that $G_W(x, 0)$ on $R^2 - B_R(0)$ is also a positive solution of minimal growth for $\frac{1}{2}\Delta$ on R^2. Let h, defined on $R^2 - B_R(0)$, denote the positive solution of minimal growth for $\frac{1}{2}\Delta$ which satisfies $h = 1$ on $\partial B_R(0)$. Since h is radially symmetric, we will denote this function by $h(r)$. We have $\frac{1}{2}h'' + \frac{1}{2r}h' = 0$ in $(1, \infty)$ and $h(1) = 1$. The two linearly independent solutions to this equation are 1 and $1 + c\log r$. Since h is a positive solution of minimal growth, we conclude that $h(r) = 1$. Comparing $G_W(x, 0)$ and h, it follows from the maximum principle in Theorem 3.6 for positive solutions of minimal growth that there exist positive constants c_1 and c_2 such that $c_1 \leq G_W(x, 0) \leq c_2$, for $|x| > 1$. This can easily be extended to

$$c_1 \leq G_W(x, y) \leq c_2, \text{ for } |x - y| > 1. \tag{3.32}$$

By Theorem 8.3.1 in Chapter 8, the Martin boundary of $L = \frac{1}{2}\Delta + W$ on R^2 consists of one point; thus $C_L(R^2)$ contains a unique function (up to positive multiples); call this function ϕ. We will show that

$$\lim_{|x|\to\infty} \frac{\phi(x)}{\log|x|} = c > 0. \tag{3.33}$$

For definiteness, normalize $\phi(x)$ by $\phi(x_0) = 1$, for some $x_0 \in R^2$ with $|x_0| = 1$. Let $\hat{W} \in C_0^{\alpha}(R^d)$ be radially symmetric, supported in the unit ball, and satisfy $\hat{W} \leq 0$ and $\hat{W} \not\equiv 0$. Then $\hat{L} \equiv \frac{1}{2}\Delta + \hat{W}$ on R^2 is subcritical and, by Theorem 8.3.1, the Martin boundary of $\hat{L}$ consists of a single point. Denote the unique function (up to positive multiples) in $C_{\hat{L}}(R^2)$ by $\hat{\phi}$. From the definition of the Green's function and the fact that two-dimensional Brownian motion never hits 0, it follows that the Green's function for $\hat{L}$ on R^2 coincides with the Green's function for $\hat{L}$ on $R^2 - \{0\}$. From this and exercise 5.5, it follows that $\hat{\phi}$ is radially symmetric. We will denote this function by $\hat{\phi}(r)$. Since $\hat{W}$ vanishes outside the unit ball, we have $\frac{1}{2}\hat{\phi}'' + \frac{1}{2r}\hat{\phi}' = 0$ in $(1, \infty)$. Normalizing by $\hat{\phi}(1) = 1$, it follows that $\hat{\phi}(r) = 1 + c \log r$, for some $c \geqslant 0$. We now show that $c > 0$. Since $\hat{W}$ vanishes outside the unit ball, it follows by Theorem 3.6 and the recurrence of two-dimensional Brownian motion that the constant function 1, restricted to the exterior of the unit ball, is a positive solution of minimal growth for $\hat{L}$. By Theorem 3.9, $\hat{\phi}$ is not a positive solution of minimal growth for $\hat{L}$, thus $\hat{\phi}(r) = 1 + c \log r$, where $c > 0$.

By the resolvent equation (Theorem 4.3.11),

$$G_{\hat{W}}(x, y) = G_W(x, y) + \int_{R^2} G_W(x, z)(\hat{W}(z) - W(z))G_{\hat{W}}(z, y)\, dz. \tag{3.34}$$

Let $\{y_n\}_{n=1}^{\infty}$ be a sequence satisfying $\lim_{n\to\infty} |y_n| = \infty$. Since the Martin boundaries of L and $\hat{L}$ consist of one point, it follows that

$$\lim_{n\to\infty} \frac{G_{\hat{W}}(x, y_n)}{G_{\hat{W}}(x_0, y_n)} = \hat{\phi}(|x|) \text{ and } \lim_{n\to\infty} \frac{G_W(x, y)}{G_W(x_0, y)} = \phi(x).$$

Now replace y by y_n in (3.34) and divide each term in the equation by $G_{\hat{W}}(x_0, y_n)$. Rewrite

$$\frac{G_W(x, y_n)}{G_{\hat{W}}(x_0, y_n)} \text{ as } \frac{G_W(x, y_n)}{G_W(x_0, y_n)} \frac{G_W(x_0, y_n)}{G_{\hat{W}}(x_0, y_n)},$$

and then let $n \to \infty$. Since all the terms but

$$\frac{G_W(x_0, y_n)}{G_{\hat{W}}(x_0, y_n)}$$

clearly converge, it follows that

$$\gamma \equiv \lim_{n\to\infty} \frac{G_W(x_0, y_n)}{G_{\hat{W}}(x_0, y_n)}$$

also exists. The limiting equation is then

$$\hat{\phi}(|x|) = \gamma\phi(x) + \int_{R^2} G_W(x, z)(\hat{W}(z) - W(z))\hat{\phi}(|z|)\, dz. \tag{3.35}$$

We will show that $\gamma > 0$. Since $\widetilde{L} = L$ and $\widetilde{\hat{L}} = \hat{L}$, it follows that $G_W(x_0, y) = G_W(y, x_0)$ and $G_{\hat{W}}(x_0, y) = G_{\hat{W}}(y, x_0)$. From this and the fact that W and $\hat{W}$ are compactly supported, it follows that $G_W(x_0, y)$ and $G_{\hat{W}}(x_0, y)$, as functions of y in a neighborhood of infinity, are positive solutions of minimal growth for $\frac{1}{2}\Delta$ on R^2. Thus, by the maximum principle in Theorem 3.6 for positive solutions of minimal growth, it follows that $\gamma > 0$.

Divide both sides of (3.35) by $\hat{\phi}(|x|) = 1 + c \log |x|$. Then, letting $|x| \to \infty$, using the fact that W and $\hat{W}$ are compactly supported, and using (3.32), we obtain (3.33).

We note that (3.33) also follows as an immediate application of Theorem 8.3.3 in Chapter 8.

7.4 Harmonic measure for exterior regions

We being by recalling the notion of weak convergence of measures. Let A be a seperable metric space. A sequence $\{\mu_n\}_{n=1}^{\infty}$ of finite measures on A is said to *converge weakly* to a (finite) measure μ on A if

$$\int_A f \, d\mu = \lim_{n \to \infty} \int_A f \, d\mu_n, \quad \text{for all } f \in C_b(A).$$

This converges is denoted by $\mu = w - \lim_{n\to\infty} \mu_n$. The space of finite measures on A under the topology of weak convergence is metrizable. A subset B of the space of probability measures on A is called *tight* if for each $\varepsilon > 0$, there exists a compact set $K_\varepsilon \subseteq A$ such that $\mu(K_\varepsilon) > 1 - \varepsilon$, for all $\mu \in B$. If B is tight, then it is precompact. If A is complete, then B is tight if and only if it is precompact. If A is compact, then, clearly, every subset B of the space of probability measures on A is tight.

We now turn to harmonic measures.

Definition. *Let $L = L_0 + V$ on $D \subseteq R^d$ be either critical or subcritical, and let $\Omega \subset\subset D$ be a subdomain. Define for each $x \in D - \Omega$, the finite measure μ_x on $\partial\Omega$ by*

$$\mu_x(A) \equiv E_x\left(\exp\left(\int_0^{\sigma_{\bar{\Omega}}} V(X(s))\, ds\right); \sigma_{\bar{\Omega}} < \tau_D, X(\sigma_{\bar{\Omega}}) \in A\right),$$

for measurable $A \subseteq \partial\Omega$; μ_x is called the harmonic measure on $\partial\Omega$ with respect to $x \in D - \Omega$ for L on D.

Remark 1. The finiteness of μ_x follows from Lemma 3.4(i).

Remark 2. Note that if $D - \bar{\Omega}$ is not connected, then μ_x and μ_y will have disjoint supports if x and y belong to different connected components of $D - \bar{\Omega}$.

Remark 3. The dependence of μ_x on the choice of subdomain Ω has been suppressed to simplify the notation.

It will be convenient to restate certain results from Section 3 in terms of harmonic measures.

Proposition 4.1. *Under the conditions of Theorem 3.6, a non-negative and not identically zero function $h \in C^2(D - \bar{\Omega}) \cap C(D - \Omega)$ is a positive solution of minimal growth at ∂D for L on D if and only if*

$$h(x) = \int_{\partial\Omega} h(y)\mu_x(dy), \quad x \in D - \Omega.$$

Proof. This is a restatement of Theorem 3.6. □

Proposition 4.2. *Let $L = L_0 + V$ satisfy Assumption $\widetilde{H}_{\text{loc}}$ on a domain $D \subseteq R^d$ and assume that L is either critical or subcritical on D. Let $\Omega \subset\subset D$ be a subdomain.*

- (i) *μ_x, the harmonic measure on $\partial\Omega$ with respect to $x \in D - \Omega$ for L on D, is weakly continuous as a function of $x \in D - \bar{\Omega}$.*
- (ii) *Let $\{\Omega_n\}_{n=1}^{\infty}$ be a sequence of domains satisfying $\Omega_1 \subset\subset D$, $\Omega_{n+1} \subset\subset \Omega_n$ and $\bigcap_{n=1}^{\infty}\Omega_n = \bar{\Omega}$. Let μ_x^n denote the harmonic measure on $\partial\Omega_n$ with respect to $x \in D - \Omega_n$ for L on D. Then $w - \lim_{n\to\infty} \mu_x^n = \mu_x$, for $x \in D - \bar{\Omega}$.*

Proof. Part (i) follows from Lemma 3.4(iii) and part (ii) is a restatement of Lemma 3.4(ii). □

As an application of Harnack's inequality, we now show that harmonic measures corresponding to different initial points within the same connected component of $D - \bar{\Omega}$ are mutually absolutely continuous with bounded Radon–Nikodym derivatives.

Theorem 4.3. *Let $L = L_0 + V$ satisfy Assumption $\widetilde{H}_{\text{loc}}$ on a domain $D \subseteq R^d$ and assume that L is either critical or subcritical on D. Let $\Omega \subset\subset D$ be a subdomain and let μ_x denote the harmonic measure on $\partial\Omega$ with respect to $x \in D - \Omega$ for L on D. Let $(D - \bar{\Omega})^{(i)}$, $i = 1, 2, \ldots, k$, denote the connected components of $D - \bar{\Omega}$. Then μ_{x_1} and μ_{x_2} are mutually absolutely continuous for all $x_1, x_2 \in (D - \bar{\Omega})^{(i)}$, $i = 1, 2,$*

..., k. Moreover, letting γ (which depends only on the coefficients of L) be as in Harnack's inequality, then for each domain $U \subset\subset (D - \bar{\Omega})^{(i)}$, there exist positive constants c_1 and c_2 depending only on γ, D, Ω, U and d such that

$$c_1 \leqslant \frac{d\mu_{x_1}}{d\mu_{x_2}}(y) \leqslant c_2, \quad \text{for } x_1, x_2 \in U \text{ and } y \in \partial(D - \bar{\Omega})^{(i)} \cap \partial\Omega.$$

Proof. The first statement of the theorem follows from the second one. To prove the second one, choose a domain U_1 with a $C^{2,\alpha}$-boundary such that $U \subset\subset U_1 \subset\subset D - \bar{\Omega}$. By Theorem 4.4.1(iii), $\lambda_c(U_1) < 0$. Thus, it follows by Theorem 3.6.5 that for each positive $\phi \in C^{2,\alpha}(\bar{U}_1)$, there exists a positive solution $u \in C^{2,\alpha}(\bar{U}_1)$ of $Lu = 0$ in U_1 and $u = \phi$ on ∂U_1. By Harnack's inequality, there exist positive constants c_1 and c_2 whose dependence is as in the statement of the proposition such that

$$c_1 \leqslant \frac{u(x_1)}{u(x_2)} \leqslant c_2 \quad \text{for all } x_1, x_2 \in U. \tag{4.1}$$

By Theorem 3.6.6, $u(x) = \int_{\partial U_1} \phi(y) \nu_x(dy)$, for $x \in U_1$, where ν_x is defined by

$$\nu_x(A) = E_x \exp\left(\int_0^{\tau_{U_1}} V(X(s))\, ds\right) 1_A(X(\tau_{U_1})),$$

for measurable $A \subseteq \partial U_1$. Thus, from (4.1), it follows that

$$c_1 \leqslant \frac{d\nu_{x_1}}{d\nu_{x_2}} \leqslant c_2, \quad \text{for } x_1, x_2 \in U. \tag{4.2}$$

The theorem follows from (4.2) and the fact that, by the strong Markov property, $\mu_x(dy) = \int_{U_1} \nu_x(dz) \mu_z(dy)$. □

Under appropriate smoothness conditions on $\partial\Omega$, the harmonic measures are in fact mutually absolutely continuous with respect to Lebesgue surface measure and the Radon–Nikodym derivative is bounded and bounded away from zero.

Theorem 4.4. *Let $L = L_0 + V$ satisfy Assumption $\widetilde{H}_{\text{loc}}$ on a domain $D \subseteq R^d$ and assume that L is either critical or subcritical on D. Let $\Omega \subset\subset D$ be a subdomain with a $C^{2,\alpha}$-boundary and let μ_x denote the harmonic measure on $\partial\Omega$ with respect to $x \in D - \Omega$ for L on D. Let $(D - \bar{\Omega})^{(i)}$, $i = 1, 2, \ldots, k$, denote the connected components of $D - \bar{\Omega}$. Let $l(dy)$ denote Lebesgue surface measure on $\partial\Omega$. Then for each $x \in (D - \bar{\Omega})^{(i)}$, $i = 1, 2, \ldots, k$, μ_x is mutually absolutely continuous with respect to Lebesgue surface measure $l(dy)$ on $\partial(D - \bar{\Omega})^{(i)} \cap$*

$\partial\Omega$. *Moreover, for each* $U \subset\subset (D - \bar{\Omega})^{(i)}$, $i = 1, 2, \ldots, k$, *there exist positive constants* c_1 *and* c_2 *such that the Radon–Nikodym derivative satisfies*

$$c_1 \leqslant \frac{d\mu_x}{dl}(y) \leqslant c_2, \quad \text{for } x \in U \text{ and } y \in \partial(D - \bar{\Omega})^{(i)} \cap \partial\Omega. \tag{4.3}$$

Proof. We will prove the theorem under the assumption that $D - \bar{\Omega}$ is connected. The proof in the general case is conceptually identical but the notation is more involved. To prove the theorem, it suffices to prove (4.3). We first prove (4.3) in the case where D is bounded with a $C^{2,\alpha}$-boundary. Choose domains $U \subset\subset U_1 \subset\subset \bar{D} - \Omega$ and let $f \in C^{\alpha}(\bar{D} - \Omega)$ satisfy $f = 1$ on U, $f = 0$ on $D - \bar{\Omega} - U_1$ and $0 \leqslant f \leqslant 1$ on $\bar{D} - \Omega$. By Proposition 3.1, L is subcritical on $D - \bar{\Omega}$; thus, by Theorem 4.3.2, $\lambda_c(D - \bar{\Omega}) < 0$. By Theorem 3.6.5, there exists a solution $\psi \in C^{2,\alpha}(\bar{D} - \Omega)$ to $\tilde{L}\psi = -f$ in $D - \bar{\Omega}$ and $\psi = 0$ on $\partial D \cup \partial\Omega$ and, for each $\phi \in C^{2,\alpha}(D - \Omega)$, there exists a solution $h \in C^{2,\alpha}(\bar{D} - \Omega)$ to $Lh = 0$ in $D - \bar{\Omega}$, $h = \phi$ on $\partial\Omega$ and $h = 0$ on ∂D. Of course, h is a positive solution of minimal growth at ∂D; thus

$$h(x) = \int_{\partial\Omega} \phi(y)\mu_x(dy). \tag{4.4}$$

By Harnack's inequality and the properties of f, there exist constants $0 < \gamma_1 < \gamma_2 < \infty$ such that for all $x \in U$,

$$\gamma_1 \int_{U_1} fh\,dy \leqslant h(x) \leqslant \gamma_2 \int_{U_1} fh\,dy. \tag{4.5}$$

Integrating by parts gives

$$\int_{U_1} fh\,dy = \int_{D-\bar{\Omega}} fh\,dy = -\int_{D-\bar{\Omega}} h\tilde{L}\psi\,dy = -\tfrac{1}{2}\int_{\partial\Omega} \phi(a\nabla\psi\cdot n)\,dl. \tag{4.6}$$

Since Ω has a $C^{2,\alpha}$-boundary, it satisfies an interior sphere condition at each point of its boundary. Thus, by the Hopf maximum principle (Theorem 3.2.5),

$$\inf_{\partial\Omega}(-\nabla\psi\cdot n) > 0. \tag{4.7}$$

From (4.4)–(4.7), it follows that there exist constants $0 < c_1 < c_2 < \infty$, independent of $\phi \in C^{2,\alpha}(\partial\Omega)$, such that for all $x \in U$,

$$c_1 \int_{\partial\Omega} \phi\,dl \leqslant \int_{\partial\Omega} \phi\,d\mu_x \leqslant c_2 \int_{\partial\Omega} \phi\,dl.$$

This proves (4.3) in the case that D is bounded with a $C^{2,\alpha}$-boundary.

To prove (4.3) in the general case, it will be helpful to assume that $V \equiv 0$. To make this reduction in the subcritical case, choose $\Omega_0 \subset\subset \Omega$ and choose a positive solution $h \in C^{2,\alpha}(D - \bar{\Omega}_0) \cap C(D - \Omega_0)$ of minimal growth at ∂D. Then the zeroth-order part of the operator L^h on $D - \bar{\Omega}_0$ vanishes and, by Theorem 4.5 below, μ_x^h, the harmonic measure on $\partial\Omega$ for L^h, satisfies

$$\mu_x^h(\mathrm{d}y) = \frac{h(y)}{h(x)}\mu_x(\mathrm{d}y);$$

thus it suffices to consider the case $V \equiv 0$. To make the reduction in the critical case, let $\Omega' \subset\subset \Omega_0 \subset\subset \Omega$. Then, by Proposition 3.1, L is subcritical on $D - \bar{\Omega}'$. Let h be as above and proceed as in the subcritical case.

For the rest of the proof, we assume that $V \equiv 0$. Let U, U_1 and D_1 be subdomains satisfying $U \subset\subset D - \bar{\Omega}$ and $\Omega \cup U \subset\subset U_1 \subset\subset D_1 \subset\subset D$, and assume that D_1 has a $C^{2,\alpha}$-boundary. Let $\tau_1 \equiv \tau_{D_1} = \inf(t \geqslant 0: X(t) \notin D_1\}$ and define by induction,

$$\sigma_n = \inf\{(t \geqslant \tau_n: X(t) \in \partial U_1\}$$

and

$$\tau_{n+1} = \inf\{t \geqslant \sigma_n: X(t) \in \partial D_1\},\ n = 1, 2, \ldots.$$

Then, for $\phi \in C(\partial\Omega)$ and $x \in U$,

$$\begin{aligned}\int_{\partial\Omega} \phi(y)\mu_x(\mathrm{d}y) &= E_x(\phi(X(\sigma_{\bar{\Omega}})); \sigma_{\bar{\Omega}} < \tau_D)\\ &= E_x(\phi(X(\sigma_{\bar{\Omega}})); \sigma_{\bar{\Omega}} < \tau_{D_1})\\ &\quad + \sum_{n=1}^{\infty} E_x(\phi(X(\sigma_{\bar{\Omega}})); \sigma_n < \sigma_{\bar{\Omega}} < \tau_{n+1}). \quad (4.8)\end{aligned}$$

By the strong Markov property,

$$\begin{aligned}&E_x(\phi(X(\sigma_{\bar{\Omega}})); \sigma_n < \sigma_{\bar{\Omega}} < \tau_{n+1})\\ &\qquad = E_x 1_{\{\sigma_n < \sigma_{\bar{\Omega}}\}} E_{X(\sigma_n)}(\phi(X(\sigma_{\bar{\Omega}})); \sigma_{\bar{\Omega}} < \tau_{D_1}). \quad (4.9)\end{aligned}$$

Since D_1 is bounded with a $C^{2,\alpha}$-boundary, applying what we have already proved to D_1, it follows that there exist positive constants c_1 and c_2, such that

$$c_1\int_{\partial\Omega} \phi\,\mathrm{d}l \leqslant E_y(\phi(X(\sigma_{\bar{\Omega}})); \sigma_{\Omega} < \tau_{D_1}) \leqslant c_2\int_{\partial\Omega} \phi\,\mathrm{d}l,$$

$$\text{for } y \in \partial U_1 \cup U \text{ and } \phi \in C(\partial\Omega). \quad (4.10)$$

Using the strong Markov property and exercise 4.5(ii), it follows that $P_x(\sigma_{n+1} < \sigma_{\bar{\Omega}}) \leqslant \rho P_x(\sigma_n < \sigma_{\bar{\Omega}})$, for some $\rho \in (0,1)$ and all $x \in U$; thus

$$\sum_{n=1}^{\infty} P_x(\sigma_n < \sigma_{\bar{\Omega}}) < \infty. \tag{4.11}$$

The theorem follows from (4.8)–(4.11). □

We now consider harmonic measures for h-transformed operators, where h is a positive solution of minimal growth at ∂D. In the sequel, given a domain $\Omega \subset\subset D$, the notation $(D - \bar{\Omega})\hat{}$ will be used to denote the one-point compactification of $D - \bar{\Omega}$ in the case where $D - \bar{\Omega}$ is a domain; in the case where $D - \bar{\Omega}$ has more than one component, $(D - \bar{\Omega})\hat{}$ will denote the union of the one-point compactifications of the components.

Definition. Let $\Omega_0 \subset\subset \Omega \subset\subset D$ and let $h \in C^{2,\alpha}(D - \bar{\Omega}_0) \cap C(D - \Omega_0)$ be a positive solution of minimal growth at ∂D for L on D. Let

$$L^h = L_0 + a\frac{\nabla h}{h} \cdot \nabla$$

denote the h-transformed operator on $D - \bar{\Omega}_0$. Let $\{P_x^h, x \in (D - \bar{\Omega}_0)\hat{}\}$ denote the solution to the generalized martingale problem for L^h on $D - \bar{\Omega}_0$ in the case where $D - \bar{\Omega}_0$ is a domain, and let it denote the collection of solutions to the generalized martingale problems for L^h on $(D - \bar{\Omega}_0)^{(i)}$, $i = 1, 2, \ldots, k$, in the case where $D - \bar{\Omega}_0$ has k ($\geqslant 2$) components. Then

$$\mu_x^h(dy) \equiv P_x^h(\sigma_{\bar{\Omega}} < \tau_D, X(\sigma_{\bar{\Omega}}) \in dy)$$

is called the *harmonic measure on $\partial\Omega$ with respect to $x \in D - \Omega$ for L^h on $D - \bar{\Omega}_0$*.

Remark. As with μ_x, the dependence of μ_x^h on Ω has been suppressed.

Theorem 4.5. *Let L satisfy Assumption $\tilde{H}_{\text{loc}}$ on a domain $D \subseteq R^d$ and let Ω_0 and Ω be subdomains satisfying $\Omega_0 \subset\subset \Omega \subset\subset D$. Let*

$$h \in C^{2,\alpha}(D - \bar{\Omega}_0) \cap C(D - \Omega_0)$$

be a positive solution of minimal growth at ∂D for L on D and let μ_x^h, $x \in D - \Omega$, be the harmonic measure on $\partial\Omega$ for L^h on $D - \bar{\Omega}_0$. Then

$P_x^h(\sigma_{\bar\Omega} < \tau_D) = 1$, for $x \in D - \Omega$; thus

$$\mu_x^h(\mathrm{d}y) = P_x^h(X(\sigma_{\bar\Omega}) \in \mathrm{d}y)$$

and μ_x^h is a probability measure on $\partial\Omega$. Furthermore,

$$\mu_x^h(\mathrm{d}y) = \frac{h(y)}{h(x)}\mu_x(\mathrm{d}y), \tag{4.12}$$

where μ_x is the harmonic measure on $\partial\Omega$ for L on D.

Proof. We can define positive solutions of minimal growth at ∂D for L^h on $D - \Omega_0$ just as we did for L on D. Call $v \in C^2(D - \bar\Omega) \cap C(D - \Omega)$ a positive solution of minimal growth at ∂D for L^h on $D - \Omega_0$ if it satisfies the following conditions:

(i) $L^h v = 0$ and $v > 0$ on $D - \bar\Omega$;
(ii) If Ω' satisfies $\Omega \subseteq \Omega' \subset\subset D$ and if $u \in C^2(D - \bar\Omega') \cap C(D - \Omega')$ satisfies $L^h u = 0$ and $u > 0$ in $D - \bar\Omega'$, and $u = v$ on $\partial\Omega'$, then $v(x) \leqslant u(x)$, for all $x \in D - \Omega'$.

The analogs of Lemma 3.4, Theorems 3.5 and 3.6, and Propositions 4.1 and 4.2 go through with virtually identical proofs. In particular, the analog of Proposition 4.1 gives

$$v(x) = \int_{\partial\Omega} v(y)\mu_x^h(\mathrm{d}y). \tag{4.13}$$

To prove the theorem, first assume that Ω satisfies an interior cone condition at each point of its boundary. Let $\phi \in C(\partial\Omega)$ and let $u \in C^{2,\alpha}(D - \bar\Omega) \cap C(D - \Omega)$ be the positive solution of minimal growth at ∂D for L on D which satisfies $u = \phi$ on $\partial\Omega$ (the solution exists by Theorem 3.5). It is easy to see then that $v \equiv u/h$ is the minimal positive solution at ∂D for L^h on $D - \bar\Omega_0$ which satisfies $v = \phi/h$ on $\partial\Omega$. By Proposition 4.1 and (4.13),

$$u(x) = \int_{\partial\Omega} \phi(y)\mu_x(\mathrm{d}y) \text{ and } v(x) = \frac{u(x)}{h(x)} = \int_{\partial\Omega} \frac{\phi(y)}{h(y)}\mu_x^h(\mathrm{d}y).$$

Thus,

$$u(x) = \int_{\partial\Omega} \phi(y)\mu_x(\mathrm{d}y) = \int_{\partial\Omega} \phi(y)\frac{h(x)}{h(y)}\mu_x^h(\mathrm{d}y), \quad \text{for all } \phi \in C(\partial D),$$

and we conclude that

$$\mu_x^h(\mathrm{d}y) = \frac{h(y)}{h(x)}\mu_x(\mathrm{d}y),$$

which is (4.12). It then follows from Proposition 4.1 that μ_x^h is a probability measure. From the definition of μ_x^h we have $1 = \mu_x^h(\partial\Omega) = P_x^h(\sigma_{\bar{\Omega}} < \tau_D)$.

In the general case, let $\{\Omega_n\}_{n=1}^{\infty}$ be a sequence of domains satisfying an interior cone condition at each point of their boundaries and such that $\Omega_1 \subset\subset D$, $\Omega_{n+1} \subset\subset \Omega_n$ and $\bigcap_{n=1}^{\infty}\Omega_n = \bar{\Omega}$. Let $\mu_x^{n,h}(\mu_x^n)$ denote the harmonic measure on $\partial\Omega_n$ with respect to $x \in D - \Omega_n$ for L^h on $D - \bar{\Omega}_0$ (L on D). By what we have already proved,

$$\mu_x^{n,h}(dy) = \frac{h(y)}{h(x)}\mu_x^n(dy).$$

The proposition now follows by applying Proposition 4.2(ii) to μ_x^n and by applying its analog to $\mu_x^{n,h}$. □

Although Theorem 4.5 could be used without any problem in the sequel, in the interest of aesthetic and unencumbered notation, we prefer to extend Theorem 4.5 to the case in which $\Omega_0 = \Omega$. The extension, though simple, requires an explanation with regard to notation. Let $(D - \bar{\Omega}_0)^{(i)}$, $i = 1, 2, \ldots, k$, denote the connected components of $D - \bar{\Omega}_0$, and let $((D - \bar{\Omega})^{(i)})\hat{}$ denote the one-point compactification of $(D - \bar{\Omega})^{(i)}$. For $x \in (D - \bar{\Omega})^{(i)}$, the harmonic measure μ_x^h in Theorem 4.5 is the exit distribution on $\partial\Omega$ for the diffusion starting at x and corresponding to L^h on $(D - \bar{\Omega}_0)^{(i)}$. Now let h be as in Theorem 4.5 and let $\Omega_0 = \Omega$. The solution to the generalized martingale problem for L^h on $(D - \bar{\Omega})^{(i)}$ is a collection $\{P_x^h, x \in ((D - \bar{\Omega})^{(i)})\hat{}\}$ of probability measures defined on the space of continuous functions $X(t)$ from $[0, \infty)$ to $((D - \bar{\Omega})^{(i)})\hat{}$. Therefore, by definition, $X(t)$ exits $(D - \bar{\Omega})^{(i)}$ into the cemetery state Δ, the point at infinity in the one-point compactification of $(D - \bar{\Omega})^{(i)}$. Thus, the notation $P_x^h(X(\sigma_{\bar{\Omega}}) \in \cdot)$ is meaningless. However, the exit distribution of $X(t)$ on $\partial\Omega$ under P_x^h can be defined by $w - \lim_{n\to\infty} P_x^h(X(\sigma_{\bar{\Omega}_n}) \in \cdot)$, if this limit exists, where $\{\Omega_n\}_{n=1}^{\infty}$ is a sequence of domains satisfying $\Omega_1 \subset\subset D$, $\Omega_{n+1} \subset\subset \Omega_n$ and $\bigcap_{n=1}^{\infty}\Omega_n = \bar{\Omega}$.

Corollary 4.6. *Let L satisfy Assumption $\tilde{H}_{\text{loc}}$ on a domain $D \subseteq R^d$. Let Ω be a subdomain satisfying $\Omega \subset\subset D$ and let $h \in C^{2,\alpha}(D - \bar{\Omega}) \cap C(D - \Omega)$ be a positive solution of minimal growth at ∂D for L on D. Let $\{\Omega_n\}_{n=1}^{\infty}$ be a sequence of domains satisfying $\Omega_1 \subset\subset D$, $\Omega_{n+1} \subset\subset \Omega_n$ and $\bigcap_{n=1}^{\infty}\Omega_n = \bar{\Omega}$, and let $\mu_x^{n,h}$ denote the harmonic measure on $\partial\Omega_n$ with respect to $x \in D - \Omega_n$ for L^h on $D - \bar{\Omega}$. Then* *for* $x \in D - \bar{\Omega}$, $\mu_x^h \equiv w - \lim_{n\to\infty} \mu_x^{n,h}$ *exists as a probability measure on* $\partial\Omega$ *and satisfies*

$$\mu_x^h(\mathrm{d}y) = \frac{h(y)}{h(x)}\mu_x(\mathrm{d}y)$$
$$= \frac{E_x(\exp(\int_0^{\sigma_{\bar{\Omega}}} V(X(s))\,\mathrm{d}s)h(X(\sigma_{\bar{\Omega}})); \sigma_{\bar{\Omega}} < \tau_D, X(\sigma_{\bar{\Omega}}) \in \mathrm{d}y)}{E_x(\exp(\int_0^{\sigma_{\bar{\Omega}}} V(X(s))\,\mathrm{d}s)h(X(\sigma_{\bar{\Omega}})); \sigma_{\bar{\Omega}} < \tau_D)},$$
$$x \in D - \bar{\Omega}, \quad (4.14)$$

where μ_x *is the harmonic measure on* $\partial\Omega$ *for* L *on* D.

Proof. By Theorem 4.5, for $x \in D - \Omega_n$, (4.12) holds with μ_x^h, μ_x and Ω replaced by $\mu_x^{n,h}$, μ_x^n and Ω_n (where μ_x^n is the harmonic measure on $\partial\Omega_n$ for L on D). The existence of μ_x^h and the lefthand equality in (4.14) now follow upon letting $n \to \infty$ by applying Proposition 4.2(ii) to μ_x^n. The righthand equality in (4.14) follows from the definition of μ_x and Proposition 4.1. □

We will call μ_x^h as in Corollary 4.6 the *harmonic measure on* $\partial\Omega$ *with respect to* $x \in D - \bar{\Omega}$ *for* L^h *on* $D - \bar{\Omega}$.

The following corollary of Theorem 4.4 and Corollary 4.6 will be required in Section 6.

Corollary 4.7. *Let the assumptions of* Corollary 4.6 *hold and assume in addition that* Ω *has a* $C^{2,\alpha}$*-boundary. Let* $(D - \bar{\Omega})^{(i)}$, $i = 1, 2, \ldots, k$, *denote the connected components of* $D - \bar{\Omega}$. *Let* $l(\mathrm{d}y)$ *denote Lebesgue surface measure on* $\partial\Omega$. *Then for* $x \in (D - \bar{\Omega})^{(i)}$, μ_x^h *is mutually absolutely continuous with respect to Lebesgue surface measure* $l(\mathrm{d}y)$ *on* $\partial(D - \bar{\Omega})^{(i)} \cap \partial\Omega$. *Moroever, for each* $\varepsilon > 0$, *there exist positive constants* $c_1^{(\varepsilon)}$ *and* $c_2^{(\varepsilon)}$ *such that the Radon–Nikodym derivative satisfies*

$$c_1^{(\varepsilon)} \leq \frac{\mathrm{d}\mu_x^h}{\mathrm{d}l}(y) \leq c_2^{(\varepsilon)}, \quad (4.15)$$

for $y \in \partial(D - \bar{\Omega})^{(i)} \cap \partial\Omega$ *and for* $x \in (D - \bar{\Omega})^{(i)}$ *satisfying* $\mathrm{dist}(x, \partial\Omega) > \varepsilon$.

Proof. Choose $\varepsilon > 0$ such that $\Omega_\varepsilon \equiv \{x \in R^d: \mathrm{dist}(x, \Omega) < \varepsilon\} \subset\subset D$ and let $x \in D - \bar{\Omega}_\varepsilon$. It follows from the strong Markov property and the fact that μ_x^h is a probability measure that there exists a probability measure ν_x on $\partial\Omega_\varepsilon$ such that $\mu_x^h(\mathrm{d}y) = \int_{\partial\Omega_\varepsilon} \mu_z^h(\mathrm{d}y)\nu_x(\mathrm{d}z)$. Equation

(4.15) follows from this, from the lefthand inequality in (4.14) and from Theorem 4.4. □

7.5 The exterior harmonic measure boundary

Let $\Omega \subset\subset D$ be a subdomain and let $\{x_n\}_{n=1}^{\infty} \subset D - \bar{\Omega}$ be a sequence with no accumulation points in D. In this section, we will study the behavior of sequences of harmonic measures $\{\mu_{x_n}^h\}_{n=1}^{\infty}$, where $h \in C^{2,\alpha}(D - \bar{\Omega}) \cap C(D - \Omega)$ is a positive solution of minimal growth at ∂D, and the behavior of sequences of ratios

$$\left\{\frac{h_1(x_n)}{h_2(x_n)}\right\}_{n=1}^{\infty},$$

where h_1 and h_2 are positive solutions of minimal growth at ∂D. This will lead to the construction, similar in style to the construction of the Martin boundary, of a boundary which we will dub the exterior harmonic measure boundary.

Proposition 5.1. *Let $L = L_0 + V$ satisfy Assumption $\tilde{H}_{\text{loc}}$ on a domain $D \subseteq R^d$ and assume that L is either critical or subcritical on D. Let $\Omega \subset\subset D$ be a subdomain and let $\{x_n\}_{n=1}^{\infty} \subset D - \bar{\Omega}$ be a sequence with no accumulation points in D. Then the three statements below satisfy the following set of implications:* (i) $\Leftrightarrow$ (ii) $\Rightarrow$ (iii). *If Ω satisfies an interior cone condition at each point of its boundary, then all three statements are equivalent. (In what follows, if $h \in C^{2,\alpha}(D - \bar{\Omega}) \cap C(D - \Omega)$, then μ_x^h denotes the harmonic measure on $\partial\Omega$ with respect to $x \in D - \bar{\Omega}$ for L on $D - \bar{\Omega}$.)*

(i) *$w - \lim_{n\to\infty} \mu_{x_n}^{h_0}$ exists for some $h_0 \in C^{2,\alpha}(D - \bar{\Omega}) \cap C(D - \Omega)$ which satisfies $h_0 > 0$ on $\partial\Omega$ and which is a positive solution of minimal growth at ∂D for L on D.*

(ii) *$w - \lim_{n\to\infty} \mu_{x_n}^h$ exists for every $h \in C^{2,\alpha}(D - \bar{\Omega}) \cap C(D - \Omega)$ which is a positive solution of minimal growth at ∂D for L on D.*

(iii) $$\lim_{n\to\infty} \frac{h_1(x_n)}{h_2(x_n)}$$ *exists for every pair $h_1, h_2 \in C^{2,\alpha}(D - \bar{\Omega}) \cap C(D - \Omega)$ of positive solutions of minimal growth at ∂D for L on D.*

Proof. (i) $\Rightarrow$ (iii) Let h_0 be as in (i) and let h_i, $i = 1, 2$, be as in (iii). It follows from Proposition 4.1 and Corollary 4.6 that

$$\frac{h_i(x)}{h_0(x)} = \int_{\partial\Omega} \frac{h_i(y)}{h_0(y)} \mu_x^{h_0}(dy), \quad \text{for } x \in D - \bar{\Omega}.$$

Since $h_0 > 0$ on $\partial\Omega$, we have

$$\frac{h_i}{h_0} \in C(\partial\Omega),$$

and it follows from the definition of weak convergence that

$$\lim_{n\to\infty} \int_{\partial\Omega} \frac{h_i(y)}{h_0(y)} \mu_{x_n}^{h_0}(\mathrm{d}y)$$

exists. Thus

$$\lim_{n\to\infty} \frac{h_i(x_n)}{h_0(x_n)}$$

exists, and (iii) follows by writing

$$\frac{h_1}{h_2} = \frac{h_1/h_0}{h_2/h_0}.$$

(i) $\Rightarrow$ (ii) Let h_0 be as in (i) and let h be as in (ii). By the implication (i) $\Rightarrow$ (iii) which we have proved above,

$$\lim_{n\to\infty} \frac{h_0(x_n)}{h(x_n)}$$

exists. The implication (i) $\Rightarrow$ (ii) now follows since, by Corollary 4.6,

$$\mu_x^h(\mathrm{d}y) = \frac{h(y)}{h(x)}\mu_x(\mathrm{d}y) = \frac{h(y)}{h_0(y)}\frac{h_0(x)}{h(x)}\mu_x^{h_0}(\mathrm{d}y).$$

(ii) $\Rightarrow$ (i) Trivial.

(iii) $\Rightarrow$ (i) Let $\phi \in C(\partial D)$. By Theorem 3.5, there exists a positive solution $h \in C^{2,\alpha}(D - \bar{\Omega}) \cap C(D - \Omega)$ of minimal growth at ∂D, which satisfies $h = \phi$ on $\partial\Omega$. Let h_0 be as in (i). By Proposition 4.1 and Corollary 4.6,

$$\frac{h(x)}{h_0(x)} = \int_{\partial\Omega} \frac{\phi(y)}{h_0(y)} \mu_x^{h_0}(\mathrm{d}y), \quad \text{for } x \in D - \bar{\Omega}.$$

By assumption,

$$\lim_{n\to\infty} \frac{h(x_n)}{h_0(x_n)}$$

exists; thus,

$$\lim_{n\to\infty} \int_{\partial\Omega} \frac{\phi(y)}{h_0(y)} \mu_{x_n}^{h_0}(\mathrm{d}y)$$

exists for all $\phi \in C(\partial D)$ or, equivalently,

$$w - \lim_{n\to\infty} \mu_{x_n}^{h_0} \quad \text{exists.}$$ □

Let $L = L_0 + V$ satisfy Assumption $\widetilde{H}_{\text{loc}}$ on a domain $D \subseteq R^d$ and assume that L is either critical or subcritical on D. Let $\Omega \subset\subset D$ be a subdomain and let $h \in C^{2,\alpha}(D - \bar{\Omega}) \cap C(D - \Omega)$ be a positive solution of minimal growth at ∂D for L on D which satisfies $h > 0$ on $\partial\Omega$. Let $\{x_n\}_{n=1}^{\infty} \subset D - \bar{\Omega}$ be a sequence with no accumulation points in D. By Corollary 4.6, μ_x^h is a probability measure on $\partial\Omega$ for each $x \in D - \bar{\Omega}$. Since $\partial\Omega$ is compact, it follows that $\{\mu_{x_n}^h\}_{n=1}^{\infty}$ is tight; thus one can always extract a weakly convergence subsequence. Call a sequence $\{x_n\}_{n=1}^{\infty} \subset D - \bar{\Omega}$ with no accumulation points in D *fundamental* if $w - \lim_{n\to\infty} \mu_{x_n}^h$ exists. By the implication (i) ⇒ (ii) in Proposition 5.1, this definition is independent of the choice of h. In the next section we will show that this definition is also independent of the choice of subdomain $\Omega \subset\subset D$. Call two fundamental sequences, $\{x_n\}_{n=1}^{\infty}$ and $\{x'_n\}_{n=1}^{\infty}$, *equivalent* if $w - \lim_{n\to\infty} \mu_{x_n}^h = w - \lim_{n\to\infty} \mu_{x'_n}^h$. We will call the collection of such equivalence classes the *exterior harmonic measure boundary for L on D*.

We note the following two characterizations of the exterior harmonic measure boundary. By Proposition 5.1, the exterior harmonic measure boundary may be characterized as the collection of equivalence classes of sequences $\{x_n\}_{n=1}^{\infty} \subset D$ with no accumulation points in D along which the ratio of every pair of positive solutions of minimal growth at ∂D converges. By Corollary 4.6, the exterior harmonic measure boundary may be characterized as the collection of equivalence classes of sequences $\{x_n\}_{n=1}^{\infty} \subset D$ with no accumulation points in D for which the exit distribution on $\partial\Omega$ for the diffusion corresponding to L^h and starting at x_n converges weakly as $n \to \infty$, where $\Omega \subset\subset D$ and $h \in C^{2,\alpha}(D - \bar{\Omega}) \cap C(D - \Omega)$ is a positive solution of minimal growth at ∂D for L on D satisfying $h > 0$ on $\partial\Omega$. One particular case of this second characterization is sufficiently important to be stated separately as a proposition.

Proposition 5.2. *Let $V \equiv 0$ and let $L = L_0$ satisfy Assumption $\widetilde{H}_{\text{loc}}$ on a domain $D \subseteq R^d$. Then a sequence $\{x_n\}_{n=1}^{\infty} \subset D$ with no accumulation points in D is fundamental for the exterior harmonic measure boundary if and only if for some subdomain $\Omega \subset\subset D$ which satisfies an interior cone condition at each point of its boundary (or, equivalently, for all such subdomains), the conditioned exit distribution $P_{x_n}(X(\sigma_{\bar{\Omega}}) \in$*

$\cdot|\sigma_{\bar{\Omega}} < \tau_D)$ *on* $\partial\Omega$ *for the diffusion corresponding to* L_0 *and, starting at* x_n, *converges weakly as* $n \to \infty$.

Proof. Let $\Omega \subset\subset D$ be as in the statement of the proposition. Then, by Theorem 3.5, it follows that $h(x) \equiv P_x(\sigma_{\bar{\Omega}} < \tau_D)$, $x \in D - \Omega$, is the positive solution of minimal growth at ∂D satisfying $h = 1$ on $\partial\Omega$. Equating the expressions on the left and the right in (4.14) gives

$$\mu_x^h(dy) = \frac{P_x(X(\sigma_{\bar{\Omega}}) \in dy, \sigma_{\bar{\Omega}} < \tau_D)}{P_x(\sigma_{\bar{\Omega}} < \tau_D)} = P_x(X(\sigma_{\bar{\Omega}}) \in dy | \sigma_{\bar{\Omega}} < \tau_D).$$

□

The exterior harmonic measure boundary is related in spirit to the theory of entrance and exit boundaries; see the notes at the end of the chapter for references.

7.6 A characterization of the Martin boundary in terms of the exterior harmonic measure boundary

In order to show that the exterior harmonic measure boundary is independent of the choice of subdomain used in its calculation, we need the following proposition.

Proposition 6.1. *Let* $L = L_0 + V$ *satisfy Assumption* $\widetilde{H}_{\text{loc}}$ *on a domain* $D \subseteq R^d$ *and assume that* L *is either critical or subcritical on* D. *If the exterior harmonic measure boundary for* L *on* D *calculated using* Ω *coincides for every choice of subdomain* $\Omega \subset\subset D$ *with a* $C^{2,\alpha}$*-boundary, then the exterior harmonic measure boundary for* L *on* D *calculated using* Ω *coincides for every choice of subdomain* $\Omega \subset\subset D$.

Proof. Let $\Omega \subset\subset D$ be a subdomain and let Ω_1 and Ω_2 be subdomains with $C^{2,\alpha}$-boundaries such that $\Omega_1 \subset\subset \Omega \subset\subset \Omega_2 \subset\subset D$. Let $h_1 \in C^{2,\alpha}(D - \bar{\Omega}) \cap C(D - \Omega_1)$ be a positive solution of minimal growth at ∂D for L on D which satisfies $h_1 > 0$ on $\partial\Omega_1$ and let $h_2 \in C^{2,\alpha}(D - \bar{\Omega}) \cap C(D - \Omega)$ be a positive solution of minimal growth at ∂D for L on D which satisfies $h_2 > 0$ on $\partial\Omega$. In what follows, we will be considering the harmonic measures on $\partial\Omega_1$ and on $\partial\Omega$ for L^{h_1} and on $\partial\Omega$ and $\partial\Omega_2$ for L^{h_2}. Therefore, we must depart from our usual notation for harmonic measures since that notation does not distinguish the support of the measure. We will employ the usual notation $\mu_x^{h_i}$ to denote the harmonic measure on $\partial\Omega$ with respect to $x \in D - \bar{\Omega}$ for L^{h_i} on $D - \bar{\Omega}$, $i = 1, 2$. We will use the notation $\mu^{h_i}_{\Omega_i,x}$ to denote the harmonic measure on $\partial\Omega_i$ with respect to $x \in D - \bar{\Omega}_i$ for L^{h_i} on $D - \bar{\Omega}_i$, $i = 1, 2$.

Let $\{x_n\}_{n=1}^{\infty} \subset D - \bar{\Omega}_2$ be a sequence with no accumulation points in D. By the strong Markov property,

$$\mu^{h_2}_{x_n}(\mathrm{d}y) = \int_{\partial\Omega_2} \mu^{h_2}_{\Omega_2,x_n}(\mathrm{d}z)\mu^{h_2}_z(\mathrm{d}y) \text{ and } \mu^{h_1}_{\Omega_1,x_n}(\mathrm{d}y)$$

$$= \int_{\partial\Omega} \mu^{h_1}_{x_n}(\mathrm{d}z)\mu^{h_1}_{\Omega_1,z}(\mathrm{d}y). \tag{6.1}$$

By Proposition 4.2(i) and (4.14), $\mu^{h_2}_z$ and $\mu^{h_1}_{\Omega_1,z}$ are weakly continuous in their dependence on z. From this and (6.1), it follows that if $w - \lim_{n\to\infty} \mu^{h_2}_{\Omega_2,x_n}$ exists, then $w - \lim_{n\to\infty} \mu^{h_2}_{x_n}$ exists, and if $w - \lim_{n\to\infty} \mu^{h_1}_{x_n}$ exists, then $w - \lim_{n\to\infty} \mu^{h_1}_{\Omega_1,x_n}$ exists. This proves the proposition. □

We can now state and prove the result for which the work of the previous three sections has been a preparation.

Theorem 6.2. *Let $L = L_0 + V$ satisfy Assumption $\widetilde{H}_{\text{loc}}$ on a domain $D \subseteq R^d$ and assume that L is subcritical on D. Then the Martin boundary for L ($\widetilde{L}$) on D coincides with the exterior harmonic measure boundary (calculated with any choice of subdomain $\Omega \subset\subset D$) for $\widetilde{L}$ (L) on D.*

Remark. It is perhaps worth while to state Theorem 6.2 exclusively in terms of partial differential equations, with no reference to harmonic measure. Using Proposition 5.1 and Theorem 6.2, the following alternative version of Theorem 6.2 is immediate.

Theorem 6.2′. *Let $L = L_0 + V$ satisfy Assumption $\widetilde{H}_{\text{loc}}$ on a domain $D \subseteq R^d$ and assume that L is subcritical on D. Let $\Omega \subset\subset D$ be a subdomain which satisfies an interior cone condition at each point of its boundary. Let $\{x_n\}_{n=1}^{\infty} \subset D - \Omega$ be a sequence with no accumulation points in D. Then $\{x_n\}_{n=1}^{\infty}$ is a Martin sequence for L ($\widetilde{L}$) on D if and only if*

$$\lim_{n\to\infty} \frac{h_2}{h_1}(x_n)$$

exists whenever $h_1, h_2 \in C^{2,\alpha}(D - \bar{\Omega}) \cap C(D - \Omega)$ are positive solutions of minimal growth at ∂D for $\widetilde{L}$ (L) on D.

The characterization of the Martin boundary in terms of the exterior harmonic measure boundary has a particular probabilistic appeal in certain special cases.

Corollary 6.3. *Let $V \equiv 0$ and let $L = L_0$ satisfy Assumption $\widetilde{H}_{\text{loc}}$ on a domain $D \subseteq R^d$. Assume that L_0 is subcritical on D. Let $\Omega \subset\subset D$ be a subdomain satisfying an interior cone condition at each point of its boundary and let $\{x_n\}_{n=1}^{\infty} \subset D$ be a sequence with no accumulation points in D. Then $\{x_n\}_{n=1}^{\infty}$ is a Martin sequence for $\widetilde{L}_0$ on D if and only if the conditioned exit distribution $P_{x_n}(X(\sigma_{\bar{\Omega}}) \in \cdot \,|\sigma_{\bar{\Omega}} < \tau_D)$ on $\partial\Omega$ for the diffusion corresponding to L_0 and starting at x_n converges weakly as $n \to \infty$. If L_0 is symmetric, that is, $b = a\nabla Q$ for some $Q \in C^{2,\alpha}(D)$, then the Martin boundaries of L_0 and $\widetilde{L}_0$ coincide; thus the above characterization also holds for L_0.*

Proof. The first statement follows from Proposition 5.2 and the second one from Proposition 1.5. □

Before proving Theorem 6.2, we present a number of results which are immediate consequences of Theorem 6.2. Although the first two in particular may seem 'obvious', we have no simple proofs without resorting to Theorem 6.2.

Theorem 6.4. *Let L_1 and L_2 be subcritical operators satisfying Assumption $\widetilde{H}_{\text{loc}}$ on a domain $D \subseteq R^d$. Assume that there exists a domain $\Omega \subset\subset D$ such that $L_1 = L_2$ on $D - \Omega$. Then the Martin boundaries of L_1 and L_2 coincide.*

Proof. By Theorem 6.2, the Martin boundaries of L_1 and L_2 can be identified with the exterior harmonic measure boundaries for $\widetilde{L}_1$ and $\widetilde{L}_2$ calculated using the subdomain Ω. Since, by construction, these exterior harmonic measure boundaries depend only on the coefficient of $\widetilde{L}_1$ and $\widetilde{L}_2$ in $D - \Omega$, it follows that they coincide and, thus, so do the Martin boundaries of L_1 and L_2. □

Theorem 6.5. *Let L satisfy Assumption $\widetilde{H}_{\text{loc}}$ on a domain $D \subseteq R^d$, $d \geq 2$, and assume that L is subcritical. Let $U \subset\subset D$ be a subdomain and assume that U is simply connected so that $D - U$ is connected. Then the Martin boundary for L on D coincides with the part of the Martin boundary for L on $D - \bar{U}$ corresponding to sequences $\{x_n\}_{n=1}^{\infty} \subset D - \bar{U}$ with no accumulation points in D.*

Proof. Choose domains U_1 and U_2 satisfying $U \subset\subset U_1 \subset\subset U_2 \subset\subset D$ and define $\Omega = U_2 - \bar{U}_1$. If h, defined on $D - \Omega$, is a positive solution of minimal growth at $\partial U \cap \partial D$ for L on $D - \bar{U}$, then its restriction to $D - U_2$ is a positive solution of minimal growth for L on D. Thus, from the construction, it is clear that a sequence $\{x_n\}_{n=1}^{\infty} \subset D - \bar{U}$ with no accumulation points in D is fundamental for the exterior harmonic measure boundary of L on D calculated with the subdomain U_2 if and only if it is fundamental for the exterior harmonic measure boundary of L on $D - \bar{U}$ calculated with the subdomain Ω. Thus, the theorem follows from Theorem 6.2. □

As we noted in Chapter 4, Section 11, the criticality theory extends to operators on non-compact, open, $C^{2,\alpha}$-Riemannian manifolds. A manifold is called a *manifold with m ends* if it can be represented in the form $D = \bar{\Omega} \cup E_1 \cup E \ldots \cup E_m$, where Ω is bounded and open, E_i is open, $\bar{E}_i \cap \bar{E}_j = \phi$, for $i \neq j$, and $\bar{\Omega} \cap E_i = \phi$. As an example of such a manifold, consider a sphere with m 'handles' protruding from it (perhaps a model of the heart if $d = 2$ and $m = 4$).

Theorem 6.6. *Let D be a non-compact, open d-dimensional $C^{2,\alpha}$-Riemannian manifold with m ends and let L on D satisfy the analog of Assumption $\bar{H}_{\text{loc}}$ and be subcritical. Then the Martin boundary for L decomposes into m components in the following sense:*

(i) *If $\{x_n\}_{n=1}^{\infty} \subset D$ is a Martin sequence, then all but a finite number of its terms lie in E_i for some $i = 1, 2, \ldots, m$.*
(ii) *If L_1 and L_2 are two subcritical operators on D and $L_1 = L_2$ on E_i, then a sequence $\{x_n\}_{n=1}^{\infty} \subset E_i$ is a Martin sequence for L_1 if and only if it is a Martin sequence for L_2.*

Proof. We may assume that Ω has a smooth boundary since otherwise we could extend Ω a bit at the expense of the E_is and obtain a smooth boundary.

(i) By Theorem 3.5, there exists a positive solution h of minimal growth at ∂D for L on D which is continuous on $D - \Omega$. Let μ_x^h, $x \in D - \bar{\Omega}$, denote the harmonic measure on $\partial\Omega$ for L^h on $D - \bar{\Omega}$, as in Corollary 4.6. If a diffusion path starting in E_i satisfies $\sigma_{\bar{\Omega}} < \infty$, then $X(\sigma_{\bar{\Omega}}) \in \partial\Omega \cap \bar{E}_i$. Since $\bar{E}_i \cap \bar{E}_j = \phi$, for $i \neq j$, it follows that a sequence with an infinite number of terms in E_i and an infinite number of terms in E_j cannot be fundamental for the exterior harmonic measure boundary. Thus, part (i) follows from Theorem 6.2.

(ii) Let $\phi \in C(\partial\Omega)$ and $\phi > 0$. By Theorem 3.5, we can construct positive solutions h_j, $j = 1, 2$, of minimal growth at ∂D for L_j on D, $j = 1, 2$, which are continuous up to $D - \Omega$ and satisfy $h_j = \phi$ on $\partial\Omega$. Let $\mu_x^{h_j}$, $x \in D - \bar{\Omega}$, denote the harmonic measure on $\partial\Omega$ for $L_j^{h_j}$ on $D - \bar{\Omega}$. Since a diffusion path which starts in E_i satisfies $X(t) \in E_i$ for all $t \in [0, \sigma_{\bar{\Omega}})$, it follows from the representation in Theorem 3.5 and the fact that $L_1 = L_2$ on E_i, that $h_1 = h_2$ on E_i and, similarly, $\mu_x^{h_1} = \mu_x^{h_2}$, for $x \in E_i$. Thus, a sequence $\{x_n\}_{n=1}^{\infty} \subset E_i$ is fundamental for the exterior harmonic measure boundary of L_1 if and only if it is fundamental for the exterior harmonic measure boundary of L_2. Part (ii) now follows from Theorem 6.2. □

A trivial example of a manifold with ends is $D = (\alpha, \beta) \subseteq R$, with $-\infty \leqslant \alpha < \beta \leqslant \infty$. Thus, we have the following corollary.

Corollary 6.7. *Let L satisfy Assumption* $\widetilde{H}_{\text{loc}}$ *and be subcritical on the domain* $D = (\alpha, \beta)$, *where* $-\infty \leqslant \alpha < \beta \leqslant \infty$. *Then the Martin boundary of L on D consists of two points. More specifically, a sequence* $\{x_n\}_{n=1}^{\infty}$ *with no accumulation points in D is a Martin sequence if and only if* $\lim_{n\to\infty} x_n = \alpha$ *or* $\lim_{n\to\infty} x_n = \beta$.

Remark. It follows from Proposition 5.1.3 that $C_L(D)$ is two-dimensional. Thus, from the Martin representation theorem (Theorem 1.2), the minimal Martin boundary must consist of exactly two points. However, this does not prove that the entire Martin boundary consists of two points, nor does it identify Martin sequences. An alternative proof of Corollary 6.7 which does not use Theorem 6.6 but, rather, is based on a direct calculation is suggested in exercise 7.6.

Theorem 6.8. *Let* $L = L_0 + V$ *satisfy Assumption* $\widetilde{H}_{\text{loc}}$ *on a domain* $D \subseteq R^d$ *and assume that L is either critical or subcritical on D. Then the exterior harmonic measure boundary is independent of the choice of subdomain* $\Omega \subset\subset D$ *used to calculate it.*

Proof. In the subcritical case, the theorem follows immediately from Theorem 6.2. Now consider the critical case. It is enough to show that if $\Omega_i \subset\subset D$, $i = 1, 2$, satisfy $\Omega_1 \subset\subset \Omega_2 \subset\subset D$, then the exterior harmonic measure boundaries obtained using Ω_1 and Ω_2 coincide. Let U_1 and U_2 be domains satisfying $U_1 \subset\subset U_2 \subset\subset \Omega_1$ and let $\Omega_i' = \Omega_i - \bar{U}_2$, $i = 1, 2$. By Proposition 3.1, L is subcritical on $D - \bar{U}_1$. By Theorem 6.2, for either $i = 1$ or $i = 2$, a sequence $\{x_n\}_{n=1}^{\infty} \subset D - \bar{U}_1$ with no accumulation points in D is fundamental for the exterior harmonic measure boundary of L on $D - \bar{U}_1$ calculated using Ω_i' if and only if $\{x_n\}_{n=1}^{\infty}$ is a Martin sequence for $\tilde{L}$ on $D - \bar{U}_1$. The theorem now follows since, as in the proof of Theorem 6.5, $\{x_n\}_{n=1}^{\infty}$ is fundamental for the exterior harmonic measure boundary of L on $D - \bar{U}_1$ calculated using Ω_i' if and only if it is fundamental for the exterior harmonic measure boundary of L on D calculated using Ω_i. □

For the proof of Theorem 6.2, we will need the following lemma; it is the uniqueness part of the Riesz decomposition theorem in potential theory.

Lemma 6.9. *Let L satisfy Assumption $\tilde{H}_{\text{loc}}$ on a domain $D \subseteq R^d$ and assume that L is subcritical on D. Let μ be a measure on D and assume that $u(x) \equiv \int_D G(x, y)\mu(dy)$ is locally bounded for $x \in D$. Then u uniquely determines μ.*

Proof. We first prove the lemma under the additional assumption that $\operatorname{supp} \mu \subset\subset D$. By a standard mollification procedure, there exists a sequence $\{\mu_n\}_{n=1}^{\infty}$ of measures with C^{∞}-densities $\{f_n\}_{n=1}^{\infty}$ such that $\mu = w - \lim_{n \to \infty} \mu_n$. Since μ is assumed to have compact support in D, we may assume that all the f_ns are supported in a domain $A \subset\subset D$. Define $u_n(x) = \int_D G(x, y) f_n(y)\, dy$ and, for $w \in C_0^{\infty}(D)$, define $v(x) = \int_D G(y, x)\tilde{L}w(y)\, dy$. By Theorem 4.3.8 and the fact that $\tilde{G}(x, y) = G(y, x)$ is the Green's function for $\tilde{L}$ on D, we have u_n, $v \in C^{2,\alpha}(D)$ and $Lu_n = -f_n$. Integrating by parts and using Fubini's theorem (which is permissible by the integrability assumption in the statement of the theorem and the fact that $\tilde{L}w$ has compact support in D), we obtain

$$\begin{aligned}\int_D w\, d\mu_n &= -\int_D wLu_n\, dx = -\int_D u_n \tilde{L}w\, dx \\ &= -\int_D\left(\int_D G(x, y) f_n(y)\, dy\right)\tilde{L}w(x)\, dx \\ &= -\int_D v\, d\mu_n.\end{aligned}$$

Since v is bounded and continuous on A, letting $n \to \infty$ and using Fubini's theorem again gives

$$\int_D w\, d\mu = -\int_D v\, d\mu = -\int_D u\tilde{L}w\, dx. \tag{6.2}$$

This shows that μ is uniquely determined by u.

In the general case, let $\{D_n\}_{n=1}^{\infty}$, be a sequence of domains satisfying $D_n \subset\subset D_{n+1}$ and $\bigcup_{n=1}^{\infty} D_n = D$. Define $\mu_n(B) = \mu(B \cap D_n)$, for measurable $B \subset D$ and define

$$u_n(x) = \int_D G(x, y)\mu_n(dy) = \int_D G(x, y)1_{D_n}(y)\mu(dy).$$

By the monotone convergence theorem, $u = \lim_{n\to\infty} u_n$. Applying (6.2) to μ_n, we have for $w \in C_0^{\infty}(D)$,

$$\int_D w \, d\mu_n = -\int_D u_n \widetilde{L} w \, dx.$$

Letting $n \to \infty$ and using the fact that Lw has compact support in D, we obtain $\int_D w \, d\mu = -\int u \widetilde{L} w \, dx$; thus μ is uniquely determined by u.

□

Proof of Theorem 6.2. We will prove that the exterior harmonic measure boundary for L on D calculated using a subdomain $\Omega \subset\subset D$ with a $C^{2,\alpha}$-boundary coincides with the Martin boundary for $\widetilde{L}$ on D. It then follows from Proposition 6.1 that the result holds for any subdomain $\Omega \subset\subset D$.

For each $x \in \Omega$, define $h_x(y) = \widetilde{G}(x, y)$, for $y \in D - \Omega$. By Theorem 4.2.5, $\widetilde{G}(x, y) = G(y, x)$. From this and Theorem 3.7, it follows that $h_x \in C^{2,\alpha}(D - \Omega)$ is a positive solution of minimal growth at ∂D for L on D and $h_x > 0$ on $\partial\Omega$. Fix $x_0 \in \Omega$. The Martin kernel is given by

$$\widetilde{k}(x, y) = \frac{\widetilde{G}(x, y)}{\widetilde{G}(x_0, y)}.$$

Thus

$$\widetilde{k}(x, y) = \frac{h_x(y)}{h_{x_0}(y)}, \quad \text{for } x \in \Omega \text{ and } y \in D - \Omega. \tag{6.3}$$

Let $\{y_n\}_{n=1}^{\infty} \subset D - \Omega$ be a sequence with no accumulation points in D. To prove the theorem, we will show that $\{y_n\}_{n=1}^{\infty}$ is fundamental for the exterior harmonic measure boundary of L on D, if and only if it is a Martin sequence for $\widetilde{L}$ on D.

Assume first that $\{y_n\}_{n=1}^{\infty}$ is fundamental for the exterior harmonic measure boundary of L on D. By Proposition 5.1 and (6.3), it follows that $\lim_{n\to\infty} \widetilde{k}(x, y_n)$ exists for all $x \in \Omega$. To show that $\{y_n\}_{n=1}^{\infty}$ is a Martin sequence for $\widetilde{L}$ on D, we must show that $\lim_{n\to\infty} \widetilde{k}(x, y_n)$ exists for all $x \in D$. Assume the contrary. Then there exist two subsequences $\{y'_n\}_{n=1}^{\infty}$ and $\{y''_n\}_{n=1}^{\infty}$ of $\{y_n\}_{n=1}^{\infty}$ such that $u_1(x) \equiv \lim_{n\to\infty} \widetilde{k}(x, y'_n)$

and $u_2(x) \equiv \lim_{n\to\infty} \tilde{k}(x, y''_n)$ exist, $u_i \in C_{\tilde{L}}(D)$, $i = 1, 2$, $u_1 = u_2$ on Ω, but $u_1 \not\equiv u_2$. Thus $v \equiv u_1 - u_2$ is a solution in D which vanishes on Ω but is not identically zero; this is impossible by the unique continuation theorem for elliptic operators [Hormander (1985), Theorem 17.2.6].

Now assume that $\{y_n\}_{n=1}^{\infty}$ is a Martin sequence for $\tilde{L}$ on D. To show that $\{y_n\}_{n=1}^{\infty}$ is fundamental for the exterior harmonic measure boundary for L on D, we will show that if $\{y'_n\}_{n=1}^{\infty}$ and $\{y''_n\}_{n=1}^{\infty}$ are subsequences of $\{y_n\}_{n=1}^{\infty}$ and $\mu_1 \equiv w - \lim_{n\to\infty} \mu_{y'_n}^{h_{x0}}$ and $\mu_2 \equiv w - \lim_{n\to\infty} \mu_{y''_n}^{h_{x0}}$ exist, then $\mu_1 = \mu_2$. By Proposition 4.1, $h_x(y) = \int_{\partial\Omega} G(z, x)\mu_y(dz)$, for $x \in \Omega$ and $y \in D - \Omega$ and, by Corollary 4.6,

$$\mu_y^{h_{x0}}(dz) = \frac{h_{x_0}(z)}{h_{x_0}(y)}\mu_y(dz), \quad \text{for } y \in D - \bar{\Omega}.$$

Thus, (6.3) can be written as

$$\tilde{k}(x, y) = \int_{\partial\Omega} G(z, x)\frac{1}{h_{x_0}(z)}\mu_y^{h_{x0}}(dz), \quad \text{for } x \in \Omega \text{ and } y \in D - \bar{\Omega}.$$

Since, by assumption, $\{y_n\}_{n=1}^{\infty}$ is a Martin sequence for $\tilde{L}$ on D, $\lim_{n\to\infty} \tilde{k}(x, y'_n) = \lim_{n\to\infty} \tilde{k}(x, y''_n)$, for $x \in D$. Thus, we obtain

$$\int_{\partial\Omega} G(z, x)\frac{1}{h_{x_0}(z)}\mu_1(dz) = \int_{\partial\Omega} G(z, x)\frac{1}{h_{x_0}(z)}\mu_2(dz), \quad \text{for } x \in \Omega. \tag{6.4}$$

We will show below that the equality in (6.4) in fact holds for all $x \in D$. Using this and applying Lemma 6.9 to $\tilde{L}$ and $\tilde{G}(x, y) = G(y, x)$, it follows that

$$\frac{1}{h_{x_0}(z)}\mu_1(dz) = \frac{1}{h_{x_0}(z)}\mu_2(dz), \quad \text{that is, } \mu_1 = \mu_2.$$

To show that (6.4) holds for all $x \in D$, recall that by Theorem 4.2.8, the singularity in $G(x, y)$ at $x = y$, which exists if $d \geqslant 2$, is of order

$$\frac{1}{|x - y|^{d-2}}, \quad \text{if } d \geqslant 3,$$

and is of order $-\log|x - y|$ if $d = 2$. By assumption, Ω has a $C^{2,\alpha}$-boundary; thus by Corollary 4.7 and the definition of μ_i, $i = 1, 2$, it follows that μ_i possesses a bounded density with respect to Lebesgue surface measure. From these two facts and the fact that $\partial\Omega$ is a $(d - 1)$-dimensional hypersurface, it follows that the expressions on the lefthand and righthand sides of (6.4) are finite for $x \in \partial\Omega$. An application of the dominated convergence theorem then shows that these two expressions are continuous over all $x \in D$. Thus

$$u_i(x) \equiv \int_{\partial\Omega} G(z, x)\frac{1}{h_{x_0}(z)}\mu_i(dz), \quad x \in D - \Omega, \quad i = 1, 2,$$

satisfies $u_i \in C^{2,\alpha}(D - \bar{\Omega}) \cap C(D - \Omega)$, $\tilde{L}u_i = 0$ in $D - \bar{\Omega}$ and $u_1 = u_2$ on $\partial\Omega$. To complete the proof that (6.4) holds for all $x \in D$, we will show that u_1 and u_2 are positive solutions of minimal growth at ∂D for $\tilde{L}$ on D. It will then follow from the maximum principle for positive solutions of minimal growth at ∂D (Theorem 3.6) that $u_1 = u_2$ on $D - \Omega$.

Let $\{D_n\}_{n=1}^{\infty}$ be a sequence of domains with $C^{2,\alpha}$-boundaries satisfying $\Omega \subset\subset D_1$, $D_n \subset\subset D_{n+1}$ and $\bigcup_{n=1}^{\infty} D_n = D$, and let $G_n(x, y)$ denote the Green's function for L on D_n. Define

$$u_{n,i}(x) = \int_{\partial\Omega} G_n(z, x)\frac{1}{h_{x_0}(x)}\mu_i(dz), \; x \in D_n - \Omega, \; i = 1, 2.$$

By Theorem 3.2 applied to the Green's function $\tilde{G}_n(x, z) = G_n(z, x)$, $u_{n,i}$ can be extended continuously to $\bar{D}_n - \Omega$ by defining $u_{n,i}(x) = 0$ for $x \in \partial D_n$. Thus $u_{n,i} \in C^2(D_n - \bar{\Omega}) \cap C(\bar{D}_n - \Omega)$, $\tilde{L}u_{n,i} = 0$ and $u_{n,i} > 0$ in $D_n - \bar{\Omega}$, and $u_{n,i} = 0$ on ∂D_n, $i = 1, 2$. Let $v \in C^{2,\alpha}(D - \bar{\Omega}) \cap C(D - \Omega)$ satisfy $\tilde{L}v = 0$ and $v > 0$ in $D - \Omega$, and $v = u_1 = u_2$ on $\partial\Omega$. Since $G_n(x, y) \leqslant G(x, y)$, for all $x \neq y$, it follows that $v \geqslant u_{n,i}$ on $\partial\Omega$. Also, $v > u_{n,i} = 0$ on ∂D_n. Thus, from Theorem 4.4.1(iii) and Theorem 3.6.5, it follows that $v \geqslant u_{n,i}$ on $D_n - \Omega$. By Theorem 4.3.7, $u_i(x) = \lim_{n\to\infty} u_{n,i}(x)$, for $x \in D - \Omega$; thus $v \geqslant u_i$ on $D - \Omega$ and it follows that u_i is a positive solution of minimal growth at ∂D for $\tilde{L}$ on D. □

Remark. In one direction of the proof of Theorem 6.2, we relied on the unique continuation theorem for elliptic PDEs. This can be avoided by defining the exterior harmonic measure boundary in a slightly different way as follows. In our construction of the exterior harmonic measure boundary, we chose a domain $\Omega \subset\subset D$ and a positive solution $h \in C^{2,\alpha}(D - \bar{\Omega}) \cap C(D - \Omega)$ of minimal growth at ∂D for L on D which satisfies $h > 0$ on $\partial\Omega$. Letting μ_x^h denote the harmonic probability measure on $\partial\Omega$ with respect to $x \in D - \bar{\Omega}$, for L^h on $D - \bar{\Omega}$, we defined a sequence $\{x_n\}_{n=1}^{\infty} \subset D$ with no accumulation points in D to be fundamental if $w - \lim_{n\to\infty} \mu_{x_n}^h$ exists. The definition was shown to be independent of the choice of Ω and h. Consider now the following alternative definition. Let $\{\Omega_m\}_{m=1}^{\infty}$ be a sequence of domains satisfying $\Omega_m \subset\subset \Omega_{m+1}$ and $\bigcup_{m=1}^{\infty} \Omega_m = D$. For $m = 1, 2, \ldots$, choose a positive solution $h_m \in C^{2,\alpha}(D - \bar{\Omega}_m) \cap$

$C(D-\Omega_m)$ of minimal growth at ∂D for L on D, which satisfies $h_m > 0$ on $\partial\Omega_m$, and let $\mu_x^{h_m}$ denote the harmonic measure on $\partial\Omega_m$ with respect to $x \in D - \bar{\Omega}_m$ for L^{h_m} on $D - \bar{\Omega}_m$. Define a sequence $\{x_n\}_{n=1}^{\infty} \subset D$ with no accumulation points in D to be fundamental if $w-\lim_{n\to\infty} \mu_{x_n}^{h_m}$ exists for $m = 1, 2, \ldots$, and define the exterior harmonic measure boundary using this new definition of a fundamental sequence. It is then easy to check that, with this new definition, the proof of Theorem 6.2 no longer requires the unique continuation theorem. Furthermore, the theorems in this chapter which relied on Theorem 6.2 and the applications in Chapter 8 which rely on Theorem 6.2 go through just as well with this new definition.

Exercises

7.1. Prove that ρ satisfies the triangle inequality and that $\rho(z_1, z_2) > 0$ if $z_1 \neq z_2$, where ρ is given by (1.1).

7.2. Let L be symmetric; that is, $L = \frac{1}{2}\nabla\cdot a\nabla + a\nabla Q\cdot\nabla + V$. Show that $\tilde{L} = L^h$, where $h(x) = \exp(-2Q(x))$. Conclude, then, that the Green's function $G(x, y)$ for L satisfies

$$\exp(2Q(x))G(x,y) = \exp(2Q(y))G(y,x).$$

7.3. Let L, D and Ω be as in Theorem 3.5. Let $A \subseteq \partial\Omega$ be open in the relative topology on $\partial\Omega$. Then there exists a positive solution $h \in C^{2,\alpha}(D-\bar{\Omega})$ of minimal growth at $\partial\Omega$ for L on D which satisfies $\lim_{x\to A} h(x) = 1$ and $\lim_{x\to\partial\Omega-\bar{A}} h(x) = 0$. In fact,

$$h(x) = E_x\left(\exp\left(\int_0^{\sigma_{\bar{\Omega}}} V(X(s))\,\mathrm{d}s\right)1_A(X(\sigma_{\bar{\Omega}}));\ \sigma_{\bar{\Omega}} < \tau_D\right).$$

In particular, if $V \equiv 0$, $h(x) = P_x(X(\sigma_{\bar{\Omega}}) \in A,\ \sigma_{\bar{\Omega}} < \tau_D)$. (*Hint*: Let $\{A_n\}_{n=1}^{\infty}$ be a sequence of open sets relative to $\partial\Omega$ satisfying $A_n \subset\subset A_{n+1}$ and $\bigcup_{n=1}^{\infty} A_n = A$. Let $\phi_n \in C(\partial\Omega)$ satisfy $\phi_n = 1$ on A_n, $\phi_n = 0$ on $\partial\Omega - A_{n+1}$ and $0 \le \phi_n \le 1$, and let h_n denote the positive solution of minimal growth at ∂D as in Theorem 3.5 with ϕ replaced by ϕ_n.)

7.4. Let L, D, Ω and A be as in exercise 7.3. Assume in addition that Ω has a $C^{2,\alpha}$-boundary and that $\bar{A} - A$ has Lebesgue surface measure zero. Let h denote the positive solution of minimal growth at ∂D from exercise 7.3. Show that Corollary 4.6 extends to this choice of h. In particular, if $V \equiv 0$, then for $x \in D - \bar{\Omega}$,

$$\begin{aligned}\mu_x^h(\mathrm{d}y) &= \lim_{n\to\infty} P_x^h(X(\sigma_{\bar{\Omega}_n}) \in \mathrm{d}y)\\ &= P_x(X(\sigma_{\bar{\Omega}}) \in \mathrm{d}y \mid X(\sigma_{\bar{\Omega}}) \in A,\ \sigma_{\bar{\Omega}} < \tau_D),\end{aligned}$$

where $\{\Omega_n\}_{n=1}^{\infty}$ is a sequence of domains satisfying $\Omega_1 \subset\subset D$, $\Omega_{n+1} \subset\subset \Omega_n$ and $\bigcap_{n=1}^{\infty}\Omega_n = \bar{\Omega}$. (*Hint*: By Theorem 4.4, $P_x(X(\sigma_{\bar{\Omega}}) \in \bar{A} - A, \sigma_{\bar{\Omega}} < \tau_D) = 0$.).

7.5. Let L, D, Ω and A be as in exercise 7.4 and assume that $V \equiv 0$. Let $h(x) = P_x(X(\sigma_{\bar{\Omega}}) \in A, \sigma_{\bar{\Omega}} < \tau_D)$. Then, for $x \in D - \bar{\Omega}$,

$$P_x(\cdot\,|X(\sigma_{\bar{\Omega}}) \in A, \sigma_{\bar{\Omega}} < \tau_D) = P_x^h(\cdot). \qquad (*)$$

To be precise, $(*)$ should be interpreted as follows: Let $\{\Omega_n\}_{n=1}^{\infty}$ be a sequence of domains satisfying $\Omega_1 \subset\subset D$, $\Omega_{n+1} \subset\subset \Omega_n$ and $\bigcap_{n=1}^{\infty}\Omega_n = \bar{\Omega}$. Then, identifying $\hat{\mathcal{F}}^{D}_{\sigma\bar{\Omega}_n}$ with $\hat{\mathcal{F}}^{D-\bar{\Omega}}_{\sigma\Omega_n}$ in the obvious way,

$$P_x(B|X(\sigma_{\bar{\Omega}}) \in A, \sigma_{\bar{\Omega}} < \tau_D) = P_x^h(B),$$
$$\text{for } B \in \hat{\mathcal{F}}^{D-\bar{\Omega}}_{\sigma\Omega_n}, \quad x \in D - \bar{\Omega} \text{ and } n = 1, 2, \ldots.$$

(*Hint*: Use Corollary 4.1.2.)

7.6. Prove Corollary 6.7 without resorting to Theorem 6.6. (*Hint*: First reduce to the case $V \equiv 0$. Then, by the strong Markov property, if $x_0 < x < y$, then $G(x_0, y) = P_{x_0}(\tau_x < \infty)G(x, y)$, where $\tau_x = \inf\{t \geq 0: X(t) = x\}$.)

7.7. Let L and $D = (\alpha, \beta)$ satisfy the assumptions of Corollary 6.7 and let h_α and h_β denote the functions in $C_L(D)$ corresponding to the Martin boundary points α and β. Let $\tau_r = \inf\{t \geq 0: X(t) = r\}$, for $r \in (\alpha, \beta)$, and let $\tau_\alpha = \lim_{r\to\alpha}\tau_r$ and $\tau_\beta = \lim_{r\to\beta}\tau_r$. Show that for $r \in (\alpha, \beta)$,

$$h_\alpha(x) = h_\alpha(r)E_x\left(\exp\left(\int_0^{\tau_r} V(X(s))\,\mathrm{d}s\right); \tau_r < \tau_\beta\right), \quad \text{for } x \in (r, \beta),$$

and

$$h_\beta(x) = h_\beta(r)E_x\left(\exp\left(\int_0^{\tau_r} V(X(s))\,\mathrm{d}s\right); \tau_r < \tau_\alpha\right), \quad \text{for } x \in (\alpha, r).$$

7.8. Let $D_{\theta_0} = \{x = (r, \theta) \in R^2: r > 0 \text{ and } 0 < \theta < \theta_0\}$, where $\theta_0 \in (0, 2\pi]$. Two-dimensional Brownian motion is transient in D_{θ_0} since it eventually leaves D_{θ_0}; thus $\frac{1}{2}\Delta$ on D_{θ_0} is subcritical. Let V, $W \in C_0^\alpha(D_{\theta_0})$ and assume that V, $W \not\equiv 0$, that $W \geq 0$ and that V changes sign. Let t_0, $t^\pm$, ϕ_c and $\phi_c^\pm$ be as in Example 3.10. Show that ϕ_c, $\phi_c^\pm \in L^2(D_{\theta_0})$ if $\theta_0 \in (0, 2\pi)$, but that ϕ_c, $\phi_c^\pm \notin L^2(D_{\theta_0})$ if $\theta_0 = 2\pi$. (*Hint*: Let $G(r, \theta, r', \theta')$ denote the Green's function for $\frac{1}{2}\Delta$ on D_{θ_0}. Using the theory of positive solutions of minimal growth, show that $G(r, \theta, 1, \theta_0/2)$ is comparable to $g(\theta)r^l$, for $r \geq 2$, where g is an appropriate function vanishing at $\theta = 0$ and $\theta = \theta_0$, and l is an appropriate exponent. Find l.)

7.9. Let L satisfy Assumption $\tilde{\text{H}}_{\text{loc}}$ on a domain $D \subseteq R^d$. Assume that L on D is subcritical and let $G(x, y)$ denote the Green's function. Fix $y_0 \in D$ and let $D_{y_0} = D - \{y_0\}$. Prove that $G(\cdot, y_0)$ is a minimal positive harmonic function for L on D_{y_0}. (*Hint*: Use the results of Section 3 and the fact (proved in Example 2.6) that y_0 may be identified with a unique Martin boundary point for L on D_{y_0}.)

Notes

The analytic theory of the Martin boundary for the Laplacian goes back to [Martin (1941)]. The probabilistic connection to the Martin boundary in the case of the Laplacian is due to [Doob (1957)]. [Kunita and Watanabe (1965)] extended the results of Martin and Doob to more general Markov processes. See also [Sur 1963)] and [Hunt (1957a, b), (1958), (1960)]. Treatments of the Martin boundary from a probabilistic approach can be found in [Doob (1984)] and [Rogers and Williams (1987)].

The notion of positive solutions of minimal growth and their connection with Green's functions and ground states is due to [Agmon (1982)]. Theorem 3.8(i) is due to [Murata (1986)] in the symmetric case and to [Pinchover (1988)] in the general case (see also [Pinchover (1990a)] for a stronger version of the result). Theorem 3.8(ii) is due to [Agmon (1982)]. Sections 3 and 4 constitute a systematic development of positive solutions of minimal growth and of the related exterior harmonic measure from a probabilistic perspective; this has not appeared in the literature before. The absolute continuity of the harmonic measure is proved in Theorem 4.4 under the assumption that everything is smooth. For absolute continuity of harmonic measures in case of less smoothness, see the references contained in the survey article [Cranston and Mueller (1988)]. The theory of entrance and exit boundaries, which is related in spirit to the exterior harmonic measure boundary, can be found in [Rogers and Williams (1987)].

Sections 3 and 4 and the construction of the exterior harmonic measure boundary in Section 5 lay the groundwork for Theorem 6.2. This theorem was proved in [R. Pinsky (1993)]; however, Sections 3–6 give a more systematic development. While the paper [R. Pinsky (1993)] was in the galley stage, I learned from John Taylor that Theorem 6.2 had already been stated and proved using the language and techniques of axiomatic potential theory. The case of a symmetric operator appears in [Constantinescu and Cornea (1963)] (this reference is in German; there does not seem to be a reference in English). The result in the non-symmetric case is stated without proof in [Taylor (1969)]; the proof appears in an unpublished manuscript of Taylor's. Theorems similar to Theorems 6.4–6.6 have been proved by Taylor using potential theoretic methods [Taylor (1970)]. Theorem 6.4 also follows from [Pinchover (1988)]. Results similar to Theorem 6.6 appear in [Cranston (1994)], [Donnelly (1986)], [Li and Tam (1987)] and [Murata (1990(a))].

8

Positive harmonic functions and the Martin boundary: applications to certain classes of operator

8.0 Introduction

For the developments in this chapter, we will need the *boundary Harnack principle*.

Theorem 0.1. *Let $L = L_0 + V$ satisfy Assumption H (defined in Chapter 3, Section 2) on a bounded domain $D \subset R^d$ with a Lipschitz boundary. Let $z \in \partial D$ and let $B_r(z)$ denote the ball of radius r centered at z. Then there exists an $r_0 > 0$ and a constant c which depend only on the numbers λ, Λ and α appearing in Assumption H, and on the Lipschitz continuity of ∂D near z such that the following property holds. If $r \in (0, r_0)$ and u_1 and u_2 are positive solutions of $Lu = 0$ in $B_{8r}(z) \cap D$ which vanish continuously on $B_{8r}(z) \cap \partial D$, then*

$$\frac{u_1(x)}{u_1(x_r)} \leqslant c\frac{u_2(x)}{u_2(x_r)}, \quad \text{for } x \in B_r(z) \cap D, \text{ where } x_r \in \partial B_r(z) \cap D.$$

Remark. Note that the boundary Harnack principle also applies to an operator L satisfying Assumption $\tilde{H}$ (defined in Chapter 3, Section 7) since such an operator also satisfies Assumption H.

8.1 The Martin boundary in the case of a Lipschitz Euclidean boundary

We begin with an extension of Theorem 7.3.2.

Theorem 1.1. *Let L satisfy Assumption $\tilde{H}_{\text{loc}}$ on a domain $D \subset R^d$ and assume that L is subcritical. Let $x_0 \in \partial D$. Assume that there exists a ball B centered at x_0 such that $\partial D \cap B$ is a Lipschitz boundary portion and such that L satisfies Assumption E (defined in Chapter 3, Section 2) on*

$D \cap B$. Then

$$\lim_{x \to x_0} G(x, y) = 0, \quad \text{for } y \in D.$$

Remark. Our proof below uses the boundary Harnack principle. However, if $V \leqslant 0$ the simpler proof of (7.3.1) can be used (see exercise 8.1).

Proof. Fix $y \in D$ and choose the ball B in the statement of the theorem so that $y \notin \bar{B}$. Let U be open and satisfy $y \in U \subset\subset D - \bar{B}$. By assumption, there exists a constant M such that

$$\sup_{x \in B \cap D} V(x) \leqslant M. \tag{1.1}$$

By shrinking B if necessary, it follows from Theorem 2.2.1 and its proof that

$$\sup_{x \in B \cap D} E_x \exp((M + 1)(\tau_B \wedge \tau_D)) < \infty, \tag{1.2}$$

where, as usual, $\tau_B = \inf\{t \geqslant 0; X(t) \notin B\}$.

Define $H(x) = E_x(\exp(\int_0^{\sigma_{\bar{U}}} V(X(s))\, ds); \sigma_{\bar{U}} < \tau_D)$, where, we recall, $\sigma_{\bar{U}} = \inf\{t \geqslant 0: X(t) \in \bar{U}\}$. Since L is subcritical on D, it follows from Lemma 7.3.4(i) that $H(x)$ is locally bounded for $x \in D - U$. By Theorem 7.3.3,

$$\begin{aligned} G(x, y) &= E_x\left(\exp\left(\int_0^{\sigma_{\bar{U}}} V(X(s))\, ds\right) G(X(\sigma_{\bar{U}}), y); \quad \sigma_{\bar{U}} < \tau_D\right) \\ &\leqslant \sup_{z \in \partial U} G(z, y) H(x), \quad \text{for } x \in D - U. \end{aligned}$$

Thus, to prove the theorem, it is enough to show that

$$\lim_{x \to x_0} H(x) = 0. \tag{1.3}$$

Using the strong Markov property, we have

$$\begin{aligned} H(x) &= E_x\left(\exp\left(\int_0^{\sigma_{\bar{U}}} V(X(s))\, ds\right); \sigma_{\bar{U}} < \tau_D\right) \\ &= E_x\left(\exp\left(\int_0^{\tau_B} V(X(s))\, ds\right) E_{X(\tau_B)}\left(\exp\left(\int_0^{\sigma_{\bar{U}}} V(X(s))\, ds\right);\right.\right. \\ &\qquad \left.\left.\sigma_{\bar{U}} < \tau_D\right); \tau_B < \tau_D\right) \\ &= E_x\left(\exp\left(\int_0^{\tau_B} V(X(s))\, ds\right) H(X(\tau_B)); \tau_B < \tau_D\right), \\ &\qquad \text{for } x \in B \cap D. \end{aligned} \tag{1.4}$$

Define

$$\nu_x(\mathrm{d}y) = \frac{E_x\left(\exp\left(\int_0^{\tau_B} V(X(s))\,\mathrm{d}s\right);\ \tau_B < \tau_D,\ X(\tau_B) \in \mathrm{d}y\right)}{E_x\left(\exp\left(\int_0^{\tau_B} V(X(s))\,\mathrm{d}s\right);\ \tau_B < \tau_D\right)},$$

$$x \in B \cap D.$$

Then, from (1.4),

$$H(x) = \left(\int_{\partial B \cap D} H(y)\nu_x(\mathrm{d}y)\right) E_x\left(\exp\left(\int_0^{\tau_B} V(X(s))\,\mathrm{d}s\right);\ \tau_B < \tau_D\right),$$

$$x \in B \cap D. \quad (1.5)$$

Since $\partial D \cap B$ is a Lipschitz boundary, D satisfies an exterior cone condition at each point of $\partial D \cap B$. Thus, by Corollary 2.3.4,

$$\lim_{x \to z \in \partial D \cap B} P_x(\tau_D > t) = 0, \quad \text{for all } t > 0. \quad (1.6)$$

From Theorem 2.2.2,

$$\lim_{t \to 0} \limsup_{x \to z \in \partial D \cap B} P_x(\tau_B < t) = 0. \quad (1.7)$$

Combining (1.6) and (1.7) gives

$$\lim_{x \to z \in \partial D \cap B} P_x(\tau_B < \tau_D) = 0. \quad (1.8)$$

Applying Holder's inequality and using (1.1), (1.2) and (1.8), we obtain

$$\lim_{x \to z \in \partial D \cap B} E_x\left(\exp\left(\int_0^{\tau_B} V(X(s))\,\mathrm{d}s\right);\ \tau_B < \tau_D\right)$$

$$= \lim_{x \to z \in \partial D \cap B} E_x\left(\exp\left(\int_0^{\tau_B \wedge \tau_D} V(X(s))\,\mathrm{d}s\right);\ \tau_B < \tau_D\right)$$

$$\leqslant \lim_{x \to z \in \partial D \cap B} \left(E_x \exp(M+1)(\tau_B \wedge \tau_D)\right)^{(M/M+1)} (P_x(\tau_B < \tau_D))^{(1/M+1)}$$

$$= 0. \quad (1.9)$$

By (1.5), it follows that

$$\int_{\partial B \cap D} H(y)\nu_x(\mathrm{d}y) < \infty, \quad \text{for } x \in B \cap D. \quad (1.10)$$

We will show below that the measures ν_x are uniformly mutually absolutely continuous for x near x_0. From this fact and (1.10), it follows that

$$\int_{\partial B\cap D} H(y)\nu_x(dy) \text{ is uniformly bounded for } x \text{ near } x_0. \tag{1.11}$$

Now (1.3) follows from (1.5), (1.9) and (1.11).

It remains to prove the uniform mutual absolute continuity of ν_x for x near x_0. For $f \in C^\infty(\bar{B})$, define

$$u_f(x) = E_x\left(\exp\left(\int_0^{\tau_B} V(X(s))\,ds\right)f(X(\tau_B));\ \tau_B < \tau_D\right), \quad x \in B\cap D.$$

By (1.9), $\lim_{x\to B\cap\partial D} u_f(x) = 0$. An argument of the type used in the proof of Theorem 7.3.5 shows that $u_f \in C^{2,\alpha}(B\cap D)$ and that $Lu_f = 0$ in $B\cap D$ (exercise 8.2). Thus, by the boundary Harnack principle, there exists a ball $B_1 \subseteq B$, centered at x_0, a point $\bar{x} \in \partial B_1 \cap D$ and a constant $c > 0$ such that

$$\frac{u_f(x)}{u_g(x)} \le c\frac{u_f(\bar{x})}{u_g(\bar{x})}, \quad \text{for } x \in B_1 \text{ and } f, g \in C^\infty(\bar{B}). \tag{1.12}$$

Choosing $g \equiv 1$ and using the definition of ν_x, (1.12) becomes

$$\int_{B\cap\partial D} f(y)\nu_x(dy) \le c\int_{B\cap\partial D} f(y)\nu_{\bar{x}}(dy), \quad \text{for } x \in B_1 \text{ and } f \in C^\infty(\bar{B}). \tag{1.13}$$

Reversing the roles of f and g in (1.12) and again choosing $g \equiv 1$ gives

$$\int_{B\cap\partial D} f(y)\nu_X(dy) \ge \frac{1}{c}\int_{B\cap\partial D} f(y)\nu_{\bar{x}}(dy), \quad \text{for } x \in B_1 \text{ and } f \in C^\infty(\bar{B}). \tag{1.14}$$

The uniform mutual absolute continuity of ν_x for x near x_0 follows from (1.13) and (1.14). □

Although they will not be used in the sequel, we note the following corollaries of Theorem 1.1.

Corollary 1.2. *Let L satisfy Assumption $\widetilde{H}_{\mathrm{loc}}$ on a bounded domain $D \subset R^d$ with a Lipschitz boundary. Assume in addition that L satisfies Assumption E (defined in Chapter 3, Section 2) on D. Then the ground state ϕ_c for $L - \lambda_c$ on D satisfies*

$$\lim_{x\to\partial D} \phi_c(x) = 0.$$

Proof. Fix a domain $\Omega \subset\subset D$ and a point $y_0 \in \Omega$. By Proposition 7.3.1, $L - \lambda_c$ is subcritical on $D - \bar{\Omega}$; let $G^{\Omega}_{\lambda_c}(x, y)$ denote its Green's function. By Theorems 7.3.6, 7.3.7 and 7.3.8 (ii), $\phi_c(x) \leqslant \gamma G^{\Omega}_{\lambda_c}(x, y_0)$, for x near ∂D and some $\gamma > 0$; now apply Theorem 1.1 to $G^{\Omega}_{\lambda_c}$. □

Corollary 1.3. *Let L satisfy Assumption $\widetilde{H}_{\text{loc}}$ on a bounded domain $D \subset R^d$ with a Lipschitz boundary. Assume in addition that L and $\widetilde{L}$ satisfy Assumption E (defined in Chapter 3, Section 2) on D. Then $L - \lambda_c$ is product L_1 critical on D; equivalently, the diffusion on D corresponding to $(L - \lambda_c)^{\phi_c}$ is positive recurrent.*

Proof. Applying Corollary 1.2 to L and $\widetilde{L}$, it follows that the ground states ϕ_c and $\widetilde{\phi}_c$ for $L - \lambda_c$ and $\widetilde{L} - \lambda_c$ on D are bounded on D; thus $\int_D \phi_c \widetilde{\phi}_c \, dx < \infty$. □

We now state the main result of the section.

Theorem 1.4. *Let L satisfy Assumption $\widetilde{H}_{\text{loc}}$ on a domain $D \subset R^d$ and assume that L is subcritical on D. Let $z_0 \in \partial D$ and assume that there exists a ball $B \subset R^d$ centered at z_0 such that $\partial D \cap B$ is a Lipschitz boundary portion and such that L satisfies Assumption E (defined in Chapter 3, Section 2) on $D \cap B$. Then there exists a minimal Martin boundary point ζ with the property that a sequence $\{x_n\}_{n=1}^{\infty} \subset D$ is a Martin sequence corresponding to ζ if and only if* $\lim_{n\to\infty} x_n = z_0$.

The particular case of Theorem 1.4 in which D is bounded with a Lipschitz boundary is important enough to merit a separate formulation.

Theorem 1.5. *Let L satisfy Assumption $\widetilde{H}_{\text{loc}}$ on a bounded domain $D \subset R^d$ with a Lipschitz boundary, and assume that L is subcritical on D. Assume in addition that L and $\widetilde{L}$ satisfy Assumption E (defined in Chapter 3, Section 2) on D. Then the Martin boundary Λ, the minimal Martin boundary Λ_0, and the Euclidean boundary ∂D coincide. More specifically, a sequence $\{y_n\}_{n=1}^{\infty} \subset D$ with no accumulation points in D is a Martin sequence if and only if there exists a $z \in \partial D$ such that $z = \lim_{n\to\infty} y_n$; furthermore, if*

$$k(x; z) \equiv \lim_{n\to\infty} \frac{G(x, y_n)}{G(x_0, y_n)},$$

then $k(x; z)$ is a minimal positive harmonic function.

The following result is essentially a corollary of Theorem 1.4.

Corollary 1.6. *Let L satisfy Assumption $\tilde{H}_{\text{loc}}$ on a domain $D \cap R^d$ and assume that L is subcritical on D. Let $\Gamma \subseteq \partial D$ be a Lipschitz boundary portion which is open in the relative topology on ∂D, and assume that L satisfies Assumption E (defined in Chapter 3, Section 2) in a D-neighborhood of each point $z \in \Gamma$. By Theorem 1.4, the Martin boundary Λ may be represented as $\Lambda = \Gamma \cup \Lambda'$, where Γ denotes the part of the Martin boundary corresponding to sequences $\{x_n\}_{n=1}^{\infty} \subset D$ satisfying $\lim_{n\to\infty} x_n \in \Gamma$, and Λ' denotes the part of the Martin boundary corresponding to $\{x_n\}_{n=1}^{\infty} \subset D$ with no accumulation points in Γ. Let $u \in C_L(D)$ and assume that u vanishes continuously on Γ. Then the measure μ_u in the Martin representation (Theorem 7.1.2) for u is supported on Λ', that is,*

$$u(x) = \int_{\Lambda'} k(x; \zeta)\mu_u(\mathrm{d}\zeta), \quad x \in D.$$

We now give the proofs of Theorem 1.4 and Corollary 1.6.

Proof of Theorem 1.4. Fix a point $x_0 \in D$ with respect to which the Martin boundary is to be calculated. Let $B_r(x)$ denote the ball of radius r centered at x. Fix an $R_0 > 0$ such that $B_{R_0}(z_0) \subset\subset B$, where B is as in the statement of the theorem. Then $\partial D \cap B_{R_0}(z_0)$ is a uniformly Lipschitz boundary segment. We begin by proving the following three statements:

Let $\{y_n\}_{n=1}^{\infty}$ be a Martin sequence for which z_0 is not an accumulation point, let ζ denote the corresponding Martin boundary point, and let $k(x; \zeta)$ denote the corresponding function in $C_L(D)$. Then

$$\lim_{x\to z_0} k(x; \zeta) = 0 \text{ and the convergence is uniform over all } \zeta$$
$$\text{for which } \liminf_{n\to\infty} |y_n - z_0| \geq \varepsilon > 0. \tag{1.15}$$

Let $\{y_n\}_{n=1}^{\infty}$ be a Martin sequence satisfying $\lim_{n\to\infty} y_n = z_0$, let ζ denote the corresponding Martin boundary point and let $k(x; \zeta)$ denote the corresponding element in $C_L(D)$. Then

$$\limsup_{x\to z_0} k(x; \zeta) > 0. \tag{1.16}$$

Let $\{y_n\}_{n=1}^{\infty}$ and $\{y'_n\}_{n=1}^{\infty}$ be Martin sequences satisfying $\lim_{n\to\infty} y_n = \lim_{n\to\infty} y'_n = z_0$. Let ζ_1 and ζ_2 denote the corresponding Martin boundary points and let $k(x; \zeta_1)$ and $k(x; \zeta_2)$ denote the corresponding functions in $C_L(D)$. Then there exists a constant C such that

$$k(x; \zeta_1) \leq Ck(x; \zeta_2). \tag{1.17}$$

Let $\{y_n\}_{n=1}^{\infty}$ be as in (1.15). By Theorem 1.1, $\lim_{x\to z} G(x, z) = 0$, for all $z \in \partial D \cap B$. Thus, by the boundary Harnack principle and the fact that z_0 is not an accumulation point of $\{y_n\}_{n=1}^{\infty}$, it follows that there exists an $r > 0$, an $x_r \in \partial B_r(z_0) \cap D$, and a $c > 0$, such that $\{y_n\}_{n=1}^{\infty} \subset D - B_{8r}(z_0)$ and such that

$$\frac{G(x, y_n)}{G(x_r, y_n)} \leqslant c\frac{G(x, y_1)}{G(x_r, y_1)}, \quad \text{for } x \in B_r(z_0) \cap D, \text{ and } n = 1, 2, \ldots. \tag{1.18}$$

By Harnack's inequality,

$$\frac{G(x_r, y_n)}{G(x_0, y_n)} \text{ is bounded in } n;$$

thus multiplying (1.18) by

$$\frac{G(x_r, y_n)}{G(x_0, y_n)},$$

letting $n \to \infty$, and then letting $x \to z_0$, we obtain $\lim_{x\to z_0} k(x; \zeta) = 0$. The uniformity claim in (1.15) also follows from the above proof.

Now let $\{y_n\}_{n=1}^{\infty}$ be as in (1.16) and assume, contrary to (1.16), that

$$\lim_{x\to z_0} k(x; \zeta) = 0. \tag{1.19}$$

By applying (1.15) with $z \in B$ in place of z_0, it follows that

$$\lim_{x\to z} k(x; \zeta) = 0, \quad \text{for all } z \in \partial D \cap B - \{z_0\}. \tag{1.20}$$

By (1.19), (1.20), Theorem 1.1, the boundary Harnack principle (applied to small neighborhoods of z for $z \in \partial D \cap B$), and Harnack's inequality, there exists an $\varepsilon > 0$, a $y_0 \in D$ and a $c > 0$ such that $B_\varepsilon(y_0) \subset\subset D \cap B_{R_0/2}(z_0)$ and such that

$$k(x; \zeta) \leqslant cG(x, y_0), \quad \text{for } x \in D \cap B_{R_0}(z_0) - B_\varepsilon(y_0). \tag{1.21}$$

Applying Theorem 1.1 to $\widetilde{G}(x, y) = G(y, x)$, it follows that $\lim_{y\to z} G(x, y) = 0$, for $x \in D$ and $z \in \partial D \cap B$. Thus, from the boundary Harnack principle applied to the operator $\widetilde{L}$ and to the function $G(x, \cdot)$, it follows that there exists an $r \in (0, R_0/2)$, a $y_r \in D \cap \partial B_r(z_0)$ and a $c > 0$ such that

$$\frac{G(x, y_n)}{G(x, y_r)} \leqslant c\frac{G(x_0, y_n)}{G(x_0, y_r)}, \quad \text{for } x \in D - B_{R_0/2}(z_0), \text{ and large } n. \tag{1.22}$$

Multiplying (1.22) by

$$\frac{G(x, y_r)}{G(x_0, y_n)}$$

and letting $n \to \infty$ gives

$$k(x; \zeta) \leqslant c\frac{G(x, y_r)}{G(x_0, y_r)}, \quad \text{for } x \in D - B_{R_0/2}(z_0). \tag{1.23}$$

By applying Harnack's inequality to $G(x, \cdot)$, it follows that

$$\frac{G(x,y_r)}{G(x, y_0)}$$

is bounded for $x \in D - B_{R_0/2}(z_0)$; thus, from (1.23), there exists a $c > 0$ such that

$$k(x; \zeta) \leqslant cG(x, y_0), \quad \text{for } x \in D - B_{R_0/2}(z_0). \tag{1.24}$$

From (1.21) and (1.24), we obtain

$$k(x; \zeta) \leqslant cG(x, y_0), \quad \text{for } x \in D - B_\varepsilon(y_0). \tag{1.25}$$

Since $k(x; \zeta) \in C_L(D)$, (1.25) contradicts Theorem 7.3.9. Thus, (1.19) in fact does not hold; this proves (1.16).

We now prove (1.17). Since $\lim_{y \to z} G(x, y) = 0$, for $x \in D$ and all $z \in \partial D \cap B$, it follows from the boundary Harnack principle that there exists an $r_0 > 0$ and a $c > 0$ such that for $r \in (0, r_0)$, there exists a $z_r \in \partial B_r(z_0) \cap D$ such that

$$\frac{1}{c}\frac{G(x_0, z)}{G(x_0, z_r)} \leqslant \frac{G(x, z)}{G(x, z_r)} \leqslant c\frac{G(x_0, z)}{G(x_0, z_r)},$$

$$\text{for } z \in B_r(z_0) \cap D \text{ and } x \in D - B_{8r}(z_0). \tag{1.26}$$

Multiplying (1.26) by

$$\frac{G(x, z_r)}{G(x_0, z)},$$

substituting alternatively y_n and y'_n for z, and letting $n \to \infty$, we obtain

$$\frac{1}{c}\frac{G(x, z_r)}{G(x_0, z_r)} \leqslant k(x; \zeta_i) \leqslant c\frac{G(x, z_r)}{G(x_0, z_r)}, \quad \text{for } x \in D - B_{8r}(z_0),$$

$$i = 1, 2. \tag{1.27}$$

Since (1.27) holds for all $r \in (0, r_0)$, we conclude that $k(x; \zeta_1) \leqslant c^2 k(x; \zeta_2)$, for all $x \in D$; thus (1.17) holds with $C = c^2$.

From (1.15) and (1.16), it follows that if $\{y_n\}_{n=1}^{\infty}$ is a Martin sequence and z_0 is an accumulation point of $\{y_n\}_{n=1}^{\infty}$, then $\lim_{n\to\infty} y_n = z_0$. To complete the proof of the theorem, then, we must show that every sequence $\{y_n\}_{n=1}^{\infty}$ satisfying $\lim_{n\to\infty} y_n = z_0$ is a minimal Martin sequence. Using (1.17) and the definition of minimality, the above statement will follow if we show that among all Martin sequences $\{y_n\}_{n=1}^{\infty}$ satisfying $\lim_{n\to\infty} y_n = z_0$, at least one of them corresponds to a minimal Martin boundary point.

Assume on the contrary that no Martin sequence $\{y_n\}_{n=1}^{\infty}$ satisfying $\lim_{n\to\infty} y_n = z_0$ corresponds to a minimal Martin boundary point. We arrive at a contradiction as follows. For each integer $m \geq 1$, let Λ_m denote the part of the Martin boundary which may be obtained via sequences $\{y_n\}_{n=1}^{\infty}$ satisfying

$$\liminf_{n\to\infty} |y_n - z_0| > \frac{1}{m}.$$

By assumption, then, Λ_0, the minimal Martin boundary, satisfies $\Lambda_0 \subseteq \bigcup_{m=1}^{\infty} \Lambda_m$. Now choose a Martin sequence $\{y_n\}_{n=1}^{\infty}$ satisfying $\lim_{n\to\infty} y_n = z_0$, let ζ denote the corresponding Martin boundary point, and let $k(x; \zeta)$ denote the corresponding function in $C_L(D)$. By the Martin representation theorem (Theorem 7.1.2), $k(x; \zeta) = \int_{\Lambda_0} k(x; \zeta')\mu(d\zeta')$, for some finite measure μ on Λ_0.

Fix an integer m_0 such that

$$\frac{3}{m_0} < R_0.$$

Let $u_{m_0}(x; \zeta) = \int_{\Lambda_0 \cap \Lambda_{m_0}} k(x; \zeta')\mu(d\zeta')$ and, for integers $m > m_0$, let

$$k_m(x; \zeta) = \int_{\Lambda_0 \cap (\Lambda_m - \Lambda_{m_0})} k(x; \zeta')\mu(d\zeta'). \tag{1.28}$$

Since $\Lambda_0 \subseteq \bigcup_{m=1}^{\infty} \Lambda_m$, it follows from the monotone convergence theorem that

$$k(x; \zeta) = u_{m_0}(x; \zeta) + \lim_{m\to\infty} k_m(x; \zeta). \tag{1.29}$$

By (1.28) and (1.15), it follows that

$$\lim_{x\to z_0} k_m(x; \zeta) = 0. \tag{1.30}$$

Applying (1.15) with $z \in \partial D \cap B_{3/m_0}(z_0) - \{z_0\}$ in place of z_0, it follows that $\lim_{x\to z} k(x; \zeta) = 0$, for $z \in \partial D \cap B_{3/m_0}(z_0) - \{z_0\}$. Since $k_m(x; \zeta) \leq k(x; \zeta)$, we obtain

$$\lim_{x\to z} k_m(x; \zeta) = 0, \quad \text{for } z \in \partial D \cap B_{3/m_0}(z_0) - \{z_0\}. \tag{1.31}$$

The same argument used to obtain (1.21) from (1.19) and (1.20) shows that there exists an $\varepsilon > 0$, a $y_0 \in D$ and a $c > 0$ such that $B_\varepsilon(y_0) \subset\subset D \cap B_{1/m_0}(z_0)$ and such that

$$k_m(x; \zeta) \leq cG(x, y_0), \quad \text{for } x \in D \cap B_{3/m_0}(z_0) - B_\varepsilon(y_0). \quad (1.32)$$

Since $k_m(\cdot\,; \zeta)$ depends on $k(\cdot\,; \zeta')$ only if $\zeta' \in \Lambda_m - \Lambda_{m_0}$, an argument such as the one leading to (1.24) shows that for some $c > 0$,

$$k_m(x; \zeta) \leq cG(x, y_0), \quad \text{for } x \in D - B_{2/m_0}(z_0). \quad (1.33)$$

From (1.32) and (1.33), it follows that for some $c > 0$,

$$k_m(x; \zeta) \leq cG(x, y_0), \quad \text{for } x \in D - B_\varepsilon(y_0). \quad (1.34)$$

Since $k_m(\cdot\,; \zeta)$ satisfies $k_m(\cdot\,; \zeta) \geq 0$ and $Lk(\cdot\,; \zeta) = 0$, it follows from (1.34) and Theorem 7.3.9 that $k_m(\cdot\,, \zeta) \equiv 0$. From (1.29), we conclude that $k(x; \zeta) = u_{m_0}(x; \zeta)$. This is a contradiction since from (1.15) it follows that $\lim_{x \to z_0} u_{m_0}(x; \zeta) = 0$, while from (1.16) we have $\limsup_{x \to z_0} k(x; \zeta) > 0$. □

Proof of Corollary 1.6. For each $z \in \Gamma$, let $k(x; z)$ denote the corresponding function in $C_L(D)$. It suffices to show that the function $u(x) \equiv \int_\Gamma k(x; z)\mu(dz)$, where μ is a finite measure on Γ, satisfies $\limsup_{x \to z} u(x) > 0$, for some $z \in \Gamma$. Assume on the contrary that

$$\lim_{x \to z} u(x) = 0, \quad \text{for all } z \in \Gamma. \quad (1.35)$$

We arrive at a contradiction as follows. Let $\{\Gamma_m\}_{m=1}^\infty \subset \Gamma$ be an increasing sequence of bounded sets satisfying $\bigcup_{m=1}^\infty \Gamma_m = \Gamma$. Let $u_m(x) = \int_{\Gamma_m} k(x; z)\mu(dz)$. Then

$$u = \lim_{m \to \infty} u_m. \quad (1.36)$$

Since $u_m \leq u$, it follows from (1.35) that

$$\lim_{x \to z} u_m(x) = 0, \quad \text{for all } z \in \Gamma.$$

Now the same type of argument used to obtain (1.34) shows that there exists an $\varepsilon > 0$, a $y_0 \in D$ and a $c > 0$ such that $B_\varepsilon(y_0) \subset\subset D$ and

$$u_m(x) \leq cG(x, y_0), \quad \text{for } x \in D - B_\varepsilon(y_0). \quad (1.37)$$

Since u_m satisfies $u_m \geq 0$ and $Lu_m = 0$ in D, it follows from (1.37) and Theorem 7.3.9 that $u_m \equiv 0$; thus, from (1.36) we conclude that $u \equiv 0$. This is a contradiction; hence (1.35) does not hold. □

The rest of this section will be devoted to proving Theorem 1.10 below which gives a probabilistic representation for the minimal positive harmonic functions $k(\cdot\,;\zeta)$ under the assumption of Theorem 1.5 or, if $V \leqslant 0$, under the assumption of Theorem 1.4.

Let L_0 satisfy Assumption $\widetilde{\mathrm{H}}_{\mathrm{loc}}$ on a domain $D \subset R^d$. Since the coefficients of L_0 are allowed to go bad at the boundary ∂D, the exit distribution of $X(t)$ on ∂D may not be well defined. Even if it is well defined, it cannot be denoted by $X(\tau_D)$ since in the framework of the generalized martingale problem, it follows by definition that $X(\tau_D) = \lim_{t\to\tau_D} X(t) = \Delta$, where Δ is the point at infinity in the one-point compactification of D. (Recall the discussion preceding Corollary 7.4.6.) We show below that, under appropriate conditions, the exit distribution is well defined.

Theorem 1.7. *Let $L = L_0 + V$ satisfy Assumption $\widetilde{H}_{\mathrm{loc}}$ on a bounded domain $D \subset R^d$ with a Lipschitz boundary and assume that L is subcritical on D. Assume in addition that L satisfies Assumption E (defined in Chapter 3, Section 2) on D. Let $\{D_n\}_{n=1}^{\infty}$ be a sequence of domains satisfying $D_n \subset\subset D_{n+1}$ and $\bigcup_{n=1}^{\infty} D_n = D$. Define $\mu_x^n(dy) = E_x(\exp(\int_0^{\tau_{D_n}} V(X(s))\,ds);\ X(\tau_{D_n}) \in dy)$, $x \in D_n$. Then for $x \in D$, $\mu_x \equiv w-\lim_{n\to\infty}\mu_x^n$ exists as a finite measure on ∂D. In fact, $\sup_n \sup_{x\in D_n} \mu_x^n(\partial D_n) = \sup_n \sup_{x\in D_n} E_x \exp(\int_0^{\tau_{D_n}} V(X(s))\,ds) < \infty$. In the case where $V = 0$, μ_x defines the exit distribution on ∂D.*

Remark. The measure μ_x in Theorem 1.7 will be called the *harmonic exit measure for L on D.*

Proof. First assume that $V = 0$. The existence of the exit distribution of $X(t)$ on ∂D follows as an easy consequence of Corollary 2.3.4 and Theorem 2.2.2. Once it exists, it must be given by $w-\lim_{n\to\infty}\mu_x^n$. This proves the theorem in the case where $V = 0$. The general case follows as soon as it is established that $\mu_x^n(\partial D)$ is bounded in x and n. This boundedness is an easy consequence of Theorem 4.7.8(i) (see the remark following that theorem). □

Proposition 1.8. *Let the conditions of Theorem 1.7 hold and let μ_x be the harmonic exit measure as in that theorem. Let $f \in C(\partial D)$ and define $u(x) = \int_{\partial D} f(y)\mu_x(dy)$. Then $u \in C^{2,\alpha}(D) \cap C(\bar{D})$, $Lu = 0$ in D and $u = f$ on ∂D.*

Proof. Choose domains A and B such that $A \subset\subset B \subset\subset D$ and such that B has a $C^{2,\alpha}$-boundary. By the strong Markov property and the definition of μ_x, it follows easily that

$$u(x) = E_x \exp\left(\int_0^{\tau_B} V(X(s))\, ds\right) u(X(\tau_B)), \text{ for } x \in B$$

(exercise 8.3). Let $\{f_n\}_{n=1}^{\infty}$ be a sequence of functions satisfying $f_n \in C^{2,\alpha}(\bar{B})$, $\|f_n\|_{0;B} \leq \|u_0\|_{0;B}$ and $\lim_{n\to\infty} f_n(x) = u(x)$, for almost all $x \in \partial B$ (with respect to Lebesgue measure on ∂B). Define $u_n(x) = E_x \exp(\int_0^{\tau_B} V(X(s))\, ds) f_n(X(\tau_B))$, $x \in \bar{B}$. Since the principle eigenvalue of L on B is negative, it follows from the boundedness of V that for some $\varepsilon > 0$, the principal eigenvalue of $L + \varepsilon V = L_0 + (1 + \varepsilon)V$ on B is also negative; thus, by Theorem 3.6.6,

$$E_x \exp\left(\int_0^{\tau_B} (1 + \varepsilon) V(X(s))\, ds\right) < \infty.$$

It follows, then, that

$$\left\{\exp\left(\int_0^{\tau_B} V(X(s))\, ds\right) f_n(X(\tau_B))\right\}_{n=1}^{\infty}$$

is uniformly integrable with respect to E_x. By Theorem 7.4.4, the distribution of $X(\tau_B)$ is absolutely continuous with respect to Lebesgue measure on ∂B. Thus, we conclude that

$$\lim_{n\to\infty} E_x \exp\left(\int_0^{\tau_B} V(X(s))\, ds\right) f_n(X(\tau_B)) =$$

$$E_x \exp\left(\int_0^{\tau_B} V(X(s))\, ds\right) u(X(\tau_B)), \text{ for } x \in B;$$

that is, $u(x) = \lim_{n\to\infty} u_n(x)$, for $x \in B$. By Theorem 3.6.6, $u_n \in C^{2,\alpha}(B)$ and $Lu_n = 0$ in B. The interior Schauder estimate (Theorem 3.2.7) gives

$$\|u_n\|_{2,\alpha;A} \leq c\|u_n\|_{0;A} \leq c\|f_n\|_{0;B} \sup_{x\in A} E_x \exp\left(\int_0^{\tau_\beta} V(X(s))\, ds\right)$$

$$\leq c\|u\|_{0;B} \sup_{x\in A} E_x \exp\left(\int_0^{\tau_\beta} V(X(s))\, ds\right),$$

where c is independent of n. A standard compactness argument then shows that $u \in C^{2,\alpha}(A)$ and $Lu = 0$ in A. Since A is arbitrary, we conclude that $u \in C^{2,\alpha}(D)$ and $Lu = 0$ in D.

We now show that $u \in C(\bar{D})$ and $u = f$ on ∂D. Letting u_x^n be as in Theorem 1.7, is enough to show that

$$\lim_{x \to x_0} \lim_{n \to \infty} \mu_x^n(S_\varepsilon) = 1, \quad \text{for } x_0 \in \partial D \text{ and } \varepsilon > 0, \tag{1.38}$$

where $S_\varepsilon = \{x \in D: |x - x_0| < \varepsilon\}$. In the case where $V \leqslant 0$, (1.38) follows from Corollary 2.3.4 and Theorem 2.2.2. In the general case, a uniform integrability condition is also needed. By Theorem 4.7.1, the generalized principal eigenvalue of L on D is negative. Thus, by the boundedness of V, $L + \varepsilon V = L_0 + (1 + \varepsilon)V$ is subcritical on D for some $\varepsilon > 0$. Applying Theorem 1.7 to $L + \varepsilon V$, it follows that

$$\sup_n \sup_{x \in D_n} E_x \exp\left(\int_0^{\tau_{D_n}} (1 + \varepsilon)V(X(s))\, ds\right) < \infty. \tag{1.39}$$

Now (1.38) follows from Corollary 2.3.4, Theorem 2.2.2, (1.39) and a Holder's inequality type of calculation as in (1.9). □

The next proposition is the analog of Theorem 1.7 and Proposition 1.8 in the case where D is unbounded and $V \leqslant 0$.

Proposition 1.9. *Let $L = L_0 + V$ satisfy Assumption $\widetilde{H}_{\text{loc}}$ on a domain $D \subset R^d$ and assume that $V \leqslant 0$. Let $z_0 \in \partial D$ and assume that there exists a ball $B \subset R^d$ centered at z_0 such that $\partial D \cap B$ is a Lipschitz boundary portion and such that L satisfies Assumption E (defined in Chapter 3, Section 2) on $D \cap B$. Let $B_1 \subset R^d$ be a ball satisfying $z_0 \in B_1 \subset\subset B$, and let $\{D_n\}_{n=1}^\infty$ be a non-decreasing sequence of bounded domains with Lipschitz boundaries satisfying $\partial D_n \cap B_1 \supseteq \partial D \cap B_1$ and $\bigcup_{n=1}^\infty D_n = D$. Let $\mu_{x;D_n}(dy)$, $x \in D_n$, denote the harmonic exit measure on ∂D_n for L on D_n as in Theorem 1.7, and let $\nu_x^{(n)}$ denote the restriction of $\mu_{x;D_n}$ to $\bar{B}_1 \cap \partial D$.*

(i) *$\nu_x \equiv w - \lim_{n\to\infty} \nu_x^{(n)}$ exists for $x \in D$.*
(ii) *Let $f \in C(\partial D)$ with compact support in $B_1 \cap \partial D$, and define $u(x) = \int_{B_1 \cap \partial D} f(y)\nu_x(dy)$. Then $u \in C^{2,\alpha}(D)$, $Lu = 0$ in D and $\lim_{x\to z} u(x) = f(z)$, for $z \in B_1 \cap \partial D$.*

Proof.
(i) $\{\nu_x^{(n)}\}_{n=1}^\infty$ is a non-decreasing sequence of measures and

$$\nu_x^{(n)}(\bar{B}_1 \cap \partial D) \leqslant 1, \text{ for all } n \text{ and all } x \in D.$$

(ii) The proof is essentially the same as the proof of Proposition 1.8. □

Theorem 1.10.
(i) *Assume that L and D satisfy the conditions of Theorem 1.5. For each $z \in \partial D$, let $k(x; z)$ denote the minimal positive harmonic function corresponding to z, as in Theorem 1.5. Normalize $k(x; z)$ by $k(x_0; z) = 1$, for*

some $x_0 \in D$. *Then for each* $z \in \partial D$,

$$k(x; z) = \frac{d\mu_x}{d\mu_{x_0}}(z), \quad \text{for } x \in D,$$

where μ_x *is the harmonic exit measure as in Theorem* 1.7.

(ii) *Let* $L = L_0 + V$, D *and* $z_0 \in \partial D$ *be as in Theorem* 1.4, *and assume that* $V \leq 0$. *Let* $k(x; z_0)$ *denote the positive harmonic function corresponding to* z_0, *as in Theorem* 1.4. *Normalize* $k(x; z_0)$ *by* $k(x_0; z_0) = 1$, *for some* $x_0 \in D$. *Then*

$$k(x; z_0) = \frac{d\nu_x}{d\nu_{x_0}}(z_0), \quad \text{for } x \in D,$$

where ν_x *is as in Proposition* 1.9.

Proof. (i) Let $z_0 \in \partial D$ and define

$$S_m = \left\{z \in \partial D: |z - z_0| > \frac{1}{m}\right\}.$$

Let $f_m \in C(\partial D)$ satisfy $f_m = 0$ on S_m and $f_m > 0$ on $\partial D - \bar{S}_m$. Let $u_m(x) = \int_{\partial D} f_m(z)\mu_x(dz)$, where μ_x is as in Proposition 1.8, and let

$$v_m(x) = \frac{u_m(x)}{u_m(x_0)}.$$

By a standard compactness argument, there exists a subsequence of $\{v_m\}_{n=1}^{\infty}$ which converges to a function $v \in C_L(D)$. Assume without loss of generality that the entire sequence converges to v. By Proposition 1.8, $\lim_{x \to S_m} v_m(x) = 0$; thus an application of the boundary Harnack principle shows that

$$\lim_{x \to z} v(x) = 0, \quad \text{for } z \in \partial D - \{z_0\}. \tag{1.40}$$

From the Martin representation theorem and Theorem 1.5, which identified the minimal Martin boundary with ∂D, v can be represented in the form

$$v(x) = \int_{\partial D - \{z_0\}} k(x; z)\mu(dz) + ck(x; z_0),$$

where $c + \mu(\partial D - \{z_0\}) = 1$. Letting $v^{(m)}(x) = \int_{S_m} k(x; z)\mu(dz)$, we have

$$v(x) = \lim_{m \to \infty} v^{(m)}(x) + ck(x; z_0). \tag{1.41}$$

By the reasoning used to obtain (1.15), we obtain

$$\lim_{x \to z_0} v^{(m)}(x) = 0. \tag{1.42}$$

Since $v^{(m)} \leqslant v$, it follows from (1.40) that

$$\lim_{x \to z} v^{(m)}(x) = 0, \quad \text{for } z \in \partial D - \{z_0\}. \tag{1.43}$$

From (1.42), (1.43), Theorem 1.1 and the boundary Harnack principle, it follows that for $y_0 \in D$, there exists an $\varepsilon > 0$ and a $c > 0$ such that $B_\varepsilon(y_0) \subset\subset D$ and

$$v^{(m)}(x) \leqslant cG(x, y_0), \quad \text{for } x \in D - B_\varepsilon(y_0). \tag{1.44}$$

Since $v^{(m)} \geqslant 0$ and $Lv^{(m)} = 0$, we conclude from (1.44) and Theorem 7.3.9 that $v^{(m)} \equiv 0$; thus, from (1.41) we have $v(x) = k(x; z_0)$.

We have now shown that

$$k(x; z_0) = v(x) = \lim_{m \to \infty} v_m(x) = \lim_{m \to \infty} \frac{u_m(x)}{u_m(x_0)}.$$

On the other hand, from a theorem in integration theory, it follows that

$$\lim_{m \to \infty} \frac{u_m(x)}{u_m(x_0)} = \frac{d\mu_x}{d\mu_{x_0}}(z_0),$$

for μ_{x_0}-almost all $z_0 \in \partial D$. Since

$$\frac{d\mu_x}{d\mu_{x_0}}$$

is only defined up to sets of μ_{x_0}-measure zero, we may identify $k(x; z_0)$ with

$$\frac{d\mu_x}{d\mu_{x_0}}(z_0).$$

(In fact it is known that $k(x; z_0)$ is continuous in z_0, but that is beyond the scope of our technique here. See [Ancona (1978)].)

(ii) The proof of (ii) is left to the reader as exercise 8.4. □

8.2 Periodic operators on R^d: positive harmonic functions and criticality, Martin boundary, behavior of λ_c under perturbations

In this section we consider operators $L = \frac{1}{2}\nabla \cdot a\nabla + b \cdot \nabla + V$ satisfying Assumption $\widetilde{H}$ (defined in Chapter 3, Section 7) on R^d with coefficients which are periodic of period one; that is, $a_{ij}(x + e_k) = a_{ij}(x)$, $b_i(x + e_k) = b_i(x)$ and $V(x + e_k) = V(x)$, for $i, j, k \in \{1, \ldots, d\}$, where e_k is the unit vector in the x_k-direction.

I Positive harmonic functions and criticality

The cone $C_L(R^d)$ of positive solutions has a special structure in the periodic case.

Theorem 2.1. *Let L satisfy Assumption $\widetilde{H}$ on R^d and assume that the coefficients of L are periodic with period one.*

(i) *If L is critical on R^d, then the ground state ϕ_c satisfies $\phi_c(x) = \exp(v \cdot x)\psi(x)$, where $v \in R^d$ and ψ is periodic of period one.*
(ii) *If L is subcritical on R^d, then every minimal positive harmonic function $u \in C_L(R^d)$ is of the form $u(x) = \exp(v \cdot x)\psi(x)$, where $v \in R^d$ and ψ is periodic of period one. Conversely, if $u(x) = \exp(v \cdot x)\psi(x) \in C_L(R^d)$, where $v \in R^d$ and ψ is periodic of period one, then u is minimal. Furthermore, if $u_i(x) = \exp(v_i \cdot x)\psi_i(x)$, $i = 1, 2$, are minimal positive harmonic functions which are not multiples of one another, then $v_1 \neq v_2$.*

Proof. (i) Let $v_i(x) = \phi_c(x + e_i)$. By the periodicity assumption, $v_i \in C_L(R^d)$. Since $C_L(R^d)$ is one-dimensional (Theorem 4.3.4), it follows that $\phi_c(x + e_i) = v_i(x) = \gamma_i\phi_c(x)$, for some $\gamma_i > 0$; the representation in (i) follows from this.

(ii) Let $u \in C_L(R^d)$ be a minimal positive harmonic function and let $v_i(x) = u(x + e_i)$. As above, $v_i \in C_L(R^d)$. Let $C(x)$ denote the unit cube in R^d centered at $x \in R^d$. By Harnack's inequality and the perodicity assumption, there exists a constant c such that

$$\sup_{y \in C(x)} u(y) \leq c \inf_{y \in C(x)} u(y), \text{ for all } x \in R^d;$$

thus $v_i(x) \leq cu(x)$, for all $x \in R^d$. It then follows from the minimality of u that $u(x + e_i) = v_i(x) = \gamma_i u(x)$, for some $\gamma_i > 0$; the representation in (ii) follows from this. The converse is now an easy consequence of the Martin representation theorem (Theorem 7.1.2).

Now let $u_i(x) = \exp(v_i \cdot x)\psi_i(x)$, $i = 1, 2$, be minimal positive harmonic functions which are not multiples of one another. If $v_1 = v_2$, then there exists a constant c such that $u_1 \leq cu_2$; this contradicts the minimality assumption. □

The next theorem describes the set of $v \in R^d$ for which a positive solution exists of the form $\exp(v \cdot x)\psi(x)$, where ψ is periodic of period one.

Theorem 2.2. *Let L on R^d be as in Theorem 2.1. Assume that L is either critical or subcritical on R^d. Define*

$$\Gamma = \{v \in R^d, \exists\, a\, u \in C_L(R^d) \text{ of the form } u(x) = \exp(v \cdot x)\psi(x), \text{ where } \psi \text{ is periodic of period one.}\}$$

and

$$K = \{v \in R^d \colon \exists\, a\, u \in C^{2,\alpha}(R^d) \text{ satisfying } Lu \leq 0 \text{ and } u > 0 \text{ of the form } u(x) = \exp(v \cdot x)\psi(x), \text{ where } \psi \text{ is periodic of period one}\}.$$

Then either

(i) *$K = \Gamma = \{v_0\}$, for some $v_0 \in R^d$,*

or

(ii) *K is a d-dimensional strictly convex, compact set and $\Gamma = \partial K$.*

For the proof of Theorem 2.2, we need a lemma. For $v \in R^d$, let $L^{(v)}$ denote the h-transform of $L = L_0 + V$ via the function $\exp(v \cdot x)$; that is,

$$\begin{aligned} L^{(v)} &= L_0 + av \cdot \nabla + \frac{L \exp(v \cdot x)}{\exp(v \cdot x)} \\ &= L + av \cdot \nabla + \tfrac{1}{2} vav + \tfrac{1}{2} v \cdot \nabla \cdot a + v \cdot b. \end{aligned} \tag{2.1}$$

Let T_d denote the d-dimensional unit torus which is, of course, a compact manifold without boundary. Since $L^{(v)}$ is periodic on R^d, it may also be considered as an operator on T_d (see Chapter 4, Section 11). However, $\exp(v \cdot x)$ is not defined on T_d since it is not periodic; thus the criticality properties and the principal eigenvalue of $L^{(v)}$ on T_d may vary with v. Let $\lambda_0(v)$ denote the principal eigenvalue of $L^{(v)}$ on T_d and let $\tilde{\lambda}_0(v)$ denote the principal eigenvalue of $\tilde{L}^{(v)}$ on T_d.

Remark. A direct calculation reveals that the formal adjoint of $L^{(v)}$ is given by $\tilde{L}^{(-v)}$. Thus, since the principal eigenvalue of an operator and of its adjoint coincide, it follows that $\lambda_0(v) = \tilde{\lambda}_0(-v)$.

Lemma 2.3. *Let L, K and Γ be as in Theorem 2.1, and let $L^{(v)}$ denote the h-transform of L via the function $\exp(v \cdot x)$. Let $\lambda_0(v)$ denote the principal eigenvalue of $L^{(v)}$ considered as an operator on the torus T_d. Then $v \in K - \Gamma$ if and only if $\lambda_0(v) < 0$ (or equivalently, if and only if $L^{(v)}$ on T_d is subcritical), and $v \in \Gamma$ if and only if $\lambda_0(v) = 0$ (or equivalently, if and only if $L^{(v)}$ on T_d is critical).*

Proof. Let $\psi > 0$ be periodic of period one and let $u(x) = \exp(v \cdot x)\psi(x)$. Clearly, $Lu \leq 0$ $(=0)$, if and only if $L^{(v)}\psi \leq 0$ $(=0)$.

The lemma now follows from Theorem 4.11.1(iii), Theorem 4.7.1 and Theorem 4.3.9. (The latter two theorems can easily be shown to hold for operators on compact manifolds without boundaries.) □

Proof of Theorem 2.2. To prove the theorem, we will assume that (i) does not hold and prove that (ii) holds. By Theorem 2.1, $\Gamma \neq \emptyset$; thus since (i) does not hold, K consists of more than one point. By Theorem 4.7.7 (which can easily be shown to hold for operators on compact manifolds without boundaries), $\lambda_0(v)$, which is defined in Lemma 2.3, is continuous in v. (Alternatively, the continuity follows from the convexity which is proved below.) Thus, by Lemma 2.3, K is a closed set, $K - \Gamma = \{v \in R^d : \lambda_0(v) < 0\}$ is the interior of K and $\Gamma = \{v \in R^d : \lambda_0(v) = 0\} = \partial K$. To show that K is d-dimensional and strictly convex, it is enough to show that if $v_0, v_1 \in K$ and $v_0 \neq v_1$, then $v_t \equiv (1-t)v_0 + tv_1 \in K - \Gamma$, for $t \in (0, 1)$. Let $v_0, v_1 \in K$ with $v_0 \neq v_1$ and let ψ_0, ψ_1 be periodic functions of period one such that $u_i(x) = \exp(v_i \cdot x)\psi_i(x) \in C_L(R^d)$, $i = 0, 1$. Let $u_t(x) = u_0^{1-t}(x)u_1^t(x) = \exp(v_t \cdot x)\psi_0^{1-t}(x)\psi_1^t(x)$, $t \in (0, 1)$. The calculation performed in the proof of Theorem 4.6.1 shows that for $t \in (0, 1)$, $L_t u_t \leqslant 0$ and $L_t u_t \not\equiv 0$; thus $v_t \in K - \Gamma$.

To complete the proof of the theorem, it remains to show that K is bounded. Assume on the contrary that K is not bounded. Then there exists a sequence $\{v_n\}_{n=1}^{\infty}$ satisfying $\lim_{n\to\infty} |v_n| = \infty$ such that $\lambda_0(v_n) \leqslant 0$, for $n = 1, 2, \ldots$. It is easy to show that $\lambda_0(v) \geqslant \lambda_c$, for all $v \in R^d$ (see Theorem 2.5(i)); thus the sequence $\{\lambda_0(v_n)\}_{n=1}^{\infty}$ is bounded. By the proof of Lemma 2.3, for each n there exists a positive function of period one ψ_{v_n} such that $\exp(v_n \cdot x)\psi_{v_n}(x) \in C_{L-\lambda_0(v_n)}(R^d)$. Since $\{\lambda_0(v_n)\}_{n=1}^{\infty}$ is bounded, it follows from Harnack's inequality that there exists a constant $c > 0$, independent of n, such that

$$\sup_{|y|<1} \exp(v_n \cdot y)\psi_{v_n}(y) \leqslant c \inf_{|x|<1} \exp(v_n \cdot x)\psi_{v_n}(x). \tag{2.2}$$

Choosing $y = \pm e_i$ and $x = 0$ in (2.2), and using the fact that ψ_{v_n} is periodic, it follows that

$$\exp(\pm v_n \cdot e_i) \leqslant c, \quad \text{for } i \in \{1, 2, \ldots, d\} \text{ and } n = 1, 2, \ldots.$$

But this is incompatible with the condition $\lim_{n\to\infty} |v_n| = \infty$; therefore we conclude that K is bounded. □

Corollary 2.4. *Let L on R^d satisfy the conditions of Theorem 2.1 and assume that L is subcritical on R^d. Let Γ be as in Theorem 2.2. Then Γ is homeomorphic to Λ_0, the minimal Martin boundary. For each $v \in \Gamma$,* *there exists a unique minimal positive harmonic function u_v (up to constant multiples) of the form $u_v(x) = \exp(v\cdot x)\psi_v(x)$, where ψ_v is periodic of period one. In fact, letting $L^{(v)}$ denote the h-transform of L via $\exp(v\cdot x)$, considering $L^{(v)}$ as an operator on the torus T_d, and letting $\lambda_0(v)$ denote its principal eigenvalue, then ψ_v is the ground state for $L^{(v)} - \lambda_0(v)$ on T_d. Every $u \in C_L(R^d)$ can be represented in the form*

$$u(x) = \int_\Gamma \exp(v\cdot x)\psi_v(x)\eta_u(dv),$$

where η_u is a unique finite measure on the compact set Γ.

Proof. The proof is an immediate consequence of Theorem 2.1, Theorem 2.2, the proof of Lemma 2.3, and the Martin representation theorem (Theorem 7.1.2). □

In the sequel, it will be useful to consider the operators $L - \lambda$, for $\lambda \in R$. Since $L - \lambda$ is periodic, Theorems 2.1 and 2.2 and Corollary 2.3 may be applied to $L - \lambda$. Let K_λ and Γ_λ correspond to the operator $L - \lambda$ as K and Γ correspond to the operator L. Then the results of Theorem 2.2 and Lemma 2.3 hold for K_λ and Γ_λ in place of K and Γ. In particular, by Lemma 2.3,

$$\left.\begin{aligned} K_\lambda &= \{v \in R: \lambda_0(v) \leq \lambda\}, \\ &\text{and} \\ \Gamma_\lambda &= \{v \in R^d: \lambda_0(v) = \lambda\}. \end{aligned}\right\} \quad (2.3)$$

Theorem 2.5. *Let L on R^d satisfy the conditions of Theorem* 2.1.

(i) $K_\lambda = \phi$ *if and only if* $\lambda < \lambda_c$;
(ii) $\Gamma_{\lambda_c} = K_{\lambda_c} = \{v_0\}$, *for some* $v_0 \in R$;
(iii) K_λ *varies continuously with* λ *and* $K_{\lambda_1} \subset\subset K_{\lambda_2}$, *for* $\lambda_2 > \lambda_1 \geq \lambda_c$;
(iv) $\lim_{\lambda\to\infty} K_\lambda = R^d$;
(v) *Let* $\tilde{K}_\lambda$ *correspond to* $\tilde{L} - \lambda$ *as* K_λ *corresponds to* $L - \lambda$. *Then* $\tilde{K}_\lambda = -K_\lambda$.
In particular, from (i)–(iii), *it follows that* $\lambda = \lambda_c$, *if and only if* Γ_λ *(or* K_λ*) consists of a single point.*

Proof.
(i) This follows immediately from Theorem 2.1.
(ii) If K_{λ_c} consists of more than one point, then by Theorem 2.2 there exists a $v \in K_{\lambda_c} - \Gamma_{\lambda_c}$ which, by (2.3), satisfies $\lambda_0(v) < \lambda_c$. Let ψ_v denote the positive eigenfunction corresponding to $\lambda_0(v)$ for $L^{(v)}$ on T_d, and let

$u(x) = \exp(\nu \cdot x)\psi(x)$. Then, on the one hand, $(L - \lambda_0(\nu))u = 0$, while on the other hand, since $\lambda_0(\nu) < \lambda_c$, $L - \lambda_0(\nu)$ is supercritical; this is a contradiction.

(iii) By Theorem 2.2, K_λ is bounded. In the proof of Theorem 2.2, it was noted that $\lambda_0(\nu)$ is continuous. (iii) follows from these facts and (2.3).

(iv) The proof of (iv) follows from (2.3) and the fact that $\lambda_0(\nu) < \infty$, for all $\nu \in R^d$.

(v) This follows from Lemma 2.3 and the remark which precedes it. □

For periodic operators, $L - \lambda_c$ may be critical and may be subcritical as the example of the Laplacian in R^2 and R^3 illustrates. In fact, for general periodic operators, it turns out that $L - \lambda_c$ is critical if $d \leqslant 2$ and subcritical if $d \geqslant 3$. The proof is much simpler in the symmetric case which we state separately.

Theorem 2.6. *Let $L = \frac{1}{2}\nabla \cdot a\nabla + b \cdot \nabla + V$ on R^d satisfy the conditions of Theorem* 2.1. *Assume that L is symmetric; that is, $b = a\nabla Q$, for some $Q \in C^{2,\alpha}(R^d)$. Then $L - \lambda_c$ on R^d is critical if $d \leqslant 2$ and subcritical if $d \geqslant 3$.*

Proof. The proof is left as exercise 8.5. □

Theorem 2.7. *Let L on R^d satisfy the conditions of Theorem* 2.1. *Then $L - \lambda_c$ on R^d is critical if $d \leqslant 2$ and subcritical if $d \geqslant 3$.*

Proof. By incorporating $-\lambda_c$ into the zeroth-order term, we may assume that $\lambda_c = 0$, and then by making an h-transform, we may assume that the zeroth-order term vanishes. Thus, we assume that $L = \frac{1}{2}\nabla \cdot a\nabla + b \cdot \nabla$ and that $\lambda_c = 0$.

We first treat the case where $d \geqslant 3$. Let $\tilde{\psi}_0$ denote the ground state of $\tilde{L} - \lambda_c$ on the torus T_d, and extend $\tilde{\psi}_0$ periodically to all of R^d. Define Q by $\exp(2Q) = \tilde{\psi}_0$. Let $\mathscr{L} = \frac{1}{2}\nabla \cdot a\nabla + a\nabla Q \cdot \nabla$ on R^d, and note that $\tilde{\mathscr{L}} \exp(2Q) = 0$. Define $\hat{b} = b - a\nabla Q$ so that $L - \lambda_c = \mathscr{L} + \hat{b} \cdot \nabla$ and $\tilde{L} - \lambda_c = \tilde{\mathscr{L}} - \hat{b} \cdot \nabla - \nabla \cdot \hat{b}$. Since $(\tilde{L} - \lambda_c)\exp(2Q) = \tilde{\mathscr{L}} \exp(2Q) = 0$, it follows that $(\hat{b} \cdot \nabla + \nabla \cdot \hat{b})\exp(2Q) = 0$; that is, $\nabla \cdot (\exp(2Q)\hat{b}) = 0$. The operator $\mathscr{L}$ on R^d is subcritical. To see this, note that the generalized principal eigenvalue of $\mathscr{L}$ on R^d must be non-positive since $\mathscr{L}1 = 0$. If the generalized principal eigenvalue is negative, then subcriticality follows trivially. If the generalized principal eigenvalue is zero, then subcriticality follows from Theorem 2.6 since $\mathscr{L}$ is symmetric and $d \geqslant 3$.

From the above analysis, we have the following situation: $\mathscr{L} = \frac{1}{2}\nabla \cdot a\nabla + a\nabla Q \cdot \nabla$ is subcritical, $L - \lambda_c = \mathscr{L} + \hat{b} \cdot \nabla$, and $\nabla \cdot (\exp(2Q)\hat{b})$

$=0$. The subcriticality of $L-\lambda_c$ on R^d now follows from Theorem 6.6.2(ii). (Recall that subcriticality is equivalent to the transcience of the corresponding diffusion in the case where the zeroth-order term vanishes.)

In the one-dimensional case, L is always symmetric; thus in this case the theorem follows from Theorem 2.6 (exercise 8.5 also suggests an alternative proof in the one-dimensional case).

It remains to treat the case $d=2$. Recall that we are assuming that $L=\frac{1}{2}\nabla\cdot a\nabla+b\cdot\nabla$ and that $\lambda_c=0$. The proof relies on a construction in [Lyons and Sullivan (1984)]. A look at both the statement and the proof of the Lyons and Sullivan construction reveals that it is possible to associate to the diffusion $X(t)$ on R^2 corresponding to L a random walk $\{Y_n\}_{n=1}^{\infty}$ on the discrete space Z^2 with the following properties:

1. The one-step increments of $\{Y_n\}_{n=1}^{\infty}$ have a finite second moment.
2. $\{Y_n\}_{n=1}^{\infty}$ on Z^2 is recurrent if and only if $X(t)$ on R^2 is recurrent.
3. It is possible to construct $X(t)$ and $\{Y_n\}_{n=1}^{\infty}$ on the same probability space such that the following conditions hold:
   (a) There exists a sequence $\{T_n\}_{n=1}^{\infty}$ of stopping times for $X(t)$ and a constant c such that $\text{Prob}\,(|X(T_n)-Y_n|\leqslant c,\text{ for } n=1,2,\ldots)=1$.
   (b) There exist positive constants c_1, c_2 such that

$$\text{Prob}\left(c_1\leqslant\liminf_{n\to\infty}\frac{T_n}{n}\leqslant\limsup_{n\to\infty}\frac{T_n}{n}\leqslant c_2\right)=1.$$

We now use property (3) above along with the fact that $\lambda_c=0$ to prove that the one-step increments of the random walk $\{Y_n\}_{n=1}^{\infty}$ have mean zero. Denote the mean of these one-step increments by $\mu\in R^2$. By the law of large numbers,

$$\text{Prob}\left(\lim_{n\to\infty}\frac{Y_n}{n}=\mu\right)=1.$$

Since $\lambda_c=0$, it follows from the remark after Theorem 2.12 below that

$$\text{Prob}\left(\lim_{t\to\infty}\frac{X(t)}{t}=0\right)=1.$$

By property 3(b) above, it then follows that

$$\text{Prob}\left(\lim_{n\to\infty}\frac{X(T_n)}{n}=0\right)=1.$$

We conclude that

$$\text{Prob}\left(\lim_{n\to\infty}\frac{X(T_n)}{n}=0\text{ and }\lim_{n\to\infty}\frac{Y_n}{n}=\mu\right)=1.$$

This is compatible with property 3(a) above only if $\mu=0$.

Thus, the one-step increments of the random walk $\{Y_n\}_{n=1}^{\infty}$ have mean zero and, by property (1) above, a finite second moment. Such a random walk in Z^2 is recurrent [Spitzer (1976)]. It then follows by property (2) above that $X(t)$ is recurrent; that is, L is critical. □

We now consider the significance of $\lambda_0(0)$.

Proposition 2.8. *Let L on R^d satisfy the conditions of Theorem 2.1 and let λ_c denote the generalized principal eigenvalue of L on R^d. Then $\lambda_0(0) \geqslant \lambda_c$ and $\lambda_0(0)$ is the unique number λ satisfying $0 \in \Gamma_\lambda$; equivalently, $\lambda_0(0)$ is the unique number λ such that $C_{L-\lambda}(R^d)$ contains a periodic function. Also, $\lambda_0(0)$ is the principal eigenvalue of L considered as an operator on the torus T_d.*

Proof. By (2.3) and Theorem 2.5, it follows that $\lambda = \lambda_0(0)$ is the unique λ such that $0 \in \Gamma_\lambda$. Since $\Gamma_{\lambda_0(0)} \neq \varnothing$, it follows from Theorem 2.5 that $\lambda_0(0) \geqslant \lambda_c$. □

Example 2.9. We give an example where $\lambda_0(0) = \lambda_c$ and an example where $\lambda_0(0) > \lambda_c$. If $L = \frac{1}{2}\nabla \cdot a\nabla + V$, then $L = \widetilde{L}$, and it follows from Theorem 2.5(ii) and (v) that $\Gamma_{\lambda_c} = \{0\}$. Thus, from Proposition 2.8, we have $\lambda_c = \lambda_0(0)$. On the other hand, if $L = L_0$, that is, $V = 0$, then by Theorem 4.11.1 $\lambda_0(0) = 0$ (and the corresponding eigenfunction is constant). Now choose, for example, $L = \frac{1}{2}\Delta + b \cdot \nabla$, where b is a non-zero constant vector. Then

$$\lambda_c = \frac{-|b|^2}{2}$$

(see exercise 8.19); thus $\lambda_0(0) > \lambda_c$.

We now study the differential structure of K_λ. If $\lambda_0(v) = \lambda$, then $v \in \Gamma_\lambda$ by (2.3). We now give an integral condition to determine whether v is the unique point in K_λ or whether K_λ is a d-dimensional strictly convex compact set. In the latter case, the integral also determines the outward normal to K_λ at v. In the former case, it follows by Theorem 2.5 that $\lambda = \lambda_c$.

Theorem 2.10. *Let $L = \frac{1}{2}\nabla \cdot a\nabla + b \cdot \nabla + V$ on R^d satisfy the conditions of Theorem 2.1. For $v \in R^d$, let $L^{(v)}$ denote the h-transform of L via the function $\exp(v \cdot x)$. Let $\lambda_0(v)$ denote the principal eigenvalue of $L^{(v)}$ considered as an operator on the torus T_d, and let ψ_v and $\widetilde{\psi}_{-v}$ denote the ground states corresponding to $L^{(v)} - \lambda_0(v)$ and $\widetilde{L}^{-(v)} - \lambda_0(v)$ on*

T_d. (Recall from the remark preceding Lemma 2.3 that $\lambda_0(v) = \tilde{\lambda}_0(-v)$.) Normalize ψ_v and $\tilde{\psi}_{-v}$ by $\int_{T_d} \psi_v \tilde{\psi}_{-v} \mathrm{d}x = 1$. *Recall from* (2.3) *that $K_\lambda = \{v \in R^d\colon \lambda_0(v) \leqslant \lambda\}$ and $\Gamma_\lambda = \partial K_\lambda = \{v \in R^d\colon \lambda_0(v) = \lambda\}$.*

(i) $\lambda_0(\cdot) \in C^2(R^d)$.

(ii) $$\nabla\lambda_0(v) = \int_{T_d}\left(b + \tfrac{1}{2}\nabla\cdot a + a\frac{\nabla\psi_v}{\psi_v} + av\right)\psi_v\tilde{\psi}_{-v}\,\mathrm{d}x, \text{ for } v \in R^d.$$

(iii) $$\sum_{i,j=1}^{d}\left(\frac{\partial^2\lambda_0}{\partial v_i\partial v_j}(v)\right)\eta_i\eta_j = \int_T\left(\eta + \nabla\left(\frac{\phi_\eta}{\psi_v}\right)\right)a\left(\eta + \nabla\left(\frac{\phi_\eta}{\psi_v}\right)\right)\psi_v\tilde{\psi}_{-v}\,\mathrm{d}x > 0,$$

for all $\eta \in R^d - \{0\}$ and $v \in R^d$, where ϕ_η is any solution of

$$(L^{(v)} - \lambda_0(v))\phi_\eta = -\eta\cdot(b + \tfrac{1}{2}(\nabla\cdot a) + a\frac{\nabla\phi_v}{\psi_v} + av - \nabla\lambda_0(v))\psi_v \text{ on } T_d. \tag{2.4}$$

(iv) *If $\lambda_0(v) = \lambda$, then $\Gamma_\lambda = K_\lambda = \{v\}$ or, equivalently, $\lambda = \lambda_c$, if and only if*

$$\int_{T_d}\left(b + \tfrac{1}{2}\nabla\cdot a + a\frac{\nabla\psi_v}{\psi_v} + av\right)\psi_v\tilde{\psi}_{-v}\,\mathrm{d}x = 0.$$

If

$$\int_{T_d}\left(b + \tfrac{1}{2}\nabla\cdot a + a\frac{\nabla\psi_v}{\psi_v} + av\right)\psi_v\tilde{\psi}_{-v}\,\mathrm{d}x \neq 0,$$

then $\lambda > \lambda_c$ or, equivalently, K_λ is a d-dimensional, strictly convex, compact set, and

$$\int_{T_d}\left(b + \tfrac{1}{2}\nabla\cdot a + a\frac{\nabla\psi_v}{\psi_v} + av\right)\psi_v\tilde{\psi}_{-v}\,\mathrm{d}x$$

is an outward normal to K_λ at $v \in \partial K_\lambda = \Gamma_\lambda$.

Corollary 2.11. *Let $L = L_0 = \frac{1}{2}\nabla\cdot a\nabla + b\cdot\nabla$ satisfy the conditions of Theorem 2.1. Then $\lambda_c = 0$ if and only if $\int_{T_d}(b + \frac{1}{2}\nabla\cdot a)\tilde{\psi}_0\,\mathrm{d}x = 0$, where $\tilde{\psi}_0$ is the invariant density for the diffusion on the torus corresponding to the solution of the martingale problem for L on T_d.*

Proof. Choose $\lambda = 0$, and $v_0 = 0$ in Theorem 2.10. Then $\psi_0 = 1$. By Theorem 4.8.6, $\tilde{\psi}_0$, the ground state for $\tilde{L}$ on T_d is the invariant density for the diffusion corresponding to L on T_d. □

Proof of Theorem 2.10. In the statement of the theorem, ψ_v and $\tilde{\psi}_{-v}$ have been normalized by $\int_{T_d}\psi_v\tilde{\psi}_{-v}\,\mathrm{d}x = 1$, for all $v \in R^d$. For the proof of the theorem, it will be convenient to normalize further by $\psi_v(x_0) = 1$, for some fixed $x_0 \in T_d$, for all $v \in R^d$. For $v \in R^d$, $\eta \in R^d - \{0\}$ and $\varepsilon \in R$, we have $(L^{(v+\varepsilon\eta)} - \lambda_0(v + \varepsilon\eta))\psi_{v+\varepsilon\eta} = 0$ in

T_d. Thus, using (2.1), we have

$$(L^{(v)} - \lambda_0(v))(\psi_{v+\varepsilon\eta} - \psi_v) = (L^{(v)} - L^{(v+\varepsilon\eta)})\psi_{v+\varepsilon\eta} + (\lambda_0(v+\varepsilon\eta) - \lambda_0(v))\psi_{v+\varepsilon\eta}$$

$$= -\varepsilon\eta\cdot\left(b + \tfrac{1}{2}\nabla\cdot a + a\frac{\nabla\psi_{v+\varepsilon\eta}}{\psi_{v+\varepsilon\eta}} + av\right)\psi_{v+\varepsilon\eta}$$

$$- \frac{\varepsilon^2}{2}(\eta a\eta)\psi_{v+\varepsilon\eta} + (\lambda_0(v+\varepsilon\eta) - \lambda_0(v))\psi_{v+\varepsilon\eta}. \tag{2.5}$$

Using the fact that $\widetilde{L}^{(-v)}$ is the formal adjoint of $L^{(v)}$, an integration by parts shows that the lefthand side of (2.5), when multiplied by $\widetilde{\psi}_{-v}$ and integrated over T_d, vanishes. The same must then be true of the righthand side; thus

$$(\lambda_0(v+\varepsilon\eta) - \lambda_0(v))\int_{T_d}\psi_{v+\varepsilon\eta}\widetilde{\psi}_{-v}\,dx$$

$$= \varepsilon\eta\cdot\int_{T_d}\left(b + \tfrac{1}{2}\nabla\cdot a + a\frac{\nabla\psi_{v+\varepsilon\eta}}{\psi_{v+\varepsilon\eta}} + av\right)\psi_{v+\varepsilon\eta}\widetilde{\psi}_{-v}\,dx$$

$$+ \frac{\varepsilon^2}{2}\int_{T_d}(\eta a\eta)\psi_{v+\varepsilon\eta}\widetilde{\psi}_{-v}\,dx. \tag{2.6}$$

By the normalization assumption, $\psi_{v+\varepsilon\eta} \to \psi_v$ in the C^2-norm as $\varepsilon \to 0$. Thus, dividing (2.6) by ε and letting $\varepsilon \to 0$ shows that the directional derivative of $\lambda_0(\cdot)$ at v in the η-direction exists and is given by

$$\eta\cdot\int_{T_d}\left(b + \tfrac{1}{2}\nabla\cdot a + a\frac{\nabla\psi_v}{\psi_v} + av\right)\psi_v\widetilde{\psi}_{-v}\,dx.$$

This proves that $\lambda_0(\cdot) \in C^1(R^d)$ and that

$$\nabla\lambda_0(v) = \int_{T_d}\left(b + \tfrac{1}{2}\nabla\cdot a + a\frac{\nabla\psi_v}{\psi_v} + av\right)\psi_v\widetilde{\psi}_v\,dx. \tag{2.7}$$

We will now show that

$$\phi_\eta = \lim_{\varepsilon\to 0}\frac{\psi_{v+\varepsilon\eta} - \psi_v}{\varepsilon} \text{ in the } C^2\text{-norm}, \tag{2.8}$$

where ϕ_η satisfies (2.4) and $\phi_\eta(x_0) = 0$. Note that a solution to (2.4) vanishing at x_0 is unique. Indeed, if there were two distinct solutions, then their difference, call it w, would satisfy $w \not\equiv 0$, $w(x_0) = 0$ and $(L^{(v)} - \lambda_0(v))w = 0$. However, by Theorem 3.5.5, the only non-trivial solution u (up to constant multiples) of $(L^{(v)} - \lambda_0(v))u = 0$ is $u = \psi_v$. Since $\psi_v(x_0) > 0$, the solution w above cannot exist.

Define

$$\phi_{\eta,\varepsilon} = \frac{\psi_{v+\varepsilon\eta} - \psi_v}{\varepsilon}.$$

If we show that $\limsup_{\varepsilon\to 0} \sup_{T_d} |\phi_{\eta,\varepsilon}| < \infty$, then (2.8) will follow from (2.5), the global Schauder estimate of Theorem 3.2.8 (with the boundary term absent since T_d is a compact manifold without boundary), a standard compactness argument, and the above-mentioned uniqueness. Let $\gamma_\varepsilon \equiv \sup_{T_d} |\phi_{\eta,\varepsilon}|$ and assume that $\limsup_{\varepsilon\to 0} \gamma_\varepsilon = \infty$. Let

$$w_{\eta,\varepsilon} = \frac{\phi_{\eta,\varepsilon}}{\gamma_\varepsilon}.$$

Then by (2.5), the global Schauder estimate and a standard compactness argument, there exists a w_η satisfying $w_\eta \not\equiv 0$, $(L^{(v)} - \lambda_0(v))w_\eta = 0$ and $w_\eta(x_0) = 0$. But, as shown in the uniqueness argument above, this is impossible. Thus $\limsup_{\varepsilon\to 0} \gamma_\varepsilon < \infty$ and we conclude that (2.8) holds. Using (2.8), it is clear from (2.7) that $\nabla\lambda_0(v)$ may be differentiated in any direction v. This shows that $\lambda_0(\cdot) \in C^2(R^d)$.

We now calculate

$$\eta D^2\lambda_0(v)\eta \equiv \sum_{i,j=1}^{d} \frac{\partial^2 \lambda_0}{\partial v_i \partial v_j}(v)\eta_i\eta_j, \text{ for } v \in R^d \text{ and } \eta \in R^d - \{0\}.$$

Since $\lambda_0(\cdot) \in C^2(R^d)$, it follows that

$$\lambda_0(v + \varepsilon\eta) - \lambda_0(v) = \varepsilon\eta \cdot \nabla\lambda_0(v) + \frac{\varepsilon}{2}(\eta D^2\lambda_0(v)\eta) + o(\varepsilon^2), \text{ as } \varepsilon \to 0. \tag{2.9}$$

In (2.6), replace $\lambda_0(v+\varepsilon\eta) - \lambda_0(v)$ by the righthand side of (2.9). Then bring the expression $\varepsilon\eta \cdot \nabla\lambda_0(v)\int_{T_d} \psi_{v+\varepsilon\eta}\tilde{\psi}_{-v}\,dx$ over to the righthand side of (2.6) and rewrite it as

$$\varepsilon\eta \cdot \nabla\lambda_0(v) + \varepsilon\eta \cdot \nabla\lambda_0(v)\int_{T_d} (\psi_{v+\varepsilon\eta} - \psi_v)\tilde{\psi}_{-v}\,dx.$$

(Recall that $\int_{T_d} \psi_v\tilde{\psi}_{-v}\,dx = 1$.) In the term $\varepsilon\eta \cdot \nabla\lambda_0(v)$, replace $\nabla\lambda_0(v)$ by the righthand side of (2.7); however, in the term

$$\varepsilon\eta \cdot \nabla\lambda_0(v)\int_{T_d} (\psi_{v+\varepsilon\eta} - \psi_v)\tilde{\psi}_{-v}\,dx,$$

leave $\nabla\lambda_0(v)$ as it is. Now divide both sides of (2.6) by ε^2 and let $\varepsilon \to 0$. Using (2.8), we obtain

$$\tfrac{1}{2}(\eta D^2\lambda_0(v)\eta) = \int_{T_d} \eta\cdot(b\phi_\eta + \tfrac{1}{2}(\nabla\cdot a)\phi_\eta + av\phi_\eta + a\nabla\phi_\eta - \nabla\lambda_0(v)\phi_\eta)\tilde{\psi}_{-v}\,dx + \tfrac{1}{2}\int_{T_d}(\eta a\eta)\psi_v\tilde{\psi}_{-v}\,dx. \quad (2.10)$$

Using (2.4), we have

$$\begin{aligned}
&\int_{T_d}\eta\cdot(b\phi_\eta + \tfrac{1}{2}(\nabla\cdot a)\phi_\eta + av\phi_\eta - \nabla\lambda_0(v)\phi_\eta)\tilde{\psi}_{-v}\,dx \\
&= \int_{T_d}\eta\cdot(b\psi_v + \tfrac{1}{2}(\nabla\cdot a)\psi_v + av\psi_v - \nabla\lambda_0(v)\psi_v)\phi_\eta\frac{\tilde{\psi}_{-v}}{\psi_v}\,dx \\
&= \int_{T_d}\eta\cdot(b\psi_v + \tfrac{1}{2}(\nabla\cdot a)\psi_v + a\nabla\psi_v + av\psi_v - \nabla\lambda_0(v)\psi_v)\phi_\eta\frac{\tilde{\psi}_{-v}}{\psi_v}\,dx \\
&\quad - \int_{T_d}\eta a\nabla\psi_v\phi_\eta\frac{\tilde{\psi}_{-v}}{\psi_v}\,dx \\
&= -\int_{T_d}[\phi_\eta(L^{(v)} - \lambda_0(v))\phi_\eta]\frac{\tilde{\psi}_{-v}}{\psi_v}\,dx - \int_{T_d}\eta a\nabla\psi_v\phi_\eta\frac{\tilde{\psi}_{-v}}{\psi_v}\,dx. \quad (2.11)
\end{aligned}$$

Using exercise 8.18 with the notation $L^{(v)}$, $\lambda_0(v)$, ψ_v and $\tilde{\psi}_{-v}$ in place of L, λ_0, ϕ_0, $\tilde{\phi}_0$, and choosing

$$f = \frac{\phi_n}{\psi_v}$$

in that exercise, we have

$$-\int_{T_d}[\phi_\eta(L^{(v)} - \lambda_0(v))\phi_\eta]\frac{\tilde{\psi}_{-v}}{\psi_v}\,dx = \tfrac{1}{2}\int_{T_d}\left(\nabla\left(\frac{\phi_\eta}{\psi_v}\right)a\nabla\left(\frac{\phi_\eta}{\psi_v}\right)\right)\psi_v\tilde{\psi}_{-v}\,dx. \quad (2.12)$$

Substituting (2.11) and (2.12) in (2.10) gives

$$\begin{aligned}
\tfrac{1}{2}(\eta D^2\lambda_0(v)\eta) &= \int_{T_d}\eta a\nabla\phi_\eta\tilde{\psi}_{-v}\,dx - \int_{T_d}\eta a\nabla\psi_v\phi_\eta\frac{\tilde{\psi}_{-v}}{\psi_v}\,dx \\
&\quad + \tfrac{1}{2}\int_{T_d}\left(\nabla\left(\frac{\phi_\eta}{\psi_v}\right)a\nabla\left(\frac{\phi_\eta}{\psi_v}\right)\right)\psi_v\tilde{\psi}_{-v}\,dx + \tfrac{1}{2}\int_{T_d}(\eta a\eta)\psi_v\tilde{\psi}_{-v}\,dx \\
&= \tfrac{1}{2}\int_{T_d}\left(\eta + \nabla\left(\frac{\phi_\eta}{\psi_v}\right)\right)a\left(\eta + \nabla\left(\frac{\phi_\eta}{\psi_v}\right)\right)\psi_v\tilde{\psi}_{-v}\,dx. \quad (2.13)
\end{aligned}$$

If

$$\eta + \nabla\left(\frac{\phi_\eta}{\psi_v}\right) \equiv 0,$$

then

$$\frac{\phi_\eta}{\psi_\nu}(x) = c - \eta x,$$

which is impossible since ϕ_η and ψ_ν are periodic. Thus, $\eta D^2\lambda_0(\nu)\eta > 0$. Since any two solutions of (2.4) differ by a multiple of ψ_ν, it follows that the righthand side of (2.13) remains unchanged if the particular solution ϕ_η satisfying $\phi_\eta(x_0) = 0$ is replaced by any solution of (2.4).

We have now completed the proof of (i)–(iii). It remains to prove (iv). If $\lambda_0(\nu) = \lambda$ and $\nabla\lambda_0(\nu) \neq 0$, then $K_\lambda - \Gamma_\lambda = \{\zeta \in R^d: \lambda_0(\zeta) < \lambda\} \neq 0$ and it follows from Theorem 2.2 that K_λ is a d-dimensional, strictly convex, compact set. Clearly,

$$\nabla\lambda_0(\nu) = \int_{T_d}\left(b + \tfrac{1}{2}\nabla \cdot a + a\frac{\nabla\psi_\nu}{\psi_\nu} + a\nu\right)\psi_\nu\tilde{\psi}_{-\nu}\,\mathrm{d}x$$

is an outward normal to $K_\lambda = \{\zeta \in R^d: \lambda_0(\zeta) \leq \lambda\}$ at $\nu \in \partial K_\lambda = \Gamma_\lambda$. On the other hand, if $\nabla\lambda_0(\nu) = 0$, then since the Hessian

$$\left\{\frac{\partial^2\lambda_0}{\partial x_i \partial x_j}(\nu)\right\}$$

is positive definite, ν must be an isolated point in $K_\lambda = \{\zeta \in R^d: \lambda_0(\zeta) \leq \lambda\}$. Since K_λ is convex by Theorem 2.2, it follows that $K_\lambda = \Gamma_\lambda = \{\nu\}$. □

II The Martin boundary

In the next theorem, again we choose a pair (λ, ν) with $\nu \in \Gamma_\lambda$. We determine the asymptotic behavior of the diffusion process corresponding to the operator $(L - \lambda)^{u_\nu}$ on R^d, where u_ν is the minimal positive harmonic function for $L - \lambda$ on R^d corresponding to $\nu \in \Gamma_\lambda$. This will then be used to investigate the Martin boundary.

Theorem 2.12. *Let L on R^d satisfy the conditions of Theorem 2.1. Let $\nu \in \Gamma_\lambda$ and let u_ν denote the minimal positive harmonic function for $L - \lambda$ on R^d corresponding to ν. Then there exists a solution $\{P_x^{u_\nu,\lambda}, x \in R^d\}$ to the martingale problem for $(L - \lambda)^{u_\nu}$ on R^d and*

$$P_x^{u_\nu,\lambda}\left(\lim_{t\to\infty}\frac{X(t)}{t} = \int_{T_d}\left(b + \tfrac{1}{2}\nabla \cdot a + a\frac{\nabla\psi_\nu}{\psi_\nu} + a\nu\right)\psi_\nu\tilde{\psi}_{-\nu}\,\mathrm{d}x\right) = 1,$$

where ψ_ν and $\tilde{\psi}_{-\nu}$, normalized by $\int_{T_d}\psi_\nu\tilde{\psi}_{-\nu}\,\mathrm{d}x = 1$, are the ground states of $L^{(\nu)} - \lambda_0(\nu)$ and $\tilde{L}^{(-\nu)} - \lambda_0(\nu)$ considered as operators on the *torus T_d, $L^{(v)}$ is the h-transform of L via* $\exp(v\cdot x)$, *and* $\lambda_0(v)$ *is the principal eigenvalue for* $L^{(v)}$ *on* T_d.

Remark. If $L = \frac{1}{2}\nabla\cdot a\nabla + b\cdot\nabla$, that is, $V = 0$, then we can choose $\lambda = 0$, $v = 0 \in \Gamma_0$, and $u_0 = 1$ in Theorem 2.12 to obtain

$$P_x\left(\lim_{t\to\infty}\frac{X(t)}{t} = \int_{T_d}(b + \tfrac{1}{2}\nabla\cdot a)\tilde{\psi}_0\,dx\right) = 1.$$

By Corollary 2.11, $\int_{T_d}(b + \frac{1}{2}\nabla\cdot a)\tilde{\psi}_0\,dx = 0$ if and only if $\lambda_c = 0$. Thus

$$P_x\left(\lim_{t\to\infty}\frac{X(t)}{t} = 0\right) = 0 \text{ or } 1$$

according to whether $\lambda_c \neq 0$ or $\lambda_c = 0$.

Proof. By Corollary 2.4, $u_v(x) = \exp(v\cdot x)\psi_v(x)$. Then

$$\begin{aligned}(L-\lambda)^{u_v} &= L_0 + a\frac{\nabla u_v}{u_v}\cdot\nabla\\ &= \tfrac{1}{2}\sum_{i,j=1}^{d} a_{ij}\frac{\partial^2}{\partial x_i\partial x_j} + \left(b + \tfrac{1}{2}(\nabla\cdot a) + a\frac{\nabla\psi_v}{\psi_v} + av\right)\cdot\nabla.\end{aligned}$$

Since all the coefficients above are bounded and uniformly Lipschitz and a is uniformly elliptic, it follows from Theorem 1.8.1 that there exists a solution $(P_x^{u_v,\lambda}, x\in R^d)$ to the martingale problem for L on R^d. As noted in Section 1.8, $X(t)$ on $(\Omega, \mathcal{F}, \mathcal{F}_t, P_x^{u_v,\lambda})$ is an Ito process; namely an

$$I_d\left(a(X(\cdot)), \left(b + \tfrac{1}{2}\nabla\cdot a + a\frac{\nabla\psi_v}{\psi_v} + av\right)(X(\cdot))\right)\text{-process.}$$

Let

$$\bar{X}(t) = X(t) - \int_0^t\left(b + \tfrac{1}{2}\nabla\cdot a + a\frac{\nabla\psi_v}{\psi_v} + av\right)(X(s))\,ds.$$

By Theorem 1.5.1(b), for any $\eta\in R^d$, $\eta\cdot\bar{X}(t)$ is a martingale under $P_x^{u_v,\lambda}$ and $E_x^{u_v,\lambda}(\eta\cdot\bar{X}(t))^2 = (\eta\cdot x)^2 + E_x^{u_v,\lambda}\int_0^t(\eta a\eta)(X(s))\,ds$. Thus, by Theorem 1.2.3,

$$E_x^{u_v,\lambda}\left(\sup_{0\le s\le t}(\eta\cdot\bar{X}(s))^2\right) \le 4\left[(\eta\cdot x)^2 + E_x^{u_v,\lambda}\int_0^t(\eta a\eta)(X(s))\,ds\right]. \quad (2.14)$$

The righthand side of (2.14) is bounded by $c(t+1)$, for some constant $c > 0$. Thus, by Chebyshev's inequality,

$$P_x^{u_v,\lambda}\left(\sup_{0\le s\le t} |\eta \cdot \bar{X}(s)| \geq \varepsilon t\right) \leq \frac{c(t+1)}{\varepsilon^2 t^2}.$$

From this it follows that

$$P_x^{u_v,\lambda}\left(\lim_{t\to\infty} \frac{\eta \cdot \bar{X}(t)}{t} = 0\right) = 1.$$

Since $\eta \in R^d$ is arbitrary, we conclude that

$$P_x^{u_v,\lambda}\left(\lim_{t\to\infty} \frac{\bar{X}(t)}{t} = 0\right) = 1. \tag{2.15}$$

Let $(\Omega_{T_d}, \mathcal{F}^{T_d})$ denote the space of continuous functions from $[0, \infty)$ to the torus T_d. Let T denote the natural projection operator from $(\Omega, \mathcal{F})$ to $(\Omega_{T_d}, \mathcal{F}^{T_d})$. For $x \in R^d$, define $Q_{Tx}^{u_v,\lambda}$ by $Q_{Tx}^{u_v,\lambda}(T\cdot) = P_x^{u_v,\lambda}(\cdot)$. Clearly, $\{Q_x^{u_v,\lambda}, x \in T_d\}$ is the solution to the martingale problem for $(L-\lambda)^{u_v}$ considered as an operator on T_d. Since h-transforms satisfy the identity $L^{h_1 h_2} = (L^{h_1})^{h_2}$, and since $\lambda = \lambda_0(v)$, it follows from the definition of u_v that $(L-\lambda)^{u_v} = (L^{(v)} - \lambda_0(v))^{\psi_v}$. Thus $\{Q_x^{u_v,\lambda}, x \in T_d\}$ is the solution to the martingale problem for $(L^{(v)} - \lambda_0(v))^{\psi_v}$ on T_d. By Theorem 4.11.1, the corresponding diffusion process on T_d is positive recurrent. Its invariant density is the ground state for the adjoint to $(L^{(v)} - \lambda_0(v))^{\psi_v}$; this ground state is $\psi_v \tilde{\psi}_{-v}$, as is easily verified. Thus, from Theorem 4.9.5 and the fact that

$$\left(b + \tfrac{1}{2}\nabla \cdot a + a\frac{\nabla \psi_v}{\psi_v} + av\right)$$

is periodic of period one, it follows that

$$P_x^{u_v,\lambda}\left(\lim_{t\to\infty} \frac{1}{t}\int_0^t \left(b + \tfrac{1}{2}\nabla \cdot a + a\frac{\nabla \psi_v}{\psi_v} + av\right)(X(s))\,ds = \int_{T_d}\left(b + \tfrac{1}{2}\nabla \cdot a + a\frac{\nabla \psi_v}{\psi_v} + av\right)\psi_v \tilde{\psi}_{-v}\,dx\right) = 1. \tag{2.16}$$

The theorem now follows from (2.15) and (2.16). □

We now investigate the Martin boundary $\Lambda(\lambda)$ for $L - \lambda$ on R^d in the case where $L - \lambda$ is subcritical on R^d. As noted in Corollary 2.4, Γ_λ is homeomorphic to $\Lambda_0(\lambda)$, the minimal Martin boundary for $L - \lambda$ on R^d. If $\lambda = \lambda_c$, then Γ_λ consists of a single point and thus $\Lambda(\lambda)$ consists of a single point. Thus, assume that $\lambda > \lambda_c$ so that $\Gamma_\lambda = \partial K_\lambda$, where K_λ is a d-dimensional, strictly convex, compact set.

Theorem 2.13. *Let L on R^d satisfy the conditions of Theorem 2.1 and let $\lambda > \lambda_0$. By Theorems 2.2 and 2.5, K_λ is a d-dimensional, strictly convex compact set and $\Gamma_\lambda = \partial K_\lambda$. By Corollary 2.4, Γ_λ is isomorphic to $\Lambda_0(\lambda)$, the minimal Martin boundary for $L - \lambda$ on R^d. Let $M_\lambda\colon \Gamma_\lambda \to \Lambda_0(\lambda)$ denote the homeomorphism between Γ_λ and $\Lambda_0(\lambda)$. Let $N_\lambda\colon \Gamma_\lambda \to S^{d-1}$ denote the map which takes $v \in \Gamma_\lambda = \partial K_\lambda$ to the outward unit normal to K_λ at v. Then for $v \in \Gamma_\lambda$, there exists a Martin sequence $\{x_n\}_{n=1}^\infty \subset R^d$ corresponding to the Martin boundary point $M_\lambda(v)$ which satisfies*

$$\lim_{n\to\infty} \frac{x_n}{|x_n|} = N_\lambda(v).$$

Remark. We conjecture that a sequence $\{x_n\}_{n=1}^\infty \subset R^d$ with no accumulation points in R^d is a Martin sequence corresponding to the Martin boundary point $M_\lambda(v)$ if and only if

$$\lim_{n\to\infty} \frac{x_n}{|x_n|} = N_\lambda(v).$$

Proof. Let $v \in \Gamma_\lambda$ and let u_v denote the minimal positive harmonic function for $L - \lambda$ on R^d corresponding to $v \in \Gamma_\lambda$. Let $\{P_x^{u_v,\lambda}, x \in R^d\}$ denote the solution to the martingale problem for $(L - \lambda)^{u_v}$ on R^d (as in Theorem 2.12). By Theorem 7.2.1, $P_x^{u_v,\lambda}(m - \lim_{t\to\infty} X(t) = M_\lambda(v)) = 1$, by Theorem 2.12

$$P_x^{u_v,\lambda}\left(\lim_{t\to\infty} \frac{X(t)}{t} = \int_{T_d}\left(b + \tfrac{1}{2}\nabla\cdot a + a\frac{\nabla\psi_v}{\psi_v} + av\right)\psi_v\tilde{\psi}_{-v}\,\mathrm{d}x\right) = 1,$$

and by Theorem 2.10,

$$N_\lambda(v) = c\int_{T_d}\left(b + \tfrac{1}{2}\nabla\cdot a + a\frac{\nabla\psi_v}{\psi_v} + av\right)\psi_v\tilde{\psi}_{-v}\,\mathrm{d}x,$$

where c is a normalization constant. Thus, letting $\{t_n\}_{n=1}^\infty$ be an increasing sequence approaching ∞, and defining $x_n(\omega) = X(t_n, \omega)$, it follows that for $P_x^{u_v,\lambda}$-almost every ω, $\{x_n(\omega)\}_{n=1}^\infty$ is a Martin sequence corresponding to $M_\lambda(v)$ and satisfying

$$\lim_{n\to\infty} \frac{x_n(\omega)}{|x_n(\omega)|} = N_\lambda(v). \qquad \square$$

III The behavior of λ_c under perturbations

Let $W \in C^\alpha(R^d)$ be periodic and define the periodic operator $L_t = \frac{1}{2}\nabla\cdot a\nabla + b\cdot\nabla + V + tW \equiv L + tW$ on R^d. Let $\lambda_c(t)$ denote the generalized principal eigenvalue for L_t on R^d. We investigate the behavior of $\lambda_c(t)$, especially for small t. It turns out, in particular, that if W averages to zero in an appropriate sense, then $\lambda_c(t) > \lambda_c$, for all $t \neq 0$. The results are more explicit in the symmetric case, that is, the case where $b = a\nabla Q$. The symmetric case is treated in Theorem 2.14 and the general case in Theorem 2.16.

Theorem 2.14. *Let* $L_t = \frac{1}{2}\nabla \cdot a\nabla + b \cdot \nabla + V + tW \equiv L + tW$ *on* R^d *satisfy the conditions of Theorem 2.1. In addition, assume that L is symmetric (that is, $b = a\nabla Q$, for some Q). Let $\lambda_c(t)$ denote the generalized principal eigenvalue for L_t on R^d. Let ν_0 be as in Theorem 2.5 so that $\lambda_c = \lambda_0(\nu_0)$, and let ψ_{ν_0} and $\tilde{\psi}_{-\nu_0}$ be as in Theorem 2.10, normalized by*

$$\int_T \psi_{\nu_0}\tilde{\psi}_{-\nu_0}\,dx = \int_{T_d} \psi_{\nu_0}^2 \frac{\tilde{\psi}_{-\nu_0}}{\psi_{\nu_0}}\,dx = 1.$$

Then $L^{(\nu_0)}$ *restricted to T_d is symmetric with respect to the density*

$$\frac{\tilde{\psi}_{-\nu_0}}{\psi_{\nu_0}}.$$

The principal eigenvalue and the corresponding principal eigenfunction have been denoted by $\lambda_0(\nu_0)$ and ψ_{ν_0}. Let $\{\lambda_j(\nu_0)\}_{j=1}^{\infty}$, satisfying $\lambda_0(\nu_0) > \lambda_1(\nu_0) \geq \lambda_2(\nu_0) \geq \ldots$, denote the rest of the eigenvalues, counted with multiplicity, and choose corresponding eigenfunctions $\{\psi_j\}_{j=1}^{\infty}$ so that ψ_{ν_0} and $\{\psi_j\}_{j=1}^{\infty}$ together form a complete orthonormal set on

$$L^2\left(T_d, \frac{\tilde{\psi}_{-\nu_0}}{\psi_{\nu_0}}\,dx\right).$$

Let $\{c_j\}_{j=0}^{\infty}$ denote the Fourier coefficients of $W\psi_{\nu_0}$; that is, $W\psi_{\nu_0} = c_0\psi_{\nu_0} + \sum_{j=1}^{\infty} c_j\psi_j$. Then

$$\lambda_c(t) = \lambda_c + c_0 t + \left(\sum_{j=1}^{\infty} \frac{c_j^2}{\lambda_0(\nu_0) - \lambda_j(\nu_0)}\right)t^2 + o(t^2), \text{ as } t \to 0.$$

Equivalently,

$$\lambda_c(t) = \lambda_c + \left(\int_{T_d} W\psi_{\nu_0}\tilde{\psi}_{-\nu_0}\,dx\right)t + \left(\frac{1}{2}\int_{T_d}(\nabla fa\nabla f)\psi_{\nu_0}\tilde{\psi}_{-\nu_0}\,dx\right)t^2 + o(t^2)$$

as $t \to 0$, where f is any solution of

$$L^{(\nu_0)}f = -\left(W - \int_{T_d} W\psi_{\nu_0}\tilde{\psi}_{-\nu_0}\,dx\right)\psi_{\nu_0} \text{ on } T_d.$$

If W is not constant, then the coefficient of t^2 in the above expansion is positive. Furthermore, if $\int_T W\psi_{v_0}\tilde{\psi}_{-v_0}\,dx = 0$ and $W \not\equiv 0$, then

(i) $\lambda_c(t) > \lambda_c$, *for all* $t \neq 0$;

(ii) $\lambda_c(t) \leqslant \lambda_c + \dfrac{t^2}{\lambda_0(v_0) - \lambda_1(v_0)}\displaystyle\int_{T_d} W^2\psi_{v_0}\tilde{\psi}_{-v_0}\,dx + o(t^2)$, *as* $t \to 0$,

and equality holds if and only if W is an eigenfunction corresponding to $\lambda_1(v_0)$;

(iii) $\lambda_c(t) \leqslant \lambda_c + \dfrac{t^2}{\lambda_0(v_0) - \lambda_1(v_0)}\displaystyle\int_{T_d} W^2\psi_{v_0}\tilde{\psi}_{-v_0}\,dx$, *for small* t.

Example 2.15. Let $L = \frac{1}{2}\Delta$ on R^d. Then $v_0 = 0$, $\lambda_c = \lambda_0(0) = 0$, $\psi_{v_0} = \tilde{\psi}_{-v_0} = 1$, and $L^{(v_0)} = \frac{1}{2}\Delta$. One can easily write down the eigenfunction expansion for W; this is left to the reader. However, note in particular that $\lambda_j(0) = 2\pi^2$, for $j = 1, 2, \ldots, 2d$, and, letting $x = (x_1, \ldots, x_d)$, the corresponding eigenfunctions are $\{\cos 2\pi x_j\}_{j=1}^d$ and $\{\sin 2\pi x_j\}_{j=1}^d$. Thus, it follows that if $\int_{T_d} W\,dx = 0$ and $W \not\equiv 0$, then $\lambda_c(t) > 0$, for all $t \neq 0$, and

$$\lambda_c(t) \leqslant \frac{t^2}{2\pi^2}\int_{T_d} W^2\,dx,$$

for small t; furthermore,

$$\lambda_c(t) = \frac{t^2}{2\pi^2}\int_{T_d} W^2\,dx + o(t^2),$$

as $t \to 0$ if and only if W is of the form $W(x) = \sum_{j=1}^d k_j \cos 2\pi x_j + \sum_{j=1}^d l_j \sin 2\pi x_j$.

Theorem 2.16. *Let* $L = \frac{1}{2}\nabla\cdot a\nabla + b\cdot\nabla + V + tW \equiv L + tW$ *on* R^d *satisfy the conditions of Theorem 2.1 and let* $\lambda_c(t)$ *denote the generalized principal eigenvalue for* L_t *on* R^d. *Let* v_0 *be as in Theorem 2.5 so that* $\lambda_c = \lambda_0(v_0)$, *and let* ψ_{v_0} *and* $\tilde{\psi}_{-v_0}$ *be as in Theorem 2.10, normalized by* $\int_{T_d}\psi_{v_0}\tilde{\psi}_{-v_0}\,dx = 1$. *Then* $\lambda_c(t) = \lambda_c + (\int_{T_d}\psi_{v_0}\tilde{\psi}_{-v_0}\,dx)t + c_W t^2 + o(t)^2$, *as* $t \to 0$, *where*

$$c_W = \inf_{v\in R^d}\frac{1}{2}\int_{T_d}\left(v + \nabla\left(\frac{f^{(v)}}{\psi_{v_0}}\right)\right)a\left(v + \nabla\left(\frac{f^{(v)}}{\psi_{v_0}}\right)\right)\psi_{v_0}\tilde{\psi}_{-v_0}\,dx$$

and $f^{(v)}$ *is any solution of*

$$(L^{(v_0)} - \lambda_0(v_0))f^{(v)} = -(L^{(v_0)} - \lambda_0(v_0))(v\cdot x)\psi_{v_0} - \left(W - \int_{T_d} W\psi_{v_0}\tilde{\psi}_{-v_0}\,dx\right)\psi_{v_0} \text{ on } T_d.$$

If W is not constant, then $c_W > 0$. *Furthermore, if* $\int_{T_d} W\psi_{v_0}\tilde{\psi}_{-v_0}\,dx = 0$, *and* $W \not\equiv 0$, *then* $\lambda_c(t) > \lambda_c$ *for all* $t \neq 0$.

For the proofs of Theorems 2.14 and 2.16, we need to establish a bit of notation and prove two propositions. Let $\lambda_0(t, v)$ denote the principal eigenvalue of $L_t^{(v)}$ restricted to T_d, and let $\psi_{t,v}$ denote the corresponding positive eigenfunction. By Theorem 2.5, $\lambda_c(t) = \inf_{v \in R^d} \lambda_0(t, v)$, and the infimum is attained uniquely at a point which will be denoted by $v_0(t)$. Note that $\lambda_0(0, v) = \lambda_0(v)$, $\psi_{0,v} = \psi_v$ and $v_0(0) = v_0$.

Proposition 2.17. (i) *$\lambda_0(t, v)$ is three times differentiable in (t, v).*

(ii) *$\psi_{t,v}$, normalized by $\psi_{t,v}(x_0) = 1$, for some fixed $x_0 \in T_d$, is twice Frechet differentiable (on $C(T_d)$) in (t, v).*

(iii) *$\lambda_c(t)$ and $v_0(t)$ are twice differentiable in t.*

Proof. By Theorem 2.10, $\lambda_0(v)$ is twice differentiable in v, and the proof of that theorem also showed that ψ_v is once Frechet differentiable in v. It can be shown *mutatis mutandis* that $\lambda_0(t, v)$ is twice differentiable in (t, v) and that $\psi_{t,v}$ is once Frechet differentiable in (t, v). It is not difficult to continue the argument in the proof of Theorem 2.10 and to conclude that $\lambda_0(t, v)$ is three times differentiable in (t, v) and that $\psi_{t,v}$ is twice Frechet differentiable in (t, v). Thus, it remains to show that $v_0(t)$ and $\lambda_c(t)$ are twice differentiable. This is left as exercises 8.20 and 8.21. □

Proposition 2.18. *If L on R^d is symmetric with respect to some reference measure, that is, if $b = a\nabla Q$, then $v_0(t) = v_0$, for all $t \in R$.*

Proof. Define $\tilde{\psi}_{t,v}$ for the operator $\tilde{L}$ analogous to $\psi_{t,v}$ for L. By Corollary 2.4, $u_t(x) \equiv \psi_{t,v_0(t)}(x) \exp(v_0(t) \cdot x)$ and $\tilde{u}_t(x) \equiv \tilde{\psi}_{t,-v_0(t)}(x) \exp(-v_0(t) \cdot x)$ are (up to positive multiples) the unique elements, respectively, of $C_{L_t - \lambda_c(t)}(R^d)$ and $C_{\tilde{L}_t - \lambda_c(t)}(R^d)$. On the other hand, a direct calculation reveals that $u_t \exp(2Q) \in C_{\tilde{L}_t - \lambda_c(t)}(R^d)$. Thus, $\tilde{u}_t = k u_t \exp(2Q)$, for some constant k; that is,

$$\exp(-2v_0(t) \cdot x) = k \exp(2Q(x)) \frac{\psi_{t,v_0(t)}(x)}{\tilde{\psi}_{t,-v_0(t)}(x)}.$$

Since $\psi_{t,v_0(t)}$ and $\tilde{\psi}_{t,-v_0(t)}$ are periodic, it follows easily that $v_0(t)$ is constant. □

Theorems 2.14 and 2.16 will be proved together.

Proof of Theorems 2.14 and 2.16. In the statement of the theorems, ψ_v and $\tilde{\psi}_{-v}$ may have been normalized by $\int_{T_d} \psi_v \tilde{\psi}_{-v} dx = 1$, for all $v \in R^d$.

For the proof of the theorems, it will be convenient to normalize further by $\psi_{t,v}(x_0) = 1$, for some fixed $x_0 \in T_d$, for all $v \in R^d$ and $t \in R$. With this normalization, Proposition 2.17(ii) holds for $\psi_{t,v}$.

By definition,

$$(L_t - \lambda_c(t)) \exp(v_0(t) \cdot x) \psi_{t,v_0(t)}(x) = 0 \text{ on } R^d \tag{2.17a}$$

$$\psi_{t,v_0(t)}(x_0) = 1. \tag{2.17b}$$

By Proposition 2.17, $\lambda_c(t)$, $v_0(t)$ and $\psi_{t,v_0(t)}$ are all twice differentiable in t. Expand them about $t = 0$ as follows:

$$\left.\begin{aligned} \lambda_c(t) &= \lambda_0(v_0) + \lambda_1 t + \lambda_2 t^2 + o(t^2), \quad \text{as } t \to 0, \\ v_0(t) &= v_0 + v_1 t + v_2 t^2 + o(t^2), \quad \text{as } t \to 0, \\ \psi_{t,v_0(t)} &= \psi_{v_0} + tf_1 + tf_2 + o(t^2), \quad \text{as } t \to 0. \end{aligned}\right\} \tag{2.18}$$

(In the symmetric case, it follows from Proposition 2.18 that $v_1 = v_2 = 0$; this fact will be used later on in the proof.) Substituting (2.18) into (2.17), multiplying (2.17) by $\exp(-v_0 \cdot x)$ and recalling the definition of $L^{(v_0)}$, and equating powers of t, we obtain

$$(L^{(v_0)} - \lambda_0(v_0))f_1 = -(L^{(v_0)} - \lambda_0(v_0))((v_1 \cdot x)\psi_{v_0}) - (W - \lambda_1)\psi_{v_0} \text{ on } T_d, \tag{2.19a}$$

$$f_1(x_0) = 0 \tag{2.19b}$$

and

$$\begin{aligned} (L^{(v_0)} - \lambda_0(v_0))f_2 = & \\ & - (L^{(v_0)} - \lambda_0(v_0))((v_2 \cdot x)\psi_{v_0}) - (L^{(v_0)} - \lambda_0(v_0))\left(\frac{(v_1 \cdot x)^2}{2}\psi_{v_0}\right), \\ & - (L^{(v_0)} - \lambda_0(v_0))((v_1 \cdot x)f_1) - (W - \lambda_1)(v_1 \cdot x)\psi_{v_0} - (W - \lambda_1)f_1 \\ & + \lambda_2 \psi_{v_0} \text{ on } T_d \end{aligned} \tag{2.20a}$$

$$f_2(x_0) = 0. \tag{2.20b}$$

The adjoint of $L^{(v_0)} - \lambda_0(v_0)$ is $\widetilde{L}^{(-v_0)} - \lambda_0(v_0)$ and the kernel of this adjoint operator is, of course, generated by $\widetilde{\psi}_{-v_0}$. Thus, by the Fredholm alternative (see Chapter 6, Section 3 following Theorem 6.3.3 where it has been stated for operators on S^{d-1}), the righthand side of (2.19a) must vanish upon integration against $\widetilde{\psi}_{-v_0}$. A direct calculation reveals that

$$(L^{(v_0)} - \lambda_0(v_0))((v_1 \cdot x)\psi_{v_0}) = v_1 \cdot \left(b + \tfrac{1}{2}\nabla \cdot a + a v_0 + a\frac{\nabla \psi_{v_0}}{\psi_{v_0}}\right)\psi_{v_0}.$$

Using the definition of v_0 for the first equality below, and Theorem 2.10(ii) for the second one, we have

$$0 = \nabla\lambda_0(v_0) = \int_{T_d}\left(b + \tfrac{1}{2}\nabla\cdot a + a\frac{\nabla\psi_{v_0}}{\psi_{v_0}} + av_0\right)\psi_{v_0}\tilde{\psi}_{-v_0}\,dx.$$

Thus, it follows that

$$\int_{T_d}(L^{(v_0)} - \lambda_0(v_0))((v_1\cdot x)\psi_{v_0})\tilde{\psi}_{-v_0}\,dx = 0. \tag{2.21}$$

Consequently, upon integrating the righthand side of (2.19a) against $\tilde{\psi}_{-v_0}$, setting the resulting expression equal to zero, and recalling the normalization $\int_{T_d}\psi_{v_0}\tilde{\psi}_{-v_0}\,dx = 1$, we obtain

$$\lambda_1 = \int_{T_d} W\psi_{v_0}\tilde{\psi}_{-v_0}\,dx. \tag{2.22}$$

Applying the Fredholm alternative to (2.20a), it follows that the righthand side of (2.20a) must vanish upon integration against $\tilde{\psi}_{-v_0}$. The argument leading up to (2.21) shows that the first term on the righthand side of (2.20a) vanishes upon integration against $\tilde{\psi}_{-v}$. Thus, we have

$$\begin{aligned}\lambda_2 = &\int_{T_d}(L^{(v_0)} - \lambda_0(v_0))\left(\frac{(v_1\cdot x)^2}{2}\psi_{v_0}\right)\tilde{\psi}_{-v_0}\,dx\\ &+ \int_{T_d}(L^{(v_0)} - \lambda_0(v_0))((v_1\cdot x)f_1)\tilde{\psi}_{-v_0}\,dx\\ &+ \int_{T_d}(W - \lambda_1)(v_1\cdot x)\psi_{v_0}\tilde{\psi}_{-v_0}\,dx\\ &+ \int_{T_d}(W - \lambda_1)f_1\tilde{\psi}_{-v_0}\,dx.\end{aligned} \tag{2.23}$$

In order to simplify the righthand side of (2.23), we analyze the integrals appearing there.

Applying the identity $(L^{(v_0)} - \lambda_0(v_0))g^2h = g^2(L^{(v_0)} - \lambda_0(v_0))h + 2gh(\frac{1}{2}\nabla\cdot a\nabla + b\cdot\nabla + av_0\nabla)g + 2g\nabla g a\nabla h + (\nabla g a\nabla g)h$, with $g = (v_1\cdot x)$ and $h = \psi_{v_0}$, and using the fact that $(L^{(v_0)} - \lambda_0(v_0))\psi_{v_0} = 0$, we have

$$(L^{(v_0)} - \lambda_0(v_0))\frac{(v_1\cdot x)^2}{2}\psi_{v_0} =$$
$$(v_1\cdot x)\psi_{v_0}\left(b + \tfrac{1}{2}\nabla\cdot a + av_0 + a\frac{\nabla\psi_{v_0}}{\psi_{v_0}}\right)\cdot v_1 + \tfrac{1}{2}(v_1 a v_1)\psi_{v_0}. \tag{2.24}$$

Applying the identity $(L^{(v_0)} - \lambda_0(v_0))gh = g(L^{(v_0)} - \lambda_0(v_0))h +$

$h(\frac{1}{2}\nabla\cdot a\nabla + b\cdot\nabla + av_0\cdot\nabla)g + \nabla g a\nabla h$, with $g = (v_1\cdot x)$ and $h = f_1$, and using (2.19a), we have

$$\begin{aligned}
&(L^{(v_0)} - \lambda_0(v_0))((v_1\cdot x)f) \\
&\quad = (v_1\cdot x)(L^{(v_0)} - \lambda_0(v_0))f_1 + f_1(\tfrac{1}{2}\nabla\cdot a\nabla + b\cdot\nabla + av_0\cdot\nabla)(v_1\cdot x) \\
&\qquad + v_1 a\nabla f_1 \\
&\quad = -(v_1\cdot x)(L^{(v_0)} - \lambda_0(v_0))((v_1\cdot x)\psi_{v_0}) - (v_1\cdot x)(W-\lambda_1)\psi_{v_0} \\
&\qquad + f_1(b + \tfrac{1}{2}\nabla\cdot a + av_0)\cdot v_1 + v_1 a\nabla f_1 \\
&\quad = -(v_1\cdot x)\psi_{v_0}\left(b + \tfrac{1}{2}\nabla\cdot a + av_0 + a\frac{\nabla\psi_{v_0}}{\psi_{v_0}}\right)\cdot v_1 \\
&\qquad - (v_1\cdot x)(W-\lambda_1)\psi_{v_0} + f_1(b + \tfrac{1}{2}\nabla\cdot a + av_0)\cdot v_1 + v_1 a\nabla f_1. \quad (2.25)
\end{aligned}$$

Using (2.19a), we have

$$\begin{aligned}
(W-\lambda_1)f_1 &= -\frac{f_1}{\psi_{v_0}}(L^{(v_0)} - \lambda_0(v_0))f_1 - \frac{f_1}{\psi_{v_0}}(L^{(v_0)} - \lambda_0(v_0))((v_1\cdot x)\psi_{v_0}) \\
&= -\frac{f_1}{\psi_{v_0}}(L^{(v_0)} - \lambda_0(v_0))f_1 - f_1\left(b + \tfrac{1}{2}\nabla\cdot a + av_0 + a\frac{\nabla\psi_{v_0}}{\psi_{v_0}}\right)\cdot v_1.
\end{aligned} \tag{2.26}$$

From (2.24)–(2.26), we have

$$\begin{aligned}
&(L^{(v_0)} - \lambda_0(v_0))\left(\frac{(v_1\cdot x)^2}{2}\psi_{v_0}\right) + (L^{(v_0)} - \lambda_0(v_0))((v_1\cdot x)f_1) \\
&\qquad\qquad + (v_1\cdot x)(W-\lambda_1)\psi_{v_0} + (W-\lambda_1)f_1 \\
&= \tfrac{1}{2}(v_1 a v_1)\psi_{v_0} + v_1 a\nabla f_1 - \frac{f_1}{\psi_{v_0}}(L^{(v_0)} - \lambda_0(v_0))f_1 - f_1 v_1 a\frac{\nabla\psi_{v_0}}{\psi_{v_0}}. \quad (2.27)
\end{aligned}$$

Applying the identity in exercise 8.18 with L, λ_0, ϕ_0 and $\tilde{\phi}_0$ replaced by $L^{(v_0)}$, $\lambda_0(v)$, ψ_{v_0} and $\tilde{\phi}_{-v_0}$, and choosing

$$\frac{f_1}{\psi_{v_0}}$$

as the arbitrary function appearing in that identity, we have

$$\int_{T_d}\frac{f_1}{\psi_{v_0}}(L^{(v_0)} - \lambda_0(v_0))(f_1)\tilde{\psi}_{-v_0}\,dx = -\tfrac{1}{2}\int_{T_d}\left(\nabla\left(\frac{f_1}{\psi_{v_0}}\right)a\nabla\left(\frac{f_1}{\psi_{v_0}}\right)\right)\psi_{v_0}\tilde{\psi}_{-v_0}\,dx. \tag{2.28}$$

From (2.23), (2.27) and (2.28), we obtain

$$\lambda_2 = \tfrac{1}{2}\int_{T_d}(v_1 a v_1)\psi_{v_0}\tilde{\psi}_{-v_0}\,dx + \int_{T_d}(v_1 a \nabla f_1)\tilde{\psi}_{-v_0}\,dx$$
$$+ \tfrac{1}{2}\int_{T_d}\left(\nabla\left(\frac{f_1}{\psi_{v_0}}\right)a\nabla\left(\frac{f_1}{\psi_{v_0}}\right)\right)\psi_{v_0}\tilde{\psi}_{-v_0}\,dx - \int_{T_d} f_1\left(v_1 a\frac{\nabla\psi_{v_0}}{\psi_{v_0}}\right)\tilde{\psi}_{-v_0}$$
$$= \tfrac{1}{2}\int_{T_d}\left(v_1 + \nabla\left(\frac{f_1}{\psi_{v_0}}\right)\right)a\left(v_1 + \nabla\left(\frac{f_1}{\psi_{v_0}}\right)\right)\psi_{v_0}\tilde{\psi}_{-v_0}\,dx. \tag{2.29}$$

We have not found λ_2 explicitly since v_1 has yet to be identified. Since the above analysis has not imposed any conditions on v_1, it follows that this analysis may be carried out with v_1 replaced by any $v \in R^d$. Let $f_1^{(v)}$ denote the solution to (2.19) with v_1 replaced by v (identify f_1 with $f_1^{(v_1)}$ and let $\lambda_2(v)$ denote the righthand side of (2.29) with f_1 replaced by $f_1^{(v)}$ and v_1 replaced by v. Let $\gamma_v(t) \equiv \lambda_0(v_0) + \lambda_1 t + \lambda_2(v)t^2 + o(t^2)$, as $t \to 0$. The above analysis has shown that, for any $v \in R^d$, if the term $o(t^2)$ is chosen appropriately in the definition of $\gamma_v(t)$, then for small t, positive solutions of the form $\psi_t \exp(v_t \cdot x)$, with $v_t \in R^d$ and ψ_t periodic, exist for $L_t - \gamma_v(t)$. By definition, for any t, positive solutions of the form $\psi_t \exp(v_t \cdot x)$, with $v_t \in R^d$ and ψ_t periodic, exist for $L_t - \lambda$ if and only if $\lambda \geq \lambda_c(t)$. Thus it follows that

$$\lambda_2 = \inf_{v \in R^d} \lambda_2(v)$$
$$= \inf_{v \in R^d} \tfrac{1}{2}\int_{T_d}\left(v + \nabla\left(\frac{f_1^{(v)}}{\psi_{v_0}}\right)\right)a\left(v + \nabla\left(\frac{f_1^{(v)}}{\psi_{v_0}}\right)\right)\psi_{v_0}\tilde{\psi}_{v_0}\,dx. \tag{2.30}$$

Since $C_{L-\lambda_c(t)}(R^d)$ is one-dimensional, it follows then that the infimum is attained uniquely at $v = v_1$. Thus,

$$\lambda_2 = \tfrac{1}{2}\int_{T_d}\left(v_1 + \nabla\left(\frac{f_1^{(v_1)}}{\psi_{v_0}}\right)\right)a\left(v_1 + \nabla\left(\frac{f_1^{(v_1)}}{\psi_{v_0}}\right)\right)\psi_{v_0}\tilde{\psi}_{-v_0}\,dx. \tag{2.31}$$

If $u^{(v)}$ solves (2.19) (a) with v_1 replaced by v, then $u^{(v)} = f_1^{(v)} + k\psi_{v_0}$. Clearly, the righthand side of (2.30) does not change if $f_1^{(v)}$ is replaced by $u^{(v)}$. Thus, (2.30) and (2.31) hold under the condition that $f_1^{(v)}$ satisfy (2.19a) alone.

We now show that $\lambda_2 = 0$ if and only if W is constant. If $\lambda_2 = 0$, then from (2.31) it follows that

$$v_1 + \nabla\left(\frac{f_1^{(v_1)}}{\psi_{v_0}}\right) \equiv 0;$$

that is,

$$v_1 \cdot x + \frac{f_1^{(v_1)}}{\psi_{v_0}}(x)$$

is constant. Since

$$\frac{f_1^{(v_1)}}{\psi_{v_0}}$$

is periodic, it follows that $v_1 = 0$ and $f_1^{(v_1)} = k\psi_{v_0}$, for some constant k. Substituting $v_1 = 0$ and $f_1^{(v_1)} = k\psi_{v_0}$ in (2.19a), we obtain $W \equiv \lambda_1$. Similarly, it is easy to show that if W is constant, then $\lambda_2 = 0$.

Since $\lambda_2 > 0$ if W is not constant, it follows that $\lambda_c(t) > \lambda_c$ for all small t if $\int_{T_d} W\psi_{v_0}\tilde{\psi}_{-v_0}\,dx = 0$ and $W \not\equiv 0$. By the convexity of $\lambda_c(t)$ (see exercise 8.21), it then follows that $\lambda_c(t) > 0$, for all $t \neq 0$. This completes the proof of Theorem 2.16.

For the proof of Theorem 2.14, we recall that, by Proposition 2.18, $v_1 = v_2 = 0$. Setting $v_1 = 0$ in (2.29) and in (2.19a), we obtain

$$\lambda_2 = \tfrac{1}{2}\int_{T_d}\left(\nabla\left(\frac{f_1}{\psi_{v_0}}\right)a\nabla\left(\frac{f_1}{\psi_{v_0}}\right)\right)\psi_{v_0}\tilde{\psi}_{-v_0}\,dx, \tag{2.32}$$

where f_1 is any solution of

$$(L^{(v_0)} - \lambda_0(v_0))f_1 = -(W - \lambda_1)\psi_{v_0} \text{ on } T_d. \tag{2.33}$$

This gives the second of the two representations for the expansion of $\lambda_c(t)$ in Theorem 2.14. We now turn to the first representation; namely, the spectral representation.

Setting $t = 0$ in the penultimate line of the proof of Proposition 2.18, it follows that

$$\frac{\tilde{\psi}_{-v_0}}{\psi_{v_0}}(x) = k\exp(2Q(x) + 2v_0\cdot x).$$

Using this and writing $L^{(v_0)} = \frac{1}{2}\nabla\cdot a\nabla + a\nabla Q\cdot\nabla + av\cdot\nabla + V = \frac{1}{2}\exp(-2Q(x) - 2v_0\cdot x)\nabla\cdot a\exp(2Q(x) + 2v_0\cdot x)\cdot\nabla + V$, it follows that $L^{(v_0)}$ on T_d is symmetric with respect to the density

$$\frac{\tilde{\psi}_{-v_0}}{\psi_{v_0}}.$$

Let $\psi_{v_0} \cup \{\psi_i\}_{i=1}^{\infty}$ denote the complete orthonormal set of eigenfunctions for $L^{(v_0)}$ on

$$L_2\left(T_d, \frac{\tilde{\psi}_{-v_0}}{\psi_{v_0}}dx\right)$$

as in the statement of Theorem 2.14, let $\lambda_0(v_0) > \lambda_1(v_0) \geq \lambda_2(v_0) \geq \ldots$ denote the corresponding eigenvalues, and let

$$W\psi_{v_0} = c_0\psi_{v_0} + \sum_{j=1}^{\infty} c_j\psi_j, \tag{2.34}$$

as in the statement of Theorem 2.14.

From (2.22) and (2.34), we have

$$\lambda_1 = c_0 \tag{2.35}$$

and

$$(W - \lambda_1)\psi_{\nu_0} = \sum_{j=1}^{\infty} c_j \psi_j. \tag{2.36}$$

From (2.33) and (2.36), it follows that

$$f_1 = -\sum_{j=1}^{\infty} \frac{c_j}{\lambda_j(\nu_0) - \lambda_0(\nu_0)} \psi_j + k\psi_{\nu_0}, \tag{2.37}$$

where k is an arbitrary constant. From (2.32), (2.33) and (2.28), we have

$$\lambda_2 = \int_{T_d} f_1 (W - \lambda)\psi_{\nu_0} \frac{\tilde{\psi}_{-\nu_0}}{\psi_{\nu_0}} \mathrm{d}x. \tag{2.38}$$

Thus, combining (2.36)–(2.38) gives

$$\lambda_2 = \sum_{j=1}^{\infty} \frac{c_j^2}{\lambda_0(\nu_0) - \lambda_j(\nu_0)}. \tag{2.39}$$

The spectral expansion for $\lambda_c(t)$ follows from (2.18), (2.34), (2.35) and (2.39).

It remains to consider the three statements appearing at the end of Theorem 2.14. Statement (i) was already proved in the proof of Theorem 2.16, and statement (ii) follows immediately from the spectral representation for $\lambda_c(t)$. It remains to consider statement (iii). By Proposition 2.18, $\nu_0(t) = \nu_0$, for all t; therefore $\lambda_c(t) = \lambda_c(t, \nu_0)$, and we conclude by Proposition 2.17 that $\lambda_c(t)$ is three times differentiable in t. In the light of statement (ii), statement (iii) will follow if we show that

$$\frac{\mathrm{d}^3 \lambda_c}{\mathrm{d}t^3}(0) < 0,$$

whenever W is an eigenfunction corresponding to $\lambda_1(\nu_0)$. This is left as exercise 8.22. □

8.3 Fuchsian operators and their adjoints

In this section, we study *Fuchsian* operators, that is, operators satisfying the following assumption.

Assumption F.

$$L=\tfrac{1}{2}\sum_{i,j=1}^{d} a_{ij}\frac{\partial^2}{\partial x_i\partial x_j}+\sum_{i=1}^{d} b_i\frac{\partial}{\partial x_i}+V(x)\text{ on }R^d$$

satisfies the following conditions. There exist constants c_1, $c_2>0$ such that

(i) $$c_1|v|^2\leqslant \textstyle\sum_{i,j=1}^{d} a_{ij}(x)v_iv_j\leqslant c_2|v|^2,\ \textit{for } x,\ v\in R^d;$$

(ii) $$|b_i(x)|\leqslant\frac{c_2}{1+|x|},\ \textit{for } i=1,2,\ldots,d,\ \textit{and } x\in R^d;$$

(iii) $$|V(x)|\leqslant\frac{c_2}{1+|x|^2},\ \textit{for } x\in R^d.$$

Remark. Note that under Assumption F, L is not in divergence form. Thus, in this section we will assume that L satisfies Assumption $\mathrm{H}_{\mathrm{loc}}$ (defined in exercise 4.16) instead of Assumption $\widetilde{\mathrm{H}}_{\mathrm{loc}}$.

In this section we will prove that if L is subcritical, then the Martin boundary of L and the Martin boundary of $\widetilde{L}$ each consist of a single point.

Theorem 3.1. *Let L satisfy Assumption F and Assumption H_{loc} on R^d. If $C_L(R^d)\neq\phi$, then $C_L(D)$ is one-dimensional. In particular, if L is subcritical, then the Martin boundary consists of a single point.*

Proof. If L is supercritical, then $C_L(R^d)=\phi$; if L is critical, then $C_L(R^d)$ is one-dimensional by Theorem 4.3.4. Thus, we may assume that L is subcritical. Clearly, the one-dimensionality of $C_L(D)$ is equivalent to the property that the Martin boundary consists of a single point. The Martin boundary consists of more than one point if and only if there are at least two minimal elements in $C_L(D)$. Thus, to prove the theorem, it is enough to show that if u_1, $u_2\in C_L(D)$, then $u_1\leqslant ku_2$, for some constant k.

Let u_1, $u_2\in C_L(D)$. For $R>0$, define the operator

$$L_R=\tfrac{1}{2}\sum_{i,j=1}^{d} a_{ij}(Rx)\frac{\partial^2}{\partial x_i\partial x_j}+\sum_{i=1}^{d} Rb_i(Rx)\frac{\partial}{\partial x_i}+R^2V(Rx)\text{ on }R^d,$$

and define $u_{i,R}(x)=u_i(Rx)$, $i=1,2$. Then $L_Ru_{i,R}=0$ in R^d. We now apply Harnack's inequality (Theorem 4.0.1) to L_R with $D=R^d$ and $D'=\{x\in R^d\colon 1<|x|<2\}$. By the assumption on L, it follows that the number γ appearing in Harnack's inequality may be chosen independently of R. Thus, it follows from Harnack's inequality that there exists a constant c, independent of R, such that $u_{i,R}(x) \leq c u_{i,R}(y)$, for $x, y \in D'$ and $i = 1, 2$. That is,

$$u_i(x) \leq c u_i(y), \text{ if } R < |x| < 2R \text{ and } R < |y| < 2R,$$
$$\text{for } i = 1, 2 \text{ and } R > 0. \quad (3.1)$$

For $n = 1, 2, \ldots$, choose $x_n \in R^d$ with $|x_n| = n$. Then either $u_1(x_n) \leq u_2(x_n)$, for infinitely many n, or $u_2(x_n) \leq u_1(x_n)$, for infinitely many n. Assume, for example, that $u_1(x_n) \leq u_2(x_n)$, for infinitely many n. By (3.1), it follows that if $u_1(x_n) \leq u_2(x_n)$, then $u_1(x) \leq c^2 u_2(x)$, for all x satisfying $|x| = |x_n|$. Thus, we conclude that there exists an increasing sequence $\{R_n\}_{n=1}^{\infty}$ satisfying $\lim_{n\to\infty} R_n = \infty$ such that

$$u_1(x) \leq c^2 u_2(x), \text{ if } |x| = R_n,\ n = 1, 2, \ldots. \quad (3.2)$$

Let $B_{R_n} = \{x \in R^d\colon |x| < R_n\}$. Define $\varepsilon_n = \sup\{\varepsilon > 0\colon u_2(x) - \varepsilon u_1(x) > 0, \text{ for } x \in B_{R_n}\}$ and $w_n = u_2 - \varepsilon_n u_1$ on B_{R_n}. Since $L w_n = 0$ and $w_n \geq 0$ in B_{R_n}, it follows from the strong maximum principle (Theorem 3.2.6) that either $w_n \equiv 0$ in B_{R_n} or $w_n > 0$ in B_{R_n}. If $w_n \equiv 0$ in B_{R_n} for all n, then $u_1 = k u_2$ on R^d and we are done. Otherwise, for all large n, we have $w_n > 0$ in B_{R_n}. By the definition of ε_n, it follows that there exists a y_n with $|y_n| = R_n$ such that $w_n(y_n) = 0$; that is,

$$\varepsilon_n = \frac{u_2(y_n)}{u_1(y_n)}.$$

From (3.2), it then follows that

$$\frac{1}{c^2} \leq \varepsilon_n, \text{ for } n = 1, 2, \ldots.$$

Thus, $u_1 \leq c^2 u_2$ on R^d. □

Theorem 3.2. *Let L satisfy Assumption F and Assumption* $\mathrm{H_{loc}}$ *on R^d. If $C_{\tilde{L}}(R^d) \neq \phi$, then, $C_{\tilde{L}}(R^d)$ is one-dimensional. In particular, if L is subcritical, then the Martin boundary of $\tilde{L}$ consists of a single point.*

Proof. As in the proof of Theorem 3.1, we may assume that L is subcritical. Let $\Omega = \{x \in R^d\colon |x| < 1\}$ and let $D = R^d$. By Theorem 7.6.2′, it suffices to show that if $h_1, h_2 \in C^{2,\alpha}(D - \bar{\Omega}) \cap C(D - \Omega)$ are positive solutions of minimal growth at ∂D for L on D, then

$$\lim_{|x|\to\infty} \frac{h_2}{h_1}(x)$$

exists. Since $h_i > 0$ on $\{|x| = 2\}$, $i = 1, 2$, it follows from the maximum principle for positive solutions of minimal growth (Theorem 7.3.6) that

$$\frac{1}{c}h_1(x) \leq h_2(x) \leq ch_1(x),$$

for $|x| > 2$ and some $c > 0$. Thus, the theorem follows from Theorem 3.3. below. □

Theorem 3.3. *Let L satisfy Assumption F and Assumption H_{loc} on R^d, and let $\Omega = \{x \in R^d: |x| < 1\}$. Let $h_1, h_2 \in C^{2,\alpha}(R^d - \bar{\Omega}) \cap C(R^d - \Omega)$ and assume that $Lh_i = 0$ and $h_i > 0$ in $R^d - \bar{\Omega}$, $i = 1, 2$. Assume that*

$$A \equiv \liminf_{|x|\to\infty} \frac{h_2}{h_1}(x) > 0.$$

Then

$$\lim_{|x|\to\infty} \frac{h_2}{h_1}(x) = A.$$

Proof. Let

$$a_n = \inf_{x=n} \frac{h_2}{h_1}(x)$$

and

$$b_n = \sup_{|x|=n} \frac{h_2}{h_1}(x), \text{ for } n = 2, 3, \ldots.$$

Let

$$w = \frac{h_2}{h_1}$$

and note that $L^{h_1}w = 0$. Since the zeroth-order term of L^{h_1} vanishes, it follows from the maximum principle (Theorem 3.2.1) that

$$\min(a_k, a_l) \leq w(x) \leq \max(b_k, b_l), \text{ for } k \leq |x| \leq l. \tag{3.3}$$

By definition, $\liminf_{k\to\infty} a_k \geq A$, and by (3.3), $\liminf_{k\to\infty} a_k \leq A$; thus $\liminf_{k\to\infty} a_k = A$. If $\limsup_{k\to\infty} a_k > A$, then it follows easily from (3.3) that $\liminf_{|x|\to\infty} w(x) > A$, a contradiction. Thus, we conclude that $\lim_{k\to\infty} a_k = A$. In the light of (3.3), to complete the proof, it suffices to show that

$$\lim_{k\to\infty} b_k = A. \tag{3.4}$$

The proof of (3.4) can be broken down into two cases. First, assume that there exists a $k_0 \geqslant 2$ such that $a_{k_0} > A$. Then, for some $j_0 \geqslant k_0$, $a_{j_0+1} < a_{j_0}$. It then follows by the maximum principle that $a_{j+1} < a_j$, for some $j \geqslant j_0$. Thus, from (3.3) and the strong maximum principle (Theorem 3.2.6), we have $h_2 - a_{j+2}h_1 > 0$ in $\{j \leqslant |x| < j+2\}$, if $j \geqslant j_0$. A proof just like that of (3.1) shows that there exists a constant c, independent of j, such that

$$(h_2 - a_{j+2}h_1)(x) \leqslant c(h_2 - a_{j+2}h_1)(y), \text{ for } |x| = |y| = j+1 \text{ and } j \geqslant j_0, \tag{3.5}$$

and

$$h_i(x) \leqslant ch_i(y), \text{ for } |x| = |y| = j+1,\ j \geqslant j_0 \text{ and } i = 1, 2. \tag{3.6}$$

Using (3.5) and the definition of b_{j+1}, we have

$$(b_{j+1} - a_{j+2}) \inf_{|x|=j+1} h_1(x) \leqslant c(a_{j+1} - a_{j+2}) \sup_{|y|=j+1} h_1(y), \text{ for } j \geqslant j_0. \tag{3.7}$$

From (3.6) and (3.7), it follows that $(b_{j+1} - a_{j+2}) \leqslant c^2(a_{j+1} - a_{j+2})$, for $j \geqslant j_0$. Since $\lim_{j\to\infty} a_j = A$, we conclude that $\lim_{j\to\infty} b_j = A$. This proves (3.4) in the first case.

Now assume that $a_k \leqslant A$, for $k \geqslant 2$. In the proof of the first case, we noted that if $a_{j_0+1} < a_{j_0}$, for some $j_0 \geqslant 2$, then in fact $a_{j+1} < a_j$, for all $j \geqslant j_0$. Thus, since $\lim_{k\to\infty} a_k = A$, we conclude that in the present case, a_k is non-decreasing in k. Thus, from (3.3), we have $w \geqslant a_k$ in $\{|x| \geqslant k\}$. By the strong maximum principle, either $w = a_k$ in $\{|x| > k\}$ or $w > a_k$ in $\{|x| > k\}$. In the former case, it is clear that (3.4) holds. Thus, assume that $w > a_k$ in $\{|x| > k\}$. Then, as in (3.5), Harnack's inequality gives

$$(h_2 - a_k h_1)(x) \leqslant c(h_2 - a_k h_1)(y), \text{ for } |x| = |y| = k+1 \text{ and } k \geqslant 2. \tag{3.8}$$

Using (3.8) and the definition of b_{k+1} gives

$$(b_{k+1} - a_k) \inf_{|x|=k+1} h_1(x) \leqslant c(a_{k+1} - a_k) \sup_{|y|=k+1} h_1(y),\ k \geqslant 2. \tag{3.9}$$

From (3.6) and (3.9), it follows that $(b_{k+1} - a_k) \leqslant c^2(a_{k+1} - a_k)$, $k \geqslant 2$; thus, as before $\lim_{k\to\infty} b_k = A$. □

8.4 Auxiliary results I: explosion inward from the boundary

The results of this section and the next one will be used to calculate Martin boundaries in Section 6; however, they are also of independent interest. Let L be defined on a domain $D \subseteq R^d$ and satisfy Assumption A which appears at the beginning of Chapter 2. Let $\{D_n\}_{n=1}^{\infty}$ be a sequence of domains satisfying $D_n \subset\subset D_{n+1}$ and $\bigcup_{n=1}^{\infty} D_n = D$. Let $\Omega \subset\subset D$ be a subdomain and define $\sigma_{\bar{\Omega}} = \inf\{t \geq 0: X(t) \in \bar{\Omega}\}$. Consider the following condition:

There exists a sequence $\{x_n\}_{n=1}^{\infty} \subset D$ with $x_n \in D - D_n$ such that

$$\liminf_{n\to\infty} P_{x_n}(\sigma_{\bar{\Omega}} \leq t) > 0, \text{ for some } t > 0. \quad (4.1)$$

It is clear that (4.1) does not depend on the choice of Ω. If (4.1) holds, we will say that the diffusion *explodes inward from the boundary*. (Although we will not need this fact, we note that the semigroup corresponding to the diffusion preserves continuous functions f 'vanishing at infinity' (that is, $\lim_{n\to\infty} f(x_n) = 0$, if $x_n \in D - D_n$) if and only if the diffusion does not explode inward from the boundary. In the probability literature, at least in the one-dimensional case when $D = (\alpha, \beta)$, with $-\infty \leq \alpha < \beta \leq \infty$, if (4.1) holds as $x_n \to \alpha$ (or as $x_n \to \beta$), then α (or β) is called an *entrance boundary*. See the notes at the end of the chapter.)

Remark. If a diffusion explodes inward from the boundary, then it is easy to see that the solution $\{P_x, x \in \hat{D}\}$ to the corresponding generalized martingale problem is not Feller continuous at $x = \Delta$, the point at infinity in the one-point compactification of D. See exercise 8.6.

We now develop an integral criterion for explosion inward from the boundary for one-dimensional diffusions. Let

$$L = \tfrac{1}{2}a\frac{\mathrm{d}^2}{\mathrm{d}x^2} + b\frac{\mathrm{d}}{\mathrm{d}x}$$

on $D = (\alpha, \beta)$, where $-\infty \leq \alpha < \beta \leq \infty$. Let $\tau_x = \inf\{t \geq 0: X(t) = x\}$. Then the diffusion explodes inward from infinity if and only if either $\lim_{x\to\alpha} P_x(\tau_y \leq t) > 0$ or $\lim_{x\to\beta} P_x(\tau_y \leq t) > 0$, for some (or, equivalently, all) $y \in D$ and some $t > 0$. Actually, it is easy to show that if this holds for some $t > 0$, then it holds for all $t > 0$. In Chapter 5, we presented Feller's test for explosion which gives an integral test at the end point α (β) to determine whether $\lim_{y\to\alpha} P_x(\tau_y \leq t) > 0$ ($\lim_{y\to\beta} P_x(\tau_y \leq t) > 0$), for some (or, equivalently, all) $x \in D$ and $t > 0$. The following theorem gives an integral test at α (β) to determine whether $\lim_{x\to\alpha} P_x(\tau_y \leqslant t) > 0$ ($\lim_{x\to\beta} P_x(\tau_y \leqslant t) > 0$). The integral test and its proof are similar in form to Feller's test and its proof.

Theorem 4.1. *Let*

$$L = \tfrac{1}{2}a\frac{d^2}{dx^2} + b\frac{d}{dx} \text{ on } D = (\alpha, \beta),$$

where $-\infty \leqslant \alpha < \beta \leqslant \infty$, *and assume that L satisfies Assumption A. Let* $x_0 \in (\alpha, \beta)$. *Then* $\lim_{x\to\alpha} P_x(\tau_y \leqslant t) > 0$, *for* $y \in (\alpha, \beta)$ *and some* $t > 0$ *if and only if the following two conditions hold:*

(i) $\displaystyle\int_\alpha^{x_0} \exp\left(-\int_{x_0}^x \frac{2b}{a}(s)\,ds\right) dx = \infty;$

(ii) $\displaystyle\int_\alpha^{x_0} dx \frac{1}{a(x)} \exp\left(\int_{x_0}^x \frac{2b}{a}(s)\,ds\right) \int_x^{x_0} dy \exp\left(-\int_{x_0}^y \frac{2b}{a}(s)\,ds\right) < \infty.$

If (i) *and* (ii) *hold, then in fact* $\lim_{x\to\alpha} E_x\tau_y < \infty$, *for* $y \in (\alpha, \beta)$, *and* $\lim_{y\to\alpha} \lim_{x\to\alpha} E_x\tau_y = 0$.

Similarly, $\lim_{x\to\beta} P_x(\tau_y \leqslant t) > 0$, *for* $y \in (\alpha, \beta)$ *and some* $t > 0$ *if and only if the following two conditions hold:*

(iii) $\displaystyle\int_{x_0}^\beta \exp\left(-\int_{x_0}^x \frac{2b}{a}(s)\,ds\right) dx = \infty;$

(iv) $\displaystyle\int_{x_0}^\beta dx \frac{1}{a(x)} \exp\left(\int_{x_0}^x \frac{2b}{a}(s)\,ds\right) \int_{x_0}^x dy \exp\left(-\int_{x_0}^y \frac{2b}{a}(s)\,ds\right) < \infty.$

If (iii) *and* (iv) *hold, then in fact* $\lim_{x\to\beta} E_x\tau_y < \infty$, for $y \in (\alpha, \beta)$, *and* $\lim_{y\to\beta} \lim_{x\to\beta} E_x\tau_y = 0$.

Proof. We will assume that $b \in C((\alpha, \beta))$. The extension to the case that b is measurable and locally bounded is similar to the extensions covered in exercise 5.1. We will prove the part of the theorem regarding the endpoint β; the method of proof regarding the endpoint α is, of course, identical.

By Theorem 5.1.1, it follows that if (iii) does not hold, then $\lim_{x\to\beta} P_x(\tau_y < \infty) = 0$; thus (iii) certainly is a necessary condition in order that $\lim_{x\to\beta} P_x(\tau_y \leqslant t) > 0$, for $y \in (\alpha, \beta)$ and $t > 0$. In exercise 8.7, the reader is asked to show that condition (iv) is equivalent to the following condition:

(iv)′ $\displaystyle\int_{x_0}^\beta dx \exp\left(-\int_{x_0}^x \frac{2b}{a}(s)\,ds\right) \int_x^\beta dy \frac{1}{a(y)} \exp\left(\int_{x_0}^y \frac{2b}{a}(s)\,ds\right) < \infty.$

We now assume that (iii) holds and show that (iv)′ holds if and only if $\lim_{x\to\beta} P_x(\tau_y \le t) > 0$, for $y \in (\alpha, \beta)$ and $t > 0$.

Fix $y \in (\alpha, \beta)$ and let $\{\beta_n\}_{n=1}^{\infty} \in (y, \beta)$ be a sequence increasing to β. Let $u_n(x) = E_x(\exp(-\tau_y); \tau_y < \tau_{\beta_n})$. By Theorem 3.6.6, u_n satisfies $(L-1)u_n = 0$ in (y, β_n), $u_n(y) = 1$ and $u_n(\beta_n) = 0$. Integrating twice and using the boundary conditions, we obtain

$$u_n(x) = 1 + c_n \int_y^x \exp\left(-\int_y^z \frac{2b}{a}(r)\,\mathrm{d}r\right)\mathrm{d}z$$
$$+ \int_y^x \mathrm{d}z \exp\left(-\int_y^z \frac{2b}{y}(r)\,\mathrm{d}r\right)\int_y^z \mathrm{d}s \frac{2u_n(s)}{a(s)} \exp\left(\int_y^s \frac{2b}{a}(r)\,\mathrm{d}r\right),$$

where

$$c_n = \frac{-1 - \displaystyle\int_y^{\beta_n} \mathrm{d}x \exp\left(-\int_y^x \frac{2b}{a}(r)\,\mathrm{d}r\right)\int_y^x \mathrm{d}z \frac{2u_n(z)}{a(z)} \exp\left(\int_y^z \frac{2b}{a}(r)\,\mathrm{d}r\right)}{\displaystyle\int_y^{\beta_n} \exp\left(-\int_y^x \frac{2b}{a}(r)\,\mathrm{d}r\right)\mathrm{d}x}.$$

Now u_n is monotone decreasing in n and $u(x) \equiv \lim_{n\to\infty} u_n(x) = E_x \exp(-\tau_y);\ \tau_y < \tau_\beta) = E_x \exp(-\tau_y)$, since by (iii) and Theorem 5.1.1, $P_x(\tau_y < \tau_\beta) = 1$. Thus, $\lim_{x\to\beta} P_x(\tau_y \le t) > 0$, for some $t > 0$, if and only if $\inf_{x\in(y,\beta)} u(x) = \lim_{x\to\beta} u(x) > 0$.

Assume first that $l = \inf_{x\in(y,\beta)} u(x) > 0$. Then, in the light of (iii), we have

$$\limsup_{n\to\infty} c_n \le -2l \int_y^\beta \mathrm{d}x \frac{1}{a(x)} \exp\left(\int_y^x \frac{2b}{a}(s)\,\mathrm{d}s\right).$$

Therefore,

$$\int_y^\beta \mathrm{d}x \frac{1}{a(x)} \exp\left(\int_y^x \frac{2b}{a}(s)\,\mathrm{d}s\right) < \infty,$$

since otherwise we would obtain $u(x) = -\infty$, for all $x \in (\alpha, \beta)$, which is impossible since $0 \le u \le 1$. Using this fact and the dominated convergence theorem, it follows that

$$\lim_{n\to\infty} c_n = -\int_y^\beta \mathrm{d}x \frac{2u(x)}{a(x)} \exp\left(\int_y^x \frac{2b}{a}(s)\,\mathrm{d}s\right),$$

and then that

$$u(x) = 1 - \int_y^x \mathrm{d}z \exp\left(-\int_y^z \frac{2b}{a}(s)\,\mathrm{d}s\right)\int_z^\beta \mathrm{d}r \frac{2u(r)}{a(r)} \exp\left(\int_y^r \frac{2b}{a}(s)\,\mathrm{d}s\right). \tag{4.2}$$

Since $\inf_{x\in(y,\beta)} u(x) > 0$, it follows from (4.2) that (iv)′ holds.

Now assume that (iv)′ holds. Let v_n solve $Lv_n = -1$ in (y, β_n) and $v_n(y) = v_n(\beta_n) = 0$. Then

$$v_n(x) = c_n \int_y^x \exp\left(-\int_y^z \frac{2b}{a}(s)\,ds\right) dz - \int_y^x dz \exp\left(-\int_y^z \frac{2b}{a}(s)\,ds\right) \int_y^z dr \frac{2}{a(r)} \exp\left(\int_y^r \frac{2b}{a}(s)\,ds\right)$$

where

$$c_n = \frac{\displaystyle\int_y^{\beta_n} dx \exp\left(-\int_y^x \frac{2b}{a}(s)\,ds\right) \int_y^x dz \frac{2}{a(z)} \exp\left(\int_y^z \frac{2b}{a}(s)\,ds\right)}{\displaystyle\int_y^{\beta_n} \exp\left(-\int_y^x \frac{2b}{a}(s)\,ds\right) dx}.$$

By Theorem 2.1.2, $v_n(x) = E_x\tau_y \wedge \tau_{\beta_n}$. Let $\tau_\beta = \lim_{n\to\infty} \tau_{\beta_n}$ and define $v(x) = \lim_{n\to\infty} v_n(x) = E_x\tau_y \wedge \tau_\beta$. By (iii) and Theorem 5.1.1, $P_x(\tau_y < \tau_\beta) = 1$; thus $v(x) = E_x\tau_y$. In the light of (iii) and (iv)′, we have

$$\lim_{n\to\infty} c_n = \int_y^\beta \frac{2}{a(x)} \exp\left(\int_y^x \frac{2b}{a}(s)\,ds\right) dx < \infty.$$

Thus,

$$E_x\tau_y = \int_y^x dz \exp\left(-\int_y^z \frac{2b}{a}(s)\,ds\right) \int_z^\beta dr \frac{2}{a(r)} \exp\left(\int_y^r \frac{2b}{a}(s)\,ds\right)$$

and, by (iv)′ it follows that $\lim_{x\to\beta} E_x\tau_y < \infty$ and that $\lim_{y\to\beta} \lim_{x\to\beta} E_x\tau_y = 0$. □

The following result concerns explosion inward from the boundary in the case of a critical operator on a smooth bounded domain.

***Theorem* 4.2.** *Let L satisfy Assumption $\tilde{H}$ on a bounded domain $D \subset R^d$ with a $C^{2,\alpha}$-boundary. Let λ_0 denote the principal eigenvalue of L on D, and let $\phi_0 \in C^{2,\alpha}(\bar{D})$ denote the corresponding positive eigenfunction. By Theorems 4.3.2 and 4.3.3, the diffusion corresponding to $(L-\lambda_0)^{\phi_0}$ on D is recurrent; let $\{P_x^{\phi_0,\lambda_0};\ x \in D\}$ denote the solution to the martingale problem for $(L-\lambda_0)^{\phi_0}$. Let $\Omega \subset\subset D$ and let $\sigma_{\bar{\Omega}} = \inf\{t \geq 0: X(t) \in \bar{\Omega}\}$. Then $\sup_{x\in D-\Omega} E_x^{\phi_0,\lambda_0} \sigma_{\bar{\Omega}} < \infty$.*

Proof. We may assume that Ω has a $C^{2,\alpha}$-boundary. Let $\lambda_0(D-\bar{\Omega})$ denote the principal eigenvalue of L on $D-\bar{\Omega}$. By the last part of

Theorem 4.3.2, $L - \lambda_0(D - \bar{\Omega})$ is critical on $D - \bar{\Omega}$. Thus, by Theorem 4.4.1 (iii), $\lambda_0(D - \bar{\Omega}) < \lambda_0$. Then, by Theorem 3.6.5, there exists a positive solution $v \in C^{2,\alpha}(\bar{D} - \Omega)$ to $(L - \lambda_0)v = -\phi_0$ in $D - \bar{\Omega}$ and $v = 0$ on $\partial\Omega \cup \partial D$. Let

$$\psi(x) = \frac{v(x)}{\phi_0(x)}, \text{ for } x \in D - \Omega.$$

By the Hopf maximum principle (Theorem 3.2.5), ψ is bounded on $D - \Omega$ and extends continuously to $\bar{D} - \Omega$. We have $(L - \lambda_0)^{\phi_0}\psi = -1$ in $D - \bar{\Omega}$ and $\psi \geq 0$ in $\bar{D} - \Omega$. Since the diffusion corresponding to $(L - \lambda_0)^{\phi_0}$ on D is recurrent, $P_x^{\phi_0,\lambda_0}(\sigma_{\bar{\Omega}} < \infty) = 1$, for $x \in D - \Omega$. Now by the method of Theorem 2.1.2, it follows that $\psi(x) \geq E_x^{\phi_0,\lambda_0}\sigma_{\bar{\Omega}}$. Thus $\sup_{x \in D - \Omega} E_x^{\phi_0,\lambda_0}\sigma_{\bar{\Omega}} \leq \sup_{x \in D - \Omega} \psi(x) < \infty$. □

8.5 Auxiliary results II: operators in skew-product form

Let $k \geq 2$ be a positive integer and for each $i \in \{1, 2, \ldots, k\}$, let $L^{(i)} = L_0^{(i)} + V^{(i)}$ satisfy Assumption $\widetilde{H}_{\text{loc}}$ on a domain $D^{(i)} \subseteq R^{d_i}$. Let $\hat{D}^{(i)} = D^{(i)} \cup \{\Delta_i\}$ denote the one-point compactification of $D^{(i)}$, let $\{P_x^{(i)}, x \in \hat{D}^{(i)}\}$ denote the solution to the generalized martingale problem for $L_0^{(i)}$ on $D^{(i)}$, and denote the canonical diffusion on $\hat{\Omega}_{D^{(i)}}$ by $X_i(t)$. Let $D = \times_{i=1}^k D^{(i)}$ and denote the points $x \in D$ by $x = (x_1, \ldots, x_k)$, where $x_i \in D^{(i)}$. Let $\hat{D} = D \cup \{\Delta\}$ denote the one-point compactification of D. In the sequel, we will identify all the points in $\times_{i=1}^k \hat{D}^{(i)} - \times_{i=1}^k D^{(i)}$ with Δ. With this convention, we may write $\hat{D} = \times_{i=1}^k \hat{D}^{(i)}$ and $\hat{\Omega}_D = \times_{i=1}^k \hat{\Omega}_{D^{(i)}}$. For $i \in \{2, \ldots, k\}$, let $M^{(i)} \in C^\alpha(D^{(1)})$ be a positive function and define $L_0 = L_0^{(1)} + \sum_{i=2}^k M^{(i)}(x_1) L_0^{(i)}$ and $L = L^{(1)} + \sum_{i=2}^k M^{(i)}(x_1) L^{(i)} = L_0 + V$, where $V(x) = V^{(1)}(x_1) + \sum_{i=2}^k M^{(i)}(x_1) V^{(i)}(x_i)$. The operator L on the domain D will be called an operator in *skew-product form*.

Remark. In order that L satisfy Assumption $\widetilde{\text{H}}_{\text{loc}}$, we would have to assume that $M^{(i)} \in C^{1,\alpha}(D^{(1)})$, whereas above we have only assumed that $M^{(i)} \in C^\alpha(D^{(1)})$. However, the purpose of Assumption $\widetilde{\text{H}}_{\text{loc}}$ is to ensure that L and $\widetilde{L}$ satisfy Assumption H_{loc}. Indeed, the entire theory we have developed for operators satisfying Assumption $\widetilde{\text{H}}_{\text{loc}}$ goes through as long as both L and $\widetilde{L}$ satisfy Assumption H_{loc}. In the present situation, both L and $\widetilde{L}$ satisfy Assumption H_{loc}.

Let $\{P_x, x \in \hat{D}\}$ denote the solution to the generalized martingale problem for L_0 on D. The following theorem gives the *skew-product decomposition* for P_x in terms of $P_{x_i}^{(i)}$, $i = 1, 2, \ldots, k$, and $\{M^{(i)}\}_{i=2}^k$.

Theorem 5.1. *Using the notation established above and, for convenience, letting* $M^{(1)}(x) \equiv 1$, *define* $\rho_i(t) = \int_0^t M^{(i)}(X_1(s))\,ds$, $i = 1, 2, \ldots, k$, *and define* $T\colon \hat{\Omega}^D \to \hat{\Omega}^D$ *by* $TX(t) = (X_1(\rho_1(t)), \ldots, X_k(\rho_k(t)))$. *Then*

$$P_x(\cdot) = \left(\mathop{\times}_{i=1}^{k} P_{x_i}^{(i)}\right)(T\,\cdot).$$

Remark. $\{X_i(\cdot)\}_{i=1}^k$ are independent under $\times_{i=1}^k P_{x_i}^{(i)}$. Thus, Theorem 5.1 can be restated as follows. Let $\{X_i(\cdot)\}_{i=1}^k$ be independent processes defined on a probability space such that $X_i(\cdot)$ is distributed according to the solution to the generalized martingale problem for $L_0^{(i)}$. Let $\rho_i(t) = \int_0^t M^{(i)}(X_1(s))\,ds$ and note that $\{\rho_i(\cdot)\}_{i=1}^k$ and $\{X_i(\cdot)\}_{i=2}^k$ are independent. Then $(X_1(\rho_1(\cdot)), \ldots, X_k(\rho_k(\cdot)))$ is distributed according to the solution of the generalized martingale problem for $L_0 = \sum_{i=1}^k M^{(i)}(x_1)L^{(i)}$.

Proof. We will prove the theorem under the assumption that $\{P_{x_i}^{(i)}, x \in D^{(i)}\}$ solves the martingale problem for $L_0^{(i)}$, $i = 1, 2, \ldots, k$. The case that for at least one $i \in \{1, 2, \ldots, k\}$ there is no solution to the martingale problem for $L_0^{(i)}$ on $D^{(i)}$, but only to the generalized martingale problem, is left as exercise 8.17. For ease of notation, we will prove the theorem for $k = 2$; however, the same method works for any k. Let $\bar{E}_x$ denote the expectation corresponding to the measure $P_{x_1} \times P_{x_2}$, where $x = (x_1, x_2)$. We must show that for each $s > 0$, $A \in \hat{\mathcal{F}}_s^D$ and $f \in C_0^2(D)$, $\bar{E}_x(f(TX(t)) - \int_0^t L_0 f(TX(r))\,dr;\ TX(\cdot) \in A)$ is constant for $t \geq s$. It follows easily from Theorem 1.13.1 and the definition of ρ that $\bar{E}_x(f(TX(t+h)) - \int_0^{t+h} L_0 f(TX(r))\,dr;\ TX(\cdot) \in A) - \bar{E}_x(f(TX(t)) - \int_0^t L_0 f(TX(r))\,dr;\ TX(\cdot) \in A) = \bar{E}_x(f(TX(t+h)) - f(TX(t)) - \int_t^{t+h} L_0 f(TX(r))\,dr;\ TX(\cdot) \in A) = \bar{E}_x 1_A(TX(\cdot))\,\bar{E}_{TX(t)}(f(TX'(h)) - f(TX(t)) - \int_0^h L_0 f(TX'(r))\,dr)$, for $t \geq s$ and $h \geq 0$, where we have used X' to denote the variable of integration in the inner expectation. Thus, to complete the proof, we will show that for each

$$y \in D, \quad \lim_{h\to 0} \frac{1}{h} \bar{E}_y\left(f(TX(h)) - f(y) - \int_0^h L_0 f(TX(r))\,dr\right) = 0.$$

It is clear that

$$\lim_{h\to 0} \frac{1}{h} E_y \int_0^h L_0 f(TX(r))\,dr = L_0 f(y);$$

thus what remains is to show that

$$\lim_{h\to 0} \frac{1}{h} \bar{E}_y(f(TX(h)) - f(y)) = L_0 f(y). \tag{5.1}$$

Since any $f \in C_0^2(D)$ can be uniformly approximated by finite sums of functions of the form $g_1 g_2$, where $g_i \in C_0^2(D^{(i)})$, it is enough to show (5.1) with $f = g_1 g_2$. Since $\bar{E}_y(g_i(X_i(t)) - \int_0^t L_0^{(i)} g_i(X(r))\, dr) = g_i(y_i)$, for $t \geq 0$, we have

$$\lim_{h\to 0} \frac{1}{h} \bar{E}_y(g_i(X_i(h)) - g_i(y_i)) = L_0^{(i)} g_i(y_i), \quad i = 1, 2. \tag{5.2}$$

Also,

$$\rho_2'(0) = M^{(2)}(y_1) \text{ a.s. } P_{y_1} \times P_{y_2}. \tag{5.3}$$

Recalling that $\hat{\mathcal{F}}^{D^{(1)}}$ and $\hat{\mathcal{F}}^{D^{(2)}}$ are independent under $P_{y_1} \times P_{y_2}$ and that $\rho_2(\cdot)$ is $\hat{\mathcal{F}}^{D^{(1)}}$-measurable, and using (5.2) and (5.3), the lefthand side of (5.1) with $f = g_1 g_2$ can be written as

$$\lim_{h\to 0} \frac{1}{h} \bar{E}_y f(X(T(h)) - f(y))$$

$$= \lim_{h\to 0} \frac{1}{h} \bar{E}_y(g_1(X_1(h)) g_2(X_2(\rho_2(h))) - g_1(y_1) g_2(y_2))$$

$$= \lim_{h\to 0} \frac{1}{h} \bar{E}_y g_1(X_1(h)) \bar{E}_y(g_2(X_2(\rho_2(h))) - g_2(y_2) | \hat{\mathcal{F}}^{D^{(1)}})$$

$$+ g_2(y_2) \lim_{h\to 0} \frac{1}{h} \bar{E}_y(g_1(X_1(h)) - g_1(y_1)) = g_1(y_1) M^{(2)}(y_1) L_0^{(2)} g_2(y_2)$$

$$+ g_2(y_2) L_0^{(1)} g_1(y_1) = L_0 f(y). \qquad \square$$

The next theorem shows that for a certain class of operators in skew-product form, the criticality problem can be appropriately 'decomposed'.

Theorem 5.2. *Let L be an operator in skew-product form as in Theorem 5.1. Assume in addition that for $i \in \{2, \ldots, k\}$, $D^{(i)}$ is bounded with a $C^{2,\alpha}$-boundary and that L satisfies Assumption $\widetilde{H}$ on D. Let $\lambda_0^{(i)}$ denote the principal eigenvalue for $L^{(i)}$ on $D^{(i)}$. Then L is subcritical (critical, supercritical) on D if and only if $L^{(1)} + \sum_{i=2}^k \lambda_0^{(i)} M^{(i)}$ is subcritical (critical, supercritical) on $D^{(1)}$.*

Proof. By the last part of Theorem 4.3.2, $L^{(i)} - \lambda_0^{(i)}$ is critical on $D^{(i)}$, $i \in \{2, \ldots, k\}$; let $\phi_0^{(i)}$ denote the corresponding positive eigenfunction. Assume first that $L^{(1)}+\sum_{i=2}^{k}\lambda_0^{(i)}M^{(i)}$ is subcritical on $D^{(1)}$. Then, by Theorem 4.3.9, there exists an $h\in C^{2,\alpha}(D^{(1)})$ such that $h>0$, $(L^{(1)}+\sum_{i=2}^{k}\lambda_0^{(i)}M^{(i)})h\leqslant 0$ and $(L^{(1)}+\sum_{i=2}^{k}\lambda_0^{(i)}M^{(i)})h\not\equiv 0$. Let $u(x)=h(x_1)\prod_{i=2}^{k}\phi_0^{(i)}(x_i)$. Then $u>0$, $Lu\leqslant 0$ and $Lu\not\equiv 0$ on D; thus, by Theorem 4.3.9, L is subcritical on D.

Now assume that $L^{(1)}+\sum_{i=2}^{k}\lambda_0^{(i)}M^{(i)}$ is supercritical on $D^{(1)}$. Then the generalized principal eigenvalue for $L^{(1)}+\sum_{i=2}^{k}\lambda_0^{(i)}M^{(i)}$ on $D^{(1)}$ is positive and, by Theorem 4.4.1(i), we can choose a domain $U\subset\subset D^{(1)}$ with a $C^{2,\alpha}$-boundary such that the principal eigenvalue for $L^{(1)}+\sum_{i=2}^{k}\lambda_0^{(i)}M^{(i)}$ on U is also positive. Let $\gamma_0>0$ denote the principal eigenvalue for $L^{(1)}+\sum_{i=2}^{k}\lambda_0^{(i)}M^{(i)}$ on U and let ψ_0 denote the corresponding positive eigenfunction. Let $v_0=\psi_0(x_1)\prod_{i=2}^{k}\phi_0^{(i)}(x_i)$. Then $v_0\in C_{L-\gamma_0}(U\times D^{(2)}\times\ldots\times D^{(k)})$ and v_0 vanishes continously on $\partial(U\times D^{(2)}\times\ldots\times D^{(k)})$. Thus, by Theorem 4.7.8(ii), $L-\gamma_0$ is critical on $U\times D^{(2)}\times\ldots\times D^{(k)}$. Now suppose that L is not supercritical on D. Let $u\in C_L(D)$. Since $(L-\gamma_0)u=-\gamma_0u<0$ on $U\times D^{(2)}\times\ldots\times D^{(k)}$, it follows from Theorem 4.3.9 that $L-\gamma_0$ is subcritical on $U\times D^{(2)}\times\ldots\times D^{(k)}$, a contradiction. Thus, L is supercritical on D.

Now assume that $L^{(1)}+\sum_{i=2}^{k}\lambda_0^{(i)}M^{(i)}$ is critical on D. Let $\phi_c^{(1)}$ denote the ground state for $L^{(1)}+\sum_{i=2}^{k}\lambda_0^{(i)}M^{(i)}$ on D and let $h(x)=\phi_c^{(1)}(x_1)\prod_{i=2}^{k}\phi_0^{(i)}(x_i)$. A simple calculation reveals that $L^h=(L^{(1)}+\sum_{i=2}^{k}\lambda_0^{(i)}M^{(i)})^{\phi_c^{(1)}}+\sum_{i=2}^{k}M^{(i)}(L^{(i)}-\lambda_0^{(i)})^{\phi_0^{(i)}}$. By Theorem 4.3.3, in order to show that L is critical, it suffices to show that the diffusion corresponding to L^h is recurrent. Let $\{P_x^h,\ x\in\hat{D}\}$ denote the solution to the generalized martingale problem for L^h on D. Let $\Omega^{(i)}\subset\subset D^{(i)}$, $i=1,2,\ldots,k$, and let $\Omega=\times_{i=1}^{k}\Omega^{(i)}\subset\subset D$. Define $\sigma_{\bar{\Omega}}=\inf\{t\geqslant 0\colon X(t)\in\bar{\Omega}\}$. By Theorem 2.8.2, the recurrence of the diffusion corresponding to L^h will follow if we show that

$$P_x^h(\sigma_{\bar{\Omega}}<\infty)=1,\quad \text{for } x\in D-\Omega. \tag{5.4}$$

The diffusion corresponding to $(L^{(1)}+\sum_{i=2}^{k}\lambda_0^{(i)}M^{(i)})^{\phi_c^{(1)}}$ is recurrent since by assumption $L^{(1)}+\sum_{i=2}^{k}\lambda_0^{(i)}M^{(i)}$ is critical. For convenience of notation, denote the solution of the martingale problem for $(L^{(1)}+\sum_{i=2}^{k}\lambda_0^{(i)}M^{(i)})^{\phi_c^{(1)}}$ on $D^{(1)}$ by $\{Q_{x_1}^{(1)},\ x_1\in D^{(1)}\}$ and for $(L^{(i)}-\lambda_0^{(i)})^{\phi_0^{(i)}}$ on $D^{(i)}$ by $\{Q_{x_i}^{(i)},\ x_i\in D^{(i)}\}$, $i=2,\ \ldots,\ k$. Let $U\subset\subset\Omega^{(1)}$. Then, by recurrence,

$$Q_{x_1}^{(1)}(X(t)\in U \text{ for arbitrarily large } t)=1,\quad \text{for } x_1\in D^{(1)}. \tag{5.5}$$

Let $\tau_{\Omega^{(1)}}=\inf\{t\geqslant 0\colon X_1(t)\notin\Omega^{(1)}\}$. By exercise 3.4,

$$\inf_{x_1\in U}Q_{x_1}^{(1)}(\tau_{\Omega^{(1)}}\geqslant t)>0,\quad \text{for all } t>0. \tag{5.6}$$

Let $\sigma_{\bar{\Omega}^{(i)}} = \inf\{t \geq 0: X_i(t) \in \bar{\Omega}^{(i)}\}$, $i = 2, \ldots, k$. By Theorem 4.2, it follows that for some $t_0 > 0$,

$$\inf_{x_i \in D^{(i)} - \Omega^{(i)}} Q_{x_i}^{(i)}(\sigma_{\bar{\Omega}^{(i)}} \leq t_0) > 0, \quad i = 2, \ldots, k. \tag{5.7}$$

By Theorem 5.1,

$$P_x^h(\cdot) = \mathop{\times}_{i=1}^{k} Q_{x_i}^{(i)}(T\cdot), \tag{5.8}$$

where $TX(t) = (X_1(\rho_1(t)), \ldots, X_k(\rho_k(t)))$, $\rho_1(t) = t$, and $\rho_i(t) = \int_0^t M^{(i)}(X_1(s))\,ds$, $i = 2, \ldots, k$. Now (5.4) follows easily from (5.5)–(5.8), the strong Markov property, and that fact that $M^{(i)}$ is bounded away from zero on U. □

8.6 The Martin boundary for a class of operators in skew-product form: statement of general theorem, applications to particular cases and comparison principles

In this section we undertake a study of the Martin boundary for a certain subclass of the class of operators in skew-product form. Because we are striving for some generality here, the assumptions and the statement of the main theorem, Theorem 6.1, are rather detailed. However, following the statement of the theorem, we state and prove some applications to particular cases and obtain a number of simply stated results which illustrate the theorem's scope. The quite technical proof of Theorem 6.1 is given in Section 7. The following assumption will be made in this section.

Assumption SP. *Let $k \geq 2$ be an integer. Let*

$$L^{(1)} = \tfrac{1}{2}a(x_1)\frac{d^2}{dx_1^2} + b(x_1)\frac{d}{dx_1} + V^{(1)}(x_1)$$

satisfy Assumption $\widetilde{H}_{\text{loc}}$ on a domain $D^{(1)} = (\alpha, \beta)$, where $-\infty \leq \alpha < \beta \leq \infty$. For $i \in \{2, \ldots, k\}$, let $L^{(i)} = L_0^{(i)} + V^{(i)}$ satisfy Assumption $\widetilde{H}$ on a domain $D^{(i)}$, where either $D^{(i)} \subset R^{d_i}$ is bounded with a $C^{2,\alpha}$-boundary or $D^{(i)}$ is a d_i-dimensional compact manifold without boundary (see Chapter 4 Section 11). Let $M^{(i)} \in C^\alpha(D^{(i)})$ satisfy $M^{(i)} > 0$, $i = 2, \ldots, k$. Define

$$L = L_0 + V = L^{(1)} + \sum_{i=2}^{k} M^{(i)}(x_1)L^{(i)} \text{ and } D = \mathop{\times}_{i=1}^{k} D^{(i)},$$

and assume that L is subcritical on D. Denote points $x \in D$ by $x = (x_1, \ldots, x_k)$, where $x_i \in D^{(i)}$. Let $\lambda_0^{(i)}$ denote the principal eigenvalue of $L^{(i)}$ on $D^{(i)}$ and let $\phi_0^{(i)}$ denote the corresponding positive eigenfunction. Define

$$\mathcal{L} = L^{(1)} + \sum_{i=2}^{k} \lambda_0^{(i)} M^{(i)}(x_1) \text{ on } D^{(1)} = (\alpha, \beta).$$

By Theorem 5.2, $\mathcal{L}$ is subcritical on $D^{(1)}$. By Corollary 7.6.7, the Martin boundary of $\mathcal{L}$ on $D^{(1)} = (\alpha, \beta)$ may be identified with $\{\alpha, \beta\}$; let g_α and g_β denote the corresponding elements of $C_{\mathcal{L}}((\alpha, \beta))$.

Remark. The functions g_α and g_β can be represented probabilistically as in exercise 7.7.

Fix $r_0 \in D_1 = (\alpha, \beta)$. For positive functions $g, M \in C((\alpha, \beta))$, define the following integrals:

$$\left.\begin{aligned}
&I_\alpha(g, M) = \\
&\quad \int_\alpha^{r_0} \mathrm{d}r \frac{M(r)}{a(r)} g^2(r) \exp\left(\int_{r_0}^{r} \frac{2b}{a}(s)\,\mathrm{d}s\right) \int_r^{r_0} \mathrm{d}t \frac{1}{g^2(t)} \exp\left(-\int_{r_0}^{t} \frac{2b}{a}(s)\,\mathrm{d}s\right) \\
&I_\beta(g, M) = \\
&\quad \int_{r_0}^{\beta} \mathrm{d}r \frac{M(r)}{a(r)} g^2(r) \exp\left(\int_{r_0}^{r} \frac{2b}{a}(s)\,\mathrm{d}s\right) \int_{r_0}^{r} \mathrm{d}t \frac{1}{g^2(t)} \exp\left(-\int_{r_0}^{t} \frac{2b}{a}(s)\,\mathrm{d}s\right) \\
&J_\alpha(g, M) = \\
&\quad \int_\alpha^{r_0} \mathrm{d}r \frac{1}{g^2(r)} \exp\left(-\int_{r_0}^{r} \frac{2b}{a}(s)\,\mathrm{d}s\right) \int_r^{r_0} \mathrm{d}t \frac{M(t)}{a(t)} g^2(t) \exp\left(\int_{r_0}^{t} \frac{2b}{a}(s)\,\mathrm{d}s\right) \\
&J_\beta(g, M) = \\
&\quad \int_{r_0}^{\beta} \mathrm{d}r \frac{1}{g^2(r)} \exp\left(-\int_{r_0}^{r} \frac{2b}{a}(s)\,\mathrm{d}s\right) \int_{r_0}^{r} \mathrm{d}t \frac{M(t)}{a(t)} g^2(t) \exp\left(\int_{r_0}^{t} \frac{2b}{a}(s)\,\mathrm{d}s\right).
\end{aligned}\right\} \tag{6.1}$$

We now state the main theorem.

Theorem 6.1. *Let $L = L^{(1)} + \sum_{i=2}^{k} M^{(i)}(x_1) L^{(i)}$ and $D = \times_{i=1}^{k} D^{(i)}$ satisfy Assumption SP. Let $\mathcal{L}$, g_α, g_β, $\phi_0^{(i)}$ and $\lambda_0^{(i)}$ be as in Assumption SP and let I_α, I_β, J_α and J_β be as in (6.1). Fix a point $x_0 = (x_{01}, \ldots, x_{0k}) \in D$ with respect to which the Martin boundary for L on D is to be calculated and let $\{x^{(n)}\}_{n=1}^{\infty} = \{(x_1^{(n)}, \ldots, x_k^{(n)})\}_{n=1}^{\infty} \subset D$ be a sequence with no accumulation points in D.*

(a) *$\{x^{(n)}\}_{n=1}^{\infty}$ is a Martin sequence for L on D if and only if it is a Martin sequence for $\tilde{L}$ on D.*

(b) *Assume that $\{x^{(n)}\}_{n=1}^{\infty}$ has at least one accumulation point in $D^{(1)} \times (\times_{i=2}^{k} \bar{D}^{(i)})$. Then $\{x^{(n)}\}_{n=1}^{\infty}$ is a Martin sequence for L on D (or, equivalently, for $\tilde{L}$ on D) if and only if $\lim_{n\to\infty} x^{(n)}$ exists on All such Martin sequences are minimal.*

(c) *Assume that $\{x^{(n)}\}_{n=1}^{\infty}$ has no accumulation points in $D^{(1)} \times (\times_{i=2}^{k} \bar{D}^{(i)})$. Define $N_{\text{fin}}^{\alpha} = N_{\text{fin}}^{\alpha}(L, D) = \{i \in \{2, \ldots, k\}: I_{\alpha}(g_{\beta}, M^{(i)}) < \infty\}$ and $N_{\text{fin}}^{\beta} = N_{\text{fin}}^{\beta}(L, D) = \{i \in \{2, \ldots, k\}: I_{\beta}(g_{\alpha}, M^{(i)}) < \infty\}$. Then $\{x^{(n)}\}_{n=1}^{\infty}$ is a Martin sequence for L on D (or, equivalently, for $\tilde{L}$ on D) if and only if*

(1) $\lim_{n\to\infty} x_1^{(n)} = \alpha$ *and* $\lim_{n\to\infty} x_i \in \bar{D}^{(i)}$ *exists for* $i \in N_{\text{fin}}^{\alpha}$

or

(2) $\lim_{n\to\infty} x_1^{(n)} = \beta$ *and* $\lim_{n\to\infty} x_i \in \bar{D}^{(i)}$ *exists for* $i \in N_{\text{fin}}^{\beta}$.

In particular, the part of the Martin boundary corresponding to sequences $\{x^{(n)}\}_{n=1}^{\infty}$ satisfying $\lim_{n\to\infty} x_1^{(n)} = \alpha (\lim_{n\to\infty} x_1^{(n)} = \beta)$ consists of a single point if $N_{\text{fin}}^{\alpha} = \phi$ ($N_{\text{fin}}^{\beta} = \phi$) and is isomorphic to $\times_{i\in N_{\text{fin}}^{\alpha}} \bar{D}^{(i)} (\times_{i\in N_{\text{fin}}^{\beta}} \bar{D}^{(i)})$ if $N_{\text{fin}}^{\alpha} \neq \phi$ ($N_{\text{fin}}^{\beta} \neq \phi$).

(d) *For $i \in \{2, \ldots, k\}$, $I_{\alpha}(g_{\beta}, M^{(i)}) < \infty$ if and only if $J_{\alpha}(g, M^{(i)}) < \infty$, for $g = c_1 g_{\alpha} + c_2 g_{\beta}$, with $c_1 > 0$ and $c_2 \geqslant 0$, and $I_{\beta}(g_{\alpha}, M^{(i)}) < \infty$ if and only if $J_{\beta}(g, M^{(i)}) < \infty$, for $g = c_1 g_{\alpha} + c_2 g_{\beta}$, with $c_1 \geqslant 0$ and $c_2 > 0$.*

(e) *All of the Martin boundary points in* (c) *are minimal Martin boundary points, and all of the corresponding minimal functions vanish continuously on* $D^{(1)} \times \partial(\times_{i=2}^{k} D^{(i)})$. *If $N_{\text{fin}}^{\alpha} = \phi$ ($N_{\text{fin}}^{\beta} = \phi$), then the minimal function in $C_L(D)$ corresponding to sequences $\{x^{(n)}\}_{n=1}^{\infty}$ satisfying $\lim_{n\to\infty} x_1^{(n)} = \alpha (\lim_{n\to\infty} x_1^{(n)} = \beta)$ is $u_{\alpha}(x) \equiv g_{\alpha}(x_1)\prod_{i=2}^{k} \phi_0^{(i)}(x_i)$ $(u_{\beta}(x) \equiv g_{\beta}(x_1)\prod_{i=2}^{k} \phi_0^{(i)}(x_i))$. If $N_{\text{fin}}^{\alpha} \neq \phi$ ($N_{\text{fin}}^{\beta} \neq \phi$), then for each $z_0 \in \times_{i\in N_{\text{fin}}^{\alpha}} \bar{D}^{(i)} (z_0 \in \times_{i\in N_{\text{fin}}^{\beta}} \bar{D}^{(i)})$, the minimal function u_{z_0} in $C_L(D)$ corresponding to sequences $\{x^{(n)}\}_{n=1}^{\infty}$ satisfying $\lim_{n\to\infty} x_1^{(n)} = \alpha$ and $\lim_{n\to\infty} \{x_i^{(n)}: i \in N_{\text{fin}}^{\alpha}\} = z_0 (\lim_{n\to\infty} x_1^{(n)} = \beta$ and $\lim_{n\to\infty} \{x_i^{(n)}: i \in N_{\text{fin}}^{\beta}\} = z_0)$ can be represented as follows. Let $\tau_r = \inf\{t \geqslant 0: X_1(t) = r\}$, for $r \in (\alpha, \beta)$ and let $\tau_{\alpha} = \lim_{r\to\alpha} \tau_r$ and $\tau_{\beta} = \lim_{r\to\beta} \tau_r$. Let $h(x) = g(x_1)\prod_{i=2}^{k} \phi_0^{(i)}(x_i) \in C_L(D)$, where $g = c_1 g_{\alpha} + c_2 g_{\beta}$ with $c_1 > 0$ and $c_2 \geqslant 0$ ($c_1 \geqslant 0$ and $c_2 > 0$), and let $\{P_x^h, x \in \hat{D})$ denote the solution to the martingale problem for L^h on D. Then for $x \in D$, $\mu_x^h(\cdot) \equiv w - \lim_{r\to\alpha} P_x^h(X_2(\tau_r) \in \cdot, \ldots, X_k(\tau_r) \in \cdot, \tau_r < \tau_{\beta})$ $(\mu_x^h(\cdot) \equiv w - \lim_{r\to\beta} P_x^h(X_2(\tau_r) \in \cdot, \ldots, X_k(\tau_r) \in \cdot, \tau_r < \tau_{\alpha}))$ exists as a non-zero measure with support $\times_{i=2}^{k} \bar{D}^{(i)}$ and the Radon–Nikodym derivative*

$$\frac{d\mu_x^h}{d\mu_{x_0}^h}(y), \quad y \in \mathop{\times}_{i=2}^{k} \bar{D}^{(i)}$$

exists. For $i \notin N_{\text{fin}}^{\alpha}$ (N_{fin}^{β}), *the* ith *marginal distribution of* μ_x^h *is independent of* x; *thus one can write*

$$\frac{d\mu_x^h}{d\mu_{x_0}^h}(z) \equiv \frac{d\mu_x^h}{d\mu_{x_0}^h}(y),$$

where z *is the projection of* y *onto* $\times_{i \in N_{\text{fin}}^{\alpha}} \bar{D}^{(i)}$ *(onto* $X_{i \in N_{\text{fin}}^{\beta}} \bar{D}^{(i)}$*). Then*

$$u_{z_0}(x) = \frac{h(x)}{h(x_0)} \frac{d\mu_x^h}{d\mu_{x_0}^h}(z_0).$$

As indicated at the beginning of this section, the proof of Theorem 6.1 will be postponed until the next section. We now give a number of applications of Theorem 6.1.

The integral conditions in part (c) of Theorem 6.1 depend on the coefficients a, b and $\{M^{(i)}\}_{i=2}^k$ of the operator L, and also on g_α and g_β, which are minimal positive harmonic functions for an auxiliary operator $\mathcal{L}$. By taking advantage of the equivalence in part (d) of the theorem, this integral condition can be given exclusively in terms of a, b and $\{M^{(i)}\}_{i=2}^k$, for a certain subclass of operators satisfying Assumption SP.

Theorem 6.2. *Let L and D satisfy the conditions of Theorem* 6.1 *and assume that* $V^{(1)} + \sum_{i=2}^k \lambda_0^{(i)} M^{(i)} \equiv 0$. *Then at least one of the two integrals*

$$\int_\alpha^{r_0} \exp\left(-\int_{r_0}^r \frac{2b}{a}(s)\,ds\right) dr$$

and

$$\int_{r_0}^{\beta} \exp\left(-\int_{r_0}^r \frac{2b}{a}(s)\,ds\right) dr$$

is finite. The integral conditions for determining the Martin boundary in Theorem 6.1(c) *may be given exclusively in terms of* a, b *and* $\{M^{(i)}\}_{i=2}^k$ *as follows:*

(i) *Assume that*

$$\int_\alpha^{r_0} \exp\left(-\int_{r_0}^r \frac{2b}{a}(s)\,ds\right) dr < \infty$$

and that

$$\int_{r_0}^{\beta} \exp\left(-\int_{r_0}^{r} \frac{2b}{a}(s)\,ds\right) dr = \infty.$$

Then $g_\alpha = 1$ *and*

$$g_\beta(x) = \int_{\alpha}^{x} \exp\left(-\int_{r_0}^{r} \frac{2b}{a}(s)\,ds\right) dr;$$

thus $N_{\text{fin}}^{\alpha} = \{i\colon J_\alpha(1, M^{(i)}) < \infty\}$ *and* $N_{\text{fin}}^{\beta} = \{i\colon I_\beta(1, M^{(i)}) < \infty\}$.

(ii) *Assume that*

$$\int_{\alpha}^{r_0} \exp\left(-\int_{r_0}^{r} \frac{2b}{a}(s)\,ds\right) dr = \infty$$

and that

$$\int_{r_0}^{\beta} \exp\left(-\int_{r_0}^{r} \frac{2b}{a}(s)\,ds\right) dr < \infty.$$

Then

$$g_\alpha(x) = \int_{x}^{\beta} \exp\left(-\int_{r_0}^{r} \frac{2b}{a}(s)\,ds\right) dr$$

and $g_\beta = 1$; *thus* $N_{\text{fin}}^{\alpha} = \{i\colon I_\alpha(1, M^{(i)}) < \infty\}$ *and* $N_{\text{fin}}^{\beta} = \{i\colon J_\beta(1, M^{(i)}) < \infty\}$.

(iii) *Assume that*

$$\int_{\alpha}^{r_0} \exp\left(-\int_{r_0}^{r} \frac{2b}{a}(s)\,ds\right) dr < \infty$$

and that

$$\int_{r_0}^{\beta} \exp\left(-\int_{r_0}^{r} \frac{2b}{a}(s)\,ds\right) dr < \infty.$$

Then

$$g_\alpha = \int_{x}^{\beta} \exp\left(-\int_{r_0}^{r} \frac{2b}{a}(s)\,ds\right) dr,$$

$$g_\beta(x) = \int_{\alpha}^{x} \exp\left(-\int_{r_0}^{r} \frac{2b}{a}(s)\,ds\right) dr,$$

and $1 = c(g_\alpha + g_\beta)$ *with* $c^{-1} = g_\alpha(\alpha) = g_\beta(\beta)$; *thus* $N_{\text{fin}}^{\alpha} = \{i\colon J_\alpha(1, M^{(i)}) < \infty\}$ *and* $N_{\text{fin}}^{\beta} = \{i\colon J_\beta(1, M^{(i)}) < \infty\}$.

Proof. Since L is subcritical on D, it follows from Theorem 5.2 and Theorem 5.1.1 that either

$$\int_{\alpha}^{r_0} \exp\left(-\int_{r_0}^{r} \frac{2b}{a}(s)\,ds\right) dr < \infty$$

or

$$\int_{r}^{\beta} \exp\left(-\int_{r_0}^{r} \frac{2b}{a}(s)\,ds\right) dr < \infty.$$

We leave it to the reader to verify that g_α and g_β are given as in the statement of the theorem. Once this is verified, the theorem follows directly from Theorem 6.1 (c) and (d). □

Remark. The condition in Theorem 6.2 will hold in particular if $V = 0$ (i.e., $V^{(i)} = 0$, $i = 1, 2, \ldots, k$) and, for $i = 2, \ldots, k$, D_i is a compact manifold without boundary. (By Theorem 4.11.1, $\lambda_0^{(i)} = 0$ in this case.)

The following theorem constitutes a particular case of Theorem 6.2.

Theorem 6.3. *Let $D = R^d - \{0\}$ and, for $x \in D$, let $r = |x|$ and*

$$\theta = \frac{x}{|x|} \in S^{d-1}.$$

Let

$$L = \tfrac{1}{2}a(r)\frac{\partial^2}{\partial r^2} + b(r)\frac{\partial}{\partial r} + M(r)\mathcal{A} \text{ on } D,$$

where $\mathcal{A}$ satisfies the analog of Assumption $\widetilde{H}$ on S^{d-1}, $\mathcal{A}1 = 0$ (i.e., the zeroth-order part of $\mathcal{A}$ vanishes), a and b are as in Assumption SP, and $0 < M \in C^\alpha((0, \infty))$. Assume that L is subcritical on D. (By Theorem 5.2 and Theorem 5.1.1, this is equivalent to the condition that at least one of the two integrals

$$\int_1^\infty \exp\left(-\int_1^r \frac{2b}{a}(s)\,ds\right) dr \text{ and } \int_0^1 \exp\left(-\int_1^r \frac{2b}{a}(s)\,ds\right) dr$$

is finite.)

(i) *Define the following conditions:*

1. $\displaystyle\int_0^1 \exp\left(-\int_1^r \frac{2b}{a}(s)\,ds\right) dr < \infty$ *and*

$$\int_0^1 dr \exp\left(-\int_1^r \frac{2b}{a}(s)\,ds\right)\int_r^1 dt \frac{M(t)}{a(t)} \exp\left(\int_1^t \frac{2b}{a}(s)\,ds\right) < \infty.$$

2. $\displaystyle\int_0^1 \exp\left(-\int_1^r \frac{2b}{a}(s)\,ds\right) dr < \infty$ *and*

$$\int_0^1 dr \exp\left(-\int_1^r \frac{2b}{a}(s)\,ds\right)\int_r^1 dt \frac{M(t)}{a(t)} \exp\left(\int_1^t \frac{2b}{a}(s)\,ds\right) = \infty.$$

3. $\displaystyle\int_0^1 \exp\left(-\int_1^r \frac{2b}{a}(s)\,ds\right) dr = \infty$ *and*

$$\int_0^1 dr \frac{M(r)}{a(r)} \exp\left(\int_1^r \frac{2b}{a}(s)\,ds\right)\int_r^1 dt \exp\left(-\int_1^t \frac{2b}{a}(s)\,ds\right) < \infty.$$

4. $\int_0^1 \exp\left(-\int_1^r \frac{2b}{a}(s)\,ds\right) dr = \infty$ *and*

$$\int_0^1 dr \frac{M(r)}{a(r)} \exp\left(\int_1^r \frac{2b}{a}(s)\,ds\right) \int_r^1 dt \exp\left(-\int_1^t \frac{2b}{a}(s)\,ds\right) = \infty.$$

If (1) or (3) holds, then a sequence $\{x_n\}_{n=1}^{\infty} \subset D$ *satisfying* $\lim_{n\to\infty} |x_n| = 0$ *is a Martin sequence if and only if*

$$\lim_{n\to\infty} \frac{x_n}{|x_n|}$$

exists on S^{d-1}. *If (2) or (4) holds, then every sequence* $\{x_n\}_{n=1}^{\infty} \subset D$ *satisfying* $\lim_{n\to\infty} |x_n| = 0$ *is a Martin sequence.*

(ii) *Define the following conditions*:

5. $\int_1^\infty \exp\left(-\int_1^r \frac{2b}{a}(s)\,ds\right) dr < \infty$ *and*

$$\int_1^\infty dr \exp\left(-\int_1^r \frac{2b}{a}(s)\,ds\right) \int_1^r dt \frac{M(t)}{a(t)} \exp\left(\int_1^t \frac{2b}{a}(s)\,ds\right) < \infty.$$

6. $\int_1^\infty \exp\left(-\int_1^r \frac{2b}{a}(s)\,ds\right) dr < \infty$ *and*

$$\int_1^\infty dr \exp\left(-\int_1^r \frac{2b}{a}(s)\,ds\right) \int_1^r dt \frac{M(t)}{a(t)} \exp\left(\int_1^t \frac{2b}{a}(s)\,ds\right) = \infty.$$

7. $\int_1^\infty \exp\left(-\int_1^r \frac{2b}{a}(s)\,ds\right) dr = \infty$ *and*

$$\int_1^\infty dr \frac{M(r)}{a(r)} \exp\left(\int_1^r \frac{2b}{a}(s)\,ds\right) \int_1^r dt \exp\left(-\int_1^t \frac{2b}{a}(s)\,ds\right) < \infty.$$

8. $\int_1^\infty \exp\left(-\int_1^r \frac{2b}{a}(s)\,ds\right) dr = \infty$ *and*

$$\int_1^\infty dr \frac{M(r)}{a(r)} \exp\left(\int_1^r \frac{2b}{a}(s)\,ds\right) \int_1^r dt \exp\left(-\int_1^t \frac{2b}{a}(s)\,ds\right) = \infty.$$

If (5) or (7) holds, then a sequence $\{x_n\}_{n=1}^{\infty} \subset D$ *satisfying* $\lim_{n\to\infty} |x_n| = \infty$ *is a Martin sequence if and only if*

$$\lim_{n\to\infty} \frac{x_n}{|x_n|}$$

exists on S^{d-1}. *If (6) or (8) holds, then every sequence* $\{x_n\}_{n=1}^{\infty} \subset D$ *satisfying* $\lim_{n\to\infty} |x_n| = \infty$ *is a Martin sequence.*

Remark. An application of this theorem is given in Example 6.8.

Proof. Apply Theorem 6.2 with $k = 2$, $\alpha = 0$, $\beta = \infty$,

$$L^{(1)} = \tfrac{1}{2}a(r)\frac{\mathrm{d}^2}{\mathrm{d}r^2} + b(r)\frac{\mathrm{d}}{\mathrm{d}r},\ L^{(2)} = \mathcal{A},\ D^{(2)} = S^{d-1},$$

$$\lambda_0^{(2)} = 0 \text{ and } M^{(2)} = M. \quad \square$$

Theorem 6.4 (Martin boundary for the Laplacian in a cone). *Let $\Sigma \subset S^{d-1}$, $d \geqslant 2$, be a domain with a $C^{2,\alpha}$-boundary and define the cone*

$$D = \left\{x \in R^d - \{0\}: \frac{x}{|x|} \in \Sigma\right\}.$$

Then a sequence $\{x_n\}_{n=1}^{\infty} \subset D$ is a Martin sequence for $\frac{1}{2}\Delta$ on D if and only if either (i) $\lim_{n\to\infty} |x_n| = 0$, (ii) $\lim_{n\to\infty} |x_n| \in (0, \infty)$ *and*

$$\lim_{n\to\infty} \frac{x_n}{|x_n|} \in \partial\Sigma$$

or (iii) $\lim_{n\to\infty} |x_n| = \infty$. *Every Martin boundary point is minimal. The cone of positive harmonic functions vanishing continuously on $\partial D - \{0\}$ is two-dimensional and is generated by $u_0(x) = r^{\gamma-}\phi_0(\theta)$ and $u_\infty(x) = r^{\gamma+}\phi_0(\theta)$, where*

$$r = |x|,\ \theta = \frac{x}{|x|},$$

ϕ_0 is the positive eigenfunction corresponding to the principal eigenvalue $\lambda_0 < 0$ for $\frac{1}{2}\Delta$ on Σ, *and $\gamma^{\pm} = \frac{1}{2}(2 - d \pm [(d-2)^2 - 8\lambda_0]^{\frac{1}{2}})$.*

Remark. See the discussion in Section 8.

Proof. Write

$$\tfrac{1}{2}\Delta = \tfrac{1}{2}\frac{\partial^2}{\partial r^2} + \frac{d-1}{2r}\frac{\partial}{\partial r} + \frac{1}{2r^2}\Delta_{S^{d-1}}$$

and $x = (r, \theta)$. Then, in the notation of Assumption SP, we have $k = 2$, $\alpha = 0$, $\beta = \infty$, $D_2 = \Sigma$,

$$L^{(1)} = \tfrac{1}{2}\frac{\partial^2}{\partial r^2} + \frac{d-1}{2r}\frac{\partial}{\partial r},\ L^{(2)} = \tfrac{1}{2}\Delta_{S^{d-1}},\ M^{(2)}(r) = \frac{1}{r^2},\ \lambda_0^{(2)} = \lambda_0$$

and

$$\mathcal{L} = \tfrac{1}{2}\frac{\mathrm{d}^2}{\mathrm{d}r^2} + \frac{d-1}{2r}\frac{\mathrm{d}}{\mathrm{d}r} + \frac{\lambda_0}{r^2},$$

where we have replaced the notation (x_1, x_2) by (r, θ). Recall that g_0

and g_∞ are the minimal positive harmonic functions for $\mathcal{L}$ corresponding to the Martin boundary points 0 and ∞. It is easy to verify that $g_0(r) = r^{\gamma-}$ and that $g_\infty(r) = r^{\gamma+}$. A direct integration reveals that

$$I_0\left(g_\infty, \frac{1}{r^2}\right) = I_\infty\left(g_0, \frac{1}{r^2}\right) = \infty.$$

The theorem now follows from Theorem 6.1, Theorem 1.4, Corollary 1.6 and a verification that u_0 and u_∞ belong to the cone of positive harmonic functions vanishing continuously on $\partial D - \{0\}$. □

The following simple lemma will be useful in the sequel.

Lemma 6.5. *Let $I_\alpha(g, M)$ and $I_\beta(g, M)$ be as in* (6.1).

(i) *$I_\alpha(g, M)$ and $I_\beta(g, M)$ are non-decreasing in their dependence on the function b appearing in their definitions.*
(ii) *$I_\alpha(g, M)$ and $I_\beta(g, M)$ are non-decreasing in their dependence on the function*

$$\frac{g'}{g}.$$

Proof. (i) is left as a simple exercise (exercise 8.14). For (ii), denote the dependence of $I_\alpha(g, M)$ and $I_\beta(g, M)$ on b by writing $I_\alpha(g, M, b)$ and $I_\beta(g, M, b)$. Since

$$g^2(r) = g^2(r_0)\exp\left(\int_{r_0}^{r} \frac{2a\frac{g'}{g}}{a}(s)\,ds\right),$$

we have

$$I_\alpha(g, M, b) = I_\alpha\left(1, M, b + a\frac{g'}{g}\right)$$

and

$$I_\beta(g, M, b) = I_\beta\left(1, M, b + a\frac{g'}{g}\right).$$

Thus (ii) follows from (i). □

Theorem 6.6 (Martin boundary for the Laplacian in a strip). *Let $\Sigma \subset R^m$, $m \geq 1$, be a bounded domain with a $C^{2,\alpha}$-boundary and define the strip $D = \{x = (y, z) \in R^{d+m}\colon\ y \in R^d,\ z \in \Sigma\}$, $d \geq 1$. Then a sequence $\{x_n\}_{n=1}^\infty = \{(y_n, z_n)\}_{n=1}^\infty \subset D$ is a Martin sequence for $\frac{1}{2}\Delta$ on D if and* *only if either* (i) $\lim_{n\to\infty} x_n \in \partial D = R^d x \partial\Sigma$ or (ii) $\lim_{n\to\infty} |y_n| = \infty$ *and*

$$\lim_{n\to\infty} \frac{y_n}{|y_n|} \in S^{d-1}$$

exists. ($S^0 \equiv \{-1, 1\}$). *Every Martin boundary point is minimal. The cone of positive harmonic functions vanishing continuously on* ∂D *is generated by* $\{u_v(x) \equiv \exp(\sqrt{-2\lambda_0}\, y \cdot v)\phi_0(z)\colon v \in R^d, |v| = 1\}$, *where* $\lambda_0 < 0$ *is the principal eigenvalue and* ϕ_0 *is the corresponding positive eigenfunction for* $\frac{1}{2}\Delta_m$, *the m-dimensional Laplacian, on* Σ.

Remark. See the discussion in Section 8.

Proof. We will assume that $d \geq 2$; in the case where $d = 1$, the proof is similar but the notation is a little different. It follows from Theorem 1.4 that a sequence $\{x_n\}_{n=1}^{\infty}$ with at least one accumulation point in ∂D is a Martin sequence if and only if $\lim_{n\to\infty} x_n \in \partial D$; furthermore, all such Martin sequences correspond to minimal Martin boundary points. We will prove the following statement:

A sequence $\{x_n\}_{n=1}^{\infty} = \{(y_n, z_n)\}_{n=1}^{\infty}$ satisfying $\lim_{n\to\infty} |y_n| = \infty$

is a Martin sequence for $\frac{1}{2}\Delta$ on D if and only if

$\lim_{n\to\infty} \frac{y_n}{|y_n|} \in S^{d-1}$ exists. Furthermore, the corresponding

Martin boundary points are all minimal. (6.2)

Using (6.2) along with Corollary 1.6, it follows that the cone of positive harmonic functions vanishing continuously on ∂D is generated by a collection of minimal elements which is homeomorphic to S^{d-1}. By inspection, it then follows that the functions $\{u_v\colon v \in R^d, |v| = 1\}$ in the statement of the theorem generate this cone.

It remains to prove (6.2). We wish to apply Theorem 6.1; however, the domain D does not satisfy Assumption SP. (If $d = 1$, then D does satisfy Assumption SP.) Let $D' = D - \{x = (y, z) \in D\colon y = 0\}$. Starting from $x = (y, z)$ with $y \neq 0$, the probability that a Brownian motion will ever hit $D - D'$ is zero since $d \geq 2$; thus the Green's functions for $\frac{1}{2}\Delta$ on D and on D' coincide and it suffices to prove (6.2) with D replaced by D'. The domain D' does satisfy Assumption SP since $D' = \{(r, \theta, z)\colon r \in (0, \infty), \theta \in S^{d-1}, z \in \Sigma\}$. We can write

$$\tfrac{1}{2}\Delta = \tfrac{1}{2}\frac{\partial^2}{\partial r^2} + \frac{d-1}{2r}\frac{d}{dr} + \frac{1}{2r^2}\Delta_{S^{d-1}} + \tfrac{1}{2}\Delta_m.$$

In the notation of Assumption SP, we have

$$k = 3,\ \alpha = 0,\ \beta = \infty,\ L^{(1)} = \tfrac{1}{2}\frac{\mathrm{d}^2}{\mathrm{d}r^2} + \frac{d-1}{2r}\frac{\mathrm{d}}{\mathrm{d}r},\ L^{(2)} = \tfrac{1}{2}\Delta_{S^{d-1}},$$

$$L^{(3)} = \tfrac{1}{2}\Delta_m,\ D^{(2)} = S^{d-1},\ D^{(3)} = \Sigma,\ M^{(2)}(r) = \frac{1}{r^2},\ M^{(3)}(r) = 1,$$

$$\lambda_0^{(2)} = 0,\ \lambda_0^{(3)} = \lambda_0 \text{ and } \mathscr{L} = \tfrac{1}{2}\frac{\mathrm{d}^2}{\mathrm{d}r^2} + \frac{d-1}{2r}\frac{\mathrm{d}}{\mathrm{d}r} + \lambda_0,$$

where we have replaced the notation (x_1, x_2, x_3) by (r, θ, z). From (6.1), we have $I_\infty(g, M) = \int_1^\infty \mathrm{d}r\, r^{d-1} M(r) g^2(r) \int_1^r \mathrm{d}t\, t^{1-d} g^{-2}(t)$. By Theorem 6.1, to prove (6.2), we must show that

$$I_\infty(g_0, M^{(2)}) = I_\infty\left(g_0, \frac{1}{r^2}\right) < \infty$$

and that $I_\infty(g_0, M^{(3)}) = I_\infty(g_0, 1) = \infty$, where g_0 is the minimal positive harmonic function for $\mathscr{L}$ corresponding to the Martin boundary point 0.

By exercise 7.7, $g_0(r) = g_0(1) E_r(\exp(\lambda_0 \tau_1);\ \tau_1 < \tau_\infty)$, for $r \geq 1$, where E_r denotes the expectation for the diffusion corresponding to the operator

$$\tfrac{1}{2}\frac{\mathrm{d}^2}{\mathrm{d}r^2} + \frac{d-1}{2r}\frac{\mathrm{d}}{\mathrm{d}r}.$$

Therefore, by the strong Markov property, g_0 is non-increasing; that is, $g_0'(r) \leq 0$. Let

$$f(r) \equiv -\frac{g_0'}{g_0}(r),\ r \geq 1.$$

Then $f \geq 0$ and, from the equation $\mathscr{L} g_0 = 0$, we obtain for f the equation

$$f' = f^2 - \frac{d-1}{r} f + 2\lambda_0,\ r \geq 1.$$

It follows that $f(r) \geq (-2\lambda_0)^{1/2}$, $r \geq 1$, since otherwise it is easy to see that f would become negative for large enough r. It also follows that $f'(r) \leq 0$, $r \geq 1$, since otherwise it is easy to see that for large r the differential equation would be governed by the term f^2 and would blow up at some finite r; thus

$$f^2 - \frac{d-1}{r} f + 2\lambda_0 \leq 0, \quad \text{for } r \geq 1.$$

The above analysis shows that there exist positive constants c_1, c_2 such that

$$c_1 \leqslant -\frac{g_0'}{g_0}(r) \leqslant c_2, \quad r \geqslant 1. \tag{6.3}$$

For $c > 0$, define $g(r) = \exp(-cr)$, $r \geqslant 1$; then

$$\frac{g'}{g}(r) = -c.$$

An easy calculation shows that

$$I_\infty\left(g, \frac{1}{r^2}\right) < \infty \text{ and } I_\infty(g, 1) = \infty.$$

(Use l'Hôpital's rule to show that

$$\lim_{r\to\infty}\left(\left(\int_1^r t^{1-d}\exp(2ct)\,\mathrm{d}t/r^{1-d}\exp(2cr) = 2c.\right)\right.$$

It then follows from Lemma 6.5(ii) and (6.3) that

$$I_\infty\left(g_0, \frac{1}{r^2}\right) < \infty \text{ and } I_\infty(g_0, 1) = \infty. \qquad \square$$

Theorem 6.7 (Picard principle). *Let $L = \frac{1}{2}\Delta - W(|x|)$ on $D = R^d - \{0\}$, $d \geqslant 2$, where $W \in C^1((0, \infty))$ and $W \geqslant 0$. Assume in addition that W satisfies the following conditions:*

(i) *If $\limsup_{r\to 0} r^2W(r) = \infty$, then $\lim_{r\to 0} r^2W(r) = \infty$, $r^4W(r)$ is non-decreasing in r for r near zero, and $r^{6-2d}W(r)$ is non-increasing in r for r near zero.*
(ii) *If $\limsup_{r\to\infty} r^2W(r) = \infty$, then $\lim_{r\to\infty} r^2W(r) = \infty$, $W(r)$ is non-increasing for large r and $r^{2d-2}W(r)$ is non-decreasing for large r.*

Then the Martin boundary for L on D is given as follows.

(a) *If*

$$\int_0^1 \frac{\mathrm{d}r}{r^2W^{1/2}(r)} = \infty,$$

then every sequence $\{x_n\}_{n=1}^\infty \subset D$ satisfying $\lim_{n\to\infty}|x_n| = 0$ is a Martin sequence. Thus, the Euclidean boundary point zero may be identified with a unique Martin boundary point.

(b) *If*

$$\int_0^1 \frac{\mathrm{d}r}{r^2W^{1/2}(r)} < \infty,$$

then a sequence $\{x_n\}_{n=1}^{\infty} \subset D$ *satisfying* $\lim_{n\to\infty} |x_n| = 0$ *is a Martin sequence if and only if*

$$\lim_{n\to\infty} \frac{x_n}{|x_n|}$$

exists on S^{d-1}. *Thus, the Euclidean boundary point zero may be identified with a collection of Martin boundary points which is homeomorphic to the sphere* S^{d-1}.

(c) *If*

$$\int_1^{\infty} \frac{dr}{r^2 W^{1/2}(r)} = \infty,$$

then every sequence $\{x_n\}_{n=1}^{\infty} \subset D$ *satisfying* $\lim_{n\to\infty} |x_n| = \infty$ *is a Martin sequence.*

(d) *If*

$$\int_1^{\infty} \frac{dr}{r^2 W^{1/2}(r)} < \infty,$$

then a sequence $\{x_n\}_{n=1}^{\infty} \subset D$ *satisfying* $\lim_{n\to\infty} |x_n| = \infty$ *is a Martin sequence if and only if*

$$\lim_{n\to\infty} \frac{x_n}{|x_n|}$$

exists on S^{d-1}.

Remark. Assumptions (i) and (ii) on W in Theorem 6.7 hold for $W(r) = r^l$, $W(r) = r^{-2}|\log r|^l$, $W(r) = r^{-2}|\log r|^2(\log|\log r|)^l$, etc., near $r = 0$ or $r = \infty$. In particular, for example, if $W(r) = r^l$ near $r = 0$, then (a) holds if $l \geqslant -2$ and (b) holds if $l < -2$; if $W(r) = r^{-2}|\log r|^l$ near $r = 0$, then (a) holds if $l \leqslant 2$ and (b) holds if $l > 2$; if $W(r) = r^{-2}|\log r|^2(\log|\log r|)^l$ near $r = 0$, then (a) holds if $l \leqslant 2$ and (b) holds if $l > 2$; etc. Similarly, if $W(r) = r^l$ near $r = \infty$, then (c) holds if $l \leqslant -2$ and (d) holds if $l > -2$; if $W(r) = r^{-2}(\log r)^l$ near $r = \infty$, then (c) holds if $l \leqslant 2$ and (d) holds if $l > 2$; if $W(r) = r^{-2}(\log r)^2(\log\log r)^l$ near $r = \infty$, then (c) holds if $l \leqslant 2$ and (d) holds if $l > 2$; etc.

Proof. We will prove (c) and (d); (a) and (b) are proved by similar techniques or, alternatively, they can be obtained from (c) and (d) by a change of variables and several calculations (see exercise 8.15). Let

$$r = |x| \text{ and } \theta = \frac{x}{|x|}.$$

We write

$$\tfrac{1}{2}\Delta - W(|x|) = \tfrac{1}{2}\frac{\partial^2}{\partial r^2} + \frac{d-1}{2r}\frac{\partial}{\partial r} - W(r) + \frac{1}{2r^2}\Delta_{S^{d-1}}$$

and apply Theorem 6.1 with

$$k = 2,\ \alpha = 0,\ \beta = \infty,\ D^{(2)} = S^{d-1},$$

$$L^{(1)} = \tfrac{1}{2}\frac{\mathrm{d}^2}{\mathrm{d}r^2} + \frac{d-1}{2r}\frac{\mathrm{d}}{\mathrm{d}r} - W(r),$$

$$L^{(2)} = \tfrac{1}{2}\Delta_{S^{d-1}},\ M^{(2)}(r) = \frac{1}{r^2},\ \lambda_0^{(2)} = 0,\ \text{and}\ \mathscr{L} = L^{(1)},$$

where we have replaced the notation (x_1, x_2) by (r, θ). Similarly, we replace the notation $(X_1(t),\ X_2(t))$ by $(r(t),\ \theta(t))$. To prove the theorem we must show that, if (c) holds, then

$$I_\infty\left(g_0, \frac{1}{r^2}\right) = \infty,$$

and that if (d) holds, then

$$I_\infty\left(g_0, \frac{1}{r^2}\right) < \infty,$$

where

$$I_\infty\left(g_0, \frac{1}{r^2}\right) = \int_1^\infty \mathrm{d}r\, r^{d-3} g_0^2(r) \int_1^r s^{1-d} g_0^{-2}(s)\, \mathrm{d}s, \qquad (6.4)$$

and where g_0 is the minimal positive harmonic function for $\mathscr{L} = L^{(1)}$ on $(0, \infty)$ corresponding to the Martin boundary point zero. By exercise 7.7, $g_0(r) = g_0(1) E_r(\exp(-\int_0^{\tau_1} W(r(s))\, \mathrm{d}s);\ \tau_1 < \tau_\infty)$, for $r \geq 1$, where E_r denotes the expectation for the diffusion corresponding to the operator

$$\tfrac{1}{2}\frac{\mathrm{d}^2}{\mathrm{d}r^2} + \frac{d-1}{2r}\frac{\mathrm{d}}{\mathrm{d}r}.$$

Therefore, by the strong Markov property, g_0 is non-increasing; that is, $g_0'(r) \leq 0$. Let

$$f(r) = \frac{-g_0'}{g_0}(r),\ r \geq 1.$$

Then $f \geq 0$ and, from the equation $\mathscr{L} g_0 = 0$, we obtain for f the equation

$$f' = f^2 - \frac{d-1}{r} f - 2W, \quad r \geq 1. \qquad (6.5)$$

We first dispense with the case in which $\limsup_{r\to\infty} r^2W(r) < \infty$; that is,

$$W(r) \leqslant \frac{c}{r^2},$$

for $r \geqslant 1$ and some $c > 0$. In this case, the integral condition in (c) holds; thus we must show that

$$I_\infty\left(g_0, \frac{1}{r^2}\right) = \infty.$$

By the comparison result of Theorem 6.11, below, it suffices to consider the case where

$$W(r) = \frac{c}{r^2}.$$

However, the proof that

$$I_\infty\left(g_0, \frac{1}{r^2}\right) = \infty$$

in the case where

$$W(r) = \frac{c}{r^2}$$

was actually given in the proof of Theorem 6.4 above (just substitute $c = -\lambda_0$ in that proof).

From now on, we assume that

$$\lim_{r\to\infty} r^2W(r) = \infty, \tag{6.6}$$

that W is non-increasing for large r, and that $r^{2d-2}W(r)$ is non-decreasing for large r. This last condition can be expressed as

$$\frac{2(d-1)}{r}W(r) + W'(r) \geqslant 0, \quad \text{for large } r. \tag{6.7}$$

It follows from (6.4) that, to prove the theorem, it suffices to show that there exist constants $c_1, c_2 > 0$ such that

$$\frac{c_1}{W^{1/2}(r)} \leqslant r^{d-1}g_0^2(r)\int_1^r s^{1-d}g_0^{-2}(s)\,ds \leqslant \frac{c_2}{W^{1/2}(r)}, \quad \text{for large } r. \tag{6.8}$$

The rest of the proof will be devoted to showing that (6.8) holds.

Using the fact that W is non-increasing, if follows that $f'(r) \leqslant 0$, for large r, since otherwise the differential equation in (6.5) would be governed by the term f^2 and f would blow up at some finite r. Thus, from (6.5) we obtain

$$f^2 - \frac{d-1}{r} f - 2W \leqslant 0,$$

for large r. Using this along with (6.6), it follows that for some $c > 0$,

$$f(r) \leqslant cW^{\frac{1}{2}}(r), \quad \text{for large } r. \tag{6.9}$$

We now show that

$$f(r) \geqslant \sqrt{2} W^{\frac{1}{2}}(r), \quad \text{for large } r. \tag{6.10}$$

Let $H(r) = \frac{1}{2} f^2(r) - W(r)$. Using (6.5), it follows that

$$\text{if } H(r_0) = 0, \text{ then } H'(r_0) = -\frac{2(d-1)}{r_0} W(r_0) - W'(r_0). \tag{6.11}$$

From (6.11) and (6.7), it follows that either (6.10) holds or $f(r) \leqslant \sqrt{2} W^{1/2}(r)$, for large r. To complete the proof of (6.10), we will assume that $f(r) \leqslant \sqrt{2} W^{1/2}(r)$, for large r, and come to a contradiction. From (6.5), we have

$$f'(r) \leqslant -\frac{d-1}{r} f,$$

for large r. Integrating, we obtain for some $c > 0$,

$$f(r) \leqslant cr^{-(d-1)}, \quad \text{for large } r. \tag{6.12}$$

Using (6.12) in (6.5) and recalling that $d \geqslant 2$, we have, for some $c > 0$,

$$f'(r) \leqslant \frac{c}{r^2} - 2W(r), \quad \text{for large } r. \tag{6.13}$$

By (6.12), $\lim_{r \to \infty} f(r) = 0$; thus integrating (6.13) gives

$$f(r) \geqslant 2\int_r^\infty W(s)\, ds - \frac{c}{r}, \quad \text{for large } r. \tag{6.14}$$

In the light of (6.6), (6.14) contradicts (6.12). Thus, we conclude that (6.10) holds.

From (6.9) and (6.10), it follows that there are constants k_1, $k_2 > 0$ such that

$$\exp\left(-k_1 \int_1^r W^{\frac{1}{2}}(s)\, ds\right) \leqslant g_0(r) \leqslant \exp\left(-k_2 \int_1^r W^{\frac{1}{2}}(s)\, ds\right), \quad \text{for } r \geqslant 1. \tag{6.15}$$

Using (6.6), (6.9), (6.10) and (6.15), it follows that

$$\lim_{r\to\infty}\int_1^r s^{1-d}g_0^{-2}(s)\,ds = \infty$$

and $\lim_{r\to\infty} r^{1-d}g_0^{-2}(r)f^{-1}(r) = \infty$; thus we may apply l'Hopitâl's rule at $r = \infty$ to $\int_1^r s^{1-d}g_0^{-2}(s)\,ds/r^{1-d}g_0^{-2}(r)f^{-1}(r)$. We have

$$\left(\int_1^r s^{1-d}g_0^{-2}(s)\,ds\right)' = r^{1-d}g_0^{-2}(r). \tag{6.16}$$

Using (6.5), we have

$$\begin{aligned}(r^{1-d}g_0^{-2}(r)f^{-1}(r))' &= r^{1-d}g_0^{-2}(r)\left[\frac{1-d}{rf(r)} + 2 - \frac{f'}{f^2}(r)\right]\\ &= r^{1-d}g_0^{-2}(r)\left[1 + \frac{2W}{f^2}\right].\end{aligned} \tag{6.17}$$

From (6.10), (6.16) and (6.17), it follows that there exist constants c_1, $c_2 > 0$ such that

$$c_1 \leqslant \frac{(\int_1^r s^{1-d}g_0^{-2}(s)\,ds)'}{(r^{1-d}g_0^{-2}(r)f^{-1}(r))'} \leqslant c_2, \quad \text{for large } r. \tag{6.18}$$

Thus, by a generalized form of l'Hopitâl's rule, there exist constants $c_1, c_2 > 0$ such that

$$c_1f^{-1}(r) \leqslant r^{d-1}g_0^2(r)\int_1^r s^{1-d}g_0^{-2}(s)\,ds \leqslant c_2f^{-1}(r), \quad \text{for large } r. \tag{6.19}$$

Now (6.8) follows from (6.9), (6.10) and (6.19). □

Example 6.8. Corollary 6.4.2 showed that a comparison principle holds with regard to transience and recurrence for diffusions corresponding to symmetric operators. Namely, if $L_i = \frac{1}{2}\nabla\cdot a_i\nabla + a_i\nabla Q_i\cdot\nabla$, $i = 1, 2$, and the matrices $a_i\exp(2Q_i)$ associated with the quadratic forms of L_i, $i = 1, 2$, satisfy $a_1\exp(2Q_1) \geqslant a_2\exp(2Q_2)$ (that is, $a_1\exp(2Q_1) - a_2\exp(2Q_2)$ is non-negative definite), then the transience of the diffusion corresponding to L_2 implies the transience of the diffusion corresponding to L_1 or, equivalently, the recurrence of the diffusion corresponding to L_1 implies the recurrence of the diffusion corresponding to L_2. Using Theorem 6.3 we can demonstrate that a comparison principle in terms of quadratic forms does not hold with regard to the size of the Martin boundary. Let L be as in Theorem 6.3 with

$\mathcal{A} = \frac{1}{2}\Delta_{S^{d-1}}$ and assume, for example, that

$$\int_1^\infty \exp\left(-\int_1^r \frac{2b}{a}(s)\,ds\right) dr < \infty.$$

The matrix (up to an arbitrary scalar multiple) associated with the quadratic form for L, resolved in polar coordinates, is

$$\begin{pmatrix} a(r) & 0 \\ 0 & M(r)A(\theta) \end{pmatrix} \exp\left(\int_1^r \frac{2b}{a}(s)\,ds\right),$$

where $A(\theta)$ is the matrix associated with the quadratic form for $\frac{1}{2}\Delta_{S^{d-1}}$ on S^{d-1}. Under the partial ordering of non-negative definite matrices, it is clear that this matrix function is increasing in M and in b. By Theorem 6.3, the part of the Martin boundary corresponding to sequences $\{x_n\}_{n=1}^\infty \subset R^d - \{0\}$ satisfying $\lim_{n\to\infty} |x_n| = \infty$ is determined by the convergence or divergence of

$$J_\infty(1, M) = \int_1^\infty dr \exp\left(-\int_1^r \frac{2b}{a}(s)\,ds\right) \int_1^r dt \frac{M(t)}{a(t)} \exp\left(\int_1^t \frac{2b}{a}(s)\,ds\right).$$

Now consider two operators, L_i, $i = 1, 2$, of the above form. First let

$$a_1 = a_2 = 1,\ b_1 = b_2 = 1,\ M_1 = \frac{1}{r^2} \text{ and } M_2 = 1.$$

The quadratic form for L_2 is larger than the quadratic form for L_1. One can check that $J_\infty(1, M_1) < \infty$ and $J_\infty(1, M_2) = \infty$. Thus, by Theorem 6.3, the part of the Martin boundary associated with sequences $\{x_n\}_{n=1}^\infty \subset R^d - \{0\}$ satisfying $\lim_{n\to\infty} |x_n| = \infty$ is larger for L_1 than for L_2. On the other hand, now assume that $a_1 = a_2 = 1$, $b_1 = 0$, $b_2 = r^\gamma$ with $\gamma > 1$, and $M_1 = M_2 = 1$. Again, the quadratic form for L_2 is larger than the quadratic form for L_1. Once can check that $J_\infty(1, M_1) = \infty$ and $J_\infty(1, M_2) < \infty$ (see Example 5.1.7). Thus, by Theorem 6.3, the part of the Martin boundary associated with sequences $\{x_n\}_{n=1}^\infty \subset R^d - \{0\}$ satisfying $\lim_{n\to\infty} |x_n| = \infty$ is larger for L_2 than for L_1.

Although Example 6.8 shows that a comparison principle for the Martin boundary in terms of quadratic forms does not hold for symmetric operators satisfying Assumption SP, it turns out that a useful comparison principle concerning the Martin boundary for operators satisfying Assumption SP may indeed be formulated.

Theorem 6.9. *Let* $O(a, b, \alpha, \beta, k)$ *denote the class of operators and domains* (L, D) *which satisfy Assumption SP for a particular choice of*

a, b, α, β, k. Let (L_1, D_1) and (L_2, D_2) belong to $\mathrm{O}(a, b, \alpha, \beta, k)$. Use the notation $M_1^{(i)}$ and $M_2^{(i)}$, $V_1^{(i)}$ and $V_2^{(i)}$ and $\lambda_{01}^{(i)}$ and $\lambda_{02}^{(i)}$ to denote for (L_1, D_1) and (L_2, D_2) the objects which are analogous to $M^{(i)}$, $V^{(i)}$ and $\lambda_0^{(i)}$ for (L, D) in Assumption SP. Let N_{fin}^{α} and N_{fin}^{β} be as in Theorem 6.1(b). *For some $i \in \{2, \ldots, k\}$, assume that the inequalities $M_1^{(i)} \leq M_2^{(i)}$ and $V_1^{(1)} + \sum_{j=2}^{k} \lambda_{01}^{(j)} M_1^{(j)} \leq V_2^{(1)} + \sum_{j=2}^{k} \lambda_{02}^{(j)} M_2^{(j)}$ hold near α (near β). Then the following comparison holds:*

If $i \in N_{\text{fin}}^{\alpha}(L_2, D_2)$ $(N_{\text{fin}}^{\beta}(L_2, D_2))$,

then $i \in N_{\text{fin}}^{\alpha}(L_1, D_1)$ $(N_{\text{fin}}^{\beta}(L_1, D_1))$.

Before proving Theorem 6.9, we single out two particular cases.

Theorem 6.10. *Let $L = L^{(1)} + \sum_{i=2}^{k} M^{(i)}(x_1) L^{(i)}$ on $D = \times_{i=1}^{k} D^{(i)}$ satisfy Assumption SP. Let $A^{(i)} \subseteq D^{(i)}$ be a subdomain with a $C^{2,\alpha}$-boundary, and let $A = D^{(1)} \times (\times_{i=2}^{k} A^{(i)})$. Then*

$$N_{\text{fin}}^{\alpha}(L, D) \subseteq N_{\text{fin}}^{\alpha}(L, A)$$

and

$$N_{\text{fin}}^{\beta}(L, D) \subseteq N_{\text{fin}}^{\beta}(L, A),$$

where N_{fin}^{α} and N_{fin}^{β} are as in Theorem 6.1.

Remark. Interpreting the word 'size' appropriately, Theorem 6.10 states that, given an operator L and a domain $D = \times_{i=1}^{k} D^{(i)}$ satisfying Assumption SP, the size of the Martin boundary for L on A, where $A = D^{(1)} \times (\times_{i=2}^{k} A^{(i)}) \subseteq D$, is non-increasing in its dependence on A. The probabilistic intuition behind Theorem 6.10 is discussed in Section 8.

Proof. Let $\lambda_0^{(i)}(D)$ and $\lambda_0^{(i)}(A)$ denote, respectively, the principal eigenvalue for $L^{(i)}$ on $D^{(i)}$ and for $L^{(i)}$ on $A^{(i)}$, $i = 2, \ldots, k$. The theorem follows immediately from Theorem 6.9 and the fact that $\lambda_0^{(i)}(D) \geq \lambda_0^{(i)}(A)$, $i = 2, \ldots, k$. □

Theorem 6.11. *Let $L_j = L_0 + V_j$, $j = 1, 2$, satisfy Assumption SP on a domain $D = \times_{i=1}^{k} D^{(i)}$. Assume that the inequality $V_1(x) \leq V_2(x)$ holds for all $x \in D$ with x_1 close to α (close to β). Then $N_{\text{fin}}^{\alpha}(L_2, D) \subseteq N_{\text{fin}}^{\alpha}(L_1, D)$ $(N_{\text{fin}}^{\beta}(L_2, D) \subseteq N_{\text{fin}}^{\beta}(L_1, D))$, where N_{fin}^{α} and N_{fin}^{β} are as in Theorem* 6.1.

Remark 1. Theorem 6.11 states that in the class of operators satisfying Assumption SP, the Martin boundary is non-increasing in its dependence on the zeroth-order term V. This turns out not to be true in general; see the first subsection in Section 9.

Remark 2. Note from Theorem 6.9 that if $\lambda_0^{(i)} \geq 0$, then the Martin boundary for L on D will be non-increasing in its dependence on $M^{(i)}$. However, if $\lambda_0^{(i)} < 0$, then Theorem 6.9 gives no information in this regard.

Proof. By assumption, we can write L_j, $j = 1, 2$, in the form

$$L_j = L_0^{(1)} + V_j^{(1)}(x_1) + \sum_{i=2}^{k} M^{(i)}(x_1)(L_0^{(i)} + V_j^{(i)}(x_i)).$$

Thus

$$V_j(x) = V_j^{(1)}(x_1) + \sum_{i=2}^{k} M^{(i)}(x_1)V_j^{(i)}(x_i),\ j = 1, 2.$$

This representation is not unique; for $i \in \{2, \ldots, k\}$, we can replace $V_j^{(i)}$ by $V_j^{(i)} + c$ and then compensate by replacing $V_j^{(1)}$ by $V_j^{(1)} - cM^{(i)}$. It is not difficult to show, though, that if $V_1(x) \leq V_2(x)$, for $x \in D$ with x_1 close to α (close to β), then one may choose the $V_j^{(i)}$s above in such a fashion that $V_1^{(1)}(x_1) \leq V_2^{(1)}(x_1)$, for x_1 close to α (close to β) and $V_1^{(i)}(x_i) \leq V_2^{(i)}(x_i)$, for $x_i \in D^{(i)}$, $i = 2, \ldots, k$. For $i = 2, \ldots, k$, and $j = 1, 2$, let $\lambda_{0j}^{(i)}$ denote the principal eigenvalue for $L_j^{(i)} = L_0^{(i)} + V_j^{(i)}$ on $D^{(i)}$. It then follows that $\lambda_{01}^{(i)} \leq \lambda_{02}^{(i)}$, $i = 2, \ldots, k$. Thus the inequality

$$V_1^{(1)} + \sum_{i=2}^{k} \lambda_{01}^{(i)} M^{(i)} \leq V_2^{(1)} + \sum_{i=2}^{k} \lambda_{02}^{(i)} M^{(i)}$$

holds for x_1 close to α (close to β). The theorem now follows from Theorem 6.9. □

The proof of Theorem 6.9 hinges on Lemma 6.5 above and the following lemma.

Lemma 6.12. *Let*

$$\mathcal{L} = \tfrac{1}{2}a(r)\frac{\mathrm{d}^2}{\mathrm{d}r^2} + b(r)\frac{\mathrm{d}}{\mathrm{d}r} + W(r)$$

be subcritical and satisfy Assumption $\widetilde{H}_{\text{loc}}$ *on* $D = (\alpha, \beta)$, *where* $-\infty \leq \alpha < \beta \leq \infty$. *Let* h_α *and* h_β *denote the elements of* $C_L(D)$ *corresponding to the Martin boundary points* α *and* β. *Then the functions*

$$\frac{h'_\alpha}{h_\alpha} \text{ and } \frac{h'_\beta}{h_\beta}$$

are non-decreasing in their dependence on W.

Proof. We will prove the lemma for h_α; the same type of proof works for h_β. Let $\{P_x, x \in \hat{D}\}$ denote the solution to the generalized martingale problem for

$$\tfrac{1}{2}a\frac{\mathrm{d}^2}{\mathrm{d}r^2} + b\frac{\mathrm{d}}{\mathrm{d}r} \text{ on } D = (\alpha, \beta).$$

For $r \in (\alpha, \beta)$, let $\tau_r = \inf\{t \geq 0\colon X(t) = r\}$, and let $\tau_\beta = \lim_{r \to \beta} \tau_r$. By exercise 7.7, up to a constant normalization factor, $h_\alpha(r) = E_r(\exp(\int_0^{\tau_{r_0}} W(X(s))\,\mathrm{d}s);\ \tau_{r_0} < \tau_\beta)$, for $\alpha < r_0 \leq r < \beta$. Using the strong Markov property, it follows that for $0 < \delta < \beta - r$, $h_\alpha(r + \delta) = h_\alpha(r) E_{r+\delta}(\exp(\int_0^{\tau_r} W(X(s))\,\mathrm{d}s);\ \tau_r < \tau_\beta)$. Thus, we have

$$\begin{aligned}\frac{h'_\alpha}{h_\alpha}(r) &= \lim_{\delta \to 0} \frac{h_\alpha(r+\delta) - h_\alpha(r)}{\delta h_\alpha(r)} \\ &= \lim_{\delta \to 0} \frac{1}{\delta}\left[E_{r+\delta}\left(\exp\left(\int_0^{\tau_r} W(X(s))\,\mathrm{d}s\right);\ \tau_r < \tau_\beta\right) - 1\right]. \qquad \square\end{aligned}$$

Proof of Theorem 6.9. Assume first that the inequalities $M_1^{(i)} \leq M_2^{(i)}$ and $V_1^{(1)} + \sum_{j=2}^k \lambda_{01}^{(j)} M_1^{(j)} \leq V_2^{(1)} + \sum_{j=2}^k \lambda_{02}^{(j)} M_2^{(j)}$ hold on the entire interval (α, β). Use the notation $g_{\alpha,1}$, $g_{\alpha,2}$, $g_{\beta,1}$ and $g_{\beta,2}$ to denote for (L_1, D_1) and (L_2, D_2) the functions which are analogous to g_α and g_β in Assumption SP. Then by Lemma 6.12,

$$\frac{g'_{\alpha,1}}{g_{\alpha,1}} \leq \frac{g'_{\alpha,2}}{g_{\alpha,2}} \text{ and } \frac{g'_{\beta,1}}{g_{\beta,1}} \leq \frac{g'_{\beta,2}}{g_{\beta,2}},$$

and by Lemma 6.5(ii), $I_\alpha(g_{\beta,1}, M) \leq I_\alpha(g_{\beta,2}, M)$ and $I_\beta(g_{\alpha,1}, M) \leq I_\beta(g_{\alpha,2}, M)$, for any M. Clearly, $I_\alpha(g, M_1^{(i)}) \leq I_\alpha(g, M_2^{(i)})$ and $I_\beta(g, M_1^{(i)}) \leq I_\beta(g, M_2^{(i)})$, for any g. Thus $I_\alpha(g_{\beta,1}, M_1^{(i)}) \leq I_\alpha(g_{\beta,2}, M_2^{(i)})$ and $I_\beta(g_{\alpha,1}, M_1^{(i)}) \leq I_\beta(g_{\alpha,2}, M_2^{(i)})$, and the theorem follows.

The extension to the case in which the above inequalities hold only near α (near β) follows form the fact that the finiteness or infiniteness of $I_\alpha(g_{\beta,l}, M_l^{(i)})$ $(I_\beta(g_{\alpha,l}, M_l^{(i)}))$, $l = 1, 2$, does not depend on the behavior of $M_l^{(i)}$ or $V_l^{(1)} + \sum_{j=2}^{k} \lambda_{0l}^{(j)} M_l^{(j)}$ outside of a neighborhood of α (β) (exercise 8.16). □

8.7 The Martin boundary for a class of operators in skew-product form: proof of general theorem

In this section, we prove Theorem 6.1. The proof is considerably simpler in the case where $\partial D^{(i)} = \phi$, $i = 2, \ldots, k$; that is, the case where $D^{(i)}$ is a compact manifold without boundary. For the proof of Theorem 6.1, we will need Propositions 7.1 and 7.2 below. These propositions are not needed in the case where $\partial D^{(i)} = \phi$, $i = 2, \ldots, k$.

Proposition 7.1 *Let L and D be as in Assumption SP. For $r \in (\alpha, \beta)$, define $\Omega_{r,\alpha} = \{x = (x_1, \ldots, x_k) \in D: \alpha < x_1 < r\}$ and $\Omega_{r,\beta} = \{x = (x_1, \ldots, x_k) \in D: r < x_1 < \beta\}$. Let $\sigma_{\bar{\Omega}_{r,\alpha}} = \inf\{t \geq 0: X(t) \in \bar{\Omega}_{r,\alpha}\}$ and let $\sigma_{\bar{\Omega}_{r,\beta}} = \inf\{t \geq 0: X(t) \in \bar{\Omega}_{r,\beta}\}$. Note that if $\sigma_{\bar{\Omega}_{r,\alpha}} < \tau_D$ ($\sigma_{\bar{\Omega}_{r,\beta}} < \tau_D$), then $X_1(\sigma_{\bar{\Omega}_{r,\alpha}}) = r$ and $X(\sigma_{\bar{\Omega}_{r,\alpha}}) \in D$ ($X_1(\sigma_{\bar{\Omega}_{r,\beta}}) = r$ and $X(\sigma_{\bar{\Omega}_{r,\beta}}) \in D$).*

(i) *The Green's function $G(x, y)$ for L on D satisfies*

$$G(x, y) = E_x\left(\exp\left(\int_0^{\sigma_{\bar{\Omega}_{r,\alpha}}} V(X(s))\, ds\right) G(X(\sigma_{\bar{\Omega}_{r,\alpha}}), y); \sigma_{\bar{\Omega}_{r,\alpha}} < \tau_D\right),$$

$$\text{for } x \in \Omega_{r,\beta} \text{ and } y \in \Omega_{r,\alpha}, \quad (7.1a)$$

and

$$G(x, y) = E_x\left(\exp\left(\int_0^{\sigma_{\bar{\Omega}_{r,\beta}}} V(X(s))\, ds\right) G(X(\sigma_{\bar{\Omega}_{r,\beta}}), y); \sigma_{\bar{\Omega}_{r,\beta}} < \tau_D\right),$$

$$\text{for } x \in \Omega_{r,\alpha} \text{ and } y \in \Omega_{r,\beta}. \quad (7.1b)$$

(ii) *Let $h_\alpha(x) = g_\alpha(x_1)\prod_{i=2}^{k}\phi_0^{(i)}(x_i)$ and $h_\beta(x) = g_\beta(x_1)\prod_{i=2}^{k}\phi_0^{(i)}(x_i)$, where g_α, g_β and $\phi_0^{(i)}$ are as in Assumption SP. Then*

$$h_\alpha(x) = E_x\left(\exp\left(\int_0^{\sigma_{\bar{\Omega}_{r,\alpha}}} V(X(s))\, ds\right) h_\alpha(X(\sigma_{\bar{\Omega}_{r,\alpha}})); \sigma_{\bar{\Omega}_{r,\alpha}} < \tau_D\right),$$

$$\text{for } x \in \Omega_{r,\beta}$$

and

$$h_\beta(x) = E_x\left(\exp\left(\int_0^{\sigma_{\bar{\Omega}_{r,\beta}}} V(X(s))\, ds\right) h_\beta(X(\sigma_{\bar{\Omega}_{r,\beta}})); \sigma_{\bar{\Omega}_{r,\beta}} < \tau_D\right),$$

$$\text{for } x \in \Omega_{r,\alpha}.$$

(iii) *Let $\{P_x^{h_\alpha}, x \in \hat{D}\}$ denote the solution to the generalized martingale problem for L^{h_α} on D, and let $\{P_x^{h_\beta}, x \in \hat{D}\}$ denote the solution to the generalized martingale problem for L^{h_β} on D. Define $\mu_x^{h_\alpha}(dy) = P_x^{h_\alpha}(X(\sigma_{\bar{\Omega}_{r,\alpha}}) \in dy)$, for $x \in \Omega_{r,\beta}$, and define $\mu_x^{h_\beta}(dy) = P_x^{h_\beta}(X(\sigma_{\bar{\Omega}_{r,\beta}}) \in dy)$, for $x \in \Omega_{r,\alpha}$. Then $\mu_x^{h_\alpha}$ and $\mu_x^{h_\beta}$ are probability measures on $D \cap \{x_1 = r\}$. Furthermore, a sequence $\{y^{(n)}\}_{n=1}^\infty$, $\{(y_1^{(n)}, \ldots, y_k^{(n)})\}_{n=1}^\infty \subset \Omega_{r,\beta}$ which has no accumulation points in D and which satisfies $\liminf_{n\to\infty} y_1^{(n)} > r$ is a Martin sequence for $\tilde{L}$ on D if and only if $w - \lim_{n\to\infty} \mu_{y^{(n)}}^{h_\alpha}$ exists, and a sequence $\{y^{(n)}\}_{n=1}^\infty = \{(y_1^{(n)}, \ldots, y_k^{(n)})\}_{n=1}^\infty \subset \Omega_{r,\alpha}$ which has no accumulation points in D and which satisfies $\limsup_{n\to\infty} y_1^{(n)} < r$ is a Martin sequence for $\tilde{L}$ on D if and only if $w - \lim_{n\to\infty} \mu_{y^{(n)}}^{h_\beta}$ exists.*

Remark. Proposition 7.1 is a generalization of Theorems 7.3.3, 7.4.5 and 7.6.2 to a case in which $\partial\Omega \cap \partial D \neq \phi$.

Proof. (i) We will prove that (7.1a); (7.1b) is proved similarly. Fix $y \in \Omega_{r,\alpha}$ and let U be a domain satisfying $y \in U \subset\subset \Omega_{r,\alpha}$. By Theorem 7.3.7,

$$G(x, y) = E_x\left(\exp\left(\int_0^{\sigma_{\bar{U}}} V(X(s))\,ds\right)G(X(\sigma_{\bar{U}}), y);\ \sigma_{\bar{U}} < \tau_D\right),$$
$$x \in D - U.$$

Using this representation for $G(x, y)$ in the first and third equalities below, and using the strong Markov property in the second equality below, we have for $x \in \Omega_{r,\beta}$,

$$\begin{aligned}G(x, y) &= E_x\left(\exp\left(\int_0^{\sigma_{\bar{U}}} V(X(s))\,ds\right)G(X(\sigma_{\bar{U}}), y);\ \sigma_{\bar{U}} < \tau_D\right)\\ &= E_x\left(\exp\left(\int_0^{\sigma_{\bar{\Omega}_{r,\alpha}}} V(X(s))\,ds\right)E_{X(\sigma_{\bar{\Omega}_{r,\alpha}})}\left(\exp\left(\int_0^{\sigma_{\bar{U}}} V(X(s))\,ds\right)\right.\right.\\ &\qquad \left.\left.\times\ G(X(\sigma_{\bar{U}}); y);\ \sigma_{\bar{U}} < \tau_D\right);\ \sigma_{\bar{\Omega}_{r,\alpha}} < \tau_D\right)\\ &= E_x\left(\exp\left(\int_0^{\sigma_{\bar{\Omega}_{r,\alpha}}} V(X(s))\,ds\right)G(X(\sigma_{\bar{\Omega}_{r,\alpha}}), y);\ \sigma_{\bar{\Omega}_{r,\alpha}} < \tau_D\right).\end{aligned}$$

(ii) We will prove the representation for h_α. Fix $r \in (\alpha, \beta)$. Let $\{D_n^{(i)}\}_{n=1}^\infty$ be a sequence of domains with $C^{2,\alpha}$-boundaries satisfying $D_n^{(i)} \subset\subset D_{n+1}^{(i)}$ and $\bigcup_{n=1}^\infty D_n^{(i)} = D^{(i)}$, for $i = 2, \ldots, k$. Let $\lambda_n^{(i)}$ denote the principal eigenvalue of $L^{(i)}$ on $D_n^{(i)}$ and let $\phi_n^{(i)}$ denote the corresponding positive eigenfunction, normalized by $\phi_n^{(i)}(x_{0i}) = \phi_0^{(i)}(x_{0i})$, for some fixed $x_{0i} \in D_1^{(i)}$. Let $\{\alpha_n\}_{n=1}^\infty$ be a decreasing sequence and let $\{\beta_n\}_{n=1}^{\infty}$ be an increasing sequence such that $\alpha_1 < r < \beta_1$, $\lim_{n\to\infty} \alpha_n = \alpha$ and $\lim_{n\to\infty} \beta_n = \beta$. Let $\mathcal{L}_n$ denote the operator obtained from $\mathcal{L}$ by replacing $\sum_{i=2}^{k} \lambda_0^{(i)} M^{(i)}$ by $\sum_{i=2}^{k} \lambda_n^{(i)} M^{(i)}$ and consider $\mathcal{L}_n$ on (α_n, β_n). Let $D_n = (\alpha_n, \beta_n) \times (\times_{i=2}^{k} D_n^{(i)})$, $\Omega_{r,\alpha,n} = \Omega_{r,\alpha} \cap D_n$ and $\Omega_{r,\beta,n} = \Omega_{r,\beta} \cap D_n$. Since $\lambda_n^{(i)} < \lambda^{(i)}$ and $M^{(i)} > 0$, it is clear that $\mathcal{L}_n$ is subcritical on (α_n, β_n). Thus, there exists a unique positive solution g_n to $\mathcal{L}_n g_n = 0$ in (α_n, β_n), $g_n(r) = 1$ and $g_n(\beta) = 0$. Let $h_n(x) = g_n(x_1) \prod_{i=2}^{k} \phi_n^{(i)}(x_i)$ on $\bar{D}_n$. Then $Lh_n = 0$ in D_n. Let U_n be a domain with a $C^{2,\alpha}$-boundary satisfying $\Omega_{r,\beta,n} \subset U_n \subset\subset D$. By Theorem 4.4.1(iii), the principal eigenvalue of L on U_n is negative. Applying Theorem 3.6.6 (with U_n and $\Omega_{r,\beta,n}$ taking the roles of D and D' in the statement of that theorem) and noting that h_n vanishes on $\partial\Omega_{r,\beta,n} \cap \{x_1 > r\}$, we obtain

$$h_n(x) = E_x\left(\exp\left(\int_0^{\sigma_{\bar{\Omega}_{r,\alpha,n}}} V(X(s))\,ds\right) h_n(X(\sigma_{\bar{\Omega}_{r,\alpha,n}}));\ \sigma_{\bar{\Omega}_{r,\alpha,n}} < \tau_{D_n}\right), \quad \text{for } x \in \Omega_{r,\beta,n}. \tag{7.2}$$

By exercise 8.8, $\lim_{n\to\infty} g_n = g_\alpha$, where g_α has been normalized by $g_\alpha(r) = 1$, and a standard compactness argument shows that $\lim_{n\to\infty} \phi_n^{(i)} = \phi_0^{(i)}$; thus $\lim_{n\to\infty} h_n = h_\alpha$. Since $\sigma_{\bar{\Omega}_{r,\alpha,n}} = \sigma_{\bar{\Omega}_{r,\alpha}}$ on $\{\sigma_{\bar{\Omega}_{r,\alpha,n}} < \tau_{D_n}\}$, and since $\{\sigma_{\bar{\Omega}_{r,\alpha,n}} < \tau_{D_n}\}$ is increasing in n, (ii) follows from (7.2) upon letting $n \to \infty$.

(iii) We will prove the claim for μ_x^β, $x \in \Omega_{r,\alpha}$. Just as in Theorem 7.4.5, we have

$$\mu_x^{h_\beta}(dy) = \frac{1}{h_\beta(x)} E_x\left(\exp\left(\int_0^{\sigma_{\bar{\Omega}_{r,\beta}}} V(X(s))\,ds\right) h_\beta(X(\sigma_{\bar{\Omega}_{r,\beta}}));\ \sigma_{\bar{\Omega}_{r,\beta}} < \tau_D;\ X(\sigma_{\bar{\Omega}_{r,\beta}}) \in dy\right), \quad \text{for } x \in \Omega_{r,\alpha}. \tag{7.3}$$

The proof is left to the reader in exercise 8.9. Thus it follows from part (ii) that $\mu_x^{h_\beta}$, $x \in \Omega_{r,\alpha}$, is a probability measure on $D \cap \{x_1 = r\}$. Writing (7.1b) in terms of (7.3) gives

$$G(x, y) = h_\beta(x) \int_{D\cap\{x_1 = r\}} \frac{G(z, y)}{h_\beta(z)} \mu_x^{h_\beta}(dz), \text{ for } x \in \Omega_{r,\alpha} \text{ and } y \in \Omega_{r,\beta}. \tag{7.4}$$

Define the Martin kernel $\tilde{k}(x, y)$ for $\tilde{L}$ on D with respect to some point $x_0 \in \Omega_{r,\beta}$, and let $\{y^{(n)}\}_{n=1}^{\infty} \subset \Omega_{r,\alpha}$ be a sequence with no accumulation points in D which satisfies $\limsup_{n\to\infty} y_1^{(n)} < r$. Then from (7.4),

$$\tilde{k}(x, y^{(n)}) = \frac{\tilde{G}(x, y^{(n)})}{\tilde{G}(x_0, y^{(n)})} = \frac{G(y^{(n)}, x)}{G(y^{(n)}, x_0)} = \frac{\displaystyle\int_{D\cap\{x_1=r\}} \frac{G(z, x)}{h_\beta(z)} \mu_{y^{(n)}}^{h_\beta}(dz)}{\displaystyle\int_{D\cap\{x_1=r\}} \frac{G(z, x_0)}{h_\beta(z)} \mu_{y^{(n)}}^{h_\beta}(dz)},$$

$$\text{for } x \in \Omega_{r,\beta}. \quad (7.5)$$

For $x \in \Omega_{r,\beta}$,

$$\frac{G(z, x)}{h_\beta(z)}$$

as a function of z is continuous on $D \cap \{x_1 = r\}$ and, by Theorem 1.1 and the boundary Harnack principle, it is also bounded there. We want to conclude that

$$\text{if } \gamma \equiv w - \lim_{n\to\infty} \mu_{y^{(n)}}^{h_\beta} \text{ exists, then } \lim_{n\to\infty} \int_{D\cap\{x_1=r\}} \frac{G(z, x)}{h_\beta(z)} \mu_{y^{(n)}}^{h_\beta}(dz)$$

$$= \int_{D\cap\{x_1=r\}} \frac{G(z, x)}{h_\beta(z)} \gamma(dz), \quad \text{for } x \in \Omega_{r,\beta}. \quad (7.6)$$

In order to establish (7.6), we need to know either that

$$\frac{G(z, x)}{h_\beta(z)}$$

is continuous on $\bar{D} \cap \{x_1 = r\}$ or that γ gives no mass to $\partial D \cap \{x_1 = r\}$. In exercise 8.10, the reader is asked to show that γ gives no mass to $\partial D \cap \{x_1 = r\}$. (The function h_β is differentiable up to the boundary $D^{(1)} \times \partial(\times_{i=2}^{k} D^{(i)})$, since $\phi_0^{(i)}$ is differentiable up to the boundary $\partial D^{(i)}$. Thus the continuity of

$$\frac{G(z, x)}{h_\beta(z)} \text{ on } \bar{D} \cap \{x_1 = r\}$$

would follow if we showed that $G\{z, y\}$ is differentiable up to the boundary $D^{(1)} \times \partial(\times_{i=2}^{k} D^{(i)})$. This is probably true, but goes beyond the techniques of this book. In particular, the corners, where the boundary ∂D is only Lipschitz, can cause some technical problems.) Once we have (7.5) and (7.6), the proof of (iii) is very similar to the proof of Theorem 7.6.2; the details are left to the reader in exercise 8.11.

□

Proposition 7.2. *For $i \in \{2, \ldots, k\}$, let $L^{(i)}$, $\lambda_0^{(i)}$, $\phi_0^{(i)}$, and $D^{(i)}$ be as in Assumption SP, and let $(L^{(i)} - \lambda_0^{(i)})^{\phi_0^{(i)}}$ denote the h-transform of*

$L^{(i)} - \lambda_0^{(i)}$ via the function $\phi_0^{(i)}$. Let $\{P_{x_i}^{(i)}, x_i \in D^{(i)}\}$ denote the solution to the martingale problem for $(L^{(i)} - \lambda_0^{(i)})^{\phi_0^{(i)}}$ on $D^{(i)}$, and let $X_i(t)$ denote the canonical path. (The solution to the martingale problem exists because, by Theorems 4.3.2 and 4.3.3, $(L^{(i)} - \lambda_0^{(i)})^{\phi_0^{(i)}}$ corresponds to a recurrent diffusion.) Let $\mathcal{D}_\alpha^{(i)} = \{u \in C^{2,\alpha}(\bar{D}^{(i)}): u = 0$ and $L^{(i)}u = 0$ on $\partial D^{(i)}\}$ and $A^{(i)} = \{g \in C^{2,\alpha}(\bar{D}^{(i)}): g\phi_0^{(i)} \in \mathcal{D}_\alpha^{(i)}\}$. Define

$$G_g(x_i, t) = \int_{D^{(i)}} g(x_i')P_{x_i}^{(i)}(X_i(t) \in dx_i'), \quad \textit{for } x_i \in D^{(i)}.$$

Let $\{x_i^{(n)}\}_{n=1}^\infty$ satisfy $\lim_{n\to\infty} x_i^{(n)} = x_i \in \bar{D}^{(i)}$. Then for $g \in A^{(i)}$,

$$H_g(x_i, t) \equiv \lim_{n\to\infty} G_g(x_i^{(n)}, t) \textit{ exists and the convergence is uniform over } t$$

$$\textit{in compact subsets of } [0, \infty). \tag{7.7}$$

Also

$$\lim_{t\to 0} H_g(x_i, t) = g(x_i). \tag{7.8}$$

Remark. The proof we give below of (7.7) uses the semigroup theory developed in Sections 3.4 and 3.5. However, if we use a result from parabolic PDE theory, then the proof of (7.7), with $[0, \infty)$ replaced by $(0, \infty)$, can be given easily as follows. (The change from $[0, \infty)$ to $(0, \infty)$ does not affect the proof of Theorem 6.1.) Let $p(t, x_i, dx_i')$ denote the transition measure for $L^{(i)}$ on $D^{(i)}$. It is known that $p(t, x_i, dx_i')$ possesses a density $p(t, x_i, x_i')$, $t > 0$, $x_i, x_i' \in \bar{D}^{(i)}$, such that $\lim_{x_i\to\partial D^{(i)}} p(t, x_i, x_i') = 0$, for $t > 0$ and $x_i' \in \bar{D}^{(i)}$ and such that the first partial derivatives in x_i exist on $\bar{D}^{(i)}$ and are continuous over x_i, $x_i', \in \bar{D}^{(i)}$ and $t > 0$. It follows from Theorem 4.1.1 and the definition of the transition measure that $P_{x_i}^{(i)}(X_i(t) \in dx_i')$, the transition measure for $(L^{(i)} - \lambda_0^{(i)})^{\phi_0^{(i)}}$, possesses the density

$$\exp(-\lambda_0^{(i)}t)\frac{\phi_0^{(i)}(x_i')}{\phi_0^{(i)}(x_i)}p(t, x_i, x_i').$$

This proves (7.7) in the case where $x_i \in D^{(i)}$. Since $\phi_0^{(i)}$ is differentiable on $\bar{D}^{(i)}$ and vanishes on $\partial D^{(i)}$, in the case where $x_i \in \partial D^{(i)}$, (7.7) follows by an application of the mean-value theorem and the bounded convergence theorem.

Proof. For convenience, we delete the superscript and subscript i. Let $\mathcal{B}_0 = \{u \in C(\bar{D}): u = 0$ on $\partial D\}$, and let T_t denote the semigroup on $\mathcal{B}_0$ as in Section 3.4. (The infinitessimal generator of T_t is an extension of $(L-\lambda_0, C_0^\infty(D))$.) As in Section 3.5, let $\mathcal{B}_\alpha = \{u \in C^\alpha(\bar{D}): u = 0$ on $\partial D\}$. By the Hille–Yoshida–Phillips theorem [Reed and Simon, Vol. 2, (1975), Theorem 10.47b] and the results of Section 3.5, the operator $(L-\lambda_0, \mathcal{D}_\alpha)$ on $\mathcal{B}_\alpha$ is the infintesimal generator of a strongly continuous semigroup S_t on $\mathcal{B}_\alpha$, where $\mathcal{D}_\alpha = \text{Ran}((L-\lambda_0)^{-1}, \mathcal{B}_\alpha) = \{u \in C^{2,\alpha}(\bar{D}) \cap \mathcal{B}_\alpha: Lu = 0 \text{ on } \partial D\}$. It is easy to deduce from the above that $T_t = S_t$ on $\mathcal{B}_\alpha$. By exercise 3.3, $S_t\mathcal{D}_\alpha \subset \mathcal{D}_\alpha$; thus $T_t\mathcal{D}_\alpha \subset \mathcal{D}_\alpha$.

By Theorem 4.1.1,

$$G_g(x,t) = \frac{(T_t g\phi_0)(x)}{\phi_0(x)}.$$

In particular, then, $G_g(x,t)$ is continuous for $x \in D$; thus the limit in (7.7) exists for $x \in D$ and $H_g(x,t) = G_g(x,t)$. Now consider $x \in \partial D$. Since $g \in A$, it follows that $T_t g\,\phi_0 \in \mathcal{D}_\alpha$. By Theorem 3.5.5, $\phi_0 \in \mathcal{D}_\alpha$. Writing

$$G_g(x^{(n)},t) = \frac{(T_t g\phi_0)(x^{(n)})}{\text{dist}\,(x^{(n)}, \partial D)}\,\frac{\text{dist}\,(x^{(n)}, \partial D)}{\phi_0(x^{(n)})}$$

and using the mean value theorem and the fact that, by the Hopf maximum principle (Theorem 3.2.5), the normal derivative of ϕ_0 on ∂D does not vanish, it follows that $H_g(x,t) \equiv \lim_{n\to\infty} G_g(x^{(n)},t)$ exists for $x \in \partial D$. To complete the proof of (7.7), we must show that this limit is uniform over t in compact subsets of $[0,\infty)$.

We will show that this limit is uniform over compact sets of $[0,\infty)$ by showing that $\|T_t g\phi_0\|_{2,\alpha;D}$ is bounded for t in compact subsets of $[0,\infty)$. It follows from the global Shauder estimate (Theorem 3.2.8) that

$$\|T_t g\phi_0\|_{2,\alpha;D} \le c(\|T_t g\phi_0\|_{0;D} + \|LT_t g\phi_0\|_{0,\alpha;D}). \tag{7.9}$$

Since $g\phi_0 \in \mathcal{D}_\alpha$, it follows from exercise 3.3 that $LT_t g\phi_0 = T_t L g\phi_0$. The boundedness of the righthand side of (7.9) for t in compact subsets of $[0,\infty)$ now follows from the fact that T_t is strongly continuous on $\mathcal{B}_\alpha$.

The proof of (7.8) is left to exercise 8.12. □

We now prove Theorem 6.1.

Proof of Theorem 6.1. (a) Let $\tilde{I}_\alpha$, $\tilde{I}_\beta$, $\tilde{g}_\alpha$ and $\tilde{g}_\beta$ correspond to $\tilde{L}$ as I_α, I_β, g_α and g_β correspond to L. Note that $\tilde{I}_\alpha$ ($\tilde{I}_\beta$) is obtained by replacing b by $a' - b$ in I_α (I_β). A straightforward calculation reveals that a function h solves $\mathcal{L}h = 0$ if and only if

$$\tilde{h}(r) \equiv \frac{h(r)}{a(r)} \exp\left(\int_{r_0}^{r} \frac{2b}{a}(s)\, ds\right)$$

satisfies $\tilde{\mathscr{L}}\tilde{h} = 0$; thus, it follows easily that

$$\tilde{g}_\alpha(r) = \frac{g_\alpha(r)}{a(r)} \exp\left(\int_{r_0}^{r} \frac{2b}{a}(s)\, ds\right)$$

and

$$\tilde{g}_\beta(r) = \frac{g_\beta(r)}{a(r)} \exp\left(\int_{r_0}^{r} \frac{2b}{a}(s)\, ds\right).$$

A simple calculation now confirms that $\tilde{I}_\alpha(\tilde{g}_\beta, M^{(i)}) = I_\alpha(g_\beta, M^{(i)})$ and $\tilde{I}_\beta(\tilde{g}_\alpha, M^{(i)}) = I_\beta(g_\alpha, M^{(i)})$. Thus, part (a) follows from part (b) and part (c).

(b) Since $D^{(1)} \times \partial(\times_{i=2}^{k} D^{(i)})$ is a Lipschitz boundary, (b) follows from Theorem 1.4.

(c) We will prove that the Martin boundary of $\tilde{L}$ behaves as in the statement of the theorem. The corresponding statement for L then follows from part (a). We will treat the endpoint α; the identical method works at β. Let $r \in (\alpha, \beta)$ and let $\Omega_{r,\alpha}$ and $\mu_x^{h_\beta}$ be as in Proposition 7.1. Let $\{x^{(n)}\}_{n=1}^{\infty} \subset \Omega_{r,\alpha}$ be a sequence satisfying $\lim_{n\to\infty} x_1^{(n)} = \alpha$. By Proposition 7.1, $\{x^{(n)}\}_{n=1}^{\infty}$ is a Martin sequence for $\tilde{L}$ if and only if $w - \lim_{n\to\infty} \mu_{x^{(n)}}^{h_\beta}$ exists. Thus, to prove (c), we must prove the following statement:

$$w - \lim_{n\to\infty} \mu_{x^{(n)}}^{h_\beta} \text{ exists if and only if}$$

$$\lim_{n\to\infty} x_i^{(n)} \in \bar{D}^{(i)} \text{ exists for all } i \in N_{\text{fin}}^{\alpha}. \tag{7.10}$$

A direct calculation shows that

$$L^{h_\beta} = \tfrac{1}{2}a(x_1)\frac{\partial^2}{\partial x_1^2} + b(x_1)\frac{\partial}{\partial x_1} + a(x_1)\frac{g'_\beta}{g_\beta}(x_1)\frac{\partial}{\partial x_1}$$
$$+ \sum_{i=2}^{k} M^{(i)}(x_1)(L^{(i)} - \lambda_0^{(i)})^{\phi_0^{(i)}},$$

where $(L^{(i)} - \lambda_0^{(i)})^{\phi_0^{(i)}}$ is the h-transform of $L^{(i)} - \lambda_0^{(i)}$ via $\phi_0^{(i)}$. Let $\{P_x^{h_\beta}, x \in \hat{D}\}$ denote the solution to the generalized martingale problem for L^{h_β} on D, and denote the canonical path by $X(t)$. Let $\{P_{x_1}^{(1)}, x \in (\alpha, \beta)\hat{}\}$ denote the solution to the generalized martingale problem for

$$\tfrac{1}{2}a(x_1)\frac{d^2}{dx_1^2} + b(x_1)\frac{d}{dx_1} + a(x_1)\frac{g'_\beta}{g_\beta}(x_1)\frac{d}{dx_1}$$

on (α, β) and denote the canonical path by $X_1(t)$; ($(\alpha, \beta)^\wedge$ denotes the one-point compactification of (α, β).) For each $i \in \{2, \ldots, k\}$, $L^{(i)} - \lambda_0^{(i)}$ is critical by the final statement in Theorem 4.3.2. Thus, by Theorem 4.3.3, the diffusion corresponding to $(L^{(i)} - \lambda_0^{(i)})^{\phi_0^{(i)}}$ is recurrent. Let $\{P_{x_i}^{(i)}, x_i \in D^{(i)}\}$ denote the solution to the martingale problem for $(L^{(i)} - \lambda_0^{(i)})^{\phi_0^{(i)}}$ on $D^{(i)}$, and let $X_i(t)$ denote the canonical path.

Define $M^{(1)}(x_1) \equiv 1$, $\rho_i(t) = \int_0^t M^{(i)}(X_1(s))\, ds$, $Q_x = \times_{i=1}^k P_{x_i}^{(i)}$ and $Z(t) = (X_1(\rho_1(t)), \ldots, X_k(\rho_k(t)))$. By Theorem 5.1, the distribution of $X(\cdot)$ under $P_x^{h_\beta}$ coincides with the distribution of $Z(\cdot)$ under Q_x. Denote expectations with respect to $P_{x_i}^{(i)}$ by $E_{x_i}^{(i)}$ and with respect to Q_x by E_x^Q. We can now express $\mu_x^{h_\beta}$ in terms of the skew-product representation. $\mu_x^{h_\beta}$, $x \in \Omega_{r,\alpha}$, is a measure on $\{r\} \times (\times_{i=2}^k D^{(i)})$; for convenience we consider it as a measure on $\times_{i=2}^k D^{(i)}$. Let $\tau_r = \inf\{t \geq 0: X_1(t) = r\}$. Using the fact that Q_x is a product measure and that $\rho_i(\tau_r)$ is measurable with respect to $\sigma(X_1(t), t \geq 0)$, we have for $x \in \Omega_{r,\alpha}$,

$$\mu_x^{h_\beta}(dx_2', \ldots, dx_k') = Q_x(X_2(\rho_2(\tau_r)) \in dx_2', \ldots, X_k(\rho_k(\tau_r)) \in dx_k')$$
$$= \int \left[\prod_{i=2}^k P_{x_i}^{(i)}(X_i(t_i) \in dx_i')\right] \nu_{x_1,r}(dt_2, \ldots, dt_k), \tag{7.11}$$

where

$$\nu_{x_1,r}(dt_2, \ldots, dt_k) = P_{x_1}^{(1)}(\rho_2(\tau_r) \in dt_2, \ldots, \rho_k(\tau_r) \in dt_k),\ x_1 \in (\alpha, r). \tag{7.12}$$

We now prove the following two statements:

$$\text{If } i \notin N_{\text{fin}}^\alpha \text{ (i.e., } I_\alpha(g_\beta, M^{(i)}) = \infty\text{), then } \lim_{x_1 \to \alpha} P_{x_1}^{(1)}(\rho_i(\tau_r) \leq t) = 0, \quad \text{for all } t > 0. \tag{7.13}$$

$$\text{If } i \in N_{\text{fin}}^\alpha \text{ (i.e., } I_\alpha(g_\beta, M^{(i)}) < \infty\text{),}$$
$$\text{then } \gamma_r^{(i)}(dt) \equiv w - \lim_{x_1 \to \alpha} P_{x_1}^{(1)}(\rho_i(\tau_r) \in dt)$$
$$\text{exists as a probability measure on } (0, \infty) \text{ and}$$
$$\lim_{r \to \alpha} \gamma_r^{(i)}(dt) = \delta_{(0)}(dt), \text{ the atom with mass one at zero.} \tag{7.14}$$

For $i \in \{2, \ldots, k\}$, let $\gamma_i(t)$, $0 \leq t \leq \rho_i(\tau_r)$, denote the inverse of $\rho_i(t)$, $0 \leq t \leq \tau_r$, and define $Y_i(t) = X_1(\gamma_i(t))$. Then $\inf\{t \geq 0: Y_i(t) = r\} = \rho_i(\tau_r)$; that is, $\rho_i(\tau_r)$ is the first hitting time of r by the process $Y_i(t)$. By the time-change formula (exercise 1.15), the distribution of $Y_i(\cdot)$ under $\{P_{x_1}^{(1)}, x_1 \in (\alpha, \beta)^\wedge\}$ coincides with the solution to the generalized martingale problem for the operator

$$M^{(i)}(x_1)\left(\tfrac{1}{2}a(x_1)\frac{d^2}{dx_1^2} + b(x_1)\frac{d}{dx_1} + a(x_1)\frac{g'_\beta}{g_\beta}(x_1)\frac{d}{dx_1}\right)$$

on (α, β). Thus, the distribution of $\rho_i(\tau_r)$ under $P_{x_1}^{(1)}$ may be identified with the distribution of the first hitting time of r for the solution to the generalized martingale problem for the above operator. Equations (7.13) and (7.14) now follow from Theorem 4.1.

Using (7.13) and (7.14), we prove (7.10). For definiteness, assume that there exists a $j \in \{2, \ldots, k\}$ such that $i \notin N_{\text{fin}}^\alpha$, for $i = 2, \ldots, j$, and $i \in N_{\text{fin}}^\alpha$, for $i = j+1, \ldots, k$. (A similar proof holds if $N_{\text{fin}}^\alpha = \phi$ or $N_{\text{fin}}^\alpha = \{2, \ldots, k\}$.) Let $\bar{\nu}_{x_1,r}(dt_{j+1}, \ldots, dt_k)$ denote the marginal distribution of $\nu_{x_1,r}(dt_2, \ldots, dt_k)$ with respect to the variables $t_{j+1}, \ldots, t_k$. Clearly, $\bar{\nu}_{x_1,r}$ is monotone non-increasing in x_1 and non-decreasing in r. Thus, it follows from (7.14) that

$$\bar{\nu}_r \equiv w - \lim_{x_1 \to \alpha} \bar{\nu}_{x_1,r} \text{ exists} \tag{7.15a}$$

and

$$\lim_{r \to \alpha} \bar{\nu}_r(dt_{j+1}, \ldots, dt_k) = \prod_{i=j+1}^{k} \delta_{(0)}(dt_i). \tag{7.15b}$$

Since finite linear combinations of functions of the form $f(x_2, \ldots, x_j)g(x_{j+1}, \ldots, x_k)$, where $f \in C(\times_{i \notin N_{\text{fin}}^\alpha} \bar{D}^{(i)})$ and $g \in C(\times_{i \in N_{\text{fin}}^\alpha} \bar{D}^{(i)})$, are dense in $C(\times_{i=2}^k \bar{D}^{(i)})$, in order to investigate the weak convergence of $\mu_{x^{(n)}}^{h_\beta}$, it is enough to study the behavior of integrals of the form $\int f(x'_2, \ldots, x'_j)g(x'_{j+1}, \ldots, x'_k)\mu_{x^{(n)}}^{h_\beta}(dx'_2, \ldots, dx'_k)$, where f and g are as above. We note for later use that in fact, of course, it is enough to study the behavior of such integrals for $f \in C(\times_{i \notin N_{\text{fin}}^\alpha} \bar{D}^{(i)})$ and g in a dense subset of $C(\times_{i \in N_{\text{fin}}^\alpha} \bar{D}^{(i)})$. Let $F_f(x_2, \ldots, x_j, t_2, \ldots, t_j) = \int f(x'_2, \ldots, x'_j) \prod_{i=2}^j P_{x_i}^{(i)}(X_i(t_i) \in dx'_i)$ and let $G_g(x_{j+1}, \ldots, x_k, t_{j+1}, \ldots, t_k) = \int g(x'_{j+1}, \ldots, x'_k) \prod_{i=j+1}^k P_{x_i}^{(i)}(X_i(t_i) \in dx'_i)$. By (7.11) and (7.12),

$$\begin{aligned}\int f(x'_2, \ldots, x'_j)g(x'_{j+1}, \ldots, x'_k)\mu_x^{h_\beta}(dx'_2, \ldots, dx'_k)\\ = \int F_f(x_2, \ldots, x_j, t_2, \ldots, t_j)G_g(x_{j+1}, \ldots, x_k, t_{j+1}, \ldots, t_k)\\ \times \nu_{x_1,r}(dt_2, \ldots, dt_k).\end{aligned} \tag{7.16}$$

Since $P_{x_i}^{(i)}$, $i \in \{2, \ldots, k\}$, corresponds to a positive recurrent diffusion with invariant density $\phi_0^{(i)}(x_i')\tilde{\phi}_0^{(i)}(x_i')$, it follows from Theorem 4.9.9 that $w - \lim_{t\to\infty} P_{x_i}^{(i)}(X_i(t) \in dx_i') = \phi_0^{(i)}(x_i')\tilde{\phi}_0^{(i)}(x_i')\,dx_i'$, and the convergence is uniform over x in any compact subset of $D^{(i)}$. A check of the proof of Theorem 4.9.9. reveals that if the compact set K appearing in (4.9.15) can be replaced by D, then the convergence in Theorem 4.9.9 is in fact uniform over all of D, not just over compact subsets. Considering (4.9.15) in the present case with $D = D^{(i)}$, it follows from Theorem 4.2 that (4.9.15) holds with $D^{(i)}$ in place of K. Thus

$$w - \lim_{t\to\infty} P_{x_i}^{(i)}(X_i(t) \in dx_i') = \phi_0^{(i)}(x_i')\tilde{\phi}_0^{(i)}(x_i')\,dx_i',$$

$$\text{and the convergence is uniform over } x \in D^{(i)}. \quad (7.17)$$

By (7.12), (7.13), (7.16), (7.17) and the definition of F_f, it follows that $\lim_{n\to\infty} \int f(x_2', \ldots, x_j')g(x_{j+1}', \ldots, x_k')\mu_{x^{(n)}}^{h_\beta}(dx_2', \ldots, dx_k')$, if it exists, does not depend on $\{(x_2^{(n)}, \ldots, x_j^{(n)})\}_{n=1}^{\infty}$. Thus, to complete the proof of (7.10), we must show that for all g in a dense subset of $C(\times_{i=j+1}^{k} \bar{D}^{(i)})$,

$$H_g(x_{j+1}, \ldots, x_k) \equiv$$

$$\lim_{n\to\infty} \int G_g(x_{j+1}^{(n)}, \ldots, x_k^{(n)}, t_{j+1}, \ldots, t_k)\bar{\nu}_{x_1^{(n)},r}(dt_{j+1}, \ldots, dt_k)$$

$$\text{exists if } \lim_{n\to\infty} x_i^{(n)} = x_i \in \bar{D}^{(i)},\ i = j+1, \ldots, k, \quad (7.18)$$

and that

$$\text{if } (x_{j+1}, \ldots, x_k) \neq (x_{j+1}', \ldots, x_k'), \text{ then there exists a } g \text{ such that}$$

$$H_g(x_{j+1}, \ldots, x_k) \neq H_g(x_{j+1}', \ldots, x_k'). \quad (7.19)$$

We first prove (7.18). Restrict g to the dense subset of $C(\times_{i=j+1}^{k} \bar{D}^{(i)})$ which consists of finite linear combinations of functions of the form $\prod_{i=j+1}^{k} g_i(x_i')$, where $g_i \in A^{(i)}$. ($A^{(i)}$ is as in Proposition 7.2.) For g as above, (7.18) follows from (7.7) in Proposition 7.2 and (7.15a).

We now consider (7.19). By Proposition 7.1, the existence of $w - \lim_{n\to\infty} \mu_{x^{(n)}}^{h_\beta}$ does not depend on the choice of $r \in (\alpha, \beta)$. Thus, it suffices to prove (7.19) with r chosen close to α. From (7.8) in Proposition 7.2, it follows that

$$\lim_{\substack{t_i\to 0\\ i=j+1,\ldots,k}} G_g(x_{j+1}, \ldots, x_k, t_{j+1}, \ldots, t_k) = g(x_{j+1}, \ldots, x_k),$$

$$\text{for } (x_{j+1}, \ldots, x_k) \in \mathop{\times}_{i=j+1}^{k} \bar{D}^{(i)}. \quad (7.20)$$

It follows from (7.15b) and (7.20) that by choosing r sufficiently close to α, $H_g(x_{j+1}, \ldots, x_k)$ can be made arbitrarily close to $g(x_{j+1}, \ldots, x_k)$. Thus, by choosing a g which satisfies $g(x_{j+1}, \ldots, x_k) \neq g(x'_{j+1}, \ldots, x'_k)$, and by choosing r sufficiently close to α, (7.19) will hold.

(d) We leave it to the reader to check that it suffices to consider the case $k = 2$. We will prove that $I_\alpha(g_\beta, M^{(2)}) < \infty$ if and only if $J_\alpha(g, M^{(2)}) < \infty$, for $g = c_1 g_\alpha + c_2 g_\beta$, with $c_1 > 0$ and $c_2 \geq 0$. The proof of the corresponding statement with the roles of α and β and the roles of c_1 and c_2 switched is obtained identically.

Fix $r_0 \in (\alpha, \beta)$ and let $\Omega_{r_0,\beta}$ be as in Proposition 7.1. Let $h(x) = g(x_1)\phi_0^{(2)}(x_2)$, where $g = c_1 g_\alpha + c_2 g_\beta$, with $c_1 > 0$ and $c_2 \geq 0$. Let $\{P_x^h, x \in \hat{D}\}$ denote the solution to the generalized martingale problem for L^h on D. For a domain $B \subseteq D^{(2)}$, define

$$u_B(x) = P_x^h(\lim_{t\to\tau_D} \operatorname{dist}^{R^d}(X_1(t), \alpha) = 0$$
$$\text{and } X_2(t) \in B \text{ for all } t \text{ sufficiently close to } \tau_D),\ x \in D.$$

(Recall that we must write $\lim_{t\to\tau_D} \operatorname{dist}^{R^d}(X_1(t), \alpha) = 0$ rather than $\lim_{t\to\tau_D} X_1(t) = \alpha$, because in the framework of the generalized martingale problem, $\lim_{t\to\tau_D} X_1(t) = \Delta$, where Δ is the point at infinity in the one-point compactification $(\alpha, \beta)\hat{}$ of (α, β).) We will prove the following two statements:

$$\text{If } J_\alpha(g, M^{(2)}) = \infty, \text{ then } u_{D^{(2)}} > 0, \text{ but } u_B \equiv 0,$$
$$\text{for every domain } B \subset D^{(2)} \text{ for which } D^{(2)} - \bar{B} \text{ is non-empty.} \quad (7.21)$$

If $J_\alpha(g, M^{(2)}) < \infty$, then $u_B > 0$ for every domain $B \subseteq D^{(2)}$. Furthermore,

$$\left.\begin{aligned} &\lim_{n\to\infty} u_B(x^{(n)}) = 1, \text{ if } \lim_{n\to\infty} x_1^{(n)} = \alpha \text{ and } \lim_{n\to\infty} x_2^{(n)} \in B, \\ &\text{and} \\ &\lim_{n\to\infty} u_B(x^{(n)}) = 0, \text{ if } \lim_{n\to\infty} x_1^{(n)} = \alpha \text{ and } \lim_{n\to\infty} x_2^{(n)} \in D^{(2)} - \bar{B}. \end{aligned}\right\} \quad (7.22)$$

Using (7.21) and (7.22), we now prove (d); then we return to prove (7.21) and (7.22).

By Proposition 7.1.4, the Martin boundaries of L on D and L^h on D coincide; thus by (c), the following dichotomy holds.

$$\text{If } I_\alpha(g_\beta, M^{(2)}) = \infty, \text{ then every sequence } \{x^{(n)}\}_{n=1}^\infty = \{(x_1^{(n)}, x_2^{(n)})\}_{n=1}^\infty$$
$$\text{satisfying } \lim_{n\to\infty} x_1^{(n)} = \alpha \text{ is a Martin sequence for } L^h \text{ on } D. \quad (7.23)$$

If $I_\alpha(g_\beta, M^{(2)}) < \infty$, then a sequence $\{x^{(n)}\}_{n=1}^\infty = \{(x_1^{(n)}, x_2^{(n)})\}_{n=1}^\infty$ satisfying $\lim_{n\to\infty} x_1^{(n)} = \alpha$ is a Martin sequence for L^h on D

$$\text{if and only if } \lim_{n\to\infty} x_2^{(n)} \text{ exists on } \bar{D}^{(2)}. \quad (7.24)$$

Assume first that $J_\alpha(g, M^{(2)}) = \infty$. By (7.21), it follows that

$$P_x^h(\lim_{t\to\tau_D} \mathrm{dist}_{R^d}(X(t), \alpha) = 0 \text{ and } \lim_{t\to\tau_D} X_2(t) \text{ does not exist}) > 0,$$
$$\text{for } x \in D. \quad (7.25)$$

By Theorem 7.2.2, the diffusion process almost surely convergences in the Martin topology; thus (7.25) is incompatible with (7.24), and we conclude that $I_\alpha(g_\beta, M^{(2)}) = \infty$.

Now assume that $J_\alpha(g, M^{(2)}) < \infty$. We will assume that $I_\alpha(g_\beta, M^{(2)}) = \infty$ and come to a contradiction as follows. For each point ζ in the Martin boundary Λ for L^h on D, let $k^h(x; \zeta)$ denote the corresponding harmonic function. Since $I_\alpha(g_\beta, M^{(2)}) = \infty$, (7.23) holds; let ζ_0 denote the Martin boundary point corresponding to sequences $\{x^{(n)}\}_{n=1}^\infty$ satisfying $\lim_{n\to\infty} x_1^{(n)} = \alpha$. Then

$$\lim_{x_1\to\alpha} k^h(x_1, x_2; \zeta) = 0, \quad \text{for } x_2 \in D^{(2)} \text{ and } \zeta \in \Lambda - \{\zeta_0\}, \quad (7.26)$$

and

$$\limsup_{x_1\to\alpha} \sup_{\zeta\in\Lambda-\{\zeta_0\}} k^h(x_1, x_2; \zeta) < \infty, \quad \text{for } x_2 \in D^{(2)}. \quad (7.27)$$

The proofs of (7.26) and (7.27) are left as exercise 8.13. By Theorem 9.1.2 in Chapter 9, $u_B \in C_{L^h}(D)$. It follows from the Martin representation theorem (Theorem 7.1.2) that for each $B \subseteq D^{(2)}$, there exists a finite measure μ_B on $\Lambda - \{\zeta_0\}$ and a constant $c_B \geqslant 0$ such that

$$u_B(x) = \int_{\Lambda-\{\zeta_0\}} k^h(x; \zeta)\mu_B(d\zeta) + c_B k^h(x; \zeta_0). \quad (7.28)$$

From (7.26) and (7.27), we have

$$\lim_{x_1\to\alpha} \int_{\Lambda-\{\zeta_0\}} k^h(x_1, x_2; \zeta)\mu_B(d\zeta) = 0, \quad \text{for } x_2 \in D^{(2)}. \quad (7.29)$$

Thus from (7.28), (7.29) and (7.22), it follows that for every $B \subseteq D^{(2)}$,

$$\lim_{n\to\infty} k^h(x^{(n)}; \zeta_0) > 0, \text{ if } \lim_{n\to\infty} x_1^{(n)} = \alpha \text{ and } \lim_{n\to\infty} x_2^{(n)} \in B,$$

and

$$\lim_{n\to\infty} k^h(x^{(n)}; \zeta_0) = 0, \text{ if } \lim_{n\to\infty} x_1^{(n)} = \alpha \text{ and } \lim_{n\to\infty} x_2^{(n)} \in D^{(2)} - \bar{B}.$$

This is, of course, impossible since $k^h(x^{(n)}; \zeta_0)$ does not depend on B; thus in fact $I_\alpha(g_\beta, M^{(2)}) < \infty$.

We now return to prove (7.21) and (7.22). The proof of (7.21) and (7.22) is similar to the proof of (7.13) and (7.14) in part (c). The difference is that there we applied the criterion for explosion inward from the boundary given in Theorem 4.1, while in the present situation, we will apply Feller's criterion for explosion in Theorem 5.1.5. We make use of the skew-product representation, using the notation introduced in the proof of part (c) with the one difference that g_β is replaced by $g = c_1 g_\alpha + c_2 g_\beta$, with $c_1 > 0$ and $c_2 \geq 0$, and $h_\beta(x) = g_\beta(x_1)\phi_0^{(2)}(x_2)$ is replaced by $h(x) = g(x_1)\phi_0^{(2)}(x_2)$. Define $\tau_\alpha = \lim_{r\to\alpha} \tau_r$ and $\tau_\beta = \lim_{r\to\beta} \tau_r$. Note that $\rho_2(\tau_\alpha \wedge \tau_\beta) = \rho_2(\tau_\alpha) \wedge \rho_2(\tau_\beta)$. Let $\gamma_2(t)$, $0 \leq t < \rho(\tau_\alpha) \wedge \rho(\tau_\beta)$, denote the inverse of $\rho_2(t)$, $0 \leq t < \tau_\alpha \wedge \tau_\beta$, and define $Y_2(t) = X_1(\gamma_2(t))$. Then $\inf\{t \geq 0: Y_2(t) = r\} = \rho_2(\tau_r)$, $\alpha < r < \beta$; that is, $\rho_2(\tau_r)$ is the first hitting time of r by the process $Y_2(t)$. By the time-change formula (exercise 1.15), the distribution of $Y_2(\cdot)$ under $\{P_{x_1}^{(1)}, x_1 \in (\alpha, \beta)\hat{}\}$ coincides with the solution to the generalized martingale problem for the operator

$$M^{(2)}(x_1)\left(\tfrac{1}{2}a(x_1)\frac{\mathrm{d}^2}{\mathrm{d}x_1^2} + b(x_1)\frac{\mathrm{d}}{\mathrm{d}x_1} + a(x_1)\frac{g'}{g}(x_1)\frac{\mathrm{d}}{\mathrm{d}x_1}\right) \quad \text{on } (\alpha, \beta).$$

Thus, the distribution of $\rho_2(\tau_r)$ under $P_{x_1}^{(1)}$ may be identified with the distribution of the first hitting time of r for the solution to the generalized martingale problem for the above operator. Applying Theorem 5.1.5 to the above operator, it follows that

$$\text{if } J_\alpha(g, M^{(2)}) = \infty, \text{ then } P_x^{(1)}(\rho_2(\tau_\alpha) < \infty) = 0, \text{ for } x \in (\alpha, \beta), \tag{7.30}$$

and

$$\text{if } J_\alpha(g, M^{(2)}) < \infty, \text{ then } \lim_{x_1\to\alpha} P_x^{(1)}(\rho_2(\tau_\alpha) < \varepsilon) = 1, \text{ for } \varepsilon > 0. \tag{7.31}$$

By Theorem 7.2.2 and the fact that $c_1 > 0$, it follows that

$$P_{x_1}^{(1)}\left(\lim_{t\to\tau_\alpha\wedge\tau_B} \operatorname{dist}_{R^d}(X_1(t), \alpha) = 0\right) > 0, \text{ for } x_1 \in (\alpha, \beta). \tag{7.32}$$

Equations (7.21) and (7.22) now follow easily from (7.30)–(7.32) and the skew-product representation.

(e) We treat the endpoint α; the identical method works for the endpoint β. First consider the case where $N_{\text{fin}}^\alpha = \phi$. The proof that the Martin boundary point corresponding to sequences $\{x^{(n)}\}_{n=1}^\infty$ satisfying

$\lim_{n\to\infty} x_1^{(n)} = \alpha$ is minimal and that the corresponding minimal positive harmonic function is given by $u_\alpha(x) = g_\alpha(x_1)\prod_{i=2}^k \phi_0^{(i)}(x_i)$ follows from Theorem 7.2.2 and that fact that

$$P_x^{u_\alpha}(\lim_{t\to\tau_D} \mathrm{dist}_{R^d}(X_1(t), \alpha) = 0) = 1. \tag{7.33}$$

To see that (7.33) holds, note that

$$P_x^{u_\alpha}(\lim_{t\to\tau_D} \mathrm{dist}_{R^d}(X_1(t), \alpha) = 0) = P_{x_1}^{(1)}(\lim_{t\to\tau_\alpha\wedge\tau_\beta} \mathrm{dist}_{R^d}(X_1(t), \alpha) = 0), \tag{7.34}$$

where $\{P_{x_1}^{(1)},\ x \in (\alpha, \beta)\hat{}\}$ denotes the solution to the generalized martingale problem on (α, β) for $\mathcal{L}^{g_\alpha}$. By definition, g_α is the minimal positive harmonic function corresponding to the Martin boundary point α for $\mathcal{L}$. Thus, the righthand side of (7.34) equals one by Theorem 7.2.1.

Now consider the case where $N_{\mathrm{fin}}^\alpha \neq \phi$. Let

$$h(x) = g(x_1)\prod_{i=2}^k \phi_0^{(i)}(x_i),$$

where $g = c_1 g_\alpha + c_2 g_\beta$, with $c_1 > 0$ and $c_2 \geqslant 0$. By part (d), $i \in N_{\mathrm{fin}}^\alpha$ if and only if $J_\alpha(g, M^{(i)}) < \infty$. Thus, the same type of argument used in part (d), along with Theorem 4.9.9, shows that μ_x^h as defined in the statement of part (e) exists as a non-zero measure and that, for $i \notin N_{\mathrm{fin}}^\alpha$, the ith marginal of $\mu_x^h(\mathrm{d}x_2', \ldots, \mathrm{d}x_k')$ is independent of $x \in D$ (and is given by $\phi_0^{(i)}(x_i')\tilde{\phi}_0^{(i)}(x_i')\,\mathrm{d}x_i'$). By Theorem 9.1.2, $u_B(x) \equiv \mu_x^h(B) \in C_L(D)$, for any Borel set $B \subseteq \times_{i=2}^k \bar{D}^{(i)}$. Thus, if follows by an application of Harnack's inequality that the Radon–Nikodym derivative

$$\frac{\mathrm{d}\mu_x^h}{\mathrm{d}\mu_{x_0}^h}$$

exists on $\times_{i=2}^k \bar{D}^{(i)}$. Since for $i \notin N_{\mathrm{fin}}^\alpha$, the ith marginal of μ_x^h is independent of $x \in D$, it follows that we can write

$$\frac{\mathrm{d}\mu_x^h}{\mathrm{d}\mu_{x_0}^h}(z) \equiv \frac{\mathrm{d}\mu_x^h}{\mathrm{d}\mu_{x_0}^h}(y),$$

where z is the projection onto $\times_{i\in N_{\mathrm{fin}}^\alpha} \bar{D}^{(i)}$ of $y \in \times_{i=2}^k \bar{D}^{(i)}$.

It remains to show that

$$\frac{\mathrm{d}\mu_x^h}{\mathrm{d}\mu_{x_0}^h}(z_0)$$

is the minimal positive harmonic function for L^h on D corresponding to sequences $\{x^{(n)}\}_{n=1}^{\infty}$ satisfying $\lim_{n\to\infty} x_1^{(n)} = \alpha$ and $\{\lim_{n\to\infty} x_i^{(n)}, i \in N_{\text{fin}}^{\alpha}\} = z_0$. (After an h-transform, it will then follow that

$$u_{z_0}(x) \equiv \frac{h(x)}{h(x_0)} \frac{\mathrm{d}\mu_x^h}{\mathrm{d}\mu_{x_0}^h}(z_0)$$

is the minimal positive harmonic function for L on D corresponding to the above Martin boundary point.) The proof is similar to the proof of Theorem 1.10 and is thus left to the reader. □

8.8 Some remarks

It is worth pointing out explicitly the probabilistic implications of the results of Sections 6 and 7. We will assume that $V = 0$; in any case, this can always be achieved after an h-transform. Let L and D be as in Theorem 6.1, and let $\Omega_{r,\alpha}$, $\Omega_{r,\beta}$, h_α and h_β be as in Proposition 7.1. In particular,

$$h_\alpha(x) = E_x(h_\alpha(X(\sigma_{\bar{\Omega}_{r,\alpha}})); \sigma_{\bar{\Omega}_{r,\alpha}} < \tau_D), \quad \text{for } x \in \Omega_{r,\beta} \tag{8.1a}$$

and

$$h_\beta(x) = E_x(h_\beta(X(\sigma_{\bar{\Omega}_{r,\beta}})); \sigma_{\bar{\Omega}_{r,\beta}} < \tau_D), \quad \text{for } x \in \Omega_{r,\alpha} \tag{8.1b}$$

Define $\hat{h}_\alpha(x) = P_x(\sigma_{\bar{\Omega}_{r,\alpha}} < \tau_D)$, $x \in \Omega_{r,\beta}$, and $\hat{h}_\beta(x) = P_x(\sigma_{\bar{\Omega}_{r,\beta}} < \tau_D)$, $x \in \Omega_{r,\alpha}$. Then $\hat{h}_\alpha$ and $\hat{h}_\beta$ are continuous up to $\partial\Omega_{r,\alpha} \cap D = \partial\Omega_{r,\beta} \cap D$ and $\hat{h}_\alpha = \hat{h}_\beta = 1$ on $\partial\Omega_{r,\alpha} \cap D = \partial\Omega_{r,\beta} \cap D$; thus

$$\hat{h}_\alpha(x) = E_x(\hat{h}_\alpha(X(\sigma_{\bar{\Omega}_{r,\alpha}})); \sigma_{\bar{\Omega}_{r,\alpha}} < \tau_D), \quad \text{for } x \in \Omega_{r,\beta} \tag{8.2a}$$

and

$$\hat{h}_\beta(x) = E_x(\hat{h}_\beta(X(\sigma_{\bar{\Omega}_{r,\beta}})); \sigma_{\bar{\Omega}_{r,\beta}} < \tau_D), \quad \text{for } x \in \Omega_{r,\alpha} \tag{8.2b}$$

Therefore, for $x \in \Omega_{r,\beta}$ and $A \subset \partial\Omega_{r,\alpha} \cap \{x_1 = r\} \cap D$,

$$\begin{aligned}
P_x^{\hat{h}_\alpha}(X(\sigma_{\bar{\Omega}_{r,\alpha}}) \in A) &= P_x^{\hat{h}_\alpha}(X(\sigma_{\bar{\Omega}_{r,\alpha}}) \in A, \sigma_{\bar{\Omega}_{r,\alpha}} < \tau_D) \\
&= \frac{1}{\hat{h}_\alpha(x)} E_x(\hat{h}_\alpha(X(\sigma_{\bar{\Omega}_{r,\alpha}})); X(\sigma_{\bar{\Omega}_{r,\alpha}}) \in A, \sigma_{\bar{\Omega}_{r,\alpha}} < \tau_D) \\
&= \frac{P_x(X(\sigma_{\bar{\Omega}_{r,\alpha}}) \in A, \sigma_{\bar{\Omega}_{r,\alpha}} < \tau_D)}{P_x(\sigma_{\bar{\Omega}_{r,\alpha}} < \tau_D)} = P_x(X(\sigma_{\bar{\Omega}_{r,\alpha}}) \in A | \sigma_{\bar{\Omega}_{r,\alpha}} < \tau_D).
\end{aligned}$$

A similar calculation holds for $P_x^{\hat{h}_\beta}$. Thus,

$$P_x^{\hat{h}_\alpha}(X(\sigma_{\bar{\Omega}_{r,\alpha}}) \in \mathrm{d}y) = P_x(X(\sigma_{\bar{\Omega}_{r,\alpha}}) \in \mathrm{d}y | \sigma_{\bar{\Omega}_{r,\alpha}} < \tau_D), \quad x \in \Omega_{r,\beta} \tag{8.3a}$$

and

$$P_x^{\hat{h}_\beta}(X(\sigma_{\bar{\Omega}_{r,\beta}}) \in dy) = P_x(X(\sigma_{\bar{\Omega}_{r,\beta}}) \in dy \mid \sigma_{\bar{\Omega}_{r,\beta}} < \tau_D), \quad x \in \Omega_{r,\alpha}. \tag{8.3b}$$

Proposition 7.1(iii) holds with h_α and h_β replaced by $\hat{h}_\alpha$ and $\hat{h}_\beta$; indeed the proof is identical. By Theorem 6.1, the Martin boundaries of L and $\tilde{L}$ coincide. Therefore, we have proved the following proposition.

Proposition 8.1. *Let L and D satisfy the conditions of Theorem* 6.1 *and let* $V = 0$. *Let* $\Omega_{r,\alpha} = \{x \in D: \alpha < x_1 < r\}$ *and* $\Omega_{r,\beta} = \{x \in D: r < x_1 < \beta\}$. *Then a sequence* $\{y^{(n)}\}_{n=1}^{\infty} = \{(y_1^{(n)}, \ldots, y_k^{(n)})\}_{n=1}^{\infty} \subset D$ *satisfying* $\lim_{n\to\infty} y_1^{(n)} = \alpha$ ($\lim_{n\to\infty} y_1^{(n)} = \beta$) *is a Martin sequence for L if and only if* $w - \lim_{n\to\infty} P_{y^{(n)}}(X(\sigma_{\Omega_{r,\beta}}) \in \cdot \mid \sigma_{\Omega_{r,\beta}} < \tau_D)$ *exists* ($w - \lim_{n\to\infty} P_{y^{(n)}}(X(\sigma_{\Omega_{r,\alpha}}) \in \cdot \mid \sigma_{\Omega_{r,\alpha}} < \tau_D)$ *exists*).

Remark. Proposition 8.1 is an extension of Corollary 7.6.3 to a case in which $\partial\Omega \cap \partial D \neq \phi$.

From the point of view of calculating Martin boundaries, Proposition 7.1(iii) is many orders of magnitude more useful than Proposition 8.1 since the operators L^{h_α} and L^{h_β} are in skew-product form (due to the fact that h_α and h_β are the separation of variables–solutions), whereas $L^{\hat{h}_\alpha}$ and $L^{\hat{h}_\beta}$ are not in skew-product form. However, the probabilistic interpretation of the Martin boundary in Proposition 8.1 is more appealing than that in Proposition 7.1(iii). Actually, the diffusions under $P_x^{h_\alpha}$ ($P_x^{h_\beta}$) and under $P_x^{\hat{h}_\alpha}(\cdot) = P_x(\cdot \mid \sigma_{\bar{\Omega}_{r,\alpha}} < \tau_D)$ ($P_x^{\hat{h}_\beta}(\cdot) = P_x(\cdot \mid \sigma_{\bar{\Omega}_{r,\beta}} < \tau_D)$) are quite similar. Indeed, from (8.1), (8.2) and the fact that $\hat{h}_\alpha = \hat{h}_\beta$ on $\partial\Omega_{r,\alpha} \cap D = \partial\Omega_{r,\beta} \cap D$, it is easy to see that

$$\left.\frac{dP_x^{h_\alpha}}{dP_x^{\hat{h}_\alpha}}\right|_{\mathcal{F}^D_{\sigma_{\bar{\Omega}_{r,\alpha}}}} = h_\alpha(X(\sigma_{\bar{\Omega}_{r,\alpha}})), \quad \text{for } x \in \Omega_{r,\beta}$$

and

$$\left.\frac{dP_x^{h_\beta}}{dP_x^{\hat{h}_\beta}}\right|_{\mathcal{F}^D_{\sigma_{\bar{\Omega}_{r,\beta}}}} = h_\beta(X(\sigma_{\bar{\Omega}_{r,\beta}})), \quad \text{for } x \in \Omega_{r,\alpha}.$$

In the sequel, we will refer to the diffusion corresponding to $P_x^{\hat{h}_\alpha}$, $x \in \Omega_{r,\beta}$ ($P_x^{\hat{h}_\beta}$, $x \in \Omega_{r,\alpha}$), as the diffusion corresponding to P_x conditioned on $\sigma_{\bar{\Omega}_{r,\alpha}} < \tau_D$ ($\sigma_{\bar{\Omega}_{r,\beta}} < \tau_D$), and we will refer to the diffusion corresponding to $P_x^{h_\alpha}$, $x \in \Omega_{r,\beta}$ ($P_x^{h_\beta}$, $x \in \Omega_{r,\alpha}$) as the diffusion corresponding to P_x conditioned on $\sigma_{\bar{\Omega}_{r,\alpha}} < \tau_D$ ($\sigma_{\bar{\Omega}_{r,\beta}} < \tau_D$) with the weight $h_\alpha(X(\sigma_{\bar{\Omega}_{r,\alpha}}))$ ($h_\beta(X(\sigma_{\bar{\Omega}_{r,\beta}}))$).

Consider Theorems 6.4 and 6.6 in light of the above discussion. In Theorem 6.6, we have $D = \{x = (y, z)\colon\ y \in R^d,\ z \in \Sigma\}$, where $\Sigma \subset R^m$, $m \geq 1$, is a bounded domain with a $C^{2,\alpha}$-boundary, and $d \geq 1$. Let $\Omega = \{x = (y, z) \in D\colon |y| < 1\}$ denote the unit cylinder in the strip D. Let $y = (r, \theta)$, $r > 0$, $\theta \in S^{d-1}$. The proof of Theorem 6.6 shows that conditioning Brownian motion in D on $\{\sigma_{\bar{\Omega}} < \tau_D\}$ (with the weight arising from the appropriate separation of variables–solution) introduces a drift in the r-direction which is bounded between two negative constants. That is, requiring that the last m components of the R^{d+m}-dimensional Brownian motion remain in a certain compact set until the radial component of the first d components reaches one, has the effect of introducing an inward radial drift on the first d components which is of the order of unity. This drift causes the diffusion to reach $\bar{\Omega}$ sufficiently quickly so that, even as the initial point approaches infinity, the diffusion at time σ_{Ω} still 'remembers' its initial θ-component (although not its initial z component). Thus, the exterior harmonic measure boundary and, consequently, the Martin boundary will depend on the θ-component.

On the other hand, consider now Theorem 6.4. We have

$$D = \left\{x \in R^d\colon \frac{x}{|x|} \in \Sigma\right\},\ d \geq 2,$$

where $\Sigma \subset S^{d-1}$ is a domain with a $C^{2,\alpha}$-boundary. Let $\Omega = \{x \in D\colon |x| < 1\}$ and let $x = (r, \theta)$, with $r > 0$ and $\theta \in S^{d-1}$. The proof of Theorem 6.4 shows that conditioning Brownian motion on $\{\sigma_{\bar{\Omega}} < \tau_D\}$ (with the weight arising from an appropriate separation of variables–solution) introduces an inward radial drift on the order $1/r$. This is a weak drift; $1/r$ is the characteristic order of the radial drift in a drift-free, uniformly elliptic multi-dimensional diffusion, that is, in a diffusion corresponding to an operator of the form

$$L = \tfrac{1}{2}\sum_{i,j=1}^{d} a_{i,j}\frac{\partial}{\partial x_i \partial x_j},\ d \geq 2,$$

where $a = \{a_{ij}\}$ is uniformly elliptic. With this weak drift, as the initial point tends to infinity, the diffusion at time $\sigma_{\bar{\Omega}}$ 'forgets' its initial angular component. Thus, the exterior harmonic measure boundary and, consequently, the Martin boundary will not depend on the θ-component.

The above discussion shows that if the domain widens at a linear rate (the case of a cone), then the Martin boundary for $\frac{1}{2}\Delta$ does not depend on the θ-component, while if the domain does not widen at all (the case of a strip), then the Martin boundary for $\frac{1}{2}\Delta$ will depend on the θ-component. This motivates the study of the Martin boundary for $\frac{1}{2}\Delta$ in domains which widen at a sublinear rate (see the second problem discussed in Section 9). The above intuition may be applied more generally to any operator satisfying Assumption SP in Section 6; Theorems 6.9–6.11 are manifestations of this intuition.

We now compare briefly the two probabilistic approaches to the Martin boundary which appear in Chapter 7. Theorems 7.2.1 and 7.2.2 characterize the minimal Martin boundary for L on D in terms of the behavior of certain diffusions as they exit D. These diffusions are the diffusions corresponding to h-transforms of L, where $h \in C_L(D)$. On the other hand, in Theorem 7.6.2 (and in its extension in Proposition 7.1), one considers the harmonic measure on a bounded subdomain Ω for the diffusion starting in $D - \Omega$ and corresponding to an h-transform of L, where h is a positive solution of minimal growth at ∂D. The Martin boundary is then characterized by the behavior of these harmonic measures as the initial point of the diffusion exits D.

Using Theorems 7.2.1 and 7.2.2, the best one can ever do is conclude that almost all paths of a certain diffusion converge in the Martin topology to a Martin boundary point; however, one can never prove by this method that a particular sequence $\{x_n\}_{n=1}^{\infty}$ is a Martin sequence. The use of this technique is illustrated by Theorem 2.13 and its proof. On the other hand, Theorem 7.6.2 or Proposition 7.1 determines every Martin sequence. The use of this technique is illustrated by Theorem 6.1(c) and its proof.

8.9 Survey of other results

In this section, we survey briefly some noteworthy results concerning positive harmonic functions and the Martin boundary which were not treated in this chapter. In the notes at the end of the chapter, references to additional results can be found.

1. Let $L = \Delta - c(x)$ in $R^2 - \{0\}$, where $c(x) \geqslant 0$ and $c(x) \not\equiv 0$. Then L is subcritical on $R^{(2)} - \{0\}$. The part of the Martin boundary corresponding to sequences $\{x_n\}_{n=1}^{\infty}$ satisfying $\lim_{n\to\infty} |x_n| = 0$ will be called the 'Martin boundary at zero'. If the Martin boundary at zero consists of a single point, the 'Picard principle' is said to hold. In the case where $c(x) = W(|x|)$, it follows from Theorem 6.1 that the Martin boundary at zero is either one point or is homeomorphic to S. Under the assumption that $r^2W(r)$ is non-increasing in r for r near 0, it was proved in [Nakai (1975)] that the Picard principle holds if and only if

$$\int_0^1 \frac{dr}{r(r^2W(r)+1)^{1/2}} = \infty .$$

This strengthens Theorem 6.7(a) and (b) in the case where $d=2$, since that theorem also required that $r^4W(r)$ be non-decreasing in r for r near 0. Without any monotonicity assumption on $r^2W(r)$, it was shown in [Nakai (1986)] that the above integral condition is necessary for the Picard principle to hold; furthermore, a counter-example was provided to show that the integral condition is not sufficient. These results may be extended to R^d.

Theorem 6.11 showed that a monotonicity result holds with regard to the Picard principle in the case where $c(x)=W(|x|)$; namely, if $W_1 \leqslant W_2$ and the Picard principle holds for W_2, then it also holds for W_1. This monotonicity breaks down in the non-radially symmetric case. In [Nakai and Tada (1988)] it was shown that for any $c(x) \geqslant 0$, one can find a C^∞-function $Q(x) \geqslant c(x)$ such that the Picard principle holds for $\Delta - Q(x)$. In [Tada (1990)], a function $c(x)=W(|x|) \geqslant 0$ and a function $Q(x)$ satisfying $0 \leqslant Q(x) \leqslant c(x)$ were found such that the Picard principle holds for $\Delta - c(x)$ but not for $\Delta - Q(x)$.

Let $D_\gamma = \{x=(r,\theta) \in R^2 - \{0\}\colon\ 0<\theta<\gamma\}$, for $\gamma \in (0,2\pi)$. Extend the definition of the Picard principle to $\Delta - c(x)$ on D_γ in the obvious way. Theorem 6.10 showed that in the radially symmetric case, the following monotonicity result holds with regard to the Picard principle: if $\gamma_1 < \gamma_2$ and the Picard principle holds for $\Delta - W(|x|)$ on D_{γ_1}, then it also holds for $\Delta - W(|x|)$ on D_{γ_2}. This monotonicity principle, like the previous one, breaks down in the non-radially symmetric case. In [Tada (1989)], a function $c(x) \geqslant 0$ was constructed such that the Picard principle holds for $\Delta - c(x)$ on D_γ, for all $\gamma \in (0,2\pi)$, but not for $\Delta - c(x)$ on $R^2 - \{0\}$.

In [Tada (1988)], it was shown that if the Picard principle holds for $\Delta - W(|x|)$, and

$$|c(x) - W(|x|)| \leqslant \frac{\text{const.}}{|x|^2},$$

then the Picard principle also holds for $\Delta - c(x)$.

2. Let $D = \{x=(y,z) \in R^{d+1}\colon y \in R^d,\ z \in R,\ |z| < a(|y|)\}$, $d \geqslant 2$, where $a\colon [0,\infty) \to (0,\infty)$, and let $L = \frac{1}{2}\Delta$. Under appropriate regularity conditions on the function a, the following result was proved in [Ioffe and Pinsky (1994)]:

(i) If

$$\int^\infty \frac{a(r)}{r^2}\, dr = \infty,$$

then every sequence $\{x_n\}_{n=1}^{\infty} \subset D$ satisfying $\lim_{n\to\infty} |x_n| = \infty$ is a Martin sequence; thus the cone of positive harmonic functions vanishing on ∂D is one-dimensional.

(ii) If

$$\int^{\infty} \frac{a(r)}{r^2}\,dr < \infty,$$

then a sequence $\{x_n\}_{n=1}^{\infty} = \{(y_n, z_n)\}_{n=1}^{\infty} \subset D$ satisfying $\lim_{n\to\infty} |y_n| = \infty$ is a Martin sequence if and only if

$$\lim_{n\to\infty} \frac{y_n}{|y_n|}$$

exists on S^{d-1}. All corresponding Martin boundary points are minimal; thus, the cone of positive harmonic functions vanishing on ∂D is generated by a collection of minimal elements which is homeomorphic to S^{d-1}.

Note that

$$\int^{\infty} \frac{a(r)}{r^2}\,dr < \infty,$$

if $a(r) \sim r^{\gamma}$, $\gamma \in [0, 1)$, as $r \to \infty$, or if

$$a(r) \sim \frac{r}{(\log r)^{l}}, \; l \in (1, \infty), \text{ as } r \to \infty.$$

However, if

$$a(r) \sim \frac{r}{(\log r)},$$

as $r \to \infty$, then

$$\int^{\infty} \frac{a(r)}{r^2}\,dr = \infty.$$

The case $a(r) = c$ was treated in Theorem 6.6; compare the case $a(r) = cr$ to Theorem 6.4.

3. Let L denote the Laplace–Beltrami operator on a d-dimensional Riemannian manifold whose sectional curvatures are bounded between two negative constants. Such an operator is subcritical (the corresponding diffusion is transient). In [Anderson and Schoen (1985)], it was shown that the Martin boundary and the minimal Martin boundary for L on D coincide and are homeomorphic to S^{d-1}. The result was extended to more general operators on such manifolds by [Ancona (1987)]. [Kifer (1986)] gave a probabilistic proof of the result of Anderson and Schoen.

4. Let $L_0 = \Delta + V_0(x)$ and $L = \Delta + V(x)$ on $R^d - \{0\}$, where $V(x) = V_0(x) + V_1(x) + V_2(x)$, V_0 is radially symmetric, V_1 is a so-called 'small' perturbation and V_2 has compact support. (For the definition of a 'small' perturbation, see [Murata (1986)].) The operator L is 'almost' radially symmetric. Assume that L_0 and L are subcritical on $R^d - \{0\}$. In [Murata (1986)], it was proved that the Martin boundaries of L_0 and L on $R^d - \{0\}$ coincide, and the integral condition in Theorem 6.1(c) for $L_0 = \Delta + V_0$ was given to determine the Martin boundary of L_0.

5. Let $V(x)$, $x \in R^d$, satisfy $V(x) = a^2$, if $x_n > 0$, and $V(x) = a^2 + b^2$, if $x_n \leq 0$, where $a, b > 0$. Let $L = \Delta + V(x)$ on R^d. In [Murata (1990b)] it was shown that Λ_0, the minimal Martin boundary, is naturally homeomorphic to $\{x \in R^d: |x| = a, x_n \geq 0\} \cup \{x \in R^d: |x| = (a^2 + b^2)^{1/2}, x_n \leq -b\}$, and that $\Lambda - \Lambda_0$, the collection of non-minimal Martin boundary points, is naturally homeomorphic to $\{(x_1, \ldots, x_{d-1}, -\gamma b): \sum_{j=1}^{d-1} x_j^2 = a^2, 0 < \gamma < 1\}$.

In a related result which appears in [Murata (1986)], the minimal Martin boundary was calculated for $\Delta + V$ in R^2, in the case where $V(x) = V(r, \theta) = r^{\beta_{2j}}$, if $\theta \in (\theta_{2j-1}, \theta_{2j})$, $j = 1, 2, \ldots, k$, and $V(x) = V(r, \theta) = 0$, otherwise, where k is a positive integer, $\beta_{2j} > -2$, and $0 = \theta_0 < \theta_1 < \ldots < \theta_{2k} = 2\pi$.

6. Let L be defined on R^2 and satisfy

$$L = \tfrac{1}{2}\Delta + r^{-\delta}\frac{\partial}{\partial r} + r^{-k-1}\frac{\partial}{\partial \theta}$$

on $\{r > 1\}$, where $\delta \in (-1, 1)$, $k \in (-\infty, \infty)$, and (r, θ) denote polar coordinates in R^2. Define

$$U(x) = U(r, \theta) = \left(\frac{r^{\delta-k}}{\delta - k} - \theta\right) \bmod 2\pi,$$

if $k \neq \delta$, and $U(x) = U(r, \theta) = (\log r - \theta) \bmod 2\pi$, if $k = \delta$. Define $V(x) = V(r, \theta) = U(r, -\theta)$. For $c \in [0, 2\pi)$, the trajectories $U(r, \theta) = c$ constitute distinct spirals approaching infinity in a counter-clockwise direction, while the trajectories $V(r, \theta) = c$ constitute distinct spirals approaching infinity in a clockwise direction. If $k > \delta$, then the spirals are degenerate; that is, they only wrap around the origin a finite number of times as $r \to \infty$. The operator L on R^2 is subcritical (the corresponding diffusion is transient). In [R. Pinsky (1993)], the following result was proved:

(i) If $k \leq \delta - \frac{1}{2}(1 - \delta)$, then the Martin boundaries of L and $\widetilde{L}$ on R^2 consist of a single point.

(ii) If $k > \delta - \frac{1}{2}(1-\delta)$, then a sequence $\{x_n\}_{n=1}^{\infty} \subset R^2$ satisfying $\lim_{n\to\infty} |x_n| = \infty$ is a Martin sequence for L on R^2 if and only if $\lim_{n\to\infty} U(x_n)$ exists; it is a Martin sequence for $\widetilde{L}$ on R^2 if and only if $\lim_{n\to\infty} V(x_n)$ exists.

In particular, note that if $k \in (\delta - \frac{1}{2}(1-\delta), \delta]$, then the Martin boundaries of L and $\widetilde{L}$ are different. We are unaware of other results in the literature which calculate the Martin boundaries explicitly for L and $\widetilde{L}$ and show that they differ.

7. Theorem 1.5 can be generalized to operators with coefficients which do not necessarily remain bounded at ∂D. Let

$$L = \sum_{i,j=1}^{d} a_{i,j} \frac{\partial^2}{\partial x_i \partial x_j} + b \cdot \nabla + V$$

on $R_+^d \equiv \{x \in R^d\colon\ x_d > 0\}$. Assume that $V \leqslant 0$, that $c_1|v|^2 \leqslant \sum_{i,j=1}^d a_{i,j}(x) v_i v_j \leqslant c_2 |v|^2$, for x, $v \in R^d$, that $x_d|b(x)|$ is bounded on R_+^d, and that $x_d^{2-\varepsilon}|V(x)|$ is bounded on R_+^d for some $\varepsilon > 0$. Let $D \subset R_+^d$ be a bounded domain with a Lipschitz boundary. In [Ancona (1980)], it was proved that Theorem 1.5 holds for L on D.

8. Let $L = \frac{1}{2}\Delta + Bx \cdot \nabla$ on R^2, where B is a constant 2×2 matrix. Assume that L is subcritical on R^2. (That is, assume that the corresponding diffusion is transient.) This is equivalent to the assumption that at least one of the eigenvalues of B is positive. Assume in addition that neither eigenvalue is zero. For simplicity, assume that B is in lower triangular form; that is,

$$B = \begin{pmatrix} \lambda_1 & 0 \\ b & \lambda_2 \end{pmatrix},$$

where λ_1 and λ_2 are the eigenvalues of B. (There is no loss of generality in this since such a form can always be achieved by an orthogonal transformation.) Define

$$\hat{B} = \begin{pmatrix} |\lambda_1| & 0 \\ b & |\lambda_2| \end{pmatrix}.$$

For each $x \in R^2$ with $|x| > 1$, let Mx denote the unique point in R^2 satisfying $|Mx| = 1$ and $x = \exp(\hat{B}t)Mx$, for some $t > 0$. In [Cranston, Orey and Rosler (1982)], it was proved that a sequence $\{x_n\}_{n=1}^{\infty} \subset R^2$ satisfying $\lim_{n\to\infty} |x_n| = \infty$ is a Martin sequence if and only if $\lim_{n\to\infty} Mx_n$ exists.

9. Let $L = \Delta$ on R^{d+1}. Denote points in R^{d+1} by $x = (y, z)$, where $y \in R^d$ and $z \in R$. Let E be a proper, closed subset of the hyperplane $\{z = 0\}$. Fix $\alpha \in (0, 1)$ and for each $y \in R^d$, let K_y denote the cube in

R^{d+1} centered at $(y, 0)$ with sides parallel to the coordinate planes of length $\alpha|y|$. Let $\Omega_y = K_y - E$, let w^y denote the solution to $\Delta w^y = 0$ in Ω_y, $w^y = 1$ on ∂K_y and $w^y = 0$ on $E \cap K_y$, and define $B_E(y) = w^y(y, 0)$. In [Benedicks (1980)] it was proved that the cone of positive harmonic functions on $R^{d+1} - E$ which vanish on E is either one- or two-dimensional; equivalently, the part of the minimal Martin boundary corresponding to sequences $\{x_n\}_{n=1}^{\infty} \subset R^{d+1} - E$ satisfying $\lim_{n\to\infty} |x_n| = \infty$ consists of either one or two points. The two cases may be distinguished as follows:

(i) If

$$\int_{|y|\geq 1} \frac{B_E(y)}{|y|^n} \mathrm{d}y = \infty,$$

then the cone of positive harmonic functions vanishing on E is one-dimensional;

(ii) If

$$\int_{|y|\geq 1} \frac{B_E(y)}{|y|^n} \mathrm{d}y < \infty,$$

then the cone of positive harmonic functions vanishing on E is two-dimensional.

A related result appears in [Cranston and Salisbury (1994)].

Exercises

8.1 In the case where $V \leq 0$, use the proof of (7.3.1) to prove Theorem 1.1 without using the boundary Harnack inequality.

8.2 Produce an argument similar to that used in Theorem 7.3.5 to show that the function u_f defined in the proof of Theorem 1.1 satisfies $u_f \in C^{2,\alpha}(B \cap D)$ and $Lu_f = 0$ in $B \cap D$.

8.3 Show that the function u defined in Proposition 1.8 satisfies $u(x) = E_x \exp\left(\int_0^{\tau_B} V(X(s))\, \mathrm{d}s\right) u(X(\tau_B))$, for $x \in B$.

8.4 Prove Theorem 1.10(ii) along the lines of the proof of Theorem 1.10(i).

8.5 Prove Theorem 2.6. (*Hint*: By making an h-transform, one may assume that the zeroth-order term in $L - \lambda_c$ vanishes; thus $L - \lambda_c$ is of the form $L - \lambda_c = \frac{1}{2}\nabla \cdot a\nabla + a\nabla Q \cdot \nabla$. Show that, up to constant multiples, $\exp(2Q)$ is the unique element in $C_{\tilde{L}-\lambda_c}(R^d)$. Since the zeroth-order term in $L - \lambda_c$ vanishes, it follows from Theorem 4.11.1 that the principal eigenvalue of $L - \lambda_c$ on T_d is zero. Use these facts to conclude that Q is periodic. Now apply Example 6.4.3, if $d \geq 2$, and Theorem 5.1.1, if $d = 1$. In the case where $d = 1$, an alternative proof can be given via Corollary 7.6.7 and Theorem 2.5.)

8.6 Show that if a diffusion explodes inward from the boundary, then the corresponding solution $\{P_x, x \in \hat{D}\}$ to the generalized martingale problem is not Feller continuous at $x = \Delta$, where Δ is the point at infinity in the one-point compactification of D.

8.7 Show that condition (iv) in Theorem 4.1 is equivalent to the condition

$$\int_{x_0}^{\beta} \mathrm{d}x \exp\left(-\int_{x_0}^{x} \frac{2b}{a}(s)\,\mathrm{d}s\right)\int_{x}^{\beta} \mathrm{d}y \frac{1}{a(y)} \exp\left(\int_{x_0}^{y} \frac{2b}{a}(s)\,\mathrm{d}s\right) < \infty.$$

8.8 Let g_n and g_α be as in the proof of Proposition 7.1(ii). Show that $g_\alpha = \lim_{n\to\infty} g_n$. (*Hint*: Use the Feynman–Kac representation for g_n (Theorem 3.6.6) and the representation for g_α in exercise 7.7.)

8.9 Prove (7.3) using the method of the proof of Theorem 7.4.5.

8.10 Let $\mu_x^{h\beta}$, $x \in \Omega_{r,\alpha}$, be as in the statement of Proposition 7.1(iii). Let $\{y^{(n)}\}_{n=1}^{\infty} = \{(y_1^{(n)}, \ldots, y_k^{(n)})\}_{n=1}^{\infty} \subset \Omega_{r,\alpha}$ be a sequence with no accumulation points in D satisfying $\limsup_{n\to\infty} y_1^{(n)} < r$. Assume that $\gamma \equiv w - \lim_{n\to\infty} \mu_{y^{(n)}}^{h\beta}$ exists. Prove that γ gives no mass to $\partial D \cap \{x_1 = r\}$. (*Hint*: Let $\{P_{x_i}^{(i)}, x \in D^{(i)}\}$ and $\phi_0^{(i)}$ be as in Proposition 7.2.) By Proposition 7.2, for $x_i \in \partial D^{(i)}$ and $t > 0$, $P_{x_i}^{(i)}(X_i(t) \in \cdot) \equiv w - \lim_{z_i\to x_i} P_{z_i}^{(i)}(X_i(t) \in \cdot)$ exists. The proof of Theorem 6.1(c) shows that if $\lim_{n\to\infty} y_1^{(n)} = \alpha$, then $\gamma \equiv w - \lim_{n\to\infty} \mu_{y^{(n)}}^{h\beta}$ exists if and only if $\lim_{n\to\infty} y_i^{(n)} \in \bar{D}^{(i)}$ exists for $i \in N_{\text{fin}}^{\alpha}$. The proof also shows that γ is of the form

$$\gamma(\mathrm{d}y_1', \ldots, \mathrm{d}y_k') =$$

$$\delta_r(\mathrm{d}y_1')\left(\prod_{i\notin N_{\text{fin}}^{\alpha}} \phi^{(i)}(y_i')\tilde{\phi}^{(i)}(y_i')\,\mathrm{d}y_i'\right)\int \prod_{i\in N_{\text{fin}}^{\alpha}} P_{y_i}^{(i)}(X_i(t_i) \in \mathrm{d}y_i')\nu_{y_1}(\mathrm{d}t),$$

where $y_i = \lim_{n\to\infty} y_i^{(n)}$, $\mathrm{d}t = \{\mathrm{d}t_i, i \in N_{\text{fin}}^{\alpha}\}$, ν_{y_1} is a probability measure on the positive orthant in R^j, and $j = \text{card}(N_{\text{fin}}^{\alpha})$. A similar proof can be obtained to show that if $\lim_{n\to\infty} y_1^{(n)} \in (\alpha, r)$, then $\gamma = w - \lim_{n\to\infty} \mu_{y^{(n)}}^{h\beta}$ exists if and only if $\lim_{n\to\infty} y_i^{(n)} \in \bar{D}^{(i)}$ exists for $i = 2, \ldots, k$, and γ is of the form

$$\gamma(\mathrm{d}y_1', \mathrm{d}y_2', \ldots, \mathrm{d}y_k') = \delta_r(\mathrm{d}y_1')\int \prod_{i=2}^{k} P_{y_i}^{(i)}(X_i(t_i) \in \mathrm{d}y_i')\nu_{y_1}(\mathrm{d}t_2, \ldots, \mathrm{d}t_k),$$

where $y_i = \lim_{n\to\infty} y_i^{(n)}$ and ν_{y_1} is a probability measure on the positive orthant in R^{k-1}.

Use the proof of Theorem 4.2 along with Theorem 4.7.4 to show that if $\{\Omega_n\}_{n=1}^{\infty}$ is an increasing sequence of domains satisfying $\bigcup_{n=1}^{\infty} \Omega_n = D^{(i)}$, then $\lim_{n\to\infty} \sup_{x \in D^{(i)} - \Omega_n} E_x^{(i)} \sigma_{\bar{\Omega}_n} = 0$, where $\sigma_{\bar{\Omega}_n} = \inf\{t \geq 0: X_i(t) \in \bar{\Omega}_n\}$. Use this and the fact that $P_{x_i}^{(i)}(X_i(t) \in \partial D^{(i)}) = 0$, for $x_i \in D^{(i)}$ and $t \geq 0$ to show that $P_{x_i}^{(i)}(X_i(t) \in \partial D^{(i)}) = 0$, for $x_i \in \partial D^{(i)}$ and $t > 0$.)

8.11 Complete the details of the proof of Proposition 7.1(iii) by mimicking the proof of Theorem 7.6.2.

8.12 Prove (7.8). (*Hint*: Since the result is a local one, by changing coordinates locally, we may assume for the proof that $D^{(i)} = \{z = (z_1, \ldots, z_{d_i}) \in R^{d_i}\colon 0 < z_1 < 1,\ |z_j| < 1,\ j = 2, \ldots, d_i\}$ and that $x_i = 0 \in \partial D^{(i)}$. Denote the diffusion process by $X_i(t) = (X_{i1}(t), \ldots, X_{id_i}(t))$, and let $Y_i(t) = (X_{i2}(t), \ldots, X_{id_i}(t))$. Recall that the diffusion $X_i(t)$ corresponds to the operator $(L^{(i)} - \lambda_c^{(i)})^{\phi_0^{(i)}}$. This operator is of the form

$$\tfrac{1}{2}\nabla \cdot a^{(i)}\nabla + b^{(i)} \cdot \nabla + a^{(i)}\frac{\nabla \phi_0^{(i)}}{\phi_0^{(i)}}\nabla,$$

where $a^{(i)}$ and $b^{(i)}$ are smooth up to the boundary of $D^{(i)}$. To prove (7.8), it suffices to show that for each $\varepsilon > 0$, there exists a $\delta = \delta(\varepsilon) > 0$ such that

$$\lim_{t \to 0} \sup_{\substack{|z| < \delta \\ z \in D^{(i)}}} P_z^{(i)}(X_{i1}(t) > \varepsilon, |X_i(t)| < \tfrac{1}{2}) = 0 \qquad (*)$$

and such that

$$\lim_{t \to 0} \sup_{\substack{|z| < \delta \\ z \in D^{(i)}}} P_z^{(i)}(|Y_i(t) - (z_2, \ldots, z_{d_1})| > \varepsilon) = 0. \qquad (**)$$

Prove (*) by using Theorem 2.2.2, the strong Markov property, and the fact that

$$\frac{|\nabla \phi_0^{(i)}|}{\phi_0^{(i)}}(y)$$

is bounded over

$$\left\{y \in D^{(i)}\colon y_1 > \frac{\varepsilon}{4}, |y| < \tfrac{1}{2}\right\}.$$

To prove (**), define $\hat{a}_{jl}(t, \omega) = a_{jl}^{(i)}(X(t, \omega))$ and

$$\hat{b}_j(t, \omega) = b_j^{(i)}(X_i(t, \omega)) + \tfrac{1}{2}\sum_{l=1}^{d_i}\left(\frac{\partial a_{lj}^{(i)}}{\partial x_l}(X_i(t, \omega)) + \left(\frac{1}{\phi_0^{(i)}} a_{jl}\frac{\partial \phi_0^{(i)}}{\partial x_j}\right)(X_i(t, \omega))\right),$$

and let $\hat{a}(t, \omega) = \{\hat{a}_{jl}(t, \omega)\}_{j,l=1}^{d_i}$ and $\hat{b}(t, \omega) = (\hat{b}_2(t, \omega), \ldots, \hat{b}_{d_i}(t, \omega))$. From the definition of the martingale problem and from the definition of Itô processes, it follows that, under $P_z^{(i)}$, $Y_i(t)$ is an $I_{d_i - 1}(\hat{a}, \hat{b})$ – Itô process. The proof of Theorem 2.2.2 reveals that this theorem holds not only for diffusions, but, more generally, for Itô processes. When Theorem 2.2.2(i) is applied to $Y_i(t)$, the rate of convergence to zero of $P_z^{(i)}(|Y_i(t) - (z_2, \ldots, z_{d_i})| > \varepsilon)$ as $t \to 0$ depends only on $\sup_y \sup_{|v|=1} \sum_{j,l=2} a_{jl}^{(i)}(y) v_j v_l$ and on

$$\sup_{j \in \{2, \ldots, d_i\}} \sup_y \left| b_j^{(i)}(y) + \tfrac{1}{2}\sum_{l=1}^{d_i}\left(\frac{\partial a_{lj}^{(i)}}{\partial x_l} + \left(\frac{1}{\phi_0^{(i)}} a_{jl}\frac{\partial \phi_0^{(i)}}{\partial x_j}\right)(y)\right)\right|,$$

where the supremum in y is over $A_\varepsilon(z) \equiv \{y \in D^{(i)}: |y_2, \ldots, y_{d_i}) - (z_2, \ldots, z_{d_i})| < \varepsilon\}$. Show that for

$$l \in \{2, \ldots, d_i\}, \left(\frac{1}{\phi_0^{(i)}(y)} \frac{\partial \phi_0^{(i)}}{\partial x_l}\right)$$

is bounded over $A_\varepsilon(z)$ if $|z|$ and ε are small.)

8.13 Prove (7.26) and (7.27). (*Hint*: We have

$$\lim_{x_1 \to \alpha} k^h(x_1, x_2; \zeta) = \lim_{n\to\infty} \frac{G^h(x_1, x_2, y_1^{(n)}, y_2^{(n)})}{G^h(x_{01}, x_{02}, y_1^{(n)}, y_2^{(n)})},$$

where $(x_{01}, x_{02}) \in D$ is an arbitrary fixed point and the sequence $\{(y_1^{(n)}, y_2^{(n)})\}_{n=1}^\infty$ corresponds to the Martin boundary point ζ. By the assumption on ζ, we may assume that $y_1^{(n)} > \alpha + 2\delta$, for all n, where $\delta > 0$. Choose $x_{01} = \alpha + \delta$ and let x_{02} be arbitrary. Using the probabilistic definition of G^h along with the strong Markov property, Harnack's inequality and the boundary Harnack principle, show that $G(x_1, x_2, y_1^{(n)}, y_2^{(n)}) \leq cP_{x_1}^{(1)}(\tau_{\alpha+\delta} < \tau_\alpha)G(x_{01}, x_{02}, y_1^{(n)}, y_2^{(n)})$, for $x_1 \in (\alpha, \alpha + \delta)$, where c is a positive constant independent of x_1 and n. It follows easily from (7.32) that $\lim_{x_1\to\alpha} P_{x_1}^{(1)}(\tau_{\alpha+\delta} < \tau_\alpha) = 0$. The proof of (7.27) is carried out similarly. (Note that in (7.27), the point $x_2 \in D^{(2)}$ is fixed.))

8.14 Prove Lemma 6.5(i).

8.15 Use the proof of Theorem 6.7(c) and (d) along with the change of variables

$$s = \frac{1}{r}$$

to prove Theorem 6.7(a) and (b). (*Hint*: One actually must be a bit careful here. To prove (a) and (b), one must show that

$$\int_0^1 \frac{dr}{r^2 W^{1/2}(r)} < \infty,$$

if and only if

$$I_0\left(g_\infty, \frac{1}{r^2}\right) < \infty, \text{ where } I_0\left(g_\infty, \frac{1}{r^2}\right) = \int_0^1 dr\, r^{d-3} g_\infty^2(r) \int_r^1 t^{1-d} \frac{1}{g_\infty^2(t)}\, dt,$$

and g_∞ is the minimal positive harmonic function for

$$\frac{1}{2}\frac{d^2}{dr^2} + \frac{d-1}{2r}\frac{d}{dr} - W(r) \text{ on } (0, \infty)$$

corresponding to the Martin boundary point ∞. Let

$$h_0(r) = g_\infty\left(\frac{1}{r}\right)$$

and show that h_0 is the minimal positive harmonic function for

$$\frac{1}{2}\frac{d^2}{dr^2} + \frac{3-d}{2r}\frac{d}{dr} - r^{-4} W\left(\frac{1}{r}\right) \text{ on } (0, \infty)$$

corresponding to the Martin boundary point 0. Let $v_0(r) = r^{2-d}h_0(r)$ and show that v_0 is the minimal positive harmonic function for

$$\frac{1}{2}\frac{d^2}{dr^2} + \frac{d-1}{2r}\frac{d}{dr} - r^{-4}W\left(\frac{1}{r}\right) \text{ on } (0, \infty)$$

corresponding to the Martin boundary point 0. Now show that

$$I_0\left(g_\infty, \frac{1}{r^2}\right) = I_\infty\left(v_0, \frac{1}{r^2}\right),$$

where

$$I_\infty\left(v_0, \frac{1}{r^2}\right) = \int_1^\infty dr\; r^{d-3}v_0^2(r)\int_1^r t^{1-d}\frac{1}{v_0^2(t)}\,dt.$$

Thus, letting

$$\hat{W}(r) = r^{-4}W\left(\frac{1}{r}\right),$$

it follows that the part of the Martin boundary for $\frac{1}{2}\Delta - W(|x|)$ on $R^d - \{0\}$ corresponding to sequences $\{x_n\}_{n=1}^\infty$ satisfying $\lim_{n\to\infty}|x_n| = 0$ is homeomorphic to S^{d-1} if and only if the part of the Martin boundary for $\frac{1}{2}\Delta - \hat{W}(|x|)$ on $R^d - \{0\}$ corresponding to sequences $\{x_n\}_{n=1}^\infty$ satisfying $\lim_{n\to\infty}|x_n| = \infty$ is homeomorphic to S^{d-1}.)

8.16 Refer to the notation in Assumption SP and (6.1). Show that for $i = 2, \ldots, k$, the finiteness or infiniteness of $I_\alpha(g_\beta, M_i)$ ($I_\beta(g_\alpha, M_i)$) depends only on $a(r)$, $b(r)$, $M_i(r)$, and $V^{(1)}(r) + \sum_{i=2}^k \lambda_0^{(i)}M^{(i)}(r)$ near $r = \alpha$ ($r = \beta$). (*Hint*: Use the representation for g_α and g_β given in exercise 7.7.)

8.17 Prove Theorem 5.1 in the case where for at least one $i \in (1, 2, \ldots, k\}$ there is no solution to the martingale problem for $L^{(i)}$ on $D^{(i)}$, but only to the generalized martingale problem. (*Hint*: Let $\{D_n\}_{n=1}^\infty$ be a sequence of domains satisfying $D_n \subset\subset D_{n+1}$ and $\bigcup_{n=1}^\infty D_n = D$. Using the notation of the proof of Theorem 5.1, one must show that for each $s > 0$, $A \in \hat{\mathcal{F}}_s^D$, $f \in C^2(D)$, and $n \geq 1$, $\bar{E}_x(f(TX(t)) - \int_0^t L_0 f(TX(r))\,dr;\ TX(\cdot) \in A)$ is constant for $t \geq s$. Follow the proof given in the case where a solution exists to the martingale problem for $L^{(i)}$ on $D^{(i)}$, for all $i \in \{1, 2, \ldots, k\}$, and use Theorem 2.2.2.)

8.18 Let $L = \frac{1}{2}\nabla\cdot a\nabla + b\cdot\nabla + V$ satisfy Assumption $\tilde{\text{H}}$ on a domain D, where either $D \subset R^d$ is bounded with a $C^{2,\alpha}$-boundary, or D is a compact $C^{2,\alpha}$-manifold without boundary. Let λ_0 denote the principal eigenvalue of L on D, let $\phi_0 > 0$ denote the corresponding eigenfunction for L and let $\tilde{\phi}_0 > 0$ denote the corresponding eigenfunction for $\tilde{L}$. Then $-\int_D f[\phi_0^{-1}(L - \lambda_0)(f\phi_0)]\phi_0\tilde{\phi}_0\,dx = \frac{1}{2}\int_D(\nabla f a \nabla f)\phi_0\tilde{\phi}_0\,dx$, for all $f \in C^2(\bar{D})$ which satisfy $f = 0$ on ∂D. (*Hint*: Let $\mathcal{A} = (L - \lambda_0)^{\phi_0}$ denote the h-transform of $L - \lambda_0$ via the function ϕ_0, and let $\tilde{\psi} = \phi_0\tilde{\phi}_0$. Then $\tilde{\mathcal{A}}\tilde{\psi} = 0$ and $\mathcal{A}$ is of the form $\mathcal{A} = \frac{1}{2}\nabla\cdot a\nabla + B\cdot\nabla$. Thus, the above identity may be written as $-\int_D f(\frac{1}{2}\nabla\cdot a\nabla\cdot f + B\nabla f)\tilde{\psi}\,dx =$

$\frac{1}{2}\int_D(\nabla f a \nabla f)\tilde{\psi}\,\mathrm{d}x$.) Using integration by parts and the fact that $\tilde{\mathscr{A}}\tilde{\psi} = 0$, prove this new form of the identity.

8.19 Let

$$L = \frac{1}{2}\sum_{i,j=1}^{d} a_{ij}\frac{\partial^2}{\partial x_i \partial x_j} + \sum_{i=1}^{d} b_i \frac{\partial}{\partial x_i}$$

be an operator on R^d with constant coefficients and assume that $a = \{a_{ij}\}$ is positive definite. Show that $\lambda_c = -\frac{1}{2}(b, a^{-1}b)$.

8.20 Show that $\lambda_0(v)$ defined in Section 2 is strictly convex in v. (*Hint*: Let v_0, $v_1 \in R^d$ and define $\lambda_s = (1-s)\lambda_0(v_0) + s\lambda_0(v_1)$ and $v_s = (1-s)v_0 + sv_1$. Calculate $(L^{v_s} - \lambda_s)\psi_{v0}^{1-s}\psi_{v1}^{s}$ and argue as in Theorem 4.6.1.)

8.21 Prove Proposition 2.17(iii). (*Hint*: Show that $\lambda_0(t, v)$ is convex in (t, v) by an argument similar to the one in exercise 8.20. This shows that $\lambda_c(t)$ is convex. Now use the convexity of $\lambda_c(t)$ and of $\lambda_0(t, v)$ along with the differentiability of $\lambda_0(t, v)$ to show that $\lambda_c(t)$ is once differentiable. By Theorem 2.10 applied to L_t, it follows that

$$\left\{\frac{\partial^2 \lambda_0}{\partial v_i \partial v_j}(t, v)\right\}_{i,j=1}^{d}$$

is positive definite. Use this, the differentiability of $\lambda_c(t)$, the fact that $\lambda_0(t, v_0(t)) - \lambda_c(t) \equiv 0$, and the implicit function theorem to show that $v_0(t)$ is once differentiable. Then use the fact that $\nabla\lambda_0(t, v_0(t)) \equiv 0$ and that

$$\left\{\frac{\partial^2 \lambda_0}{\partial v_i \partial v_j}(t, v)\right\}_{i,j=1}^{d}$$

is positive definite to show that $v_0(t)$ is twice differentiable. Since $\lambda_c(t) = \lambda_0(t, v_0(t))$, it follows that $\lambda_c(t)$ is twice differentiable.)

8.22 Using the notation and assumptions of Theorem 2.14, prove that

$$\frac{\mathrm{d}^3\lambda_c}{\mathrm{d}t^3}(0) < 0,$$

whenever W is an eigenfunction corresponding to $\lambda_1(v_0)$. (*Hint*: By Proposition 2.18, $v_0(t) \equiv v_0$. Write $\lambda_c(t) = \lambda_0(v_0) + \lambda_1 t + \lambda_2 t^2 + \lambda_3 t^3 + o(t^3)$ and show that $\lambda_3 < 0$ via a perturbation analysis similar to the one used in the proof of Theorems 2.14 and 2.16.)

Notes

The boundary Harnack principle as it appears in Theorem 0.1 is due to [Ancona (1978)]. An extension to the case of uniformly elliptic operators in divergence-form with measurable coefficients is due to [Cafferelli, Fabes, Mortola, and Salsa (1981)]. An extension to the case of a uniformly elliptic non-divergence-form operator with a continuous diffusion matrix $\{a_{ij}\}$, a bounded, measurable drift $\{b_i\}$ and $V = 0$ is due to [Ballman (1984)]. The equivalence of the Martin boundary and the Euclidean boundary in the case where the Euclidean boundary is Lipschitz goes back to [Hunt and Wheeden (1970)], where Theorem 1.5 and Theorem 1.10 (i) were proved in the case where $L = \Delta$. The general case of Theorem 1.5 is due to [Ancona (1978)]; it is also proved in [Taylor (1978)] with the restriction $V \leqslant 0$. Theorem 1.4, which extends Theorem 1.5 to unbounded domains, is due to [Murata (1990a)]; the proof is somewhat elliptic (no pun intended) and I had trouble following some of the missing steps, but the author has sent me the supplementary details which have convinced me of the proof's veracity. I was unable to find a reference in the literature for Theorem 1.10 (ii). The proofs of the results in Section 1 are quite different from the proofs given in the articles cited above.

Theorem 2.1 is due to Agmon and to Pinchover; see [Agmon (1984)]. Theorems 2.2 and 2.5 are due to [Agmon (1984)], where they are stated without proof. The rest of the chapter starting with Theorem 2.6 is due to [R. Pinsky (1994)]. See [Schroeder (1988)] for related work concerning the asymptotic behavior of the Green's function for operators with periodic coefficients.

Theorem 3.1 was originally proved by [Gilbarg and Serrin (1955/1956)] in the case where $V \equiv 0$; see also [R. Pinsky (1991)] for a probabilistic proof using coupling. In the general case, Theorem 3.1 was proved by [Murata (1990a)]. The proof in the text follows [Pinchover (1992)]. Theorem 3.2 is due to [Pinchover (1994)] and Theorem 3.3 is due to [Berestycki and Nirenberg (1991)].

The concept of *explosion inward from the boundary* is due to Azencott, although he did not call it by that name. He developed the connection between semigroups that leave invariant the space of continuous functions vanishing at ∂D and diffusions which do not explode inward from the boundary. These results and Theorem 4.1 are in [Azencott (1974)]. In the one-dimensional case, the concept of an entrance boundary goes back to [Feller (1952)].

The skew-product construction as in Theorem 5.1 goes back to [Itô and McKean (1965)]. Theorem 5.2 is new; however, in the case where the operators $L^{(i)}$, $i = 2, \ldots, k$, are symmetric, it was proved in [Murata (1990a)]. It is easy to show that if $L^{(1)} + \sum_{i=2}^{k} \lambda_0^{(i)} M^{(i)}$ is subcritical (supercritical) on $D^{(1)}$, then L is subcritical (supercritical) on D. It is more difficult to show that if $L^{(1)} + \sum_{i=2}^{k} \lambda_0^{(i)} M^{(i)}$ is critical on $D^{(1)}$, then L is critical on D; note that we have given a probabilistic proof of this.

Theorem 6.1 is new. In the case where L_i is symmetric for $i = 2, \ldots, k$, [Murata (1990a)] has proved a result similar in spirit to Theorem 6.1 (b) and (c), but less useful from the point of view of calculations. Theorems 6.2 and 6.3 are new. Theorem 6.4 has been proved with the Laplacian replaced by a more general operator of the form $\nabla \cdot a\nabla$ or

$$\sum_{i,j=1}^{d} a_{ij} \frac{\partial^2}{\partial x_i \partial x_j}$$

in [Landis and Nadirashvili (1986)]. Theorem 6.6 was proven by [Aikawa

(1986)] who allowed for $\Sigma \subset R^d$ to have a Lipschitz boundary; the proof in the text, which is an application of Theorem 6.1, is completely different. The Picard principle (Theorem 6.7) was proved by Picard in the case where $W \equiv 0$. The case of non-zero W was first considered by Brelot in the 1930s. The first subsection in Section 9 gives a sketch of more recent results concerning the Picard principle and makes reference to a few articles. A look at the references in these articles will supply the interested reader with a long list of references. Theorems 6.9–6.11 are apparently new. A special case of Theorem 6.11 appears in [Nakai (1975)]. Proposition 8.1 is new.

9

Bounded harmonic functions and applications to Brownian motion and the Laplacian on a manifold of non-positive curvature

9.1 Probabilistic characterization of bounded harmonic functions

In this chapter we assume that $V \equiv 0$. In this case, it is of interest to study the class of bounded harmonic functions. If u is bounded and satisfies $Lu = 0$, then by the strong maximum principle, $v(x) \equiv u(x) - \inf_{z \in D} u(z)$ is a bounded positive harmonic function. Thus, there is no loss of generality in restricting our considerations to bounded positive harmonic functions. Define

$$BC_L = BC_L(D) = \{u \in C_L(D)\colon u \text{ is bounded}\}.$$

Recall the shift operator $\theta_s : \hat{\Omega}_D \to \hat{\Omega}_D$ defined by $X(t, \theta_s \omega) = X(t+s, \omega)$. A set $A \in \hat{\mathcal{F}}^D$ is called *invariant* if $\theta_s^{-1} A = A$, for all $s > 0$. It is easy to show that the collection of invariant sets is a σ-algebra. It is called the *invariant σ-algebra* and will be denoted by $\mathcal{J}$.

Lemma 1.1. Let $L = \frac{1}{2}\nabla \cdot a\nabla + b \cdot \nabla$ satisfy Assumption $\tilde{\mathrm{H}}_{\mathrm{loc}}$ on a domain $D \subseteq R^d$. Let $\mathcal{U}$ be a non-negative bounded $\mathcal{J}$-measurable function on $\hat{\Omega}_D$ and let $u(x) = E_x \mathcal{U}$.

(i) *For any domain $B \subset\subset D$,*

$$u(x) = E_x u(X(\tau_B)), \text{ for } x \in D, \text{ where } \tau_B = \inf\{t \geq 0 : X(t) \notin B\}.$$

(ii) *If $u(x) = 0$, for some $x \in D$, then $u(x) = 0$, for all $x \in D$.*

Proof.

(i) Since any bounded $\mathcal{J}$-measurable $\mathcal{U}$ is the bounded pointwise limit of finite linear combinations of indicator functions of invariant sets, it suffices to prove (i) for $\mathcal{U} = 1_A$, where $A \in \mathcal{J}$. For $t > 0$, $\theta_t^{-1}A = A$ and $\theta_t^{-1}A \in \hat{\mathcal{F}}^D(X(s), s \geq t)$; thus by the strong Markov property,

$$u(x) = E_x 1_A = E_x 1_{\theta_t^{-1}A} = E_x E_x(1_{\theta_t^{-1}A} | \mathcal{F}_{t \wedge \tau_B}) = E_x E_{X(t\wedge\tau_B)} 1_{\theta_t^{-1}A}$$
$$= E_x E_{X(t\wedge\tau_B)} 1_A = E_x u(X(t \wedge \tau_B)).$$

Since $P_x(\tau_B < \infty) = 1$, we now obtain (i) by letting $t \to \infty$ and using the boundedness of u, which follows from the boundedness of $\mathcal{U}$.

(ii) The proof of (ii) follows from (i) and exercise 4.7. □

The invariant σ-algebra is called *trivial* if $P_x(A) = 0$ or 1, for each $A \in \mathcal{J}$. By Lemma 1.1, this definition is independent of $x \in D$.

Theorem 1.2. *Let $L = \frac{1}{2}\nabla \cdot a\nabla + b \cdot \nabla$ satisfy Assumption $\widetilde{\mathrm{H}}_{\mathrm{loc}}$ on a domain $D \subseteq R^d$. Then $u \in BC_L(D)$ if and only if there exists a bounded, non-negative $\mathcal{J}$-measurable function $\mathcal{U}: \hat{\Omega}_D \to \hat{\Omega}_D$ satisfying $P_x(\mathcal{U} > 0) > 0$, for $x \in D$, such that*

$$u(x) = E_x\mathcal{U}, \quad x \in D. \tag{1.1}$$

In particular, $BC_L(D)$ contains only the positive constants if and only if the invariant σ-algebra $\mathcal{J}$ is trivial under P_x for some (or, equivalently, for all) $x \in D$; that is, if and only if $P_x(A) = 0$ or 1, for each $A \in \mathcal{J}$.

Proof. Let $\mathcal{U}$ be as in the statement of the theorem and let $u(x) = E_x\mathcal{U}$. By the assumptions on $\mathcal{U}$, it follows that u is positive. Fix a domain $B \subset\subset D$ with a $C^{2,\alpha}$-boundary and let $\{f_n\}_{b=1}^{\infty}$ be a bounded sequence of $C^{2,\alpha}(\bar{B})$ functions on ∂B such that

$$\lim_{n\to\infty} f_n(x) = u(x), \quad \text{for almost all } x \in \partial B$$
$$\text{with respect to Lebesgue measure on } \partial B. \tag{1.2}$$

Define

$$u_n(x) = E_x f_n(X(\tau_B)), \quad \text{for } x \in B, \tag{1.3}$$

where $\tau_B = \inf\{t \geq 0 : X(t) \notin B\}$. By Theorems 2.1.2 and 3.6.5,

$$u_n \in C^{2,\alpha}(B) \quad \text{and} \quad Lu_n = 0 \text{ in } B. \tag{1.4}$$

By Theorem 7.4.4, the distribution of $X(\tau_B)$ is absolutely continuous with respect to Lebesgue surface measure on ∂B. Thus, it follows from (1.2), (1.3) and Lemma 1.1(i) that $u(x) = \lim_{n\to\infty} u_n(x)$, for $x \in B$. Using this along with (1.4) and a standard compactness argument, it follows that $u \in C^{2,\alpha}(B)$ and $Lu = 0$ in B. Since B is arbitrary, we conclude that $u \in BC_L(D)$.

Now assume that $u \in BC_L(D)$. We will prove that u has the form given in (1.1) under the condition that the diffusion corresponding to L does not explode; that is, under the condition that $P_x(\tau_D = \infty) = 1$, for some (or equivalently, all) $x \in D$. Without this assumption, the proof is a bit more technical, and is left to exercise 9.1. Since $P_x(\tau_D = \infty) = 1$, $\{P_x, x \in D\}$ solves the martingale problem for L on D; thus, we work with $(\Omega, \mathcal{F}, \mathcal{F}_t)$ rather than with $(\hat{\Omega}_D, \hat{\mathcal{F}}^D, \hat{\mathcal{F}}_t^D)$. Since $Lu = 0$ and u is bounded, it follows from Proposition 2.0.1 that $u(X(t))$ is a martingale with respect to $(\Omega, \mathcal{F}, \mathcal{F}_t, P_x)$. By the martingale convergence theorem (Theorem 1.2.4), for each $x \in D$,

$$\mathcal{U} \equiv \lim_{t\to\infty} u(X(t)) \text{ exists a.s. } P_x,\ x \in D. \tag{1.5}$$

In fact, a standard argument, which we leave to the reader, shows that there exists a set $N \subset \Omega$ such that $P_x(N) = 0$, for all $x \in D$, and $\mathcal{U} \equiv \lim_{t\to\infty} u(X(t))$ exists for all $\omega \in \Omega - N$. Extend $\mathcal{U}$ to Ω by defining $\mathcal{U} = 1$ on N. Since $u(X(t))$ is a martingale, $u(x) = E_x u(X(t))$, for all $t > 0$; thus from the bounded convergence theorem, we obtain $u(x) = E_x\mathcal{U}$, $x \in D$. By construction, $\mathcal{U}$ if $\mathcal{I}$-measurable and non-negative. Since $u(x) = E_x\mathcal{U} > 0$, it follows that $P_x(\mathcal{U} > 0) > 0$. □

We now give a similar, but more diffusion-path-oriented, characterization of $BC_L(D)$. For each measurable set $B \subseteq D$, define $u_B(x) = P_x(X(t) \in B$, for all $t \in (s, \tau_D)$, for some $s < \tau_D)$. (If the diffusion corresponding to L does not explode, then u_B may be defined more simply as $u_B(x) = P_x(X(t) \in B$, for all large t).) A function u_B, $B \subseteq D$, will be called an *exit probability function* for L. Since the event $\{X(t) \in B$, for all $t \in (s, \tau_D)$, for some $s < \tau_D\}$ belongs to the invariant σ-algebra $\mathcal{I}$, it follows from Theorem 1.2 that $u_B \in BC_L(D)$.

We note the following lemma.

Lemma 1.3. *Let L satisfy Assumption $\tilde{\mathrm{H}}_{\mathrm{loc}}$ on a domain $D \subseteq R^d$ and let u_B, $B \subseteq D$, be an exit probability function. If $u_B \equiv c$, then $c = 0$, or 1.*

Proof. The proof uses the proof of Theorem 1.2. We proved Theorem 1.2 under the assumption that the diffusion does not explode, leaving the explosive case to exercise 9.1. Thus, we will assume here also that the diffusion does not explode; the explosive case will follow by using exercise 9.1. Since $u_B \in BC_L(D)$, it follows from the proof of Theorem 1.2 that $I_A = \lim_{t\to\infty} u_B(X(t))$ a.s. P_x, where $A = \{X(t) \in B$, for all large $t\}$. Since I_A only takes on the values 0 and 1, it follows that $c = 0$ or 1. □

Theorem 1.4. *Let $L = \frac{1}{2}\nabla \cdot a\nabla + b \cdot \nabla$ satisfy Assumption $\widetilde{\mathrm{H}}_{\mathrm{loc}}$ on a domain $D \subseteq R^d$. Then each function $u \in BC_L(D)$ is the bounded pointwise limit of finite linear combinations of exit probability functions. In particular, $BC_L(D)$ contains only the positive constants if and only if every exit probability function is equal to* 0 *or* 1.

Proof. We will prove the theorem under the assumption that the diffusion does not explode; use exercise 9.1 for the explosive case. Thus, $\{P_x, x \in D\}$ solves the martingale problem for L on D and we work with $(\Omega, \mathcal{F}, \mathcal{F}_t)$ rather than with $(\hat{\Omega}_D, \hat{\mathcal{F}}^D, \hat{\mathcal{F}}_t^D)$. Let $u \in BC_L(D)$. As in the proof of Theorem 1.2, there exists a bounded non-negative random variable $\mathcal{U}$ on $(\Omega, \mathcal{F})$ such that (1.5) holds and such that

$$u(x) = E_x \mathcal{U}, \quad x \in D. \tag{1.6}$$

Fix $a \in R$. If $P_x(\mathcal{U} = a) = 0$, for some $x \in D$, then in fact $P_x(\mathcal{U} = a) = 0$, for all $x \in D$. This follows by Lemma 1.1(ii) (substitute $1_{\{a\}}(\mathcal{U})$ for the $\mathcal{U}$ appearing in that lemma). Thus, there exists a countable set $C \subset R$ such that $P_x(\mathcal{U} = a) = 0$, for all $x \in D$ and all $a \in R - C$. For simplicity of notation only, we will assume that

$$\frac{j}{2^n} \notin C,$$

for all positive integers j and n. That is,

$$P_x\left(\mathcal{U} = \frac{j}{2^n}\right) = 0, \quad \text{for } x \in D \text{ and } j, n = 1, 2, \ldots. \tag{1.7}$$

Choose $M > \sup \mathcal{U}$ and define

$$B_{n,j} = \left\{x \in D : \frac{j-1}{2^n} M \leqslant u(x) < \frac{j}{2^n} M\right\},$$

for $n = 1, 2 \ldots$ and $j = 1, 2, \ldots, 2^n$. Let $u_{B_{n,j}}$ denote the exit probability function for the set $B_{n,j}$ and define

$$u_n(x) = M \sum_{j=1}^{2^n} \frac{j}{2^n} u_{B_{n,j}}(x), \quad x \in D.$$

From (1.5)–(1.7), it follows that $u(x) = \lim_{n\to\infty} u_n(x)$, for $x \in D$. By construction, $\{u_n\}_{n=1}^{\infty}$ is bounded by M. This proves the first statement of the theorem. The second statement of the theorem follows from the first statement and Lemma 1.3. □

Theorems 1.2 and 1.4 gave two very similar characterizations for the existence or non-existence of non-constant bounded harmonic functions. We now give another such characterization in terms of the Martin boundary.

Proposition 1.5. *Let $L = \frac{1}{2}\nabla \cdot a\nabla + b \cdot \nabla$ satisfy Assumption $\tilde{\mathrm{H}}_{\mathrm{loc}}$ on a domain $D \subseteq R^d$. Assume that L is subcritical; that is, assume that the diffusion corresponding to L is transient. Then $BC_L(D)$ contains only the positive constants if and only if the function $u(x) \equiv 1$ is minimal in $C_L(D)$.*

Remark. There is no loss of generality in considering only subcritical operators since in the critical case, it follows from Theorem 4.3.4 that $C_L(D)$ contains only the positive constants.

Proof. Assume first that 1 is minimal. Let $u \in BC_L(D)$. Then $\varepsilon u \leqslant 1$, for some $\varepsilon > 0$; thus, it follows from the minimality of 1 that εu is a multiple of 1; that is, u is constant. Now assume that 1 is not minimal and let μ_1 denote the measure on the minimal Martin boundary Λ_0 in the Martin representation theorem for the function 1 (Theorem 7.1.2). By the assumption of non-minimality, the support of the measure μ_1 contains more than one point. Choose $A \subset \Lambda_0$ such that $\mu_1(A) > 0$ and $\mu_1(\Lambda_0 - A) > 0$, and let $u_A(x) = \int_A k(x;\zeta)\mu_1(d\zeta)$. Then $u_A \in C_L(D)$, and u_A is bounded since $u_A(x) \leqslant \int_{\Lambda_0} k(x;\zeta)\mu_1(d\zeta) = 1$. It remains to show that u_A is not constant. Since $u_A(x) = \int_{\Lambda_0} k(x;\zeta)1_A(\zeta)\mu_1(d\zeta)$, it follows that μ_A, the unique measure in the Martin representation theorem for u_A, satisfies $\mu_A(d\zeta) = 1_A(\zeta)\mu_1(d\zeta)$. For any constant $c > 0$, $c = \int_{\Lambda_0} k(x;\zeta)c\mu_1(d\zeta)$; thus μ_c, the unique measure in the Martin representation theorem for c, satisfies $\mu_c(d\zeta) = c\mu_1(d\zeta)$. Since $\mu_A(\Lambda_0 - A) = 0$ and $\mu_c(\Lambda_0 - A) > 0$, it follows that $\mu_c \not\equiv \mu_A$ and thus, by the uniqueness of the measure in the Martin representation theorem, $u_A \not\equiv c$. □

9.2 The connection between bounded harmonic functions and bounded solutions in exterior domains

In this section, we characterize the non-existence of non-constant bounded harmonic functions in terms of the behavior of bounded solutions in exterior domains. We begin with a simple proposition.

Proposition 2.1. *Let $L = \frac{1}{2}\nabla \cdot a\nabla + b \cdot \nabla$ satisfy Assumption $\widetilde{\mathrm{H}}_{\mathrm{loc}}$ on a domain $D \subseteq R^d$, $d \geq 2$. Let $\Omega \subset\subset D$ be a subdomain satisfying an interior cone condition at each point of its boundary and let $\phi \in C(\partial\Omega)$. Consider the equation*

$$\left.\begin{aligned} &Lu = 0 \text{ in } D - \bar{\Omega}, \\ &u = \phi \text{ on } \partial\Omega, \\ &u \text{ is bounded.} \end{aligned}\right\} \tag{2.1}$$

(i) *Assume that L is critical; that is, assume that the diffusion corresponding to L is recurrent. Then there exists a unique solution u_0 to (2.1). The solution is given by $u_0(x) = E_x\phi(X(\sigma_{\bar{\Omega}}))$, for $x \in D - \Omega$, where $\sigma_{\bar{\Omega}} = \inf\{t \geq 0 : X(t) \in \bar{\Omega}\}$. If $\phi \geq 0$ and $\phi \not\equiv 0$, then u is a positive solution of minimal growth at ∂D.*

(ii) *Assume that L is subcritical; that is, assume that the diffusion corresponding to L is transient. Let $u_0(x) = E_x(\phi(X(\sigma_{\bar{\Omega}})); \sigma_{\bar{\Omega}} < \tau_D)$ and let $v_0(x) = P_x(\sigma_{\bar{\Omega}} = \infty)$, for $x \in D - \Omega$. Then u_0, $v_0 \in C^{2,\alpha}(D - \bar{\Omega}) \cap C(D - \Omega)$, $v_0 > 0$ on $D - \bar{\Omega}$ and for every $c \in R$, $u(x) = u_0 + cv_0$ is a solution to (2.1). If $\phi \geq 0$ and $\phi \not\equiv 0$, then u_0 is a positive solution of minimal growth at ∂D.*

Proof. By linearity it suffices to consider the case $\phi \geq 0$.

(i) If $\phi \equiv 0$, then $u_0 \equiv 0$ is a solution of (2.1). By the recurrence assumption $P_x(\sigma_{\bar{\Omega}} < \tau_D) = 1$, for $x \in D - \Omega$. Thus, if $\phi \geq 0$ and $\phi \not\equiv 0$, then it follows from Theorem 7.3.5. that $u_0 \in C^{2,\alpha}(D - \bar{\Omega}) \cap C(D - \Omega)$, that u_0 solves (2.1), and that u_0 is a positive solution of minimal growth at ∂D. To prove uniqueness, let $\{D_n\}_{n=1}^{\infty}$ be a sequence of domains satisfying $\Omega \subset\subset D_1$, $D_n \subset\subset D_{n+1}$ and $\bigcup_{n=1}^{\infty} D_n = D$. Let $u \in C^{2,\alpha}(D - \bar{\Omega}) \cap C(D - \Omega)$ be a solution of (2.1). Then $u(X(t \wedge \tau_{D_n} \wedge \sigma_{\bar{\Omega}}))$ is a martingale; thus $u(x) = E_x u(X(t \wedge \tau_{D_n} \wedge \sigma_{\bar{\Omega}}))$, for $x \in D_n - \Omega$, $t \geq 0$, and $n = 1, 2, \ldots$. By the recurrence assumption, $\lim_{n\to\infty} \lim_{t\to\infty} t \wedge \tau_{D_n} \wedge \sigma_{\bar{\Omega}} = \sigma_{\bar{\Omega}}$ a.s. P_x. Thus, letting $t \to \infty$ and $n \to \infty$ and using the bounded convergence theorem and the boundary condition gives $u(x) = E_x\phi(X(\sigma_{\bar{\Omega}})) = u_0(x)$. This proves uniqueness.

(ii) If $\phi \equiv 0$, then $u_0 \equiv 0$ is a solution of (2.1). If $\phi \geq 0$ and $\phi \not\equiv 0$, then it follows from Theorem 7.3.5 that $u_0 \in C^{2,\alpha}(D - \bar{\Omega}) \cap C(D - \Omega)$, that u_0 solves (2.1) and that u_0 is a positive solution of minimal growth at ∂D. By the transience assumption, $v_0(x) > 0$ for $x \in D - \bar{\Omega}$. By arguments similar to those given in the proof of Theorem 7.3.5, $v_0 \in C^{2,\alpha}(D - \bar{\Omega}) \cap C(D - \Omega)$, $Lv_0 = 0$ in $D - \bar{\Omega}$, and $v_0 = 0$ on $\partial\Omega$. Thus $u = u_0 + cv_0$ solves (2.1) for any $c \in R$. □

Theorem 2.2. *Let* $L = \frac{1}{2}\nabla \cdot a\nabla + b \cdot \nabla$ *satisfy Assumption* $\widetilde{\mathrm{H}}_{\mathrm{loc}}$ *on a domain* $D \subseteq R^d$, $d \geq 2$. *Assume that* L *is subcritical on* D*; that is, assume that the diffusion corresponding to* L *is transient. Let* $\Omega \subset\subset D$. *Define* $v_0(x) = P_x(\sigma_{\bar{\Omega}} < \infty)$, *where* $\sigma_{\bar{\Omega}} = \inf\{t \geq 0: X(t) \in \bar{\Omega}\}$, *and define*

$$BC_L(D - \Omega) = \{v \in C^{2,\alpha}(D - \bar{\Omega}) \cap C(D - \Omega): Lv = 0 \text{ and } v > 0 \text{ in } D - \Omega \text{ and } v \text{ is bounded}\}.$$

Then there are no non-constant bounded harmonic functions for L *on* D *if and only if for each* $v \in BC_L(D - \Omega)$, *there exists a constant* c_v *such that* $\lim_{n\to\infty} v(x_n) = c_v$, *whenever* $\{x_n\}_{n=1}^{\infty} \subset D - \Omega$ *is a sequence satisfying* $\lim_{n\to\infty} v_0(x_n) = 0$.

Proof. We will prove the theorem under the condition that the diffusion does not explode; that is, $P_x(\tau_D = \infty) = 1$; use exercise 9.1 for the explosive case. Then $\{P_x, x \in D\}$ solves the martingale problem for L on D and we can work with $(\Omega, \mathcal{F}, \mathcal{F}_t)$ rather than with $(\hat{\Omega}_D, \hat{\mathcal{F}}^D, \hat{\mathcal{F}}_t^D)$. (The notation Ω is now unfortunately being used in two different roles in this proof; however, the context will clarify any ambiguity.) First, assume that there are no non-constant bounded harmonic functions. Let $v \in BC_L(D - \Omega)$. Extend v boundedly so that it is defined on all of D, and define $\mathcal{U}$ on Ω by

$$\mathcal{U} = \begin{cases} \lim_{t\to\infty} v(X(t)), & \text{if the limit exists} \\ 0 & \text{otherwise.} \end{cases}$$

Clearly, $\mathcal{U}$ is non-negative, $\mathcal{U}$ is measurable with respect to the invariant σ-algebra and $\mathcal{U} \leq \sup_{x\in D-\Omega} v(x)$. Thus, by Theorem 1.2, $h(x) \equiv E_x\mathcal{U}$ is a bounded harmonic function. By assumption then, there exists a constant c_v such that $h(x) = E_x\mathcal{U} = c_v$. Define

$$\mathcal{V} = \begin{cases} \lim_{t\to\infty} v(X(t \wedge \sigma_{\bar{\Omega}})), & \text{if the limit exists} \\ 0 & \text{otherwise.} \end{cases}$$

By Proposition 2.0.1, $v(X(t \wedge \sigma_{\bar{\Omega}}))$ is a bounded martingale with respect to $(\Omega, \mathcal{F}, \mathcal{F}_t, P_x)$; thus, by the martingale convergence theorem (Theorem 1.2.4), $\lim_{t\to\infty} v(X(t \wedge \sigma_{\bar{\Omega}}))$ exists a.s. P_x and

$$v(x) = E_x\mathcal{V} = E_x(\mathcal{V}; \sigma_{\bar{\Omega}} < \infty) + E_x(\mathcal{V}; \sigma_{\bar{\Omega}} = \infty), \quad x \in D. \tag{2.2}$$

But $\mathcal{U} = \mathcal{V}$ on $\{\sigma_{\bar{\Omega}} = \infty\}$; thus

$$c = h(x) = E_x\mathcal{U} = E_x(\mathcal{U}; \sigma_{\bar{\Omega}} < \infty) + E_x(\mathcal{V}; \sigma_{\bar{\Omega}} = \infty), \quad x \in D. \tag{2.3}$$

Substituting (2.3) into (2.2) gives

$$v(x) = c - E_x(\mathcal{U}; \sigma_{\bar{\Omega}} < \infty) + E_x(\mathcal{V}; \sigma_{\bar{\Omega}} < \infty). \tag{2.4}$$

Let $\{x_n\}_{n=1}^{\infty} \subset D - \Omega$ be a sequence satisfying $\lim_{n\to\infty} v_0(x_n) = \lim_{n\to\infty} P_{x_n}(\sigma_{\bar{\Omega}} < \infty) = 0$. Since $\mathcal{U}$ and $\mathcal{V}$ are bounded, it follows from (2.4) that $\lim_{n\to\infty} v(x_n) = c$.

Now assume that for each $v \in BC_L(D - \Omega)$, there exists a constant c_v such that $\lim_{n\to\infty} v(x_n) = c_v$, whenever $\{x_n\}_{n=1}^{\infty} \subset D - \Omega$ is a sequence satisfying $\lim_{n\to\infty} v_0(x_n) = 0$. Define $A_t = \{X(s) \in \Omega$, for some $s \geq t\}$, $t > 0$, and define $A = \{X(t_n) \in \Omega$ for some sequence $t_n \to \infty\}$. For each $t_0 > 0$, $M(t) \equiv E_x(1_{A_{t_0}} | \mathcal{F}_t)$ is a martingale with respect to $(\Omega, \mathcal{F}, \mathcal{F}_t, P_x)$ and thus, by the martingale convergence theorem,

$$\lim_{t\to\infty} E_x(1_{A_{t_0}} | \mathcal{F}_t) = 1_{A_{t_0}} \text{ a.s. } P_x.$$

By the Markov property and the definition of v_0,

$$v_0(X(t)) = E_x(1_{A_t} | \mathcal{F}_t) \text{ a.s. } P_x, \quad \text{for } x \in D \text{ and } t \geq 0.$$

Since A_t is non-increasing in t, it follows that for each $t_0 > 0$,

$$\lim_{t\to\infty} v_0(X(t)) = \lim_{t\to\infty} E_x(1_{A_t} | \mathcal{F}_t) \leq \lim_{t\to\infty} E_x(1_{A_{t_0}} | \mathcal{F}_t) = 1_{A_{t_0}} \text{ a.s. } P_x. \tag{2.5}$$

Since $A_t \downarrow A$ and since, by the transience assumption, $P_x(A) = 0$, it follows upon letting $t_0 \to \infty$ in (2.5) that

$$\lim_{t\to\infty} v_0(X(t)) = 0 \text{ a.s. } P_x. \tag{2.6}$$

Let $u \in BC_L(D)$. Then by Proposition 2.0.1, $u(X(t))$ is a bounded martingale with respect to $(\Omega, \mathcal{F}, \mathcal{F}_t, P_x)$, and by the martingale convergence theorem,

$$u(x) = \lim_{t\to\infty} E_x u(X(t)). \tag{2.7}$$

Since $u \in BC_L(D - \Omega)$, it follows from the assumption on $BC_L(D - \Omega)$ and from (2.6) that

$$c \equiv \lim_{t\to\infty} u(X(t)) \text{ exists a.s. } P_x. \tag{2.8}$$

From (2.7), (2.8) and the bounded convergence theorem, it follows that $u \equiv c$. □

9.3 Examples

Using the results of Section 1, we analyze the structure of $BC_L(D)$ in two particular cases.

Proposition 3.1. *Let $L = \frac{1}{2}\nabla \cdot a\nabla + b \cdot \nabla$ on R^d be a periodic operator as in Theorem* 8.2.1. *Then there are no non-constant bounded harmonic functions.*

Proof. By Corollary 8.2.4, it follows easily that $u \in BC_L(D)$ if and only if $u = c\psi_0$, for some $c > 0$, where ψ_0 is the ground state of L on the torus T_d. Clearly, $\psi_0 \equiv 1$. □

Theorem 3.2. *Let*

$$L = \tfrac{1}{2}a(r)\frac{\partial^2}{\partial r^2} + b(r)\frac{\partial}{\partial r} + M(r)\mathcal{A}$$

on $D = R^d - \{0\}$ *be as in Theorem* 8.6.3. *Define*

$$A_0 = \left\{\lim_{t\to\tau_D}|X(t)| = 0,\ \lim_{t\to\tau_D}\frac{X(t)}{|X(t)|}\ \textit{exists}\right\},$$

$$A_\infty = \left\{\lim_{t\to T_D}|X(t)| = \infty,\ \lim_{t\to\tau_D}\frac{X(t)}{|X(t)|}\ \textit{exists}\right\},$$

$B_0 = \{\lim_{t\to T_D}|X(t)| = 0\}$ *and* $B_\infty = \{\lim_{t\to T_D}|X(t)| = \infty\}$. *Let*

$$\mathcal{U}_0 = \lim_{t\to\tau_D}\frac{X(t)}{|X(t)|}1_{A_0},$$

$$\mathcal{U}_\infty = \lim_{t\to\tau_D}\frac{X(t)}{|X(t)|}1_{A_\infty},$$

$\mathcal{U}_0 = 1_{B_0}$ *and* $\mathcal{V}_\infty = 1_{B_\infty}$.

(i) *Assume that integral conditions* (1) *and* (5) *hold in Theorem* 8.6.3. *Then for each open $U \subseteq S^{d-1}$, $P_x(\mathcal{U}_0 \in U)$ and $P_x(\mathcal{U}_\infty \in U)$ are not constant in x. Up to sets of P_x-measure zero, the invariant σ-algebra is generated by $\mathcal{U}_0$ and $\mathcal{U}_\infty$.*

(ii) *Assume that integral conditions* (1) *and* (6) *hold in Theorem* 8.6.3. *Then $E_x\mathcal{V}_\infty$ is not constant in x and, for each open $U \subseteq S^{d-1}$, $P_x(\mathcal{U}_0 \in U)$ is not constant in x. Up to sets of P_x-measure zero, the invariant σ-algebra is generated by $\mathcal{U}_0$ and $\mathcal{V}_\infty$.*

(iii) *Assume that the integral conditions* (1) *and either* (7) *or* (8) *hold in Theorem* 8.6.3. *Then for each non-empty, open $U \subsetneqq S^{d-1}$, $P_x(\mathcal{U}_0 \in U)$ is not constant in x. Up to sets of P_x-measure zero, the invariant σ-algebra is generated by $\mathcal{U}_0$.*

(iv) *Assume that the integral conditions* (2) *and* (5) *hold in Theorem* 8.6.3. *Then $E_x\mathcal{V}_0$ is not constant in x and for each open $U \subseteq S^{d-1}$, $P_x(\mathcal{U}_\infty \in U)$ is not constant in x. Up to sets of P_x-measure zero, $\mathcal{V}_0$ and $\mathcal{U}_\infty$ generate the invariant σ-algebra.*

(v) *Assume that the integral conditions* (2) *and* (6) *hold in Theorem* 8.6.3.

Then $E_x \mathcal{V}_0$ and $E_x \mathcal{V}_\infty$ are not constant in x. Up to sets of P_x-measure zero, $\mathcal{V}_0$ and $\mathcal{V}_\infty$ generate the invariant σ-algebra.

(vi) *Assume that the integral conditions (2) and either (7) or (8) hold in Theorem 8.6.3. Then the invariant σ-algebra is trivial under P_x.*

(vii) *Assume that the integral conditions (3) or (4) and the integral condition (5) in Theorem 8.6.3 hold. Then for each non-empty, open $U \subsetneqq S^{d-1}$, $P_x(\mathcal{U}_0 \in U)$ is not constant. Up to sets of P_x-measure zero, $\mathcal{U}_0$ generates the invariant σ-algebra.*

(viii) *Assume that the integral condition (3) or (4) and the integral condition (6) hold in Theorem 8.6.3. Then the invariant σ-algebra is trivial under P_x.*

(ix) *Assume that the integral condition (3) or (4) and the integral condition (7) or (8) hold in Theorem 8.6.3. Then the invariant σ-algebra is trivial under P_x.*

Remark. In the light of Theorem 1.2, Theorem 3.2 gives an explicit probabilistic representation for $BC_L(R^d - \{0\})$. In particular, there are no non-constant bounded harmonic functions if and only if either (vi), (viii) or (ix) holds.

Proof. We leave it to the reader to verify that the proof follows almost directly from Theorem 8.6.1(e), Theorem 8.6.3 and the proof of Theorem 8.6.1(d). (In the present case, in the notation of Theorem 8.6.1, we have $\alpha = 0$, $\beta = \infty$, $k = 2$, $\phi_0^{(2)} = 1$ and $\lambda_0^{(2)} = 0$. Also,

$$g_0(x_1) = \int_{x_1}^{\infty} \exp\left(-\int_1^r \frac{2b}{a}(s)\, ds\right) dr,$$

for $x_1 \in (0, \infty)$ if (5) or (6) holds, and $g_0(x_1) = 1$, for $x_1 \in (0, \infty)$ if (7) or (8) holds;

$$g_\infty(x_1) = \int_0^{x_1} \exp\left(-\int_1^r \frac{2b}{a}(s)\, ds\right) dr,$$

for $x_1 \in (0, \infty)$ if (1) or (2) holds, and $g_\infty(x_1) = 1$, for $x_1 \in (0, \infty)$ if (3) or (4) holds.) □

9.4 Brownian motion on a manifold of non-positive curvature

In this section, we recall some fundamental results in differential geometry. The reader is directed to the notes at the end of the chapter for references. Let M be a complete, simply connected, d-dimensional Riemannian manifold, $d \geq 2$, with non-positive sectional curvature. Denote the Riemannian distance between x and y in M by $\text{dist}_M(x, y)$. Fix $x_0 \in M$. Identify M_{x_0}, the tangent manifold to M at x_0, with R^d. Since M is complete, it follows by the Hopf–Rinow theorem that for each $v \in R^d$, there exists a unique constant speed geodesic $\gamma_v(t)$ defined for $t \in (-\infty, \infty)$ and satisfying $\gamma_v(0) = x_0$ and $\gamma_v'(0) = v$. Define $\exp_{x_0}: R^d \to M$, the *exponential map* at x_0, by $\exp_{x_0}(v) = \gamma_v(1)$. Using polar coordinates, $r = |v|$ and

$$\theta = \frac{v}{|v|} \in S^{d-1},$$

we will write $\exp_{x_0}(r, \theta) = \gamma_v(1)$. Since the sectional curvatures are all non-positive, it follows by a theorem of Cartan and Hadamard that geodesics are distance minimizing over arbitrary time intervals and that $\exp_{x_0}$ is a diffeomorphism from R^d onto M. Since $\gamma_v(t)$ is a constant speed geodesic, $\text{dist}_M(x_0, \gamma_v(t)) = t|v|$. Thus, M may be given a global system of geodesic coordinates (r, θ), where $r(x) = \text{dist}_M(x_0, x)$ and $\theta(x)$ is defined by $\exp_{x_0}(r(x), \theta(x)) = x$, for $x \in M$.

The *Laplace–Beltrami* operator, Δ_M on M, is defined intrinsically as the trace of the second covariant derivative on M, where covariant differentiation is defined with respect to the Levi–Civita connection, the unique symmetric connection which is compatible with the metric. Let $G(r, \theta)$ denote the square root of the determinant of the metric at $x = \exp_{x_0}(r, \theta)$; $G(r, \theta)$ is the natural volume element for the manifold. Then Δ_M can be written in (r, θ)-coordinates as

$$\Delta_M = \frac{\partial^2}{\partial r^2} + \frac{G_r}{G}(r, \theta)\frac{\partial}{\partial r} + \Delta_{x_0,r}, \tag{4.1}$$

where $\Delta_{x_0,r}$ is the Laplace–Beltrami operator on the sphere in M of radius r centered at x_0 with the metric induced by the metric of M. We will assume enough smoothness so that Δ_M in (4.1) satisfies Assumption $\widetilde{\text{H}}_{\text{loc}}$ on $R^d - \{0\}$.

Brownian motion on the manifold M may be defined in several equivalent ways. In particular, it may be defined intrinsically on the manifold; however, we will define it as the solution to the generalized martingale problem for $\frac{1}{2}\Delta_M$ on $R^d - \{0\}$ as in (4.1). Using the diffeomorphism provided by the exponential map, the Brownian motion may be mapped back to M. Let $X_M(t)$ denote the Brownian motion on M and let $\hat{r}(t) = r(X_M(t))$ and $\hat{\theta}(t) = \theta(X_M(t))$. Then $(\hat{r}(t), \hat{\theta}(t))$ denotes the Brownian motion on $R^d - \{0\}$. Denote the lifetime of the process by τ_M. Denote probabilities by P_x or by $P_{r,\theta} = P_{r(x),\theta(x)}$, for $x \in M - \{x_0\}$.

Actually, by using global geodesic polar coordinates, we have created a singularity artificially at $r = 0$; the operator Δ_M in (4.1) and the Brownian motion as defined above are only defined on $R^d - \{0\}$. The point $0 \in R^d$ corresponds to the reference point $x_0 \in M$, called the pole of the manifold. Since x_0 was chosen arbitrarily, it is clear that the singularity is removable. Indeed, it is a standard fact that

$$\frac{G_r}{G}(r, \theta) = \frac{d-1}{r} + O(1),$$

as $r \to 0$, uniformly over $\theta \in S^{d-1}$. Using this, it is not difficult to show that for (r, θ) with $r > 0$, $P_{r,\theta}(\lim_{t\to\tau_D} \hat{r}(t) = 0) = 0$ (exercise 9.2); in particular, then, $P_x(X_M(t)) = x_0$, for some $t \geqslant 0) = 0$, for $x \neq x_0$. Therefore, from the definition of the Green's function, it follows that the Green's function for the Laplace–Beltrami operator on M coincides with the Green's function for the Laplace–Beltrami operator on $M - \{x_0\}$. Thus, in the sequel, when using the representation (4.1) to study the transience and recurrence properties of Brownian motion on M, or to study the Martin boundary of Δ_M on M, we may ignore the boundary point $r = 0$.

In the rotationally symmetric case, that is, the case where $G(r, \theta) = G(r)$, the Riemannian distance s satisfies $ds^2 = dr^2 + g^2(r)\, d\theta^2$, where $g(0) = 0$ and $g'(0) = 1$. The square root of the determinant of this metric is $g^{d-1}(r)$; thus $G(r) = g^{d-1}(r)$. The sectional curvatures of course all coincide and depend only on r; this curvature $k(r)$ is given by

$$k(r) = \frac{-g''}{g}(r).$$

Since $k(r) \leqslant 0$, $g(0) = 0$ and $g'(0) = 1$, it follows that

$$g'(r) \geqslant 1. \tag{4.2}$$

In geodesic polar coordinates, the operator Δ_M takes the simple form

$$\Delta_M = \frac{\partial^2}{\partial r^2} + (d-1)\frac{g'}{g}(r) + \frac{1}{g^2(r)}\Delta_{S^{d-1}}, \tag{4.3}$$

where $\Delta_{S^{d-1}}$ is the usual Laplace–Beltrami operator on S^{d-1}; that is, the Laplace–Beltrami operator corresponding to S^{d-1} with the metric induced from the Euclidean metric on R^d. In order that Δ_M satisfy Assumption $\widetilde{\mathrm{H}}_{\mathrm{loc}}$, we will assume that $g \in C^{2,\alpha}((0, \infty))$. Manifolds with rotationally symmetric metrics are called *model manifolds*; using curvature comparison theorems in differential geometry, results for general manifolds can sometimes be obtained by studying model manifolds.

9.5 Brownian motion on a model manifold of non-positive curvature: transience, recurrence, bounded harmonic functions, Martin boundary

In this section, we make the following assumption.

Assumption I. *Let M be a simply connected, d-dimensional, model Riemannian manifold of non-positive curvature whose metric is given in global geodesic coordinates by $\mathrm{d}s^2 = \mathrm{d}r^2 + g^2(r)\,\mathrm{d}\theta^2$, where $d \geq 2$, $g \in C^{2,\alpha}((0,\infty))$, $g(0) = 0$, $g'(0) = 1$ and $g(r) > 0$, for $r > 0$. Let Δ_M denote the Laplace–Beltrami operator; in geodesic polar coordinates, Δ_M is given by* (4.3), *where g is the Riemannian metric. Let*

$$k(r) = -\frac{g''}{g}(r) \leq 0$$

denote the curvature. Denote the Brownian motion on M by $X_M(t)$ and let $\hat{r}(t) = r(X_M(t))$, $\hat{\theta}(t) = \theta(X_M(t))$ denote its representation in geodesic polar coordinates. Denote its lifetime by τ_M. Denote probabilities by P_x or by $P_{r,\theta} = P_{r(x),\theta(x)}$.

Theorem 5.1. *Let M be a simply connected d-dimensional model Riemannian manifold of non-positive curvature as in Assumption I.*

(i) *If $\int_1^\infty g^{1-d}(r)\,\mathrm{d}r = \infty$, then Brownian motion on M is recurrent; equivalently, Δ_M on M is critical.*
(ii) *If $\int_1^\infty g^{1-d}(r)\,\mathrm{d}r < \infty$, then Brownian motion on M is transient and $\lim_{t\to\infty} \hat{r}(t) = \infty$, a.s. P_x; equivalently, Δ_M on M is subcritical.*

Proof. The Brownian motion is transient (recurrent) if and only if its radial part is transient (recurrent) (see exercise 5.5, for example). The radial part corresponds to the operator

$$\frac{\mathrm{d}^2}{\mathrm{d}r^2} + (d-1)\frac{g'}{g}(r)\frac{\mathrm{d}}{\mathrm{d}r} \text{ on } (0,\infty).$$

Since $g(0) = 0$ and $g'(0) = 1$, it follows that

$$\int_0^1 g^{1-d}(r)\,\mathrm{d}r = \infty. \tag{5.1}$$

If (i) holds, then it follows from Theorem 5.1.1 and (5.1) that $\hat{r}(t)$ is recurrent on $(0,\infty)$. If (ii) holds, then it follows from Theorem 5.1.1 and (5.1) that $\hat{r}(t)$ is transient on $(0,\infty)$ and $\lim_{t\to\tau_D} \hat{r}(t) = \infty$ a.s. $P_{r,\theta}$. □

***Theorem* 5.2.** *Let M be a simply connected, d-dimensional model Riemannian manifold of non-positive curvature as in Assumption I. Assume that Δ_M on M is subcritical; that is, $\int_1^\infty g^{1-d}(r)\,dr < \infty$.*

(a) *Assume that* $\int_1^\infty dr\, g^{1-d}(r) \int_1^r dt\, g^{d-3}(t) = \infty$.
Then
   (i) *the Martin boundary for Δ_M on M consists of a single point;*
   (ii) $P_x(\lim_{t\to\tau_D} \hat{\theta}(t)$ *does not exist*$) = 1$, *for* $x \in M$;
   (iii) *the invariant σ-algebra is trivial under* P_x, *for* $x \in M$.
(b) *Assume that* $\int_1^\infty dr\, g^{1-d}(r) \int_1^r dt\, g^{d-3}(t) < \infty$.
Then
   (i) *the Martin boundary for Δ_M on M is homeomorphic to S^{d-1}; more specifically, a sequence $\{x_n\}_{n=1}^\infty \subset M$ satisfying $\lim_{n\to\infty} r(x_n) = \infty$ is a Martin sequence if and only if $\lim_{n\to\infty} \theta(x_n) \in S^{d-1}$ exists;*
   (ii) *for* $x \in M$, $P_x(\mathcal{U} \equiv \lim_{t\to\tau_D} \hat{\theta}(t)$ *exists*$) = 1$; *furthermore, for each non-empty, open* $U \subsetneqq S^{d-1}$, $P_x(\mathcal{U} \in U)$ *is not constant in x. Thus, for each non-empty, open* $U \subsetneqq S^{d-1}$, $u(x) \equiv P_x(\mathcal{U} \in U)$ *is a non-constant bounded harmonic function for* Δ_M *on* M;
   (iii) *up to sets of P_x-measure zero, the invariant σ-algebra is generated by* $\mathcal{U}$.

Remark. If $d = 2$, then the conclusion of part (b) must necessarily hold. This is because the subcriticality condition $\int_1^\infty g^{-1}(r)\,dr < \infty$ is equivalent to $\int_1^\infty dr\, g^{-1}(r) \int_1^r dt\, g^{-1}(t) < \infty$.

Proof. As was noted in the previous section, when using (4.3) to study the Martin boundary for Δ_M on M, we may ignore the artificial singularity at $r = 0$. The theorem follows as a particular case of Theorem 8.6.3(ii) and Theorem 3.2. □

***Lemma* 5.3.** *For* $i = 1, 2$, *let g_i satisfy the conditions on g in Assumption I and let*

$$k_i(r) = -\frac{g_i''}{g_i}(r).$$

Assume that $k_1(r) \leqslant k_2(r)$, *for* $r > 0$. *Then*

(i) $\dfrac{g_1'}{g_1}(r) \geqslant \dfrac{g_2'}{g_2}(r)$, for $r > 0$
and
(ii) $g_1(r) \geqslant g_2(r)$, for $r \geqslant 0$.

Proof. Let

$$\mathcal{L}_i = \frac{d^2}{dr^2} + k_i(r) \text{ on } (0, \infty).$$

Then it is easy to see that g_i is the minimal positive harmonic function for $\mathcal{L}_i$ on $(0, \infty)$ corresponding to the Martin boundary point ∞. Thus, (i) is actually a particular case of Lemma 8.6.12. Using the fact that $g_1(0) = g_2(0)$ and $g_1'(0) = g_2'(0) = 1$, (ii) follows from (i) upon integration. □

Theorem 5.4. *Let M_i be a simply connected, d-dimensional model Riemannian manifold with non-positive curvature as in Assumption I with metric $g_i(r)$ and curvature*

$$k_i(r) = \frac{-g_i''}{g_i}(r),\ i = 1, 2.$$

Assume that Δ_{M_i} is subcritical on M_i, i = 1, 2, and that $k_1(r) \leq k_2(r)$, for large r. If the conclusion of Theorem 5.2(a) holds for M_1, then it also holds for M_2. Equivalently, if the conclusion of Theorem 5.2(b) holds for M_2, then it also holds for M_1.

Proof. The integral appearing in Theorem 5.2 can be written as

$$\int_1^\infty dr\, g^{1-d}(r) \int_1^r dt\, g^{d-3}(t) =$$

$$c\int_1^\infty dr \exp\left(-\int_1^r (d-1)\frac{g'}{g}(s)\, ds\right) \int_1^r dt \frac{1}{g^2(t)} \exp\left(\int_1^t (d-1)\frac{g'}{g}(s)\, ds\right),$$

for an appropriate constant c. Define

$$J(b, M) = \int_1^\infty dr \exp\left(-\int_1^r b(s)\, ds\right) \int_1^r dt \frac{1}{M(t)} \exp\left(\int_1^t b(s)\, ds\right).$$

Then in terms of $J(b, M)$, we have

$$\int_1^\infty dr g^{1-d}(r) \int_1^r dt g^{d-3}(t) = cJ\left((d-1)\frac{g'}{g}, \frac{1}{g^2}\right). \tag{5.2}$$

We first prove the theorem in the case where $k_1(r) \leq k_2(r)$, for all $r > 0$. A proof just like that of Lemma 8.6.5(i) (which was left to exercise 8.14) shows that $J(b, M)$ is non-increasing in its dependence on b; clearly $J(b, M)$ is non-decreasing in its dependence on M. Thus, it follows from (5.2) and Lemma 5.3 that

$$\int_1^\infty dr g_2^{1-d}(r) \int_1^r dt g_2^{d-3}(t) = \infty \text{ whenever } \int_1^\infty dr g_1^{1-d}(r) \int_1^r dt g_1^{d-3}(t) = \infty.$$

We now treat the general case. Define $k^+(r) = \max(k_1(r), k_2(r))$, for $r > 0$ and $k^-(r) = \min(k_1(r), k_2(r))$, for $r > 0$; it follows that $k^\pm \in C^\alpha((0, \infty))$. Let $u_\pm$ solve $u_\pm'' + k^\pm u_\pm = 0$ in $(0, \infty)$, $u_\pm(0) = 0$ and

$u'_\pm(0) = 1$ ($u_\pm$ is the minimal positive harmonic function corresponding to the Martin boundary point ∞ for the operator

$$\frac{\partial^2}{\partial r^2} + k^\pm \text{ on } (0, \infty)).$$

Choose R such that $k_1(r) \leqslant k_2(r)$, for $r \geqslant R$. Since (4.2) holds with g replaced by g_i, $i = 1, 2$, or by $u_\pm$, it is clear that there exist positive constants c_1 and c_2 such that $c_1u_+(R) \leqslant g_2(R) \leqslant c_2u_+(R)$, $c_1u'_+(R) \leqslant g'_2(R) \leqslant c_2u'_+(R)$, $c_1u_-(R) \leqslant g_1(R) \leqslant c_2u_-(R)$, and $c_1u'_-(R) \leqslant g'_1(R) \leqslant c_2u'_-(R)$. Using this and the fact that $k^+(r) = k_2(r)$ and $k^-(r) = k_1(r)$, for $r \geqslant R$, it follows that $c_1u_+(r) \leqslant g_2(r) \leqslant c_2u_+(r)$ and $c_1u_-(r) \leqslant g_1(r) \leqslant c_2u_-(r)$, for $r \geqslant R$. From these two-sided inequalities, it follows that $\int_1^\infty dr g_2^{1-d}(r)\int_1^r dt g_2^{d-3}(t) = \infty$ if and only if $\int_1^\infty dr\, u_+^{1-d}(r)\int_1^r dt\, u_+^{d-3}(t) = \infty$, and $\int_1^\infty dr g_1^{1-d}(r)\int_1^r dt g_1^{d-3}(t) = \infty$ if and only if $\int_1^\infty dr u_-^{1-d}(r)\int_1^r dt u_-^{d-3}(t) = \infty$. Since $k^-(r) \leqslant k^+(r)$; for $r > 0$, it follows from the case already proved that $\int_1^\infty dr u_+^{1-d}(r)\int_1^r dt\, u_+^{d-3}(t) = \infty$ whenever $\int_1^\infty dr\, u_-^{1-d}(r)\int_1^r dt u_-^{d-3}(t) = \infty$; thus $\int_1^\infty dr g_2^{1-d}(r)\int_1^r dt g_2^{d-3}(t) = \infty$ whenever $\int_1^\infty dr g_1^{1-d}(r)\int_1^r dt g_1^{d-3}(t) = \infty$. □

***Corollary* 5.5.** *Let M be a simply connected d-dimensional model Riemannian manifold of non-positive curvature as in Assumption I. Assume that Δ_M is subcritical on M.*

$$\textit{Let } c_d = \begin{cases} 1, & \textit{if } d = 2 \\ \frac{1}{2}, & \textit{if } d \geqslant 3 \end{cases}.$$

(i) *If the curvature $k(r)$ satisfies*

$$k(r) \geqslant \frac{-c}{r^2 \log r},$$

for large r and some constant $c < c_d$, then the conclusion of Theorem 5.2(a) *holds.*

(ii) *If the curvature $k(r)$ satisfies*

$$k(r) \leqslant \frac{-c}{r^2 \log r},$$

for large r and some constant $c > c_d$, then the conclusion of Theorem 5.2(b) *holds.*

Proof. Let $\phi(r) = r(\log r)^\gamma$, $\gamma > 0$, and choose $a > 1$ such that $\phi'(a) > 0$ and $\phi''(a) > 0$. Let

$$\hat{g}(r) = \frac{\phi(r + a) - \phi(a)}{\phi'(a)}.$$

Then $\hat{g}(0) = 0$, $\hat{g}'(0) = 1$, and

$$\hat{k}(r) \equiv -\frac{\hat{g}''}{\hat{g}}(r) \leqslant 0.$$

Let $\hat{M}$ denote the manifold with metric $\hat{g}$. Since

$$-\frac{\phi''}{\phi}(r) = -\frac{\gamma}{r^2 \log r}\left(1 - \frac{\gamma - 1}{\log r}\right),$$

and since

$$\frac{\hat{g}''}{\hat{g}}(r) \sim \frac{\phi''}{\phi}(r), \text{ as } r \to \infty,$$

it follows that for any $\varepsilon > 0$,

$$\frac{-\gamma - \varepsilon}{r^2 \log r} \leqslant \hat{k}(r) \leqslant \frac{-\gamma + \varepsilon}{r^2 \log r}, \quad \text{for large } r. \tag{5.3}$$

Since $\hat{g} \sim \phi$, as $r \to \infty$, it follows that

$$\int_1^\infty \mathrm{d}r\, \hat{g}^{1-d}(r) \int_1^r \mathrm{d}t\, \hat{g}^{d-3}(t) = \infty \text{ if and only if}$$
$$\int_a^\infty \mathrm{d}r\, \phi^{1-d}(r) \int_a^r \mathrm{d}t\, \phi^{d-3}(t) = \infty. \tag{5.4}$$

We will show below that

$$\int_a^\infty \mathrm{d}r\, \phi^{1-d}(r) \int_a^r \mathrm{d}t\, \phi^{d-3}(t) = \infty \text{ if and only if } \gamma \leqslant c_d. \tag{5.5}$$

The theorem then follows by comparing $k(r)$ to $\hat{k}(r)$ and using (5.3)–(5.5) and Theorem 5.4.

It remains to prove (5.5). First consider the case $d = 2$. As was noted in the remark following Theorem 5.2, the infiniteness of $\int_a^\infty \mathrm{d}r\,\phi^{-1}(r) \int_a^r \mathrm{d}t\,\phi^{-1}(t)$ is equivalent to the infiniteness of $\int_a^\infty \phi^{-1}(r)\,\mathrm{d}r$; clearly $\int_a^\infty \phi^{-1}(r)\,\mathrm{d}r = \infty$ if and only if $\gamma \leqslant 1 = c_2$.

Now consider the case where $d \geqslant 3$. We apply l'Hôpital's rule to the quotient $\int_a^r \phi^{d-3}(r)\,\mathrm{d}r / \psi(r)\phi^{d-1}(r)$, where $\psi(r) = r^{-1}(\log r)^{-2\gamma}$. One finds that

$$\lim_{r\to\infty} \left(\int^r \phi^{d-3}(s)\,\mathrm{d}s\right)' / (\psi(r)\phi^{d-1}(r))' = \frac{1}{d-2};$$

thus

$$\left(\int_a^r \phi^{d-3}(s)\,\mathrm{d}s / \phi^{d-1}(r)\right) \sim \frac{1}{d-2}\psi(r), \text{ as } r \to \infty,$$

and it follows that $\int_a^\infty dr\phi^{1-d}(r)\int_a^r dt\phi^{d-3}(t) = \infty$ if and only if $\gamma \leq \frac{1}{2} = c_d$. □

9.6 Bounded harmonic functions on manifolds with curvature bounded between two negative constants

In this section, we will prove the following result.

Theorem 6.1. *Let M be a simply connected d-dimensional Riemannian manifold with sectional curvatures bounded between two negative constants. Let $X_M(t)$ denote the Brownian motion on M with lifetime τ_M and let $\hat{r}(t) = r(X_M(t))$, $\hat{\theta}(t) = \theta(X_M(t))$ denote its representation in geodesic polar coordinates with respect to some $x_0 \in M$. Then*

(i) $\tau_M = \infty$ *a.s.* P_x;
(ii) $\lim_{t\to\infty} \hat{r}(t) = \infty$ *a.s.* P_x; *thus, the Brownian motion on M is transient or, equivalently, Δ_M on M is subcritical*;
(iii) $\mathcal{U} \equiv \lim_{t\to\infty} \hat{\theta}(t)$ *exists a.s.* P_x;
(iv) *for each non-empty, open $U \subsetneqq S^{d-1}$, $P_x(\mathcal{U} \in U)$ is not constant in x. Thus, $u(x) \equiv P_x(\mathcal{U} \in U)$ is a non-constant bounded harmonic function for each non-empty, open $U \subsetneqq S^{d-1}$*;
(v) $\mathcal{U}$ *under P_x converges to the atom at $\theta_0 \in S^{d-1}$ as $r(x) \to \infty$ and $\theta(x) \to \theta_0$.*

The following corollary is important.

Corollary 6.2. *Let M be as in Theorem 6.1. Then for any $f \in C(S^{d-1})$, there exists a unique solution to the following asymptotic Dirichlet problem*:

$$\Delta_M u = 0 \text{ in } M$$

$$\lim_{\substack{r(x)\to\infty \\ \theta(x)\to\theta_0}} u(x) = f(\theta_0),\ \theta_0 \in S^{d-1}.$$

The solution u is given by $u(x) = E_x f(\mathcal{U})$, where $\mathcal{U}$ is as in Theorem 6.1.

Proof. The corollary is an immediate consequence of Theorem 6.1(v) and Theorem 1.2. □

For the proof of the theorem, we need a comparison result from differential geometry. First, consider the case of a model manifold of constant curvature $k(r) = -\gamma^2$. We have

$$g(r) = \frac{\exp(\gamma r) - \exp(-\gamma r)}{2\gamma} \text{ and } G(r) = g^{d-1}(r).$$

Thus,

$$\frac{G_r}{G}(r) = (d-1)\gamma \coth \gamma r.$$

A classical calculation for manifolds of constant negative curvature shows that

$$\operatorname{dist}_{S^{d-1}}(\theta(x), \theta(y)) = c \operatorname{dist}_M(x, y) \exp(-\gamma(r(x) \wedge r(y)),$$

$$\text{for } x, y \in M,$$

where $\operatorname{dist}_M(\cdot, \cdot)$ is the Riemannian distance on the manifold M, $\operatorname{dist}_{S^{d-1}}(\cdot, \cdot)$ is the distance on the sphere S^{d-1} with its usual metric, and $c > 0$. We state without proof the following comparison theorem [Bishop and Crittendon (1970), p. 250].

Theorem 6.3. *Let M be a simply connected, d-dimensional, Riemannian manifold with sectional curvatures bounded between $-b^2$ and $-a^2$, where $0 < a < b < \infty$. Then*

(i) $(d-1)a \coth(ar) \leqslant \dfrac{G_r}{G}(r, \theta) \leqslant (d-1)b \coth(br)$, *for* $r > 0$.
(ii) $\operatorname{dist}_{S^{d-1}}(\theta(x), \theta(y)) \leqslant c \operatorname{dist}_M(x, y) \exp(-a(r(x) \wedge r(y)))$, *where* $\operatorname{dist}_M(\cdot, \cdot)$ *is the Riemannian distance on* M, $\operatorname{dist}_{S^{d-1}}(\cdot, \cdot)$ *is the distance on the sphere with its usual metric, and* $c > 0$.

Remark. The corresponding lower bound also holds in (ii); however it is not necessary for the proof of Theorem 6.1.

Proof of Theorem 6.1. $(\hat{r}(t), \hat{\theta}(t))$ is a diffusion on $R^d - \{0\}$ corresponding to the operator $\frac{1}{2}\Delta_M$, where Δ_M is as in (4.1). By Theorem 6.3, the drift term

$$\frac{G_r}{G}$$

is bounded for r away from zero; thus, by exercise 6.8, the process $\hat{r}(t)$ cannot explode to ∞. By exercise 9.2, $\hat{r}(t)$ does not explode to 0. Thus, $\tau_M = \infty$ a.s. P_x. This proves (i).

Applying Itô's formula (Theorem 1.6.3) to $f(r) = r$ gives

$$\hat{r}(t) = r(x) + B(t) + \int_0^t \frac{G_r}{G}(\hat{r}(s), \hat{\theta}(s))\, ds \text{ a.s. } P_x, \qquad (6.1)$$

where $B(t)$ is a standard one-dimensional Brownian motion under P_x.

(Actually, since

$$\frac{G_r}{G}$$

is not bounded, one must use a standard localization argument in applying Theorem 1.6.3.) By the law of the iterated logarithm [Breiman (1968 or 1992), chapter 12],

$$\limsup_{t\to\infty} \frac{|B(t)|}{(2t\log\log t)^{1/2}} = 1 \text{ a.s. } P_x. \tag{6.2}$$

From Theorem 6.3(i), we have

$$\frac{G_r}{G}(r, \theta) \geq (d-1)a \coth(ar) \geq (d-1)a. \tag{6.3}$$

Fix $\delta \in (0, 1)$. It follows from (6.1)–(6.3) that for almost P_x-almost all ω,

$$\exists \text{ a } t_0 = t_0(\omega) \text{ such that } \hat{r}(t) \geq (1-\delta)(d-1)at, \quad \text{for } t \geq t_0. \tag{6.4}$$

This proves (i).

We also have from Theorem 6.3(i) that

$$\frac{G_r}{G}(r, \theta) \leq (d-1)b \coth(br) \leq \left(1 + \frac{\delta}{2}\right)(d-1)b, \quad \text{for large } r. \tag{6.5}$$

From (6.1)–(6.5), it follows that for P_x-almost all ω,

$$\exists \text{ a } t_0 = t_0(\omega) < \infty \text{ such that } \hat{r}(t) \leq (1+\delta)(d-1)bt, \quad \text{for } t \geq t_0. \tag{6.6}$$

Thus, for P_x-almost all ω, we can choose a $t_0 = t_0(\omega)$, for which (6.4) and (6.6) hold. For any $s \in (0, 1]$, it follows from the triangle inequality and (6.6) that

$$\begin{aligned} &\text{dist}_M (X_M(t_0 + n), X_M(t_0 + n + s)) \\ &\quad \leq r(X_M(t_0 + n)) + r(X_M(t_0 + n + s)) \\ &\quad = \hat{r}(t_0 + n) + \hat{r}(t_0 + n + s) \\ &\quad \leq (1+\delta)(d-1)b(2t_0 + 2n + 1); \; n = 0, 1, \ldots. \end{aligned} \tag{6.7}$$

Using (6.4) and (6.7) in Theorem 6.3(ii) gives

$$\text{dist}_{S^{d-1}}(\hat{\theta}(t_0+n), \hat{\theta}(t_0+n+s)) \leqslant$$
$$c(1+\delta)(d-1)b(2t_0+2n+1)\exp(-a^2(1-\delta)(d-1)(t_0+n)),$$
$$\text{for } s \in (0,1] \text{ and } n = 0, 1, \ldots. \quad (6.8)$$

Now (iii) follows since the righthand side of (6.8) is convergent as a series in n.

(iv) follows from (v) and Theorem 1.2. It remains to prove (v). We need to refine (6.4) and (6.6) a bit. Recall that under P_x, $B(t)$ is a standard Brownian motion starting from 0. We note the following equality [Breiman (1968 or 1992), chapter 13]:

$$P_x(B(t) \geqslant at + b, \text{ for some } t \geqslant 0) =$$
$$P_x(B(t) \leqslant -at - b, \text{ for some } t \geqslant 0) = \exp(-2ab),\ a, b \geqslant 0. \quad (6.9)$$

It follows from (6.1), (6.3) and (6.9) that

$$\lim_{r(x)\to\infty} P_x\left(\hat{r}(t) \geqslant \frac{r(x)}{2} + (1-\delta)(d-1)at, \quad \text{for all } t \geqslant 0\right) = 1. \quad (6.10)$$

Similarly, it follows from (6.1), (6.5) and (6.9) that

$$\lim_{r(x)\to\infty} P_x(\hat{r}(t) \leqslant 2r(x) + (1+\delta)(d-1)t, \quad \text{for all } t \geqslant 0) = 1. \quad (6.11)$$

Let ω be such that

$$\frac{r(x)}{2} + (1-\delta)(d-1)at \leqslant \hat{r}(t) \leqslant 2r(x) + (1+\delta)(d-1)bt,$$
$$\text{for all } t \geqslant 0. \quad (6.12)$$

Then for any $s \in (0, 1]$, we obtain from the triangle inequality,

$$\text{dist}_M(X_M(n), X_M(n+s))$$
$$\leqslant r(X_M(n)) + r(X_M(n+s)) = \hat{r}(n) + \hat{r}(n+s)$$
$$\leqslant 4r(x) + (1+\delta)(d-1)b(2n+1), \text{ for } n = 0, 1, \ldots. \quad (6.13)$$

Using (6.12) and (6.13) in Theorem 6.3(ii) gives

$$\text{dist}_{S^{d-1}}(\hat{\theta}(n), \hat{\theta}(n+s)) \leqslant c[4r(x) + (1+\delta)(d-1)b(2n+1)]$$
$$\times \exp\left(-a\left[\frac{r(x)}{2} + (1-\delta)(d-1)an\right]\right),$$
$$\text{for } s \in (0,1] \text{ and } n = 0, 1, \ldots. \quad (6.14)$$

Thus, we conclude that if (6.12) holds, then

$$\operatorname{dist}_{S^{d-1}}(\hat{\theta}(0), \lim_{t\to\infty}\hat{\theta}(t)) \leqslant \sum_{n=0}^{\infty} c[4r(x) + (1+\delta)(d-1)b(2n+1)] \times \exp\left(-a\left[\frac{r(x)}{2} + (1-\delta)(d-1)an\right]\right). \tag{6.15}$$

(v) follows from (6.10), (6.11) and the fact that the righthand side of (6.15) converges to zero as $r(x) \to \infty$. □

With the curvature assumption of Theorem 6.1, it has been proved that the Martin boundary for Δ_M on M is homeomorphic to S^{d-1}. (See subsection 3 of Chapter 8, Section 9.) From this and Theorem 6.1, it follows that $\mathcal{U} = \lim_{t\to\infty} \hat{\theta}(t)$ generates the entire invariant σ-algebra up to sets of P_x-measure zero. The results of Theorem 6.1 have been proved with weaker assumptions on the curvature; see the notes at the end of the chapter.

Exercises

9.1 Let $u \in BC_L(D)$. Show that $\mathcal{U} \equiv \lim_{t\to\tau_D} u(X(t))$ exists a.s. P_x, for $x \in D$. Use this in place of (1.5) to prove Theorem 1.2, Lemma 1.3 and Theorem 1.4 without the assumption that $P_x(\tau_D = \infty) = 1$. (*Hint*: Let $\{D_n\}_{n=1}^{\infty} \subset D$ be an increasing sequence of domains satisfying $D_n \subset\subset D_{n+1}$ and $\bigcup_{n=1}^{\infty} D_n = D$. Let $V(n, t) = u(X(t \wedge \tau_{D_n}))$. Show that for fixed t, $V(n, t)$ is a discrete parameter martingale in n. Since it is bounded, by Theorem 1.2.4, $V(t) \equiv \lim_{t\to\infty} V(n, t)$ exists a.s. P_x. Show that the limit is independent of the choice of $\{D_n\}_{n=1}^{\infty}$. This shows that $\lim_{t\to\tau_D} u(X(t))$ exists a.s. P_x on $\{\tau_D \leqslant s\}$, for any $s \in (0, \infty)$. Now show that $V(t)$ is a martingale. Since $V(t)$ is bounded, $V \equiv \lim_{t\to\infty} V(t)$ exists a.s. P_x. This shows that $\lim_{t\to\infty} u(X(t))$ exists a.s. P_x on $\{\tau_D = \infty\}$.)

9.2 Let $\hat{r}(t)$ denote the radial part of the process corresponding to the operator Δ_M in (4.1). Assume that

$$\frac{G_r(r, \theta)}{G(r, \theta)} = \frac{d-1}{r} + O(1), \text{ as } r \to 0$$

uniformly over $\theta \in S^{d-1}$. Then $P_{r,\theta}(\lim_{t\to\tau_D} \hat{r}(t) = 0) = 0$. (*Hint*: By the strong Markov property, it is enough to show that $\lim_{\varepsilon\to 0} P_{r,\theta}(\tau_\varepsilon < \tau_{1/2}) = 0$, where $\tau_a = \inf\{t \geqslant 0: \hat{r}(t) = a\}$. Let

$$u(r) = \log\log\frac{1}{r}, \text{ for } r \leqslant \tfrac{1}{2}.$$

Then $\Delta_M u \leqslant 0$ on $0 < r \leqslant \frac{1}{2}$; thus $u(\hat{r}(t \wedge \tau_\varepsilon \wedge \tau_{1/2}))$ is a bounded submartingale and $u(r, \theta) = E_{r,\theta} u(\hat{r}(t \wedge \tau_\varepsilon \wedge \tau_{1/2}))$. Now let $t \to \infty$ and $\varepsilon \to 0$. (If $d \geqslant 3$, one can use the test function $u(r) = r^{-l}$, with $0 < l < d - 2$, instead of $u(r) = \log\log \frac{1}{r}$.).)

9.3 (Seven ways to prove the original Liouville theorem). As a parting shot, we suggest seven ways to prove the original Liouville's theorem:

If $\Delta u = 0$ *and* u *is bounded in* R^d, *then* u *is constant.*

Five of the methods will in fact prove the stronger result that there are no non-constant positive harmonic functions; two of the methods only work on R^2 and one of the methods only works in R^d, $d \geqslant 3$.

(i) (Via complex analysis). This is the original proof of Liouville for R^2. Liouville proved that every bounded entire function is constant. (For a proof, consult any textbook on complex analysis.) Now let u be a bounded harmonic function. Using the Cauchy–Riemann equations and integrating, one obtains a v such that $w \equiv u + iv$ is an entire function. Then $\exp(w)$ is also entire and since $|e^w| = \exp(u)$, it follows by assumption that $\exp(w)$ is bounded. Thus, $\exp(w)$ is constant and, consequently, so is u.

(ii) (Via Martin's theory). Assume that $d \geqslant 3$. The Green's function (up to a constant multiple) for Δ on R^d is given by $G(x, y) = |x - y|^{2-d}$. Fix $x_0 \in R^d$ and let

$$k(x; y) = \frac{|x - y|^{2-d}}{|x_0 - y|^{2-d}}.$$

Since $\lim_{|y| \to \infty} k(x; y) = 1$, it follows that the Martin boundary is trivial. Therefore, $C_\Delta(R^d)$ is one-dimensional and contains only the positive constants.

(iii) (Via the method of periodic operators). Since $\lambda_c = 0$ for Δ on R^d, it follows by Theorem 8.2.5 that $C_\Delta(R^d)$ is one-dimensional, thus it contains only the positive constants.

(iv) (Via the method of Fuchsian operators). By Theorem 8.3.1, there are no non-constant positive harmonic functions.

(v) (Via recurrence if $d = 2$). Δ is critical on R^2, that is, the Brownian motion is recurrent. Thus, by Theorem 4.3.4, $C_\Delta(R^2)$ is one-dimensional and contains only the positive constants.

(vi) (Via the mean value theorem). Let $B_r(x)$ denote the ball of radius r centered at x. First, let $u \in BC_\Delta(R^d)$. By the martingale formulation, we have $u(x) = E_x u(X(\tau_{B_r(x)}))$. Since this holds for any r, we obtain the mean value property:

$$u(x) = \frac{1}{|B_r(x)|} \int_{B_r(x)} u(y)\, dy, \text{ for } r > 0.$$

Now write

$$u(x) - u(0) = \frac{1}{|B_r(0)|} \int_{B_r(x)} u(y)\, dy - \frac{1}{|B_r(0)|} \int_{B_r(0)} u(y)\, dy. \quad (*)$$

Use the boundedness of u and the fact that

$$\lim_{r\to\infty} \frac{B_r(x) \cap B_r(0)}{B_r(0)} = 1$$

to show that the righthand side of (*) goes to zero as $r \to \infty$. To extend to $u \in C_\Delta(R^d)$, let $r > |x|$ and use the fact that $(B_r(x) - B_r(0)) \cup (B_r(0) - B_r(x)) \subset B_{r+|x|}(0) - B_{r-|x|}(0)$ and that

$$\int_{B_{r+|x|}(0)} u(y)\,\mathrm{d}y = u(0)|B_{r+|x|}(0)|$$

and

$$\int_{B_{r-|x|}(0)} u(y)\,\mathrm{d}y = u(0)|B_{r-|x|}(0)|.$$

(vii) (Via coupling). Let x, $y \in R^d$ with $x \neq y$ and let L_{xy} denote the hyperplane in R^d which is perpendicular to the line segment joining x and y, and which contains the point

$$\frac{x+y}{2}.$$

Define $T_{xy}(z)$ = the mirror image of z with respect to L_{xy}. Let P_x denote Wiener measure and let $X(t)$ denote the canonical realization. Let $\tau_{xy} = \inf\{t \geq 0\colon\ X(t) \in L_{xy}\}$. It is clear that $P_x(\tau_{xy} < \alpha) = 1$. Define

$$Y(t) = \begin{cases} T_{xy}\,(X(t)), \text{ if } t \leq \tau_{xy} \\ X(t), \text{ if } t > \tau_{xy} \end{cases}$$

It is easy to see that under P_x, $Y(t)$ is a Brownian motion starting at y. Since $P_x(\tau_{xy} < \infty) = 1$, it follows that

$$\lim_{t\to\infty} P_x(X(t) = Y(t)) = 1. \qquad (**)$$

If $u \in BC_\Delta(R^d)$, then by the martingale formlation and the boundedness of u, it follows easily that $u(x) = E_x u(X(t))$ and $u(y) = E_x u(Y(t))$. Letting $t \to \infty$ and using (**) and the boundedness of u gives $u(x) = u(y)$. This proof appears in [Lindvall and Rogers (1986)].

Notes

The connection between bounded harmonic functions and the invariant σ-algebra goes back to [Blackwell (1955)] and [Feller (1956)]. The theory is developed in [Dynkin (1965)]. For other results concerning bounded harmonic functions and the invariant σ-algebra, see [Cranston (1983)], [R. Pinsky (1987)] and [Cranston (1992)]. Theorem 2.2 is due to [R. Pinsky (1991)].

For texts on differential geometry, see, for example, [Bishop and Crittendon

(1970)], [Carmo (1992)], [Chavel (1993)] or [Spivak (1979)]. See also [Milnor (1963), part II] for a quick introduction. See [Cheeger and Ebin (1975)] and [Greene and Wu (1979)] for comparison theorems in differential geometry. For an introduction to stochastic differential geometry, see [Elworthy (1982)], [M. Pinsky (1978)] and [Kendall (1987)].

The study of Brownian motion on manifolds of negative curvature and the connection to bounded harmonic functions was initiated by [Dynkin (1965)]. Theorem 5.2, a(ii), a(iii), b(ii), and b(iii) is due to [March (1986)], although there is a gap in the proof of a(iii). Theorem 5.2, a(i) and b(i) is due to [Murata (1986)] by purely analytic means and by [R. Pinsky (1993)] using the combination of analytic and probabilistic methods that appear in the proof given in the text. Theorem 5.4 and Corollary 5.5 are due to [March (1986)].

Theorem 6.1(i)–(iii) is due to [Prat (1975)]. [Kifer (1976)] proved Theorem 6.1 and Corollary 6.2. Kifer's proof of Corollary 6.2 is by probabilistic methods; [Anderson (1983)] gave an analytic proof of Corollary 6.2 and [Sullivan (1983)] gave a geometric proof. Under the assumptions of Theorem 6.1, it has been shown that the Martin boundary is homeomorphic to S^{d-1}; for references, see Subsection 3 of Chapter 8, Section 9.

In certain cases, Theorem 6.1 has been proved under relaxed curvature assumptions. In the two-dimensional case, [Kendall (1984)] proved Theorem 6.1 with no lower bound on the curvature and with a negative constant upper bound. In the case of a d-dimensional manifold, $d \geqslant 3$ [Hsu and March (1985)] proved Theorem 6.1 under the condition that for each x outside of a compact set, all the sectional curvatures at x are bounded below by $-\gamma(r(x))^{2\beta}$ and above by $-c(r(x))^{-2}$, where γ and c are positive constants, $0<\beta<1$, and β and c are related by the condition $\alpha(1-\beta)>2$, with $\alpha=\alpha(c)>1$ defined by $c=\alpha(\alpha-1)$. In [Hsu and Kendall (1992)], Theorem 6.1 was proved in the two-dimensional case with no lower bound on the curvature and with upper bound

$$\frac{-c}{r(x)}$$

for all x off a compact set, where c is a positive constant.

For other related results, see, for example, [Cheeger and Yau (1981)], [Cheng, Li and Yau (1981)], [Debiard, Gaveau and Mazet (1976)], [Ichihara (1982)], [Varopoulus (1983)] and [Yau (1975)].

References

Agmon, S. (1982) On positivity and decay of solutions of second order elliptic equations on Riemannian manifolds, in *Methods of Functional Analysis and Theory of Elliptic Equations*, D. Greco, (ed.) Liguori, Naples, pp. 19–52.

Agmon, S. (1984) On positive solutions of elliptic equations with periodic coefficients in R^n, spectral results and extensions to elliptic operators on Riemannian manifolds, in *Proceedings of the International Conference on Differential Equations*, I. W. Knowles and R. T Lewis (eds.), pp. 7–17, North Holland.

Aikawa, H. (1986) On the Martin boundary of Lipschitz strips, *J. Math. Soc. of Japan*, **38**, 527–41.

Ancona, A. (1978) Principe de Harnack à la frontière et théorème de Fatou pour un opérateur elliptique dans un domaine Lipschitzien, *Ann. Inst. Fourier (Grenoble)*, **28**, 169–213.

Ancona, A. (1980) Théorème de Fatou et frontière de Martin pour une classe d'opérateurs elliptiques dans un demi-espace, *C.R. Acad. Sci. Paris Ser. A–B*, **290**, A401–4.

Ancona, A. (1987) Negatively curved manifolds, elliptic operators, and the Martin boundary, *Ann. Math.*, **125**, 495–536.

Anderson, M. (1983) The Dirichlet problem at infinity for manifolds of negative curvature, *J. Diff. Geom.*, **18**, 701–21.

Anderson, M. and Schoen, R (1985) Positive harmonic functions on complete manifolds of negative curvature, *Ann. Math.*, **121**, 429–61.

Azencott, R. (1974) Behavior of diffusion semigroups at infinity, *Bull. Soc. Math. France*, **102**, 193–240.

Ballman, P. (1984) Positive solution of elliptic equations in nondivergence form and their adjoints, *Arkiv för Matematik*, **22**, 153–73.

Benedicks, M. (1980) Positive harmonic functions vanishing on the boundary of certain domains in R^n, *Arkiv för Matematik*, **18**, 53–72.

Berestycki, H. and Nirenberg, L. (1991) Asymptotic behaviour via the Harnack inequality, in *Nonlinear Analysis, a tribute in honor of Giovanni Prodi*, pp. 135–44, Scuola Normale Superiure Pisa.

Berestycki, H., Nirenberg, L. and Varadhan, S. R. S. (1994) The principal eigenvalue and maximum principle for second-order elliptic operators in general domains, to appear in *Comm. Pure App. Math.*, **47**, 47–92.

Bhattacharya, R. N. (1978) Criteria for recurrence and existence of invariant measures for multidimensional diffusions, *Ann. Prob.*, **6**, 541–53.

Bishop, R. L. and Crittendon, R. J. (1970) *Geometry of Manifolds*, Academic Press.

Blackwell, D. (1955) On transient Markov processes with a countable number of states and stationary transition probabilities, *Ann. Math. Stat.*, **26**, 654–8.

Breiman, L. (1968) *Probability*, Addison-Wesley; republished (1992), SIAM.

Brown, A. L. and Page, A. (1970) *Elements of Functional Analysis*, Van Nostrand Reinhold.

Brown, L. D. (1971) Admissible estimators, recurrent diffusion and insoluble boundary value problems, *Ann. Math Statist.*, **42**, 855–903.

Cafferelli, L., Fabes, E., Mortola, S. and Salsa, S. (1981) Boundary behavior of nonnegative solutions of elliptic operators in divergence form, *Indiana Univ. Math. J.*, **30**, 621–40.

Carmo, M. P. (1992) *Riemannian Geometry*, Birkhauser.

Chavel, I. (1993) *Riemannian Geometry: A Modern Introduction*, Cambridge University Press.

Chavel, I., Karp, L. (1991) Large time behavior of the heat kernel: the parabolic λ-potential alternative, *Comment. Math. Helv.*, **66**, 541–56.

Cheeger, J. and Ebin, D. B. (1975) *Comparison Theorems in Differential Geometry*, North Holland.

Cheeger, J. and Yau, S.-T. (1981) A lower bound for the heat kernel, *Comm. Pure Appl. Math.*, **34**, 465–80.

Cheng, S. Y., Li, P. and Yau, S.-T. (1981) On the upper estimate of the heat kernel of a complete Riemannian manifold, *Amer. J. Math.*, **103**, 1021–63.

Chung, K. L. (1974) *A course in Probability Theory*, Academic Press.

Chung, K. L. and Rao K. M. (1981) Feynman–Kac functional and the Schrödinger equation, *Seminar on Stochastic Processes*, E. Cinlar, K. L. Chung and R. K. Getoor eds., Birkhauser.

Chung, K. L. and Williams, R. (1990) *Introduction to Stochastic Integration*, 2nd edition, Birkhauser.

Constantinescu, C. and Cornea, A. (1963) *Ideale Ränder Riemannsher Flächen*, Springer Verlag.

Cranston, M. (1983) Invariant σ-fields for a class of diffusions, *Z. Wahr. verw. Geb.*, **65**, 161–80.

Cranston, M. (1992) On specifying invariant σ-fields, *Sem. on Stoch. Proc. 1991, Prog. Probab.*, **29**, Birkhauser.

Cranston, M. (1994) A probabilistic approach to Martin boundaries for manifolds with ends, preprint.

Cranston, M., Fabes, E. and Zhao, Z. (1988) Conditional gauge and potential theory for the Schrödinger operator, *Trans. Amer. Math. Soc.*, **307**, 171–94.

Cranston, M. and Mueller, C. (1988) A review of recent and older results on the absolute continuity of harmonic measure, in *Geometry of Random Motion, Contemporary Mathematics*, Vol. 73, pp. 9–19, AMS.

Cranston, M., Orey S. and Rosler, U. (1982) The Martin boundary of two-dimensional Ornstein–Uhlenbeck processes, *London Math. Soc. Lec. Notes* **Ser. no. 79**: *Probability, Statistics and Analysis*, J. F. C. Kingman and G. E. H. Reuter (eds.), pp. 63–78, Cambridge University Press.

Cranston, M. and Salisbury, T. (1994) Martin boundaries of sectorial domains, preprint.

Debiard, A., Gaveau, B. and Mazet, E. (1976) Théorèmes de comparaison en géométrie Riemannienne, *Publ. RIMS, Kyoto Univ.*, **12**, 391–425.

Donnelly, H. (1986) Bounded harmonic functions and positive Ricci curvature, *Math. Z.*, **191**, 559–65.

Donsker, M. and Varadhan, S. R. S. (1976) On the principal eigenvalue of second-order elliptic differential operators, *Comm. Pure Appl. Math.*, **29**, 595–621.

Doob, J. L. (1953) *Stochastic Processes*. Wiley and Sons.

Doob, J. L. (1957) Conditional Brownian motion and the boundary limits of harmonic functions, *Bull. Soc. Math. France*, **85**, 431–58.

Doob, J. L. (1984) *Classical Potential Theory and its Probabilistic Counterpart*, Springer Verlag.

Durrett, R. (1984) *Brownian Motion and Martingales in Analysis*, Wadsworth.

Durrett, R. (1991) *Probability: Theory and Examples*, Wadsworth.

Dynkin E. (1965) *Markov Processes*, Vols. I and II, Springer Verlag.

Elliott, R. (1982) *Stochastic Calculus and Applications*, Springer Verlag.

Elworthy, K. D. (1982) *Stochastic Differential Equations on Manifolds*, Cambridge University Press.

Englander, J. and Pinsky, R. (1994) The asymptotic behavior of the principal eigenvalue for small perturbations of critical one-dimensional Schrödinger operators with $V(x) = l_{\pm}/x^2$ for $x^{\pm} \gg 1$, submitted.

Ethier, S. and Kurtz, T. (1986) *Markov Processes: Characterization and Convergence*, Wiley-Interscience.

Falkner, N. (1983) Feynman–Kac functionals and positive solutions of $1/2\Delta u + qu = 0$, *Z. Wahr. verw. Geb.*, **65**, 19–33.

Feller, W. (1952) The parabolic differential equations and the associated semi-groups of transformations, *Ann. Math.*, **55**, 468–519.

Feller, W. (1956) Boundaries induced by nonnegative matrices, *Trans. Amer. Math. Soc.*, **83**, 19–54.

Freidlin, M. (1985) *Functional Integration and Partial Differential Equations*, Princeton University Press.

Freidman, A. (1964) *Partial Differential Equations of Parabolic Type*, Prentice-Hall.

Friedman, A. (1975) *Stochastic Differential Equations and Applications*, Vol. 1, Academic Press.

Gihman, I. I and Skorohod, A. V. (1972) *Stochastic Differential Equations*, Springer Verlag.

Gilbarg, D. and Serrin, J. (1955/1956) *J. D'Analyse Math.*, **4**, 309–40.

Gilbarg, D. and Trudinger, N. (1983) *Elliptic Partial Differential Equations of Second Order*, 2nd edition, Springer Verlag.

Greene, R. and Wu, H. (1979) Function theory of manifolds which possess a pole, *Lecture Notes in Mathematics*, Vol. 699, Springer Verlag.

Hasminskii, R. Z. (1960) Ergodic properties of recurrent diffusion processes and stabilization of the solution of the Cauchy problem for parabolic equations, *Theor. Prob. Appl.*, **5**, 179–96.

Hasminskii, R. Z. (1980) *Stochastic Stability of Differential Equations*, Sijthoff and Noordhoff.

Helms, L. (1969) *Introduction to Potential Theory*, Wiley-Interscience.

Hormander, L. (1985) *The Analysis of Linear Partial Differential Operators*, Vol. III, Springer Verlag.

Hsu, P. and Kendall, W. (1992) Limiting angle of Bownian motion in certain two-dimensional Cartan–Hadamard manifolds, *Ann. Fac. Sci. Toulouse Math.*, *Ser. 6*. **1**, 169–86.

Hsu, P. and March, P. (1985) The limiting angle of certain Riemannian Brownian motions, *Comm. Pure Appl. Math.*, **38**, 755–68.

Hunt, G. A. (1957a) Markoff processes and potentials, I, *Illinois J. Math.*, **1**, 44–93.

Hunt, G. A. (1957b) Markoff processes and potentials, II, *Illinois J. Math.* **1**, 316–69.

Hunt, G. A. (1958) Markoff processes and potentials, III, *Illinois J. Math.* **2**, 151–213.

Hunt, G. A. (1960) Markoff chains and Martin boundaries *Illinois J. Math.* **4**, 313–340.

Hunt, G. A. and Wheeden, R. L. (1970) Positive harmonic functions on Lipschitz domains, *Trans. Amer. Math. Soc.*, **132**, 307–22.

Ichihara, K. (1978) Some global properties of symmetric diffusion processes, *Publ. RIMS, Kyoto Univ.*, **14**, 441–486.

Ichihara, K. (1982) Curvature, geodesics and the Brownian motion on a Riemannian manifold I: recurrence properties, *Nagoya Math. J.*, **87**, 101–14.

Ikeda, N. and Watanabe, S. (1981) *Stochastic Differential Equations and Diffusion Processes*, North Holland.

Ioffe, D. and Pinsky, R. (1994) Positive harmonic functions vanishing on the boundary for the Laplacian in unbounded horn-shaped domains, *Trans. Amer. Math. Soc.*, **342**, 773–91.

Ito, K. and McKean, H. (1965) *Diffusion Processes and their Sample Paths*, Springer Verlag.

Kac, M. (1949) On distributions of certain Wiener functionals, *Trans. Amer. Math. Soc.*, **65**, 1–13.

Kakutani, S. (1944a) On Brownian motion in n-space, *Proc. Imp. Acad. Tokyo*, **20**, 648–52. (Reprinted in *Shizuo Kakutani: Selected Papers*, Vol. 2, Birkhauser, 1986.)

Kakutani, S. (1944b) Two-dimensional Brownian motion and harmonic functions, *Proc. Imp. Acad. Tokyo*, **20**, 706–14. (Reprinted in *Shizuo Kakutani: Selected Papers*, Vol. 2, Birkauser, 1986.)

Kakutani, S. (1945) Markoff process and the Dirichlet problem, *Proc. Japan Acad.*, **21**, 227–33. (Reprinted in *Shizuo Kakutani: Selected Papers*, Vol. 1, Birkhauser, 1986.)

Karatzas, A. and Shreve, S. F. (1988) *Brownian motion and Stochastic Calculus*, Springer-Verlag.

Kendall, W. (1984) Brownian motion on 2-dimensional manifolds of negative curvature, *Seminaire de Probabilités*, Vol. 18, *Lecture Notes in Mathematics*, **1059**, 70–6, Springer Verlag.

Kendall, W. (1987) Stochastic differential geometry, in: *Proceedings of the First World Congress of the Bernoulli Society*, Yu. Prohorov and V. Sazonov (eds.), Vol. 1, 515–24, VNU Press.

Kifer, Yu. (1976) Brownian motion and harmonic functions on manifolds of negative curvature, *Theor. Prob. Appl.*, **21**, 81–95.

Kifer, Yu. (1986) Brownian motion and positive harmonic functions on complete manifolds of non-positive curvature, *Pitman Research Notes in Math.*, Vol. 150, pp. 187–232.

Klaus, M. (1977) On the bound state of Schrödinger operators in one dimension, *Ann. Phys.*, **108**, 288–300.

Krasnosel'skii, M. A., Lifshits, Je. A. and Sobolev, A. V. (1989) *Positive Linear Systems – The Method of Positive Operators*, Heldermann Verlag.

Krein, M. G. and Rutman, M. A. (1948) Linear operators leaving invariant a cone in a Banach space, *AMS Translations* (1962), Series 1, **10**, 199–324.

Krylov, N. V. and Safanov, M. V. (1979) An estimate of the probability that a diffusion process hits a set of positive measure, *Soviet Math. Dokl.*, **20**, 253–5.

Krylov, N. V. and Safanov, M. V. (1980) Certain properties of solutions of parabolic equations with measurable coefficients, *Izvestia Akad. Nauk. SSSR*, **40**, 161–80 [Russian].

Kunita, H. and Watanabe, T. (1965) Markov processes and Martin boundaries, part I, *Illinois J. Math.*, **9**, 485–526.

Lamperti, J. (1977) *Stochastic Processes*, Springer Verlag.

Landis, E. M. and Nadirashvili, N. S. (1986) Positive solutions of second order elliptic equations in unbounded domains, *Math. USSR Sbornik*, **54**, 129–34.

Li, P. and Tam, L. F. (1987) Positive harmonic functions on complete manifolds with nonnegative curvature outside a compact set, *Ann. Math.*, **125**, 171–207.

Lindvall, T. and Rogers, L. C. G. (1986) Coupling of mutlidimensional diffusions by reflection, *Ann. Prob.*, **14**, 860–72.

Lyons, T. and Sullivan, D. (1984) Function theory, random paths and covering spaces, *J. Diff. Geom.*, **19**, 299–323.

March, P. (1986) Brownian motion and harmonic functions on rotationally symmetric manifolds, *Ann. Prob.*, **14**, 793–801.

Martin, R. S. (1941) Minimal positive harmonic functions, *Trans. Amer. Math. Soc.*, **49**, 137–72.

Milnor, J. (1963) *Morse Theory*, Princeton Univ. Press.

Miranda, C. (1970) *Partial Differential Equations of Elliptic Type*, 2nd edition, Springer Verlag.

Murata, M. (1984) Positive solutions and large time behavior of Schrödinger semigroups, Simon's problem, *J. Func. Anal.*, **56**, 300–10.

Murata, M. (1986) Structure of positive solutions to $(-\Delta + V)u = 0$ in R^n, *Duke Math. J.*, **53**, 869–943.
Murata, M. (1990a) On construction of Martin boundaries for second order elliptic equations, *Publ. RIMS, Kyoto Univ.*, **26**, 585–627.
Murata, M. (1990b) Martin compactification and asymptotics of Green functions for Schrödinger operators with anisotropic potentials, *Math. Ann.*, **288**, 211–30.
Nakai, M. (1975) A test of Picard principle for rotation free densities, *J. Math. Soc. Japan*, **27**, 412–31.
Nakai, M. (1986) A test of Picard principle for rotation free densities, II, *J. Math. Soc. Japan*, **28**, 323–42.
Nakai, M. and Tada, T. (1988) Extreme nonmonotoneity of the Picard principle, *Math. Ann.*, **281**, 279–93.
Nussbaum, R. and Pinchover, Y. (1992) On variational principles for the generalized principal eigenvalue of second order elliptic operators and some applications, *J. D'Analyse Math.*, **59**, 161–77.
Persson, A. (1960) Bounds for the discrete part of the spectrum of a semi-bounded Schrödinger operator, *Math. Scand.*, **8**, 143–53.
Pinchover, Y. (1988) On positive solutions of second order elliptic equations, stability results and classification, *Duke Math, J.*, **57**, 955–80.
Pinchover Y. (1989) Criticality and ground states for second-order elliptic equations, *J. Diff. Equs.*, **80**, 237–50.
Pinchover, Y. (1990a) On criticality and ground states of second order elliptic equations II, *J. Diff. Equs.*, **87**, 353–64.
Pinchover, Y. (1990b) Large scale properties of multiparameter oscillation problems, *Comm. Partial Diff. Equs.*, **15**, 647–73.
Pinchover, Y. (1992) Large time behavior of the heat kernel and the behavior of the Green function near critically for nonsymmetric elliptic operators, *J. Func. Anal.*, **104**, 54–70.
Pinchover, Y. (1994) On positive Liouville theorems and asymptotic behavior of solutions of Fuchsian type elliptic operators, to appear in *Annales de l'I.H.P. Analyse Non-Linéaire.*
Pinsky, M. (1978) Stochastic Riemannian geometry, in *Probabilistic Analysis and Related Topics*, Vol. 1, pp. 199–236, Academic Press.
Pinsky, R. (1987) Recurrence, transience and bounded harmonic functions for diffusions in the plane, *Ann. Prob.*, **15**, 954–84.
Pinsky, R. (1988a) A generalized Dirichlet principle for second order non-selfadjoint elliptic operators with boundary conditions, *SIAM J. Math. Anal.*, **19**, 204–13.
Pinsky, R. (1988b) A mini-max variational formula giving necessary and sufficient conditions for the transience or recurrence of multidimensional diffusion processes, *Ann. Prob.*, **16**, 662–71.

Pinsky, R. (1988c) Transience and recurrence for multidimensional diffusions: a survey and a recent result, in *Geometry of Random Motion, Contemporary Mathematics Series*, Vol. 73, pp. 273–85.

Pinsky, R. (1991) A probabilistic approach to a theorem of Gilbarg and Serrin, *Israel Jour. Math.*, **74**, 1–12.

Pinsky, R. (1993) A new approach to the Martin boundary via diffusions conditioned to hit a compact set, *Ann. Prob.*, **21**, 453–81.

Pinsky, R. (1994) Second order elliptic operators with periodic coefficients: criticality theory, perturbations and positive harmonic functions, to appear in *J. Func. Anal.*

Prat, J. J. (1975) Étude asymptotique et convergence angulaire du mouvement brownian sur une variété à courbure négative, *C.R. Acad. Sc. Paris*, **280**, serie A, 1539–42.

Protter, M. and Weinberger, H. (1966) On the spectrum of general second order operators, *Bull. Amer. Math. Soc.*, **72**, 251–5.

Protter, M. and Weinberger, H. (1984) *Maximum Principles in Differential Equations*, Springer Verlag.

Reed, M. and Simon, B. (1975) *Methods of Modern Mathematical Physics*, II, *Fourier Analysis*, *Self-Adjointness*, Academic Press.

Reed, M. and Simon, B. (1978) *Methods of Modern Mathematical Physics*, IV, *Analysis of Operators*, Academic Press.

Revuz, D. and Yor, M. (1991) *Continuous Martingales and Brownian Motion*, Springer Verlag.

Rogers, L. C. G. and Williams, D. (1987) *Diffusions, Markov Processes and Martingales*, Vol. II, Wiley.

Safanov, M. V. (1980) Harnack inequalities for elliptic equations and Hölder continuity of their solutions, *Zap. Nauk. Sem. Leningrad, Otdel. Mat. Inst. Steklov. (LOMI)*, **96**, 271–87 [Russian].

Schroeder, C. (1988) Green's functions for the Schrödinger operator with periodic potential, *J. Func. Anal.*, **77**, 60–87.

Serrin, J. (1955/56) On the Harnack inequality for linear elliptic equations, *J. D'Analyse Math.*, **4**, 292–308.

Simon, B. (1976) The bound state of weakly coupled Schrödinger operators in one and two dimensions, *Ann. Phys.*, **97**, 279–88.

Simon, B. (1980) Brownian motion, L^p properties of Schrödinger operators and the localization of binding, *J. Func. Anal.*, **35**, 215–29.

Simon, B. (1981) Large time behavior of the L^p norm of Schrödinger semigroups, *J. Func. Anal.*, **40**, 66–83.

Simon, B. (1982) Schrödinger semigroups, *Bull. Amer. Math. Soc.*, **7**, 447–526.

Simon, B. (1993) Large time behavior of the heat kernel: On a theorem of Chavel and Karp, *Proc. Amer. Math. Soc.*, **118**, 513–4.

Spitzer, F. (1976) *Principles of Random Walk*, Springer Verlag.

Spivak, M. (1979) *A Comprehensive Introduciton to Differential Geometry*, five volumes, Publish or Perish Press.

Stroock, D. (1971) On the growth of stochastic integrals, *Z. Wahr. verw. Gebiete*, **18**, 340–4.

Stroock, D. (1982) On the spectrum of Markov semigroups and the existence of invariant measures, in *Functional Analysis in Markov Processes*, *Lecture Notes in Mathematics*, Vol. 923, Springer Verlag.

Stroock, D. and Varadhan, S. R. S. (1970) On the support of diffusion processes with applications to the strong maximum principle, *Proc. 6th Berkeley Symp. Math. Stat. and Prob.*, **3**, 333–60.

Stroock, D. and Varadhan, S. R. S. (1979) *Mulitdimensional Diffusion Processes*, Springer Verlag.

Sullivan, D. (1983) The Dirichlet problem at infinity for a negatively curved manifold, *J. Diff. Geom.*, **18**, 723–32.

Sullivan, D. (1986) Related aspects of positivity: λ-potential theory on manifolds, lowest eigenstates, Hausdorff geometry, renormalized Markov processes . . . . *Aspects of Mathematics and its Applications*, *North Holland Math. Library*, **34**, 747–79, North Holland.

Sullivan, D. (1987) Related aspects of positivity in Riemannian geometry, *J. Diff. Geom.*, **25**, 327–51.

Sur, M. G. (1963) The Martin boundary for a linear elliptic second-order operator, *Izv. Acad. Nauk. SSSR Ser. Math.*, **27**, 45–60.

Tada, T. (1988) Martin compactifications of the punctured disk with close to rotation free densities, *Proc Amer. Math. Soc.*, **103**, 483–6.

Tada, T. (1989) A note on Martin boundary of angular regions for Schrödinger operators, *J. Math. Soc. Japan*, **41**, 285–90.

Tada, T. (1990) Nonmonoteneity of Picard principle for Schrödinger operators, *Proc. Japan Acad.*, **66**, Ser. A, 19–21.

Taylor, J. (1969) The Martin boundary and adjoint harmonic functions, in *Contributions to Extension Theory of Topological Structures*, VEB Deutscher Verlag.

Taylor, J. (1970) The Martin boundaries of equivalent sheaves, *Ann. Inst. Fourier (Grenoble)*, **20**, 433–56.

Taylor, J. (1978) On the Martin compactification of a bounded Lipschitz domain in a Riemannian manifold, *Ann. Inst. Fourier (Grenoble)*, **28**, 25–52.

Varadhan, S. R. S. (1980) *Lectures on Diffusion Problems and Partial Differential Equations*, Tata Institute.

Varopoulus, N. (1983) Potential theory and diffusion on Riemannian manifolds, *Conference on Harmonic Analysis in Honor of Antoni Zygmund*, Vols. I, II, 821–37, Wadsworth.

Williams, D. (1991) *Probability with Martingales*, Cambridge University Press.

Yau, S.-T. (1975) Harmonic functions on complete Riemannian manifolds, *Comm. Pure Appl. Math.*, **28**, 201–28.

Yosida, K. (1965) *Functional Analysis*, Springer Verlag.

Zhao, Z. (1983) Conditional gauge and unbounded potential, *Z Wahr. verw. Gebiete*, **65**, 13–18.

Zhao, Z. (1986) Green function for Schrödinger operator and conditioned Feynman–Kac gauge *J. Math. Anal. Appl.*, **116**, 309–34.

Index